50 Years of CFD in Engineering Sciences

Akshai Runchal
Editor

50 Years of CFD in Engineering Sciences

A Commemorative Volume in Memory of D. Brian Spalding

 Springer

Editor
Akshai Runchal
ACRI, The CFD Innovators
Los Angeles, CA, USA

ISBN 978-981-15-2669-5   ISBN 978-981-15-2670-1   (eBook)
https://doi.org/10.1007/978-981-15-2670-1

This Springer imprint is published by the registered company Springer Nature Singapore Pte Ltd.
The registered company address is: 152 Beach Road, #21-01/04 Gateway East, Singapore 189721, Singapore

Foreword

This book, entitled *50 Years of CFD in Engineering Sciences: A Commemorative Volume in Memory of D. Brian Spalding*, is a significant addition to the literature in *Computational Fluid Dynamics* (CFD). It brings together experts in the field from around the world to discuss their areas of research, presenting the current state-of-the-art and future trends and needs. Thus, it establishes an important timeline in the field and provides guidance for future research. As the name suggests, it is a commemorative volume to honor one of the leading pioneers in Computational Fluid Dynamics and Computational Heat Transfer (CHT), Prof. D. Brian Spalding. Brian contributed extensively to strategies and methods to solve complex problems, while meeting a wide variety of challenges to obtain accurate, realistic and valid solutions. Besides providing important basic insight and solution approaches, Professor Spalding was a mentor to large number of students, who went on to establish their own significant research programs and thus expand the work undertaken by Brian and his research group. It is worth noting that Brian's work was instrumental in establishing the acceptability of the computational approach to solve complex problems that were generally studied only experimentally at considerably higher cost and time. His focus on a wide range of problems indicated the versatility of this approach and the importance of using computational results in studying various processes, as well as employing these for system design and optimization.

Many of the chapters in the book are contributed by students and colleagues of Professor Spalding, as well as their students. Also, some chapters are written by researchers like me, who were not directly involved with Prof. Spalding and his group but were strongly influenced by his work. Brian's impact was felt worldwide, his approaches were universally applied and the environment created by him for researchers in CFD influenced virtually all practitioners of the art. The diversity of the topics covered in this book and that of the authors is a testament to Prof. Spalding's wide-ranging impact on this field.

The book is an outcome of the dedication, love and respect that Dr. Akshai Runchal has for his former mentor. Dr. Runchal is a well-known, distinguished, researcher and engineer himself, who has contributed extensively to CFD/CHT as an educator and an entrepreneur. He has followed in the footsteps of Prof. Spalding in that he has applied his knowledge of CFD to a wide variety of practical problems and obtained results that lead to better designs with lower costs or improved performance.

Overall, this book represents a timely contribution to the field of Computational Fluid Dynamics and also honors one of the giants in the field, Prof. D. Brian Spalding. The Editor, Authors and the Publisher must be congratulated on accomplishing such a significant achievement in an area of considerable importance to science and engineering.

Yogesh Jaluria
Board of Governors Professor &
Distinguished Professor
Rutgers, the State University of New Jersey, USA

Preface

Brian Spalding observed that "Man inhabits fluid and fluid inhabits Man". This book commemorates the life and seminal contributions of Spalding in advancing the science and practice of fluid dynamics.

The mathematical foundation of fluid dynamics was laid by the eighteenth Century Swiss mathematician, Leonhard Euler—one of the most prolific mathematicians in history. He developed the Euler Equations that describe the flow of an ideal frictionless fluid. However, most flows of engineering interest stayed outside the purview of these equations since friction plays a very important role in real fluids. In 1821, the French engineer Claude-Louis Navier introduced the concept of viscosity, or friction, to generalize the Euler equations to deal with the flow of real fluids. Though Navier realized that there were forces between the molecules of fluids, he had no formal concept of shear stress or its relation to the state of the fluid. The next major breakthrough was in the middle of the nineteenth century by British physicist and mathematician Sir George Gabriel Stokes. He derived a formal mathematical expression that related the stress due to friction (viscosity) to the strain of the fluid. These developments led to the Navier–Stokes Equations that put fluid dynamics on a firm and formal mathematical basis. One result of these theoretical advances was that whereas the Euler equations were, in many cases, somewhat amenable to analytic solutions, the Navier–Stokes equations were intractable except for trivial and highly idealized one- or two-dimensional flows in simple geometrical configurations. Real-life flows that occur in three- or four-dimensional space-time frame in complex geometries evaded all attempts of an analytic or even, in most cases, approximate semi-analytic mathematical solution.

Solution of the Navier–Stokes equations is now recognized to be one of the most intractable mathematical problems. Despite the equations' broad applicability, no proof is available that a solution always exists in three dimensions and, if it does exist, that it is smooth and unique. The Clay Mathematics Institute (2000) has called this one of the seven most important open problems in mathematics and has offered a prize of US$1 million for a solution or a counterexample.

It is in this context that Computational Fluid Dynamics arrived on the scene. In spite of some breakthroughs such as the Boundary Layer Theory of Prandtl or long-wave approximations of Stokes, most problems of practical interest stayed intractable. This changed drastically with the availability of commercial electronic computers in the early 1950s. Manual computations of fluid dynamics problems based on the Finite Difference Method (FDM) had existed for a while (e.g. Thom, 1932) but it was the advent of commercial electronic machines that revolutionized fluid dynamics practice. This led to modern Computational Fluid Dynamics, or CFD as we know it today.

The first dedicated exploration of fluid dynamics by electronic computations was by a group led by Frank Harlow at Los Alamos National Laboratory starting in the early 1950s. Their workhorse was an IBM 701 computer that arrived in 1953. Their primary interest was in otherwise intractable fluid dynamics problems related to the US atomic weapons program. Most of the early work appeared as internal reports of the Laboratory and it was not until the late 1950s that publications started appearing in the open literature. One of the first hints of their work to outside research community was the paper on the particle-in-cell method (Harlow, 1957).

Unaware of the work at Los Alamos, Brian Spalding at Imperial college had been trying to get a handle on complex practical problems faced by practicing engineers. Frank Harlow was by training a physicist whereas Spalding was an engineer. This led to a difference in their approach that permeated throughout their work. Most of the processes that impact practical problems—turbulence, combustion, radiation, wall interactions, multi-phase flows, etc.—were (and are) poorly understood. There is no generally agreed unique mathematical formulation of such processes. Approximations, called "models", are therefore used to capture some essential aspects of the impact of such processes. Spalding boldly ventured into this field—with the philosophy "if it's good enough for design, it's good enough for me"—and derived methodologies that enabled an engineer to gain some insight into complex problems.

Brian Spalding did not invent CFD. He did not even coin the name. The name was coined much later—in 1972—by Roache (1972). But in one sense, more than anyone else, he created the *engineering practice* of CFD—its application to problems of interest to engineers. He was the first to start a consulting company to provide CFD services. IMARC (2019) estimates that the global commercial CFD market will surpass US\$ 3.0 billion in the next 5 years. Commercial CFD is only a small part of the total CFD activities. The actual CFD market is at least an order of magnitude larger since most of the CFD research and problem solving occurs in government organizations and private R&D departments of large corporations and research organizations.

This book commemorates the life and contributions of D. Brian Spalding. The editor was associated with, and was an integral part of, the team led by Prof. Spalding that developed the basic engineering practice that came to be known as the Imperial College (IC) approach to "CFD". This is the formulation that came to be known as the Finite Volume Method (FVM). Many other methods exist to solve the governing equations. These include the Finite Element Method (FEM), the Finite Difference Method (FDM), the Spectral Method, etc. But none of these have had the impact that FVM has had on engineering problem solving. Most of today's commercially available CFD software tools trace their origin to the work done by the IC group in the decade spanning the mid-60s and mid-70s.

This book has been assembled by former students and colleagues of Spalding who were deeply influenced by his work. Most of the authors knew him personally. So, its scope is principally directed to the impact of his work. Along with the technical material on various aspects of CFD and its applications, the book contains two appendices. The first of these is a brief biography and the second contains personal tributes from his former students and associates.

Los Angeles, USA   Akshai Runchal

Introduction

Abstract Computational Fluid Dynamics (CFD) has come a long way from its start in the late 1950s. In the process it has revolutionized engineering design practice. This book provides a window into the history and the state-of-the-art of CFD since it first became established as a generally available engineering tool in the late 60s. The early applications of CFD were largely in the field of defense and aerospace. However, since then it has impacted practically all areas of human economic and social activities. CFD has spurred advances in health physics, electronics, energy, power, propulsion, transportation, food processing, and many other human endeavors. In many cases, CFD analysis has resulted in significant improvements in efficiency, performance, and environmental impact. This book is written by internationally renowned experts who were students of, or closely associated with, Spalding. It captures the essence of what has gone on in the last 5 decades since the Finite Volume Method (FVM) of CFD was developed at Imperial College by Spalding and his students. It also provides a personal glimpse of Spalding's life story and tributes from his students and close associates whose lives were impacted by him. It is the aim of this book to serve as a reference source for practicing engineers, educators, and research professionals.

Keywords Computational fluid dynamics, CFD, Finite volume method, Energy, Environment, Turbulence, Combustion, Aerospace, Heat Transfer, Mass transfer, Fuel cells

Man inhabits fluid and fluid inhabits Man. That was Spalding's dictum. All human endeavors involve motion of some quantity. Fluid Dynamics, the science of a continuum that moves, is thus pervasive in our social and economic life. Computational Fluid Dynamics (CFD), a mix of science and technology with multi-disciplinary aspects, is the most general tool that enables us to understand the behavior of fluids.

In its early days, CFD was used primarily for defense, aerospace, and traditional industrial applications involving flow and heat transfer. However, since then it has impacted practically all areas of human economic and social activities. CFD tools today are integral components of aerospace, defense, energy, power generation, transport, electronics, food processing, environmental management, fire safety, computational chemistry, particle physics, genetics, architecture and building design, and life, biomedical, and pharmaceutical sciences. CFD is helping us design better computer chips. It is helping us understand the flow of blood in the heart, that of oxygen in the lungs, and the diffusion of drugs in the brain. It is helping us develop better neurological treatments and better utilization of CAT SCAN devices by exploring the behavior of matter and fluids in the human brain. The movement of a school of fish and the growth and decay of bio-populations can be studied with CFD. In the field of sports, CFD is beginning to play an important role, e.g., in the design and analysis of America's Cup (where the movement of a boat in water was simulated) and the Tour de France (simulation of the impact of wind on a cyclist). Even the entertainment industry is benefitting from the use of CFD in film making such as cost-effective simulation of a sinking ship. The traffic flow on a congested highway can be studied by CFD. In geosciences, CFD is being used to simulate the movement of pollutants in air, the development of weather patterns and forecasts, the flow of surface and ground water, and the spread of an oil slick in the ocean. It is helping us understand the movement of tectonic plates and their impact on the behavior of volcanos and earthquakes. CFD is helping us tackle the difficult problems of environmental and disaster management and safe disposal of hazardous and nuclear waste. CFD tools play a crucial role in astronomy in helping us understand the evolution, movement, and fate of stars and other astronomical bodies. In summary, CFD has impacted and transformed all aspects of human endeavor and industry. By rough estimates, the global CFD, both commercial and in-house R&D by government and private industry, is today worth tens of billions of dollars and is growing at a fast pace.

Computational Fluid Dynamics had its origin in the late 1950s. Early applications of CFD were based on the age-old method of finite differences (FDM) that employed Taylor Series. Starting in the early 1940s, the Finite Element Method (FEM) was developed to solve structural engineering problems. The hallmark of the FEM was the minimization of a functional that implied the existence of a Hamiltonian and was used for systems governed by linear differential equations. Gupta & Leek (1996) provide an interesting glimpse into the history of development of the FEM. The general set of fluid dynamics equations, the Navier–Stokes Equations, are inherently non-linear. No proof of unique solutions or a Hamiltonian exists except for those cases where the equations reduce to a linear system. The equations are known to possess multiple and chaotic solutions and their solution has now been recognized as one of the seven most intractable mathematical problems (Clay Mathematics Institute, 2000). It was not until the late 1960s that attempts were made to extend the FEM to fluid dynamics problems. Till then FDM remained the method of choice for non-linear Navier–Stokes equations.

In the mid-1950s, a group led by Harlow at the Los Alamos National Laboratory had used the FDM to tackle some theoretically intractable problems of fluids. Their work then was confined mostly to weapons related problems and was not generally available in the open literature. In the early 1950s, Spalding became interested in developing a general theory of combustion. By the mid-1960s, he had already developed a comprehesive theoretical framework for coupled hydrodynamics, heat and mass transfer and a Unified Theory for boundary layers, jets and wakes. When the first generally available commercial computer, the IBM 7090/7094, arrived at Imperial College in the early 1960s, the group led by Spalding was well-placed to explore its use for engineering and industrial applications related to fluid dynamics, heat and mass transfer, and combustion.

The group led by Spalding at Imperial College also started with the application of the FDM to problems of fluid dynamics. However, Spalding, with an engineer's insight, quickly realized the advantages of thinking in terms of, and working with, a methodology based on the physics of the problems rather than the mathematics. Both FDM and FEM are primarily "mathematical" methods. FDM is based on expressing the "derivatives" in terms of Taylor series. The FEM is based on expressing the variation of a governing quantity (dependent variable) in terms of a linear or non-linear mathematical (basis) function. The method developed by Spalding and his students stressed the physics of a problem and is now known as the Finite Volume Method (FVM). Its conceptualization is in terms of "fluxes" and "amounts". The "amount" refers to the inventory of a property in a given volume and the "flux" is the rate at which the quantity is transferred between adjacent volumes. Early conceptualization by the Imperial College group was in terms of a fluid flowing through a series of tanks (grid cells or elements) interconnected by a series of tubes (grid lines)—named the tank-and-tube method. Each tank held a given amount and the tubes conveyed the fluxes. Runchal (1969) realized that FVM represented an integral rather than a differential approach to the set of governing equations. For his dissertation, he integrated the Navier–Stokes equations by applying the Gauss Theorem and showed that the resulting algebraic equations were equivalent to those derived by FDM. The formal foundation was thus laid to derive the FVM formulation by application of the Gauss Theorem to the Navier–Stokes equations. One major advantage of the FVM over FDM is that, like the FEM, it can be applied to domains with arbitrary boundaries. The control volumes, or elements, can also be arbitrary in shape with structured or unstructured grids. The primary advantages that FVM possesses over the FEM is that mass and other properties can be conserved over each finite volume rather than only globally. This becomes important for problems, such as those in combustion, where local errors in one quantity (say, temperature) can lead to orders of magnitude larger errors in another quantity (such as reaction rate). Another significant advantage is that FVM being integral in nature, reduces the order of continuity required by one degree over the differential approach. Thus, sharp interfaces and discontinuities, such as shocks, require only first order smoothing. It is to be noted that most of the current commercial codes employ the FVM formulation developed at Imperial College.

The main focus of a practicing engineer is to design reliable, efficient, and cost-effective systems with a low footprint on the environment. The analytical tools have benefited from advances in computational sciences, with more powerful computing systems and significant improvements in our abilities to make use of evolving computer architecture with multiple processors, and parallel and distributed computing. From meager beginnings in the 1960s, computational sciences have revolutionized our theoretical analysis capabilities, and CFD tools have now become ubiquitous in academia and industry.

Scientific computing in general, and CFD in particular, appear to be poised on the threshold of rapid advances based on Machine Learning, Artificial Neural Networks (ANN), and AI (Artificial Intelligence). Furthermore, software tools are now being augmented with hardware components such as FPGA (field-programmable gate array), ASIC (application specific integrated circuits), and SoC (system-on-a-chip). These, combined with Virtual Reality, will improve the speed, accuracy, and user friendliness of CFD software. This will revolutionize engineering design and result in robust and reliable tools for practicing engineers that might be called Engineering Virtual Reality (EVR) where the engineer can see what is being designed before any metal is cut.

This book is the result of a collaboration by internationally acclaimed experts who were associated with Brian Spalding. It is divided into sections that deal with various aspects of CFD engineering practice. The first part deals with the fundamentals of CFD and provides a glimpse of Spalding's contributions and the basic technology underlying CFD. Turbulence is still one of the poorly understood processes that plays a very important role in practical flows of engineering interest. The second part of this book provides historical perspective and reviews of specific aspects of turbulence. The third part of this book deals with combustion and reactive flows. Most of the energy generation today depends on reactive flows. These also play a very important role in transportation and in the chemical processing industry. This part of the book captures some salient aspects of energy generation by combustion and fuel cell technology. The fourth part discusses computational models of liquid films and pool boiling that play a very important role in many engineering applications. This is followed by a section on validation and specific applications of CFD and a discussion of the comparison between the FEM and FVM methods. The last contribution is a speculation of what might lie ahead for CFD and how the technology may develop. The volume concludes with two appendices. The first is a short biographical sketch of Brian Spalding and the second contains tributes to Spalding from his former students and colleagues.

We hope this book provides a glimpse into the current state of CFD and encourages further R&D efforts worldwide to make CFD an essential component of engineering design tools. There are limitless novel applications yet to be developed and extended that will further contribute to efficient and sustainable social and economic activities within the human biosphere.

I would like to express my gratitude to all the authors for contributing to this commemorative volume to celebrate an exceptional life. Professor Spalding had a deep and profound impact on the field of fluid dynamics and helped to revolutionize the engineering practice that led to unimaginable advances in the design of systems that are at the core of human endeavors.

Los Angeles, USA & McLeod Ganj, India   Akshai Runchal
                                                                                      runchal@ACRiCFD.com

References

1. Clay Mathematics Institute. (2000). *Millennium problems.* https://www.claymath.org/millennium-problems.
2. Gupta, K. K., & Meek, J. L. (1996). A brief history of the beginning of the finite element method. *International Journal for Numerical Methods in Engineering, 39,* 3761–3774.
3. Harlow, F. H. (1957). Hydrodynamics problem involving large fluid distortions. *Journal of the ACM (JACM)* 137.
4. IMARC Group. (2019). *Computational fluid dynamics market: Global industry trends, share, size, growth, opportunity and forecast 2019–2024.* https://www.imarcgroup.com/computational-fluid-dynamics-market.
5. Roache, P. J. (1972). *Computational fluid dynamics* (p. 87801). Socorro, NM: Hermosa Publishers.
6. Thom, A. (1932). *Arithmetical solution of problems in steady viscous flow* (p. 1475). A.R.C., R&M.

Contents

CFD and Turbulence

CFD and Reactive Flows

Multiphase Flows

Applications and Validation

About the Editor

Dr. Akshai Runchal is the founder and Director of CFD Virtual Reality Institute and ACRI group of companies. His expertise is in Computational Fluid Dynamics (CFD). He obtained his Ph.D. in 1969 from Imperial College (London, U.K.) under the guidance of Prof. D. B. Spalding and was a key member of the 3-person team led by Spalding that invented the Finite Volume Method (FVM) in the mid 1960's. Since 1979, he has consulted widely on flow, heat and mass transfer, combustion, environmental impact, hazardous and nuclear waste, ground water and decision analysis to over 200 clients in 20 countries. He is the principal author of PORFLOW, TIDAL, ANSWER and RADM simulation models that are used worldwide. He received his Bachelor's in Engineering from PEC in1964. He has taught as regular or adjunct faculty at a number of leading institutes including IIT (Kanpur), Imperial College (London), University of California (Los Angeles) and Cal State (Northridge). He has authored or co-authored 7 books and over 200 technical publications. He has received many honors and awards, and given many invited contributions at conferences.

Dr. Runchal grew up in the scenic hill town of McLeod Ganj, Dharamsala that is now on the world map as the hometown of His Holiness the Dalai Lama. Since 2006 he has been deeply engaged in reviving the lost heritage of Kangra Miniature Paintings through Kangra Arts Promotion Society – an NGO based in McLeod Ganj. He divides his time between Los Angeles and McLeod Ganj and is actively engaged in promoting education, training and R&D in CFD and related disciplines.

,

Fundamentals of CFD

Brian Spalding: Some Contributions to Computational Fluid Dynamics During the Period 1993 to 2004

Michael R. Malin ⓘD

1 Introduction

Professor Brian Spalding and his colleagues at both Imperial College and CHAM made tremendous contributions to the field of CFD. In particular, they pioneered its development and use for industrial applications, and one of the most distinguished achievements in this respect was the development of the first general-purpose CFD code, PHOENICS [1]. Much has been written elsewhere about these contributions, as well as Brian's important works in the fields of fluid dynamics, combustion and heat and mass transfer [2–7]. This paper will discuss some novel contributions that stem from the author's association with Brian at CHAM on particular CFD contracts and code developments during the period 1993–2004. This was a period when computer resources were significantly inferior to those on offer today, a fact which should help provide further perspective to the requirements and challenges faced by Brian in finding solutions to the practical CFD problems discussed here.

There is much that could be written about pioneering work carried out at CHAM, but I have limited the discussion here to include only topics with which I was directly involved with Brian during the period mentioned, and sometimes this was done in collaboration with other colleagues, as described in the text, and listed as co-authors in the references. In the remainder of this article, attention will be focused on the technical aspects of Brian's work in the following areas: differential-equation wall-distance calculation; the LVEL turbulence model; the IMMERSOL model of thermal radiation; virtual mass modelling in Eulerian–Eulerian descriptions of two-phase flow; a space-marching method for hyperbolic and transonic flow and an automatic convergence-promoting algorithm.

M. R. Malin (✉)
Concentration Heat and Momentum Limited (CHAM), Bakery House, 40 High Street, Wimbledon, London SW19 5AU, UK
e-mail: mrm@cham.co.uk

© Springer Nature Singapore Pte Ltd. 2020
A. Runchal (ed.), *50 Years of CFD in Engineering Sciences*,
https://doi.org/10.1007/978-981-15-2670-1_1

2 Differential-Equation Wall-Distance Calculator

2.1 Introduction

Most CFD codes today employ a differential-equation wall-distance calculator as an economical means of computing approximate values of the normal wall distance for use in turbulence models. The origin of this now ubiquitous method can be traced back to 1993 when CHAM was working as the CFD partner in a R&D Collaborative project funded by the European Commission under the BRITE-EURAM programme. Specifically, the author and Fred Mendonca implemented a low-Reynolds-number two-equation turbulence model into PHOENICS, but in order to handle generic complex geometries, a means had to found to compute the nearest wall distances required by the turbulence model. Consequently, Fred implemented a generic search-based procedure to compute these distances for each mesh cell in the solution domain.

2.2 Equations

When we briefed Brian on the turbulence-model implementation, his immediate response was that the search-based wall-distance method was far too expensive and cumbersome for use in industrial applications with arbitrary complex geometries. A day later he produced a technical note describing an ingenious and generic Poisson-based differential-equation method for calculating approximate values of the nearest wall distance [8]. This method solved the following Poisson-type equation for a wall-distance-related scalar variable ϕ

$$\nabla^2 \phi = -1 \tag{1}$$

where $\phi = 0$ in all solids and at wall boundaries. The variable ϕ was converted into a wall distance δ through the auxiliary equation

$$\delta = \sqrt{\left(|\nabla\phi|^2 + 2\phi\right)} - |\nabla\phi| \tag{2}$$

Brian also derived the following equation for evaluating the distance separating the two nearest walls

$$G = 2\sqrt{\left(|\nabla\phi|^2 + 2\phi\right)} \tag{3}$$

During the derivation of this method Brian was thinking physically in terms of the turbulent length scale rather than mathematically in terms of the wall distance, as emphasised in his original technical note by the comment: *That two distances are required for the computation of the turbulent length scale is revealed by consideration of the Nikuradze formula for the turbulent length scale in a channel. His formula expresses the turbulent length scale as a quartic function of the distance from the wall, and the distance across the channel.* As we shall see later, Brian had in mind radiation modelling as an application area for using the distance between walls.

Brian also wrote: *I have devised a method of computing the turbulence length scale which will be more economical and more realistic than any other which we know of.* This assertion was realised when the method was implemented immediately in PHOENICS by Brian, Dr. Igor Poliakov and Dr. Slava Semin; and then put to use on a number of commercial contracts. The generic benefits of the method were obvious because the nearest wall distance is a key parameter in a large number of turbulence models, and also in some radiation models. Moreover, the ease of implementation and computational economy of the method becomes especially useful for aeroelastic, adaptive and transient mesh problems where there is a need to recompute the nearest wall distance.

2.3 Further Developments

The method was published obscurely by Brian in a poster session at the 1994 International Heat Transfer Conference [9], although Professor Paul Tucker, then at the University of Dundee, acquired the details of the method through an explanation given to his colleague Dr. Desmond Adair, who was using PHOENICS at the time. There can be little doubt that Brian opened up a new area of research with his invention of the Poisson wall-distance calculator, because thereafter, the method was vigorously worked and elaborated on by Tucker [10, 12–15, 20] and other researchers [11, 16–19]; and eventually the method or some variant found its way into most commercial and open-source CFD codes.

Nowadays wall-distance calculation methods are still classified into three groups: search algorithms [21–23], advancing surface-front methods [24] and differential-equation approaches based on either the Eikonal, Hamilton–Jacobi or Poisson equation. The Eikonal equation ($|\nabla\delta| = 1$) produces the exact wall distance [13] and so it has attracted a lot of attention, but it is challenging to implement and is less numerically robust than the Poisson equation. The Hamilton–Jacobi equation [13, 14] appends a Laplacian term to the Eikonal equation; and the relationship between the various differential equations, and their respective merits and demerits have been discussed in detail by Tucker [25]. Interestingly, Brian's development [8, 9] was a combination of intuition and analytical flair; but its arrival initiated a large number of refinements or alternatives based on mathematical derivation. Tucker [10] explored Brian's original approach, whereas Fares and Schroder [11] derived what is essentially an Eikonal-type equation that uses an inverse distance function. Tucker [12] and Tucker et al. [13, 14] compared approaches based on solving Eikonal, Hamilton–Jacobi and Poisson equations. Later Tucker [15] developed a hybrid Hamilton–Jacobi/Poisson approach, which was also investigated by Jefferson-Loveday [19].

Xu et al. [16] also derived a wall-distance transport equation from the Eikonal equation, and Wukie and Orkwis [18] investigated the *p*-Poisson approach proposed by Belyaev and Fayolle [17]. This method is a generalisation of the Poisson equation which reduces to Brian's method when the *p*-Laplace operator $p = 2$. As p is increased further, the approximate distances were found [18] to approach the exact distance fields. More recently, Watson et al. [20] investigated a hyperbolic form of the Poisson equation for unsteady aerodynamics applications on unstructured meshes, and reported improved convergence rates over the original method.

Despite providing only an approximate estimate of the wall distance, Brian's Poisson-based method remains the method of choice for established industrial CFD codes because it is the easiest to implement, and the most numerically stable and economical. It also scales well on larger meshes, and is efficient for parallelisation.

3   The LVEL Turbulence Model

3.1   *Introduction*

The invention of the differential-equation wall-distance calculator emerged almost simultaneously with Brian's proposal for a zero-equation low-Reynolds-number turbulence model named LVEL [26], which is targeted for situations in which fluid flows through spaces cluttered with many solid objects. The motivation for this model stemmed from a contract placed by a large electronics company, which was concerned with using PHOENICS for thermal and fluid-flow simulations of various electronic packages and systems. The challenges here are the need to mesh large numbers of electronic components, and because the spaces between them are small and low velocity, the turbulence is often transitional in the cooling airflow. This means that the standard k-ε model is not strictly valid and a low-Reynolds-number turbulence model is required. Also, the grid density between nearby solids is often too coarse to allow accurate computation of the turbulence energy generation rate. Therefore, two-equation turbulence models cannot be meaningfully employed, which is the reason other CFD practitioners favoured using zero-equation effective-viscosity models at that time. These very simple models were based on empirical data and knowledge of the local values of the bulk velocity and characteristic flow dimension.

3.2   *Model Equations*

The key elements of the LVEL model [26] are the differential-equation wall-distance calculator, which was mentioned earlier, and Brian's law of the wall [27] which allows the computation of a turbulent viscosity v_t that covers the entire laminar and turbulent regimes of the flow, and is given by:

$$v_t^+ = \frac{\kappa}{E}\left[exp(\kappa u^+) - \sum_{i=0}^{3} \frac{(\kappa u^+)^i}{i!}\right] \qquad (4)$$

where $v_t^+ = v_t/v$, and v is the kinematic laminar viscosity, κ (=0.41) is von Karman's constant, E (=8.6) is the roughness parameter, and $u^+ = u/u^*$ where u is the local velocity, and u^* is the friction velocity defined by $\sqrt{(\tau_w/\rho)}$ where τ_w is the wall shear stress, and ρ the fluid density. The local value of $u+$ is computed from Brian's law of the wall

$$y^+ = u^+ + \frac{1}{E}\left[exp(\kappa u^+) - \sum_{i=0}^{4} \frac{(\kappa u^+)^i}{i!}\right] \qquad (5)$$

by noting that $Re = u\delta/v = u^+y^+$ and then substituting this expression into (5) for $y+$ to yield the equation

$$x^2 + x\frac{\kappa}{E}\left[exp(x) - \sum_{i=0}^{4} \frac{(x)^i}{i!}\right] = Re\kappa^2 \qquad (6)$$

where $x = \kappa u^+$ and δ is the distance to the nearest wall. Equation (6) may be solved iteratively for x by using Newton-Raphson iteration.

3.3 Applications

The model was implemented in PHOENICS by Brian and Dr. John Worrell, and it was then used to great practical advantage in electronics-cooling applications, so much so, that almost inevitably the model began to appear in other commercial CFD codes, and especially ones dedicated to the thermal design of electronic components and systems. Certainly, the usefulness of the model has been demonstrated by many other workers in the field of electronics cooling [28–35], as well as in other applications areas such as buildings [36–42], as will now be discussed by considering a selection of these publications.

Dhinsa et al. [28] tested seven different turbulence models for a wall-mounted heated cube in a channel under conditions relevant to electronics-cooling applications. These models included the standard k-ε model, the LVEL model, the low-Reynolds-number k-ε and k-ω models and the k-ω-SST model. An important finding was that even though the LVEL model was the simplest model, its performance was as effective as the more advanced turbulence models and better than the standard k-ε model. These workers also encountered the well-known problem of the LVEL model overestimating the eddy viscosity remote from walls, as for example in the centre of a channel where the mixing length implied by the model will be too large. Therefore, the LVEL-CAP model [28, 29] was proposed where the eddy viscosity was limited to $0.01*V_b*L_{max}$, where V_b is the bulk viscosity, and L_{max} is taken as 1/6th of the channel width. Interestingly, Dhinsa [29] also reported on the use of a two-layer turbulence model that employed the LVEL model in the near-wall viscosity-effected region, and the standard k-ε model in the fully turbulent region away from the wall. Following a well-established practice for two-layer models, the transition between the inner and outer layers was achieved by multiplying the eddy viscosity by a damping function dependent on the dimensionless wall distance, $y+$. This form of the model will feature later when we consider applications to flow inside buildings.

For practical electronics applications, Eveloy [31] reported that *despite its limitations, the LVEL model predictions are on order within 6% of those for the k-ε model, when comparing predictions of the steady-state junction temperatures for a PCB package design, and similar trends were reported by* Rodgers et al. [33] *for another PCB topology. The LVEL model produced slightly larger prediction errors than the k-ε model, but the model maintained its predictive accuracy using lower grid densities, unsuitable for the k-ε model, indicating "the greater applicability of the LVEL model for system-level turbulent flow analysis, where grid density is constrained by computational limits.*

Later, Choi et al. [34] also reported that the LVEL model was as effective as the more advanced two-equation models for simulating data-centre rack-mounted servers, with the big advantage being that *significant computation time can be saved (factor of three or higher), especially when conducting transient CFD simulations or when making steady-state runs with different rack settings.* In more recent work related to electronics-cooling applications, Marchi Neto and Altemani [35] used the LVEL model to investigate the air cooling by two impinging jets of an array of heaters mounted on a substrate in a duct. Favourable comparison was reported with measurements over jet Reynolds numbers ranging from 2000 to 7000.

Over the last thirty to forty years, CFD has become well established as an important design and analysis tool for modelling airflow in and around buildings. Many indoor airflows involve forced, natural and mixed convection, and compared to more complex turbulence models, the LVEL model offers low-Reynolds-number modelling with the benefits of reduced solution times and robust performance, while still delivering acceptable engineering accuracy for initial design studies. Dhoot et al. [36] wrote *Agonofer et al. (1996) proposed the popular LVEL zero-equation model which utilizes a wall function at all points in the flow; this model is particularly good for confined flows, like small-scale electronics thermal applications, but is less appropriate for typical room-scale applications.* Nevertheless, both the original LVEL model and its derivative the two-layer LVEL k-ε model have been used successfully for building design and analysis. For example, the LVEL model has been used in CFD studies aimed at improving air-conditioning systems in industrial buildings [37] and optimising solar wall-heating systems in residential buildings [43]. The two-layer LVEL k-ε model was applied to study natural ventilation for cooling in commercial buildings [38], modelling roof-mounted cowls in hybrid ventilation systems [39], assessing thermal comfort for occupants in rooms using different space-heating methods [40], investigating thermal comfort in an office using traditional and ventilation radiators [41], analysing building-integrated photovoltaic (BIPV) systems in building facades

[42] and investigating the effectiveness of ventilation in reducing human exposure to indoor airborne pollutants [44].

The simplicity and robustness of the LVEL model also makes it a candidate for complex CFD applications involving multiple physical models [45–54]. For example, Tchouvela and co-workers [46, 47] employed the model in their hazard assessment studies for the accidental release of high-pressure hydrogen jets when using a real-gas-thermodynamics CFD model. Wang et al. [48, 49] also used the model in their complex CFD modelling of the tilt-casting process of titanium alloys. This application involved complex transient, three-phase flow with heat transfer and solidification, and the dynamic evolution of a free surface (the metal/gas interface) during the filling of the mould when heat transfer takes place between the metal, mould and resident gas. The LVEL model was adapted to handle the moving solid boundaries by setting the wall-distance scalar variable $\phi = 0$ in the solidifying regions of the advancing solid front.

For a continuous flow reactor (tundish), the performance of the LVEL model was compared against measurements and more complex turbulence models for the prediction of residence times of chemical tracer solutions in water [50]. The predictions of the LVEL model and LES were reported to show better agreement with the experimental residence-time-distribution curves than the results of a number of two-equation models and an explicit algebraic Reynolds-stress model. It should be mentioned that this research group had previously used the LVEL model to study successfully slag-steel entrapment phenomena in a tundish [51, 52].

In the field of high-temperature superconductivity for power transmission systems, Artemov et al. [53] reported that experimental data for liquid-Nitrogen flow in a corrugated tube were reproduced with greater accuracy by the LVEL model and the k-ω turbulence model than the low-Reynolds-number k-ε model.

Finally, Tucker and Liu [54] investigated the performance of a large number of turbulence models for flow in the stagnation and wake regions of a small cylinder (wire) at a Reynolds number of 2,300. Despite its simplicity, the LVEL model produced broadly similar results to all the other models, and unlike some two-equation turbulence models, it more correctly showed relatively little cylinder influence on turbulence production in the wake region.

3.4 Discussion

Brian [26] emphasised that *The LVEL model is to be regarded as providing a practical solution to heat-transfer-engineers' problems, and not as one providing new scientific insight.* In summary, the model is intended for use in wall-bounded cluttered regions where the between-solid distances are often too small for fine grid resolution. The model is valid over the laminar, transitional and fully turbulent regimes, and gives the well-known experimental results for simple cases, such as for example, flow between parallel plates; and it gives plausible results for more complex cases. It is particularly useful for making economic simulations of heat transfer in electronics-cooling applications, because of the excessive grid requirements of low-Reynolds-number two-equation-model extensions. As with all zero-equation models, no account is taken of the transport of turbulence quantities, and so it cannot be expected to perform better than more sophisticated models in cases where these effects are important. As a final note, Brian once suggested that further development of LVEL might be to allow the length scale to be diminished by velocity gradients. This would offer a more aesthetic solution than putting an upper limit on the eddy viscosity in regions remote from walls [28, 29].

4 The IMMERSOL Model of Thermal Radiation

4.1 Introduction

Brian's ability to recognize the need for engineers to get an adequate solution within reasonable computing time also led to the development of the IMMERSOL radiation model in 1994 [55–57]. Here, IMMERSOL stands for IMMERsed SOLids. The history is that the author was engaged on a contract for a large US Corporation, who wanted an economical radiation model for use on curvilinear meshes in a participating medium with turbulent combustion and complex geometry. The client had previous experience of using the discrete transfer radiation model [58] in another commercial CFD code, but they found the model prohibitively expensive. The author proposed we implement Rosseland's diffusion model [59], but Brian expressed the view that whilst this was a reasonable suggestion, he had something much more generic in mind.

4.2 The Radiosity Model

A few days later Brian sent me a technical note outlining his proposal for a diffusional model to represent radiation in arbitrary geometries for both optically thick and thin media [55]. This model involved the solution of a diffusion equation for the radiosity R with a diffusivity proportional not only to the medium's absorption and scattering coefficients, but also to the local distance between solid walls, which was needed for transparent media. The radiosity equation was derived by simplifying the six-flux model of Hamaker [60] and Schuster [61] (see also Spalding [62]) by ignoring the directional aspects of radiation; and the inter-wall distances were computed from Brian's Poisson equation wall-distance calculator, which was introduced earlier.

The author implemented the radiosity model in PHOENICS in the following optically thick form, but without the inter-wall distance term for transparent media because of the pressing needs of the contract:

$$\nabla\left(\frac{4}{3k_e}\nabla R\right) + 4a(E - R) = 0 \tag{7}$$

where E is the emissive power, defined by $E = \sigma T^4$, with T the thermodynamic temperature and σ (=5.6678 × 10^{-8} W m^{-2} K^{-4}) Stefan–Boltzmann's constant. The extinction coefficient k_e is given by:

$$k_e = a + s \tag{8}$$

where a and s are the absorption and scattering coefficients per unit length, respectively. The diffusion term in Eq. (7) indicates that the radiative heat flux vector is given by

$$q_r = -\left(\frac{4}{3k_e}\right)\nabla R \tag{9}$$

The physical meaning of Eq. (7) is that it represents the net rate of loss or gain of radiant energy per unit volume. The term aE represents the local rate of emission, while the term aR represents the local rate of absorption of radiation per unit volume.

For those conversant with radiation modelling, it is easy to see that Eq. (7) has the same form as for the irradiance I in the P-1 spherical-harmonics approximation [63] to the general radiative transfer equation:

$$\nabla\left(\frac{1}{3k_e}\nabla I\right) + a(4E - I) = 0 \tag{10}$$

where anisotropic scattering has been ignored, and so $I = 4R$. The spherical-harmonics approach, also known as the Eddington approximation [64], is one of the oldest approximate methods for radiative transfer. The idea is to express the radiation intensity function in terms of a Fourier series and to truncate this series to the lowest order to obtain the P-1 approximation, but as explained above, Brian arrived at an equivalent model by a different route.

The diffusion approximation is not valid at a boundary, but at the time Brian was not forthcoming about the wall boundary conditions for the radiosity model. I can recall Brian saying *I have done as much as I have time for, so I leave the rest to you.* Consequently, the author adopted a Marshak [65] boundary condition:

$$S_{R,w} = \frac{2\varepsilon_w}{(2 - \varepsilon_w)}(E_w - R) \tag{11}$$

where ε_w is the wall emissivity and E_w is the emissive power at the wall temperature T_w, i.e. $E_w = \sigma T_w^4$. This type of boundary condition is valid only for optically thick media, and it has been used in the past with both effective thermal conductivity models [59, 66] and the P-1 model [67].

4.3   The Immersed Solids Model

Soon after the successful completion of the contract, Brian delivered a lecture at VKI in Belgium [68] where he proposed the development of a model for radiative heat transfer in the congested spaces encountered in electronics-cooling situations. Here, the medium is transparent; *so the conduction* [59] *and Eddington* [64] *approximations are invalid*; and the use of more sophisticated models, such as for example, the Zone and Monte Carlo [63] methods, would be computationally very expensive. Brian again suggested the use of inter-wall distances as a basis for evaluating radiant fluxes in a transparent media via an *effective conductivity*. The following year, further technical notes arrived from Brian [56, 57] elaborating on how the radiosity model could be employed in conjugate-heat-transfer situations by reformulating the radiosity equation in terms of a radiant temperature T_r, where $R = \sigma T_r^4$. This resulted in Brian and Dr. Igor Poliakov implementing the final form of IMMERSOL into PHOENICS:

$$\nabla \left(\frac{4\sigma T_r^3}{\left(\frac{3}{4} k_e + \frac{1}{G} \right)} \nabla T_r \right) + 4a\sigma \left(T^4 - T_r^4 \right) = 0 \tag{12}$$

where G is the distance between the two nearest walls, which is computed from Eq. (3). The radiative heat flux is now given by

$$q_r = -\frac{4\sigma T_r^3}{\left(\frac{3}{4} k_e + \frac{1}{G} \right)} \nabla T_r \tag{13}$$

For optically thick media, where the mean free path $1/k_e \gg G$, the expression for the heat flux vector reduces to Eq. (9), which is Rosseland's formula. For optically thin media, the radiative heat flux vector reduces to:

$$q_r = -G\sigma \nabla T_r^4 = -G\nabla R \tag{14}$$

so that $1/G$ is the geometrical resistance to radiation.

Brian [69] proposed the use of the following boundary condition at a solid wall,

$$S_{r,w} = \sigma \beta \left(T_w^4 - T_r^4 \right) \tag{15}$$

with the wall resistance term β given by

$$\beta = \left[\delta \left(\frac{3}{4} k_e + \frac{1}{G} \right) + \frac{(1 - \varepsilon_w)}{\varepsilon_w} \right]^{-1} \tag{16}$$

so as to account for both transparent and participating media.

From the outset, Brian anticipated the possibility of slow convergence when the absorptivity of the medium is large, or when it is small and the solid temperatures are strongly influenced by radiation. For the resolution of this problem, Brian adapted the Partial Elimination Algorithm (PEA) [70] from his work on two-phase flows, but this was not implemented until several years later [71]. The PEA accelerates convergence by rearranging the discretisation of the energy and radiation equations so that they are more implicit and take directly into account the coupling between T^4 and T_r^4.

Subsequent application of IMMERSOL in CHAM showed that whilst it does not necessarily procure close agreement with experiments, it always produces physically reasonable predictions with computer times that are only a fraction of those expended by more sophisticated radiation models. Even so, apart from the work of Rasmussen [72] and Osenbroch [73], the model appears to have made little impact outside of CHAM and its base of PHOENICS users.

Brian [69] proposed that since the computational expense of IMMERSOL was so small, it could be used when the wavelength dependency is too great to be ignored, as for example, when short-wave solar radiation provides a source of heat that is redistributed by way of long-wavelength low-temperature infrared radiation between the terrestrial objects on which it impinges. In a similar vein, for combustion applications, Rasmussen [72] suggested extending the model to use a number of wavelength bands, and the weighted-sum-of-gray-gases model to compute the total emissivity and absorptivity of the medium.

4.4 Applications

Successful use of the model across a broad range of combustion applications has been reported by a number of workers [72–80]. Rasmussen [72] implemented IMMERSOL in the commercial CFD code, STAR-CD, and then used it in his CFD modelling of flow, combustion and heat transfer in a self-recuperative radiant burner, a burner tube for liquid heating, and a glass furnace. Osenbroch [73] implemented the model in the FLEXSIM (Fire Leak Explosion Simulator) CFD code to predict horizontally released jet fires for scenarios relevant to accidental gas releases in both onshore and offshore installations. Similar applications were considered by Aloqaily and co-workers [77–79] to assess the thermal hazard resulting from the impact of horizontal jet flames on portable building structures. Yang et al. [74, 75] made use of IMMERSOL when simulating the transient operation of heat treatment furnaces with turbulent flow, combustion and conjugate heat transfer. Later, Zhubrin [76] used IMMERSOL in combination with his discrete reaction model to investigate sooting flames in a carbon black furnace, a laboratory cylindrical furnace, a lead-smelting furnace and an experimental compartment fire. Agranat and Perminov [80] validated their multiphase CFD model against an experimental representation of a wildland fire. The forest was considered as a chemically reactive multiphase medium, and IMMERSOL was used to model the radiation in the presence of turbulent gas flow and heat transfer, drying, pyrolysis and the combustion of wood char and volatiles.

Building applications have been the subject of a large number of simulations. For example, Corbin and Zhai [81] investigated the performance of a building-integrated photovoltaic cell thermal collector, and reported good agreement between predictions and measurements. Chiang et al. [82] investigated various design studies for radiant cooling ceilings and mechanical ventilation systems. The focus here was on thermal comfort in an office in a hot, humid region; and indoor air temperatures were predicted to within 2% of the measurements. The use of IMMERSOL to study the influence of solar radiation and outdoor airflow on buildings, the so-called Urban Heat Island (UHI) effect, has been reported by Radhi et al. [83, 85], Maragkogiannis et al. [84] and Zhang et al. [86]. Radhi et al. [87] also used the model to study the impact of climate-interactive façade systems on the cooling energy in fully glazed buildings. For other building studies with solar radiation, Zhang et al. [88] studied green-roof systems for different building layouts and ambient conditions; whilst Hien and Istiadji [89] used the model to investigate the effects of external-shading devices on natural ventilation in a residential building.

IMMERSOL has also seen applications in the nuclear industry. Vaidya et al. [90] investigated the refuelling operation of a nuclear reactor during the dry period. Steady and transient simulations of turbulent convection and radiative heat transfer were carried out under different decay power conditions. For natural convection and radiation in a square enclosure, the CFD model was also compared and matched to benchmark data over a wide range of Rayleigh numbers and wall emissivities. Similarly, for thermofluid studies related to the cooling of a very high-temperature reactor, Kuriyama et al. [91] simulated separate cases of forced and natural convection with radiation in a rectangular channel containing porous inserts in the form of copper wire.

In studies related to electronics cooling and energy storage devices, Zamora [92–95] used IMMERSOL to model radiation and buoyancy-driven turbulent airflows in cavities and enclosures. Finally, Budiyanto et al. [96] performed CFD studies to investigate the energy consumption of a refrigerated cargo container under wind flow and solar radiation in a container port.

4.5 Discussion

IMMERSOL provides an economically realisable approximation to the radiative transfer equation that gives the exact solution for thermal radiation between two infinitely long parallel plates; and in more complex situations, its predictions are always plausible, and of the right order of magnitude. Brian [57] commented that *IMMERSOL is like the LVEL model, with which it shares the wall-distance and inter-wall distance calculations based on the Poisson equation. Its use can be justified in part by reference to the facts that: (a) it is rare for the absorption and scattering coefficients to be known with great precision; and (b) the neglect of wavelength dependencies, which is commonly regarded as acceptable, is probably no less serious cause of error. Its practical realisability is its greatest asset; but it should NOT be represented or defined as providing more than plausible predictions, at least until it has been subjected to rigorous tests. Like LVEL, and the radiosity model, IMMERSOL is based upon intuition rather than rigorous analysis. It should therefore be treated with both respect and circumspection.*

5   Eulerian Two-Phase Flow Simulations with Virtual Mass Modelling

5.1   Introduction

In 1996, CHAM was commissioned by a European oil company to develop a PHOENICS-based CFD model to simulate the sudden release of natural gas from a break in a subsea pipeline. An underwater gas release produces a so-called bubble plume comprising a swarm of rising bubbles within a column of rising water, the flow is driven by the buoyancy of the gas. The CFD model allowed for the interaction of the bubble plume with the sea surface, including the deformation of the surface. Also, the transport of natural gas into the atmosphere was modelled, which is important for hazard considerations.

The liquid and gas phases were represented by a Eulerian–Eulerian approach and the two-phase system of equations was solved using IPSA [70]. Specifically, conservation equations were solved for phase continuity and momentum, the mass fraction of natural gas, as well as additional liquid-phase transport equations for the turbulence parameters k and ε, where k is the turbulent kinetic energy and ε is its dissipation rate. Computations were also made using Large eddy simulation (LES) to solve directly for the large-scale fluctuating motion. With LES, Smagorinsky's sub-grid-scale model was employed for the unresolved small-scale motion. The phase momentum equations included the additional effects of lift, interfacial pressure and virtual mass forces.

The physical significance of the virtual mass (also called added mass) force in the momentum equations for dispersed two-phase flow is that it represents the force required to accelerate the apparent mass of the surrounding continuous phase in the immediate vicinity of the dispersed phase. These effects are significant if the dispersed phase density ρ_d is much smaller than the density ρ_c of the carrier phase, as in the motion of gas bubbles through liquids. Specifically, in the present application, the added mass will influence the prediction of the plume rise time to the surface.

The CFD model produced results that were in agreement with laboratory observations of a steady-source bubble plume. The main features were a conical-shaped plume undergoing fluctuating horizontal excursions (with LES) before striking the surface to form a radially outflowing surface flow with a fountain (rise of water surface) in the central boil area; and the emergence and dispersion of the gas plume into the atmosphere. As anticipated, the influence of the virtual mass sources on the solution was to increase the time taken for the gas plume to reach the surface. However, it was discovered that good monotonic convergence could be achieved only in the absence of virtual mass forces, whereas with them, typically, the whole field residuals would oscillate downwards to convergence in triple the number of iterations per time step. This led to a significant increase in computer time for the entire transient run, which was carried out in Parallel mode using MPI on a 4-CPU SGI Origin 2000. Consequently, a more economical remedy was needed than simply reducing the time step, or increasing the linear relaxation applied to the explicitly coded source terms for the virtual mass. It was here that Brian was consulted, who in turn proposed [98] and implemented himself an approximate formulation of the virtual mass source terms amenable to implicit linearisation in the discretised momentum equations. This treatment of the source terms will now be described by replicating below Brian's original analysis.

5.2 Momentum Equations

So as to provide context for Brian's virtual mass analysis, the phase momentum equations are reproduced below:

$$\frac{\partial}{\partial t}(\rho_k r_k U_k) + \nabla.(\rho_k r_k \mathbf{U_k U_k}) = \nabla.\left(\rho_k r_k \nu_{e,k} \nabla U_k\right) + \nabla \cdot \left(\rho_k r_k \frac{\nu_t}{\sigma_r} U_k \nabla r_k\right)$$
$$-r_k \nabla p + r_k \rho_k g + M_{d,k} + M_{vm,k} + M_{l,k} + M_{p,k} \tag{17}$$

where: ρ_k, r_k and $\mathbf{U_k}$ denote the density, volume fraction and velocity vector of phase k, with $k = c$ for the continuous phase (liquid) and $k = d$ for the dispersed phase (gas); σ_r is the turbulent Schmidt number for volume fractions; p is the static pressure; $\mathbf{g}$ is the gravitational acceleration; $\nu_{e,k}$ is the effective viscosity for phase k; and $\mathbf{M_{d,k}}$, $\mathbf{M_{vm,k}}$, $\mathbf{M_{l,k}}$ and $\mathbf{M_{p,k}}$ are the interfacial forces per unit volume due to drag, virtual mass, lift and interfacial pressure.

5.3 Virtual Mass Terms

The virtual mass terms take the following form:

$$\mathbf{M}_{vm,c} = \rho_c C_{vm} r_d \mathbf{a}_{vm}; \ \mathbf{M}_{vm,d} = -\mathbf{M}_{vm,c} \tag{18}$$

where C_{vm} is the virtual mass coefficient, which describes the volume of displaced fluid that contributes to the effective mass of the dispersed phase; r_d is the volume fraction of phase d; and $\mathbf{a}_{vm}$ is the virtual mass acceleration vector, given by:

$$\mathbf{a}_{vm} = \frac{D}{Dt}(\mathbf{U}_d - \mathbf{U}_c) = \frac{\partial}{\partial t}(\mathbf{U}_d - \mathbf{U}_c) + (\mathbf{U}_d.\nabla\mathbf{U}_d - \mathbf{U}_c.\nabla\mathbf{U}_c) \tag{19}$$

where D/Dt is the material derivative; and $\mathbf{U}_d$ and $\mathbf{U}_c$ are the velocity vectors of phase d and c, respectively. The coefficient C_{vm} is likely to be a function of r_d, but most workers use $C_{\text{vm}} = 0.5$, as for spherical bubbles.

5.4 The Term to Be Implemented

From Eqs. (18) and (19), the virtual mass source term to be introduced in the momentum equations is of the following form:

$$T = C_{\text{vm}} \int \left\{ r_d \rho_c \left[\frac{D}{Dt} (U_d - U_c) \right] dV \right\} \tag{20}$$

where the integral refers to integration over the cell volume. This term must be subtracted from the net source in the dispersed phase (light) momentum equation, and added to the net source in the continuous-phase (heavy) momentum equation.

5.5 The Term to Be Calculated

During the formulation of the coefficients and contributions to the momentum imbalances, convection (through space and time) coefficients (c) are computed, and $\sum c_{d,n}(U_{d,n} - U_d)$ is added into the Su source term of the finite-volume equation, which is solved in correction form in PHOENICS. Here, the subscript n stands for 'neighbour' in space and time. The term $\sum c_{d,n}(U_{d,n} - U_d)$ has the significance of

$$- \int \left\{ r_d \rho_d \frac{D}{Dt} U_d dV \right\} \tag{21}$$

and $\sum c_{c,n}(U_{c,n} - U_c)$ has the corresponding significance for the continuous phase. Thus, the virtual mass term T can be expressed approximately as:

$$T = C_{\text{vm}} \left[-\frac{\rho_c}{\rho_d} \sum c_{d,n}(U_{d,n} - U_d) + \frac{r_d}{r_c} \sum c_{c,n}(U_{c,n} - U_c) \right] \tag{22}$$

The word approximately is appropriate because the volume fractions and densities used in formulating the $c_{d,n}$'s and $c_{c,n}$'s are neighbour cell rather than in-cell values. This approximation is insignificant in comparison with the uncertainty regarding the proper value of the virtual mass coefficient C_{vm}.

The finite-volume form of the phase momentum equations can be expressed in correction form as:

$$U'_d = [\text{Su}_d + f(U_c - U_d) - T]/A_{p,d} \tag{23}$$

$$U'_c = [Su_c + f(U_d - U_c) + T]/A_{p,c} \tag{24}$$

where f is the interphase-friction coefficient, $A_{p,d}$ and $A_{p,c}$ are the central coefficients, Su_d and Su_c are all other terms in the corresponding phasic momentum equations, and the $'$ denotes the correction to be applied.

5.6   Linearisation

In the interests of convergence, T is expressed as

$$T = T^* + T' \tag{25}$$

where the $*$ denotes the value based upon current values, and the correction T' is given by

$$T' = gU'_d - GU'_C \tag{26}$$

with

$$g = C_{vm}(\varrho_c/\varrho_d)\sum c_{d,n} \tag{27}$$

and

$$G = C_{vm}(r_d/r_c)\sum c_{c,n} \tag{28}$$

If Eqs. (25) and (26) are substituted in Eqs. (23) and (24), the following linearised form of the momentum equations is obtained:

$$U'_d = \frac{\left[Su_d + f(U_c - U_d) - T' + (f + G)U'_c\right]}{\left[A_{p,d} + f + g\right]} \tag{29}$$

$$U'_c = \frac{\left[Su_c + f(U_d - U_c) + T' + (f + g)U'_d\right]}{\left[A_{p,c} + f + G\right]} \tag{30}$$

5.7   The Partial Elimination Algorithm (PEA)

In PHOENICS, Eqs. (29) and (30) are combined and then rearranged for the solution by the PEA. First, Eqs. (29) and (30) are rewritten as:

$$U'_d = \left[B_d + (f + G)U'_c\right]/A_d \tag{31}$$

$$U'_c = \left[B_c + (f + g)U'_d\right]/A_c \tag{32}$$

and after substitution of Eq. (31) for U_d' into Eq. (32), one obtains:

$$U'_c = \frac{[B_c A_d + (f + g) B_d]}{[A_d A_c - (f + g)(f + G)]} \tag{33}$$

and

$$U'_d = \frac{[B_d A_c + (f + G) B_c]}{[A_d A_c - (f + g)(f + G)]} \tag{34}$$

as the final form of the phasic momentum equations in correction form.

5.8 Discussion

Brian's formulation of the virtual mass terms produced effectively the same results as the exact explicit implementation, but with the following benefits; the size of coding was reduced; no relaxation was required; convergence was promoted rather than threatened and the execution time was faster. Specifically, for the bubble plume simulations, the oscillatory-downward behaviour of the residual sums was eliminated; and the number of iterations per time step was reduced to roughly the same number as for runs without the virtual mass terms.

6 Extensions of PHOENICS Parabolic Solver to Hyperbolic and Transonic Flow

6.1 Introduction

During 1998 CHAM was awarded a contract by an Engineering Consultancy firm to extend the PHOENICS parabolic solver to handle both purely supersonic flow (hyperbolic) and supersonic jets issuing into a subsonic free stream (transonic) [99]. The motivation for this work was to effect massive savings in computer time for those rocket exhaust plume applications that would qualify from switching from elliptic computations to the space-marching integration procedure offered by the parabolic solver. Of course, marching integration, when permissible (high Reynolds number without flow reversal), also allows fine grid solutions with modest memory. The transonic parabolic option is useful when a supersonic jet emerges into a subsonic atmosphere, but it was recognised that applications involving embedded subsonic regions, such as those resulting from Mach-disc formation in highly under-expanded jets, could not be handled rigorously by the modified parabolic solver. Therefore, accurate simulations of these cases would remain the remit of the elliptic solver.

For the purposes of this contract, Brian devised and implemented very rapidly several algorithmic changes to the PHOENICS parabolic solver. The author's role was twofold, first to create numerous cases to test and validate the modified solver, and then to report the results to the client. These developments finally put the 'H' into PHOENICS, in the sense that the hyperbolic solution for a purely supersonic flow could now be obtained in a single marching-integration sweep, rather than by making repeated sweeps of the solution domain with the elliptic solver. The nature of the contract was such that the hyperbolic solver would be validated for relevant analytical cases, but for the transonic parabolic option, sufficient funding was available only to demonstrate capability by testing on a number of non-reacting 2d and 3d cases relevant to rocket exhaust plumes. Before discussing this work in more detail, a brief review will be given of Brian's earlier pioneering work in this field.

6.2   Earlier Work

It is well known that Brian and his group at Imperial College (IC) began working on the solution of the parabolic Navier–Stokes equations right at the dawn of CFD in the 1960s. A number of space-marching parabolic solvers were developed for both two- [100–103] and three-dimensional boundary layers [104]. The key developments were the non-iterative 2d parabolic boundary-layer procedure of Patankar and Spalding [101], which was later developed into GENMIX [103]; and the SIMPLE algorithm of Patankar and Spalding [104], which was applied first to 3d parabolic flows, but was soon adapted to elliptic problems [105]. In the 2d parabolic procedure, the primary transport equations were transformed into a von Mises coordinate system, with the longitudinal distance and dimensionless stream function as independent variables. This facilitated the automatic growth and contraction of the integration domain to match precisely the evolving boundary layer. Brian called this the 'Bikini' method, and it was achieved by way of suitable prescriptions of the entrainment rate at free boundaries. Also, the cross-stream momentum equation wasn't solved in the 2d procedure, but rather the cross-stream velocity was calculated from continuity. In the 3d parabolic procedure, both cross-stream momentum equations were solved, but the streamwise momentum equation was decoupled from them by using a uniform cross-sectional mean pressure. For unconfined flows, the streamwise variation of the mean pressure was obtained from the external free stream, whereas in confined flows, the mean pressure was adjusted automatically to satisfy overall mass conservation.

Soon after these developments and throughout the 1970s, parabolic/hyperbolic variants appeared for handling supersonic boundary-layer flows [106–110]. The impetus for most of this high-speed work stemmed from contracts placed with IC and CHAM for modelling either scramjet-combustor flow fields (as discussed for example by Drummond [111]) or rocket exhaust plumes. Specifically, the hyperbolic extensions of the parabolic solver involved solving for both the cross-stream pressure distribution and velocities so as to allow for the formation of shock and expansion waves. A notable code from this period was the rocket exhaust plume code REP3 [106] developed for the UK's RPE at Westcott. The modified versions of this code were still in use at PERME, Westcott during the 1980s [112] and DERA, Westcott in the late 1990s. Later, embedded transonic regions, without flow reversal were handled by adapting the partially parabolic method of Pratap and Spalding [113]. For example, Singhal and Spalding [114] computed 2d transonic flow through axial turbomachinery cascades, and Jennions simulated 2d axisymmetric supersonic under-expanded jets [115].

As the decade came to a close, Brian reviewed the various CHAM and IC codes then in current use for rocket plume applications [116]. These codes included the parabolic/hyperbolic code PAM2, which was an extension of GENMIX for 2d axisymmetric plumes; and PAM3 simulated 3d plumes on a cylindrical-polar mesh which adapted to contain the curved shape of the plume. PAM2 solved the radial momentum equation so that expansion and shock waves appeared as part of the solution. Both codes were equipped with several turbulence models, and there were physical models coded for the chemical kinetics, eddy-break-up turbulent combustion, particle growth and diminution and radiation with wavelength dependence.

In parallel with the development of these parabolic and hyperbolic codes, elliptic CFD codes for rocket exhaust plumes were also developed by Brian and co-workers so as to handle the region of flow separation behind the rocket base wall [105, 117–120].

6.3 PHOENICS Parabolic Solver

Brian's earlier concern with parabolic codes led to the slab-centred architecture of PHOENICS, which can work in both elliptic and parabolic modes [1]. The independent variables are x, y and z, and in parabolic flows, the solution is achieved by marching through space in the z-direction, starting from the slab of finite-volume cells at the low-z boundary. There is a considerable saving in memory requirement because storage need only be provided for the current and upstream slab of cells. For elliptic problems, when operating in the slabwise mode of solution, repeated sweeps are made in the z-direction through the entire solution domain, from the low-z to high-z boundary, until convergence. Brian [1] equipped PHOENICS with a parabolic solver capable of handling both 2d and 3d confined or unconfined flows in planar or cylindrical polar coordinates. Provision was made for users to vary the lateral dimensions of the solution domain with downstream distance z so as to contain the growing extent of the diffusion-influenced region; and to expand or contract according to whether the main flow velocity is decreasing or accelerating. The solution algorithm was similar to the original Patankar–Spalding method [104] except that SIMPLEST [121, 122] was used with iteration at each forward step in the marching-integration procedure.

6.4   *Hyperbolic-Flow Extension*

Brian's modifications to the parabolic solver for handling wholly supersonic flow will now be described. So as to allow the streamwise velocity w to be influenced by lateral pressure variations, the upstream streamwise pressure gradient was used for the solution of the w-momentum equation at the current slab. In addition, the sign of the pressure-correction influence coefficient $\partial w/\partial p$ was changed so as to obtain the correct velocity-pressure dependence in purely supersonic flow.

For expanding and contracting grids, the author identified the need to modify the cross-stream convection fluxes because under these conditions, the default fluxes ignored the contribution arising from the streamwise velocity, w. An important consequence was the ability to capture the formation of waves due to the turning of a supersonic flow. Specifically, the convection velocity through, say, the north face of a cell, v_c, must be computed from:

$$v_c = v_m - w \tan \alpha \tag{35}$$

where v_m is the y-direction velocity resulting from the solution of the momentum equation, w is the streamwise velocity and α is the local grid inclination of the cell face.

As in earlier studies [103, 107, 108, 123], the small-wave theory was applied at supersonic free boundaries to avoid wave reflection. The boundary condition was specified in terms of a prescribed mass efflux, so as to allow any outgoing waves to pass through the boundary. The mass efflux was derived from the theory of simple wave flow, which shows that there must be a unique relation between the turning angle of the flow θ; the approach velocity w; the Mach number M and the static pressure p.

6.5   *Transonic Extension to the Hyperbolic Solver*

The following extensions were made by Brian to the hyperbolic option so as to allow for supersonic jets to spread into subsonic regions:

1. Store and compute a Mach number variable M_z, based on the streamwise velocity w alone.
2. Multiply the influence coefficients of the pressure-correction equation, $\partial u/\partial p$ and $\partial v/\partial p$, by a constant β, where $\beta > 1$, wherever $M_z < 1$.
3. Use the local axial pressure-gradient $\partial p/\partial z$ in the w-momentum equation when $M_z > 1$, otherwise use the mean axial pressure gradient dp/dz as with subsonic flow.
4. Eliminate lateral-momentum convection in subsonic regions, i.e. wherever $M_z < 1$.

5. Wherever $M_z < 1$, set the influence coefficient $\partial w/\partial p = 0$. Otherwise if $M_z \geq 1$, retains the hyperbolic practice of changing the sign of $\partial w/\partial p$, so as to obtain the correct velocity-pressure dependence.

It can be seen that the transonic extension involves multiplying the cross-stream influence coefficients of the pressure-correction equation by a factor β. Basically, a test is made at each cell location as to whether the velocity is supersonic or subsonic; and then the multiplier is introduced so as to maintain uniformity of pressure across the flow in the subsonic region. For turbulent flow calculations with supersonic injection, a value of $\beta = 100$ was found suitable for the cases simulated in the contract. For sonic injection, a value of unity was recommended for β. For convergence at each forward step, it was necessary to solve the momentum equations in a point-by-point manner.

6.6 *Hyperbolic Applications*

The hyperbolic-flow solver was validated by considering the following analytical two-dimensional inviscid cases: inclined supersonic flow in a duct; supersonic flow in a diffuser and the merging of two supersonic free streams behind a splitter plate. For all of these cases, the conservation equations and equation of state were solved in dimensionless form by introducing the following reference values [123, 124]: length L; density ρ_o; velocity $a_o/\sqrt{\gamma}$; pressure p_o; enthalpy h_o $(\gamma - 1)/\gamma$ and temperature T_o. Here, L is a characteristic flow dimension; γ is the specific heat ratio, taken as 1.4; a is the acoustic velocity, and the subscript o denotes total conditions at an inlet boundary.

2d Inclined Supersonic Flow in a Duct. The first case considered was isentropic, supersonic flow at Mach 3 entering a plane duct of unit height L at an angle of $6°$ to the horizontal. The simulation employed 40 uniform cross-stream mesh cells, and the marching integration was carried out for one-duct height downstream using 40 uniform forward steps. For this type of flow, a compression wave is formed from the top corner of the duct inlet, and an expansion wave runs from the bottom corner of the duct inlet. For conditions downstream of these two waves, Table 1 shows that the results of the hyperbolic solver were in good agreement with the analytical results.

Table 1 Inclined supersonic flow in a duct: comparison of hyperbolic solver and analytical solution

	Ma	p/p_0	ρ/ρ_0
Expansion wave			
Hyperbolic solver	3.3	0.0166	0.1047
Analytical	3.3	0.0167	0.1045
Compression wave			
Hyperbolic solver	2.7	0.0424	0.0537
Analytical	2.7	0.0425	0.0542

Table 2 Supersonic diffuser flow: comparison of hyperbolic solver and analytical solution

	Ma_1	Ma_2	Ma_3	p_1/p_o	p_2/p_o	p_3/p_o
Analytical	2.0	1.71	1.42	0.128	0.197	0.293
Hyperbolic solver	2.0	1.75	1.45	0.128	0.192	0.293

Table 3 Merging of two supersonic streams: comparison of the hyperbolic solver and analytical solution

	Upper expansion wave		Lower compression wave	
	M_1	M_3	M_2	M_3
PHOENICS	3.0	3.26	2.0	1.82
Analytical	3.0	3.28	2.0	1.80

2d Supersonic Diffuser Flow. The second case considered was isentropic super-sonic flow in a converging plane duct with a diffuser angle of $8°$ to the horizontal. The flow entered axially into a duct of unit inlet height with a Mach number of 2. The simulation employed 80 uniform mesh cells across the duct, and the marching integration was carried out for 2.3 duct heights downstream using 320 uniform forward steps. For this flow situation, a weak oblique shock wave (of angle $37°$) is generated at the top corner of the inlet, which is reflected from the bottom of the duct to arrive at the top corner of the exit plane. The comparison between the PHOENICS hyperbolic solver and analytical results for the conditions downstream of the two waves was very good, as indicated in Table 2.

2d Merging of Two Supersonic Free Streams. The third case considered was the inviscid merging of two supersonic streams downstream of a splitter plate. The top and bottom streams entered with Mach numbers of $M_1 = 3$ and $M_2 = 2$, respectively. The inlet streams had the same total temperature, but different static pressures p_1 and p_2, with $p_1/p_2 = 2$. The interaction of the two streams proceeds such that their pressures are equalised at p_3 far downstream of the plate. This results in the propagation of a shock wave through the low-pressure stream, and an expansion wave through the high-pressure stream. As a consequence of this process, both streams turn through an equal angle. Since both streams extend to infinity, the outgoing waves should continue to travel outwards, unhindered without reflection. These features were well predicted by the PHOENICS hyperbolic solver. The simulation employed 20 uni-form mesh cells above and below the splitter plate, and the marching integration was carried out for 1.5 heights downstream with 60 uniform steps. The ideal gas law was used in the simulations, and the energy equation was not needed because the total enthalpy is uniform everywhere.

The PHOENICS hyperbolic solver predicted a downstream deflection angle of $5°$ below the horizontal and $p_3/p_2 = 1.317$, which is in excellent agreement with the analytical result. The comparison between the predictions and the analytical results for the downstream Mach numbers on either side of the slip line was also very good, as can be seen from Table 3. The mean Mach line of the expansion fan is located 18.5° above the horizontal, and the compression wave is located 35° below the horizontal.

6.7   Transonic Applications

The satisfactory working of the modified parabolic solver for transonic flow was verified by considering a number of 2d and 3d under-expanded jets; and a typical 2d supersonic, rocket exhaust plume. The testing of under-expanded jets covered invis-cid and turbulent flow, sonic and supersonic injection, stagnant and moving subsonic and supersonic free streams; as well as cases involving expanding grids. There was insufficient contract funding to cover the costs of validation and grid-dependency studies, the requirement being to demonstrate capability by producing physically plausible solutions for comparison, where possible, with the corresponding results of the PHOENICS elliptic solver and the earlier CHAM code, REP3 [106, 112].

Some sample results are discussed here briefly for an axisymmetric, supersonic, under-expanded rocket exhaust plume discharging into a subsonic moving air stream, as defined by the client, but simplified by representing the multispecies exhaust gas as a single conserved scalar; and using temperature-independent specific heats. The energy equation was solved in terms of the stagnation enthalpy, and the density was determined from the ideal gas law. The turbulence was represented by means of the standard two-equation eddy viscosity k-ε model.

The exhaust gas had a specific heat ratio of 1.21, a Mach number of 2.39, a total pressure of 57.86 bar and a total temperature of 3462 K. The moving air stream had a specific heat ratio of 1.4, a Mach number of 0.029, a static temperature of 288 K and a static pressure of 1.01325 bar. The nozzle-to-ambient static pressure ratio was 3.8, and the nozzle diameter was 0.07265 m. The radial extent of the solution domain was 2.5 diameters, and the marching integration was carried out for a downstream distance of 12.5 diameters using a uniform forward step size of 1% of the domain radius. The calculations employed a mesh of 40 uniform cells across the nozzle, and a further 20 cells in the external stream concentrated towards the nozzle using geometric progression.

The computed Mach number contours were comparable with those produced by the PHOENICS elliptic solver and REP3. The predicted flow field contained two shock cells, and the flow structure was typical of an under-expanded jet. The exhaust gas was expanded rapidly by the rarefaction fan emanating from the nozzle; and the flow was turned away from the jet axis before being pushed back by the higher free-stream pressures. A converging conical shock wave (incident shock) was formed, which reflected back from the flow axis to form a reflected shock wave.

In later work, Smith and Taylor [125] applied the modified parabolic solver to simulate under-expanded rocket exhaust plumes for practical applications. For non-reacting plumes with weak to moderate (but not strong) shock waves, they reported that the parabolic solver provided results that matched those of the elliptic solver, but at a much shorter computer time—typically 8 times faster than the elliptic solver when predicting 3d plumes.

7   An Automatic Convergence-Promoting Algorithm

7.1   Introduction

For those engaged at the coalface of CFD, convergence difficulties with the solution procedure on new applications is a common problem; and Brian once humorously commented that he could tell from the worried looks on their faces in the corridor, which engineers had a convergence problem. In my experience, Brian certainly had a remarkable ability to resolve difficult convergence problems by devising novel algorithmic changes or relaxation devices, such as for example implicit relaxation through linearisation of challenging source terms. Examples of the latter include the virtual mass forces for two-phase flow, which was discussed earlier; the drift flux source terms for the multiphase algebraic slip model; source terms for chemically reacting flow; and explicit source and sink terms of any type that do not depend directly on the solved for dependent variable. This ability to address convergence problems was evident with Brian's embryonic automatic convergence-promoting algorithm for PHOENICS, which he named CONWIZ (Convergence Wizard) [126].

The concept of an automatic convergence-promoting algorithm was aired first by Brian at the Basel World User Days CFD Conference in 1992 [127]. It was part of his proposal that general-purpose CFD codes should be equipped with an expert system designed to procure economically, converged and reliable solutions for code users without specialist CFD knowledge. Since the CFD solution algorithm is iterative in nature, the idea was that the CFD code should become self-steering by making expert choices on the numerical inputs, which include: the type of meshing; the mesh fineness and distribution; the most appropriate algorithm for solving the equations; the type of solver for each variable; the initial guesses of field values for steady-state problems; the relaxation parameters; the solver settings and the iteration-termination (convergence cut-off) criteria. The convergence of the iteration process can be assisted greatly by careful selection of the various relaxation and solver options and their parameters, but even CFD specialists find numerical inputs hard to choose optimally. Moreover, such settings tend to be problem dependent, and so Brian argued that the general-purpose CFD code itself should choose the convergence-promoting settings on its own for any given flow simulation; and ideally there should be an in-flight adjustment of the relaxation parameters to increase the rate of convergence, and reduce the computer time.

Soon after the Basel Conference, Brian implemented his so-called EXPERT system into PHOENICS as an option. This made automatic 'in-flight' adjustments to the relaxation parameters in order to speed up the convergence of the solution procedure. This was a fairly rudimentary system being concerned mainly with monitoring the variation of the residual errors with iteration, and then adjusting the false time step and/or using over-relaxation in the solver. Like its predecessor SARAH (Self-Adjusting-Relaxation AlgoritHm), which also used false-time-step adjustment at runtime, EXPERT did not always improve the rate of convergence in all circumstances. Like all self-adaptive devices, sometimes it suffered from 'over-shoots' by making excessive changes in the effort to increase speed. Brian also felt that the use of false-time-step relaxation was undesirable because of its dimensional character, and therefore any further developments should concentrate on linear relaxation.

Despite these early explorations, it was not until 2004 that Brian devised and implemented CONWIZ into PHOENICS as a default option for setting numerical parameters for all flow simulations. This algorithm was designed to ensure convergence provided that the input data supplied by the code user were sufficient and self-consistent, thereby removing the need for users to set any numerical inputs associated with promoting convergence. This was an ambitious objective given the expansive range of applications that might be presented to a general-purpose CFD code.

CONWIZ will now be described, and liberal use will be made of Brian's lecture notes [126]. Although the discussion will focus on PHOENICS, CONWIZ should be transferable to any CFD methodology based on the finite-volume method with iterative solution by SIMPLE or one of its variants. The reader is reminded that PHOENICS is based on SIMPLEST [121, 122].

7.2 The Main Features of CONWIZ

The main features of CONWIZ can be summarized as follows:

1. It starts by estimating reference values of length L_r, velocity V_r, density ρ_r and temperature T_r.

2. From these, it deduces and sets some initial values of variables, including the '∂vel/∂p's', i.e. the rates of change of velocity with pressure difference at every point. For example, $(0.1V_r)^2$ is used for the turbulent kinetic energy k, and $C_d(0.1V_r)^3/(0.1L_r)$ for its dissipation rate ε.

3. It sets linear under-relaxation factors for all variables, including the '∂vel/∂p's', e.g. a factor of 0.5 is used for the pressure and velocities, and 0.25 for the energy equation.

4. It sets maximum values to the increments ($\Delta\phi_m$) per iteration for some variables, as well as the upper and lower limits for variables (ϕ_{max} and ϕ_{min}). For example, $\Delta\phi_m = 10V_r$ is used for the momentum equations, $\Delta\phi_m = (0.001V_r)^2$ for k, and V_r^3/L_r for ε.

5. It invokes the whole field solution of the momentum equations. The PHOENICS default is slabwise solution, unless performing parallel computations.

6. For reasons to be explained in the next section, it applies some background false-time-step relaxation of $\Delta t_f = L_r/V_r$ on the momentum equations.

If the code user wants to intervene and make their own relaxation settings, most of these will be respected by CONWIZ, but not false time steps for velocities.

7.3   The Ideas Underlying CONWIZ

Brian listed the following ideas as having provided guidance for his development of the algorithm:

1. SIMPLE-type CFD algorithms differentiate the momentum equations to deduce the influence coefficients of the pressure-correction equation as: $\partial\mathrm{vel}/\partial p = A/[\sum a_c + M/(\Delta t + \Delta t_f)]$ where A is the flow area, a_c the convection coefficient, M the mass in the cell and Δt the real time step.
2. In steady-flow problems the true time step, $\Delta t = 0$; and at the start of the simulation, the convection coefficients a_c may be 0; so, a finite false time step, Δt_f, is needed to keep $\partial\mathrm{vel}/\partial p$ finite; and so, prevent small pressure differences from producing enormous velocities; but these are not needed once the a_c's are finite because motion has started.
3. *The much-used constant false-transient relaxation (Δt_f) is thus now recognised as intrinsically bad; it should no longer be used.*
4. *An initial Δt_f should be based on an initial estimate of what the a_c sum will be; then linear-under-relaxation allows the momentum-equation-based values to take over gradually.*
5. *The fact that PHOENICS users have been successfully using constant-Δt_f strategies for a long time demonstrates the essential sturdiness of SIMPLEST.*
6. *Nevertheless, if any Δt_f is to remain in use, it should be deduced (by CONWIZ, of course) from local flow conditions and be based on monitored convergence behaviour. The same is true of other variables than velocity, except that the 'initial estimates' are harder for CONWIZ to make. For this reason, limitation of the size of the correction which may be made in any sweep provides a safer strategy.*

These then, were Brian's thoughts behind the strategies employed by CONWIZ, although he did indicate that further development would be needed because his thinking was far from complete. The discussion will now move on to other aspects and details of the convergence-promoting algorithm.

7.4   Other Features of CONWIZ

Correction Form of the Finite-Volume Equations. PHOENICS, and hence CONWIZ, has always used the correct form of the finite-volume equations, which for any variable ϕ may be written as

$$a_p\phi'_p = \sum a_i \phi'_p + R_\phi \tag{37}$$

where

$$\emptyset_P = \emptyset_p^* + \emptyset_p' \tag{38}$$

and ϕ_p^* is the previous iteration value of ϕ_p, and ϕ_p' is the correction needed to satisfy the original finite-volume equation. In Eq. (37), $a_p = \sum a_i$ with $\sum$ indicating summation over all neighbouring grid nodes i. The a_i are the neighbouring coefficients representing the combined effects of convection (including time) and diffusion for the neighbour node i. The residual error in Eq. (37) is defined by:

$$R_\phi = \sum a_i\left(\phi_i^* - \phi_p^*\right) + S_\phi \tag{39}$$

The source term S_ϕ in the above also includes contributions from boundary conditions; and it is represented in the following generic, linearised form:

$$S_\emptyset = T\left(C_\phi + \max\langle S_m, 0\rangle\right)\left(V_\phi - \emptyset_P\right) \tag{40}$$

where T is a multiplier (often geometrical), C_ϕ and V_ϕ are the coefficient and value for ϕ, and S_m represents a mass-flow source, and ϕ_p is the value of the variable at the node p. Linearisation of S_ϕ proceeds in the usual way by adding $T\left(C_\phi + \max < S_m, 0 >\right)$ to the central coefficient a_p of the finite-volume equation.

The advantage of solving equation (37) is that the coefficients need be only approximate, the sources are replaced by the residual errors R_ϕ in the original equation, and the corrections tend to zero as convergence is approached, reducing the possibility of round-off errors affecting the solution.

Linearisation of the Turbulence Model Source Terms. For engineering applications, turbulence effects are most commonly represented by the two-equation k-ε model in conjunction with wall functions. In the early days of PHOENICS, three different linearisation practices for the source terms of these equations were in current usage, but these were soon superseded by a very robust formulation proposed by Brian. This was based on a Newton–Raphson linearisation using the presumption that the local turbulence length scale changes only slowly. This option is used by CONWIZ, and the source terms per unit mass are linearised as follows:

$$S_k = \upsilon_t E - \varepsilon_p = C_k\left(V_k - k_p\right) + \upsilon_t E \tag{41}$$

$$S_\varepsilon = \frac{\varepsilon}{k}(C_{1\epsilon}\upsilon_t E - C_{2\epsilon}\varepsilon) = C_\varepsilon\left(V_\varepsilon - \varepsilon_p\right) + C_{1\epsilon}\upsilon_t Ef \tag{42}$$

with

$$C_k = \frac{3\varepsilon}{2k}; \quad V_k = \frac{k}{3}; \quad C_\varepsilon = 4C_{2\epsilon}\frac{\varepsilon}{3k}; \quad V_\varepsilon = \frac{\varepsilon}{4} \tag{43}$$

In the foregoing, E is the mean rate of strain, v_t is the kinematic turbulent viscosity, $C_{1\varepsilon}$ and $C_{2\varepsilon}$ are model coefficients, and f $(=\varepsilon/k)$ is the turbulent frequency.

Linearisation of Explicit Source Terms. Explicit source terms of relatively large magnitude have the potential to generate numerical instabilities. Examples are fires represented as volumetric heat sources, or solar loads applied as heat sources on surfaces. For positive or negative sources of magnitude Q, CONWIZ applies implicit relaxation through the following linearisation:

For $Q > 0$

$$C_\phi = \frac{|Q|}{\Delta\emptyset_m}; \quad V_\emptyset = \emptyset + \Delta\emptyset_m \tag{44}$$

For $Q < 0$

$$C_\phi = \frac{|Q|}{\Delta\emptyset_m}; \quad V_\emptyset = \emptyset - \Delta\emptyset_m \tag{45}$$

where $\Delta\phi_m$ is the maximum allowable increment of ϕ per iteration.

Relaxation of the Influence Coefficients. CONWIZ uses harmonic under-relaxation of the $\partial\mathrm{vel}/\partial\mathrm{p}$'s, as follows:

$$\Phi = \frac{\Phi_n \Phi_o}{(\Phi_n + \alpha[\Phi_o - \Phi_n])} \tag{46}$$

where Φ signifies $\partial\mathrm{vel}/\partial p$, α is the relaxation factor set to a default value of 0.5, and the subscripts o and n denote the old and latest iterate values. Provision was also made to set upper and lower limits on the $\partial\mathrm{vel}/\partial p$'s.

7.5 Monitoring Convergence

As part of the CONWIZ initiative, Brian introduced two additional measures for monitoring convergence. The first monitored the change of the minimum and maximum values of each field variable with sweep. The second monitored the maximum absolute corrections to each variable. The latter has proved particularly useful, not only for monitoring convergence, but also for diagnosing the cause of convergence problems. In effect, the maximum absolute corrections represent the maximum change in a variable over the whole domain. Once the largest correction falls to zero, or at least to a negligible fraction of the value being corrected, it is reasonable to assume that convergence has been achieved, even if the sum of the residuals has not fallen below any specific level.

Whilst developing CONWIZ, Brian made extensive use of the maximum-absolute-correction mode of monitoring convergence, as opposed to relying on the long-established practices of monitoring the sum of absolute residuals, selected spot values and overall imbalances with sweep. Brian argued that the correction mode was much more informative than the 'spot-value' mode because one cannot know in advance which are the important cells to monitor during the course of the solution. He also expressed the view that the 'residual error' plots may be misleading, especially towards the end of a run. Specifically, Brian commented: *It has become apparent that the computed sum of residuals may include a large contribution from cells for which the balance is already as good as the round-off error the computer allows. This practice has two ill-effects, namely: it causes a solution which has truly converged to appear to the user as though it has not; and it causes the calculation process to continue to produce never-diminishing corrections which have no more significance than 'noise'.* Almost a decade later, Brian [128] again emphasized that the residuals were imperfect measures of convergence, and that the corrections were more physically meaningful. As an example, the deficiencies of the former were illustrated for situations where fixed-value boundary conditions are implemented through a source term with a large coefficient.

At CHAM, the maximum absolute corrections have proved to be a powerful tool for investigating convergence difficulties, especially for large, complex engineering applications. This supports Brian's observation that *watching the maximum absolute corrections is of especial value when one finds that some of them do not fall to acceptably low values. For this is a sign that something may be wrong with the problem set-up (or, of course, with the CFD code). PHOENICS has means of printing out not only how big are the maximum absolute corrections for each variable, but also where they occur. Its use points out where the investigation of the (convergence) problem can usefully begin.*

7.6 Discussion

For the last fifteen years, CONWIZ has been used extensively for CFD applications across a wide range of industries; and experience has shown that whilst it doesn't always procure monotonic convergence, it often performs pretty well. It has also been known to procure convergence in some cases when all else has failed. The convergence is somewhat slower than when using manual settings established by CFD specialists for particular types of application, but more rapid convergence is not to be expected of CONWIZ in these circumstances. This is because CONWIZ makes conservative rather than optimal choices of the relaxation settings—its purpose being to secure a converged solution for non-CFD experts. For cases where CONWIZ experiences convergence difficulties not related to code and/or user problems, tighter relaxation, sometimes by use of Brian's then new maximum-absolute-correction limitation has proved successful.

Brian never returned to continue his work on CONWIZ, but he did suggest that there was plenty of scope to improve its performance, for example as follows:

a. *By systematically inspecting all geometry-affecting settings, and all boundary and (physically significant initial) conditions, so as to make better estimates of reference lengths, velocities and densities.*
b. *By increasing horizontal-direction vertical-velocity coefficients in regions where un-balanced horizontal gradients of gravity are likely to provoke large vertical-velocity corrections.*
c. *By detecting regions in which corrections exhibit periodicity, and applying appropriate damping.*
d. *Allowing better for low-Reynolds-number effects.*

Although Brian included two-phase Eulerian–Eulerian flows in CONWIZ, no attention was paid to Eulerian–Lagrangian descriptions, other than stabilisation of the interphase-source terms along particle trajectories via the explicit source linearisation offered by CONWIZ.

8 Concluding Remarks

This article has reviewed some of the contributions to CFD made by Professor Brian Spalding whilst at CHAM during the period 1993–2004. The following topics were covered and reviewed: the differential-equation wall-distance calculator; the LVEL model of turbulence; the IMMERSOL model of thermal radiation; the implementation of virtual mass terms in Eulerian–Eulerian descriptions of two-phase flow; the provision of a space-marching method for hyperbolic and transonic flow in the general-purpose CFD code PHOENICS and an automatic convergence-promoting algorithm. It is hoped that these descriptions of Brian's lesser-known contributions will serve to illustrate further his impressive technical creativity over a wide range of subject matter.

Brian was an original thinker and a remarkable engineer and scientist, but first and foremost Brian thought like an engineer. The emphasis was always on getting an adequate solution within reasonable computing time. In essence, his approach was to seek generic and economical solutions to practical engineering problems by making approximations based on a mixture of mathematical flair and impressive physical insight. As we have seen with the work on the virtual mass sources, parabolic-solver extensions and the automatic convergence algorithm, Brian's approach could also be very hands-on; by which is to say he was not content with just inventing and formulating new models and features, and then instructing others engaged in their implementation, because often he would code and test the entire method himself.

References

1. Spalding, D. B. (1981). A general-purpose computer program for multi-dimensional one- and two-phase flow. *Mathematics and Computers in Simulation, 23,* 267–276.
2. Patankar, S. V., Pollard, A., Singhal, A. K., & Vanka, S. P. (1983). *Numerical prediction of flow, heat transfer, turbulence and combustion: selected works of Professor D. Brian Spalding.* Oxford: Pergamon Press.
3. Artemov, V., Escudier, M. P., Fueyo, N., Launder, B. E., Leonardi, E., Malin, M. R., et al. (2009). A tribute to D.B. Spalding and his contributions in science and engineering. *International Journal of Heat and Mass Transfer, 52,* 3884–3905.
4. Runchal, A. K. (2009). Brian Spalding: CFD and reality—A personal recollection. *International Journal of Heat and Mass Transfer, 52,* 4063–4073.
5. Runchal, A. K. (2013). Emergence of computational fluid dynamics at Imperial College (1965–1975): A personal recollection. *ASME Journal of Heat Transfer, 135*(1), 011009-1.
6. Runchal, A. K.(2017). Origins and development of the finite volume CFD method at Imperial College. In *CHT-17,* 29 May–2 June, Naples, Italy.
7. Launder, B. E., Patankar, S. V., & Pollard, A. (2019). Dudley Brian Spalding. 9 January 1923–27 November 2016. *Biographical Memoirs of Fellows of the Royal Society, 66,* Article ID: 20180024.
8. Spalding, D. B. (1993). A turbulence length-scale formulation. CHAM Technical Note 4/9/93, CHAM, Wimbledon, London.
9. Spalding, D. B. (1994). *Calculation of turbulent heat transfer in cluttered spaces.* Presented at the 10th International Heat Transfer Conference, Brighton, UK.
10. Tucker, P. G. (1998). Assessment of geometric multilevel convergence robustness and a wall distance method for flows with multiple internal boundaries. *Applied Mathematical Modelling, 22,* 293–311.
11. Fares, E., & Schroder, W. A. (2002). Differential equation to determine the wall distance. *International Journal for Numerical Methods in Fluids, 39,* 743–762.
12. Tucker, P. G. (2003). Differential equation-based wall distance computation for DES and RANS. *Journal of Computational Physics, 190*(1), 229–248.
13. Tucker P. G, Rumsey, C. L, Bartels R. E., & Biedron R. T. (2003). Transport equation-based wall distance computations aimed at flows with time-dependent geometry. NASA TM-2003-212680, December.
14. Tucker, P. G., Rumsey, C. L., Spalart, P. R., Bartels, R. E., & Biedron, R. T. (2005). Computations of wall distances based on differential equations. *AIAA Journal, 43*(3), 539–549.
15. Tucker, P. G. (2011). Hybrid Hamilton–Jacobi–Poisson wall distance function model. *Computers & Fluids, 44,* 130–142.
16. Xu, J., Yan, C., & Fan, J. (2011). Computations of wall distances by solving a transport equation. *Applied Mathematics and Mechanics, 32*(2), 141–150.
17. Belyaev, A. G., & Fayolle, P. A. (2015). On variational and PDE-based distance function approximations. *Computer Graphics Forum, 34*(8), 104–118.
18. Wukie, N. A., & Orkwis, P. D. (2017). A p-Poisson wall distance approach for turbulence modelling. In *AIAA 2017-3945, 23rd AIAA CFD Conference,* 5–9 June, Denver, Colorado, USA.
19. Jefferson-Loveday, R. J. (2017). Differential-equation based specification of turbulence integral length scales for cavity flows. *Journal of Engineering for Gas Turbines and Power, 139*(6).
20. Watson, R. A., Trojak, W., & Tucker, P. G. (2018). A simple flux reconstruction approach to solving a Poisson equation to find wall distances for turbulence modelling. In *2018 Fluid Dynamics Conference, AIAA Aviation Forum (AIAA 2018-4261),* Atlanta, Georgia.
21. Boger, D. A. (2001). Efficient method for calculating wall proximity. *AIAA Journal, 39*(12), 2404–2406.

22. Van der Weide, E., Kalitzin, G., Schluter, J., & Alonso, J. J. (2006). Unsteady turbomachinery computations using massively parallel platforms. In *44th AIAA Aerospace Sciences Meeting and Exhibit, AIAA Paper 2006-0421*, Reno, NV.

23. Roget, B., & Sitataman, J. (2012). Wall distance search algorithm using voxelised marching spheres. In *7th International Conference on CFD (ICCFD7)*, Big Island, Hawaii, July 9–13.

24. Lohner, R., Sharov, D., Luo, H., & Ramamurthi, R. (2001). Overlapping unstructured grids. In *AIAA 2001-0439*, Reno, NV.

25. Tucker, P. G. (2016). Section 7.6.3 Nearest wall distance. In *Advanced computational fluid and aerodynamics*. Cambridge: Cambridge University Press.

26. Agonafer, D., Gan-Li, L., & Spalding, D. B. (1996). The LVEL turbulence model for conjugate heat transfer at low Reynolds numbers. In *Proceedings of the EEP Application of CAE/CAD to Electronic Systems, ASME International Mechanical Engineering Congress and Exposition*, Atlanta, GA.

27. Spalding, D. B. (1961). A single formula for the law of the wall. *ASME Journal of Applied Mechanics, 28*(3), 455–458.

28. Dhinsa, K. K., Bailey, C. J., & Pericleous, K. A. (2004). Turbulence modelling and its impact on CFD predictions for cooling of electronic components. In *Proceedings of 9th Intersociety Conference Thermal and Thermomechanical Phenomena in Electronic Systems*.

29. Dhinsa, K. K. (2006). *Development and application of low Reynolds number turbulence models for air-cooled electronics*. Ph.D. thesis, University of Greenwich, London, UK.

30. Rodgers, P., Lohan, J., Eveloy, V., Fager, C. M., & Rantala, J. (1999). Validating numerical predictions of component thermal interaction on electronic printed circuit boards in forced convection air flows by experimental analysis. *Advanced Electronic Packaging, 1*, 999–1008.

31. Eveloy, V. C. (2003). *An experimental assessment of CFD predictive accuracy for electronic component operational temperatures*. Ph.D. thesis, Dublin City University, Ireland.

32. Eveloy, V., Rodgers, P., & Hashmi, M. S. J. (2003). An experimental assessment of computational fluid dynamics predictive accuracy for electronic component operational temperature. In *Proceedings of the ASME Heat Transfer Conference*, Las Vegas, Nevada, USA, Paper Number HT2003-47282.

33. Rodgers, P., Eveloy, V., & Davies, M. (2003). An experimental assessment of numerical predictive accuracy for electronic component heat transfer in forced convection: Parts I and II. *Transactions of the ASME, Journal of Electronic Packaging, 125*(1), 67–83.

34. Choi, J., Kim, Y., Sivasubramaniam, A., Srebic, J., Wang, Q., & Lee, J. (2008). A CFD-based tool for studying temperatures in rack-mounted servers. *IEEE Transactions on Computers, 57*(8) 1129–1142.

35. De Marchi Neto, I., & Altemani, C. A. C. (2017). A matrix to evaluate the conjugate cooling of a heaters' array. *International Journal of Thermal Sciences, 118*, 278–291.

36. Dhoot, P., Healey, C. M., Pardey, Z., & van Gilder, J. W. (2017). Zero-equation turbulence models for large electrical and electronics enclosure applications. LV-17-C078. In *ASHRAE Winter Conference*, Las Vegas, NV, USA.

37. Wang, S., & Zu, D. (2003). Application of CFD in retrofitting air-conditioning systems in industrial buildings. *Energy and Buildings, 35*, 893–902.

38. Favarolo, P. A., & Manz, H. (2005). Temperature-driven single-sided ventilation through a large rectangular opening. *Building and Environment, 40*, 689–699.

39. Pfeiffer, A., Dorer, V., & Weber, A. (2008). Modelling of cowl performance in building simulation tools using experimental data and computational fluid dynamics. *Building and Environment, 43*, 1361–1372.

40. Myhren, J. A., & Holmberg, S. (2008). Flow patterns and thermal comfort in a room with panel, floor and wall heating. *Energy and Buildings, 40*, 524–536.

41. Myhren, J. A., & Holmberg, S. (2009). Design considerations with ventilation-radiators: Comparisons to traditional two-panel radiators. *Energy and Buildings, 41*, 92–100.

42. Yoo, S.-H., & Manz, H. (2011). Available remodelling simulation for a BIPV as a shading device. *Solar Energy Materials and Solar Cells, 95*, 394–397.

43. Wang, F., Manzanares-Bennett, A., Tucker, J., Roaf, S., & Heath, N. (2012). Feasibility study on solar-wall systems for domestic heating—An affordable solution for fuel poverty. *Solar Energy, 86,* 2405–2415.

44. Jurelionis, A., Gagytea, L., Seduikytea, L., Prasauskas, T., Ciuzas, D., & Martuzevicius, D. (2016). Combined air heating and ventilation increases risk of personal exposure to airborne pollutants released at the floor level. *Energy and Buildings, 116,* 263–273.

45. Mathioulakis, E., Karathanos, V. T., & Belessiotis, V. G. (1998). Simulation of air movement in a dryer by computational fluid dynamics: Application for the drying of fruits. *Journal of Food Engineering, 36,* 183–200.

46. Tchouveleva, A. V., Cheng, Z., Agranat, V. M., & Zhubrin, S. V. (2007). Effectiveness of small barriers as means to reduce clearance distances. *International Journal of Hydrogen Energy, 32,* 1409–1415.

47. Hourri, A., Angers, B., Benard, P., Tchouvelev, A., & Agranat, V. (2011). Numerical investigation of the flammable extent of semi-confined hydrogen and methane jets. *International Journal of Hydrogen Energy, 36,* 2567–2570.

48. Wang, H., Djambazov, G., Pericleous, K. A., Harding, R. A., & Wickins, M. (2011). Modelling the dynamics of the tilt-casting process and the effect of the mould design on the casting quality. *Computers & Fluids, 42,* 92–101.

49. Wang, H., Wang, S., Wang, X., & Li, E. (2015). Numerical modelling of heat transfer through casting–mould with 3D/1D patched transient heat transfer model. *International Journal of Heat and Mass Transfer, 81,* 81–89.

50. Chen, C., Jonsson, L. T. I., Tilliander, A., Cheng, G., & Jönsson, P. G. (2015). A mathematical modelling study of the influence of small amounts of KCl solution tracer son mixing in water and residence time distribution of tracers in a continuous flow reactor-metallurgical tundish. *Chemical Engineering Science, 137,* 914–937.

51. Solhed, H., Jonsson, L., & Jönsson, P. (2002). A theoretical and experimental study of continuous-casting tundishes focusing on slag-steel interaction. *Metallurgical and Materials Transactions, B33B*(2), 173–185.

52. Solhed, H., Jonsson, L., & Jönsson, P. (2008). Modelling of the steel/slag interface in a continuous casting tundish. *Steel Research International, 79*(5), 348–357.

53. Artemov, V. I., Minko, K. B., & Yankov, G. G. (2015). Numerical simulation of fluid flow in an annular channel with outer transversally corrugated wall. *International Journal of Heat and Mass Transfer, 90,* 743–751.

54. Tucker, P. G., & Liu, Y. (2007). Turbulence modelling for flows around convex features giving rapid eddy distortion. *International Journal of Heat and Fluid Flow, 28,* 1073–1091.

55. Spalding, D. B. (1994). Proposal for a diffusional radiation model for attachment to PHOENICS. CHAM Technical Note 18/10/94, CHAM, Wimbledon, London, UK.

56. Spalding, D. B. (1996). Radiation in PHOENICS HOTBOX, FLAIR, etc. CHAM Technical Note 4/9/96, CHAM, Wimbledon, London, UK.

57. Spalding, D. B. (1996). Immersed-solid heat transfer. CHAM Technical Note 11/9/96, CHAM, Wimbledon, London, UK.

58. Lockwood, F. C., & Shah, N. G. (1981). A new radiation method for incorporation in general combustion prediction procedures. In *Proceedings of the 18th International Symposium on Combustion* (pp. 1405–1414). London: The Combustion Institute.

59. Rosseland, S. (1936). *Theoretical astrophysics: Atomic theory and the analysis of stellar atmospheres and envelopes.* Clarendon Press.

60. Hamakar, H. C. (1947). Radiation and heat conduction in a light-scattering material. *Philips Research Reports, 2,* 55–67.

61. Schuster, A. (1905). Radiation through a foggy atmosphere. *Astrophysical Journal, 21,* 1–22.

62. Spalding, D. B (1980). Lecture 9, Idealisations of radiation. In *Mathematical modelling of fluid-mechanics, heat-transfer and chemical-reaction processes: A lecture course.* HTS/80/1, Mech. Eng. Dept., Imperial College, University of London.

63. Siegel, R., & Howell, J. R. (1992). *Thermal radiation heat transfer* (3rd ed.). Washington DC, USA: Hemisphere Publishing Corporation.

64. Eddington, A. (1916). On the radiative equilibrium of the stars. *Monthly Notices of the Royal Astronomical Society, 77*, 16–35.
65. Marshak, R. E. (1947). Note on the spherical harmonics methods as applied to the Milne problem for a sphere. *Physical Review, 71*, 443–446.
66. Deissler, R. G. (1964). Diffusion approximation for thermal radiation in gases with jump boundary condition. *ASME Journal Heat Transfer*, 240–246.
67. Liu, F. M., & Swithenbank, J. (1990). Modelling radiative heat transfer in pulverised coal-fired furnaces. In M. G. Carvalho, F. Lockwood, & J. Taine (Eds.), *Heat transfer in radiating and combusting systems. Proceedings of the EUROTHERM* (Vol. 17, pp. 358–373). Cascais, Portugal: Springer.
68. Spalding, D. B. (1995). Modelling convective, conductive and radiative heat transfer. Lecture LE3-1 in Industrial Computational Fluid Dynamics. Lecture Series 1995-03, Von Karman Institute for Fluid Dynamics, Belgium, April 3–7.
69. Spalding, D. B. (2013). Chapter 1, trends, tricks, and try ons in CFD/CHT, Section 3.1. The IMMERSOL radiation model. In E. M. Sparrow, Y. I. Cho, J. P. Abraham, & J. M. Gorman (Eds.), *Advances in heat transfer* (Vol. 45, pp. 1–78). Burlington: Academic Press.
70. Spalding, D. B. (1980). Numerical computation of multi-phase flow and heat transfer. In C. Taylor & K. Morgan (Eds.), Recent advances in numerical methods in fluids (pp. 139–167). Swansea: Pineridge Press.
71. Spalding, D. B. (2002). PEA for IMMERSOL. CHAM Technical Notes 23/7/02 & 24/07/02, CHAM, Wimbledon, London, UK.
72. Rasmussen, N. B. K. (2002). The composite radiosity and gap (CRG) model of thermal radiation. In: *Proceedings of the 6th European Conference on Industrial Furnaces and Boilers (INFUB-6) 2002 Conference*, Estoril, Lisbon, Portugal, 2002. (Also published as Danish Gas Technology Centre Report No. CO201, Hørsholm, Denmark.)
73. Osenbroch, J. (2006). *CFD study of gas dispersion and jet fires in complex geometries*. Ph.D. Thesis, The Faculty of Engineering and Science, Aalborg University, Denmark.
74. Yang, Y., de Jong, R. A., & Reuter, M. (2005). Use of CFD to predict the performance of a heat treatment furnace, In *Proceedings of the 4th International Conference on CFD in the Oil and Gas, Metallurgical and Process Industries*, Trondheim, Norway (pp. 1–9).
75. Yang, Y., de Jong, R. A., & Reuter, M. (2007). CFD prediction for the performance of a heat treatment furnace. *Progress in Computational Fluid Dynamics, An International Journal, 7*(2–4), 209–218.
76. Zhubrin, S. V. (2009). Discrete reaction model for composition of sooting flames. *International Journal of Heat and Mass Transfer, 52*, 4125–4133.
77. Aloqaily, A. M., & Chakrabarty, A. (2010). Jet flame length and thermal radiation: Evaluation with CFD simulations. In *Global Congress on Process Safety*. San Antonio, TX: AIChE.
78. Chakrabarty, A., & Aloqaily, A. (2011). Using CFD to assist facilities comply with thermal hazard regulations such as new API RP-752 recommendations. Hazards XXII, AICheE. Symp. Series No. 156.
79. Chakrabarty, A., Edel, M., Raibagkar, A., & Aloqaily, A. (2011). Thermal hazard evaluation for process buildings using CFD analysis techniques. In *AIChE Annual Meeting, Conference Proceedings* (Vol. 29).
80. Agranat, V., & Perminov, V. (2016). Multiphase CFD model of wildland fire initiation and spread. In *Proceedings of the 5th International Fire Behavior and Fuels Conference*, April 11–15, Portland, Oregon, USA.
81. Corbin, C. D., & Zhai, Z. J. (2010). Experimental and numerical investigation on thermal and electrical performance of a building integrated photovoltaic–thermal collector system. *Energy and Buildings, 42*, 76–82.
82. Chiang, W. H., Wang, C. Y., & Huang, J. S. (2012). Evaluation of cooling ceiling and mechanical ventilation systems on thermal comfort using CFD study in an office for subtropical region. *Building and Environment, 48*, 113–127.
83. Radhi, H., Fikiry, F., & Sharples, S. (2013). Impacts of urbanisation on the thermal behaviour of new built up environments: A scoping study of the urban heat island in Bahrain. *Landscape and Urban Planning, 113*, 47–61.

84. Maragkogiannis, K., Kolokotsa, D., Maravelakis, E., & Konstantara, A. (2014). Combining terrestrial laser scanning and computational fluid dynamics for the study of the urban thermal environment. *Sustainable Cities and Society, 13,* 207–216.

85. Radhi, H., Sharples, S., & Assem, E. (2015). Impact of urban heat islands on the thermal comfort and cooling energy demand of artificial islands—A case study of AMWAJ Islands in Bahrain. *Sustainable Cities and Society, 19,* 310–318.

86. Zhang, L., Zhang, L., Jin, M., & Liu, J. (2017). Numerical study of outdoor thermal environment in a university campus in summer. *Procedia Engineering, 205,* 4052–4059.

87. Radhi, H., Sharples, S., & Fikiry, F. (2013). Will multi-facade systems reduce cooling energy in fully glazed buildings? A scoping study of UAE buildings. *Energy and Buildings, 56,* 179–188.

88. Zhang, L., Jin, M., Liu, J., & Zhang, L. (2017). Simulated study on the potential of building energy saving using the green roof. *Procedia Engineering, 205,* 1469–1476.

89. Hien, H. N., & Istiadji, A.D. (2003). Effects of external shading devices on daylighting and natural ventilation. In *Proceedings of the 8th International IBPSA Conference*, Eindhoven, The Netherlands (pp. 475–482).

90. Vaidya, A. M., Maheshwari, N. K., & Vijayan, P. K. (2010). Estimation of fuel and clad temperature of a research reactor during dry period of de-fueling operation. *Nuclear Engineering and Design, 240,* 842–849.

91. Kuriyama, S., Takeda, T., & Funatani, S. (2015). Study on heat transfer characteristics of the one side-heated vertical channel with inserted porous materials applied as a vessel cooling system. *Nuclear Engineering and Technology, l47,* 534–545.

92. Zamora, B., & Kaiser, A. S. (2012). Influence of the variable thermophysical properties on the turbulent buoyancy-driven airflow inside open square cavities. *Heat and Mass Transfer, 48,* 35–53.

93. Zamora, B., & Kaiser, A. S. (2016). Radiative effects on turbulent buoyancy-driven airflow in open square cavities. *International Journal of Thermal Sciences, 100,* 267–283.

94. Zamora, B., & Kaiser, A. S. (2017). Radiative and variable thermophysical properties effects on turbulent convective flows in cavities with thermal passive configuration. *International Journal of Heat and Mass Transfer, 109,* 981–996.

95. Zamora, B. (2018). Heating intensity and radiative effects on turbulent buoyancy-driven airflow in open square cavities with a heated immersed body. *International Journal of Thermal Sciences, 126,* 218–237.

96. Budiyanto, M. A., Shinoda, T., & Nasruddin, N. (2017). Study on the CFD simulation of refrigerated container. *IOP Conference Series: Materials Science and Engineering, 257*(1), 012042.

97. Baltas, N., & Malin, M. R. (1997). The sudden release of gas from undersea pipelines. CHAM 2938/3, CHAM, Wimbledon, London.

98. Spalding, D. B. (1997). The virtual mass force in two-phase flow. CHAM Technical File Note: IPSA.

99. Malin, M. R., & Spalding, D. B. (1998). Extensions to the PHOENICS parabolic solver for under-expanded jets. CHAM C/4366/1 & C/4366/2, CHAM, London.

100. Patankar, S. V., & Spalding, D. B. (1966). A calculation procedure for heat transfer by forced convection through two-dimensional uniform-property turbulent boundary layers on smooth impermeable walls. In *Proceedings of the 3rd International Heat Transfer Conference*, Chicago (Vol. 2, pp. 50–63).

101. Patankar, S. V., & Spalding, D. B. (1967). A finite-difference procedure for solving the equations of the two-dimensional boundary layer. *International Journal of Heat and Mass Transfer, 10,* 1339.

102. Patankar, S. V., & Spalding, D. B. (1970). *Heat and mass transfer in boundary layers* (2nd ed.). London: Intertext Books.

103. Spalding, D. B. (1977). *GENMIX: A general computer program for two-dimensional parabolic phenomena* (1st ed.). Oxford: Pergamon Press.

104. Patankar, S. V., & Spalding, D. B. (1972). A calculation procedure for heat, mass and momentum transfer in three-dimensional parabolic flows. *International Journal of Heat and Mass Transfer, 15*, 787.

105. Spalding, D. B., & Tatchell, D. G. (1973). A prediction procedure for flow, combustion and heat transfer close to the base of a rocket. HTS/73/42, Imperial College, London, UK.

106. Issa, R. I., Spalding, D. B., & Tatchell, D. G. (1974). Guide to the computer program REP3. CHAM Report 631/2, CHAM, London, UK.

107. Elgobashi, S., & Spalding, D. B. (1977). Equilibrium chemical reaction of supersonic hydrogen-air jets (The ALMA computer program). NASA CR-2725.

108. Markatos, N. C., Spalding, D. B., & Tatchell, D. G. (1977). Combustion of hydrogen injected into a supersonic air stream. NASA-CR 2802.

109. Spalding, D. B. (1977). The PAM2 code: An introduction. CHAM/TR/40, CHAM, Wimbledon, London, UK.

110. Jennions, I. K., Ma, A. S. C., & Spalding, D. B. (1977). A prediction procedure for 2-dimensional steady, supersonic flows (The GENMIX-H computer program). HTS/77/24, Imperial College, London.

111. Drummond, J. P. (2014). Methods for prediction of high-speed reacting flows in aerospace propulsion. *AIAA Journal, 52*(3), 465–485.

112. Cousins, J. M. (1981). Calculation of conditions in an axisymmetric rocket exhaust plume: The REP3 computer program. PERME Technical Report No.218, Westcott, UK.

113. Pratap, V. S., & Spalding, D. B. (1975). Numerical computations of flow in curved ducts. *Aeronautical Quarterly, 26*, 219–228.

114. Singhal, A. K., & Spalding, D. B. (1978). A 2d partially-parabolic procedure for turbomachinery cascades. ARC R & M No. 3807, London, UK.

115. Jennions, I. K. (1980). *The impingement of axisymmetric supersonic jets on cones.* Ph.D. thesis, Imperial College, University of London, UK.

116. Spalding, D. B. (1978). Computer codes for rocket-plume analysis. CHAM TR/38, CHAM, Wimbledon, London.

117. Spalding, D. B., & Tatchell, D. G. (1973). The rocket base-flow computer program—BAFL, CHAM/640/1. CHAM, Wimbledon, London.

118. Jensen, D. E., Spalding, D. B., Tatchell, D. G., & Wilson, A. S. (1979). Computations of structures of flames with recirculating flow and radial pressure gradients. *34*, 309–26.

119. Markatos, N. C., Spalding, D. B., Tatchell, D. G., & Mace, A. C. H. (1982). Flow and combustion in the base-wall region of rocket exhaust plumes. *Combustion Science and Technology, 28*, 15–29.

120. Markatos, N. C., Mace, A. C. H., & Tatchell, D. G. (1982). Analysis of combustion in recirculating flow for rocket exhausts in supersonic streams. *Journal of Spacecraft and Rockets, 19*(6), 557–563.

121. Spalding, D. B. (1980). Lecture 25, Improved procedures for hydrodynamic problems. In *Mathematical modelling of fluid-mechanics, heat-transfer and chemical-reaction processes: A lecture course.* HTS/80/1, Imperial College, University of London.

122. Spalding, D. B. (1982). Lecture 2, 4.2 SIMPLEST. In *Four lectures on the PHOENICS computer code.* CFD/82/5, Imperial College, University of London.

123. Palacio, A., Malin, M. R., Proumen, N., & Sanchez, L. (1990). Numerical computations of steady transonic and supersonic flow fields. *International Journal of Heat and Mass Transfer, 33*(6), 1193–1204.

124. Malin, M. R., & Sanchez, L. (1988). One-dimensional steady transonic shocked flow in a nozzle. *PHOENICS Journal, 1*(2), 214–246 (CHAM, Wimbledon, London, UK).

125. Smith, A. G., & Taylor, K. (2000). Modelling of two-phase rocket exhaust plumes and other plume prediction development. *PHOENICS Journal, 13*(1) (CHAM, Wimbledon, London).

126. Spalding, D. B. (2004). The 'convergence-promoting wizard' for PHOENICS. In *PHOENICS User Conference*, Melbourne, Australia. http://www.cham.co.uk/phoenics/d_polis/d_lecs/d_conwiz/conwiz.htm.

127. Spalding, D. B. (1992). The expert-system CFD code; problems and partial solutions. In *Conference Proceedings. Basel World User Days CFD 1992*, May 24–28.
128. Spalding, D. B. (2013). Chapter 1 trends, tricks, and try ons in CFD/CHT, Section 2.2.2 General remarks about linear-equation solvers. In E. M. Sparrow, Y. I. Cho, J. P. Abraham, & J. M. Gorman (Eds.), *Advances in heat transfer* (Vol. 45, pp. 1–78). Burlington: Academic Press.

.

Some Observations on Thermodynamic Basis of Pressure Continuum Condition and Consequences of Its Violation in Discretised CFD

A. W. Date

Nomenclature

A_k, AP	Coefficients in Discretised Equations
D	Mass Diffusivity
F	Volume Fraction
Fr	Froude Number
F_{st}	Surface Tension Force
g	Gravity Acceleration
k	Thermal Conductivity
Kn	Knudsen Number
Pc	Peclet Number
p	Pressure
q	Continuum-Preserving Stress/Pressure
R	Residual or Gas Constant
Re	Reynolds Number
Ra	Rayleigh Number
t	Time
u_i	Velocity in x_i, i = 1, 2, 3 direction
V	Volume
$\vec{V}$	Total Velocity Vector
We	Weber Number

A. W. Date (✉)
Mechanical Engineering Department, IIT Bombay, Mumbai 400076, India
e-mail: awdate@me.iitb.ac.in

© Springer Nature Singapore Pte Ltd. 2020
A. Runchal (ed.), *50 Years of CFD in Engineering Sciences*,
https://doi.org/10.1007/978-981-15-2670-1_2

Greek Symbols

α, β Under-Relaxation Factors
γ Second viscosity coefficient
μ Dynamic viscosity
ρ Density
σ Surface Tension Coefficient or Stress
τ Stress

Suffixes

a Refers to Heavier Fluid
b Refers to Lighter Fluid or to Boundary Node
cont   Refers to continuum
disc   Refers to discretised space
f Refers to CV Face Location
i In i-direction
m Refers to mass conservation or to mixture
n Normal to the Interface
sm Refers to Smoothing
th Thermodynamic
x_i Refers to $x_i, i = 1, 2, 3$ directions

Superscripts

l Iteration Number
o Refers to old time
$u_i, u_{f,i}$ Refers to Momentum Equation
$-$ Refers to Multidimensional Average
$'$ Refers to Correction

Acronyms and Short Forms

CV Control Volume
CFD Computational Fluid Dynamics
$F(\ )$ Function of
LS Level-Set Method
VOF Volume of Fluid Method

1 Introduction

1.1 Navier–Stokes Equations

In the study of transport in moving fluids, the fundamental laws of motion (conservation of mass and Newton's second law of motion) are applied to an elemental fluid. Two approaches are possible: (a) Particle approach and (b) Continuum approach.

In the *particle approach*, the fluid is assumed to consist of particles (molecules, atoms) and the laws are applied to study particle motion. Fluid motion is then described by statistically averaged motion of a group of particles. For most applications arising in engineering and the environment, however, this approach is too cumbersome[1] because the significant dimensions of the flow are considerably bigger than the mean free path length between molecules. In the *continuum approach*, therefore, statistical averaging is assumed to have been *already performed* and the fundamental laws are applied to portions of fluid (or *control volumes*) that contain a large number of particles. The information *lost in averaging* must, however, be *recovered*. This is done by invoking some further *auxiliary laws* and by empirical specifications of *transport properties* such as viscosity (μ), thermal conductivity (k) and mass diffusivity (D). The transport properties are typically determined from experiments. Notionally, the continuum approach is very attractive because one can now speak of temperature, pressure or velocity *at a point* and relate them to what is *measured* by most practical instruments.

Guidance for deciding whether particle or continuum approach is to be used can be had from *Knudsen number $Kn = l/L$*, where l is the mean free path length between molecules and L is a characteristic dimension (say, the radius of a pipe) of the flow. When Kn is very small ($<10^{-4}$), continuum approach is considered valid. In macro-engineering and environmental flows, therefore, continuum approach is adopted.

Control Volume (CV): *The CV may be defined as a region in space across the boundaries of which matter, energy and momentum may flow, and it is a region within which source or sink of the same quantities may prevail. Further, it is a region on which external forces may act.*

The Navier–Stokes equations are derived by applying the law of conservation of mass and Newton's second law of motion to a CV shown in Fig. 1. The CV having dimensions Δx_1, Δx_2 and Δx_3 is located at (x_1, x_2, x_3) from a fixed origin. The statements of the laws yield *algebraic equations* of mass and momentum conservation. These statements are then *converted* to *partial differential equations* by letting Δx_1, Δx_2 and $\Delta x_3 \to 0$ followed by invoking the mathematical definition of a derivative in a continuum. Thus,

---

[1]This can be appreciated from the Avogadro's number which specifies that at normal temperature and pressure, a gas will contain 6.022×10^{26} molecules per kmol. Thus in air, for example, there will be 10^{16} molecules per mm^3.

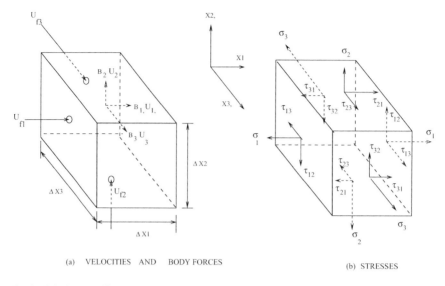

(a) VELOCITIES AND BODY FORCES   (b) STRESSES

Fig. 1 Eulerian specifications of the control volume (CV)

$$\frac{\partial(\rho_m)}{\partial t} + \frac{\partial(\rho_m\, u_{fj})}{\partial x_j} = 0 \tag{1}$$

$$\frac{\partial(\rho_m\, u_i)}{\partial t} + \frac{\partial}{\partial x_j}(\rho_m\, u_{fj}\, u_i) = \frac{\partial \sigma_{xi}}{\partial x_i} + \frac{\partial}{\partial x_j}\left\{\tau_{ij}\,(1 - \delta_{ij})\right\} + \rho_m\, B_i \tag{2}$$

$$\tau_{ij} = \mu\left[\frac{\partial u_i}{\partial x_j} + \frac{\partial u_j}{\partial x_i}\right] \tag{3}$$

where ρ_m is fluid (or mixture) density, u_{fj} are CV face velocities, u_j are representative CV velocities and B_i are volumetric body forces such as Buoyancy or Centrifugal or Coriolis force. Equation 3 expresses Stokes's law connecting surface stress τ_{ij} to the co-planar strain rate via fluid property μ. Finally, σ_{xi} are *total surface-normal (tensile) stresses* and are modelled as [27]

$$\sigma_{x_i} \equiv -p + \sigma'_{xi} = -(p - q) + \tau_{ii} = -(p - q) + 2\,\mu\,\frac{\partial u_i}{\partial x_i}$$
$$= -(p - q) + 2\,\mu\,\nabla\,.\,\vec{V}_f \tag{4}$$

where pressure p is compressive, normal stress τ_{ii} are tensile and σ' is called the deviatoric stress. The significance of the newly introduced quantity q in its definition requires further elaboration.

1.2 Stokes's Continuum Condition

In Date [6], a quantity $\overline{p}$ is defined[2] as

$$\overline{p} \equiv -\frac{1}{3} \left(\sigma_{x_1} + \sigma_{x_2} + \sigma_{x_3} \right) \tag{5}$$

Now, an often overlooked *requirement* of the Stokes's relations (with or without variable properties and *in the absence of relaxation processes at the molecular level* [27]) is that $\overline{p}$ must equal the point value of pressure p and the latter, in turn, must equal thermodynamic pressure p_{th}. Thus, using Eq. 4, it follows that

$$\overline{p} = p = p_{th} = (p - q) - \frac{2}{3} \mu \nabla \cdot \vec{V}_f \tag{6}$$

Now, to obey the above equality, q must be appropriately chosen in continuum as well as in discretised space. We now consider the following three cases:

1. Case 1: ($\vec{V} = 0$) In this *hydrostatic case*,

$$\overline{p} = p - q \tag{7}$$

But in this case, p can only vary linearly with x_1, x_2, x_3 and, therefore, the point value of p exactly equals its space averaged value $\overline{p}$ in both continuum and discretised space and hence

$$q = q_{cont} = q_{disc} = 0 \quad \rightarrow \quad \text{exactly} \tag{8}$$

2. Case 2: ($\mu = 0$ or $\nabla \cdot V_f = 0$)

   Clearly when $\mu = 0$ (inviscid flow) or $\nabla \cdot V_f = \partial u_{fi}/\partial x_i = 0$ (constant density incompressible flow) $p = \overline{p}$ (Eq. 6) in a continuum, and hence $q_{cont} = 0$ exactly.[3] But, in this case, since fluid motion is considered, p can vary arbitrarily with x_1, x_2, x_3 and, therefore, p may not equal $\overline{p}$ in a *discrete space*. To understand this matter, consider a case in which pressure varies arbitrarily in x_1 direction whereas its variation in x_2 and x_3 directions is constant or linear (as in a hydrostatic case). Such a variation is shown in Fig. 2. Now consider a point P. According to Stokes's requirement, p_P must equal $\overline{p}_P$ in a continuum. But, in a discretised space, the values of pressure are available at points E and W only, and if these points are

---

[2]In [27], symbol $\overline{\sigma} = (\sum_{i=1}^{3} \sigma_{xi})/3$ is used. Here, $\overline{p} = -\overline{\sigma}$ is preferred. Both $\overline{p}$ and q are newly introduced to serve a pedagogic purpose.

[3]It is important to recognise that in discretised CFD, the incompressible condition ($\nabla \cdot V_f = 0$) is defined in terms of CV face velocities u_{fi} as shown in Fig. 1. In fact, when this definition is explicitly implemented, there results the SIMPLE staggered grid procedure of Patankar and Spalding [19]. Further, u_{fi} must *satisfy momentum equations*. In a continuum, u_{fi} and u_i fields coincide but in a discretised space, it is important to distinguish them. This will become apparent in the next section.

Fig. 2 1D variation of pressure and Stokes's requirement

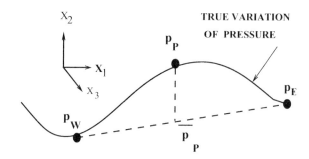

equi-distant from P then $\overline{p}_P = 0.5\,(p_W + p_E)$. Now, this $\overline{p}_P$ *will not equal* p_P as seen from the figure and, therefore, the requirement of the Stokes's relations is not met.

However, without violating the continuum requirement, we may set

$$q = q_{\text{cont}} = q_{\text{disc}} = \lambda\,(p - \overline{p}) \tag{9}$$

where λ is an arbitrary constant. In most textbooks, where continuum is assumed, λ is trivially set to zero because $p = \overline{p}$ (Eq. 6) in a continuum.

3. Case 3: ($\mu \neq 0$ or $\nabla \cdot V_f \neq 0$)

This case represents either compressible flow where density is a state function of both temperature and pressure or an incompressible flow with density dependent on temperature or any other scalar (e.g. void or volume fraction). Thus, in this case, Stokes's requirement will be satisfied (see Eq. 6) if we set

$$q = \lambda\,(p - \overline{p}) + \gamma\,\nabla \cdot V_f \quad \rightarrow \quad \gamma = -\frac{2}{3}\mu \tag{10}$$

where γ is the well-known second viscosity coefficient whose value is routinely set to $-(2/3)\mu$ *even in a continuum*. It is instructive to note the reason for this setting. Because, if this was not done then by combining Eqs. 10 and 6, it can be shown that

$$(1 - \lambda)\,(p - \overline{p})\,\nabla \cdot V_f = \left(\gamma + \frac{2}{3}\mu\right)(\nabla \cdot V_f)^2 \tag{11}$$

Clearly, this equation suggests that the system will experience *dissipation* (or reversible work done at finite rate since $\nabla \cdot V_f$ is associated with the *rate of volume change*) even in an *isothermal flow* [27, 36]. This is, of course, highly improbable.[4]

---

[4]In passing we note that in all three cases, it can be verified that the quantity q is invariant under rotation of the coordinate system or interchange of axes. This property ensures isotropy [27].

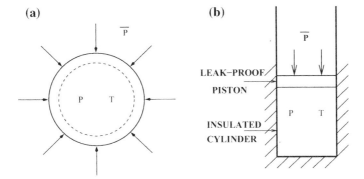

Fig. 3 Consequences of violating Stokes's condition **a** isolated spherical mass of fluid [27], **b** insulated piston cylinder

1.3 Thermodynamic Explanations

This improbability has been explained by Schlichting [27] by considering the case of an isolated spherical mass of *isothermal fluid* subjected to uniform normal stress $\overline{\sigma} = -\overline{p}$ (see Fig. 3a). Now, if γ is not set to $-(2/3)\mu$, the sphere will undergo radial oscillations of compression and expansion. Here, however, we consider an alternative arrangement that will yield one further interpretation.

Figure 3b shows the piston-cylinder arrangement typically used in undergraduate thermodynamics. The system fluid is held in a leakproof adiabatic cylinder at temperature T and pressure p. The system is in equilibrium. Now, suppose the unlikely circumstance in which external pressure $\overline{p}$ exceeds system pressure p. It is obvious that the piston will move downwards compressing the fluid. But, if we now require that the temperature of the fluid must remain constant at T (this is analogous the Schlichting's isothermal flow), then clearly, from first law of thermodynamics, no change in internal energy ΔU is permitted in an adiabatic cylinder, and hence there can only be two consequences.

1. The piston will instantly *bounce back* to its original position to restore equilibrium. The process may repeat resulting in oscillations of the piston *in time*. However, in a *steady-state* problem, these oscillations will manifest as *spatially zig-zag* pressure.
2. Alternatively, to maintain constant internal energy, some fluid must *somehow leak out* although the piston is leakproof.

It is obvious that both these occurrences are improbable but are nonetheless encountered in discretised CFD when Stokes's continuum condition (6) is violated as will be shown in Sects. 2 and 3. Incidentally, the second circumstance above is a new interpretation associated with the violation of the Stokes's condition.

Thus, as shown in Eq. 10, the Stokes's relations for normal stress require modifications *even in a continuum* when compressible flow is considered (requiring introduction of the so-called second viscosity coefficient) and a physical explanation for this modification is found from Thermodynamics. Now, the same interpretation can be afforded to $\lambda \, (p - \overline{p})$ part of q in Eqs. 9 or 10. This term represents a necessary modification in a discretised space. This is an important departure from the forms of normal stress expressions given in standard textbooks on fluid mechanics.

1.4   Appropriate Forms of N-S Equations

Finally, from the above discussion, it is clear that the Navier–Stokes equations written out for finite-volume discretisation should preferably read as

$$\frac{\partial (\rho_m)}{\partial t} + \frac{\partial (\rho_m \, u_{fj})}{\partial x_j} = 0 \tag{12}$$

$$\frac{\partial (\rho_m \, u_i)}{\partial t} + \frac{\partial}{\partial x_j} (\rho_m \, u_{fj} \, u_i) = -\frac{\partial (p - q)}{\partial x_i} + \frac{\partial}{\partial x_j} \left\{ \tau_{ij} \right\} + \rho_m \, B_i \tag{13}$$

where, in the most general case, q stands for

$$q = \lambda \, (p - \overline{p}) - \frac{2}{3} \, \mu \times \nabla \, . \, \vec{V}_f \tag{14}$$

Of course, $\lambda = 0$ in a continuum but finite[5] in a discretised space. Note, however, that in the latter, as mesh size is reduced, $p \to \overline{p}$, and hence $q_{\text{disc}} \to q_{\text{cont}}$. Also, it is important to note that since $\overline{p}$ must equal point value of pressure p in a continuum (see Eq. 6), the former must essentially correspond to the *hydrostatic or spatially linear* variation of pressure (in a local sense) irrespective of the flow considered. Mathematically, therefore, we may define $\overline{p}$ as

$$\overline{p} = -\frac{1}{3} \sum_{i=1}^{3} \sigma_{x_i} = \frac{1}{3} \sum_{i=1}^{3} \overline{p}_{x_i} \tag{15}$$

where $\overline{p}_{x_i}$ are each a solution[6] to

$$\frac{\partial^2 p}{\partial x_i^2} = 0 \tag{16}$$

This manner of evaluation of $\overline{p}$ can be implemented on both structured and unstructured meshes [4–7, 22, 23] in discretised CFD.

---

[5]Analysis of the discretised equations presented in the next section shows that $\lambda = 0.5$.
[6]Equation 16 is validated in Eqs. 39–43 for a two-dimensional flow.

2 Computations on Colocated Grids

2.1 Pressure-Correction Equation on Colocated Grids

Since it has been known that in the SIMPLE algorithm [19], zig-zag pressure predic-tion is avoided by the use of staggered grid arrangement of pressure and the velocity variables (see Fig. 4a), it is obvious that the pressure-correction equation applicable to colocated grids must mimic the main features of the staggered grid practice. Thus, we begin by stating that in a fully implicit iterative procedure, the cell-face velocity will be calculated from

$$u_{fi}^{l+1} = \frac{\alpha}{AP^{u_{fi}}} \left[\sum_k A_k u_{fi,k}^{l+1} - \Delta V \frac{\partial p^{l+1}}{\partial x_i} \right] + (1 - \alpha) u_{fi}^{l} \tag{17}$$

where $AP^{u_{fi}} = \sum A_k + \rho_m^0 \, \Delta V / \Delta t$. Now, this velocity field must satisfy mass con-servation equation (1). Thus

$$\frac{\partial(\rho_m^{l+1})}{\partial t} + \frac{\partial(\rho_m^{l+1} u_{fi}^{l+1})}{\partial x_i} = 0 \tag{18}$$

After substituting Eq. 17 in Eq. 18, we make use of following representations:

$$u_{fi}^{l+1} = u_{fi}^{l} + u_{fi}^{'} \qquad \text{and} \qquad p^{l+1} = p^l + p_m^{'} \tag{19}$$

The above operations result[7] in

$$\frac{\partial}{\partial x_i} \left[\rho_m^{l+1} D_i \frac{\partial p_m^{'}}{\partial x_i} \right] = \frac{\partial(\rho_m^{l+1})}{\partial t} + \frac{\partial(\rho_m^{l+1} u_{fi}^{l})}{\partial x_i}$$

$$- \frac{\partial}{\partial x_i} \left[\rho_m^{l+1} D_i R_{ufi} \right] \quad \rightarrow D_i = \frac{\alpha \Delta V}{AP^{u_{fi}}} \tag{20}$$

where R_{ufi} is the residual per unit volume and is given by

$$R_{ufi} = \frac{AP^{ufi} u_{fi}^{l} - \sum A_k u_{fi,k}^{l}}{\Delta V} + \frac{\partial p^l}{\partial x_i} \tag{21}$$

---

[7]In deriving Eq. 20, it is assumed that $\sum A_k u_{fi,k}^{'} = 0$. This is consistent with the staggered grid practice [19].

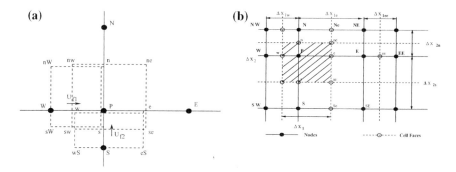

Fig. 4 **a** Staggered grid arrangement of variables and associated CVs, **b** colocated grids—cell faces are midway between the adjacent nodes

2.2 Analysis of Discretised Equation

To understand further developments of this paper, following comments are now pertinent:

1. On both staggered and colocated grids, the pressure is stored at node P and the mass conservation equation is solved over the control volume surrounding node P. Therefore, with reference to Fig. 4b, the discretised version of Eq. 20 in two dimensions will read as

$$
AP \, p'_{m,P} = AE \, p'_{m,E} + AW \, p'_{m,W} + AN \, p'_{m,N} + AS \, p'_{m,S} \\
- \dot{m}_P + \dot{m}_R \tag{22}
$$

where

$$
AE = \frac{\rho_m^{l+1} \, \alpha \, \Delta x_2^2}{A P^{uf1}} \bigg|_e, \quad AW = \frac{\rho_m^{l+1} \, \alpha \, \Delta x_2^2}{A P^{uf1}} \bigg|_w \\
AN = \frac{\rho_m^{l+1} \, \alpha \, \Delta x_1^2}{A P^{uf2}} \bigg|_n, \quad AS = \frac{\rho_m^{l+1} \, \alpha \, \Delta x_1^2}{A P^{uf2}} \bigg|_s \\
AP = AE + AW + AN + AS \tag{23}
$$

Further, the mass source $\dot{m}_P$ and residual source $\dot{m}_R$ will be given by

$$
\dot{m}_P = (\rho_m^{l+1} \, u'_{f1} \big|_e - \rho_m^{l+1} \, u'_{f1} \big|_w) \, \Delta x_2 \\
+ (\rho_m^{l+1} \, u'_{f2} \big|_n - \rho_m^{l+1} \, u'_{f2} \big|_s) \, \Delta x_1 \\
+ (\rho_{m,P}^{l+1} - \rho_{m,P}^{o}) \, \frac{\Delta V}{\Delta t} \tag{24}
$$

$$
\dot{m}_R = AE \, R_{uf1} \, \Delta x_1 \big|_e - AW \, R_{uf1} \, \Delta x_1 \big|_w \\
+ AN \, R_{uf2} \, \Delta x_2 \big|_n - AS \, R_{uf2} \, \Delta x_2 \big|_s \tag{25}
$$

2. On staggered grids, momentum equations are solved at the cell faces and, therefore, residuals R_{uf1} and R_{uf2} must vanish at full convergence rendering $\dot{m}_R = 0$. Although this state of affairs will prevail only at convergence, one may ignore $\dot{m}_R$ *even during iterative solution*. Thus, effectively, Eq. 20 applicable to staggered grid arrangement is

$$\frac{\partial}{\partial x_i} \left[\rho_m^{l+1} \, D_i \, \frac{\partial p_m'}{\partial x_i} \right] = \frac{\partial(\rho_m^{l+1})}{\partial t} + \frac{\partial(\rho_m^{l+1} \, u_{fi}')}{\partial x_i} \tag{26}$$

The above equation is solved with the boundary condition [19]

$$\frac{\partial p_m'}{\partial n} \Big|_b = 0 \tag{27}$$

If the boundary pressure p_b is specified then, of course, $p_{m,b}' = 0$.

3. When computing on colocated grid, however, cell-face velocities must be evaluated by interpolation to complete evaluation of $\dot{m}_P$ because only nodal velocities $u_{i,P}$ are computed through momentum equations. Thus, $\dot{m}_P$ in Eq. 24 is evaluated as

$$\begin{aligned}
\overline{\dot{m}}_P &= (\rho_m^{l+1} \, \overline{u}_1^l \mid_e - \rho_m^{l+1} \, \overline{u}_1^l \mid_w) \, \Delta x_2 \\
&+ (\rho_m^{l+1} \, \overline{u}_2^l \mid_n - \rho_m^{l+1} \, \overline{u}_2^l \mid_s) \, \Delta x_1 \\
&+ (\rho_{m,P}^{l+1} - \rho_{m,P}^o) \, \frac{\Delta V}{\Delta t}
\end{aligned} \tag{28}$$

where the mean velocities $\overline{u}_i$ are evaluated by one-dimensional averaging although multidimensional averaging can also be preferred. Thus, since the cell faces are midway between the nodes, we may write

$$\begin{aligned}
\overline{u}_{1,e} &= \frac{1}{2} \, (u_{1,P} + u_{1,E}) & \overline{u}_{1,w} &= \frac{1}{2} \, (u_{1,P} + u_{1,W}) \\
\overline{u}_{2,n} &= \frac{1}{2} \, (u_{2,P} + u_{2,N}) & \overline{u}_{2,s} &= \frac{1}{2} \, (u_{2,P} + u_{2,S})
\end{aligned} \tag{29}$$

Replacing $u_{f1,e}$ by $\bar{u}_{1,e}$, etc. in the above manner, of course, does not guarantee that $\dot{m}_P$ will vanish *even at convergence.*[8]

4. Similarly, to evaluate $\dot{m}_R$ from Eq. 25, we reconsider Eq. 21 for cell-face location e, for example, and write it as

$$R_{u_{f1,e}} = \frac{\overline{A\,P^{u_{f1}}\,u'_{f1} - \sum A_k\,u'_{f1,k}}}{\Delta V}\,|_e + \frac{\partial p'}{\partial x_1}\,|_e \tag{33}$$

In this equation, the net-momentum-transfer terms are now *multidimensionally averaged.* This is necessary because when computing on colocated grids, coefficients A_k are not available at the cell-face locations. Thus, again using Eq. 21, we have

$$\frac{\overline{A\,P^{u_{f1}}\,u'_{f1} - \sum A_k\,u'_{f1,k}}}{\Delta V}\,|_e = \overline{R}_{u_{f1,e}} - \overline{\frac{\partial p'}{\partial x_1}}\,|_e \tag{34}$$

Effectively, therefore

$$R_{u_{f1,e}} = \overline{R}_{u_{f1,e}} - \overline{\frac{\partial p'}{\partial x_1}}\,|_e + \frac{\partial p'}{\partial x_1}\,|_e \tag{35}$$

5. Multidimensionally averaged $\overline{R}_{u_{f1,e}}$ is evaluated as

---

[8]Incidentally, in the literature, several different types of interpolations have been proposed. Some of these are given below by way of example.

- Rhie and Chow [24] (1D Pressure gradient interpolation)

$$u_{f1,e} = \bar{u}_{1,e} - \frac{\Delta V}{A\,P^u}\left[\frac{\partial p}{\partial x_1}\,|_e - \overline{\frac{\partial p}{\partial x_1}}\,|_e\right]$$

$$\text{where}\quad \overline{\frac{\partial p}{\partial x_1}}\,|_e = \frac{1}{2}\left[\frac{\partial p}{\partial x_1}\,|_P + \frac{\partial p}{\partial x_1}\,|_E\right] \tag{30}$$

- Peric [8] (1D Mom-Outflow interpolation)

$$u_{f1,e} = \frac{1}{2}\left[\frac{\sum A_k\,u_{1,k}}{A\,P^{u_1}}\,|_P + \frac{\sum A_k\,u_{1,k}}{A\,P^{u_1}}\,|_E\right] - \frac{\Delta V}{A\,P^u}\frac{\partial p}{\partial x_1}\,|_e \tag{31}$$

- Thiart [34] (Power Law Scheme [20])

$$u_{f1,e} = \theta\,u_{1,P} + (1-\theta)\,u_{1,E}\quad\text{where}$$
$$\theta(Pc_e) = [Pc_e - 1 + \max(0, -Pc_e)]\,/\,Pc_e$$
$$+ \max\left\{0, (1 - 0.1|Pc_e|)^5\right\}\,/\,Pc_e \tag{32}$$

where cell-face Reynolds/Peclet number $Pc_e = (\rho_m\,u_{f1}\Delta x_1/\mu)_e$.

$$\overline{R}_{u_{f1,e}} = \frac{1}{2} \left[\frac{1}{2} (R_{u1,P} + R_{u1,E}) + \frac{\Delta x_{2,n} R_{u1,se} + \Delta x_{2,s} R_{u1,ne}}{\Delta x_{2,n} + \Delta x_{2,s}} \right]$$

$$R_{u1,se} = \frac{1}{4} (R_{u1,P} + R_{u1,E} + R_{u1,S} + R_{u1,SE})$$

$$R_{u1,ne} = \frac{1}{4} (R_{u1,P} + R_{u1,E} + R_{u1,N} + R_{u1,NE}) \tag{36}$$

This representation shows that effectively residuals at nodal locations P, E, N, S, NE and SE only are involved. These residuals will, of course, vanish at full convergence because momentum equations are being solved at the nodal positions. Therefore, effectively $\overline{R}_{u_{f1,e}} = 0$ at convergence and Eq. 35 can be written as

$$R_{u_{f1,e}} = \frac{\partial p^l}{\partial x_1} |_e - \overline{\frac{\partial p^l}{\partial x_1}} |_e \tag{37}$$

6. Now, to evaluate multidimensionally averaged pressure gradient in the above equation, we write

$$\overline{\frac{\partial p^l}{\partial x_1}} |_e = \frac{1}{2} \left[\frac{1}{2} \left(\frac{\partial p^l}{\partial x_1} |_P + \frac{\partial p^l}{\partial x_1} |_E \right) \right.$$
$$+ \left. \frac{\Delta x_{2,n} \partial p^l / \partial x_1 |_{se} + \Delta x_{2,s} \partial p^l / \partial x_1 |_{ne}}{\Delta x_{2,n} + \Delta x_{2,s}} \right]$$
$$= \frac{1}{4} \left[\frac{p^l_E - p^l_W}{\Delta x_{1,e} + \Delta x_{1,w}} + \frac{p^l_{EE} - p^l_P}{\Delta x_{1,e} + \Delta x_{1,w}} \right]$$
$$+ \frac{1}{4} \frac{\Delta x_{2,s}}{\Delta x_{2,n} + \Delta x_{2,s}} \left[\frac{p^l_E + p^l_{NE} - p^l_P - p^l_N}{\Delta x_{1,e}} \right]$$
$$+ \frac{1}{4} \frac{\Delta x_{2,n}}{\Delta x_{2,n} + \Delta x_{2,s}} \left[\frac{p^l_E + p^l_{SE} - p^l_P - p^l_S}{\Delta x_{1,e}} \right] \tag{38}$$

To simplify the above evaluation further, following definitions are introduced allowing for the non-uniform grid spacing:

$$\overline{p^l}_{x_1,P} \equiv \frac{\Delta x_{1,w} p^l_E + \Delta x_{1,e} p^l_W}{\Delta x_{1,w} + \Delta x_{1,e}} \qquad \text{(solution to } \frac{\partial^2 p^l}{\partial x_1^2}|_P = 0) \tag{39}$$

$$\overline{p^l}_{x_2,P} \equiv \frac{\Delta x_{2,s} p^l_N + \Delta x_{2,n} p^l_S}{\Delta x_{2,s} + \Delta x_{2,n}} \qquad \text{(solution to } \frac{\partial^2 p^l}{\partial x_2^2}|_P = 0) \tag{40}$$

$$\overline{p^l}_P = \frac{1}{2} (\overline{p^l}_{x_1,P} + \overline{p^l}_{x_2,P}) \tag{41}$$

$$\overline{p'}_{x_1,E} \equiv \frac{\Delta x_{1,e}\, p'_{EE} + \Delta x_{1,ee}\, p'_P}{\Delta x_{1,e} + \Delta x_{1,ee}} \qquad \text{(solution to } \frac{\partial^2 p'}{\partial x_1^2}|_E = 0) \qquad (42)$$

$$\overline{p'}_{x_2,E} \equiv \frac{\Delta x_{2,s}\, p'_{NE} + \Delta x_{2,n}\, p'_{SE}}{\Delta x_{2,s} + \Delta x_{2,n}} \qquad \text{(solution to } \frac{\partial^2 p'}{\partial x_2^2}|_E = 0) \qquad (43)$$

$$\overline{p'}_E = \frac{1}{2}\,(\overline{p'}_{x_1,E} + \overline{p'}_{x_2,E}) \qquad (44)$$

Substituting the above definitions[9] in Eq. 38 and replacing p'_{EE} (Eq. 42) and p'_W (Eq. 39), respectively, in favour of p'_E and p'_P, it can be shown that

$$\frac{\overline{\partial p'}}{\partial x_1}|_e = \frac{1}{2}\left[\frac{p'_E - p'_P}{\Delta x_{1,e}} + \frac{\overline{p'}_E - \overline{p'}_P}{\Delta x_{1,e}}\right]$$

$$= \frac{1}{2}\frac{\partial(p' + \overline{p'})}{\partial x_1}|_e \qquad (45)$$

and, therefore, from Eq. 37

$$R_{u_{f1,e}} = \frac{\partial p'}{\partial x_1}|_e - \frac{1}{2}\frac{\partial(p' + \overline{p'})}{\partial x_1}|_e = \frac{1}{2}\frac{\partial(p' - \overline{p'})}{\partial x_1}|_e = \frac{\partial p'_{sm}}{\partial x_1}|_e \qquad (46)$$

where

$$p'_{sm} = \frac{1}{2}\,(p' - \overline{p'}) \qquad \text{(Smoothing Pressure Correction)} \qquad (47)$$

Note that the analysis of the discretised equations has yielded a value $\lambda = 0.5$ and $q = p'_{sm}$ (see Eq. 9).

7. Repeating items 4, 5 and 6 at other cell faces, it can be shown that

$$R_{u_{f1,w,e}} = \frac{\partial p'_{sm}}{\partial x_1}|_{w,e} \qquad R_{u_{f2,s,n}} = \frac{\partial p'_{sm}}{\partial x_2}|_{s,n} \qquad (48)$$

Thus, substituting the above equations in Eq. 25, it follows that

$$\dot{m}_R = AE\,\frac{\partial p'_{sm}}{\partial x_1}\,\Delta x_1|_e - AW\,\frac{\partial p'_{sm}}{\partial x_1}\,\Delta x_1|_w$$

$$+ AN\,\frac{\partial p'_{sm}}{\partial x_2}\,\Delta x_2|_n - AS\,\frac{\partial p'_{sm}}{\partial x_2}\,\Delta x_2|_s \qquad (49)$$

8. In evaluating coefficients AE, AW, AN and AS, we need AP coefficients at the cell faces (see Eq. 23). But, these can be evaluated by one-dimensional averaging as

---

[9]Equations 39–43 justify the assertion made in Eqs. 15 and 16 for a two-dimensional flow.

$$AP_e^{uf1} = \frac{1}{2}(AP_P^{u_1} + AP_E^{u_1}) \qquad AP_w^{uf1} = \frac{1}{2}(AP_P^{u_1} + AP_W^{u_1})$$

$$AP_n^{uf2} = \frac{1}{2}(AP_P^{u_2} + AP_N^{u_2}) \qquad AP_s^{uf2} = \frac{1}{2}(AP_P^{u_2} + AP_S^{u_2}) \qquad (50)$$

where the AP coefficients at the nodal locations P, N, E, S, W are known on colocated grids.

9. The above derivations show that Eqs. 24 and 25 can be replaced by Eqs. 28 and 49, respectively. Thus, the mass-conserving pressure-correction Eq. 20 appropriate for colocated grids can effectively be written as

$$\frac{\partial}{\partial x_i}\left[\rho_m^{l+1} D_i \frac{\partial p'_m}{\partial x_i}\right] = \frac{\partial(\rho_m^{l+1})}{\partial t} + \frac{\partial(\rho_m^{l+1}\, \overline{u}'_i)}{\partial x_i}$$

$$- \frac{\partial}{\partial x_i}\left[\rho_m^{l+1} D_i \frac{\partial p'_{sm}}{\partial x_i}\right] \qquad (51)$$

2.2.1 Further Simplification

It is possible to further simplify Eq. 51. To understand this simplification, consider, for example, the grid disposition near the west boundary as shown in Fig. 5. When computing at the near-boundary node $P(2,j)$, the pressure gradient $\partial p/\partial x_1 |_P$ must be evaluated in the momentum equation for velocity $u_{1,P}$. This will require knowledge of boundary pressure $p_b = p\,(1, j)$. On colocated grids, this pressure is not known and, therefore, is evaluated by *linear* extrapolation from interior flow points. Thus,

$$p_b = \frac{L_{bE}}{L_{PE}}\, p_P - \frac{L_{bP}}{L_{PE}}\, p_E \qquad (52)$$

where L denotes length. The same procedure is adopted at Nb and Sb. Now, assuming that the pressure variation near a boundary is *locally* linear in both x_1 and x_2 directions, it follows that

$$p_b - \overline{p}_b = p_P - \overline{p}_P \qquad \text{or} \qquad p'_{sm,b} = p'_{sm,P} \qquad (53)$$

and, therefore,

$$\frac{\partial p'_{sm}}{\partial x_1}\,|_b = \frac{\partial p'_{sm}}{\partial n}\,|_b = 0 \qquad (54)$$

The same condition is also applicable to p'_m (see Eq. 27). Now, Eq. 51 shows that multipliers of gradients of p'_m and p'_{sm} are identical and since the boundary conditions for these two variables are also identical, we may write the mass-conserving pressure correction in the following form:

$$\frac{\partial}{\partial x_i} \left[\rho_m^{l+1} \ D_i \ \frac{\partial p'}{\partial x_i} \right] = \frac{\partial (\rho_m^{l+1})}{\partial t} + \frac{\partial (\rho_m^{l+1} \ \overline{u}_i')}{\partial x_i} \tag{55}$$

where p' is called *total pressure correction* and is expressed as

$$p' = p'_m + p'_{sm} \tag{56}$$

Here, p'_{sm}, of course, is evaluated from Eq. 47. Following Eqs. 54 and 27, Eq. 55 must be solved with boundary condition

$$\frac{\partial p'}{\partial x_1} \Big|_b = \frac{\partial p'}{\partial n} \Big|_b = 0 \tag{57}$$

The discretised form of Eq. 55 is

$$AP \ p'_P = AE \ p'_E + AW \ p'_W + AN \ p'_N + AS \ p'_s - \overline{\dot{m}}_P \tag{58}$$

where $\overline{\dot{m}}_P$ is given by Eq. 28.

Finally, note that Eq. 55 has similarity with the staggered grid equation (26). Both are Poisson's equations with similar boundary conditions (57) and (27).

2.2.2 Modification for Compressible Flow

So far it has been assumed that ρ_m = constant as in incompressible flow. However, in compressible flow ρ_m is connected to pressure p via equation of state such as $p = \rho_m \ R_g \ T$. Thus,

Fig. 5 West boundary $I = 1$

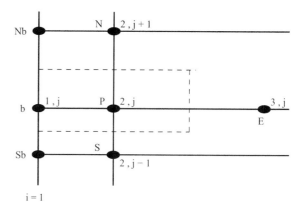

$$\rho_m^{l+1} = \rho_m^l + \rho_m^{'} = \rho_m^l + \frac{p_m^{'}}{R_g T} = \rho_m^l + \frac{p^{'} - p_{sm}^{'}}{R_g T} \tag{59}$$

With this substitution, Eq. 55 will read as

$$\frac{\partial}{\partial x_i} \left[\rho_m^l D_i \frac{\partial p^{'}}{\partial x_i} - U_i^* \frac{p^{'}}{R_g T} \right] = \frac{\partial(\rho_m^{'})}{\partial t} + \frac{\partial}{\partial x_i} \left[\rho_m^l \overline{u}_i^l - U_i^* \frac{p_{sm}^{'}}{R_g T} \right] \tag{60}$$

$$U_i^* = u_i^l - D_i \frac{\partial p_{sm}^{'}}{\partial x_i} \tag{61}$$

The left-hand side of Eq. 60 contains diffusion $(D_i \partial p^{'} / \partial x_i)$ as well as convection $(U_i^* p^{'} / R_g T)$ terms. Thus, for compressible flow, the pressure-correction equation is a transport equation for $p^{'}$ and not a Poisson's equation. Also, note that if $U_i^* = 0$, the incompressible form (see Eq. 51) is recovered.

2.2.3 Overall Calculation Procedure on Colocated Grids

The sequence of calculations on colocated grids is as follows:

1. At a given time step, guess pressure field p^l. This may be the pressure field from the previous time step.
2. Solve momentum equations once for each u_i with problem-dependent boundary conditions. Thus

$$AP^{u_i} u_{i,P}^l = \sum A_k^{u_i} u_{i,k}^l - \Delta V \frac{\partial p^l}{\partial x_i}|_P + \frac{\rho_{m,P}^o \Delta V}{\Delta t} u_{i,P}^o \tag{62}$$

where $AP^{u_i} = \sum A_k^{u_i} + \rho_m^0 \Delta V / \Delta t$.
3. Using the u_i^l distribution, solve Eq. 55 (or, 60) with boundary condition (57) to yield the total pressure correction $p_{i,j}^{'}$ field. This implies iterative solution of Eq. 58. The number of iterations typically may not exceed 10.
4. Recover mass-conserving pressure correction via Eq. 56. Thus

$$p_{m,P}^{'} = p_P^{'} - p_{sm,P}^{'} = p_P^{'} - \frac{1}{2} (p_P^{'} - \overline{p}_P^{'}) \tag{63}$$

where $\overline{p}_P^{'}$ is evaluated from Eqs. 15 and 16. Using this $p_m^{'}$ field, the mass conservation error is evaluated from[10]

$$R_{\text{mass},P} = AP \, p_{m,P}^{'} - (AE p_{m,E}^{'} + AW p_{m,W}^{'} + AN p_{m,N}^{'} + AS p_{m,S}^{'}) \tag{64}$$

---

[10]This is unlike the staggered grid practice in which the mass error is estimated from discretised version of Eq. 1.

5. Correct pressure and velocity fields according to

$$p_P^{l+1} = p_P^l + \beta \, p_{m,P}' \qquad \text{where} \qquad 0 < \beta < 1 \qquad (65)$$

$$u_{i,P}^{l+1} = u_{i,P}^l - D_i \, \frac{\partial p_m'}{\partial x_i}|_P \qquad (66)$$

6. Using this new velocity field, solve discretised forms of scalar transport equations relevant to the problem at hand.
7. Evaluate residuals R_Φ of momentum and scalar (Φ) equations. The mass residual R_m evaluated from Eq. 64. When maximum residual as per the L2-norm is $<10^{-5}$, convergence is declared.
8. If this convergence criterion is not satisfied, treat $p^{l+1} = p^l$, $u_i^{l+1} = u_i^l$ and $\Phi^{l+1} = \Phi^l$ and return to step 2.
9. To execute the next time step, set $u_i^o = u_i^{l+1}$, $\Phi^o = \Phi^{l+1}$ and return to step 1.

2.3   Some Illustrative Problems

In all problems, computations are carried out with ($\lambda = 0.5$) and without ($\lambda = 0$) application of smoothing pressure correction.

2.3.1   Essentially Parabolic Flows

We consider 2D laminar developing flow between two parallel plates $2b$ apart ($D_h = 4b$). Although the flow is parabolic, here it is treated as being governed by 2D elliptic equations.

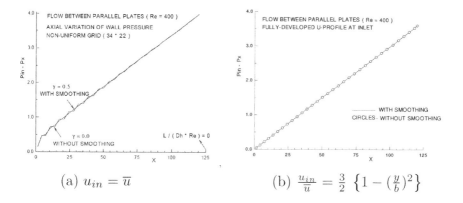

$$(a) \; u_{in} = \overline{u} \qquad\qquad (b) \; \frac{u_{in}}{\overline{u}} = \frac{3}{2} \left\{ 1 - \left(\frac{y}{b}\right)^2 \right\}$$

Fig. 6   Entrance region flow between parallel plates

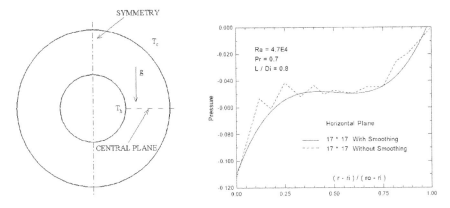

Fig. 7 Natural convection in horizontal concentric annulus [4]

Figure 6a shows predicted axial variation of wall pressure for the case when the fluid enters the channel with uniform axial velocity $u_{in} = \bar{u}$. As the boundary layers develop along the channel walls, the true pressure variation is expected to be non-linear with x in the entrance region but, at large x, it is expected to be linear because the flow is now fully developed. It is seen that with $\lambda = 0$, the predicted pressure is zig-zag near the entrance, whereas at large x, pressure zig-zagness disappears.

In contrast, Fig. 6b shows predictions for the case in which the fluid enters with fully developed parabolic axial velocity profile. Then, it is expected that true axial pressure variation will be linear right from the $x = 0$. Figure 6b confirms this expectation even when $\lambda = 0$.

2.3.2   Essentially Elliptic Flows

Natural Convection: Figure 7 (left) shows concentric cylinders in which the inner cylinder (dia D_i) is hotter (T_h) than the outer (T_c) one (dia D_o). The annulus gap width $L = (D_o - D_i)/2$. Figure 7 (right) shows predicted pressure variation along the horizontal plane in the presence of natural convection. Here again, since the true pressure variation is non-linear, the predicted pressure with $\lambda = 0$ shows zig-zagness. But, with smoothing pressure correction ($\lambda = 0.5$), zig-zagness disappears. The annulus natural convection flow is fully elliptic that involves simultaneous solution of the energy equation along with the flow equations. Further, the domain of computation is mapped by unstructured grid of triangular elements [4].

Mixed Convection: We now consider the problem of mixed convection flow in a right-angled corner. As shown in Fig. 8 (top), flow enters the domain of computation at the north boundary, turns anticlockwise around the corner and leaves through the east boundary. For this problem, an exact solution is developed by Shih and Ren [28] by specifying *artificial boundary conditions*. The unique feature of the exact solution is that the pressure variation in Y-direction is linear at all constant X-planes.

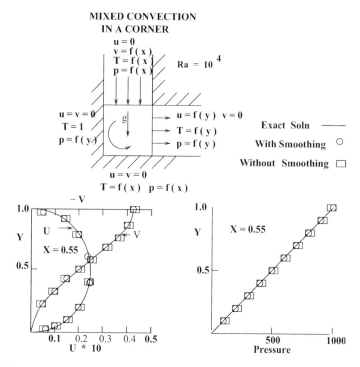

Fig. 8 Mixed convection in a corner—artificial boundary conditions

Figure 8 (left) shows comparison of predicted and exact profiles of velocity (u and v). Similar comparison for pressure is shown in Fig. 8 (right). It is seen that since the exact pressure variation is linear in Y, no zig-zagness is predicted for both $\lambda = 0$ and 0.5.

2.3.3 Compressible Flow in 2D Plane Nozzle

Figure 9 (top) shows the computational domain of a 2D convergent-divergent plane nozzle. The bottom boundary represents the axis (centreline) of the nozzle whereas the top boundary is a wall. The flow enters the left boundary and leaves through the right boundary. The total nozzle length $L = 11.56$ cm and the *throat* is midway. The half-heights of the nozzle at entry, throat and exit are 3.52 cm, 1.37 cm and 2.46 cm, respectively. The inlet Mach number is $M_{in} = 0.232$ and the exit *static* pressure is $p / p_0 = 0.1135$ where p_0 is the stagnation pressure. The stagnation enthalpy is assumed constant. For these specifications, experimental data are available [16]. This flow has been computed by Karki and Patankar [13] using curvilinear grids and Upwind Difference Scheme (UDS) for the convective terms and assuming $\mu = 0$

(that is, Euler equations are solved). Here, the flow is computed using unstructured mesh and Total Variation Diminishing (TVD) scheme [14, 17] again with $\mu = 0$.

At inflow plane, since M_{in} is known, u_{in}, T_{in} and p_{in} are specified using standard isentropic relationships [10]. In the exit plane, except for pressure (which is fixed), all other variables are linearly extrapolated from the near-exit boundary node values. At the upper nozzle wall, *tangency* condition is applied. The pressure distribution is determined by discretising a compressible flow version of the total pressure-correction equation (60). For velocities, equations for u_1, u_2 are solved and temperature is recovered from definition of stagnation enthalpy. Finally, density is determined using equation of state $p = \rho_m R_g T$. Computations are performed using 570 elements.

Figure 9 (middle) shows the predicted variations of pressure (dashed line) and Mach numbers (solid line) at the upper wall and the centreline. The experimental data (open circles) for pressure have been read from a figure in [13]. It is seen that the agreement between experiment and predictions is satisfactory. Note that the predicted Mach number at the upper wall passes through $M = 1$ exactly at the throat ($X/L = 0.5$) and reaches supersonic state $M = 2.01$ at exit. At the centreline, however, $M = 1$ location is *downstream* of the throat. Finally, Fig. 9 (bottom) shows the iso-Mach contours. Notice that the iso-Mach lines are slanted which have been found to be in agreement with computations of [13].

2.4 Main Findings

The above examples confirm our theoretical deductions (see Sect. 1.2) on importance of including simple algebraic smoothing pressure correction p'_{sm} on colocated grids to avoid prediction of zig-zag pressure. The efficacy of the smoothing correction has been shown in respect of

1. Incompressible as well as compressible flows.
2. Problems governed by parabolic and elliptic equations.
3. Problems with/without body forces.
4. Structured as well as unstructured grids.
5. Applicability to any boundary condition.

Several other applications of more complex flows can be found in [7, 11, 22, 23, 30] by way of example.

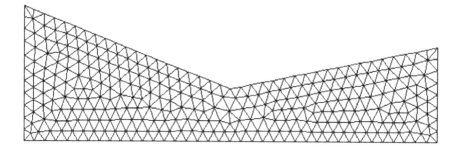

Computational Grid

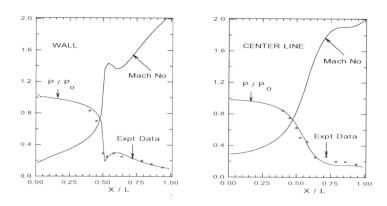

Comarison with Experimental Data [16]

Fig. 9 2D plane nozzle: computational domain (top), variation of p and M (middle), Mach number contours 0.2 (0.1) 2.0 (bottom)

3 Interfacial Flows—Volume Error

3.1 Single-Fluid Formalism

In interfacial unsteady flows of two immiscible fluids, a problem of loss of fluid volume/mass is encountered when computing in discretised space. This problem is particularly severe when the density ratio of the two fluids is large (as, for example, in air and water). Such interfacial problems are often solved by employing *single-fluid formalism* in which the flow of two fluids is treated as that of a *single fluid* whose properties change abruptly across the interface. In such flows, the governing equations are [17]

$$\frac{\partial u_{f,j}}{\partial x_j} = \nabla \cdot \vec{V}_f = 0 \qquad \text{(Volume Conservation)} \qquad (67)$$

$$\frac{\partial (\rho_m u_i)}{\partial t} + \frac{\partial}{\partial x_j}(\rho_m u_{f,j} u_i) = \frac{\partial}{\partial x_j}\left[\mu_m \frac{\partial u_i}{\partial x_j}\right] - \frac{\partial p}{\partial x_i}$$

$$+ \rho_m g_i + F_{st,i} + \frac{\partial}{\partial x_j}\left[\mu_m \frac{\partial u_j}{\partial x_i}\right] \qquad (68)$$

$$\frac{\partial \rho_m}{\partial t} + \frac{\partial}{\partial x_j}(\rho_m u_{f,j}) = 0 \qquad \text{(Mass Conservation)} \qquad (69)$$

The above equations are special in that the pressure is determined from the continuity equations (67) (or volume conservation equation) but the *superficial density* ρ_m is determined from the conserved scalar equation (69) (or mass conservation equation). Thus, the equations carry characteristics of both an incompressible flow and a compressible flow. As such, incorporation of q_{disc} becomes important. Then, following from Eq. 10 and combining Eqs. 67 and 69, it can be shown that

$$p'_{sm} = q_{\text{disc}} = \frac{1}{2}(p - \overline{p}) + \gamma \nabla \cdot V_f$$

$$= \frac{1}{2}(p - \overline{p}) - \frac{\gamma}{\rho_m}\left\{\frac{\partial \rho_m}{\partial t} + u_{f,j}\frac{\partial \rho_m}{\partial x_j}\right\}$$

$$= \frac{1}{2}(p - \overline{p}) - \gamma \frac{D}{Dt}(\ln \rho_m) \qquad (70)$$

Notice that the last term will become significant (in the neighbourhood of the interface in discretised CFD) when density ratio of the two fluids is large. Most authors (see, for example [9, 25, 31–33]) instead of solving Eq. 69, solve an equation for *Volume Fraction F* of the heavier fluid a (say). *F* is defined as

$$F \equiv \frac{\rho_m - \rho_b}{\rho_a - \rho_b} \qquad (71)$$

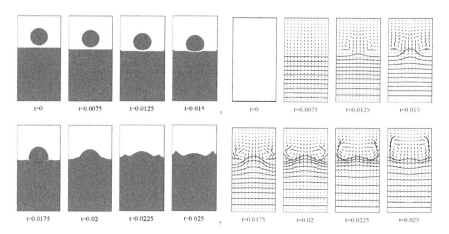

Fig. 10 Splashing of a water drop on a surface: $F = 0.5$ contours (left), pressure and velocity vectors (right)

where suffix b stands for the lighter fluid.[11] The interface is now notionally identified with $F = 0.5$. The predicted smeared (about $F = 0.5$) F-distribution at a time step is then *corrected* to conserve volume in the two-fluid CVs (in which $0 < F < 1.0$ as shown in Fig. 13-left) on the basis of *geometric* considerations. The correction procedures become extremely complex in 3D flows [25]. Other authors identify the interface with *Level Set* (usually zero) and define a level-set distance function [31]. The level-set equation is same as the F-equation but the correction procedure is invoked such that $| \bigtriangledown \cdot F | = 1$. In both procedures, apart from the additional work requirement, volume/mass balance errors arise [33].

Use of smoothing pressure correction indicated in Eq. 70, on the other hand, is very simple to implement and avoids volume/mass errors (within discretisation errors). To illustrate this, we consider few problems[12] with and without effects of surface tension.

3.2   Problems with $F_{st} = 0$

3.2.1   Splashing of a Drop on a Liquid Surface

Consider a two-dimensional rectangular enclosure (Fig. 10) of 7 mm × 14 mm dimensions. The enclosure is filled with water to a height of 8.75 mm. Initially, a cylindrical

---

[11]Incidentally, the superficial viscosity is now evaluated as $\mu_m = F \, \mu_a + (1 - F) \, \mu_b$.

[12]In all problems, the convective terms are discretised using a Total Variation Diminishing (TVD) scheme [14] to minimise interface smearing around $F = 0.5$. Implementation details are given in [17].

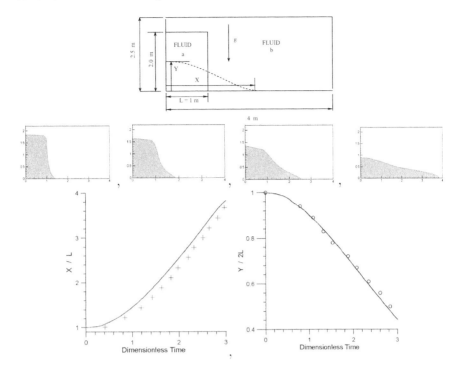

Fig. 11 Collapse of a water column

water drop of radius $r_d = 1.4$ mm is placed above the water surface in air at the centre plane $x_1 = 3.5$ mm and height $x_2 = 10.55$ mm. For $t > 0$, the drop falls under the action of gravity ($g = 9.81$ m/s^2) and splashes on the water surface creating ripples and merges with the body of water.

Computations are performed with 64×128 grid without exploiting symmetry. 50 iterations per time step (10^{-5} s) are required with p' under-relaxed for obtaining convergence. Computations are continued up to $t = 0.025$ s and the interface profiles ($F = 0.5$) are shown in Fig. 10 (left). The present results had maximum volume error[13] 0.002% at the last time step used in the computations. Notice also the smooth pressure prediction and the velocity vectors in Fig. 10 (right). Velocities in air have higher magnitude than in water. These motions cause development of pressure variations that deviate from pure hydrostatic pressure variation.

---

[13] Volume error is defined as

$$\text{Error} (t) = \left(\sum F_{i,j} \, \Delta V_{i,j} \right) / \left(\sum F^0_{i,j} \, \Delta V_{i,j} \right) \tag{72}$$

where F^0 is the initial F-distribution at $t = 0$ and $\Delta V_{i,j}$ is the volume of the cell surrounding node (i, j).

3.2.2   Collapse of a Water Column

Figure 11 (top) shows the problem configuration and the domain of computation. Initially, water (fluid a) column $H = 2$ m high and $L = 1$ m wide is kept at rest by means of a dam. Fluid b is air. At $t = 0$, the dam breaks resulting in collapse of water column followed by horizontal spread of water. Computations are performed with $40(x_1) \times 22(x_2)$ uniform grid. Taking $g = 9.81$, fixed time step $\Delta \tau = 0.001$ is used where $\tau = t \times (2 \times g/L)^{0.5}$. Computations are carried out till $\tau = 3$. At this time, the maximum error in volume balance was found to be 0.432%.

Figure 11 (middle) shows the $F = 0.5$ contours at $\tau = 0.9, 1.4, 2.0$ and 3.0. These contours mimic those computed by Jun and Spalding [12] using the explicit van Leer scheme [35]. Thus, the TVD scheme [14] used in present computations succeeds in sharp interface capturing. The time variations of horizontal spread X and of vertical fall Y (solid lines) are compared in Fig. 11 (bottom) with experimental data (open circles) of Martin and Moyce [15]. The comparison is reasonable notwithstanding the experimental difficulties mentioned by them.

3.2.3   Sloshing in a Tank

Figure 12 (top) shows the problem specification. The tank (40×25 cm) is moved with a horizontal displacement $x_1(t) = A \{\sin(2 \pi f_1 t) - \sin(2 \pi f_2 t)\}$ where $A = 7.5 \times 10^{-3}$, $f_1 = 1.598$ Hz and $f_2 = 1.307$ Hz. Computations are performed with 60×60 cells and time step $\Delta t = 0.001$ s. On all boundaries, $u_1(t) = dx_1/dt$, $u_2 = 0$ was specified at each time step. Nearly 500 iterations per step are required to obtain convergence. The pressure-correction equation was under-relaxed in the first 100

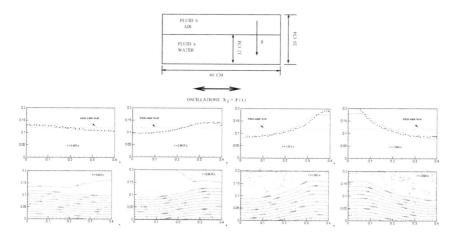

Fig. 12 Sloshing in a tank (top), interface ($F = 0.5$) locations (middle), pressure contours and velocity vectors (bottom) arrow size: 1 cm = 13 cm/s

time steps to procure convergence. At $t = 2.004$ s, the volume error (see Eq. 72) was 0.45%.

In Fig. 12 (middle), the predicted interface profiles (lines) are compared with experimental data (dots) as read from paper by Andrillon and Alessandrini [1]. Again, the agreement is seen to be reasonable. Figure 12 (bottom) shows dimensionless pressure $p^* = (p - p_{min})/(p_{max} - p_{min})$ contours and velocity vectors at different times. The velocity in the air greatly exceeds that in the water. The fluid re-circulations due to interface movement accord with the expectation and the pressure contours are indeed smooth.

3.3 Problems with Finite F_{st}

In computation of many interfacial flows involving merger and splits of the interfaces, the surface tension force F_{st} is included in the momentum equations. The force acts tangent to the interface. However, the interface within a control volume is taken to be *locally* spherical. As such, the *net force* acts *normal to the interface* while the net forces normal to the interface normal cancel out. Thus, the net force per unit volume in direction i is given by [17]

$$F_{st,i} = -\sigma \kappa \frac{\partial F}{\partial x_i} = -\sigma \kappa \frac{\partial F^*}{\partial x_i} \tag{73}$$

$$\text{where} \quad F^* \equiv 0.5 \left\{ 1 + \frac{(F - 0.5)}{|(F - 0.5)|} \right\} \tag{74}$$

where F is the volume fraction of the heavier fluid a and κ is the *interface curvature*. The replacement of F by F^* simply ensures that the surface tension force is evaluated at the interface ($F = 0.5$) only even when the F-distribution is smeared. Of course, $F_b^* = 0$ and $F_a^* = 1$.

3.3.1 Geometric Evaluation of κ

Most authors using VOF or LS methods, evaluate κ from reconstructed F-distribution as

$$\kappa = -\frac{1}{A} \left[\frac{1}{A} \frac{\partial F}{\partial x_i} \frac{\partial A}{\partial x_i} - \frac{\partial^2 F}{\partial x_i^2} \right] \quad \text{summation} \tag{75}$$

$$\text{where} \quad A = \sqrt{(\frac{\partial F}{\partial x_1})^2 + (\frac{\partial F}{\partial x_2})^2 + (\frac{\partial F}{\partial x_3})^2} \tag{76}$$

Following comments are now considered pertinent:

1. Evaluation of κ according to Eq. 75 is complex and is known to introduce discretisation errors. This has been shown by Takahira et al. [33] where a computation of a static bubble surrounded by static liquid generates spurious velocities.
2. Further, it is important to point out that many authors [2, 9, 26] study effect of surface tension coefficient σ by keeping the density and viscosity values of two fluids unchanged.[14] As such, $F_{st,i}$ calculated using Eq. 75 produces different magnitudes of the force (see Eq. 73) for the same fluid pair. However, in the literature, no real two fluid pairs having same values of density and viscosity but different values of σ are found (see, for example [21]).
3. The source of the difficulty mentioned above, however, can be traced to non-dimensionalisation of momentum equation (68). Many authors (see [33], for example) use reference velocity U, reference length L and reference properties ρ_a and μ_a to non-dimensionalise equation (68). The dimensionless equation then reads

$$\frac{\partial \rho_m^* u_i^*}{\partial t^*} + \frac{\partial \rho_m^* u_{f,j}^* u_i^*}{\partial x_j^*} = \frac{1}{Re} \frac{\partial}{\partial x_j^*} \left[\mu_m^* \frac{\partial u_i^*}{\partial x_j^*} \right] - \frac{\partial p^*}{\partial x_i^*}$$

$$+ \rho_m^* g_i^* - \frac{\kappa^*}{We} \frac{\partial F^*}{\partial x_i^*} + \frac{1}{Re} \frac{\partial}{\partial x_j^*} \left[\mu_m^* \frac{\partial u_j^*}{\partial x_i^*} \right] \quad (77)$$

where the dimensionless terms are

$$Re = \frac{\rho_a \, U \, L}{\mu_a} \qquad \text{Reynolds number} \qquad (78)$$

$$g^* = \frac{g \, L}{U^2} \qquad \text{Froude number} \qquad (79)$$

$$\kappa^* = \kappa \, L \qquad \text{Dimensionless Curvature} \qquad (80)$$

$$We = \frac{\rho_a \, U^2 \, L}{\sigma} \qquad \text{Weber number} \qquad (81)$$

$$\rho_m^*, \mu_m^* = \frac{\rho_m}{\rho_a}, \frac{\mu_m}{\mu_a} \qquad \text{Dimensionless properties} \qquad (82)$$

$$u_i^*, p^* = \frac{u_i}{U}, \frac{p}{\rho_a \, U^2} \qquad \text{Dimensionless Velocity and Pressure} \qquad (83)$$

$$x_i^*, t^* = \frac{x_i}{L}, \frac{t}{L/U} \qquad \text{Dimensionless coordinates and Time} \qquad (84)$$

Equation 77 thus shows that since κ^* is evaluated geometrically from F-distribution, the Weber number (We) now *appears* to be an *independent parameter* of the flow system. This interpretation leads to investigation of effect of surface tension coefficient σ (or We) for the same fluid pair. In the discussion below, we show that this is misleading and that *Weber number is not an independent parameter of the flow system.*

---

[14]This ignores the fact that σ is essentially a property of a specified fluid pair (a, b).

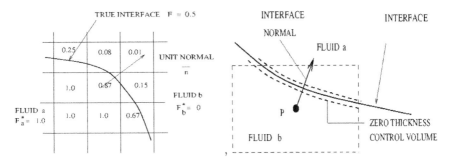

Fig. 13 Illustrative F-distribution and interface normal (left), zero-thickness CV surrounding an interface (right)

3.3.2 Fluid Dynamic Evaluation of κ

In view of the last comment, we consider an alternative approach to evaluation of κ and $F_{st,i}$. Thus, we assume that the interface is a surface of *zero thickness having no physical properties* as shown in Fig. 13 (right). Then, taking the dot product of momentum equation (68) (along with Eq. 73) with interface normal $\vec{n}$ will result in

$$\frac{\partial}{\partial n}(p - \tau_{nn}) = -\sigma\,\kappa\,\frac{\partial F^*}{\partial n} \qquad \rightarrow \qquad \tau_{nn} = 2\,\mu\,\frac{\partial V_n}{\partial n} \qquad (85)$$

and V_n is the velocity normal to the interface. Note that the unsteady, convective and gravity terms disappear because the interface has zero thickness and no mass. However, invoking the Stokes's continuum requirement (with $q = 0$ for a control volume of zero thickness normal to the interface), we have

$$p - \tau_{nn} = \overline{p} = -\sigma_n \qquad (86)$$

where σ_n is total normal stress. Hence, Eq. 85 will read as

$$\frac{d\,\overline{p}}{d\,n} = -\sigma\,\kappa\,\frac{d\,F^*}{d\,n} \qquad (87)$$

This equation is same as the familiar Young–Laplace equation for the equilibrium of a static bubble in a static fluid in which $\tau_{nn} = 0$ [29]. Here, τ_{nn} is finite. Now, upon discretisation along the normal to the interface, it follows that (see Fig. 13 (right))

$$\sigma\,\kappa = -\frac{d\,\overline{p}/d\,n}{d\,F^*/d\,n} = -\frac{\overline{p}_b - \overline{p}_a}{F_b^* - F_a^*} = -\frac{\overline{p}_b - \overline{p}_a}{0 - 1} = \overline{p}_b - \overline{p}_a \qquad (88)$$

and hence

$$F_{st,i} = -\,(\overline{p}_b - \overline{p}_a)\,\frac{\partial F}{\partial x_i} \tag{89}$$

This *Fluid Dynamic expression* with mean pressure difference $(\overline{p}_b - \overline{p}_a)$ on either side of the interface is again a new result. On both structured and unstructured grids, the difference of average pressures can be evaluated in the following Eq. 15:

$$\sigma\,\kappa = (\overline{p}_b - \overline{p}_a)_P = \frac{1}{3}\sum_{i=1}^{3}(\overline{p}_b - \overline{p}_a)_{x_i,P} \tag{90}$$

$$(\overline{p}_b - \overline{p}_a)_{x_i,P} = \text{solution of} \quad \frac{\partial^2}{\partial x_i^2}\left[p\,(1 - 2\,F^*)\right]_P = 0 \tag{91}$$

This manner of evaluation ensures that $\sigma\,\kappa$ is calculated at the interface ($F = 0.5$) only.

Thus, Eq. 89 evaluates the surface tension force directly from flow variables without evaluating the interface curvature. As such, the remarkable feature of this expression is that it *does not require knowledge of surface tension coefficient* σ.

3.3.3 Bursting of an Air Bubble Through Water

To demonstrate validity of two different methods of evaluating the surface tension force, we consider the bubble-burst problem (see Fig. 14). This problem has been solved by Takahira et al. [33] using the level-set method using $60 \times 60 \times 120$ grid and the surface tension force is evaluated from geometric considerations (which

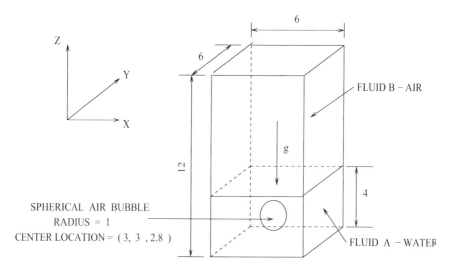

Fig. 14 Bursting of a bubble

(a) (b) (c)

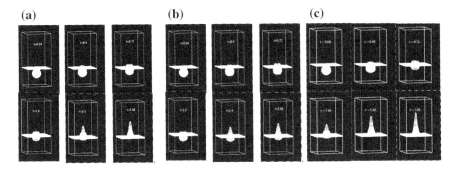

Fig. 15 Interface evolutions during bursting: **a** geometric evaluation of F_{st}, **b** fluid dynamic evaluation of F_{st}, **c** $F_{st} = 0$

requires knowledge of the value of σ). Here, in order to save computer time, the same problem is solved on coarser grid of dimensions $30 \times 30 \times 60$. Maximum of 50 iterations per time step ($\Delta \tau = 0.001$) are required to procure convergence. The reference velocity U_{ref}, properties ρ_a, μ_a and surface tension coefficient σ are chosen such that

$$Re = \frac{\rho_a \, U_{ref} \, D}{\mu_a} = 474 \qquad We = \frac{\rho_a \, U_{ref}^2 \, D}{\sigma} = 1.0$$

$$Fr = \frac{U_{ref}^2}{g \, D} = 0.64 \qquad \frac{\rho_b}{\rho_a} = 0.001 \qquad \frac{\mu_b}{\mu_a} = 0.01$$

The problem has been computed in three different ways as follows:

1. F_{st} from Eqs. 73, 75 and 76 → Fig. 15a.
2. F_{st} from Eqs. 89, 90 and 91 → Fig. 15b.
3. Ignoring surface tension force. That is, $F_{st} = 0$ → Fig. 15c.

Figure 15a, b show that predictions with both types of evaluations of F_{st} are nearly identical. The predictions also accord with the fine grid solutions obtained by [33] although the burst heights at large times are somewhat smaller due to coarseness of the grid used here. Nonetheless the formation of ripples on the liquid surface is clearly seen at $\tau = 0.4$ and 0.72. Likewise, at $\tau = 1.4$ and 1.68, formation of a *neck* in the entrained liquid suggests that a detached liquid drop is about to form.

Since predictions with both types of evaluations are nearly identical, computations are repeated with $F_{st} = 0$ (see Fig. 15c). It is seen that the burst heights are now greater indicating that the absence of surface tension force fails to minimise the interface surface during bursting, as expected.

Finally, the time variation of volume error (see Eq. 72) evaluated on coarse grid is found to increase with time in an oscillatory manner [18]. At $\tau = 1.68$, the maximum error with fluid dynamic evaluation of F_{st} is 0.078%. The same was found to be 0.093% with geometric evaluation. These errors are much smaller than the ones reported by Takahira et al. [33] on a finer grid.

Incidentally, solution to the problem of merger of two asymmetrically placed air bubbles rising in a box of square cross-section filled with water has been reported in [18]. These solutions further support the deduction that Weber number is not an independent parameter of the system.

4   Conclusions

1. Continuum requirement of Stokes's stress laws is obeyed to

    a. eliminate problem of zig-zagness pressure prediction in computation of incompressible flow on colocated grids.
    b. eliminate (within discretisation errors) the problem of loss of mass/volume at large times encountered in computation of unsteady interfacial flows.
    c. evaluate surface tension force in interfacial flows using dynamic flow variables and *without requiring knowledge of surface tension coefficient* σ.

2. Thermodynamic explanation for validity of Stokes's continuum requirement is presented with new interpretations that follow from its violations.

Acknowledgements It is a privilege for me to contribute to this volume in honour of Prof. D. B. Spalding to whom, in 2004, I sent the initial draft of my book [6] to seek his approval of the contents of the book. He not only read the draft but also gave me a task to solve a 1D problem of highly resisted flow through a porous medium. He solved the problem himself by using what he called *Date-Colocated* procedure using the PHOENICS code. Over exchange of four emails, he was satisfied that my solutions and his were in agreement. I have included with pride that 1D problem in my book. Material of section 3 was developed during sponsorship of Project No: 2005/36/47/BRNS by the Board of Research in Nuclear Science, Department of Atomic Energy, Govt of India. Contribution of Dr. Kausik Nandi, Scientist F, BARC is gratefully acknowledged.

References

1. Andrillon, Y., & Alessandrini, B. (2004). A 2D+T VOF fully coupled formulation for the calculation of breaking free-surface flow. *Journal of Marine Science and Technology, 8*, 159–168.
2. Daly, B. J. (1969). Numerical study of the effect of surface tension on interface instability. *Physics of Fluids, 17*(7), 1340–1354.
3. Date, A. W. (1998). Solution of Navier-Stokes equations on non-staggered grids at all speeds. *Numerical Heat Transfer, Part B, 33*, 451.
4. Date, A. W. (2002). SIMPLE procedure on structured and unstructured meshes with collocated variables. In *Proceedings of the 12th International Heat Transfer Conference*, Grenoble, France.
5. Date, A. W. (2004). Fluid dynamic view of pressure checker-boarding problem and smoothing pressure correction on meshes with collocated variables. *International Journal of Heat and Mass Transfer, 48*, 4885–4898.
6. Date, A. (2005). *Introduction to computational fluid dynamics*. New York: Cambridge University Press.

7. Date, A. W. (2008). Computational fluid dynamics. In *Kirk-Othmer encyclopedia of chemical technology*, NY, USA.
8. Ferziger, J. H., & Peric, M. (1999). *Computational methods for fluid dynamics* (2nd ed.). Springer.
9. Gerlach, D., Tomar, G., Biswas, G., & Durst, F. (2005). Comparison of volume-of-fluid methods for surface tension dominant two phase flows. *International Journal of Heat and Mass Transfer, 49,* 740–754.
10. Holman, J. P. (1980). *Thermodynamics* (3rd ed.). Tokyo: McGraw-Hill Kogakusha.
11. Jagad, P. I., Puranik, B. P., & Date, A. W. (2015). A novel concept of measuring mass flow rates using flow induced stresses. *SADHANA, 40*(Part 5), 1555–1566.
12. Jun, L., & Spalding, D. B. (1988). Numerical simulation of flows with moving interfaces. *Physico-Chemical Hydrodynamics, 10,* 625–637.
13. Karki, K. C., & Patankar, S. V. (1988). A pressure based calculation procedure for viscous flows at all speeds in arbitrary configurations. In *AIAA 26th Aerospace Science Meeting*, Paper No AIAA-88-0058, Nevada, USA.
14. Lin, C. H., & Lin, C. A. (1997). Simple high-order bounded convection scheme to model discontinuities. *AIAA Journal, 35,* 563–565.
15. Martin, J. C., & Moyce, W. J. (1952). An experimental study of collapse of liquid columns on a rigid horizontal plane. *Philosophical Transactions of the Royal Society A: Mathematical, Physical and Engineering Sciences, 244,* 312–324.
16. Mason, M. L., Putnam, L. E., & Re, R. J. (1980). The effect of throat contouring on two-dimensional converging-diverging nozzles at static conditions. NASA Technical Paper, 1704.
17. Nandi, K., & Date, A. W. (2009). Formulation of fully implicit method for simulation of flows with interfaces using primitive variables. *International Journal of Heat and Mass Transfer, 52,* 3217–3224.
18. Nandi, K., & Date, A. W. (2009). Validation of fully implicit method for simulation of flows with interfaces using primitive variables. *International Journal of Heat and Mass Transfer, 52,* 3225–3234.
19. Patankar, S. V., & Spalding, D. B. (1971). A calculation procedure for heat mass and momentum transfer in three-dimensional parabolic flows. *International Journal of Heat and Mass Transfer, 15,* 1787.
20. Patankar, S. V. (1981). *Numerical fluid flow and heat transfer*. New York: Hemisphere Publ Co.
21. Perry, R. H., & Chilton, C. H. *Chemical engineers handbook* (5th ed.). Tokyo: McGraw-Hill Kogakusha.
22. Pimpalnerkar, S., Kulkarni, M., & Date, A. W. (2005). Solution of transport equations on unstructured meshes with cell-centered collocated variables. Part II: Applications. *International Journal of Heat and Mass Transfer, 48,* 1128–1136.
23. Ray, S., & Date, A. W. (2003). Friction and heat transfer characteristics of flow through square duct with twisted tape insert. *International Journal of Heat and Mass Transfer, 46,* 889–902.
24. Rhie, C. M., & Chow, W. L. (1983). A numerical study of the turbulent flow past an isolated airfoil with trailing edge separation. *AIAA Journal, 21,* 1525.
25. Rudman, M. (1997). Volume-tracking methods for interfacial flow calculations. *International Journal for Numerical Methods in Fluids, 24,* 671–691.
26. Salih, A., & Ghosh Moulic, S. (2006). Simulation of Rayleigh-Taylor instability using level-set method. In *33rd National and 3rd International Conference on Fluid Mechanics and Fluid Power* (p. 2006), Paper no 1303. India: IIT Bombay.
27. Schlichting. (1968). *Boundary layer theory* (English trans. Kestin J.). McGraw-Hill.
28. Shih, T. M., & Ren, A. L. (1984). Primitive variable formulations using non-staggered grid. *Numerical Heat Transfer, 7,* 413–428.
29. Shyy, W. (1994). *Computational modeling for fluid flow and interfacial transport*. Amsterdam: Elsevier.
30. Soni, B., & Date, A. W. (2011). Prediction of turbulent heat transfer in radially outward flow in twisted-tape inserted tube rotating in orthogonal mode. *Computational Thermal Science, 3,* 49–61.

31. Sussman, M., Smereka, P., & Osher, S. (1994). A level set approach for capturing solutions to incompressible two-phase flow. *Journal of Computational Physics, 114*, 146–159.
32. Sussman, M., Smith, K. M., Hussaini, M. Y., Ohta, M., & Zhi-Wei, R. (2007). A sharp interface method for incompressible two-phase flows. *Journal of Computational Physics, 221*, 469–505.
33. Takahira, H., Horiuchi, T., & Banerjee, S. (2004). An improved three dimensional level set method for gas-liquid two-phase flows. *Transaction of the ASME Journal of Fluids Engineering, 126*, 578–585.
34. Thiart, G. D. (1990). Improved finite-difference formulation for convective-diffusive problems with SIMPLEN algorithm. *Numerical Heat Transfer, Part B, 18*, 81–95.
35. van Leer, B. (1977). Towards the ultimate conservative difference scheme IV, a new approach to numerical convection. *Journal of Computational Physics, 23*, 276–283.
36. Warsi, Z. U. A. (1993). *Fluid dynamics—Theoretical and computational approaches*. London: CRC Press.

The SUPER Numerical Scheme for the Discretization of the Convection Terms in Computational Fluid Dynamics Computations

N. C. Markatos and D. P. Karadimou

1 Introduction

Numerical diffusion is a significant source of error in numerical solution of conservation equations and can be separated into two components, namely, cross-stream and streamwise numerical diffusion. The former occurs when gradients in a convected quantity exist perpendicular to the flow and the direction of the flow is oblique to the finite volumes, i.e. due to the multidimensional nature of the flow. The latter happens when gradients in a convected quantity exist parallel to the flow even in one-dimensional situations [1–3].

According to Vahl Davis et al. [4] a theoretical approximation for the false diffusion term in two-dimensional geometries is

$$\Gamma_{\text{false}} = \frac{\rho U \Delta x \Delta y \sin 2\theta}{4 \left(\Delta y \sin^3 \theta + \Delta x \cos^3 \theta \right)}, \tag{1.1}$$

where U is the resultant velocity and θ is the angle, which lies between $0°$ and $90°$, made by the velocity vector with the x-direction. Expression (1.1) reveals that there is no false diffusion present for $\theta = 0°$ and $\theta = 90°$, and that it is at its maximum at $\theta = 45°$.

The first-order accurate upwind difference scheme is more stable than the second-order accurate central-difference scheme or higher order accurate schemes, e.g.

N. C. Markatos (✉) · D. P. Karadimou
School of Chemical Engineering, National Technical University
of Athens, Zografou Campus, 15780 Athens, Greece
e-mail: n.markatos@ntua.gr; nicholas.markatos@qatar.tamu.edu

D. P. Karadimou
e-mail: dkaradimou@gmail.com

N. C. Markatos
Texas A&M University at Qatar, Education City, Doha, Qatar

© Springer Nature Singapore Pte Ltd. 2020
A. Runchal (ed.), *50 Years of CFD in Engineering Sciences*,
https://doi.org/10.1007/978-981-15-2670-1_3

QUICK at high grid Peclet numbers. Non-linear schemes, e.g. van LEER scheme appear highly stable but they do not deal with the flow direction inclination at all [5–8].

One way to overcome diffusion errors is to use an upwind approximation, which essentially follows the streamlines. This approach, originally derived by Raithby [9], is formally called the skew upwind differencing scheme. The Corner UPwInDing (CUPID) scheme retains the general objectives of the Raithby's approach but uses an entirely different 2D formulation that eliminates the shortcomings of the original scheme. The Skew Upwind Corner Convection Algorithm (SUCCA) scheme is based on a simplified formulation of the CUPID scheme [10, 11]. Among the numerical schemes that are evaluated is the SUPER (Skew UPwind and cornER algorithm) numerical scheme that takes into account the 3D flow orientation and follows all the rules for the successful solution of the equations [12].

Numerical dispersion comes about when the convective scheme used is unstable and large gradients are present [3]. The use of the upwind or the hybrid numerical scheme of Spalding [6] ensures the stability of the calculations but the first-order accuracy makes them prone to streamwise numerical diffusion errors. Higher order schemes involve more neighbour points and reduce the streamwise false diffusion by bringing in a wider influence. These schemes are based on one-dimensional formulations that do not take into account the flow direction, which makes them unstable at high grid Peclet numbers [13]. Furthermore, numerical schemes that take into account the flow direction become unstable at high grid Peclet numbers in regions of high convection discontinuities. The phenomenon of the numerical dispersion causes spatial oscillations that lead to significant numerical errors and non-physical results [1, 2]. Generally, the numerical schemes should follow some basic rules in order to be able to calculate a successful solution [13]:

(a) The formulation of the scheme should ensure the flux consistency at the cell faces in adjacent control volumes satisfying the conservative property.

(b) The Conservation principle should be retained for all the variables everywhere and over the whole domain.

(c) The transportiveness property of a fluid flow is the effect of convection and diffusion from the nearby nodes to the central node. The non-dimensional Peclet number is a measure of the relative strengths of convection and diffusion. The scheme should recognize the relationship between the directionality of influencing and the flow direction and the magnitude of the Peclet number.

(d) The influence coefficients in the finite volume equations should be always positive.

(e) The neighbouring coefficients for the solution control volume of central node P should obey the following relation: $\sum a_{nb} = a_p$, so as to be consistent with the differential transport equation.

(f) Satisfying the Scarborough criterion $\frac{\sum |a_{nb}| \leq 1 \text{ at all nodes}}{|a'_p| < 1 \text{ at least one node}}$ ensures that the resulting matrix of coefficients is diagonally dominant which is a sufficient condition for a converged iterative method.

(g) Linearizing the source term with a negative slope, so that the incidence of unbounded solutions is reduced.
(h) The formulation of the scheme should be easy to implement without increasing the computational cost.

The method employed in this study follows the so-called 'control volume' approach, which is developed by formulating the governing equations on the basis that mass, heat and momentum fluxes are balanced over control volumes. If there are two phases the control volume can be regarded as containing a volume fraction of each phase (r_i) that obey the following relation:

$$r_1 + r_2 = 1$$

Each phase is treated as a continuum in the control volume under consideration. The phases share the control volume and they may, as they move within it, interpenetrate [14, 15].

2 Mathematical Modelling

The general form of the differential transport equations in the Cartesian coordinate system is described as follows [14, 15]:

$$\frac{\partial}{\partial t}(r_i \rho_i \phi_i) + div\left(r_i \rho_\iota \vec{u}_i \, \phi_\iota - r_i \Gamma_{\phi,\iota} grad\phi_\iota\right) = S_{\phi,i}, \tag{2.1}$$

where

ϕ_i the dependent variable of each phase i
r_i the volume fraction of each phase, (m^3/m^3)
ρ_i the density of each phase, (kg/m^3)
u_i the velocity vector of each phase, (m/s)
$\Gamma_{\phi,i}$ within-phase diffusion coefficient
$S_{\phi,i}$ within-phase volumetric sources, (kg/m^3 s)

The dependent variables solved for are

(a) Pressure that is shared by the two phases, P (Pa)
(b) Volume fractions of each phase, r_i (m^3/m^3)
(c) Three components of velocity for each phase, .., v_i, w_i (m/s)
(d) Turbulence kinetic energy (the kinetic energy of the fluctuating velocity components) and dissipation rate of turbulence (the rate of transformation of turbulence energy to internal energy) for the first phase (air), k and ε (m^2/s^2 and m^2/s^3)
(e) Temperature T (K)
(f) Enthalpy h (J/kg)

The phase volume—fraction equation is obtained from the continuity equation:

$$\frac{\partial(r_i\rho_i)}{\partial t} + div(r_i\rho_i\vec{u}_i) = S_{\phi,i}, \tag{2.2}$$

where

r_i phase volume fraction, (m^3/m^3).
ρ_i phase density, (kg/m^3).
$\vec{u}_i$ phase velocity vector, (m/s).
$S_{\phi,i}$ net rate of mass entering phase i from phase j (kg/m^3 s), if there is phase change.

The system of discretized, by the finite volume method, conservation equations is solved by the SIMPLEST and IPSA algorithms [14] embodied in the Computational Fluid Dynamics (CFD) code PHOENICS [16].

3   Discretization Schemes for Convection: Previous Contributions and the Present Proposal

A wide variety of schemes has been proposed to reduce false diffusion errors. An assessment of how interactions between false diffusion, numerical stability and the source formulation affect the accuracy and convergence of various schemes has been carried out by Patel et al. [10]. It is worth pointing out that all the schemes, which essentially involve discretization improvements on the simple upwind representation of the convection term, have severe restrictions on their utility. Most higher order schemes fail to converge easily at high grid Peclet numbers. For flows inclined to the grids, performance is not significantly improved, because although the truncation/discretization errors are reduced in the cell, the multidimensional false diffusion problem is not being dealt at all [1, 2, 10]. Raithby [9] described a class of schemes, which allow for grid/flow direction inclination. These schemes tend to involve (in 2D) the diagonal neighbouring cells as well as those opposite to each cell face. The SKEW schemes based upon Raithby's idea suffer from a number of shortcomings [1, 17] that can be summarized as follows:

(a) The influence coefficients in the finite volume equations are not constrained to be non-negative. Consequently, there are occasions when the scheme will produce non-physical oscillatory solutions or fail to converge at all.
(b) Even though the formulation was carried out in the context of the control volume framework (which normally guarantees conservation) the scheme may actually be non-conservative.
(c) The scheme is complex to implement; also, it is slow to converge when it does so, and thus, expensive in computer time.

Patel et al. [10] retained the general objectives of Raithby's scheme and derived a new discretization scheme, the so-called CUPID (Corner UPwInDing) scheme, in a finite difference-based approach, using an entirely different formulation to ensure that

(a) The influence coefficients are always positive, so that non-physical oscillations are avoided and convergence is enhanced.
(b) Conservation is retained.
(c) The scheme is easy to implement and efficient in computer time.

The central idea of the scheme was to concentrate attention at the corners of the finite volume cells rather than Raithby's approach of concentrating at the cell faces. This approach was afterwards followed and simplified by Carey et al. [11], who thus created the SUCCA scheme; but the analysis was restricted to 2D cases.

The method described in this paper retains the objectives of the CUPID and SUCCA schemes and reformulates the convection terms in the momentum and scalar conservation equations in a novel way to treat 3D flows. Scanlon et al. [18], who analysed the transport of a scalar variable in a three-dimensional cavity test case, have carried out an example of implementation of the SUCCA scheme in 3D flows. The new, proposed here, numerical scheme formally called SUPER (Skew UPwind and cornER Algorithm) is validated against experimental data from the literature [19], in the case of dilute indoor aerosol airflow in a model room geometry. Moreover, the numerical results of the SUPER scheme are compared with those obtained by applying the unidirectional upwind scheme in the case of inclined inflow, at an angle of ($\theta = 45°$), when false diffusion is expected to be at its maximum.

As mentioned above, the equations in the present work are discretized by the finite volume method and solved by the IPSA algorithm [14, 15], embodied in the PHOENICS Computational Fluid Dynamics program [16]. A fully implicit method is employed for the numerical solution. The convection terms of all the conservation equations of the first phase are discretized by (a) the new SUPER scheme and (b) the conventional upwind, hybrid and Van Leer schemes, for comparison. The convection terms of the conservation equations of the second phase are discretized by the upwind scheme. The diffusion terms are discretized by the central-differencing scheme. All results presented in Sect. 4 below are grid-independent and time-step independent.

3.1 The Novel SUPER Scheme Applied to the Momentum Equations in 3D Coordinate Systems

The numerical scheme proposed and evaluated in this study has been formally titled SUPER (Skew UPwind and Corner Algorithm) and is an extension of the CUPID scheme of Patel et al. [10], as simplified by Carey et al. [11], but extended here to three dimensions and applied also to the momentum equations. In this study, the SUPER scheme is applied to analyse the momentum transport as well as the scalar transport, in a finite volume-based approach in x, y and z directions, so that it becomes appropriate for 3D flows. It aims at minimizing the numerical false diffusion errors,

Fig. 1 Segment of the
calculation domain
(seven-cell grid cluster)

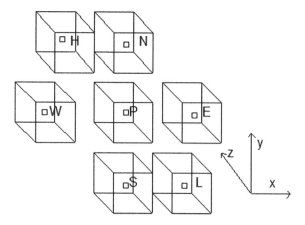

especially in regions where the local flow direction is inclined at an angle to the
solution cell.

The cell grid cluster on which the discretized conservation equations are based,
within the PHOENICS code, comprises seven cells (control volumes) for a three-
dimensional problem. The seven-cell cluster includes one central grid cell (current
solution cell), where the value of the central coefficient of the discretized conservation
equations is saved at the grid node P, and six neighbouring cells, situated at the North,
South, East, West, High, Low sides of the central grid cell, for saving the values of
the neighbouring coefficients of the discretized conservation equations (Fig. 1) [12].

Improvement of the accuracy applying the SUPER numerical scheme stems from
taking into account the influence of additional neighbouring cells when formulating
the discretized conservation equations for each cell. In a 3D flow problem two more
coefficients (high and low coefficients) need to be included in the discretized equations, furthermore, the influence of high and low flow directions needs to be added
when formulating each coefficient.

The possible additional neighbouring cells in Y-X, Z-X and Y-Z planes that are
used by the implementation of the SUPER numerical scheme in a 3D flow problem
are illustrated in Fig. 2 [12].

The overall number of the possible additional neighbouring cells in a x-y-z coordinate system is illustrated in Fig. 3. The names of the central grid nodes of the
cells, where the values of the possible neighbouring coefficients of the discretized
momentum equations are saved, when the SUPER scheme is employed, are based
on the position of each coefficient relative to the position of the central grid cell. The
SUPER scheme is implemented based on a staggered grid for the velocity components. Scalar variables, such as pressure, temperature and density are evaluated at
the central grid nodes and velocity components are calculated on the cell faces.

In Fig. 3 eight additional neighbouring cells are distinguished which are possible
flow paths [12]. The distances between adjacent cells have been exaggerated in the
drawing to highlight the position of the coefficients while, in reality, cell faces of
neighbouring coefficients are common, of course.

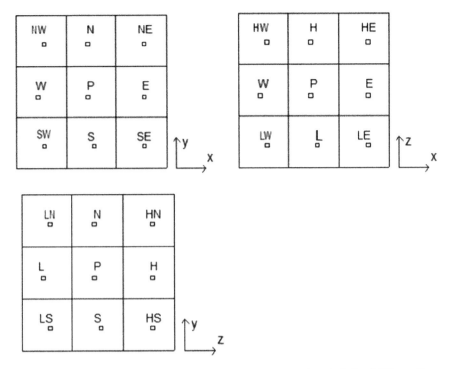

Fig. 2 Segment of the calculation domain for the SUPER scheme in *Y-X*, *Z-X* and *Y-Z* coordinate planes

The central idea of the scheme formulation lies in considering the neighbouring control volume velocities as situated at the corner of the central grid cell and distributing the mass flow rates and transported properties according to the local flow angle. The form of the coefficients is based on linear interpolation of lengths that are formed using the corner velocity vector, extrapolated from the current solution cell grid node *P*. This is done in a way such that it intersects the horizontal line connecting the two immediate nodes (one corner and one face cell grid node). Specifically, the velocity vector extrapolated from the central grid node *P* intersects the *HS* and *S* cell grid nodes at the point *C* and forms the lengths $S - C$ and $C - HS$ as shown in Fig. 4 in the *Z-Y* plane. The ratio of the SUPER coefficients is seen to be directly proportional to the ratio of lengths $S - C$ and $C - HS$ (Fig. 4) [12, 19].

$$\text{length ratio} = \frac{S - C}{C - HS} = \frac{c_{hs}}{c_s} = \text{coefficient ratio}.$$

Considering the High and South corner cell inflow from the East side to the central grid cell *P* (Fig. 1) the SUPER algorithm may be written for the convective transport of variable ϕ ($\phi = V_{ix}, V_{iy}, V_{iz}$) as (the dot denotes fluxes):

Fig. 3 Segment of the
calculation domain for the
SUPER scheme in three
directions (27-cell grid
cluster)

$$c_p \phi_P = \left(\left(\dot{m}_s - \frac{(\dot{m}_h)^2}{\dot{m}_s} \right) \phi_S + \left(\dot{m}_h + \frac{(\dot{m}_h)^2}{\dot{m}_s} \right) \phi_{HS} + 0 \cdot \phi_H \right) \text{ for } 45 < \theta < 90,$$

$$(3.1)$$

i.e. the form of the relevant finite volume equation is $c_p \phi_P = c_s \phi_S + c_{hs} \phi_{HS}$,
where $c_s = \dot{m}_s - \frac{(\dot{m}_h)^2}{\dot{m}_s}$ and $c_{hs} = \dot{m}_h + \frac{(\dot{m}_h)^2}{\dot{m}_s}$

Fig. 4 Illustration of the SUPER scheme concept (45 < θ < 90) in Z-Y plane, when considering the High-South corner inflow

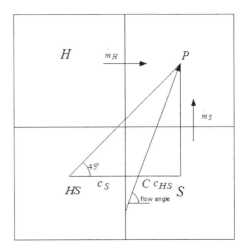

$$c_p \phi_P = \left(\left(\dot{m}_h - \frac{(\dot{m}_s)^2}{\dot{m}_h} \right) \phi_H + \left(\dot{m}_s + \frac{(\dot{m}_s)^2}{\dot{m}_h} \right) \phi_{HS} + 0 \cdot \phi_S \right) \text{ for } 0 < \theta \leq 45,$$

(3.2)

i.e. the form of the relevant finite volume equation is $c_p \phi_P = c_h \phi_H + c_{hs} \phi_{HS}$, where $c_h = \dot{m}_h - \frac{(\dot{m}_s)^2}{\dot{m}_h}$ and $c_{hs} = \dot{m}_s + \frac{(\dot{m}_s)^2}{\dot{m}_h}$.

Moreover, in a 3D flow problem the influence of eight more additional neighbouring cells is taken into account, when the incoming cell mass fluxes from neighbouring face cells are equal ($\theta = 45$), in the one 2D coordinate system, and different ($\theta \neq 45$) in the other two 2D coordinate systems. These additional neighbouring cells are situated at the High-North and East (HNE), High-North and West (HNW), High-South and East (HSE), High-South and West (HSW), Low-North and East (LNE), Low-North and West (LNW), Low-South and East (LSE), Low-South and West (LSW) sides of the central grid cell of the 27-cell grid cluster illustrated in Fig. 3 [12].

$$c_p \phi_P = \left(\left(\dot{m}_h - \frac{(\dot{m}_s)^2}{\dot{m}_h} \right) \phi_H + \left(\dot{m}_s + \frac{(\dot{m}_s)^2}{\dot{m}_h} \right) \phi_{HSE} + 0 \cdot \phi_S \right) \text{ for } \theta = 45^\circ$$

(3.3)

in the X-Y plane and $\theta \neq 45$ in the Y-Z and X-Z planes,

i.e. the form of the relevant finite volume equation is $c_p \phi_P = c_h \phi_H + c_{hse} \phi_{HSE}$, where $c_h = \dot{m}_h - \frac{(\dot{m}_s)^2}{\dot{m}_h}$ and $c_{hse} = \dot{m}_s + \frac{(\dot{m}_s)^2}{\dot{m}_h}$.

Furthermore, in the case of zero skew flow angle the normal hybrid upwind scheme is obtained by default (Fig. 5).

The SUPER scheme should comply with all the basic rules that a general numerical scheme must obey and must overcome the limitations of earlier alternative algorithms

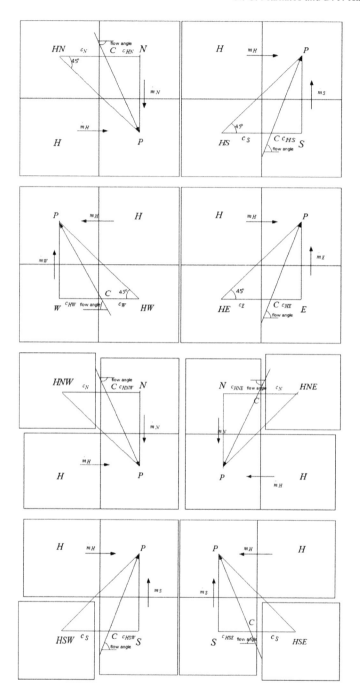

Fig. 5 Illustration of possible mass inflow sources when forming the High coefficient within the SUPER scheme concept

Fig. 6 Momentum control
volumes in the Z-Y plane

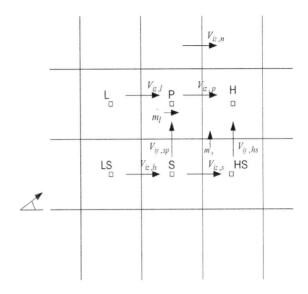

(e.g. SKEW scheme, Raithby [9], QUICK[1] scheme, Leonard [7]) constructed to reduce the effects of multidimensional false diffusion.

The following three-dimensional steady-state convection–diffusion example, illustrated in Fig. 6 in the Z-Y plane, shows how the SUPER scheme is applied to the transport of momentum when the Low-South corner inflow is taken into account [12].

The solution variable is the velocity component in the z-direction, V_{iz} and the local flow is from the Low and South corner cell to the central grid cell with the local flow direction lying within the range $0 < \theta < 45$. The mass fluxes $\dot{m}_l$, $\dot{m}_s$ from the Low and South grid cells to the solution cell respectively, are

$$\dot{m}_l = \rho_p A_p \frac{\left(V_{iz.p} + V_{iz.l}\right)}{2} \tag{3.4}$$

$$\dot{m}_s = \frac{\left(\dot{m}_{sp} + \dot{m}_{hs}\right)}{2}, \tag{3.5}$$

where ρ_p is the value of density at the central node of the solution cell, A_p the face area of the solution cell, $V_{iz.p}$ is the value of the velocity component in the z-direction at the central node of the solution cell, $V_{iz.l}$ is the value of the velocity component in the z-direction at the low face node of the solution cell, $\dot{m}_{sp}$ is the mass flux from the South grid cell to the solution cell and $\dot{m}_{hs}$ is the mass flux from the cell situated at the High and South sides of the central grid cell to the solution cell, $\dot{m}_s$ is the mass flux from the South grid cell to the solution cell as modified by the SUPER scheme.

---

[1]Quadratic-Upwind Differencing Scheme.

On the first iterative sweep through the calculation domain, the default (hybrid) discretization scheme is used. This is done to achieve initial values for the mass fluxes, which are to be modified within the SUPER scheme from the first sweep onwards.

For the normal hybrid discretization scheme, the discretized momentum equation in the IPSA algorithm is of the form:

$$a_p V_{iz,p} = a_s V_{iz,s} + a_w V_{iz,w} + a_e V_{iz,e} + a_n V_{iz,n}$$
$$+ \alpha_h V_{iz,h} + \alpha_l V_{iz,l} + b + (P_P - P_E)A_p, \tag{3.6}$$

where

b source term
P_p value of pressure at the central node of the solution cell
P_E value of pressure at the central node of the cell of the East coefficient
A_p face area of the solution cell

The coefficients α_s, α_w, α_e, α_n, α_l, α_h of the discretized momentum equation in the above example contain as follows:

α_s Convection plus diffusion terms
α_w Convection term only
α_e Diffusion term only
α_n Diffusion term only
α_l Convection term only
α_h Diffusion term only

It should be noted that the downwind convection coefficients within a_n, a_e, a_h have been negated due to the upwinding nature of the hybrid scheme and that the mass fluxes are contained within the momentum coefficients a_s, a_w, and α_l.

The influence of the additional neighbouring cells is taken into account by including it in the source term of the equation and not by increasing the number of terms (thus retaining the tri-diagonal character of the matrices).

As it stands, this discretization process will promote the numerical diffusion of momentum normal to the streamlines. The SUPER scheme is implemented on the second sweep by modifying the mass fluxes (convection coefficients) as in Eq. (3.1), but considering the Low-South corner inflow.

The discretized equation for the SUPER scheme takes the form:

$$a_p V_{iz,p} = a_s V_{iz,s} + a_w V_{iz,w} + a_e V_{iz,e} + a_n V_{iz,n} + \alpha_h V_{iz,h}$$
$$+ \alpha_l V_{iz,l} + b + (P_P - P_E)A_p, \tag{3.7}$$

where the SUPER scheme sets the following parameters:

(a) $c_s = 0$ (i.e. the convective part of coefficient a_s)

(b) $a_l = c_l =$ modified convection coefficient $= \left(\dot{m}_l - \frac{(\dot{m}_s)^2}{\dot{m}_l} \right)$

(c) b_s = created convection coefficient = $\left(\dot{m}_s + \frac{(\dot{m}_s)^2}{\dot{m}_l} \right) = c_{ls}$

The term $b_s V_{iz,ls}$ is added to the 'main' source term b and the series of these linear algebraic finite volume equations is solved within PHOENICS using the IPSA algorithm.

It is concluded that in the example described above the value of the Low coefficient is calculated by subtracting from its conventional hybrid upwind value the mass flow rate-weighted term that stems from the flow crossing the corner cell. The South coefficient is assumed zero and its value is increased by the mass flow-rate weighted term that corresponds to the corner cell inflow and is included as a source term in place of the Low-South coefficient.

The highlights of the steps followed to embody the SUPER scheme in the calculation domain (Fig. 3) when the Low-South corner inflow is taken into account are summarized below:

(a) Negation of the convection coefficient c_s.
(b) Modification of the convection coefficient c_l.
(c) Creation of the convection coefficient b_s.
(d) Identification of the Low and South cell as the upwind corner cell.
(e) Implementation of the value of V_{iz} in the Low and South corner cell for the solution of momentum.

The same procedure is followed for taking into account the influence of all the local corner inflows and all the velocity components (V_{ix}, V_{iy}, V_{iz}). In the cases of corner flows from the High-North and East (HNE), High-North and West (HNW), High-South and East (HSE), High-South and West (HSW), Low-North and East (LNE), Low-North and West (LNW), Low-South and East (LSE), Low-South and West (LSW) grid cells to the solution grid cell, the same steps are followed except for the last two steps, because the value of the transported variable is implemented in a neighbouring cell.

4 Test Cases for Numerical Validation

In this section, the performance of various numerical schemes is compared as far as the accuracy of their predictions is concerned. Various cases of physical problems that include one- and two-phase flows, heat and mass transfer, geometry of two and three dimensions, grid of Cartesian and curvilinear coordinate system, straight and inclined inflows are investigated and the conclusions are presented.

4.1 Inclined Flow of a Scalar Quantity

The physical problem presented here describes the transport of a scalar quantity ($C1$) in a two-dimensional geometry. The flow of $C1$ is inclined to the grid lines ($\theta = 45°$) and the natural diffusion is assumed equal to zero. This physical problem is considered as a benchmark for comparing the performance of various numerical schemes.

The boundary conditions applied to the grid are the following:

Air enters the domain from the west side with constant mass flow rate ($\rho \cdot u$) kg/m²s, uniform values of the two components of velocity ($u_1 = v_1 = 1$ m/s) and uniform value of the convected quantity, say concentration ($C1$) $\left(1 \, \text{kg/m}^3\right)$.

Air enters the domain also from the south side with constant mass flow rate ($\rho \cdot u$) kg/m²s, uniform values of the two components of velocity ($u_1 = v_1 = 1$ m/s) and uniform value of the convected quantity ($C1$) $\left(1 \, \text{kg/m}^3\right)$.

At the two outlets (north and east side) of the domain, the air is supposed to exhaust at an environment of fixed uniform pressure.

The dimensions of the domain are 1 by 1 m. The numerical results presented are based on the grid-independent grid of 33×33 cells. For the discretization of the convection term in the transport equation of the scalar quantity $C1$ the numerical schemes: (a) UPWIND, (b) van LEER, (c) SUCCA, (d) SUPER are applied [6, 8, 11, 12].

In Fig. 7 the velocity vector distribution predicted by the SUPER numerical scheme is presented [19]. In Fig. 8 the vertical (concentration) distribution of the scalar quantity $C1$ in the middle of the domain is presented.

The vertical distribution of the scalar quantity $C1$ that is transferred by air with inclined direction is predicted more accurately, as being more abrupt, by the numerical schemes (SUCCA, SUPER) that take into account the flow direction than the conventional numerical scheme (UPWIND) and the non-linear numerical scheme (van LEER). It is concluded that the numerical diffusion errors are minimized when the discretization schemes (SUCCA, SUPER) are applied to predict the transport of the scalar quantity $C1$, when the airflow direction appears at the major inclination angle 45° to the grid lines. The performance of the van LEER numerical scheme is also satisfactory in predicting the scalar concentration distribution. The conventional UPWIND numerical scheme that does not take into account the phenomenon of false diffusion presents the poorest accuracy.

4.2 Tubular Flow of a Scalar Quantity and Heat Conduction in the Radial Direction

A physical problem used in this study for assessing the numerical schemes is the numerical prediction of the energy transfer in a triple tube heat exchanger [18]. The

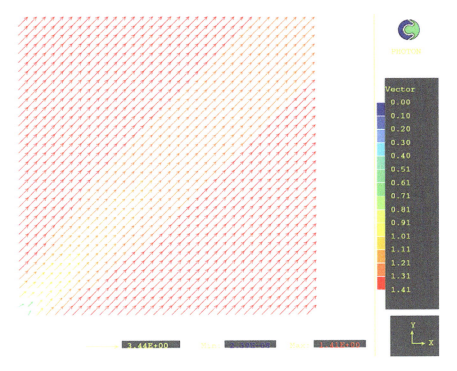

Fig. 7 Air velocity vector distribution predicted by the SUPER scheme

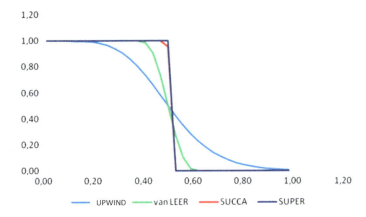

Fig. 8 Vertical distribution of the scalar quantity $C1$ (kg/m^3) in the middle of the domain applying the numerical schemes: **a** UPWIND, **b** van LEER, **c** SUCCA, **d** SUPER

heat exchanger has a three-dimensional curvilinear geometry similar to a circular cylinder. Three fluids are being considered to flow inside the domain that is separated by solid. Chilled water (10 °C) flows in the inner tube, hot water (70 °C) flows in the inner annulus and normal tap water (18 °C) flows in the outer annulus. The three fluids follow a co-current flow. The numerical model is first validated and then used to compare the performance of various numerical schemes.

The boundary conditions applied for solving the above physical problem are the following:

Water with uniform properties (temperature, density, thermal capacity, kinematic viscosity) enters the three different inlets of the heat exchanger. The heat exchanger is separated into three components that are filled with water of different properties and the solid parts that prevent the water vertical flow. Heat is exchanged between the fluids through conduction. The solid is steel at 27 °C of uniform properties (density, viscosity, specific heat, thermal conductivity, and thermal expansion coefficient). At the three outlets of the heat exchanger, the boundary condition of zero mass flow is applied.

In Fig. 9 [19] the numerical prediction of the temperature distribution along the heat exchanger is validated against experimental data [20]. The average % relative error of the numerical prediction with respect to the experiments does not exceed the value of 20%.

In Fig. 10 [19] the numerical results of the enthalpy distribution at the lateral plane in the middle of the heat exchanger applying three different numerical schemes (HYBRID, van LEER, SUPER) [6, 8, 12] is presented.

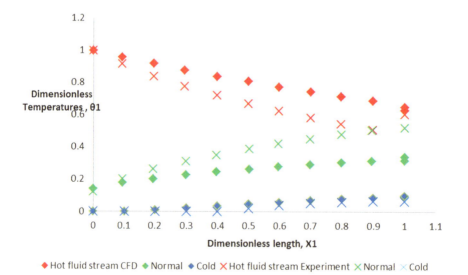

Fig. 9 Local temperature variation of the co-current three fluid streams along the length of the heat exchanger

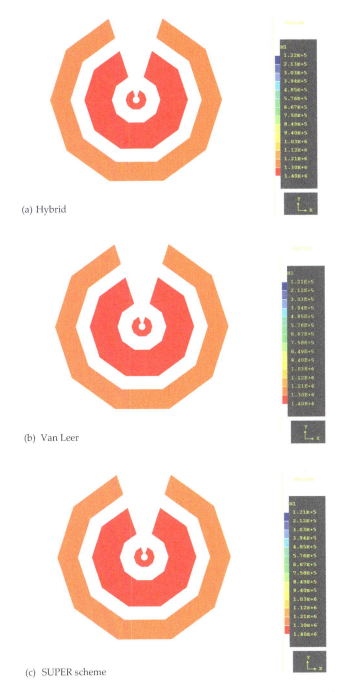

(a) Hybrid

(b) Van Leer

(c) SUPER scheme

Fig. 10 Enthalpy distribution at the lateral plane of the heat exchanger applying: **a** the HYBRID, **b** the van Leer, **c** the SUPER numerical scheme

Comparing the performance of the numerical schemes in predicting the transfer of the scalar quantity (the enthalpy), it is concluded that the numerical prediction is equally satisfactory for the three discretization schemes (nearly coinciding). This is to be expected as the flow is virtually unidirectional.

4.3   Airflow in a Backward Facing Step

Another physical problem appropriate for assessing the accuracy of various numerical schemes is the prediction of the airflow velocity vectors distribution in a backward facing step.

The separation of the airflow, the recirculation zone formed forward of the step and the reattachment length are the main characteristics of the flow in this geometry. The numerical solution of the airflow field in a backward facing step is of major concern due to the difficulty in accurately predicting these complex phenomena. The flow field may become more complex in the presence of solid particles.

In this study, the two-dimensional backward facing step of Benavides and Wachem [21] is investigated.

In Figs. 11 and 12 [19] the velocity flow field is presented applying various numerical schemes (UPWIND, HYBRID, van LEER, SUPER) [6, 8, 12] and validated against experimental data [21].

According to the numerical results that agree well with the experimental data, the performance of the numerical schemes (UPWIND, HYBRID, van LEER, SUPER) is equally satisfactory. The same conclusion is derived for the case of the two-phase air-particles flow in a backward facing step, where the velocity fields are calculated nearly coinciding.

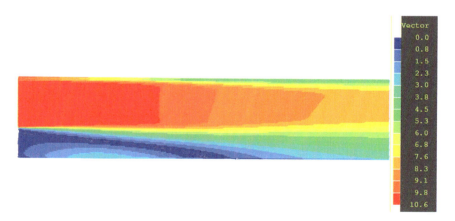

Fig. 11 Velocity flow field applying various numerical schemes (upwind, hybrid, van Leer, super)

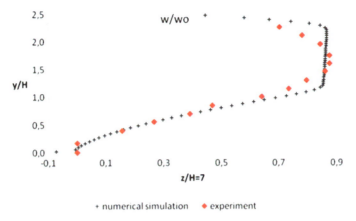

Fig. 12 Vertical velocity distribution applying various numerical schemes (UPWIND, HYBRID, van LEER, SUPER, nearly coinciding predictions) and comparison with the experimental data

4.4 Inclined Air-Particles Flow

A physical problem that has been used for assessing the performance of various numerical schemes is the dispersed two-phase air-particles flow in a model room geometry (Fig. 13) [22]. A two-phase Euler–Euler mathematical model is applied to calculate the oblique inflow in the interior of the internal space. The accuracy of the four numerical schemes: (a) the conventional first-order UPWIND numerical scheme, (b) the second-order HYBRID scheme, (c) the non-linear van LEER and (d) the flow-oriented SUPER scheme [6, 8, 12] are compared in the case of inclined inflow ($\theta = 45°$).

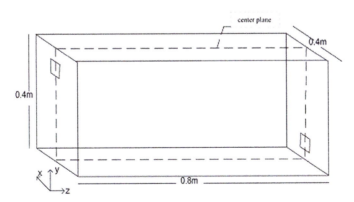

Fig. 13 Geometry of the model room

4.4.1 Boundary Conditions

At the inlet, the total mass flow rate is multiplied by each phase volume fraction. The dispersed turbulent air-particles flow enters the room with uniform velocity (0.225 m/s) in an inclined direction. Turbulence is modelled by the RNG k-ε model [23]. Particles are assumed to be transported and dispersed due to turbulence of the carrier fluid (air).

The turbulence kinetic energy of the air-phase that is applied at the inlet is defined as [13, 24] $k_{\mathrm{in}} = \frac{3}{2}\left(U_{\mathrm{avg}}T_i\right)^2$ where U_{avg} the mean air inlet velocity and T_i the turbulence intensity, considered as 6%. The dissipation rate is given by $\varepsilon = C_\mu^{3/4}\frac{k^{3/2}}{\ell}$, where ℓ, the turbulence length scale is taken as $\ell = 0.07d_h$ (d_h hydraulic diameter of the duct) and $C_\mu = 0.0845$ an empirical constant of the turbulence model [25]. The mathematical model uses the logarithmic 'wall functions' of Spalding near the solid surfaces.

At the outlet, both phases are supposed to exhaust at an environment of fixed uniform pressure. At the walls, the non-slip and non-penetration condition is applied for both phases.

4.4.2 Results

Figure 14a–c presents the vertical w_1 velocity distribution at the longitudinal plane of the domain at distances 0.2, 0.4, 0.6 m from the supply inlet [19, 24].

As far as the accuracy of the four numerical schemes in the case of inclined inflow ($\theta = 45°$) is concerned the higher order and non-linear schemes (hybrid and van Leer numerical scheme) present a similar performance with the first-order upwind numerical scheme and a different performance than the flow-oriented scheme (SUPER scheme). The vertical w_1 velocity distribution predicted by the SUPER scheme presents a more abrupt and intuitively more accurate profile due to the successful minimizing of the false diffusion errors, that tend to smear abrupt changes.

4.5 Heat and Mass Transfer

A physical problem that has been used to evaluate the performance of the numerical accuracy is the water vapour condensation of humid air in the three-dimensional geometry of a real-scale indoor space [19]. A two-phase flow Euler–Euler mathematical model has been used, wherein the humid air and water droplets are being treated as separate phases. The two phases exchange momentum and energy and, as the temperature drops below the dew point of humid air, mass transfer, and phase change of water vapour to liquid takes place. The flow of humid air inside the room is buoyancy-driven in the temperature range of 290–303 K. The properties of humid air (enthalpy, relative humidity, concentration of water vapour, saturation vapour

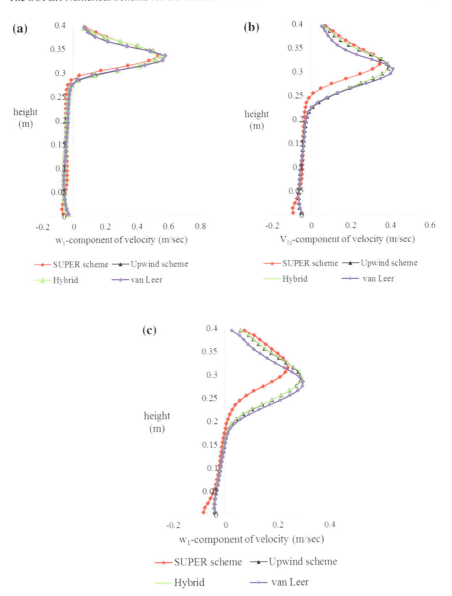

Fig. 14 **a** Vertical w1 velocity distribution at the longitudinal plane of the domain and distance 0.2 m from the inlet. **b** Vertical w_1 velocity distribution at the longitudinal plane of the domain and distance 0.4 m from the inlet. **c** Vertical w_1 velocity distribution at the longitudinal plane of the domain and distance 0.6 m from the inlet

pressure) vary with the temperature [26]. The dimensions of the domain are width $(X) \times$ height $(Y) \times$ length $(Z) = 4.0 \times 4.0 \times 8.0$ m.

4.5.1 Boundary and Initial Conditions

The interior of the domain is filled with humid air of 303 K and no liquid phase. The temperature on the surface of the walls is 303 K and on the surface of the cold floor is 290 K. The density of humid air is 1.16 kg/m^3 and the water droplets have a constant density (996 kg/m^3) and specific heat (4190.0 J/kg K) at the dew point temperature. The initial pressure inside the domain is equal to the atmospheric pressure. The initial relative humidity condition tested is 90%. The overall water vapour mass of humid air at the initial temperature 303 K is taken from the psychrometric chart [27].

4.5.2 Results

In Fig. 15 the vertical temperature distribution in the middle of the domain at the height (0–4 m) is presented [19].

The temperature of humid air near the floor is below the dew point (301.2 K) and water phase humidity covers the whole surface. The hot air close to the floor that comes into contact with the cold surface reduces its temperature and flows down due to the gravity. The remaining hot air flows up to the roof.

In Fig. 16 [19] the vertical temperature distribution in the region close to the floor calculated by three different numerical schemes (HYBRID, van LEER, SUPER) [6, 8, 12] is presented.

Temperature profile in the region of major gradient near the floor surface is predicted with a smaller slope by the SUPER scheme.

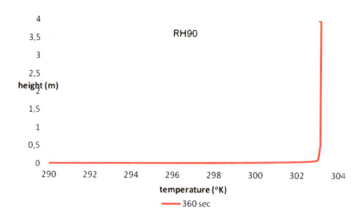

Fig. 15 Vertical temperature distribution at time 360 s for initial humidity condition 90%

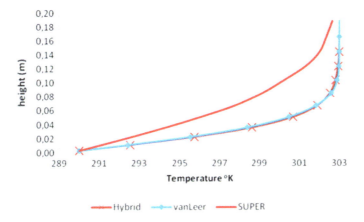

Fig. 16 Vertical temperature distribution at time 360 s for initial humidity condition 90% in the middle of the domain at the height (0–0.2 m) calculated by three different numerical schemes

In Fig. 17 the vertical absolute humidity ratio (kg H_2O/kg of dry air) predicted by the three different numerical schemes (HYBRID, van LEER, SUPER) is presented at time 360 s [19].

Mass transfer and phase change of humid air accompany heat convection. The larger gradient of temperature profile predicted by the SUPER scheme [12] leads to the formation of larger amount of water phase. Comparing the performance of the discretization schemes a more intuitively accurate solution of the condensation procedure is observed when applying the SUPER scheme. Of course, the lack of

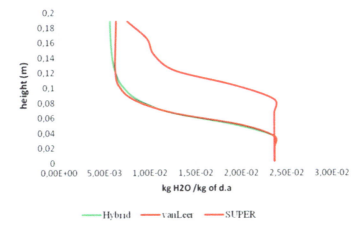

Fig. 17 Vertical absolute humidity ratio (kg H_2O/kg of dry air) distribution at time 360 s for initial humidity condition 90% applying the van LEER, the HYBRID, the SUPER numerical schemes at height 0–0.2 m

experimental data does not allow for a quantitative criterion to be used for a more concrete conclusion.

4.6   RANS and LES Modelling

In this section, the numerical results of the Reynolds averaged Navier–Stokes (RANS)-RNG [23] mathematical model are compared with those obtained by Large Eddy Simulation [28] in terms of particle concentration evolution (in the model room described in Sect. 4.4 above).

Figures 18 and 19 [29] present the vertical particle concentration distribution normalized by the inlet concentration as predicted by both mathematical models.

Comparing the performance of LES and RANS model, a similar numerical solution is obtained by both mathematical models and the agreement with the experimental data is satisfactory. Additionally, it is observed that both aforementioned mathematical two-phase models predict a higher rate of particle concentration evolution than the one-phase model of Chen et al.

4.6.1   Discretization Schemes

In this section, the accuracy of four numerical schemes (a) the conventional first-order upwind numerical scheme, (b) the second-order hybrid scheme, (c) the non-linear van Leer and (d) the flow-oriented SUPER scheme applying the LES mathematical model in a steady-state calculation are compared in the case of inclined inflow for θ

Fig. 18 Vertical particle concentration distribution at distance 0.2 m from the inlet along the centerline applying **a** RANS model and **b** LES model with the van Leer numerical scheme

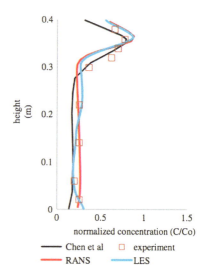

Fig. 19 Vertical particle
concentration distribution at
distance 0.4 m from the
inlet along the centerline
applying **a** RANS model and
b LES model with the van
Leer numerical scheme

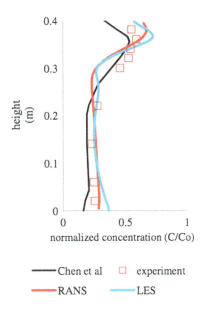

$= 45°$. Figure 20 [29] presents the vertical $w1$ velocity distribution at the longitudinal
plane of the domain at distance 0.2 m from the supply inlet.

As far as the accuracy of the four numerical schemes in the case of inclined inflow
($\theta = 45°$) is concerned the higher order and non-linear schemes (hybrid and van
Leer numerical scheme) present a similar performance with the first-order upwind
numerical scheme and different from the flow-oriented scheme (SUPER scheme).

Fig. 20 Vertical $w1$ velocity
distribution at the
longitudinal plane of the
domain at distance 0.4 m
from the inlet along the
centerline in the case of
inclined inflow ($\theta = 45°$)

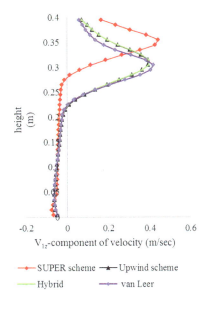

Additionally, in Figs. 21 and 22 [29] some flow features such as flow separation and subsequent recirculation predicted at the corners of the domain are clearly distinguished in the flow pattern obtained by applying the SUPER scheme and thus, it raises the expectation that it is better in resolving them.

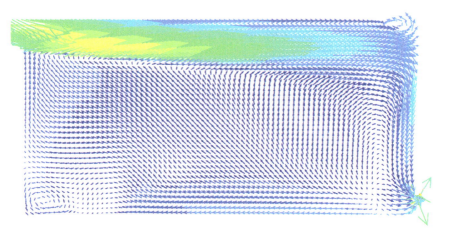

Fig. 21 Velocity vectors at the longitudinal plane of the domain applying the SUPER numerical scheme in the case of inclined inflow ($\theta = 45°$)

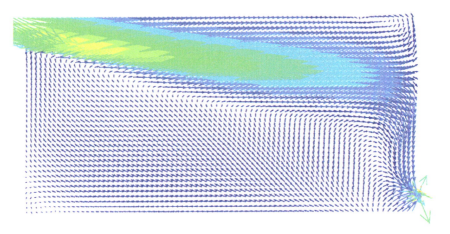

Fig. 22 Velocity vectors at the longitudinal plane of the domain applying the upwind numerical scheme in the case of inclined inflow ($\theta = 45°$)

4.7 Water Vapour Condensation on the Walls of Indoor Spaces

The numerical simulation of the water vapour condensation on the walls of a real-scale interior space, for various cases of humidity conditions, is presented below. The two-phase Euler–Euler flow model of Spalding [14] is used within the computer program PHOENICS [16], which considers the phases as interpenetrating continua. The study focused on the mass transfer interaction between liquid droplets and humid air on the wall surfaces, for the discomfort case of the absence of natural ventilation. Several discretization schemes were used. Results are presented for several humidity conditions at 303 K, while the surrounding walls are kept at temperature 290 K. The evolution of water vapour condensation is studied for a real-time interval of 10 min. According to the numerical results, when the temperature drops below the dew point of the water vapour, humidity droplets start to form on the surface of the walls, as expected, and the humid air temperature in the interior becomes uniform and equal to the temperature of the wall surfaces. The process of condensation starts to take place later when the initial humidity value is lower.

4.7.1 The Physical Problem Considered

Humidity and condensation can compromise building occupants' health and comfort, damage interior furniture, and raise heating costs. High indoor humidity may cause condensation on windows, deterioration of building materials and, in some cases, mould and mildew growth, as possible by-products of condensation, related to allergic symptoms [30]. Because of its significance and because this moisture problem is less researched than the previously discussed, equally important physical problems, some more details are given here than before.

Moisture problems are normally related to three conditions: the indoor temperature and humidity, the outside climate, and the design and construction of the building envelope [31, 32]. Indoor humidity follows the trend of outdoor humidity, especially in summer because of the increased ventilation, but also it is influenced by the seasonal variational habits of occupants and the moisture storage of hygroscopic material in space. In winter, the difference between indoor and outdoor humidity reaches a maximum due to increased activities and reduced ventilation [30]. Therefore, regardless of the climate and construction factors, properly operating the building can be an effective alternative in alleviating moisture problems. This involves regulating indoor temperature and moisture levels [31]. To do that we need predictions of the moisture behaviour in time.

Moisture can be generated into the space by the occupants and their households operations, infiltration, mechanical ventilation, evaporating damp surfaces, interior surfaces through desorption, as well as by indoor moisture-generating operations [30]. The largest generation of moisture in residential buildings may stem from the bathroom and cooking areas [32]. Moisture may be lost from the space by exfiltration, surface condensation, mechanically exhausting air and absorption by interior hygroscopic materials [30]. The main factors affecting the moisture distribution are its initial distribution, the supply air relative humidity, the location of the vents, the moisture generation rate, the location of the sources, water vapour condensation, temperature variation, and others. Acceptable indoor conditions are considered those when the indoor temperature lies between 23 and 25 °C and the relative humidity in the range of 20–60% [31].

The prediction and control of the indoor moisture distribution, particularly of wall condensation, is very important for a healthy and energy-efficient environment. Condensation usually occurs at local wall areas in most rooms, a thermal bridge or a roof above vapour sources [33]. Air-conditioning systems cause condensation on the exposed cold surface of their indoor terminals.

Surface condensation is caused by a larger air humidity ratio near the boundary, relative to the saturated humidity ratio corresponding to the temperature of the boundary, and occurs when the surface temperature is lower than the dew point temperature of the moist air. There is moisture mass transfer from the indoor air to the walls surface. The moisture mass flux varies with time during the condensation process, because the humidity ratio varies with the moisture dispersion. At the boundary: (a) the water vapour pressure at the surface is equal to the saturated vapour pressure at the same temperature as the boundary when condensation occurs, (b) the heat released during condensation has limited effect on surface temperature and the moisture has a negligible effect on the airflow field. Consequently, the dew point temperature and the saturated humidity ratio at the wall surface can be considered to be constant. Accordingly, the saturated humidity ratio equals that at the critical wall condensation condition [33]. Pressure level in the interior is strongly conditioned by both the temperature and the water vapour content of humid air. Liquid droplets form along the wall at nucleation sites. Vapour condensation on droplets makes them grow. Once the droplet diameter reaches a critical value, gravitational forces compensate the surface tension force and then the droplets slide over the wall. Droplets can also join the neighbouring droplets and form a film layer [34].

Indoor air humidity processes include moisture absorption/desorption, surface condensation, air movement across enclosure boundaries, indoor evaporation, indoor moisture generation [30, 35]. The majority of the available models are based on the mass balance between gained and lost moisture. The simplest model predicts the indoor humidity as a function of occupancy and ventilation rate. Other models have considered more time-dependent mechanisms of moisture transport in building cavities with and without condensation. Moisture absorption has also been studied [30, 36]. In [30], humidity dynamics within an enclosure is described by a linear differential equation, which includes most of the humidity processes. In [37, 38], a convection–diffusion transport equation of a scalar quantity is solved, so that the humidity of atmospheric air is taken into account for the prediction and optimization of thermal comfort conditions. Methods for calculating surface condensation are based on the well-known mass transfer equations [30–32, 36, 38, 39] and on a lumped parameter theory, in which the indoor air is assumed to be well-mixed.

Numerical simulation and prediction of a three-dimensional heat and mass transfer problem, such as the condensation process, must satisfy two main conditions [40]: (a) the knowledge of temperature and moisture distribution that explains why condensation happens at certain locations in space; most of the available prediction models deal with the average relative humidity of a space, and do not take into account the temperature and humidity distributions within the space, and (b) the transient evolution of heat and mass transfer and the time-dependent condensation rate that explain whether and when it will happen and how much the condensation will be. In the present work, a two-phase mathematical model [12] is applied for the prediction of humid air–water droplets distribution and water vapour condensation processes; it calculates the distributions of pressure, temperature and absolute humidity, as well as the transient heat and mass transfer to the water phase. To the authors' best knowledge it is the first time a full two-phase flow model is applied for the cases under consideration.

4.7.2 The Test Case Considered

The two-phase flow Eulerian–Eulerian mathematical model of Spalding [14] is used, as adapted in [12], wherein the humid air and water droplets are being treated as separate phases. Specifically, the mixture of dry air and water vapour is considered as the primary gaseous phase and water droplets as the secondary dispersed liquid phase. The two phases exchange momentum and energy and, as the temperature drops below the dew point of humid air, mass transfer and phase change of water vapour to liquid take place. The condensation process is assumed to be at equilibrium, therefore, as soon as saturation conditions are reached, the condensation process starts immediately. Furthermore, the process is assumed homogeneous; hence, it is assumed that the water droplets are created at constant rate, at a prescribed pressure and temperature. The water droplets are considered to be spherical with diameter of $d_{wd} = 10^{-5}$ m, that is much greater than the critical size for growth. Therefore, at the saturated state the liquid droplets reach the critical size. The condensation process is accompanied by the release of latent heat that may not be as significant as to change the temperature appreciably [41]. The flow of humid air inside the room is buoyancy-driven at a low Rayleigh number.

The geometry examined is a real-scale hypothetical room with dimensions: width $(X) \times$ height $(Y) \times$ length $(Z) = 4 \times 4 \times 8$ m. The case of worst discomfort humidity conditions, in the absence of natural ventilation, is investigated, so that the process of condensation takes place within the temperature range of 290–303 K in a typical residence.

4.7.3 Mathematical Modelling

Mathematical modelling consists of the time-dependent Navier–Stokes (N-S) equations and the continuity equation for a 3D, two-phase, fluid flow (Eqs. 2.1, 2.2 above).

The assumptions made for the problem are the following: (a) both phases, gaseous phase (humid air) and liquid phase (water droplets), are treated as incompressible fluids, (b) each phase is a continuum, so that the derivatives are uniquely defined, (c) both fluids are Newtonian, interdispersed and they exchange momentum, mass and energy as they move along, (d) phase change takes place, (e) homogeneous condensation process takes place at equilibrium, (f) spherical monodisperse droplets of one size group (mean AED $10\,\mu$m) are considered, (g) laminar viscosity is assumed constant and equal to the dry air viscosity, (h) the moisture storage capacity of the wall surfaces is not taken into account.

4.7.4 The Constitutive Expressions

– The mass of water vapour that changes phase to become water droplets is calculated from the following equation:

$$m_{wv} = 0.62 \cdot 1.0E - 05 \cdot p \cdot \rho_{\text{dry air}} \cdot V_{\text{cell}} \cdot R_1 \tag{4.1}$$

where p the water vapour pressure, $\rho_{\text{dry air}}$ the density of dry air, V_{cell} the volume of the cell, R_1 the volume fraction of the first phase (humid air).

– The frictional force $F_{f,ip}$ per unit volume at the humid air–water droplets interphase, due to the differing phase velocities is [42]:

$$F_{f,ip} = 0.5C_D A_{pr} \rho_g |V_g - V_w| (V_g - V_w) \equiv C_{f,ip}(V_g - V_w), \tag{4.2}$$

where C_D the interface friction coefficient, A_{pr} is the total projected area of droplets per unit volume, ρ_g density of the gaseous phase (humid air), V_g the gaseous phase velocity, V_{wd} the water droplet phase velocity, $C_{f,ip}$ the interphase momentum transfer coefficient.

– The empirical correlation used for C_D is Stoke's drag law [43]:

$$C_D = \frac{24}{\text{Re}_{wd}} \tag{4.3}$$

where Re_{wd} is the water droplet Reynolds number defined as [44]:

$$\text{Re}_{wd} = \frac{d_{wd}\rho_g}{\mu_g} V_{slip} \tag{4.4}$$

where d_{wd} is the droplet diameter, ρ_g is the density of the gaseous phase, μ_g is the viscosity of the gaseous phase, and V_{Slip} is the relative velocity between the two phases.

– The interphase heat transport rate is described by Eq. (4.5):

$$\dot{q}_{int} = C_{q,\,int} \cdot (T_1 - T_2) \tag{4.5}$$

where $C_{q.int}$ the interphase thermal transfer coefficient and T_1, T_2 the bulk temperatures of each phase.

– The buoyancy force F_b per unit volume in the vertical direction is computed according to the Boussinesq approximation.

– The properties of humid air are described by the following equations [26, 27]:

(a) The relationship between vapour pressure and concentration is defined by the following equation:

$$p = \frac{n R_a T}{V} = C R_a T. \tag{4.6}$$

where p is the vapour pressure (Pa), V is the volume (m^3), T is the temperature K, n is the molar mass (kg), R_a is the gas constant 0.000831 kJ/(kmole K).

Concentration of water vapour is usually defined as kg water vapour/kg dry air. Dry air has a molar mass of 0.029 kg and is denser than water vapour, which has a molar mass of 0.018 kg. If the total pressure is P and the water vapour pressure is p, the partial pressure of the dry air component is $P - p$. The weight ratio of the two components, water vapour and dry air is

$$\text{kg water vapour/kg dry air} = 0.018 \cdot p/0.029(P - p)$$

At room temperature $P - p$ is nearly equal to P, which at ground level is close to atmospheric $1.0E^{+05}$ (P_a) so, approximately:

$$\text{kg water vapour/kg dry air} = 0.62 \cdot 1.0E^{-05} \cdot p.$$

(b) *Saturation vapour pressure* (p_s) obeys the following equation:

$$p_s = 610.78 \cdot \exp(T/T + 238.3) \cdot 17.2694, \tag{4.7}$$

where T is the temperature in degrees Celsius.

(c) The *Relative Humidity* (RH) is the ratio of the actual water vapour pressure p to the saturation water vapour pressure p_s at the prevailing temperature:

$$\text{RH} = \frac{p}{p_s}. \tag{4.8}$$

Table 1 Condensation on walls of indoor spaces-initial relative-humidity conditions tested	Temperature of humid air	303 K		
	Wall temperature	290 K		
	Humidity of air	90%	60%	30%
	Dew point temperature	301.2 K	294.4 K	283.5 K

(d) *Enthalpy of humid air* (kJ/kg) follows the equation:

$$h = (1.007 \cdot T - 0.026) + b \cdot (2501 + 1.84T), \tag{4.9}$$

where b the water content in kg water vapour/kg dry air.

4.7.5 Initial and Boundary Conditions

At time zero, inside the room, there is humid air at rest. The volume fraction of gaseous phase (humid air) at initial time $t = 0$ s is 1 (the whole of space is occupied by it). The volume fraction of liquid phase at $t = 0$ s is zero and increases as the process of condensation proceeds. The initial temperature of humid air inside the room is 303 K, three cases of temperature boundary conditions are investigated: (a) low temperature 290 K on the surface of the walls (Case 1), (b) lower temperature 290 K on the surface of the floor (Case 2), (c) lower temperature 290 K on the surface of the roof (Case 3). The second phase, which is not formed yet, will be at the initial temperature of its dew point relevant to the surface temperature and the relative humidity of the interior, when formed. The density of humid air depends on the conditions (pressure, temperature, relative humidity) of the case studied and ranges between (1.14–1.17 kg/m^3). The water droplets have a constant density (996 kg/m^3) and specific heat (4190.0 J/kg K) at the dew point temperature. The initial pressure inside the room is equal to the atmospheric pressure. Four cases of initial relative humidity conditions are tested (Table 1) [27, 45]. The overall water vapour mass of humid air at the initial temperature 303 K for the various values of relative humidity is taken from the psychrometric chart.

4.7.6 Numerical Simulation

The equations are discretized by the finite volume method and solved by the IPSA algorithm of Spalding [14–16], properly modified to include the extra physical processes. A fully implicit method is employed for the numerical solution. The convection terms of all the conservation equations are discretized by three different numerical schemes, i.e. the hybrid, van Leer, SUPER numerical schemes [6, 8, 12]. The first-order fully implicit scheme is used for time discretization. The overall real time of the condensation procedure simulated is 10 min, enough time for the water

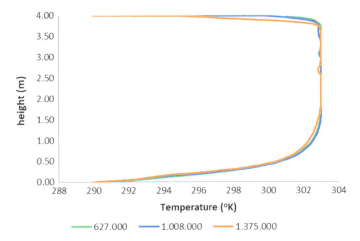

Fig. 23 Vertical temperature distribution at the centre line of the room predicted by three grids of different size: **a** 627.200 cells, **b** 1.008.000 cells, **c** 1.375.296 cells at time 100 s for the case of initial relative humidity 90% applying the SUPER scheme

vapour to meet the saturation conditions and to start changing its phase to water droplets. The initial relative humidity and the overall water vapour mass inside the room, as well as the dew point temperature, are all embodied in the algorithm and their values are defined as an input at time $t = 0$ s.

Due to different temperatures between the ambient air and the wall surface there is heat transfer from the humid air to the wall. When the humid air touches the cold surface and the temperature drops below the dew point, mass transfer from water vapour to water droplets is activated in the algorithm and the process of condensation starts.

For the numerical solution, a non-uniform structured grid is constructed. Spatial independency is tested by repeating the transient simulation for a gradually increased grid cell density. Four grid sizes have been tested consisting of 396.720, 627.200, 1.008.000 and 1.375.296 cells and compared, in order to ensure spatial independency of the results. Time-step independency is tested for each grid by repeating the transient simulation for smaller time steps. As an example, Figs. 23 and 24 reveal that the optimum spatial discretization is that of the third grid (1.008.000), which combines accuracy and economy. The same procedure is repeated for four different time steps. The results presented are thus grid- and time-step independent and correspond to a grid of 1.008.000 nodes and a time-step of 0.02 s.

4.7.7 Typical Results and Discussion

The numerical results of the moisture distribution inside the room are presented, for three different initial humidity conditions of moist air (Table 1) at real time 100 s (Case 1). The results are presented in terms of the absolute humidity ratio (kg H_2O/kg

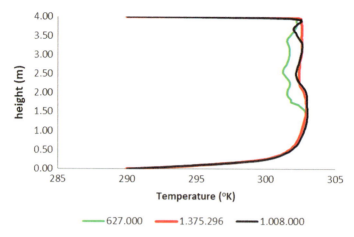

Fig. 24 Vertical temperature distribution at the centre line of the room predicted by three grids of different sizes: **a** 627.200 cells, **b** 1.008.000 cells, **c** 1.375.296 cells at time 100 s for the case of initial relative humidity 60% applying the Hybrid scheme

of dry air), as well as the relevant temperature and velocity distribution of the humid airflow field. The saturation limit is 0.024 kg H_2O/kg of dry air.

For the case of initial relative humidity 90% the initial temperature of humid air inside the room is 303 K near the dew point temperature (301.2 K). The process of condensation of water vapour takes place for the first time mainly in the region close to the floor, where the temperature drops for the first time below the dew point (Fig. 25). Humid air of initial high temperature 303 K that flows upwards, delays the cooling of the region near the cold (290 K) surface of the roof. Humid air that is cooled in the region close to the roof, condenses to water phase and flows downwards due to gravity.

Comparing Figs. 25 and 26 in regions where the humid air temperature is higher than the dew point temperature (301.2 K) no humidity has been formed, while near the floor where the humid air temperature is below 301.2 K condensation of gaseous humidity has taken place.

The humid air of higher temperature (Fig. 27) flows up to the roof, where it encounters the colder wall surface, its temperature drops and then moves down towards the floor, where its temperature drops further and below the dew point, as expected. Cold dry air close to the floor moves downwards due to gravity, while convection moves air towards the upper layers. There appear therefore two convective cells in the flow.

For the case of initial relative humidity 60% the dew point temperature is 294.4 K, thus the process of condensation starts at a later time, when the temperature of humid air drops below the dew point temperature, and as expected at time 100 s, the water phase covers less part of the region close to the floor surface.

Comparing Figs. 28 and 29 in regions, near the floor, where the humid air temperature falls below the dew point temperature (294.4 K) water phase humidity has been

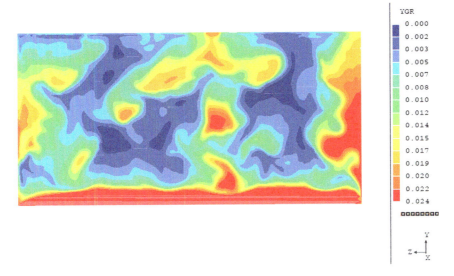

Fig. 25 Absolute humidity distribution (kg H$_2$O/kg of dry air) formed at the centre plane of the room at time 100 s (initial relative humidity 90%)

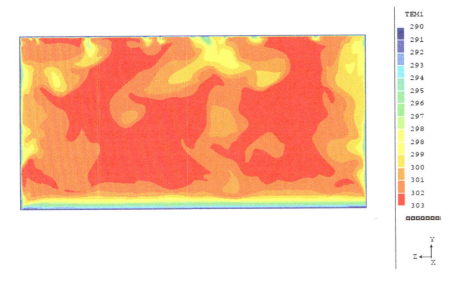

Fig. 26 Humid air temperature distribution at time 100 s (initial relative humidity 90%)

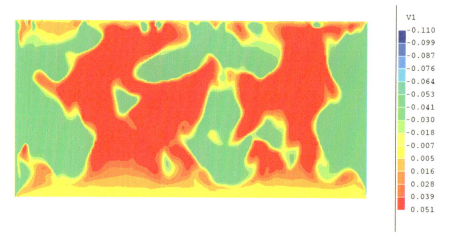

Fig. 27 *V*1-component of humid air velocity (initial relative humidity 90%) at time 100 s

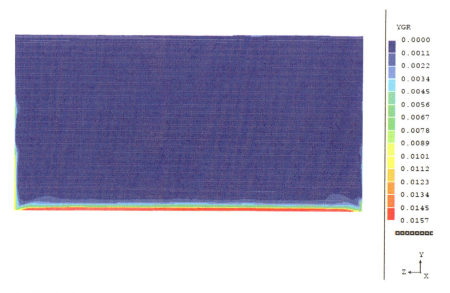

Fig. 28 Absolute humidity distribution (kg H$_2$O/kg of dry air) formed on the walls at time 100 s (initial relative humidity 60%)

formed. Humid air that encounters the cold surface of the roof, flows downwards due to gravity.

In Fig. 30a–c we observe that water phase humidity is formed firstly near the floor, where the maximum concentration of water vapour has changed phase (24.5 gr/kg d.a for RH 90% and 16.5 gr/kg of d.a for RH 60%). At time 360 s the process

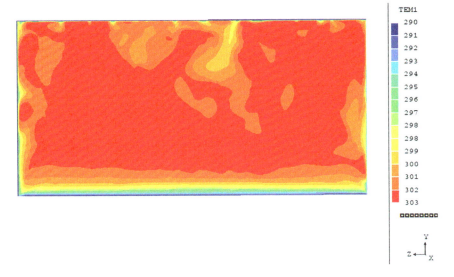

TEM1
290
291
292
293
294
295
296
297
298
298
299
300
301
302
303

Fig. 29 Humid air temperature distribution at time 100 s (initial relative humidity 60%)

of condensation has taken also place near the roof for the initial relative humidity 90%.

At time 360 s the process of condensation has taken place almost everywhere in the interior of the room for the case of initial relative humidity 90% (Fig. 30). For the case of initial relative humidity 60% water vapour maintains its gaseous phase in the internal of the room, where the temperature is above its due point value (294.4 K) (Fig. 31).

In Fig. 32, we observe that at time 360 s water phase humidity is formed near the floor, where the maximum concentration of water vapour has changed phase (24.5 gr/kg d.a for RH 90% and 16.5 gr/kg of d.a for RH 60%) (Case 2). Although the process of condensation has taken place in the interior, it maintains its gaseous phase for both the initial relative humidities 90, 60%.

As far as the temperature distribution is concerned it depends on the concentration of water vapour (kg H_2O/kg dry air) in humid air and on whether the process of condensation has taken place, or not. Therefore, for the former case (initial relative humidity 90%) a temperature drop, as well as the formation of liquid phase is observed as the time proceeds. It is interesting to add that the results of the numerical schemes of different sophistications are more or less the same.

For the case of initial relative humidity 30% the lowest initial temperature of humid air inside the room is the wall surface temperature 290 K, which is higher than the dew point temperature (283.5 K), thus it is expected that the process of condensation of water vapour will not take place and water phase will not be produced. Indeed according to the numerical results, the process of condensation does not take place.

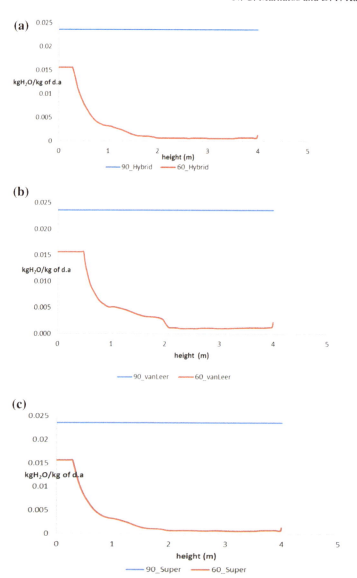

Fig. 30 **a** Vertical absolute humidity ratio (kg H$_2$O/kg of dry air) distribution at time 360 s applying the Hybrid scheme (Case 1). **b** Vertical absolute humidity ratio (kg H$_2$O/kg of dry air) distribution at time 360 s applying the van Leer scheme (Case 1). **c** Vertical absolute humidity ratio (kg H$_2$O/kg of dry air) distribution at time 360 s applying the Super scheme (Case 1)

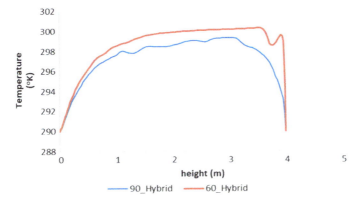

Fig. 31 Vertical temperature distribution at time 360 s for initial humidity conditions 90 and 60% (Case 1)

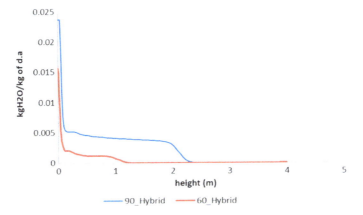

Fig. 32 Vertical absolute humidity ratio (kg H_2O/kg of dry air) distribution at time 360 s calculated by the hybrid scheme (Case 2)

According to the numerical results obtained, some of which are presented in Figs. 25, 26, 27, 28, 29, 30, 31, 32, 33, 34 and 35, the following statements can be summarized:

(a) The water phase humidity is formed due to the phase change of the water vapour at its saturated conditions when the humid air temperature drops below its dew point, as expected. As the process of condensation proceeds, more water vapour changes phase to water droplets that cover larger part of the interior. The process of condensation starts to take place later when the initial humidity value is lower.

(b) For the case of the humid, non-ventilated room, investigated in this study, the high-temperature humid air flows up to the roof, where it encounters the colder wall surface. The low-temperature humid air travels downwards to the floor, where its temperature drops further and for the first time below the dew point,

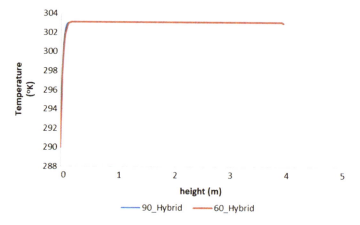

Fig. 33 Vertical temperature distribution at time 360 s for initial humidity conditions 90 and 60% (Case 2)

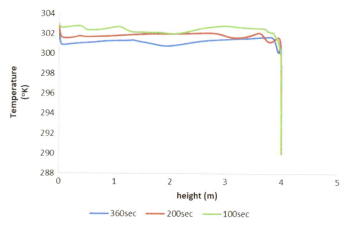

Fig. 34 Vertical temperature distribution at times 100, 200, 360 s for initial humidity conditions 90% (Case 3)

thus the condensation process starts to take place on the floor, as expected. Temperature distribution depends on the concentration of water vapour in humid air and its amount of mass transfer to the water phase. Finally, after a long time (~10 min for the case of initial relative humidity 90%) the whole water vapour mass of humid air has changed phase to water droplets that fill the air inside the room. Additionally, the humid air temperature in the interior has become uniform and equal to the initial low temperature of the wall surfaces (290 K).

(c) When the temperature inside the room is higher than the dew point temperature the condensation process does not take place, thus no humidity (water phase) is formed.

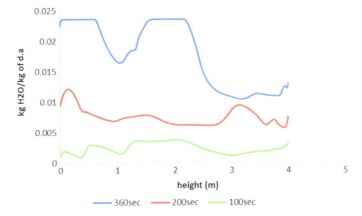

Fig. 35 Vertical absolute humidity distribution at times 100, 200, 360 s for initial humidity conditions 90% (Case 3)

5 Conclusions

Numerical diffusion is the main source of errors in computational calculations. In this study, the performance of various numerical schemes is evaluated in order to find the limitations of the convection formulations. The first-order UPWIND scheme presents stability and overcomes the streamwise diffusion but, as the higher order schemes as well, it does not deal at all with the cross-stream numerical diffusion. The second-order hybrid numerical schemes take into account the influence of more neighbouring points at low Peclet numbers but appear unstable at high Peclet numbers. The non-linear van LEER numerical scheme is highly stable and appears to reduce satisfactorily the false diffusion errors. The flow-oriented SUPER scheme appears to overcome the phenomenon of the numerical diffusion in most of the cases investigated without increasing significantly the computational cost. It is satisfying that the scheme works well under any modelling conditions (laminar, turbulent-RANS, turbulent-LES, single-phase, multiphase, multiphase with phase change, steady state, transient, two-dimensional, three-dimensional). Furthermore, it produces reasonable results, which are as accurate as any other scheme tested, when there is minor inclination of the flow direction to the grid lines, and intuitively more accurate in cases of strong inclinations as it leads to more abrupt profiles not smeared by false diffusion. Indeed, as for example Fig. 12 indicates, SUPER produces for some cases at least nearly identical results with the classical numerical schemes, (which are also close to the experimental measurements for this case); and certainly superior when false diffusion is prominent as shown in Fig. 8, for example. Unfortunately, due to lack of experimental data that would provide quantitative criteria for further comparisons, the conclusions cannot be more than tentative.

Acknowledgements The second author (D. P. Karadimou) gratefully acknowledges the financial support from the State Scholarships Foundation of Greece through the "IKY Fellowships of Excellence for Postgraduate studies in Greece-SIEMENS" Program.

References

1. Patel, M. K., Markatos, N. C., & Cross, M. (1985). Technical note-method of reducing false diffusion errors in convection diffusion problems. *Applied Mathematical Modelling, 9,* 302–306.
2. Patel, M. K., & Markatos, N. C. (1986). An evaluation of eight discretization schemes for two-dimensional convection-diffusion equations. *International Journal for Numerical Methods, 6,* 129–154.
3. Darwish, M., & Moukalled, F. (1996). A new approach for building bounded skew-upwind schemes. *Computer Methods in Applied Mechanics and Engineering, 129,* 221–233.
4. De Vahl Davis, G., & Mallinson, G. D. (1976). An evaluation of upwind and central difference approximations by a study of recirculating flows. *Computers & Fluids, 4*(1), 29–43.
5. Fromm, J. E. (1968). A method for reducing dispersion in convective difference schemes. *Journal of Computational Physics, 3,* 176–189.
6. Spalding, D. B. (1972). A novel finite-difference formulation for different expressions involving both first and second derivatives. *International Journal for Numerical Methods in Engineering, 4,* 551–559.
7. Leonard, B. P. (1979). A stable and accurate convective modelling procedure based on quadratic upstream interpolating. *Computational Mechanics and Applied Mechanical Engineering, 4,* 557–559.
8. Van Leer, B. (1985). Upwind-difference methods for aerodynamics problems governed by the Euler equations. *Lectures in Applications of Mathematics, 22,* 327–336.
9. Raithby, G. D. (1976). Skew upstream differencing schemes for problems involving fluid flow. *Computer Methods in Applied Mechanics and Engineering, 9,* 151–156.
10. Patel, M. K., Markatos, N. C., & Cross, M. (1988). An assessment of flow oriented schemes for reducing 'false diffusion'. *International Journal for Numerical Methods in Engineering, 26,* 2279–2304.
11. Carey, C., Scanlon, T. J., & Fraser, S. M. (1993). SUCCA-an alternative scheme to reduce the effects of multidimensional false diffusion. *Applied Mathematical Modelling, 17*(5), 263–270.
12. Karadimou, D. P., & Markatos, N. C. (2012). A novel flow oriented discretization scheme for reducing false diffusion in three-dimensional (3D) flows: An application in the indoor environment. *Atmospheric Environment, 61,* 327–339.
13. Versteeg, H. K., & Malalasekera, W. (1995). *An introduction to computational fluid dynamics—The finite volume method.* Essex, England: Longman Group Ltd.
14. Spalding, D. B. (1978). Numerical computation of multiphase flow and heat-transfer. In C. Taylor, & K. Morgan (Eds.), *Contribution to recent advances in numerical methods in fluids* (pp. 139–167). Pineridge Press.
15. Markatos, N. C. (1983). Modelling of two-phase transient flow and combustion of granular propellants. *International Journal of Multiphase Flow, 12*(6), 913–933.
16. Spalding, D. B. (1981). A general-purpose computer program for multi-dimensional one or two-phase flow. *Mathematics and Computers in Simulation, XII,* 267–276.
17. Huang, P. G., Launder, B. E., & Leschziner, M. A. (1985). Discretization of nonlinear convection processes: A broad-range comparison of four schemes. *Computer Methods in Applied Mechanics and Engineering, 48,* 1–24.
18. Scanlon, T. J., Carey, C., & Fraser, S. M. (1993). SUCCA3D—an alternative scheme to false-diffusion in 3D flows. *Proceedings of the Institution of Mechanical Engineers, Journal of Mechanical Engineering Science, 207,* 307–313.

19. Karadimou, D. P., & Markatos, N. C. (2018). *Study of the numerical diffusion in computational calculations* (pp. 65–78). London, UK: Intechopen Publisher.

20. Gomma, A., Halim, M. A., & Elsaid, A. M. (2016). Experimental and numerical investigations of a triple-concentric tube heat exchanger. *Applied Thermal Engineering, 99,* 1303–1315.

21. Benavides, A., & Van Wachem, B. (2009). Eulerian-Eulerian prediction of dilute turbulent gas particle flow in a backward facing step. *International Journal of Heat and Fluid Flow, 30,* 452–461.

22. Chen, F., Yu, S., & Lai, A. (2006). Modeling particle distribution and deposition in indoor environments with a new drift-flux model. *Atmospheric Environment, 40,* 357–367.

23. Yakhot, V., Orszag, S. A., Thangam, S., Gatski, T. B., & Speziale, C. G. (1992). Development of turbulence models for shear flows by a double expansion technique. *Physics of Fluids A, 4*(7), 1510–1520.

24. Markatos, N. C. (1986). The mathematical modelling of turbulent flows. *Applied Mathematical Modelling, 10,* 190–220.

25. Argyropoulos & Markatos. (2015). Recent advances on the numerical modelling of turbulent flows. *Applied Mathematical Modelling, 39*(2), 693–732.

26. Padfield, T. *Conservation physics. An online textbook in serial form.* http://www.conservationphysics.org/atmcalc/atmoclc2.pdf.

27. Wexler, A. (1983). *ASHRAE Handbook: Thermodynamic properties of dry air, moist air and water and SI psychrometric charts* (p. 360). New York.

28. Smagorinsky, J. (1963). General circulation experimental with the primitive equations. *Monthly Weather Review, 93*(3), 99.

29. Karadimou, D. P., & Markatos, N. C. (2013). Two-phase transient mathematical modelling of indoor aerosol by means of a flow-oriented discretization scheme. In *CD Proceedings of 10th HSTAM International Congress on Mechanics*, Chania, Crete, May 2013.

30. El Diasty, R., Fazio, P., & Budaiwi, I. (1992). Modelling of indoor air humidity: The dynamic behavior within an enclosure. *Energy and Buildings, 19,* 61–73.

31. Lu, X. (2002). Modelling of heat and moisture transfer in buildings II, Applications to indoor thermal and moisture control. *Energy and Buildings, 34,* 1045–1054.

32. Lu, X. (2003). Estimation of indoor moisture generation rate from measurement in buildings. *Building and Environment, 38,* 665–675.

33. Ma, X., Li, X., Shao, X., & Jiang, X. (2013). An algorithm to predict the transient moisture distribution for wall condensation under a steady flow field. *Building and Environment, 67,* 56–68.

34. Mimouni, S., Foissac, A., & Lavieville, J. (2011). CFD modeling of wall steam condensation by a two-phase flow approach. *Nuclear Engineering and Design, 241,* 4445–4455.

35. Argyropoulos, et al. (2017). Mathematical modelling and computing simulation of toxic gas building infiltration. *Process Safety and Environmental Protection, 111,* 687–700.

36. Isetti, C., Laurenti, L., & Ponticiello, A. (1988). Predicting vapour content of the indoor air and latent loads for air-conditioned environments: Effect of moisture storage capacity of the walls. *Energy and Buildings, 12,* 141–148.

37. Teodosiu, C., Hohota, R., Rusaouen, G., & Woloszyn, M. (2003). Numerical prediction of indoor air humidity and its effects on indoor environment. *Building and Environment, 38,* 655–664.

38. Stavrakakis, G. M., Karadimou, D. P., Zervas, P. L., Sarimveis, H., & Markatos, N. C. (2011). Selection of window sizes for optimizing occupational comfort and hygiene based on computational fluid dynamics and neural networks. *Building and Environmnet, 46,* 298–314.

39. Pu, L., Xiao, F., Li, Y., & Ma, Z. (2012). Effects of initial mist conditions on simulation accuracy of humidity distribution in an environmental chamber. *Building and Environment, 47,* 217–222.

40. Liu, J., Aizawa, H., & Yoshino, H. (2004). CFD prediction of surface condensation on walls and its experimental validation. *Building and Environment, 39,* 905–911.

41. Karabelas, S. J., & Markatos, N. C. (2008). Water vapor condensation in forced convection flow over an airfoil. *Aerospace and Technology, 12,* 150–158.

42. Ishii, M., & Mishima, K. (1984). Two-fluid model and hydrodynamic constitutive relations. *Nuclear Engineering and Design*, (82), 107–126.
43. Lee, S. L. (1987). Particle drag in a dilute turbulent two-phase suspension flow. *International Journal of Multiphase Flow*, 2(13), 247–256.
44. Hetsroni, G. (1989). Particles-turbulence interaction. *International Journal of Multiphase Flow*, 5(15), 735–746.
45. ASHRAE handbook-Fundamentals, 2013.

Examples of Decompositions for Time and Space Domains and Discretization of Equations for General Purpose Computational Fluid Dynamics Programs and Historical Perspective of Some Key Developments

Milorad B. Dzodzo [iD]

Nomenclature

a	Coefficient in the FDE
a^C	Convection coefficient for FDE
a^d	Diffusion coefficient for FDE
a'	Coefficients for Φ' in FDE
a^*	Coefficients for Φ^* in FDE
A	Surface area of the relevant cell area
c	Concentration
$\langle c \rangle$	Concentration in the polluted cell at the beginning
C	Courant number
$C' = c/\langle c \rangle$	Dimensionless concentration
d	Distance between the two neighbor grid nodes
D	Tube diameter or enclosure width
D	Diffusion coefficient
f	Friction factor
g	Acceleration of gravity
$Gr = g\beta(\overline{T}_w - T_b)D^3/\nu^2$	Grashof number
$Gr^+ = \frac{g\beta Q'D^3}{(k\nu^2 8)} = \frac{Gr\overline{Nu}\pi}{16}$	Modified Grashof number
$h = q/(T_w - T_b)$	Local heat transfer coefficient
$\bar{h} = q/(\overline{T}_w - T_b)$	Average heat transfer coefficient
k	Thermal conductivity

Presented work had been performed at CFD Unit, ME Building, Imperial College, Exhibition Road, London, U.K.

M. B. Dzodzo (✉)
Westinghouse Electric Company, Cranberry, USA
e-mail: dzodzomb@westinghouse.com

© Springer Nature Singapore Pte Ltd. 2020
A. Runchal (ed.), *50 Years of CFD in Engineering Sciences*,
https://doi.org/10.1007/978-981-15-2670-1_4

L Dimension of the cell, see Fig. 9
$Nu = hD/k$ Local Nusselt number
$\overline{Nu} = \bar{h}D/k$ Average Nusselt number
Nu_0 Forced convection value of $\overline{Nu}$
p Pressure
p^* Reduced pressure
P Dimensionless pressure
$Pe = \rho u(\delta x)/\Gamma$ Local (grid) Peclet number
$Pe = Re\,Pr = u\,D/\Gamma$ Enclosure Peclet number
$Pr = \nu/\Gamma$ Prandtl number
$q = \bar{h}\big(\overline{T}_w - T_b\big) = 2Q'/(\pi D)$ Rate of heat transfer per unit area
$Q' = \bar{h}\big(\overline{T}_w - T_b\big)\pi D/2$ Rate of heat transfer per unit length
r Radial coordinates
R Tube radius
$Re = \bar{w}D/\nu$ Reynolds number
S_T Source term for the cells near the heated boundary
t Time
$T = t/\Delta t$ Dimensionless time
T Temperature
T_b Bulk temperature
T_w Local wall temperature
u, v, w Velocity components in x, y, z directions
U, V, W Dimensionless velocities
$\bar{w}$ Mean axial velocity
$\overline{W}$ Mean value of W
x, y, z, X, Y, Z Cartesian coordinates
$Z = \frac{z}{L}$ Dimensionless distance

Greek Symbols

α Angle of gravitation action, see Fig. 1
β Thermal expansion coefficient
Δr Distance between the boundary and the boundary grid node, see Fig. 4
Δt Time increment
$\Delta\theta$ Corresponding angle for the boundary cell, see Fig. 4
θ Angular coordinate
ν Kinematic viscosity
ρ Density
Γ Diffusion coefficient
$\Phi = (T - T_b)/\big(Q'/k\big)$ Dimensionless temperature

Φ	Value at grid node
Φ'	Correction to Φ
Φ^*	Stored values of the Φ
Φ_h, Φ_l	Values at higher and lower cell faces, see Fig. 10
Φ_T	Value at grid point P at previous time step
ψ	Stream function

1 Introduction

This paper presents two examples from mid-eighties related to decomposition for time and space domains and discretization of equations for the general purpose CFD programs. In both cases the implementation of linearized source terms for various equations is used to allow regrouping and adding new terms in equations without the need for major changes to the general purpose CFD programs. Presented examples provide a historical perspective of some key developments. They are based on the well-planned code architecture that set the foundation for today's general purpose CFD programs and their ability to address more complex numerical and physics models through implementation of source terms. These developments are contrasted with other selected historical developments and current practices.

The pressure- and velocity-based Marker And Cell (MAC) numerical algorithm by [1, 2], and Particle In Cell (PIC) [1, 3], developed in Los Alamos Laboratory, allowed later more general approaches and development of various algorithms for turbulent and two-phase flows. However, due to the limited computational capabilities and numerical stability issues, the implementations were restricted to two-dimensional transient cases. In early 1970s Semi-Implicit Method for Pressure Linked Equations (SIMPLE) [4], and other Finite Volume Methods based algorithms were developed at Imperial College by several Prof. Spalding's students. This allowed three-dimensional steady-state implementations for parabolic [4] and later elliptical [5] flows which utilized limited hardware available at that time. This was followed by the development and benchmarking of various turbulence and multiphase models and algorithms and resulted in the first general purpose CFD program PHOENICS (Parabolic, Hyperbolic, Or Elliptic Numerical Integration Code Series) in 1981 (see [6]). Details related to the numerical aspects of the applied Finite Volume Methods were presented by [7]. More details about history and emergence of CFD and development of general purpose CFD program modules (numerical algorithms, turbulence models, multiphase models, etc.) at Imperial College are presented by [8].

General purpose CFD program platform underwent further improvements of flow simulations and numerical algorithms. Implementation of numerical concepts without the need to access the entire source code features specific challenges and requires preplanning program organization and capabilities in order to accept user scripts. It also demands that user adjust input files and scripts to the general purpose CFD code input format, connects scripts with adequate steps in numerical procedure, and applied solver options. The implementation of linearized source terms, as explained in [7], p. 143, for various equations allows regrouping and adding new terms in equations without need to change the major parts of the general purpose CFD programs. This allows organized communication of the new ideas and implementations to the other users of the program as well as better quality control of the further developments.

After initial proofs of the concepts and benchmarking based on the comparisons with analytical or numerical solutions for simple cases, further development was possible and some of the tested concepts were later included as standard options in the general purpose CFD code.

This paper presents examples of initial tests and implementations documented in [9, 10].

The first example is related to the implementation of rectangular coordinates to simulate flow and heat transfer in the arbitrarily shaped domains with different heat transfer conditions at boundaries.

The second example demonstrates capabilities to introduce and test implicit and explicit higher order numerical schemes by comparing results with an analytical solution for three-dimensional transient pollutant-cloud transport.

This paper presents a portion of the historical developments that lead to current implementations of CFD general purpose programs.

In the early 1980s, considering and modeling the complex shaped domains based on Finite Difference Methods [11], and Finite Volume Methods was challenging due to the limited available memory, application of staggered grids, and non-existence of the reliable body fitted meshing and solver algorithms. The non-staggered mesh approach [12, 13], and development of the reliable body fitted meshing and solver algorithms was just emerging. Further developments and applications of Finite Difference Methods and Finite Volume Methods are presented by [14].

However, the size and complexity of domains is very often the limiting factor for industrial applications. In this case, the fine body fitted mesh is usually used only locally to resolve flow and heat transfer in the boundary layers, or in the portions of the domain special interest. The rest of domain is usually decomposed with a coarse grid. In some cases, depending on the size and complexity of the domain of interest for industrial practice, only coarse grid might be achieved and special adjustments for the cells near the boundaries are needed. Sometimes the combination of sub-channel codes (used in nuclear industry) and CFD codes is used. One possibility is the Coarse Grid CFD approach. An overview of new trends and Coarse Grid CFD is presented by [15, 16].

Other methods, such as Finite Element Methods, are offering straightforward treatment of the arbitrarily shaped domains because the computational mesh is formed by arbitrarily shaped subdomains called finite elements. All dependent variables and parameters are presented within each finite element on a local basis by N degree approximation polynomials. These approximation basis functions reduce to zeros and ones at certain locations (nodes) within or on the boundary of each finite element. Thus, the boundary of the computational domain does not need to match the global coordinate planes. More details related to the initial applications of Finite Element Methods for CFD are available in [17, 18]. Later, Boundary Element Methods [19], Spectral Element Methods [20–22], and Meshfree Methods such as Smoothed Particle Hydrodynamics [23, 24], were developed. A comparison of Finite Element Methods, Boundary Element Methods, and Meshfree Methods and their relation to the method of weighted residuals is presented in [25]. More details about Meshfree Methods and recent progress are available in [26–28]. The most promising feature of Meshfree Methods is that adaptive analyses for CFD problems are possible. However, the computing costs increase as flexibilities of methods expand.

It is interesting to note that Particle In Cell [1, 3], with individual particles (or fluid elements) tracked in a Lagrangian frame and velocities and densities computed on Eulerian (stationary) mesh can capture fluid–solid, or liquid–gas interfaces, or shock waves, and can be considered as predecessor of Meshfree Methods. Other ways to overcome Eulerian based methods limitations to track free surfaces, or zones with steep gradients, is to introduce deformable meshes, use two-grids as in Arbitrary Lagrange–Eulerian coupling [29], or to develop higher order numerical schemes tailored to do it.

However, at that time (early 1980s) the emerging general purpose Finite Volume Methods based CFD programs did not have options to utilize the higher order numerical schemes (like QUICK, QUICKEST [30], Leith [31], Van Lear [32], MUSCL as in [33] based on [34], Total Variation Diminishing (TVD) [35], and Essentially Non-Oscillatory (ENO) [36]).

In the early 1980s, the first-order upwind [37] and the second order hybrid scheme [38], and power law scheme [7] (based on the exponential scheme [39]), were used in the emerging Finite Volume Methods CFD programs. However, both schemes, hybrid and power law, switch to the first-order upwind if the absolute values of local (grid) Peclet numbers are higher than $|Pe| \geq 2$ and $|Pe| \geq 10$, respectively.

Due to the coarse meshes used at that time these numerical schemes were almost always acting as the first-order schemes. In some cases, the solutions were improved by mesh refinement (and reaching $|Pe| \leq 2$). Based on the performed benchmarks it was concluded that they can produce physically realistic solutions for some practical (elliptic and parabolic, steady state, or time averaged) flows.

Additional comparisons and evaluations of upwind, central, hybrid, and power law schemes are available in [7, 40, 41].

Next, some examples of development and applications of higher order numerical schemes based on Finite Difference Methods derivations are presented by [42–44].

However, for some hyperbolic and transient cases where shock capturing or tracking of steep gradients is needed, the first-order numerical schemes, and combinations with higher order numerical schemes, based on Finite Difference Methods derivations, are not adequate. After several time steps the solutions are not useful. Also, if the mesh is too coarse some flow details like local wiggles will not be captured even for steady-state parabolic and elliptic cases. Based on that the authors of higher order schemes, like [45], were advocating that an approach of creating the general purpose CFD program (based on the lower order numerical schemes) does not have a chance to become an engineering tool if this issue is not solved.

However, the implementation of higher order schemes introduces complexity in the general purpose CFD program organization (introduction of additional cells for boundary conditions and solvers) and again does not guarantee the useful solution for an engineering application. They sometimes produce unrealistic and unacceptable 'solutions' (like negative concentrations—presented in the second example), or additional local wiggles—not observed experimentally. Also, sometimes the converged solution cannot be achieved due to the solver instabilities.

So, additional developments and improvements, as in [33, 35, 36, 46–48], and some other listed and implemented in [48], were needed to provide better engineering tools.

All higher order numerical schemes bring additional complexity of the program organization by introducing more surrounding cells needed to calculate values for the central cells and by introducing additional steps and operations between the start and the end of time steps computations. Also, the implementation of boundary and initial conditions and partitioning of the computational domain needed for parallelization and exchange of data between partitions is more complex.

An evaluation and comparison of the computational costs due to the mentioned additional complexities for various numerical schemes, such as [49], helps to decide which numerical schemes are the best candidates to be implemented in the general purpose CFD program.

Usually all new higher order schemes are developed and tested on simple one- or two-dimensional domains with a uniform mesh. However, implementations of the same numerical scheme for arbitrarily shaped three-dimensional domains decomposed with a nonuniform body fitted mesh needs additional developments and testing. One example for a particular problem is presented in [50].

The final implementation in the general purpose CFD program demands additional adjustments and tests taking into account already established overall organization of the program, solver options, established boundary condition options, existing models (like turbulence and multiphase models) and corresponding equations and ways to introduce new developments.

Two presented cases in this paper are examples of activities in middle 1980s and need to be viewed while taking into account the CFD numerical analysis state of the art at that time and historical perspective of the CFD general purpose program developments. The implementation of linearized source terms was used in both cases to regroup and add new terms in equations and establish connections with the solver without the need to change major parts of the general purpose CFD program, including the solver. This weak connection is adequate for initial implementation and testing of new numerical analysis concepts. Decrease of the computing costs (by decreasing the number of needed iterations, or sweeps) might be achieved later on by modifying the general purpose CFD programs closer to the solver, or the solver as well.

Currently, almost all general purpose CFD programs based on Finite Volume Methods have options to utilize higher order numerical schemes introduced by [30] and Total Variation Diminishing concepts of [34] and [33]. Also, they all have an option to utilize linear source terms and, if necessary, test some new numerical analysis concepts if they are needed for a specific application.

Today, most of the general purpose CFD programs are based on Finite Volume Methods. They are playing an important role for research and industry while supporting new designs and developments. However, there are still some types of problems where more adaptive CFD analysis is needed. Even after all Finite Volume Methods improvements (such as introduction of higher order and more complex numerical schemes, more adaptive meshing and more robust solvers) the Finite Volume Methods solutions for some challenging cases do not match experimental results after several, or several hundred, time steps. The new CFD methods like Meshfree Methods, or combination of Meshfree Methods and Finite Volume Methods (based on Lagrange–Eulerian coupling approaches), might provide an option and tools for more adaptive CFD analyses if needed.

2 Application of Rectangular Coordinates to Simulate Flow and Heat Transfer in Arbitrarily Shaped Domains

2.1 The Objective and Description of the Phenomenon Simulated

The objective was to test and demonstrate the application of the general purpose CFD computer program to utilize rectangular coordinates to simulate flow and heat transfer in the arbitrarily shaped domains with different conditions at boundaries. The effects of nonuniform heating on laminar mixed convection in a straight horizontal tube of circular cross section were used as a test case (see Fig. 1). The one half of tube circumference is heated while the other half is adiabatic. The cases of top heating and bottom heating have been analyzed with the polar coordinates first in a paper by [51]. In contrast, the implementation of rectangular coordinates offers more cases of cross sections to be considered but the special consideration needs to be paid to the grid points near the boundaries.

Fig. 1 A straight horizontal tube of circular cross section the one half of circumference of which is heated while the rest is adiabatic

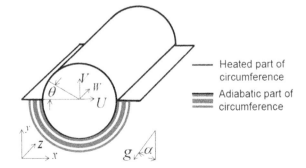

Heated part of circumference

Adiabatic part of circumference

2.2 *Mathematical Description of the Phenomenon Simulated*

Equations. The Eqs. (1–5) are for rectangular coordinates expressing conservation of mass momentum and energy in dimensionless form for steady, fully developed, laminar mixed convection in a horizontal tube. The assumptions are that viscous heat dissipation is negligible, and all fluid properties are constant except the density in the buoyancy term of momentum equation.

$$\frac{\partial U}{\partial X} + \frac{\partial V}{\partial Y} = 0 \tag{1}$$

$$U\frac{\partial U}{\partial X} + V\frac{\partial U}{\partial Y} = \frac{\partial^2 U}{\partial X^2} + \frac{\partial^2 U}{\partial Y^2} + 8Gr^+ \Phi \sin\alpha - \frac{\partial P}{\partial X} \tag{2}$$

$$U\frac{\partial V}{\partial X} + V\frac{\partial V}{\partial Y} = \frac{\partial^2 V}{\partial X^2} + \frac{\partial^2 V}{\partial Y^2} + 8Gr^+ \Phi \cos\alpha - \frac{\partial P}{\partial Y} \tag{3}$$

$$U\frac{\partial W}{\partial X} + V\frac{\partial W}{\partial Y} = \frac{\partial^2 W}{\partial X^2} + \frac{\partial^2 W}{\partial Y^2} + 1 \tag{4}$$

$$U\frac{\partial \Phi}{\partial X} + V\frac{\partial \Phi}{\partial Y} + \frac{4W}{\pi \, Pr \overline{W}} = \frac{1}{Pr}\left(\frac{\partial^2 \Phi}{\partial X^2} + \frac{\partial^2 \Phi}{\partial Y^2}\right) + \frac{S_T}{Pr} \tag{5}$$

The dimensionless variables are

$$X = \frac{x}{D},\, Y = \frac{y}{D},\, Z = \frac{z}{D},\, U = \frac{uD}{\nu},\, V = \frac{\nu D}{\nu},\, W = \frac{w}{\left(-\frac{dp}{dz}\right)\left(\frac{D^2}{\rho \nu}\right)},$$

$$\Phi = \frac{T - T_b}{\left(\frac{Q'}{k}\right)},\, P = \frac{p^*}{\rho\left(\frac{\nu}{D}\right)^2} \tag{6}$$

based on a modified Grashof number

$$Gr^+ = \frac{g\beta Q' R^3}{k\nu^2} = \frac{g\beta Q' D^3}{k\nu^2 8} \tag{7}$$

the reduced pressure

$$p^* = p + g\rho_b x \sin\alpha + g\rho_b y \cos\alpha \tag{8}$$

and the dimensionless heat source term for the cells near the heated boundary

$$S_T = s_T \frac{D^2}{Q'} = \frac{\Delta\theta}{\pi} \frac{Q' 1}{\Delta x \Delta y 1} \frac{D^2}{Q'} = \frac{\Delta\theta}{\pi \Delta X \Delta Y} \tag{9}$$

Boundary Conditions. The boundary conditions for the momentum equations imply that the velocity at the wall is zero. When using the rectangular coordinates with the finite volume method in the case of irregular boundary contours, special attention must be paid to the boundary grid points taking into account three staggered grids and the complicated situation near the boundary. The approach of implementing rectangular coordinates and boundary conditions for the grid points near the boundary is as in [52] which utilizes the real boundary shape for diffusion terms (well known in Finite Difference Methods, see [11], p. 86) and an approximate boundary for convection terms.

The true boundary for the diffusion terms and approximate one for the convection terms and the boundary grid points for the central and staggered grids are presented in Fig. 2. The approximate boundary applied for convection terms is marked with dotted line. Practically the convection fluxes are prescribed equal to zero at the central cell faces (cells for pressure, temperature, and velocities w in axial direction) near the boundary (facing the tube walls), while the central cells and staggered cells for velocities u and v in the cross-sectional plane are informed about distances between the grid points (cell centers) and the real boundary (tube wall). This appeared to be a good combination. The balance calculation for the boundary cells was not complicated while the error in the overall flow (in z-direction) was only 0.3%.

The details of the true and approximate boundaries for the central and staggered grids are presented in Fig. 3. The different distances between true boundary and cell centers for each grid (central grid and staggered grids) are used to calculate diffusion effects on the boundary control volume cells as in [52],

Fig. 2 The true boundary for the diffusion terms and approximate one for the convection terms with the boundary grid points for the central and staggered grids

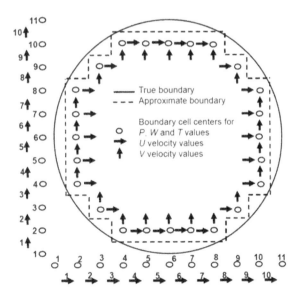

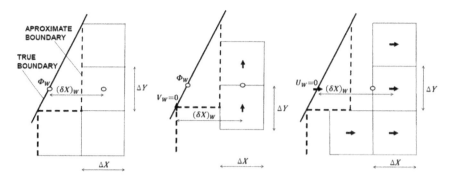

Fig. 3 Central and staggered grids near the true and approximate boundaries

$$a_W = \frac{\Gamma_{\Phi_W} \Delta Y \Delta X}{(\delta X)_W}, \quad a_P = \sum_m a_m, \quad a_P \Phi_P = \sum_m a_m \Phi_m + C_P,$$
$$m = E, W, N, S, H, L \tag{10}$$

The same effect might be introduced in general purpose CFD program by implementing linearized source terms for the boundary control volume cells as explained in [7], p. 143,

$$S_\Phi = S_\Phi' - S_\Phi'' \Phi_P, \quad S_\Phi' = \frac{\Gamma_{\Phi_W} \Delta Y \Delta X}{(\delta X)_W} \Phi_W, \quad S_\Phi'' = \frac{\Gamma_{\Phi_W} \Delta Y \Delta X}{(\delta X)_W} \tag{11}$$

and specifying diffusion and convection terms near the boundaries to be equal to zero:

$$a_W = 0, \quad a_P = \sum_m a_m + S_\Phi'', \quad a_P \Phi_P = \sum_m a_m \Phi_m + C_P + S_\Phi',$$
$$m = E, W, N, S, H, L \tag{12}$$

In the top heating case, there is an uniform heat flux q on the upper half of the tube wall, so that the boundary condition for the energy equation reads

$$\frac{\partial T}{\partial r} = \frac{Q'}{k \pi R} \quad \text{for} \quad 0 < \theta < \pi \quad \text{and} \quad r = R \tag{13}$$

The lower half is insulated resulting in

$$\frac{\partial T}{\partial r} = 0 \quad \text{for} \quad \pi < \theta < 2\pi \quad \text{and} \quad r = R \tag{14}$$

Also, the solution must be compatible with the definition of the bulk temperature, the dimensionless form of which is

Fig. 4 Calculation of the heat source term and local wall temperature for the cell near the heated boundary

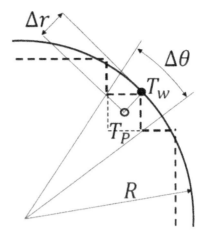

$$\iint \Phi W \, dX \, dY = 0 \tag{15}$$

The details of the boundary condition implementations for the energy equation are presented in Fig. 4. The heat source terms for the boundary cells are calculated based on the portion of the heated circumference facing them (see explanations in Fig. 4). The local heated portion of the wall temperatures and average wall temperature are calculated by using the same segments of the circumference.

2.3 Presentation and Discussion of Results

Dependence of the average and local Nusselt numbers on the modified Grashof number. The average Nusselt number $\overline{Nu} = \bar{h}D/k$ is calculated with average heat transfer coefficient $\bar{h} = q/(\overline{T}_w - T_b)$ based on the average wall temperature

$$\overline{T}_w = \frac{1}{\pi} \int_0^\pi T \, d\theta \tag{16}$$

of the heated part of the wall. For small Gr^+ numbers one can compare $\overline{Nu}$ with analytical solution Nu_0 [53] for forced convection and note that the agreement of results is good. The ratio of $\overline{Nu}/Nu_0$ is presented in Fig. 5 (left side) and the results for top heating agree with [51]. The discrepancy for bottom heating between the results obtained here and in [51] is greater. The average Nusselt number for $Gr^+/\pi = 1$ obtained by using the grid with 11×11 nodes in [9, 52] is $\overline{Nu} = 3.144$. By using rectangular coordinates and the grid with 41×41 nodes the average Nusselt number is $\overline{Nu} = 3.088$, which is closer to the analytically obtained value for the average Nusselt number $Nu_0 = 3.058$ for the forced convection in [53]. The discrepancy for

Fig. 5 Average Nusselt numbers for top and bottom heating cases as a function of modified Grashof number

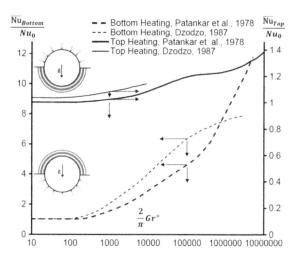

the greater values of Gr^+ numbers in the case of bottom heating would probably be smaller when using a finer grid.

The local Nusselt numbers $Nu = hD/k$ are calculated with the local wall temperatures

$$T_w = T_P + \frac{Q'}{2\pi k} \ln\left(\frac{R}{R - \Delta r}\right), \text{ or in dimensionless form } \Phi_w = \Phi_P + \frac{1}{2\pi} \ln\left(\frac{R}{R - \Delta r}\right)$$

$$(17)$$

(see Fig. 4) and the bulk temperature in the cross section, where $h = q/(T_w - T_b)$.

Dependence of the friction factor on the modified Grashof number. The ratio $(f\, Re)/(f\, Re)_0$ for various values of Gr^+ numbers for the considered case of the top heating is almost equal unity where the definition of the friction factor is

$$f = \frac{\left(-\frac{dp}{dz}\right)D}{\left(\frac{1}{2}\right)\rho \bar{w}^2} \quad \text{and then} \quad f\, Re = \frac{2}{W}.$$

$$(18)$$

The ratio $(f\, Re)/(f\, Re)_0$ for various values of Gr^+ numbers for the case of the top and bottom heating is presented in Fig. 6. The agreement with results in [51] for the case of the bottom heating is better than for the average Nusselt numbers.

Temperature and velocity fields and their comparison with the previous results. In Fig. 7 there are presented the dimensionless temperature functions $\frac{T-T_b}{T_w-T_b}$ and the stream functions $\frac{\psi}{\nu} = -\int_0^1 U dY$ for top heating case and one half of the cross section for $2Gr^+/\pi = 10,000$. The temperature and velocity fields are in good agreement with those in [51, 52].

Fig. 6 Friction factors

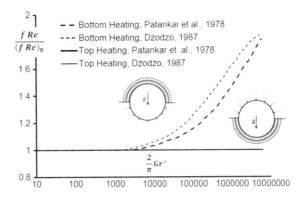

Fig. 7 Isotherm and streamline maps for top heating

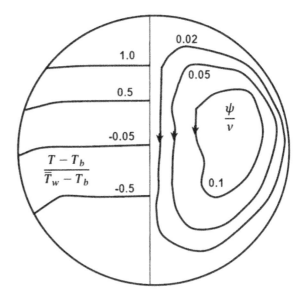

2.4 Additional Validations

Further mesh sensitivity is studied and validations of the approach were performed and reported later on [54–57]. The results were compared with experimental results for natural convection in three different rhomb-shaped three-dimensional enclosures, and with corresponding numerical results using non-orthogonal body fitted coordinates as in [58] and by applying collocated (non-staggered) grid as in [13]. The fluid thermophysical properties were constant (and defined based on the mean temperature of the hot and cold walls) except for density (linearly dependent on the temperature) in the buoyancy force term where Boussinesq approximation [59] was used. The first-order upwind scheme was used and majority of simulations were performed for 2D domain (vertical symmetry plane of the enclosure). Limited number of half and full 3D domains simulations were performed using 3D solver as in [5] and body fitted coordinates as in [58]. Good agreements of the numerical and experimental results were achieved. One exception was that the velocities maximums and minimums near the hot and cold vertical walls were under-predicted and over-predicted, respectively. This was the consequence of applying only Boussinesq approximation for density in the buoyancy force term and not modeling lower and higher fluid viscosity near the hot and cold walls, respectively.

To improve agreements between experimental and numerical results, the uniform rectangular mesh and central difference scheme with the 'deferred correction' as in [60] were used for the square enclosure [61–64]. Good agreement of results with the previous benchmark solutions [65, 66] for air ($Pr = 0.71$) is achieved [63]. However, it is worth to note that good agreement between experimental and numerical results was achieved only after introduction of the measured nonlinear thermophysical properties as a function of temperature in the numerical model. This was especially important for the measured values of viscosity of the applied mixture of glycerol with small concentration of un-encapsulated, or capsulated, liquid crystals present.

It can be concluded that in this case (viscous fluid and enclosure Peclet numbers in the range $Pe < 160$) the role of applied order of numerical schemes was secondary. It was more important to use nonlinear thermophysical properties of the used fluid as a function of temperature instead of applying as usual (as in the majority of validation cases) only Boussinesq approximation and allowing only linear density dependence on the temperature in the buoyancy force term. The limitations of Boussinesq approximation applications were presented in [67]. However, in the majority of CFD validation cases for different fluids and various (sometimes big) temperature differences Boussinesq approximation was used.

Another important cause of the discrepancy between experimental results and numerical validations for natural convection in enclosures is that experiments are performed in the real three-dimensional enclosures and although some 3D domain experimental results [68] and numerical models results, like [41, 69] were available, due to the mesh size limits of the numerical models are most of the time validated and compared among each other (at that time—early 1980s), as in [65, 70] based on the 2D domains (not allowing modeling of the flow restructuring near the front and back walls and symmetry plane). Later on, the same 2D benchmark case was calculated with 640×640 cells [66], instead 41×41 and 81×81 cells in [65]. However, this trend continues and most of the new numerical analysis approaches, like Meshfree Methods ([26], see p. 379 and p. 384) and ([25], see p. 236) are compared and validated based on this [65] idealized 2D benchmark case.

The third cause of the discrepancy between experimental results and numerical validations is that only constant (different) temperatures of the opposite vertical walls are well-defined boundary conditions in the experiments. The other four walls (front, back, bottom, and top) are usually transparent to allow the flow visualization, and their temperatures (or temperature distributions) are not measured or reported. The numerical analysts are usually using adiabatic walls boundary conditions for the bottom and top walls to compare results with the idealized 2D benchmark case as in [65]. The influence of some additional boundary conditions, like heat conducting top and bottom walls, were considered and presented in [71].

Thus, experimental results need to be reported for several vertical planes and complete walls boundary conditions information and compared with the numerical results obtained for three-dimensional domains. Limited number of comparisons based on the numerical models for three-dimensional domains is presented in [55, 57].

Also, in the real applications, very rarely, the ideal textbooks boundary conditions exist (like constant temperature, constant heat flux, or prescribed convective heat transfer conditions). The temperature distributions (used to define boundary conditions) are usually measured at the outside wall surfaces, or inside the walls, but not at the interface between walls and fluid. Thus, more benchmark cases with numerically challenging (due to the slow convergence) conjugated heat transfer examples like in [25, 64] are needed.

Well-documented three-dimensional experimental results for various enclosures are rare. Usually the complete information for temperature boundary conditions (at fluid–solid interfaces, or inside the walls, or on the outer wall surfaces), as well as applied fluid thermophysical properties are very often not available. Thus, the CFD community will most likely continue utilizing simpler benchmarks based on two-dimensional square-shaped enclosures with constant temperatures at walls and Boussinesq approximation for density variation. Simpler benchmarks can compare various numerical analysis approaches but are less useful for real applications needed in industry. Very often it is necessary to evaluate various three-dimensional natural circulation and stratification effects on the transient responses of passively cooled systems.

2.5 Summary and Current Practices

A proof of principle for general approach by applying rectangular coordinates (instead of polar coordinates) to the problem of the top and bottom heating of a horizontal tube with a circular cross section is obtained. A good agreement with the previous analytical and numerical results is obtained. Based on that, various tubes with irregular cross sections and nonuniform peripheral heating can be treated. In the case of energy equation, the more specific treatment of the linear source terms based on each boundary cell corresponding wall surface area was needed to be able to specify the constant heat flux boundary condition. In the case of non-staggered (collocated) cells the cells configuration near the boundary are presented in Fig. 8a. The true boundary can be used for the diffusion terms and an approximate boundary for the convection terms. This combination provides better agreement with the results in [51], obtained with the polar coordinates, than use of the approximate boundary for all terms.

Another possible approach is to utilize two meshes: the rectangular mesh in the core of the domain and an overlapping mesh near the boundary, which follows the shape of the boundary, as presented in Fig. 8b. The data transfer between two meshes

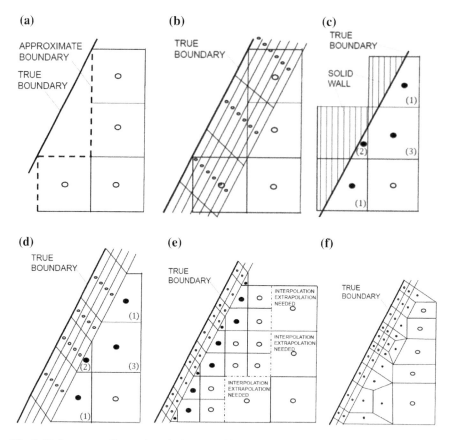

Fig. 8 Various ways of space domain decompositions and mesh configurations near the boundary walls

is organized by interpolation and extrapolation. However, this approach, often used for the turbomachinery applications, like in [72], demands special adjustments for each case (shape of the boundary, sizes of and distribution of the meshes, etc.) and cannot be easily generalized. One of the first examples is utilization of the rectangular coordinates for the whole field and an overlapping polar coordinates mesh near the cylinder surface used for the case of the flow around cylinder as in [73].

The cut-cell approach, (see Fig. 8c), computes the areas and volumes for the cells near the boundaries and employs special procedures for computing advection and diffusion near solid boundaries to calculate wall shear stresses and heat fluxes. In [74], twelve possible cut cells are identified and presented (based on [75]) for the two-dimensional cases of flows around solid bodies. If the three types presented in Fig. 8c are rotated 90° three times, all possible twelve types are obtained. Additional possible arrangements of cut cells for the flows around bodies with sharp tips are presented in [76] (see Appendix B in [76]). Nowadays, the cut-cell method for the cells near the moving solid body boundaries or free-surface interfaces provides an alternative to the body fitted coordinates methods or the need for excessive local mesh refinements in Volume Of Fluids (VOF) [77] methods. Similar concepts are used by several general purpose CFD program developers to introduce so cold 'cut-cell' and 'trimmed-cell' methods.

The trim-cell approach can take into consideration variously shaped cells which are result of hexahedral cell with one part of them removed (in fact cut). The trim-cells are usually used in combination with the body fitted coordinates methods. To resolve the boundary layer velocity and temperature profiles the fine mesh based on body fitted coordinates is applied near the walls and coarse mesh based on the rectangular coordinates is in the core of the domain. The trim-cells provide transition and connection between these two parts of the computational domain (see Fig. 8d). This approach results in an appropriate distribution of cells and an acceptable total number of cells. This arrangement is useful when the size of the mesh is limiting but both: flow in the boundary layers near walls and flow patterns in the core of the domain need to be modeled so that good agreement with the experimental results can be achieved (see for example [78]).

Further improvement of this approach is to refine the rectangular mesh near the boundary (presented in Fig. 8e). If necessary, several layers of rectangular mesh refinements can be inserted in between the coarse rectangular mesh in the core of the domain and fine non-orthogonal body fitted mesh near the boundary walls. This way, the very fine mesh can be produced in the boundary layers zone. However, the drawback of this approach is that interpolations and extrapolations need to be performed at the interfaces between the coarse and fine rectangular cells (see dotted lines in Fig. 8e). This drawback can be removed if the coarser cells are divided into tetrahedral or prism cells so that all cell faces are connected.

Another way to remove this drawback is to use structured coarse hexahedral cells (but based on non-orthogonal mesh) in the core of the domain and structured very fine non-orthogonal mesh with hexahedral cells near the boundary walls with an unstructured layer of hexahedral cells in between of them (as presented in Fig. 8f). In this case, all cells are hexahedral and all cell faces are connected. There is no need for interpolations and extrapolations in the mesh refinement zones. To achieve very fine mesh near the boundary walls several layers of refinements with unstructured hexahedral cells (or blocks of cells) can be applied (as in [79]).

If the blocks of the structured hexahedral cells are applied instead of the hexahedral cells presented in Fig. 8f, an additional mesh refinement might be achieved. This mesh arrangement is suitable for Spectral Element Methods (where all cells need to be hexahedral with 20 or 27 nodes). The arrangement of unstructured blocks and the number of structured cells inside them in each direction need to be adjusted so that the same number and distribution of cell faces are present at the two touching block faces. The same arrangement (unstructured blocks with internal structured cells) and Finite Volume Methods were used in mid-nineties to simulate flow inside complex three-dimensional domains representing portions or complete devices of interest to industry. Some examples are presented in [80–82].

Moreover, various combinations of the meshing approaches presented in Fig. 8 can be combined. For example, depending on the size of the domain and the purpose of an analysis, the local mesh refinement as in Fig. 8e can be applied without applying trim-cells and non-orthogonal body fitted mesh and combined with the approach as in Fig. 8a. Accordingly, an acceptable solution for an engineering application can be obtained with low computational costs. Some Adaptive Mesh Refinement developments and applications are presented by [83–85].

Further discussion and additional meshing arrangements for Finite Volume Methods and complex geometries are presented in [14] (in Chap. 8, pp. 217–259).

During late nineties the Finite Volume Methods solvers developed for unstructured meshes become available in general purpose CFD programs. They provide flexibility to use combinations of the body fitted hexahedral, tetrahedral, prism or polyhedral meshes. The most often used combinations are very fine hexahedral, or prism meshes near the walls with other types of coarser meshes further from the walls.

3 Test of FIP, QUICK, QUICKEST, and LEITH's Formulations

3.1 The Objective and Description of the Phenomenon Simulated

The objective was to demonstrate the application of the general purpose CFD program to the prediction of the three-dimensional transient pollutant-cloud transport by convection and diffusion in a steady stream. For the purpose of testing, the problem was simplified so that the velocity of the stream is uniform and aligned with z-coordinate direction. Prior to the start of calculation, the concentration of the pollutant is zero through the domain except for one cell in which the pollutant is uniformly distributed.

The same problem has been treated using FIP (fully-implicit upwind), QUICK, QUICKEST, and LEITH formulations and comparisons are made between the results. In particular, the problem of developing and incorporating the discretized equations for each method in the general purpose CFD program solver is discussed. The details of implementations with commented input file and developed subroutines are available in [10].

3.2 Mathematical Description of the Phenomenon Simulated

Equation. For the flow in z-direction the equation governing described phenomenon is, in Cartesian coordinates:

$$\frac{\partial c}{\partial t} + w\frac{\partial c}{\partial z} = D\left(\frac{\partial^2 c}{\partial x^2} + \frac{\partial^2 c}{\partial y^2} + \frac{\partial^2 c}{\partial z^2}\right) \tag{19}$$

Fig. 9 3D Computational domain, initial, and boundary conditions applied for the FIP, QUICK, LEITH's, and QUICKEST methods

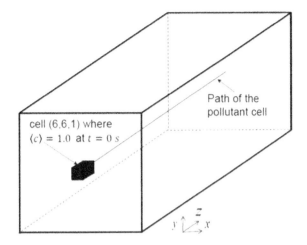

where c is the pollutant concentration; w is the velocity in the z-direction; t is the time, and D is a diffusion coefficient.

Initial and boundary conditions and fluid properties. A finite part of the considered domain presented in Fig. 9 was divided into a number of uniformly distributed cubical cells. In z-direction 18 cells and in x- and y-directions 11 cells are applied. Each cell has dimension $L = 0.1$ m. The initial pollutant concentration is zero throughout the whole infinite domain except for one cell, a cube whose dimension is L (see Fig. 9—cell (6, 6, 1), where $\langle c \rangle = 1.0$ at $t = 0$ s). The flow stream entering the computational domain with velocity $w = 1.732$ m/s has a zero concentration of pollutant. All other boundaries have a zero concentration gradient across their planes. Diffusion coefficient $D = 0.001$ m^2/s was used.

Because of the constant and uniform velocity of the stream, the equation for a coordinate system which moves with the stream (Lagrangian approach) becomes

$$\frac{\partial c}{\partial t} = D \left(\frac{\partial^2 c}{\partial x^2} + \frac{\partial^2 c}{\partial y^2} + \frac{\partial^2 c}{\partial z^2} \right) \tag{20}$$

For the infinite region where only one cell is polluted at the beginning an analytical solution is given as in [86].

$$\frac{c}{\langle c \rangle} = \frac{1}{8} \left[\text{erf}\left(\frac{\frac{L}{2} - x}{l} \right) + \text{erf}\left(\frac{\frac{L}{2} + x}{l} \right) \right] \left[\text{erf}\left(\frac{\frac{L}{2} - y}{l} \right) + \text{erf}\left(\frac{\frac{L}{2} + y}{l} \right) \right]$$
$$\left[\text{erf}\left(\frac{\frac{L}{2} - z}{l} \right) + \text{erf}\left(\frac{\frac{L}{2} + z}{l} \right) \right] \tag{21}$$

where $\langle c \rangle$ is the initial concentration in the polluted cell; L is the dimension of the cell; $l = 2\sqrt{Dt}$; and x; y; z are distances from the center of the pollutant cloud.

3.3 The Implementation of the FIP, QUICK, QUICKEST, and LEITH's Formulations

The FIP method (fully-implicit upwind) is well known and documented in many references, see for example [7]. In the calculation of the convection terms, only one node values upstream of the cell faces are taken into account. The QUICK method is different from FIP method in the quadratic upstream interpolation used for the convection terms calculation. The QUICK method for a two-dimensional flow was described by [87]. Details of how it is implemented in the present three-dimensional case can be found in the following subsections. Both, the FIP and QUICK methods are implicit methods.

The LEITH and QUICKEST methods are explicit. Both methods take into account the unsteadiness of the variables at the cell face. Both explicit methods are the same for the special case when the Courant number is equal to one. Both methods are then as FEX (fully explicit upwind), and for this special case solution is very close to the analytical solution.

In the QUICKEST method, which was presented for a one-dimensional flow by [30], diffusion was calculated as a weighted average of the central difference approximations for diffusion at the central node and the neighboring upstream grid node. The procedure adopted here for applying this scheme to a three-dimensional flow is presented in following subsections.

The LEITH method [31], was used originally for convection situations, only; but diffusion was included in the same manner as for the QUICKEST method.

The finite difference equations for QUICK, LEITH, and QUICKEST methods and the method they were implemented in the general purpose CFD program equation solver procedure are explained in the following subsections.

Derivation of the Finite Domain Equations for the QUICK, LEITH and QUICKEST methods. The finite domain equation (FDE) is obtained from integrating the differential equation over control volumes of finite and different sizes in various directions, but the calculation was performed for uniform cells, see Fig. 10.

The final form of the FDE is

$$
\begin{aligned}
a_T(\Phi_P - \Phi_T) &+ a_h^C \Phi_h - a_l^C \Phi_l \\
&= a_H^d \Phi_H + a_L^d \Phi_L + a_{LL}^d \Phi_{LL} + a_N^d \Phi_N + a_S^d \Phi_S + a_E^d \Phi_E + a_W^d \Phi_W \\
&+ a_{LN}^d \Phi_{LN} + a_{LS}^d \Phi_{LS} + a_{LE}^d \Phi_{LE} + a_{LW}^d \Phi_{LW} + a_P^d \Phi_P
\end{aligned} \tag{22}
$$

where

Φ_T is the value at grid point P at previous time step;

Φ_P is the value at grid point P;

Φ_H, Φ_N, Φ_S, Φ_E, Φ_W, Φ_L, Φ_{LN}, Φ_{LS}, Φ_{LE}, Φ_{LW}, Φ_{LL} are the values at grid points (see Fig. 10);

Φ_h, Φ_l are the values at higher and lower cell faces (see Fig. 10);

$a_h^C = a_l^C = A_z w$ are the coefficients expressing convection;

Fig. 10 Locations around a
control volume enclosing the
node P

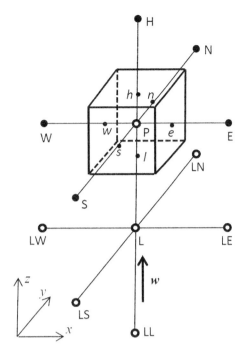

$a_H^d = \frac{DA_z(1-C)}{d_z}$

$a_N^d = a_S^d = \frac{DA_y(1-C)}{d_y}$

$a_E^d = a_W^d = \frac{DA_x(1-C)}{d_x}$ are the coefficients expressing diffusion in the P slab;

$a_{LL}^d = \frac{DA_z C}{d_z}$

$a_{LN}^d = a_{LS}^d = \frac{DA_y C}{d_y}$

$a_{LE}^d = a_{LW}^d = \frac{DA_x C}{d_x}$ are the coefficients expressing diffusion in the L slab;

$a_L^d = \frac{DA_z(1-C)}{d_z} - \frac{2DA_x C}{d_x} - \frac{2DA_y C}{d_y} - \frac{2DA_z C}{d_z}$ is the coefficient expressing diffusion
in the P and L slabs;

$a_P^d = \frac{DA_z C}{d_z} - \frac{2DA_x(1-C)}{d_x} - \frac{2DA_y(1-C)}{d_y} - \frac{2DA_z(1-C)}{d_z}$ is the coefficient expressing
diffusion in the L and P slabs;

$a_T = \frac{V}{\Delta t}$ is the coefficient representing the influence of the past on the present;

V is volume of the cell;

Δt is time increment;

$A_x = A_y = A_z$ is the surface area of the relevant cell area which is normal in the
x-, y- and z-directions;

D is the diffusion coefficient;

$d_x = d_y = d_z$ is the distance between the two neighbor grid nodes in the x-, y-
and z-directions;

$C = 0$ for QUICK method and

$C = w\frac{\Delta t}{d_z}$ is Courant number for QUICKEST and LEITH's methods.

This equation is derived for flow in the positive z-direction, which is a special case. If the flow is in the negative z-direction the derived equation (22) needs to be modified and includes node in HH slab (above the node H) and all surrounding nodes (HW, HE, HS and HN) in H slab. For an arbitrary flow direction (general case) nodes WW, EE, SS, and NN and surrounding nodes in W, E, S, and N slabs are needed in Eq. (22) and all other are presented in Eqs. (23–34).

FDE equation for the QUICK method. In the QUICK method, the diffusion coefficients were calculated using $C = 0$. This means that only the influence of diffusion surrounding the grid node P is taken into account. Concentration values of the higher and lower cell face were calculated as

$$\Phi_h = \frac{1}{2}(\Phi_H + \Phi_P) + \frac{1}{8}(2\Phi_P - \Phi_H - \Phi_L) + \frac{1}{24}(\Phi_E + \Phi_W - 2\Phi_P)$$
$$+ \frac{1}{24}(\Phi_N + \Phi_S - 2\Phi_P) \tag{23}$$

$$\Phi_l = \frac{1}{2}(\Phi_P + \Phi_L) + \frac{1}{8}(2\Phi_L - \Phi_P - \Phi_{LL}) + \frac{1}{24}(\Phi_{LE} + \Phi_{LW} - 2\Phi_L)$$
$$+ \frac{1}{24}(\Phi_{LN} + \Phi_{LS} - 2\Phi_L) \tag{24}$$

Then the difference between cell face values can be expressed as

$$\Phi_h - \Phi_l = \frac{5}{24}\Phi_P - \frac{1}{24}(17\Phi_L + \Phi_{LE} + \Phi_{LW} + \Phi_{LN} + \Phi_{LS} - 9\Phi_H$$
$$- 3\Phi_{LL} - \Phi_E - \Phi_W - \Phi_N - \Phi_S) \tag{25}$$

The coefficients for the FDE equation have finally the following form:

$$a_P = a_T + \frac{5}{24}a^C - a_P^d$$

$$a_H = -\frac{9}{24}a^C + a_H^d$$

$$a_L = \frac{17}{24}a^C + a_L^d$$

$$a_{LL} = -\frac{3}{24}a^C$$

$$a_{LN} = a_{LS} = a_{LW} = a_{LE} = \frac{1}{24}a^C$$

$$a_N = a_S = -\frac{1}{24}a^C + a_N^d$$

$$a_E = a_W = -\frac{1}{24}a^C + a_E^d \tag{26}$$

FDE equation for the LEITH method. In the LEITH method, the influence of diffusion from the surroundings of the grid node L (in the L slab) was taken into account as well as the influence from the surroundings of the grid node P (in the P slab). Which influence will prevail depends on the Courant number (or on the time step used for the same grid arrangement and on the velocity of the stream).

The concentration values at higher and lower cell face are calculated as

$$\Phi_h = \frac{1}{2}(\Phi_P + \Phi_H) - \frac{C}{2}(\Phi_H - \Phi_P) \tag{27}$$

$$\Phi_l = \frac{1}{2}(\Phi_L + \Phi_P) - \frac{C}{2}(\Phi_P - \Phi_L) \tag{28}$$

The difference between cell face values can be expressed as

$$\Phi_h - \Phi_l = \Phi_P C + \Phi_H\left(\frac{1}{2} - \frac{C}{2}\right) + \Phi_L\left(-\frac{1}{2} - \frac{C}{2}\right) \tag{29}$$

And coefficients for the final form of the FDE are

$$a_P = a_T + a^C C - a_P^d$$

$$a_H = -a^C\left(\frac{1}{2} - \frac{C}{2}\right) + a_H^d$$

$$a_L = -a^C\left(-\frac{1}{2} - \frac{C}{2}\right) + a_L^d$$

$$a_{LL} = a_{LL}^d = \frac{D A_z C}{d_z}$$

$$a_{LN} = a_{LS} = a_{LN}^d = \frac{D A_y C}{d_y}$$

$$a_{LE} = a_{LW} = a_{LE}^d = \frac{D A_x C}{d_x}$$

$$a_N = a_S = a_N^d = \frac{D A_y (1 - C)}{d_y}$$

$$a_E = a_W = a_E^d = \frac{D A_x (1 - C)}{d_x} \tag{30}$$

FDE equation for the QUICKEST method. In QUICKEST, the formulation of the transient term produces additional "time difference curvature" terms. These terms for the x- and y-directions cancel, while the term in z-direction changes the corresponding terms in the convection expression, and that is why the expressions for the cell face values

$$\Phi_h = \frac{1}{2}(\Phi_P + \Phi_H) - \frac{C}{2}(\Phi_H - \Phi_P) - \frac{(1 - C^2)}{6}(\Phi_H - 2\Phi_P + \Phi_L) \tag{31}$$

$$\Phi_l = \frac{1}{2}(\Phi_L + \Phi_P) - \frac{C}{2}(\Phi_P - \Phi_L) - \frac{(1 - C^2)}{6}(\Phi_P - 2\Phi_L + \Phi_{LL}) \qquad (32)$$

are simpler than those for the QUICK method.

Then the difference between the cell face values can be expressed as

$$\Phi_h - \Phi_l = \Phi_P\left(\frac{1}{2} + C - \frac{C^2}{2}\right) + \Phi_H\left(\frac{1}{3} - \frac{C}{2} + \frac{C^2}{6}\right)$$

$$+ \Phi_L\left(-1 - \frac{C}{2} + \frac{C^2}{2}\right) + \Phi_{LL}\left(\frac{1}{6} - \frac{C^2}{6}\right) \qquad (33)$$

And the coefficients for the final form of the FDE equation are

$$a_P = a_T + a^C\left(\frac{1}{2} + C - \frac{C^2}{2}\right) - a_P^d$$

$$a_H = -a^C\left(\frac{1}{3} - \frac{C}{2} + \frac{C^2}{6}\right) + a_H^d$$

$$a_L = -a^C\left(-1 - \frac{C}{2} + \frac{C^2}{2}\right) + a_L^d$$

$$a_{LL} = -a^C\left(\frac{1}{6} - \frac{C^2}{6}\right) + a_{LL}^d$$

$$a_{LN} = a_{LS} = a_{LN}^d = \frac{DA_yC}{d_y}$$

$$a_{LE} = a_{LW} = a_{LE}^d = \frac{DA_xC}{d_x}$$

$$a_N = a_S = a_N^d = \frac{DA_y(1 - C)}{d_y}$$

$$a_E = a_W = a_E^d = \frac{DA_x(1 - C)}{d_x} \qquad (34)$$

FDE equation for the FEX (fully explicit upwind) method. If the Courant number is equal to 1, $C = w\Delta t/d_z = 1$, the concentration value propagates from one node to the next because $w\Delta t = d_z$. For this special case the derived Eqs. (27–30) for LEITH and Eqs. (31–34) for QUICKEST produce the same values and coefficients (35–37) as for FEX (fully explicit upwind) method. Only nodes in slabs LL, L, and P, marked as white circles in Fig. 10 contribute to the solution, because

$$\Phi_h = \Phi_P \qquad (35)$$

$$\Phi_l = \Phi_L \qquad (36)$$

and the coefficients for the final form of the FDE equation are:

$$a_P = a_T + a^C - a_P^d = \frac{V}{\Delta t} + A_z w - \frac{DA_z}{d_z}$$

$$a_L = a^C + a_L^d = A_z w + \left(-\frac{2DA_x}{d_x} - \frac{2DA_y}{d_y} - \frac{2DA_z}{d_z} \right)$$

$$a_{LL} = a_{LL}^d = \frac{DA_z}{d_z}$$

$$a_{LN} = a_{LS} = a_{LN}^d = \frac{DA_y}{d_y}$$

$$a_{LE} = a_{LW} = a_{LE}^d = \frac{DA_x}{d_x}$$

$$a_H = a_N = a_S = a_E = a_W = 0 \tag{37}$$

The form of FDE (22) is then

$$\frac{V}{\Delta t}(\Phi_P - \Phi_T) + A_z w(\Phi_P - \Phi_L)$$

$$= \left(-\frac{2DA_x}{d_x} - \frac{2DA_y}{d_y} - \frac{2DA_z}{d_z} \right)\Phi_L + \frac{DA_z}{d_z}\Phi_{LL}$$

$$+ \frac{DA_y}{d_y}\Phi_{LN} + \frac{DA_y}{d_y}\Phi_{LS} + \frac{DA_x}{d_x}\Phi_{LE} + \frac{DA_x}{d_x}\Phi_{LW} + \frac{DA_z}{d_z}\Phi_P \tag{38}$$

Implementation of the new numerical methods using the general purpose CFD software equation solver. If FDE's are expressed in the 'correction' form:

$$a_P \Phi_P' = a_H' \Phi_H' + a_L' \Phi_L' + a_N' \Phi_N' + a_S' \Phi_S' + a_E' \Phi_E' + a_W' \Phi_W'$$

$$+ \left\{ a_H^* \Phi_H^* + a_L^* \Phi_L^* + a_N^* \Phi_N^* + a_S^* \Phi_S^* + a_E^* \Phi_E^* + a_W^* \Phi_W^* + a_T \Phi_T + b \right.$$

$$\left. - a_P^* \Phi_P^* \right\} \tag{39}$$

The Φ^* values are the in-store values of the Φ values and Φ' are the corrections to Φ values. When the solution has been attained, Φ_P' as well as the summation of all terms in brackets $\left\{ a_H^* \Phi_H^* + a_L^* \Phi_L^* + \cdots + b - a_P^* \Phi_P^* \right\}$ are equal to zero. This allows for the generation of the coefficients a^* independently of whether or not the solver is still generating the a' according to the in-built numerical scheme.

The advantage of this is that the coefficients a^* multiplying the Φ^* values may be different from those a' for the Φ' values. This can be done by changing the linearized source terms where $S_{\Phi'}' = \left\{ a_H^* \Phi_H^* + a_L^* \Phi_L^* + \cdots + b - a_P^* \Phi_P^* \right\}$ and $S_{\Phi'}'' = a_P$.

Thus, the new numerical methods can be implemented using linearized source terms. This advantage has been used for the QUICK method where a' coefficients corresponded to the diffusion coefficients a^d (defined in Eq. (22) for $C = 0$) calculated by the solver and the other a^* coefficients were calculated according to QUICK method (defined in Eq. (26) where all diffusion coefficients are set $a^d = 0$).

Setting the new linearized source terms back in the solver and using the equation solver is necessary for the implicit schemes only. Several sweeps (iterations) are then needed because the sweeps are then in fact iterations.

The solution for the explicit schemes, after recalculating the linear source terms, can be obtained without using the equation solver, because the iterations are not necessary.

The final form of the FDE equation for the implicit methods. The final form of the FDE equation for the implicit methods is adjusted as

$$
\begin{aligned}
a_P \Phi'_P = {}& a'_H \Phi'_H + a'_L \Phi'_L + a'_N \Phi'_N + a'_S \Phi'_S + a'_E \Phi'_E + a'_W \Phi'_W \\
& + \{ a_T \Phi_T + a^*_H \Phi^*_H + a^*_L \Phi^*_L + a^*_{LL} \Phi^*_{LL} + a^*_N \Phi^*_N + a^*_S \Phi^*_S + a^*_E \Phi^*_E \\
& + a^*_W \Phi^*_W + a^*_{LN} \Phi^*_{LN} + a^*_{LS} \Phi^*_{LS} + a^*_{LE} \Phi^*_{LE} + a^*_{LW} \Phi^*_{LW} - a^*_P \Phi^*_P \}
\end{aligned} \tag{40}
$$

All values on the right side of the equation are from the previous iteration except for Φ_T which is from the previous time step. The solution procedure is iterative and after rearranging the linear source terms ($S''_{\Phi'} = a_P$ and $S'_{\Phi'} = \{ a_T \Phi_T + \ldots + a^*_{LW} \Phi^*_{LW} - a^*_P \Phi^*_P \}$) and setting them back in the solver, the equation solver procedure is applied.

The final form of the FDE equation for the explicit methods. The final form of the FDE equation for the explicit methods is

$$
\begin{aligned}
a_T \Phi_P = {}& a_T \Phi_T - (a_P - a_T) \Phi_T + a_H \Phi_H + a_L \Phi_L + a_{LL} \Phi_{LL} + a_N \Phi_N + a_S \Phi_S \\
& + a_E \Phi_E + a_W \Phi_W + a_{LN} \Phi_{LN} + a_{LS} \Phi_{LS} + a_{LE} \Phi_{LE} + a_{LW} \Phi_{LW}
\end{aligned} \tag{41}
$$

All the values on the right hand side of the equation are from the previous time step. The linear source terms are rearranged according to the explicit method ($S''_{\Phi'} = a_T$ and $S'_{\Phi'} =$ right hand side of equation). The values for the new time step are calculated by dividing $S'_{\Phi'}$ by $S''_{\Phi'}$.

The details of implementations of both, implicit and explicit methods are available in [10].

3.4   Presentation and Discussion of Results

For all runs, only the values in the polluted cloud path at the last time step are presented. Comparison of the analytical solution with results from the various numerical methods is presented in Fig. 11, where C' stands for dimensionless concentration, c for concentration, $\langle c \rangle$ for concentration in the polluted cell at the beginning of the first time step, C is the Courant number ($C = w \Delta t / d_z$), T is the dimensionless time ($T = t / \Delta t$), and Z is dimensionless distance ($Z = z / L = wt / L$) from the initial to the current center of the polluted cloud. The presented results are for the last time step; therefore, T is 10, and Z is 8.66 (because the Courant number is 0.866 for $w = 1.732$ m/s, $\Delta t = 0.05$ s, and $d_z = L = 0.1$ m).

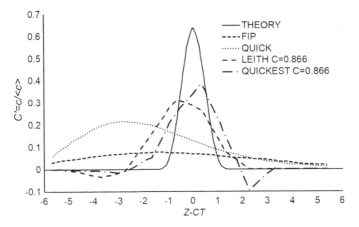

Fig. 11 Comparison of various numerical methods at $CT = 8.66$ and $Pe = 173.2$

As it is shown in Fig. 12, explicit methods, LEITH, and QUICKEST, predict correctly the speed of the polluted cloud, but the calculated peak value is lower. It should be emphasized that the exact prediction of the speed of the polluted cloud in the above schemes is due to the possibility of assuming unsteady concentration profiles at the cell faces during each time step.

The QUICK method underestimates the peak value and the cloud is slower. All three higher order methods have physically unrealistic regions with negative concentrations. However, FIP does not produce such overshoots, but the peak value is well underestimated and the cloud spreading is much over-predicted.

Further results for LEITH and QUICKEST methods for the same conditions but with Courant number values of 0.866, 0.433, and 0.1732 are displayed in Fig. 12.

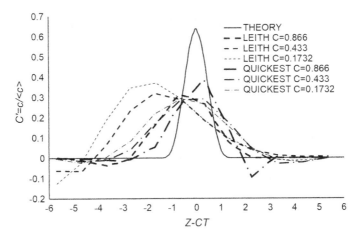

Fig. 12 Effects of Courant number at $CT = 8.66$ and $Pe = 173.2$

The changes in Courant number have been effected through appropriate changes in time steps.

Figure 12 shows that the predicted pollutant-cloud center (with explicit methods) is the closest to the theoretical one when the Courant number is the highest. In LEITH method, speed of cloud is lower with smaller Courant number. The same effect is not so outstanding for QUICKEST method.

In the QUICKEST method, the highest Courant number gives the best peak in comparison with LEITH method, where the effects of Courant number on the peak value cannot be clearly identified.

Unphysical values of concentration have been produced by both methods. In LEITH method, the negative values are produced behind the cloud center in comparison with QUICKEST method, where the unphysical values are behind and in front of peak value. This difference can be explained by analyzing the coefficients for finite domain equations.

In Fig. 13 the comparison between theoretical solution and LEITH, QUICKEST, and FEX solutions for Courant number of unity is presented. The changes in Courant number have been effected here through appropriate changes in the velocity ($w = 0.2$ m/s instead 0.1732 m/s).

During one time step a concentration value propagates from one node to the next. Consequently, under these conditions the methods of LEITH, QUICKEST, and fully explicit upwind (FEX) are all identical as presented in Eqs. (35–37).

For this case, the results are close to the analytical solution based on Lagrangian approach (see Eq. (21)).

In Table 1 a quantitative summary of results, the peak values, position of the cloud center, and errors revealed by the various methods are presented. It is obvious that the behavior of the explicit schemes for Courant number near unity is the best one.

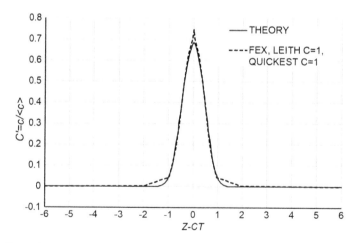

Fig. 13 Comparison of various explicit numerical methods at $C = 1$, $CT = 10$ and $Pe = 200$

Table 1 Comparison of analytical results and FIP, FEX, QUICK, LEITH, and QUICKEST results

Method	Peak value	Solution/Analytical solution	Position of the cloud center (IX, IY, IZ)
Analytical	0.6959		
FIP C = 0.866	0.0866	0.1244	6, 6, 8
QUICK C = 0.866	0.2380	0.3420	6, 6, 7
LEITH C = 0.866	0.3477	0.4996	6, 6, 9
LEITH C = 0.433	0.3280	0.4713	6, 6, 8
LEITH C = 0.1732	0.3775	0.5425	6, 6, 8
QUICKEST C = 0.866	0.4240	0.6093	6, 6, 10
QUICKEST C = 0.433	0.2907	0.4176	6, 6, 9
QUICKEST C = 0.1732	0.3014	0.4331	6, 6, 9
FEX = LEITH = QUICKEST for C = 1	0.7426	1.0776	

3.5 Summary and Current Practices

A detailed mathematical description has been given for implementation of three numerical formulations (QUICK, LEITH, and QUICKEST) for the three-dimensional transient scalar transport problem. Incorporation of the implicit and explicit methods into the general purpose CFD program by using the linearized source terms, as explained in [7], has been described. The same problem has been treated using FIP (implicit upwind) as well, and comparisons are made between the results.

It was shown that results for QUICKEST and LEITH explicit schemes are strongly dependent on Courant number. In the special case when the Courant number is equal to 1, both methods are the same as FEX (fully explicit upwind) and results are very close to the analytical solution.

Results indicate that the QUICKEST and LEITH methods could be acceptable, if the physically unrealistic negative concentrations appearance can be prevented. Only flow aligned with z-direction was considered. For the flow not aligned with any coordinate direction the additional terms need to be included in the finite domain equations.

The presented test case is an example where Eulerian approach, even if more complex numerical schemes are applied, will not likely be able to provide satisfactory results for general applications after a certain number of time steps. As discussed in the introduction, combined Lagrange–Eulerian approach like Arbitrary Lagrange–Eulerian coupling, combined Finite Volume Methods and Meshfree Methods, or Meshfree Methods alone are good candidates for these types of problems.

It is worth nothing that some activities in this direction were present at that time (1980s) as well. In parallel with testing various higher orders numerical schemes (some of them presented here), the work based on the particle tracking routines and combining Lagrange–Eulerian approach and implementing it in the general purpose CFD program was being carried out as well [88–90].

Nowadays, some of the tested formulations and their improvements based on TVD and ENO are the standard options of some general purpose CFD programs. The unphysical undershoots and overshoots are typically handled through TVD or ENO schemes. They can locally change the order of the method for the convective flux calculation and mitigate these problems.

The insights obtained from various benchmarks and similar testings and analyses such as the one presented, resulted in the development of these new classes of convective formulations.

4 Conclusions

This paper presents examples of initial tests and implementations of two numerical concepts in a general purpose CFD program.

The first example is related to the implementation of rectangular coordinates to simulate flow and heat transfer in the arbitrarily shaped domains with different heat transfer conditions at boundaries.

By applying the rectangular coordinates (instead of the polar ones) to the problem of the top and bottom heating of a horizontal tube with a circular cross section, a general approach to the problem of various (and irregular) cross sections and nonuniform peripheral heating is obtained. This is achieved with specific adjustments of linear source terms for boundary cells, by considering distances from the walls, corresponding wall surfaces and boundary cells volumes.

Current cut-cell and trim-cell methods for the cells near the boundaries provide an alternative to the body fitted coordinate methods. They are also used to provide a connection between fine prism or hexahedral cells near the walls and a coarse hexahedral, tetrahedral, or polyhedral cells in the core of the domain.

The second example demonstrates capabilities to introduce and test implicit and explicit higher order numerical schemes in the general purpose CFD program by comparing results with an analytical solution for three-dimensional transient pollutant-cloud transport.

It was shown that results for QUICKEST and LEITH explicit schemes are strongly dependent on Courant number and that they could be acceptable if the physically unrealistic negative concentration appearance can be prevented.

Currently, some of the tested formulations and their improvements which are not producing physically unrealistic values, as well as other higher order numerical schemes, are the standard options of some, but not all, general purpose CFD programs.

Both presented examples are based on the implementation of linearized source terms which allow regrouping and adding new terms in various equations without need to change the major parts of the general purpose CFD programs. The presented examples demonstrate a historical perspective of some key developments that set the foundation for today's general purpose CFD codes and their ability to address more complex numerical and physics models.

The implementation of linearized source terms option is currently available in most of the general purpose CFD programs and allows for similar modifications in case that users need to apply some new numerical concepts before they become standard options. This allows the refinement of models by adjusting inlet conditions, properties of the porous media subdomains, or the shape and effects of multiple internal obstacles.

Two presented examples demonstrate planning and organizing of the general purpose CFD program approach to be not only user friendly but also easily adapted for developing and applying new numerical concepts and modeling methods.

Presently, general purpose CFD programs play an important role for research and industry and support new designs and developments. However, even after all Finite Volume Methods improvements, there are still some types of problems where more adaptive CFD analysis and continuation of developments and improvements are needed. The new developments in numerical analyses, such as improving Finite Volume Methods, Finite Element Methods, Meshfree Methods, and Adaptive Mesh Refinements will most likely be applied in future general purpose CFD programs. Also, adjustments to the new hardware platforms (such as transition toward graphic processing units instead of central processing units), continuation of parallelization and decrease of overall computing costs and time to complete analyses (including time needed for meshing) will have an impact on industrial applications.

Acknowledgements The author would like to thank Professor Brian D. Spalding for guidance throughout the specialization at Imperial College, and for the privilege to experience a unique educational approach resulting in step-by-step gradual increase of the complexity and challenge of the problems to be solved. At the same time, giving me the freedom and time to follow intuition, search literature, and explore alternative ways instead of just recommending certain courses. The author would like to thank Dr. W. M. Pun for his encouragement and guidance throughout the work on the second example which was performed in connection with U.K. Department of Environment contract (PECD 7/9/322). The author is thankful to the British Council for providing the fellowship for study at the Imperial College in London, England, U.K., and to the Community of Science and Education of the Socialist Republic of Serbia for the financial support of my family in Belgrade, Serbia, Yugoslavia, during the 1985/86 school year.

References

1. Harlow, F. H. (1955). A machine calculation method for hydrodynamic problems. LAMS-1956, Los Alamos.
2. Harlow, F. H., & Welch, J. E. (1965). Numerical calculation of time-dependent viscous incompressible flow of fluid with a free surface. *Physics of Fluids, 8,* 2182–2189. https://doi.org/10.1063/1.1761178.
3. Harlow, F. H. (1964). The particle-in-cell computing method for fluid dynamics. In B. Alder (Ed.), *Methods in computational physics* (pp. 319–343). New York: Academic Press.

4. Patankar, S. V., & Spalding, D. B. (1972). A calculation procedure for heat, mass and momentum transfer in three-dimensional parabolic flows. *International Journal of Heat and Mass Transfer, 15,* 1787–1806. https://doi.org/10.1016/0017-9310(72)90054-3.

5. Pratap, V. S., & Spalding, D. B. (1976). Fluid flow and heat transfer in three-dimensional duct flows. *International Journal of Heat and Mass Transfer, 19,* 1183–1188. https://doi.org/10.1016/0017-9310(76)90152-6.

6. Rosten, H. I., & Spalding, D. B. (1985). PHOENICS-84 Reference Handbook. CHAM TR/100, CHAM, London, UK.

7. Patankar, S. V. (1980). *Numerical heat transfer and fluid flow.* USA: Hemisphere Publishing Corp. https://doi.org/10.1201/9781482234213.

8. Runchal, A. K. (2012). Emergence of computational fluid dynamics at Imperial College (1965–1975): A personal recollection. *Journal of Heat Transfer, 135,* 011009-1–011009-9. https://doi.org/10.1115/1.4007655.

9. Dzodzo, M. B. (1987). Application of rectangular co-ordinates to the problem of laminar combined convection in a straight horizontal tube whose circumference is heated nonuniformly. PDR/CFDU IC/33 Report, Computational Fluid Dynamics Unit, Imperial College of Science and Technology, London.

10. Dzodzo, M. B., & Spalding D. B. (1986). Three-dimensional transient pollutant-cloud transport as a comparison test for the FIP, QUICK, QUICKEST and LEITH'S formulations. PDR/CFDU IC/28 Report, Computational Fluid Dynamics Unit, Imperial College of Science and Technology, London.

11. Noye, B. J. (1978). an introduction to finite difference technologies. In B. J. Noye (Ed.), *Numerical simulation of fluid motion.* North-Holland Publishing Company.

12. Rhie, C. M. (1981). A numerical study of the flow past an isolated airfoil with separation. Ph.D. thesis, Dept. of Mech. and Ind. Eng., University of Illinois at Urbana-Champaign.

13. Rhie, C. M., & Chow, W. L. (1983). Numerical study of the turbulent flow past an airfoil with trailing edge separation. *AIAA Journal, 21,* 1525–1532. https://doi.org/10.2514/3.8284.

14. Ferziger, J. H., & Peric, M. (2002). *Computational methods for fluid dynamics* (3rd ed.). Springer. https://doi.org/10.1007/978-3-642-56026-2.

15. Roelofs, F., Gopala, V. R., Van Tichelen, K., Cheng, X., Merzari, E., & Pointer, W. D. (2013). Status and future challenges of CFD for liquid metal cooled reactors. In *IAEA fast reactor conference.*

16. Viellieber, M., & Class, A. (2015). Investigating reactor components with the coarse-grid-methodology. In *16th International Topical Meeting on Nuclear Reactor Thermal Hydraulics, NURETH-16,* Chicago, IL, August 30–September 4, 2015 (pp. 2788–2801).

17. Zienkiewicz, O. C. (1975). Why finite elements? In R. H. Gallagher, J. T. Oden, C. Taylor, & O. C. Zienkiewicz (Eds.), *Finite elements in fluids—Vol. 1 Viscous flow and hydrodynamics,* Chapter 1 (pp. 1–23). Wiley.

18. Baker, A. J. (1983). *Finite element computational fluid mechanics.* Hemisphere Publishing Corporation.

19. Brebbia, C. A., Telles, J. C. F., & Wrobel, L. (1984). *Boundary element techniques in engineering: Theory & application in engineering.* New York: Springer. https://doi.org/10.1007/978-3-642-48860-3.

20. Patera, A. T. (1984). A spectral element method for fluid dynamics: Laminar flow in a channel expansion. *Journal of Computational Physics, 54,* 468–488. https://doi.org/10.1016/0021-9991(84)90128-1.

21. Fischer, P. T., & Patera, A. T. (1991). Parallel spectral element solution of the Stokes problem. *Journal of Computational Physics, 92,* 380–421. https://doi.org/10.1016/0021-9991(91)90216-8.

22. Deville, M. O., Fischer, P. F., & Mund, E. H. (2002). High-order methods for incompressible fluid flow. *Cambridge University Press.* https://doi.org/10.1017/cbo9780511546792.

23. Lucy, L. (1977). A numerical approach to testing the fission hypothesis. *The Astronomical Journal, 82,* 1013–1024. https://doi.org/10.1086/112164.

24. Gingold, R. A., & Monaghan, J. J. (1982). Kernel estimates as a basis for general particle methods in hydrodynamics. *Journal of Computational Physics, 46,* 429–453. https://doi.org/10.1016/0021-9991(82)90025-0.

25. Pepper, D. W., Kassab, A., & Divo, E. (2014). Introduction to finite element, boundary element, and Meshless methods. *ASME Press.* https://doi.org/10.1115/1.860335.

26. Liu, G. R. (2003). Mesh free methods: Moving beyond the finite element method. In *Chapter 9: Mesh free methods for fluid dynamics problems* (pp. 345–389). CRC Press. https://doi.org/10.1201/9781420040586.ch9.

27. Liu, M. B., & Liu, G. R. (2010). Smoothed particle hydrodynamics (SPH): An overview and recent developments. *Archives of Computational Methods in Engineering, 17,* 25–76. https://doi.org/10.1007/s11831-010-9040-7.

28. Chen, J. S., Hillman, M., & Chi, S. W. (2017). Meshfree methods: Progress made after 20 years. *Journal of Engineering Mechanics, 143,* 04017001. https://doi.org/10.1061/(asce)em.1943-7889.0001176.

29. Shin, Y. S., & Chisum, J. E. (1997). Modeling and simulation of underwater shock problems using a coupled Lagrangian-Eulerian analysis approach. *Shock and Vibration, 4,* 1–10. https://doi.org/10.1155/1997/123617.

30. Leonard, B. P. (1979). A stable and accurate convective modelling procedure based on quadratic upstream interpolation. *Computer Methods in Applied Mechanics and Engineering, 19,* 59–98. https://doi.org/10.1016/0045-7825(79)90034-3.

31. Leith, C. E. (1965). Numerical simulation of the Earth's atmosphere. *Methods Computational Physics, 4,* 1–28.

32. Van Lear, B. (1974). Towards the ultimate conservative difference scheme. II. Monotonicity and conservation combined in a second-order scheme. *Journal of Computational Physics, 14,* 361–370. https://doi.org/10.1016/0021-9991(74)90019-9.

33. Van Lear, B. (1979). Towards the ultimate conservative difference scheme. V. A second-order sequel to Godunov's method. *Journal of Computational Physics, 32,* 101–136. https://doi.org/10.1016/0021-9991(79)90145-1.

34. Godunov, S. K. (1959). A difference scheme for numerical solution of discontinuous solution of the equations of hydrodynamic (in Russian). *Matematicheskii Sbornik, 47*(89), Number 3, 271–306. www.mathnet.ru/php/archive.phtml?wshow=paper&jrnid=sm&paperid=4873&option_lang=rus.

35. Harten, A. (1983). High resolution schemes for hyperbolic conservation laws. *Journal of Computational Physics, 49,* 357–393. https://doi.org/10.1016/0021-9991(83)90136-5.

36. Harten, A., Engquist, B., Osher, S., & Chakravarthy, S. R. (1987). Uniformly high order accurate essentially non-oscillatory schemes, III. *Journal of Computational Physics, 71,* 231–303. https://doi.org/10.1016/0021-9991(87)90031-3.

37. Courant, R., Isaacson, E., & Rees, M. (1952). On the solution of nonlinear hyperbolic differential equations by finite differences. *Communications on Pure and Applied Mathematics, 5,* 243–255. https://doi.org/10.1002/cpa.3160050303.

38. Spalding, D. B. (1972). A novel finite-difference formulation for differential expression involving both first and second derivatives. *International Journal for Numerical Methods in Engineering, 4,* 551–559. https://doi.org/10.1002/nme.1620040409.

39. Raithby, G. D., & Torrance, K. E. (1974). Upstream-weighted differencing schemes and their application to elliptic problems involving fluid flow. *Computers & Fluids, 2,* 191–206. https://doi.org/10.1016/0045-7930(74)90013-9.

40. Runchal, A. K. (1972). Convergence and accuracy of the three finite difference schemes for a two-dimensional conduction and convection problem. *International Journal for Numerical Methods in Engineering, 4,* 541–550. https://doi.org/10.1002/nme.1620040408.

41. De Vahl Davis, G., & Mallinson, G. D. (1976). An evaluation of upwind and central difference approximations by study of recirculating flow. *Computers and Fluids, 4,* 29–43. https://doi.org/10.1016/0045-7930(76)90010-4.

42. Agrawal, R. K. (1981). A third-order-accurate upwind scheme for Navier-Stokes solutions in three dimensions. In K. N. Ghia, T. J. Mueller, & Patel (Eds.), *Computers in flow predictions*

and fluid dynamics experiments. Winter Annual Meeting of the ASME, ASME, Washington, D.C., November 15–20, 1981 (pp. 73–82).

43. Kawamura, T., & Kuwahara, K. (1984). Computation of high Reynolds number flow around a circular cylinder of high reynolds number flow around a circular cylinder with surface roughness. In *AIAA-84–0340, AIAA 22nd Aerospace Sciences Meeting*, Reno, Nevada, USA, January 9–12, 1984. https://doi.org/10.2514/6.1984-340.

44. Kawamura, T., & Kuwahara, K. (1985). Direct simulation of a turbulent inner flow by finite-difference method. In *AIAA-85-0376, AIAA 23rd Aerospace Sciences Meeting*, Reno, Nevada, USA, January 14–17, 1985. https://doi.org/10.2514/6.1985-376.

45. Leonard, B. P. (1979). A survey of finite differences of opinion on numerical muddling of the incomprehensible defective confusion equation. In T. J. R. Hughes (Ed.), *Finite element methods for convection dominated flows*. Presented at The Winter Annual Meeting of the American Society of Mechanical Engineers, sponsored by: The Applied Mechanics Division, ASME, AMD, New York, NY, USA, December 2–7, 1979 (Vol. 34, pp. 1–17).

46. Leonard, B. P. (1991). The ULTIMATE conservative difference scheme applied to unsteady one-dimensional advection. *Computer Methods in Applied Mechanics and Engineering, 88*, 17–74. https://doi.org/10.1016/0045-7825(91)90232-u.

47. Darwish, M. S. (1993). A new high-resolution scheme based on the normalized variable formulation. *Numerical Heat Transfer, Part B: Fundamentals, 24*, 353–371. https://doi.org/10.1080/10407799308955898.

48. Malin, M. R., & Waterson, N. P. (1999). Schemes for convection discretization in PHOENICS. *PHOENICS Journal, 12*(2), 173–201.

49. Jakobsen, H. A. (2003). Numerical convection algorithms and their role in Eulerian CFD reactor simulations. *International Journal of Chemical Reactor Engineering, 1*, Article A1, 1–15. https://doi.org/10.2202/1542-6580.1006.

50. Hou, J., Liang, Q., Zhang, H., & Hinkelmann, R. (2015). An efficient unstructured MUSCL scheme for solving the 2D shallow water equations. *Environmental Modelling and Software, 66*, 131–152. https://doi.org/10.1016/j.envsoft.2014.12.007.

51. Patankar, S. V., Ramadhyani, S., & Sparrow, E. M. (1978). Effect of circumferentially nonuniform heating on laminar combined convection in a horizontal tube. *Journal of Heat Transfer, 100*, 63–70. https://doi.org/10.1115/1.3450505.

52. Dzodzo, M. B. (1983). Effect of non-uniform heating on laminar mixed convection in a straight horizontal tube (in Serbian). M.Sc., Dept. of Mechanical Engineering, The University of Belgrade.

53. Reynolds, W. C. (1960). Heat transfer to fully developed laminar flow in a circular tube with arbitrary circumference heat flux. *Journal of Heat Transfer, 82*, 108–112. https://doi.org/10.1115/1.3679887.

54. Dzodzo, M. (1991). Application of rectangular coordinates to the problem of laminar natural convection in enclosures of arbitrary cross-section. In *Proceedings of the 1st ICHMT International Numerical Heat Transfer Conference and Software Show*, Part II, Guildford, Surrey, July 22–26, 1991 (pp. 1–11).

55. Dzodzo, M. B. (1991). Laminar natural convection in some enclosures of arbitrary cross sections (in Serbian). Ph.D. thesis, Dept. of Mechanical Engineering, The University of Belgrade.

56. Dzodzo, M. B. (1993). Visualization of laminar natural convection in romb-shaped enclosures by means of liquid crystals. In S. Sideman, K. Hijikata, & W. J. Yang (Eds.), *Imaging in transport processes* (pp. 183–193). Begell House Publishers.

57. Dzodzo, M. B. (2013). Natural convection in cubic and rhomb-shaped enclosures. In *Proceedings of the ASME 2013 Heat Transfer Summer Conference*, Paper No. HT2013-17724, Minneapolis, MN, USA, July 14–19, 2013 (pp. V003T21A004, 10 pp.). https://doi.org/10.1115/ht2013-17724.

58. Peric, M. (1985). A finite volume method for the prediction of three dimensional fluid flow in complex ducts. Ph.D. thesis, Imperial College, London, U.K.

59. Boussinesq, J. (1903). *Théorie Analitique de la Chaleur (in French)* (Vol. 2, p. 172). Paris: Gauthier-Villars.

60. Khosla, P. K., & Rubin, S. G. (1974). A diagonally dominant second-order accurate implicit scheme. *Computers and Fluids, 2*, 207–209. https://doi.org/10.1016/0045-7930(74)90014-0.
61. Cuckovic-Dzodzo, D. M. (1996). Effects of heat conducting partition on laminar natural convection in an enclosure (in Serbian). M.Sc. Thesis, Department of Mechanical Engineering, The University of Belgrade.
62. Cuckovic-Dzodzo, D. M., Dzodzo, M. B., & Pavlovic, M. D. (1996). Visualization of laminar natural convection in a cubical enclosure with partition. In H. W. Coleman (Ed.), *Proceedings of the ASME Fluids Engineering Division Summer Meeting, 1996.* Presented at the 1996 ASME Fluids Engineering Division Summer Meeting, FED, San Diego, California, July 7–11, 1996 (Vol. 239, pp. 225–230).
63. Cuckovic-Dzodzo, D. M., Dzodzo, M. B., & Pavlovic, M. D. (1998). A mathematical model and numerical solution for the conjugated heat transfer in a fully partitioned enclosure containing the fluids with nonlinear thermophysical properties. *Theoretical and Applied Mechanics, Yugoslav Society of Mechanics, 24,* 29–54.
64. Cuckovic-Dzodzo, D. M., Dzodzo, M. B., & Pavlovic, M. D. (1999). Laminar natural convection in a fully partitioned enclosure containing fluid with nonlinear thermophysical properties. *International Journal of Heat and Fluid Flow, 20,* 614–623. https://doi.org/10.1016/s0142-727x(99)00053-3.
65. De Vahl Davis, G. (1983). Natural convection of air in a square cavity: A bench mark numerical solution. *International Journal for Numerical Methods in Fluids, 3,* 249–264. https://doi.org/10.1002/fld.1650030305.
66. Hortmann, M., Peric, M., & Scheurerer, G. (1990). Finite volume multigrid prediction of laminar natural convection: Bench-mark solutions. *International Journal for Numerical Methods in Fluids, 11,* 189–207. https://doi.org/10.1002/fld.1650110206.
67. Gray, D. D., & Giorgini, A. (1976). The validity of the Boussinesq approximation for liquids and gases. *International Journal of Heat and Mass Transfer, 19,* 545–551. https://doi.org/10.1016/0017-9310(76)90168-x.
68. Graham, A. D., & Mallinson, G. D. (1977). Three-dimensional convection in an inclined differentially heated box. In *6th Australasian Hydraulics and Fluid Mechanics Conference,* Adelaide, Australia, 5–9 December 1977 (pp. 467–476).
69. Mallinson, G. D., & De Vahl Davis, G. (1977). Three-dimensional natural convection in a box: a numerical study. *Journal of Fluid Mechanics, 83,* 1–31. https://doi.org/10.1017/s0022112077001013.
70. De Vahl Davis, G., & Jones, I. P. (1983). Natural convection in a square cavity: A comparison exercise. *International Journal for Numerical Methods in Fluids, 3,* 227–248. https://doi.org/10.1002/fld.1650030304.
71. Ciofalo, M., & Karayiannis, T. G. (1991). Natural convection heat transfer in a partially or completely portioned vertical rectangular enclosure. *International Journal of Heat and Mass Transfer, 34,* 167–179. https://doi.org/10.1016/0017-9310(91)90184-g.
72. Rai, M. M. (1985). Navier-stokes simulations of rotor-stator interaction using patched and overlaid grids. In *AIAA 7th Computational Fluid Dynamics Conference,* July 15–17, Cincinnati, Ohio, USA (pp. 282–289). https://doi.org/10.2514/6.1985-1519.
73. Thoman, D. C., & Szewczyk, A. A. (1969). Time dependent viscous flow over a circular cylinder. *The Physics of Fluids, Supplement II, 12,* 76–87. https://doi.org/10.1063/1.1692472.
74. Belotserkovskii, O. M., & Davydov, Yu M. (1982). *The large-particle method in gas dynamics—A computational experiment.* Moscow: Izdatel'stvo Nauka. (in Russian).
75. Davydov, Yu. M. (1971). Calculation by the "coarse particle" method of the flow past a body of arbitrary shape. *USSR Computational Mathematics and Mathematical Physics, 11,* 241–271. https://doi.org/10.1016/0041-5553(71)90026-7.
76. Kelly, D. M., Chen, Q., & Zang, J. (2015). PICIN: A particle-in-cell solver for incompressible free surface flows with two-way fluid-solid coupling. *SIAM Journal on Scientific Computing, 37,* B403–B424. https://doi.org/10.1137/140976911.
77. Hirt, C. W., & Nichols, B. D. (1981). Volume of fluid (VOF) method for the dynamics of free boundaries. *Journal of Computational Physics, 39*(1), 201–225. https://doi.org/10.1016/0021-9991(81)90145-5.

78. Smith, L. D., Conner, M. E., Liu, B., Dzodzo, M. B., Paramonov, D. V., Beasley, D. E., et al. (2002). Benchmarking computational fluid dynamics for application to PWR fuel. In *10th International Conference on Nuclear Engineering*, Paper No. ICONE10–22475, Arlington, VA, USA, April 14–18, 2002 (Vol. 3, pp. 823–830). https://doi.org/10.1115/icone10-22475.

79. Carrilho, L. A., & Dzodzo, M. B. (2018). Conjugated heat transfer model for ribbed surface convection enhancement and solid body temperature fluctuations. In A. P. Silva Freire, K. Hanjalic, K. Suga, D. Borello, M. Haziabdic (Eds.), *Turbulence, heat and mass transfer. Proceedings of the Ninth International Symposium on Turbulence, Heat and Mass Transfer*, Rio de Janeiro, Brazil, 10–13 July 2018 (pp. 383–386). New York, Wallingford: Begell House Inc.

80. Dzodzo, M. B. (1995). A multiblock procedure for the prediction of the fluid flow inside the complex three dimensional domains with specified pressures on the open boundaries. In D. Hui & S. Michaelides (Eds.), *SES'95 Society of Engineering Science 32nd Annual Technical Meeting*, University of New Orleans, New Orleans, Louisiana, USA, October 29–November 2, 1995 (pp. 729–730).

81. Braun, M. J., & Dzodzo, M. B. (1997). Three-dimensional flow and pressure patterns in a hydrostatic journal bearing pocket. *Journal of Tribology, 119*, 711–719. https://doi.org/10.1115/1.2833875.

82. Dzodzo, M. B., & Braun, M. J. (1996). A three dimensional model for a hydrostatic bearing. In *AIAA 96-3104, 32nd AIAA/ASME/SEA/ASEE Joint Propulsion Conference*, Lake Buena Vista, Florida, USA, July 1–3 1996. https://doi.org/10.2514/6.1996-3104.

83. Cumber, P. S., Fairweather, M., Falle, S. A. E. G., & Giddings, J. R. (1997). Predictions of impacting sonic and supersonic jets. *Journal of Fluids Engineering, 119*, 83–89. https://doi.org/10.1115/1.2819123.

84. Cumber, P. S., Fairweather, M., Falle, S. A. E. G., & Giddings, J. R. (1998). Body capturing in impacting supersonic flows. *International Journal of Heat and Fluid Flow, 19*, 23–30. https://doi.org/10.1016/s0142-727x(97)10011-x.

85. Olivieri, D. A., Fairweather, M., & Falle, S. A. E. G. (2010). Adaptive mesh refinement applied to the scalar transported PDF equation in a turbulent jet. *International Journal for Numerical Methods in Engineering, 84*, 434–447. https://doi.org/10.1002/nme.2899.

86. Carslaw, H. C., & Jaeger, J. C. (1959). *Conduction of heat in solids* (2nd ed.). Oxford University Press.

87. Leonard, B. P. (1980). The QUICK algorithm: A uniformly third-order finite-difference method for highly convective flows. In K. Morgan, C. Taylor, & C. A. Brebbia (Eds.), *Computer methods in fluids* (pp. 159–195). Pentec Press.

88. Castrejon, A. (1983). Particle tracking subroutines for numerical flow visualization. PDR/CFDU IC/10 Report, Computational Fluid Dynamics Unit, Imperial College of Science and Technology, London.

89. Castrejon, A., & Andrews, M. J. (1986). A procedure for calculating moving-interface flows with Phoenics-84. In N. C. Markatos, M. Cross, D. G. Tatchell, & N. Rhodes (Eds.), *Numerical simulation of fluid flow and heat/mass transfer processes. Lecture notes in engineering* (Vol. 18, pp. 433–443). Berlin, Heidelberg: Springer. https://doi.org/10.1007/978-3-642-82781-5_34.

90. Castrejon, A., & Spalding, D. B. (1988). An experimental and theoretical study of transient free-convection flow between horizontal concentric cylinders. *International Journal of Heat and Mass Transfer, 31*, 273–284 (1988). https://doi.org/10.1016/0017-9310(88)90010-5.

Accurate Numerical Modeling of Complex Thermal Processes: Impact of Professor Spalding's Work

Yogesh Jaluria

1   Introduction

Mathematical and numerical modeling of thermal processes that arise in important applications such as those related to energy, manufacturing, transportation, aerospace, heating, cooling, and the environment is crucial in a detailed study of the phenomena and in the design and optimization of the relevant systems. Most of these practical circumstances are too complicated to be investigated by analytical methods. Also, relatively limited data are usually available from existing processes and from experimental studies, which are often expensive and time consuming. Therefore, in most cases, mathematical models of the processes are developed and solved by numerical simulation. The models are validated by means of analytical and experimental results available on simpler and similar systems. Then, the numerical simulation is used to provide the extensive inputs needed for understanding and characterizing the processes, as well as for design, control, and optimization [1–3].

Practical thermal processes and systems typically involve complex, coupled, transport mechanisms and interacting components, and subsystems that constitute the overall system. Therefore, several challenges are commonly encountered in obtaining accurate results from the numerical simulation of these systems. Some of these are variable material properties, accurate and realistic imposition of boundary conditions, model validation, combined mechanisms, complex phenomena, multiple scales, multi-objective optimization, uncertainties, and other additional effects. This paper considers some of these aspects, presents examples where these considerations are important, and discusses possible approaches to meet these challenges. The work done by Professor Spalding and his group is indispensable in meeting many of the challenges discussed here. This will be pointed out at several places in this paper.

Y. Jaluria (✉)
Department of Mechanical & Aerospace Engineering,
Rutgers University, Piscataway, NJ 08854, USA
e-mail: jaluria@soe.rutgers.edu

© Springer Nature Singapore Pte Ltd. 2020   155
A. Runchal (ed.), *50 Years of CFD in Engineering Sciences*,
https://doi.org/10.1007/978-981-15-2670-1_5

Professor Spalding contributed to a wide range of problems in heat and mass transfer and in fluid mechanics. Among the major areas that he contributed substantially to are the following:

- Combustion
- Multiphase transport
- Separated flows
- Turbulent flows
- Boundary layers; parabolic flows
- Recirculating flows
- Complex domains; grid generation; boundary conditions
- Complicated Practical Processes and Systems
- Computational Fluid Dynamics (CFD)
- Computer codes or programs such as Phoenics, Simple and Genmix

Clearly, he tackled some very complex problems and developed techniques to solve them through computational methods. This legacy has been of great importance to the fluids and heat transfer communities. His methods and basic approaches to various problems have paved the way to the solution of many challenging problems today.

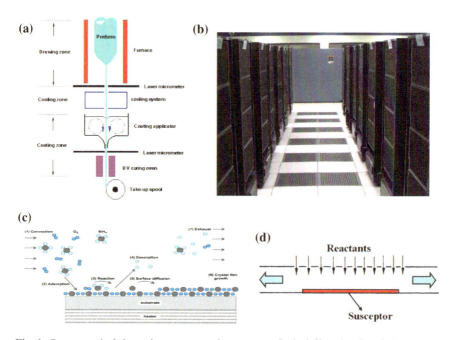

Fig. 1 Some practical thermal processes and systems: **a** Optical fiber drawing; **b** data center; **c** horizontal chemical vapor deposition (CVD) reactor for thin-film fabrication; **d** vertical impinging CVD reactor

A few practical thermal systems are shown as examples in Fig. 1. Seen here are the fabrication process for an optical fiber, a data center being cooled by chilled airflow, and two configurations, horizontal and vertical flow, of chemical vapor deposition (CVD) reactors for thin-film fabrication. Many of the complexities mentioned above are encountered in these systems. For instance, material properties of glass in optical fiber drawing are strong functions of temperature, combined modes of radiation, conduction and convection operate at various stages in the process, polymers that are non-Newtonian are generally used for the fiber coating process and large changes in preform/glass diameter occur in the draw furnace [4]. The airflow in data center cooling is generally turbulent, the geometry is complicated, and the flow is coupled with the conduction transport in the electronic system [5]. Chemical reactions and varying species concentrations arise in CVD. This involves chemical kinetics, which vary strongly with temperature and concentration [6]. The boundary conditions are generally quite complicated in all cases and combined transport mechanisms have to be considered. Other practical processes have similar complexities and challenges with respect to computational modeling.

2 Numerical Modeling

The various thermal processes and systems mentioned earlier and many others similar to these may be modeled and simulated to obtain the results needed for better insight into the basic processes and for system design and optimization. The equations that describe the flow and convective transport are based on the conservation of mass and energy and the force-momentum balance that give rise to the well-known equations for fluid flow and heat transfer. These may be written for a general three-dimensional, time-dependent process with variable properties, as [1, 5]

$$\frac{\partial \rho}{\partial t} + \nabla \cdot (\rho \bar{V}) = 0 \tag{1}$$

$$\rho \left(\frac{\partial \bar{V}}{\partial t} + \bar{V} \cdot \nabla \bar{V} \right) = \bar{F} - \nabla p + \nabla \cdot \left[\mu \left(\nabla \bar{V} + \nabla \bar{V}^T \right) \right] - \frac{2}{3} \nabla (\mu \nabla \cdot \bar{V}) \tag{2}$$

$$\rho C_p \left(\frac{\partial T}{\partial t} + \bar{V} \cdot \nabla T \right) = \nabla \cdot (k \nabla T) + \dot{Q} + \mu \Phi + \beta T \left(\frac{\partial p}{\partial t} + \bar{V} \cdot \nabla p \right) \tag{3}$$

where ρ is density, T is temperature, t is time, $\bar{V}$ is the velocity vector, $\bar{F}$ is body force, p is pressure, μ is dynamic viscosity, C_p is specific heat at constant pressure, β is coefficient of volumetric thermal expansion, $\mu \Phi$ is viscous dissipation, and $\dot{Q}$ is a volumetric heat source arising from chemical reactions, absorbed radiation, etc. The viscous dissipation and pressure work effects are included in the energy equation, Eq. (3), with the last two terms multiplied by βT representing the pressure effect. The bulk viscosity is taken as zero, giving the second viscosity coefficient as $-(2/3) \mu$

and Stokes' relationships are used for the viscous forces in the momentum equation, Eq. (2).

The preceding equations may be written for different processes and systems, retaining the relevant terms. For instance, the following forms of the equations apply to a wide range of steady flows:

Mass:

$$\nabla \cdot \left(\rho \bar{V} \right) = 0 \tag{4}$$

Momentum:

$$\rho \bar{V} \cdot \nabla \bar{V} = -\nabla p + \nabla \cdot \left(\mu \nabla \bar{V} \right) \tag{5}$$

Energy:

$$\rho C_p \bar{V} \cdot \nabla T = \nabla \cdot (k \nabla T) \tag{6}$$

where $\bar{V}$ is the velocity vector. For a solid region, the conduction equation is used, given in terms of the summation notation, with thermal conductivity of the solid k, as

$$\frac{\partial}{\partial x_i} \left(k \frac{\partial T}{\partial x_i} \right) = 0 \tag{7}$$

For conjugate problems, the heat conduction in the solid region and the convection flow in the fluid region are generally solved separately and then coupled at the solid–fluid interface through the boundary conditions, though other approaches have also been developed.

Recirculating flow is of interest in many problems of practical interest, such as environmental flows, fires, ventilation, and cooling systems. The case of data centers was mentioned earlier. Figure 2 shows some of these circumstances, including heat rejection from power plants to a water body or a cooling tower and a room fire. In the case of heat rejection to water bodies like lakes, rivers, cooling ponds and oceans, the recirculation that may short circuit the hot discharge from the power plant to the inlet is of particular concern since it would lower the efficiency of the plant. Also, of considerable importance is the thermal effect of heat rejection on the thermal cycle of the water body itself. This includes temperature rise and variation in the yearly cycle due to heat rejection. These effects could adversely affect the micro-organisms and life existing in the water body.

Depending on the fluid, dimensions, and operating conditions, the continuum approximation may be inapplicable, leading to slip flow or molecular flow [7]. Such a situation arises, for instance, in an electronic component cooled by the microchannel single-phase flow of a coolant [8]. The Knudsen number, which is the ratio of the mean free path of the particles λ to the physical length dimension L, determines the flow regime and thus the solution strategy. The overall system, on the other

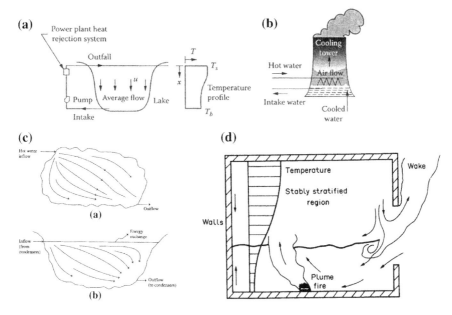

Fig. 2 **a** Heat rejection to a water body; **b** cooling tower; **c** recirculating flow due to thermal discharge into a lake; **d** room fire

hand, is usually at engineering, or macro-scale, and can be modeled using the usual conservation equations and boundary conditions.

For an important practical process, let us consider the basic characteristics of the optical fiber drawing process, shown in Fig. 1a. The flow of the glass and that of the aiding purge gas in a cylindrical furnace are generally assumed to be axisymmentric. Glass flow is laminar due to the high viscosity and the gas flow is also generally kept laminar to avoid disturbance to the free surface. Then the equations for the glass and the gas are given as [1, 4]

$$\frac{\partial v}{\partial z} + \frac{1}{r}\frac{\partial (ru)}{\partial r} = 0 \tag{8}$$

$$\frac{\partial v}{\partial t} + u\frac{\partial v}{\partial r} + v\frac{\partial v}{\partial z} = -\frac{1}{\rho}\frac{\partial p}{\partial z} + \frac{1}{r}\frac{\partial}{\partial r}\left[r v\left(\frac{\partial v}{\partial r} + \frac{\partial u}{\partial z} \right) \right] + 2\frac{\partial}{\partial z}\left(v\frac{\partial v}{\partial z} \right) \tag{9}$$

$$\frac{\partial u}{\partial t} + u\frac{\partial u}{\partial r} + v\frac{\partial u}{\partial z} = -\frac{1}{\rho}\frac{\partial p}{\partial r} + \frac{2}{r}\frac{\partial}{\partial r}\left(r v\frac{\partial u}{\partial r} \right) + \frac{\partial}{\partial z}\left[v\left(\frac{\partial v}{\partial r} + \frac{\partial u}{\partial z} \right) \right] - \frac{2vu}{r^2} \tag{10}$$

$$\rho C_p\left(\frac{\partial T}{\partial t} + u\frac{\partial T}{\partial r} + v\frac{\partial T}{\partial z} \right) = \frac{1}{r}\frac{\partial}{\partial r}\left(rk\frac{\partial T}{\partial r} \right) + \frac{\partial}{\partial z}\left(k\frac{\partial T}{\partial z} \right) + \Phi + S_r \tag{11}$$

Here, u and v are the velocity components in the axial and radial directions, z and, r, respectively, p is the local pressure, T the temperature, t the time, υ the kinematic viscosity, $\varPhi$ is the viscous dissipation term, and S_r is the radiation source term. The gravitational force term is not included in these equations since its effect is negligible in glass flow. However, the gravitational force is considered for determining the neck-down or free surface profile and for obtaining the buoyancy force in the gas flow, resulting from a combination of the gravity force and the pressure term, as discussed later. For glass, the material properties are strong functions of the temperature, composition, and microstructure. Radiation is the dominant mechanism in the heating of the glass preform for fiber drawing.

The variation in the viscosity has a substantial effect on the flow since it varies dramatically with temperature. An equation based on the curve fit of available data for kinematic viscosity υ is written for silica, in S.I. units, as

$$\upsilon = 4545.45 \ \exp\left[32\left(\frac{T_{\text{melt}}}{T} - 1\right)\right] \tag{12}$$

This indicates the strong, exponential variation of υ with temperature. Here, T_{melt} is the glass softening temperature, being around 1900 K for silica glass. The radiative source term S_r in Eq. (11) is nonzero for the glass preform/fiber because glass is a participating medium. The variation of the absorption coefficient α with wavelength λ can often be approximated in terms of bands with constant absorption over each band. Usually, a two- or three-band model is adequate to determine the radiative transport. Even though the fiber diameter is small, being around 125 μm, the high temperature dependence of the viscosity makes it necessary to calculate the temperature variation across the fiber using a large number of grid points, typically around 50.

The fiber coating process is another important step in the fabrication of optical fibers and must be modeled and simulated accurately to obtain results that can be used to design the coating applicator and choose the operating conditions. Typical coating thicknesses are on the order of 40–50 μm and are applied to the uncoated fiber or as a secondary coating to a coated fiber. The basic process involves drawing the optical fiber through a reservoir of coating fluid, which is usually a polymer such as an acrylate, with inlet and outlet dyes. The curing process of the acrylate polymer coating material is carried out by ultraviolet radiation immediately after applying the liquid coating. A balance between the forces, such as surface tension, viscous, gravitational, and pressure forces, gives rise to an upstream meniscus at the entrance, as well as a downstream meniscus at the die exit. At high draw speeds, the upper meniscus breaks down, and the air is entrained into the coating. These entrapped bubbles are undesirable since they can lead to stripping of the coating. Pressurized applicators are often used to reduce the shear at the fiber surface and help in establishing a stable free surface flow. The control of the coating characteristics has been of major concern to the optical fiber industry. These aspects are particularly important at high speeds, ranging beyond 20 m/s to enhance productivity, and also for specialty fibers and fibers of different materials, including polymer fibers. The physical properties of the polymer coating materials, particularly the viscosity, are of primary importance in the coating process. Surface tension has a significant effect on the shape, stability, and other characteristics of the interface, as discussed later.

Chemical kinetics play a critical role in many important thermal processes such as food extrusion, chemical bonding, and chemical vapor deposition of material from the gas phase. The temperature and concentration of the chemical species in the CVD reactor affect the chemical kinetics, which in turn affects the deposition. In some cases, the process is limited by the chemical kinetics, implying that the transport processes are quite vigorous, and the deposition is restricted largely by the kinetics. It is limited by the transport processes in some other cases. The chemical kinetics for depositing several materials is available in the literature. For instance, the chemical kinetics for the deposition of silicon from silane (SiH_4) with hydrogen as the carrier gas in a CVD reactor is given by the expression [9]

$$\hat{K} = \frac{K_o\,p_{SiH_4}}{1 + K_1\,p_{H_2} + K_2\,p_{SiH_4}} \tag{13}$$

where the surface reaction rate $\hat{K}$ is in mole of Si/m^2s. The parameter $K_o = A \exp(-E/RT)$, E being the activation energy, and A, K_1, and K_2 are constants which are obtained experimentally. The p's are the partial pressures of the two species in the reactor. However, the chemical processes are typically much more complicated, with several intermediate reactions in the gaseous phase and several at the surface. This is particularly true for the deposition of SiC and GaN in metal-organic reactors (MOCVD), as seen later.

3 Typical Results and Discussions

Several practical processes have been described in the preceding, along with some of the major complexities faced in obtaining an accurate simulation. These challenges are considered in greater detail here, along with examples and results.

3.1 Variable Material Properties and Characteristics

The accuracy of the results obtained from numerical simulation is a strong function of the material properties used. As mentioned earlier, in most practical processes, properties vary with the local conditions like temperature, pressure, and changes in the material during the process. Unfortunately, accurate property data are often not available at the conditions of the process. This is particularly true for optical fiber drawing where the process is strongly influenced by the temperature-dependent physical properties of silica glass. The radiation properties, such as the variation of the absorption coefficient α with wavelength λ, have been experimentally obtained for certain compositions and glasses. But these data are often available only at room temperature, whereas the process occurs at much higher temperatures. Also, data may not be available for the particular glass or composition that is being simulated. Dopants such as rare earth materials are often used to modify the transmission characteristics and for specialized applications. Even though accurate models may be developed for the process, considering different materials and configurations [10], the data on the effect of the dopants on viscosity and on radiation properties are very limited [11]. Some typical data are shown in Fig. 3 and results obtained on optical fiber drawing using these property data are presented later.

Similarly, properties of the coating fluid are important and their variation with temperature is needed for an accurate simulation. The fluids are generally non-Newtonian and large material property changes occur with temperature [12], as shown qualitatively in Fig. 4. The fluid viscosity is often taken as

$$\mu = \mu_o (\dot{\gamma}/\dot{\gamma}_o)^{n-1} \exp(b/T) \qquad (14)$$

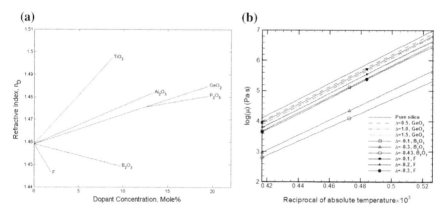

Fig. 3 Effect of various dopants on the refractive index and viscosity of silica glass in the optical fiber drawing process

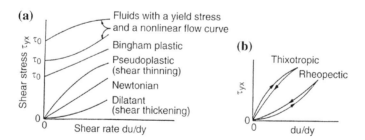

Fig. 4 Shear stress variation with shear rate for purely viscous, nonelastic, non-Newtonian fluids. **a** Time-independent, and **b** time-dependent fluids [12]

where $\dot{\gamma}$ is the total strain rate, b the temperature coefficient of viscosity, subscript o indicates reference conditions, and n is the power-law index of the fluid. The coating material is thus treated as a generalized Newtonian fluid [13]. Other rheological models may also be used, depending on the fluid. Similar models are used for polymers such as polyethylene and polystyrene in extrusion, injection molding, die flow, and other polymer processing systems. For food extrusion, chemical conversion occurs and the dependence of the viscosity on the degree of conversion is included [14]. Viscoelastic effects, which involve elastic behavior of the material, are also considered in most food processing applications.

Similarly, chemical kinetics play a critical role in chemical vapor deposition and food processing. Simple equations like the one given earlier for Silicon are generally not available or applicable for the wide variety of materials of practical interest. A large number of chemical reactions in the gases and at the deposition surface have to be solved in most cases. The results are strongly affected by the material properties and by the chemical kinetics employed. Lack of accurate property data is clearly a major hurdle in obtaining accurate simulation results in this case.

Besides temperature, pressure, and concentration, the material properties are also often sensitive to the conditions under which the material is stored, as well as the fabrication process, age, and raw materials used. The properties can also change with time, resulting in different values for experiments done at different times. This is of particular concern with biological materials, polymers, and chemicals. Therefore, it is important to know what material and under what conditions it is being employed so that the appropriate properties can be used in the simulation. If possible, properties may be measured as part of the experiment or the simulation for more accurate inputs. Interpolation is employed with available data to obtain the best estimate of the properties under the operating conditions.

Since variable properties arise in many of the processes mentioned here, it is necessary to develop strategies to take these variations into account. In many cases, the variation with temperature, pressure, or other local conditions is so strong that very fine grids are needed to accurately model the process. In transient problems, properties at the previous time step or extrapolated from previous known values are used to avoid iteration at a given time step. Similarly, in steady-state transport, the values at the previous iteration or extrapolated ones are used. The extrapolation approach is similar to the one used by Patankar and Spalding [15] for simulating boundary layer flows.

3.2 Verification and Validation

It is necessary to use simplifications and idealizations in the modeling of practical thermal processes to enable numerical predictions with available or affordable computational resources. Therefore, it is important to verify and validate the mathematical and numerical models to ensure that the results obtained are applicable, realistic, and accurate. It is critical to ensure that the numerical method accurately solves the equations obtained from the mathematical model. This aspect, known as *verification*, involves grid refinement and varying the arbitrarily chosen parameters, such as convergence criteria and computational domain boundaries, to ensure that the results are essentially independent of the values chosen. *Validation* involves ensuring that the mathematical model is an accurate representation of the physical problem [16]. Among the approaches used is a consideration of the physical behavior of the system, comparisons with available analytical and numerical results, including benchmark solutions, and comparisons with available experimental data on existing similar systems.

Because of the critical importance of verification and validation, extensive efforts are often made to obtain experimental data, whenever possible, for use as checks on numerical predictions. In some cases, a separate, well-designed, experimental setup may be needed to achieve this. For instance, in the modeling and simulation of single and twin-screw polymer extruders, a specially designed cam-driven thermocouple system was developed to obtain the temperature profile in the rotating screw. Also, two rotating cylinders were used to study the mixing phenomena and thus validate the model for twin-screw extrusion. Further details on the various strategies used to validate models for polymer extrusion are given in Ref. [17].

As mentioned earlier and shown in Fig. 1a for the manufacture of optical fibers, a polymer coating is applied for protection against abrasion and to increase strength. The basic coating process involves moving the fiber through a reservoir of coating fluid, followed by a curing process. At the die exit, the coating material is drawn out with the fiber, forming a downstream meniscus, which influences the coating thickness, uniformity, concentricity, and other characteristics. This meniscus is a free surface whose profile is determined by the forces of gravity, surface tension, shear due to the moving fiber, and external shear due to air. These forces are determined from the calculated flows and the forces acting on a guessed profile. The force imbalance is used to generate an iteration scheme, starting with this guessed profile, till the force balance is satisfied, yielding a converged meniscus [18]. Figure 5 shows the numerical results and compares these with experimental data obtained on an actual, practical, fiber coating system. Glycerin was used as the fluid in the experiment for convenience. As seen here, a good agreement was obtained, indicating the validity and accuracy of the model. Several other similar validation studies in materials processing are presented in Ref. [19]. In the simulation of complex practical processes, it is necessary to make all possible efforts to validate the mathematical/numerical models, even if it means spending considerable time and effort in developing an experimental arrangement to obtain the data needed.

3.3 Numerical Imposition of Realistic Boundary Conditions

The accuracy of the results obtained from any numerical simulation is strongly dependent on the accuracy of the numerical treatment of the boundary conditions. In mathematical modeling, isothermal and uniform heat flux conditions, as well as uniform

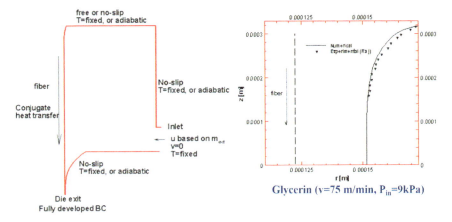

Fig. 5 Sketch of an optical fiber coating applicator and calculated meniscus at the exit of the die, along with experimental measurements of the profile, for glycerin at a fiber speed of 75 m/min and applicator pressure of 9 kPa

or developed flow conditions at the inlet and outlet, are commonly used to study the basic aspects of the problem. However, such conditions are seldom applicable to practical circumstances. For example, in the modeling of the casting process, with solidification occurring in an enclosed region, the conduction in the walls of the mold is coupled with the transport processes in the solidifying fluid. It is an important consideration that strongly affects the results obtained. The conditions at the inner surface of the mold are not known and the conjugate problem has to be solved to obtain the temperature distribution in the mold as well as that in the solid and the liquid [20]. Numerical solution methods are developed for the solid and liquid regions and continuity of heat flux and temperature at the mold inner surface is employed as the boundary condition. Figure 6a shows the effect of conduction in the mold on the resulting temperature and velocity distributions, as well as on the solidification process. For the casting of metals, alloys, polymers, and other materials, it has been shown in several studies that it is important to model the conjugate transport in the mold walls and in any insulation that may be used in order to obtain realistic and accurate results.

Similarly, in the thermal management of electronic systems, isolated heat sources that approximate components like electronic chips and devices are located on substrates that are conducting. Imposing adiabatic or isothermal conditions on the surfaces is thus not a valid representation of the practical situation. The conduction in the walls distributes the heat input over a larger area of the surface and a concentrated heat source is not a realistic assumption. This conjugate transport results in a substantial effect on the flow and the heat transfer in the electronic system [21]. Figure 6b shows the calculated isotherms and streamlines in an enclosed region with multiple, isolated, heat sources that approximate electronic devices. Clearly, it is seen that the conducting walls play a very significant role in the heat transfer process and must be included in the numerical imposition of the boundary conditions.

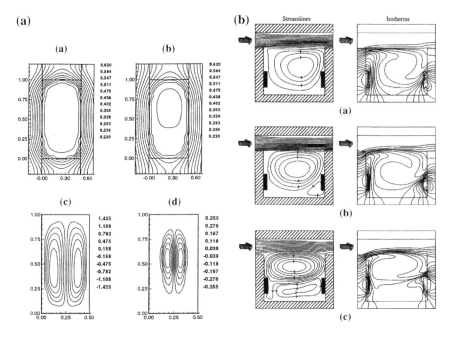

Fig. 6 Effect of conjugate boundary conditions on the flow and heat transfer in **a** solidification process in casting; and **b** cooling of heat sources representing electronic devices located in an enclosed region

In some practical situations, the boundary conditions are not accurately known. It may be possible to model the entire system, with all its components, to obtain the relevant boundary condition. But this is often difficult in terms of cost and effort. An example of such a circumstance is the optical fiber drawing furnace, where the wall temperature distribution is a critical input needed for simulating the process. This distribution is not easily obtained experimentally because of limited access to the furnace and typically high temperatures that arise. Modeling of the entire system is complicated by the presence of many control and traverse subsystems. To solve this problem, a graphite rod, with thermocouples located at the central axis, was immersed in the furnace, as shown in Fig. 7a. The temperature data thus obtained was employed to solve the inverse problem of determining the furnace wall temperature distribution that would yield the measured rod temperature [22]. Figure 7b shows the measured temperature data in the graphite rod, along with the wall temperature distribution obtained from the inverse calculation. The ends of the heater are water cooled. It is seen that the wall temperature distribution is not significantly affected by the rod size. Inverse calculations do not yield a unique solution. But optimization procedures can be used to minimize the uncertainty in the results obtained, as was also done here. This wall temperature distribution can then be used to accurately simulate the draw process. Other such approaches have been used in practice to obtain the relevant boundary conditions and thus accurately simulate the process.

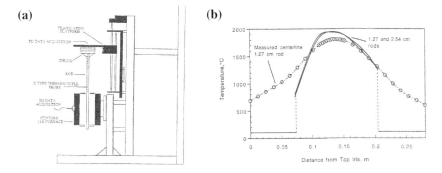

Fig. 7 **a** Schematic of an experimental system for measuring the temperature distribution in a graphite rod located in an optical fiber drawing furnace; **b** computed furnace wall temperature distributions (solid line) from graphite rod data

3.4 Simulation of Recirculating Flows

There are many other circumstances where recirculating flows such as those shown in Fig. 6 are encountered. The flow in a room due to a fire and the recirculating flow due to heat rejection to a water body were mentioned earlier, see Fig. 2. Because of the importance of these flows, a further discussion on the relevant boundary conditions and on the simulation is needed. Conjugate transport arises in many cases and the conditions similar to the ones outlined in the preceding section are employed. The inlet and outlet flow, as well as temperature distributions are particularly important and must be imposed correctly. Frequently developed conditions are employed at the outlet and uniform conditions at the inlet. The computational domain may also be extended to impose the conditions correctly by considering the region beyond the enclosed space. Spalding and his group carried out extensive simulations of recirculating flows and this work formed the basis for much of the work done on this topic in later years [23].

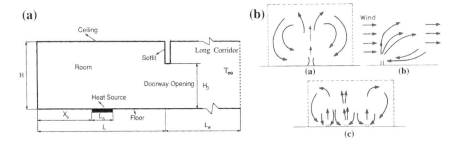

Fig. 8 **a** Computational domain extension to apply the boundary conditions for flow due to a heat source in a room; **b** recirculating flows in the environment

Figure 8a shows the extension of the computational domain that may be used to apply the velocity and temperature boundary conditions away from the flow region for the circumstance of a heat source in a room. The heat source could be a fire in a room with a doorway opening. Then the boundary conditions are applied at the end of the extended corridor. Typically, developed flow conditions are used, with gradients normal to the opening taken as zero. The temperature of the fluid entering the domain is taken at the outside or ambient temperature, and the temperature gradient of the exiting fluid as zero to ensure negligible diffusion compared to advection (the so-called outflow condition). The extension needed is determined numerically by varying it until the effect on the results is negligible [24]. Figure 8b shows several recirculating environmental flows. The computational boundaries are chosen far from the sources and fluid is allowed to enter and leave the domain, with developed conditions used in most cases.

Figure 9 shows the computed results for two circumstances: flow due to heat rejection into a water body and flow in a channel with two isolated heat sources and a passive vortex generator to enhance the heat transfer. In the first case, the parameter Gr/Re^2, where Gr is the Grashof number and Re the Reynolds number, indicates the relative importance of the buoyancy effect. Both Gr and Re are based on inlet

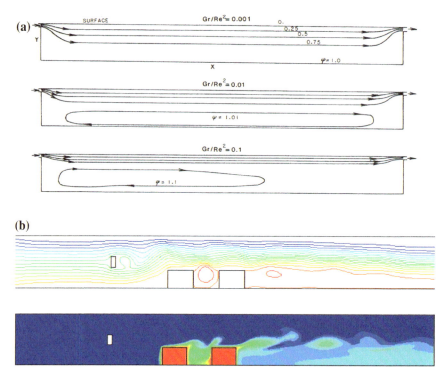

Fig. 9 a Recirculating flow due to thermal discharge into a water body; **b** streamlines and isotherms for flow in a channel with heat sources and a vortex generator

conditions. Therefore, $Gr/Re^2 = g\beta\Delta T d/U^2$, where ΔT is the temperature difference between the inlet and the ambient, U is the inlet velocity, and d is the inlet height. As this parameter increases due to increase in temperature difference ΔT or decrease in inlet flow velocity, the buoyancy effects increase and reduce the depth of the flow close to the surface and strengthen the recirculation below this top layer. Similarly, in the second case, Fig. 9b, the inflow and outflow conditions may be imposed. The inlet flow and temperature may be taken as uniform and the outflow condition may be imposed at the outlet plane. The inlet conditions may also be applied upstream by employing a domain extension to ensure that the thermal and flow conditions are more realistic. In actual practice, fluid flow at uniform temperature and velocity is generally available not at the inlet but at an upstream location.

3.5 Combined and Complex Transport Mechanisms

Single transport modes have been studied extensively in the literature. However, most practical thermal processes and systems involve coupled transport mechanisms that complicate the modeling and simulation. Similarly, complex mechanisms like chemical reactions, non-Newtonian fluids, surface tension effects, and free surfaces arise and need to be modeled accurately. Conjugate boundary conditions that involve coupled conduction and convection were considered earlier. Similarly, in the furnace for optical fiber drawing, thermal radiation and convection arise as coupled mechanisms, with conduction in the solid walls, as shown in Fig. 10a. Convection arises

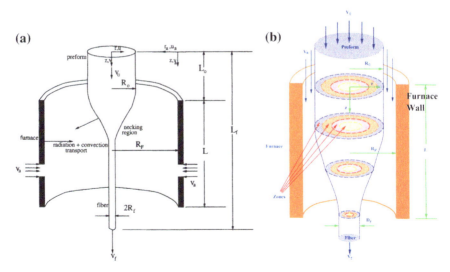

Fig. 10 **a** Transport mechanisms in the drawing furnace for an optical fiber; **b** axisymmetric finite volume zones for the calculation of radiation in the glass preform and fiber

both in the inert gas environment and in the glass. Beyond the softening point, T_{melt}, the glass, which is a subcooled liquid, is treated as a highly viscous liquid, with the viscosity obtained from an equation such as Eq. (12). Below the softening point, the viscosity is quite large, and the behavior is almost like that of a solid. Thermal radiation is the dominant mode of transport for heating up the glass. Radiation models such as the zonal method and the discrete ordinates method may be used to determine the radiation transport in the glass, as well as in the furnace, to obtain the energy absorbed by the preform/fiber, see Fig. 10b. The temperature distribution in the perform/fiber depends on the combined radiation and convection transport and on the viscous dissipation in the glass. The viscous dissipation is particularly important near the end of the neck-down region where the diameters are small and approaching 125 μm [19].

Similarly, forced flow and buoyancy effects arise in different regions of the furnace and the fiber in the drawing of hollow fibers, as shown in Fig. 11. The flow in the fiber core is largely driven by buoyancy and the shear imparted by the moving fiber.

Fig. 11 Sketch of the optical fiber drawing process for hollow fibers

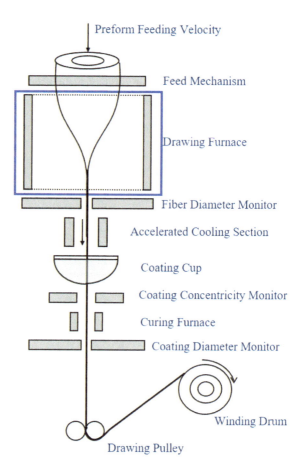

Preform Feeding Velocity

Feed Mechanism

Drawing Furnace

Fiber Diameter Monitor

Accelerated Cooling Section

Coating Cup

Coating Concentricity Monitor

Curing Furnace

Coating Diameter Monitor

Winding Drum

Drawing Pulley

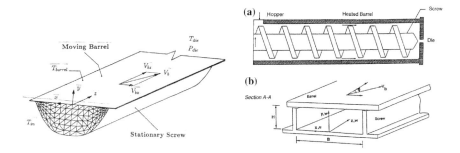

Fig. 12 A practical single-screw extruder, an idealized extruder; and a simplified model of the resulting channel flow

Thus, the models must include the combined mechanisms to determine the resulting transport, temperature variation, and the flow, as well as the free surface profile as the fiber is drawn from a cylindrical preform of several centimeters in diameter to the fiber diameter of 125 μm.

Another important practical system that may be considered is the screw extrusion process for plastics, including reactive polymers like food materials. Figure 12 shows a sketch of a practical single screw extruder, a simplified version with a rectangular screw profile and a mathematical model, with the coordinate axes located on the rotating screw and the curvature effects being neglected, to yield a simple channel flow. The barrel is then represented by a lid moving at the screw angle. The transport processes involve convective combined heat and mass transfer, conduction in the walls and the screw, and chemical reactions for a reactive material. The resulting product depends on the inlet conditions and imposed concentration C and the temperature T. The basic equations thus involve the flow equations, along with the energy and mass transfer equations. Chemical reactions occur and give rise to source terms in the energy and mass conservation equations. The properties vary with concentration and the temperature. The viscosity varies with the shear rate for the non-Newtonian materials typically encountered, as given by Eq. (14).

It is important to model the different transport mechanisms accurately and to ensure that the coupling effects are not neglected. For instance, in combined heat and mass transfer, the Soret and Dufour effects may not be negligible and may need to be included. A typical set of results in the channel flow of a single screw extruder is shown in Fig. 13. The Reynolds numbers are typically much less than one due to the large fluid viscosity so that the diffusion terms are much larger than the convection terms. However, viscous dissipation is important and heats up the polymer as it moves through the channel and the die, as seen from the temperature distributions. It is seen that there is very little mixing going on in the flow. Consequently, various strategies such as gaps in the screw, reverse elements, and different screw profiles are used to enhance mixing. Also, twin-screw extruders have been developed and are increasingly being used because of their superior mixing characteristics.

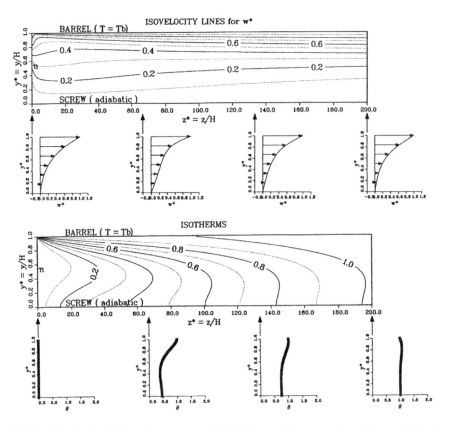

Fig. 13 Calculated velocity and temperature fields in the extruder at power-law index $n = 0.5$ and dimensionless throughput $qv = 0.3$

Several significant additional effects, such as buoyancy, surface tension, chemical conversion, complex domains, and free surfaces, that considerably complicate the transport phenomena being modeled are encountered in various practical thermal processes. The free surfaces that arise during fiber drawing and the menisci obtained in the coating process have been considered earlier. In these cases, the resulting shape of the free surfaces is governed by a balance of the forces due to shear, tension, gravity, and surface tension. Similarly, a force balance is used at interfaces in multilayered fibers, along with the conservation principles, to determine the resulting profiles.

In the drawing of hollow optical fibers, which are used for applications such as power delivery, sensors, and infrared radiation transmission, a major concern is the collapse of the central core. A collapse ratio C is defined as

$$C(z) = 1 - (R_1(z)/R_2(z))/(R_{10}/R_{20}) \tag{15}$$

where R_1 and R_2 are given in Fig. 14b. Thus, C is zero when the radius ratio of the final fiber equals the initial radius ratio, and C is 1 when the central cavity is

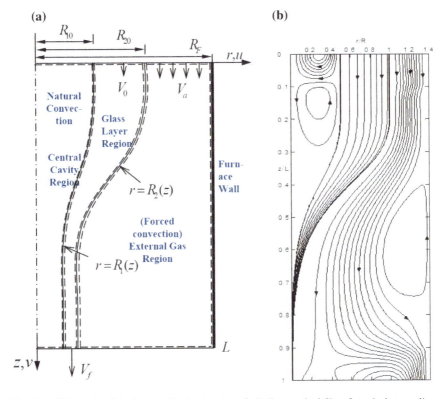

Fig. 14 a Schematic of the furnace for the drawing of a hollow optical fiber; **b** typical streamlines in hollow optical fiber drawing

closed. The effects of the feeding speed of the glass preform, fiber drawing speed, and the furnace temperature on collapse ratio have been studied in detail [25, 26]. Figure 14b shows the calculated streamlines in a typical drawing process, indicating the outer and inner free surfaces of the hollow fiber. Because of the small dimensions of the core, surface tension effects are important and play a very significant role in the collapse. Pressurizing the inner core can also be used to affect the collapse. Figure 15 shows the variation of the collapse ratio C with pressure in the core and with the furnace temperature. It is found that in order to avoid the collapse of the central cavity, we can increase the drawing and feeding speeds, decrease the furnace temperature, or increase the preform outer to inner radius ratio. It is seen that the collapse ratio decreases with a decrease in the pressure difference. This is because higher pressure in the central cavity tends to prevent a collapse of the central cavity.

Thus, a wide variety of additional effects often arise in practical processes and complicate the transport phenomena being modeled. It is necessary to consider all the additional effects that arise and to carry out a detailed scale analysis to determine which ones need to be retained. Then the complex process, with the appropriate additional effects, can be modeled. It must also be noted that Professor Spalding and

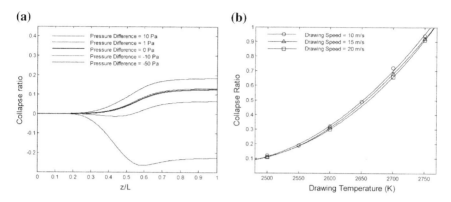

Fig. 15 Variation of the collapse ratio C in the hollow optical fiber drawing with **a** pressure in the inner core and **b** with furnace temperature

his group studied a wide range of practical problems with concerns similar to those discussed here, paving the way for further research on new and emerging complex problems.

3.6 Multiple Length and Time Scales

A common challenge that arises in the modeling of thermal systems is the presence of different length and time scales in the transport processes. The equations that describe the processes for different scales in the problem are generally different and thus the numerical approaches used for different regions and times may be quite different [26]. For example, consider the numerical simulation of pressure-driven nitrogen flow in long microchannels. Depending on the dimensions of the microchannel, slip flow at the boundaries needs to be considered. Also, conjugate heat transfer arises due to conduction in the walls [8]. Figure 16a shows a sketch of this problem, which is important in many practical circumstances such as those related to the cooling of electronic devices, energy delivery, diagnostics and localized heat input in materials processing systems. For the gas phase, the two-dimensional, axisymmetric or three-dimensional momentum and energy equations are solved, considering variable properties, rarefaction, which involves velocity slip, thermal creep and temperature jump, compressibility, and viscous dissipation. For the solid region, the energy equation is solved with variable properties. Therefore, the two regions are treated with different equations and solution strategies. Figure 16b, c show typical results obtained for different solid, or substrate, materials, including commercial bronze, silicon nitride, pyroceram, and fused silica. The effects of substrate conduction, thermal conductivity, and thickness are clearly seen. It was found that substrate conduction leads to a flatter bulk temperature profile along the channel length, lower maximum temperature, and lower Nusselt number. The effect of substrate thickness

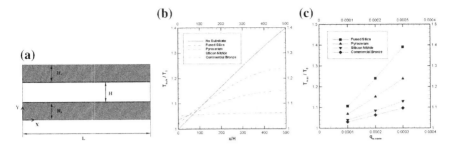

Fig. 16 Nitrogen flow in a microchannel, with conjugate transport in the walls. **a** Sketch of the microchannel; **b** bulk temperature distribution for different wall materials; **c** maximum temperature as a function of the heat input for different wall materials

is quite similar to that of the material thermal conductivity. That implies that, in terms of thermal resistance, an increase in substrate thickness has a similar effect as that caused by an increase in its thermal conductivity.

The characteristics and quality of the material being processed in manufacturing are generally determined by the transport processes that occur at the micro- or nanometer scale in the material. Examples of this are regions close to the solid–liquid interface in casting or crystal growing, over molecules involved in chemical reactions in chemical vapor deposition and reactive extrusion, or at sites where defects are formed in an optical fiber. On the other hand, engineering aspects are generally concerned with the macroscale, involving much larger dimensions, systems and appropriate boundaries. The operating conditions, for instance, are imposed on boundaries that are typically in centimeter or meter scale. Therefore, different length scales arise and need to be solved by different methodologies, ultimately coupling the two to obtain the overall behavior. Similarly, different time scales arise. For instance, chemical reactions, fluid flow, and thermal diffusion occur at very different time scales.

Changes at the molecular level are important in the generation of thermally induced defects in optical fiber drawing. One such defect is the E' defect, which is a point defect generated at high temperatures during the drawing process and which causes transmission loss and degradation of mechanical strength in the fiber. The differential equation for the time-dependent concentration of these defects was formulated by Hanafusa et al. [27] based on the theory of thermodynamics of lattice vacancies in crystals. It was assumed that the E' defects are generated through the breaking of the Si–O band, and, at the same time, a portion of the defects recombine to form Si–O again. Then the net concentration of the defects is the difference between the generation and the recombination. The equation for E' defect concentration was given as [27],

$$v\frac{dn_d}{dz} = n_p(0)v\exp\left(-\frac{E_p}{KT}\right) - n_d v\left[\exp\left(-\frac{E_p}{KT}\right) + \exp\left(-\frac{E_d}{KT}\right)\right] \quad (16)$$

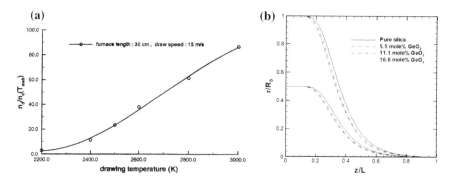

Fig. 17 **a** Dependence of the average concentration of E' defects on furnace wall temperature in optical fiber drawing; **b** neck-down profile in optical fiber drawing for various GeO_2 concentrations in a doped fiber

where n_d and E_d represent the concentration and activation energy of the E' defect; n_p and E_p represent those of the precursors. The initial values and constants are defined by Ref. [27]. Figure 17a shows the dependence of the average concentration of E' defects on the drawing temperature, indicating an increase with temperature, as expected from the higher breakage of the Si–O bond. Figure 17b shows the results on the neck-down profile for a doped fiber, considering different concentrations of the dopant GeO_2. Therefore, the defects can be controlled by doping and by varying the operating conditions, particularly the furnace temperature. Several other dopants and operating conditions have been considered in the literature. It was also found that the concentration of the defects is reduced if the drawing is followed by slow cooling, which anneals the fiber and allows the broken bonds to recombine [19].

Another example is provided by reactive thermal processing, such as food and reactive polymer extrusion. In these cases, the microscopic changes in the material are linked with the operating conditions that are imposed on the system. The chemical conversion process is then quantified by chemical kinetics, which depends on the temperature and the concentration [28]. These microscale conversion mechanisms may be coupled with the flow and heat transfer in a screw extruder to obtain the conversion, pressure, and other important quantities. It was shown in Ref. [28] that chemical conversion occurs due to thermal as well as shear effects and the two could be varied to obtain different product characteristics. If the conversion occurs due to shear at room temperature, the product is quite different from that obtained largely by thermal conversion. Thus, by linking microscale behavior with the transport processes, the product and the overall process can be varied.

Multi-scale transport is of interest in environmental flows as well, as shown for a couple of cases in Fig. 18. The heat transfer near the source in a fire or in a polluting system involves much smaller length scales than the transport far downstream. In thermal discharges from power plants and industries, the length scale is of the order of a few meters at the source and of the order or several kilometers far downstream [29]. In room fires, the source may only be a few centimeters, with the room itself

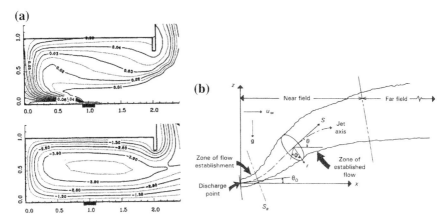

Fig. 18 Multiple length scales in **a** a room fire and **b** in a thermal discharge into the environment

being several meters in scale. The scales are further apart in large-scale fires as in forest fires. In such cases, the modeling involved, as well as the numerical grid, near the source could be quite different from those far away. For instance, radiation is particularly important near the fire source. But as one moves far away, the flow is dominated by turbulence and buoyancy. Similar considerations apply for heat and material discharges into the environment. Much of the numerical simulation work on such environmental flows has benefited from the extensive research done by Professor Spalding and his group on the modeling of turbulent flow.

3.7 Optimization

Optimization in terms of the operating conditions and the physical design of the system is very important in most practical systems due to global competition. Optimization is carried out with respect to a chosen objective, such as efficiency, cost, production rate, heat transfer rate, and product quality, which is maximized or minimized. For a given physical design, the optimal conditions that yield the best results may be determined. Similarly, the system design may be optimized for the best performance under given conditions. However, multiple objective functions are often of interest. For instance, in the cooling of electronic circuitry, typical design objectives are maximizing the heat removal rate from the components and minimizing the pressure drop. These two objectives can often be treated by considering the two objective functions separately. But the simple approach of combining the objectives into a single function simplifies the problem. Weighted sums may be employed in the development of the combined objective function [30]. However, such a composite objective function is arbitrary, and the results obtained will depend on the way it is formulated.

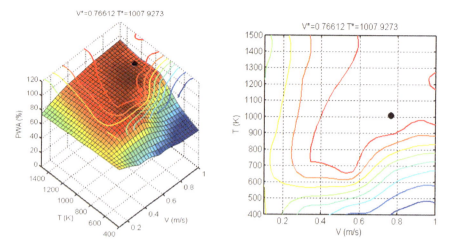

Fig. 19 Response surface model for PWA and the optimal operating point in a typical CVD process for depositing a silicon film

Lin et al. [31] carried out a detailed study on the design and optimization of chemical vapor deposition (CVD) systems. The main objectives are the deposition rate, which is to be maximized, and the film uniformity, which affects the percentage working area (PWA) that can be used for device fabrication. Results from several computational simulations were obtained to determine the effect of the operating conditions and examine the system performance. Response Surface Method (RSM) models were developed to study the behavior or the response of the system with respect to these variables. The optimization problem was formulated in terms of the RSM models, which were utilized to provide the operating conditions for higher productivity and quality of the film deposited, with a typical result shown in Fig. 19 for PWA in terms of the inlet velocity V and susceptor temperature T. The optimum point is shown on the response surface as well as on the contours of constant PWA. In this case, PWA is maximized. However, we could take the deposition rate as the objective function as well and obtain the resulting optimum. Another possibility is to take the acceptable PWA as a constraint and maximize the deposition rate or use an acceptable value of the latter as a constraint and maximize PWA. Some of these results are given in Ref. [19].

If the two or more objective functions that are of interest in a given problem are considered separately and a strategy is developed to trade-off one objective function in comparison to the others, a set of non-dominated designs, termed the *Pareto Set*, which represents the best collection of designs is generated [30, 32, 33]. Then, for any design in the Pareto Set, one objective function can be improved, at the expense of the other objective function. The set of designs that constitute the Pareto Set represent the formal solution in the acceptable design space to the multi-objective optimization problem. The selection of a specific design from the Pareto Set is left to the design engineer. A large literature exists on utility theory, which seeks to provide additional insight into the decision-making process to assist in selecting a specific design. Many multi-objective optimization methods are available that can be used to generate Pareto solutions. Various quality metrics are often used to evaluate the *goodness* of a Pareto solution obtained and thus improve the method as well as the optimal solution.

The use of this multi-objective optimization approach was demonstrated by Zhao et al. [34] for an electronic system cooling problem, which involved multiple isolated heat sources that represented electronic devices in a channel. Figure 20 shows the response curves separately for the heat transfer rate, given in terms of the Stanton number, $St = Nu/(Re \cdot Pr)$, and the dimensionless pressure drop DP. The former is to be maximized and the latter minimized. Also, shown is the Pareto frontier that may be used to obtain an appropriate design by trading off between the heat transfer and the pressure drop. A higher heat transfer rate is associated with higher pressure drop as well, which is not desirable. Therefore, a designer could choose the best solution for a given application by choosing an acceptable pressure drop for a desired heat transfer rate.

Figure 21 shows another example of multi-objective optimization. Microchannel liquid flows are used for cooling electronic chips. As before, the pressure difference and the heat transfer rate are the main objectives. A detailed simulation is carried out and the non-dominant solutions for the two objectives are used to obtain the Pareto front, given in terms of pumping power and thermal resistance. If a lower

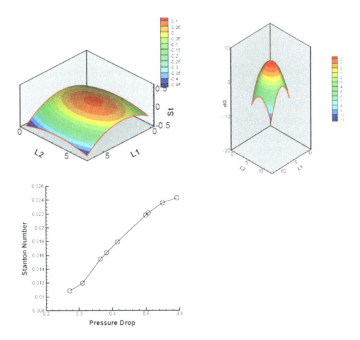

Fig. 20 Multi-objective optimization with response surfaces for two objective functions St and DP; Pareto Front for selection of optimal design by trade-off between the two objectives

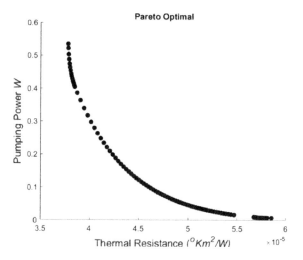

Fig. 21 Pareto Front for a system for the thermal management of electronic devices by the use of microchannel liquid flow

thermal resistance is desired for higher heat transfer, the pumping power increases. Similarly, for low pumping power the thermal resistance is high and heat transfer is low. From this curve, the best solution may be obtained for a given circumstance using the information on available pumps and costs.

3.8 Uncertainties

Uncertainties arise in the operating conditions and in the hardware in thermal systems. Even if an optimal design is obtained using deterministic conditions and parameters, the uncertainties can lead to unacceptable or unreliable designs. For example, the compositions of the chemical species entering a CVD reactor could have variations on the order of 15%. Similarly, uncertainties exist in inflow rate, susceptor temperature, and rotational speed. Errors may also be encountered in material properties and system dimensions. The rate constants of the chemical reactions may also have significant uncertainties. Several researchers have estimated the randomness of the operating parameters in different systems and processes [35–37].

It can be shown that, due to the existence of the design uncertainties, the traditional deterministic optimization formulation is no longer reliable to generate safe designs because it may lead to a design with a high risk of system failure. Reliability-Based Design Optimization (RBDO) evaluates the probabilities of the system failures and provides a more conservative design which reaches optimality as the failure probabilities that are subject to some acceptable level. Typically, normal distributions of the probabilistic conditions are taken, and an optimal design generated. The productions of the thermal systems are executed based on the optimal design variables. If any design uncertainties are found in the experiments, the simulations, or the mass productions, the information on the uncertainties is fed back to the formulation of

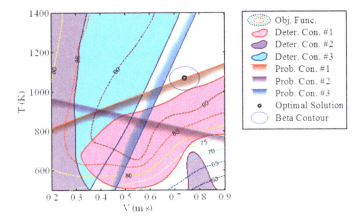

Fig. 22 Optimization with deterministic design variables and operating conditions, as well as with uncertainties associated with these

the RBDO problems and new optimal conditions can be generated by the proposed strategy. Figure 22 shows the results of the optimization of a CVD system with some deterministic and some probabilistic or uncertainty conditions. The probabilistic constraints are shown as a spread, rather than a deterministic line. The optimum point moves away from that obtained under deterministic conditions [35]. Typically, a failure rate of 0.13% is taken as the acceptable level in reliability-based design. Uncertainties form an important consideration in the design of thermal systems. However, not much has been done so far on this aspect, despite its importance in obtaining realistic and practically useful designs.

4   Concluding Remarks

This paper presents several important challenges faced in accurate numerical simulations of practical thermal systems and processes. The paper is a tribute to the extensive work done by Professor D. B. Spalding and his group. This work has provided the basis for meeting many of the challenges discussed in this paper. Among these challenges are variable material properties, that are often treated by employing methods developed by Professor Spalding's group. Similarly, validation of models is a well-known and critical concern in accurate modeling. Accurate imposition of boundary conditions, combined and complex transport mechanisms, uncertainties in the design variables and operating conditions, additional complexities and multiple scales are other challenges that commonly arise and must be addressed. These aspects are considered, along with practical examples where they are of particular importance. Possible approaches to meet these challenges are discussed. Accurate simulation results are needed for studying system behaviour and for design, control, and optimization. Though only a selected number of practical systems are outlined here, the basic considerations are applicable to a much wider range of problems.

It must be noted that to design and optimize thermal systems and processes, extensive numerical simulations are needed, along with experimental data for validation and physical insight. To obtain accurate simulation results, verified and validated mathematical models and numerical solution methods must be developed, as outlined here. Many books on computational fluid dynamics and heat transfer such as those referenced here, as well as those by Professor Spalding and his group [38, 39] among others, may be used for selecting the algorithm and developing the numerical code. However, a variety of computational software, some of which is based on Professor Spading's work, is commercially available and may be used advantageously to generate the desired numerical results.

Acknowledgements The author acknowledges the support of the National Science Foundation, through several grants, and of the industry for the work reported here. The author also acknowledges the interactions with several collaborators and the work done by several students that made it possible to present this review. Finally, the inspiration provided by Professor Spalding is gratefully acknowledged.

References

1. Jaluria, Y. (2003). Thermal processing of materials: From basic research to engineering. *ASME Journal of Heat Transfer, 125,* 957–979.
2. Jaluria, Y. (2008). *Design and Optimization of Thermal Systems* (2nd ed.). Boca Raton, FL: CRC Press.
3. Bejan, A., Tsatsaronis, G., & Moran, M. (1996). *Thermal design and optimization*. New York: Wiley.
4. Paek, U. C. (1999). Free drawing and polymer coating of silica glass optical fibers. *ASME Journal of Heat transfer, 121,* 774–788.
5. Zhang, J., & Jaluria, Y. (2017). Steady and transient behavior of data centers with variations in thermal load and environmental conditions. *International Journal of Heat and Mass Transfer, 108,* 374–385.
6. Mahajan, R. L. (1996). Transport phenomena in chemical vapor-deposition systems. *Advances in Heat Transfer, 28,* 339–425.
7. Gad-el-Hak, M. (1999). The fluid mechanics of microdevices—The Freeman scholar lecture. *Journal of Fluids Engineering, 121,* 5–33.
8. Sun, Z., & Jaluria, Y. (2010). Unsteady two-dimensional nitrogen flow in long microchannels with uniform wall heat flux,". *Numerical Heat Transfer, 57,* 625–641.
9. Eversteyn, F. C., Severin, P. J. W., Brekel, C. H. J., & Peek, H. L. (1970). A stagnant layer model for the epitaxial growth of silicon from silane in a horizontal reactor. *Journal of the Electrochemical Society, 117,* 925–931.
10. Chen, C., & Jaluria, Y. (2009). Effects of doping on the optical fiber drawing process. *International Journal of Heat and Mass Transfer, 52,* 4812–4822.
11. Izawa, T., & Sudo, S. (1987). *Optical fibers: Materials and fabrication*. Tokyo: KTK Scientific Publishers.
12. Skelland, A. H. O. (1967). *Non-Newtonian flow and heat transfer*. New York: Wiley.
13. Tadmor, Z., & Gogos, C. (1979). *Principles of polymer processing*. New York: Wiley.

14. Kokini, J. L., Ho, C.-T., & Karwe, M. V. (Eds.). (1992). *Food extrusion science and technology*. New York: Marcel Dekker.

15. Patankar, S. V., & Spalding, D. B. (1970). *Heat and mass transfer in boundary layers* (2nd ed.). Intertext: Taylor & Francis, Oxfordshire, UK.

16. Roache, P. J. (1998). *Verification and validation in computational science and engineering*. Albuquerque, New Mexico: Hermosa Publishers.

17. Jaluria, Y. (1996). Heat and mass transfer in the extrusion of non-newtonian materials. *Advances in Heat Transfer, 28*, 145–230.

18. Yoo, S. Y., & Jaluria, Y. (2008). Numerical simulation of the meniscus in the non-isothermal free surface flow at the exit of a coating die. *Numerical Heat Transfer, 53A*, 111–131.

19. Jaluria, Y. (2018). *Advanced materials processing and manufacturing*. Cham, Switzerland: Springer.

20. Viswanath, R., & Jaluria, Y. (1995). Numerical study of conjugate transient solidification in an enclosed region. *Numerical Heat Transfer, 27*, 519–536.

21. Papanicolaou, E., & Jaluria, Y. (1993). Mixed convection from a localized heat source in a cavity with conducting walls: A numerical study. *Numerical Heat Transfer, 23*, 463–484.

22. Issa, J., Yin, Z., Polymeropoulos, C. E., & Jaluria, Y. (1996). Temperature distribution in an optical fiber draw tower furnace. *Journal of Materials Processing and Manufacturing Science, 4*, 221–232.

23. Gosman, A. D., Pun, W. M., Spalding, D. B., & Wolfshtein, M. (1969). *Heat and mass transfer in recirculating flows*. New York: Academic Press.

24. Abib, A., & Jaluria, Y. (1988). Numerical simulation of the buoyancy-induced flow in a partially open enclosure. *Numerical Heat Transfer, 14*, 235–254.

25. Fitt, A. D., Furusawa, K., Monro, T. M., & Please, C. P. (2001). Modeling the fabrication of hollow fibers: Capillary drawing. *Journal of Lightwave Technology, 19*, 1924–1931.

26. Jaluria, Y. (2009). Microscale transport phenomena in materials processing. *ASME Journal of Heat Transfer, 131*, 033111-1–17.

27. Hanafusa, H., Hibino, Y., & Yamamoto, F. (1985). Formation mechanism of drawing-induced E′ centers in silica optical fibers. *Journal of Applied Physics, 58*(3), 1356–1361.

28. Wang, S. S., Chiang, C. C., Yeh, A. I., Zhao, B., & Kim, I. H. (1989). Kinetics of phase transition of waxy corn starch at extrusion temperatures and moisture contents. *Journal of Food Science, 54*, 1298–1301.

29. Gebhart, B., Hilder, D. S., & Kelleher, M. (1984). The diffusion of turbulent jets. *Advances in Heat Transfer, 16*, 1–57.

30. Deb, K. (2002). *Multi-objective optimization using evolutionary algorithms*. New York, NY: Wiley.

31. Lin, P. T., Gea, H. C., & Jaluria, Y. (2009). Parametric modeling and optimization of chemical vapor deposition process. *Journal of Manufacturing Science and Engineering, 131*, 011011-1–7.

32. Miettinen, K. M. (1999). *Nonlinear multi-objective optimization*. Boston, MA: Kluwer Acad. Press.

33. Ringuest, J. L. (1992). *Multiobjective optimization: Behavioral and computational considerations*. Boston, MA: Kluwer Acad. Press.

34. Zhao, H., Icoz, T., Jaluria, Y., & Knight, D. (2007). Application of data driven design optimization methodology to a multi-objective design optimization problem. *Journal of Engineering, 18*, 343–359.

35. Lin, P. T., Gea, H. C., & Jaluria, Y. (2010). Systematic strategy for modeling and optimization of thermal systems with design uncertainties. *Frontiers Heat Mass Transfer, 1*(013003), 1–20.

36. Tu, J., Cho, J., & Park, Y. H. (1999). A new study on reliability-based design optimization. *Journal of Mechanical Design, 121*, 557–564.

37. Youn, B. D., & Choi, K. K. (2004). An investigation of nonlinearity of reliability-based design optimization approaches. *Journal of Mechanical Design, 126*, 403–411.

38. Spalding, D. B. (1977). *Genmix: A general computer program for two-dimensional parabolic phenomena*. Oxford, UK: Pergamon Press.
39. Patankar, S. V. (1980). *Numerical heat transfer and fluid flow*. Boca Raton, FL: CRC Press.

Numerical Predictions of Temporally Periodic Fluid Flow and Heat Transfer in Spatially Periodic Geometries

Alexandre Lamoureux and Bantwal R. (Rabi) Baliga

1 Introduction

The methods of computational fluid dynamics and heat transfer (CFDHT) have become an integral and critically important part of the procedures used today for numerical predictions of the behavior of thermofluid systems in engineering and the environment. This development is mainly due to the many seminal contributions of Professor D. B. Spalding (individually and in collaboration with his former students and colleagues) to CFDHT over the last 50 years. Some examples of such contributions put forward in the late 1960s and in the 1970s include the works described in Patankar and Spalding [1, 2], Runchal [3], Spalding [4], Patankar and Spalding [5], Launder and Spalding [6], Pratap and Spalding [7], Pollard and Spalding [8], and Patankar [9]. Elaborations of these and many other seminal contributions of Professor D. B. Spalding and his colleagues are available in Patankar et al. [10], Artemov et al. [11], and Runchal [12].

In this chapter, some observations on numerical solutions of the mathematical models of temporally periodic fluid flow and heat transfer in spatially periodic geometries, in both spatially developing and fully developed regions, are presented and discussed. Such fluid flow and heat transfer phenomena are encountered in a wide variety of engineering devices and systems: examples include shell-and-tube and compact heat exchangers [13–16], cooling arrangements for gas turbine blades [17–20] and electronics [21], and porous media [22]. Detailed discussions of the numerical simulations of such phenomena, for laminar and turbulent flow conditions, are available, for example, in the works of Sparrow et al. [23], Patankar et al. [24], Acharya et al.

A. Lamoureux
Hatch Limited, 5 Place Ville Marie, Suite 1400, Montreal, QC H3B 2G2, Canada
e-mail: alexandre.lamoureux@hatch.com

B. R. (Rabi) Baliga (✉)
Heat Transfer Laboratory, Department of Mechanical Engineering, McGill University, 817 Sherbrooke St. W., Montreal, QC H3A 0C3, Canada
e-mail: bantwal.baliga@mcgill.ca

© Springer Nature Singapore Pte Ltd. 2020
A. Runchal (ed.), *50 Years of CFD in Engineering Sciences*,
https://doi.org/10.1007/978-981-15-2670-1_6

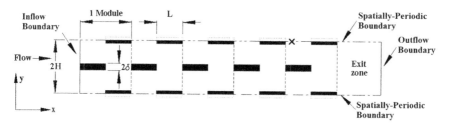

Fig. 1 Schematic representation of a calculation domain used in the simulations of developing, two-dimensional, laminar flow and heat transfer in uniform arrays of staggered rectangular plates. The symbol "**x**" indicates the location where the flow oscillation frequency was monitored in the computer simulations

[25], Wang and Vanka [26], Murthy and Mathur [27], Beale and Spalding [28, 29], Bahaidarah et al. [30, 31], Fullerton and Anand [32, 33], Hutter et al. [34], Labbé [35], Iacovides et al. [36], Ruck and Arbeiter [37], Hasis et al. [38], and Dezan et al. [39]. There are numerous other published investigations (experimental and numerical) of fluid flow and heat transfer in spatially periodic geometries: representative examples include the works described in [40–66].

With regard to mathematical models and numerical predictions of the above-mentioned fluid flow and heat transfer phenomena, special attention is given in this chapter to several issues that have not been fully resolved in earlier publications. The key points pertaining to these issues are presented and discussed in this chapter in the context of computationally convenient numerical solutions of the mathematical models of temporally periodic, two-dimensional, laminar, constant-property New-tonian fluid flow and forced convection heat transfer in uniform arrays of staggered rectangular plates, in both spatially developing and fully developed regions akin to those illustrated schematically in Figs. 1 and 2, respectively. The focus in this chapter is on numerical predictions obtained using finite volume methods that are extensions of the abovementioned seminal works of Professor D. B. Spalding and his colleagues [1–12].

The aforementioned published investigations have established that in spatially periodic (interrupted surface, wavy, or zig-zag) geometries, the augmentations of the rate of heat transfer and the related pressure drop (or pumping power) for a given fluid flow rate are due to one or more of the following aspects: interrupted boundary layers; secondary flows (in cross-sections that are transverse to the main-flow direction); increased surface area (in plate-fin ducts); and, in some cases, unsteady features of the flow (including vortex shedding). These investigations have also shown that the designers of devices with such spatially periodic geometries are faced with a thermofluid optimization problem, in which the objective is usually to maximize the rate of heat transfer for a fixed pumping power (subject to some constraints), over the full range of required operating conditions. However, despite all of the aforementioned publications, several issues related to the mathematical modeling and numerical predictions of fluid flow and heat transfer in spatially periodic geometries have not yet been fully resolved. Some of these issues are elaborated and addressed in

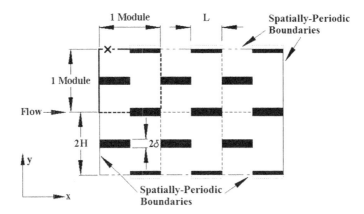

Fig. 2 Schematic representation of a calculation domain used in the simulations of spatially periodic, two-dimensional, laminar flow and heat transfer in uniform arrays of staggered rectangular plates. The symbol "**x**" indicates the location where the flow oscillation frequency was monitored in the computer simulations

the remainder of this chapter, in the context of spatially periodic interrupted-surface fluid flow passages.

In the following section, the nomenclature used in this paper is presented. After that, some issues pertaining to numerical predictions of the abovementioned problems of interest are elaborated, and followed by an overview of the specific problems considered in this work. The mathematical models of the problems of interest are then presented and discussed. Following that, the formulation, implementation, and testing of a finite volume method for solving the problems of interest are described concisely. Then the results of this investigation are presented and discussed, and followed by a brief review of the main findings of this work. Finally, some acknowledgements and a list of references are presented.

2 Nomenclature

$A_{c\text{-}s\,\min}$	Minimum cross-sectional area per unit depth
A_{wetted}	Wetted module surface area per unit depth
c_p	Specific heat at constant pressure
D_h	Hydraulic diameter
f	Time-mean modular friction factor
$\bar{h}^{[H]}$, $\bar{h}^{[T]}$	Modular heat transfer coefficient, for $[H]$ and $[T]$ thermal boundary conditions
H, L	Module half-height and plate length
$j^{[H]}$, $j^{[T]}$	Time-mean modular Colburn factor, for $[H]$ and $[T]$ thermal boundary conditions

k	Thermal conductivity
L^*, δ^*	Dimensionless plate length and thickness
$\dot{m}_{\text{module}}$	Instantaneous mass flow rate per unit depth in a module
n_x, n_y	Integer periodicity indices in the x- and y-directions
p	Instantaneous reduced pressure
p^*	Dimensionless instantaneous reduced pressure
Pr	Prandtl number
q_{module}	Modular rate of heat addition per unit depth
q_w''	Heat flux at plate–fluid interface
Re_{D_h}	Reynolds number, based on time-mean $U_{c\text{-}s\,\text{min}}$ and D_h
$St^{[H]}, St^{[T]}$	Stanton number, for $[H]$ and $[T]$ thermal boundary conditions
S	Strouhal number
t	Time
t^*	Dimensionless time
T	Instantaneous temperature
T_b	Instantaneous bulk temperature at a given x position
$T_{c\text{-}s\,\text{ref}}$	Instantaneous cross-sectional reference temperature
T_w	Plate surface temperature
u, v	Instantaneous velocity components in the x- and y-directions, respectively
u^*, v^*	Dimensionless instantaneous velocity components
$U_{c\text{-}s\,\text{min}}$	Instantaneous spatial average of u at minimum cross-sectional area
x, y	Cartesian coordinates (see Figs. 1 and 2)
X, Y	Dimensionless Cartesian coordinates
$-\beta$	Modular x-direction gradient of the time-mean reduced pressure
$-\beta^*$	Dimensionless modular x-direction gradient of the time-mean reduced-pressure
δ	Half-thickness of the plates
$\Delta x, \Delta y$	Control-volume widths in the x- and y-directions, in the fine-grid regions
γ	Modular x-direction temperature gradient
μ	Dynamic viscosity
ρ	Mass density
$\theta^{[H]}, \theta^{[T]}$	Dimensionless temperature, for $[H]$ and $[T]$ thermal boundary conditions
ω_v	Frequency of self-sustained oscillations at a monitoring point
ξ_{surf}	Coordinate along the local unit vector normal to the plate surface, pointing into the fluid

3 Some Issues Pertaining to Numerical Predictions of Fluid Flow and Heat Transfer in Spatially Periodic Interrupted-Surface Passages

In many of the published numerical investigations of laminar fluid flow and heat transfer in interrupted-surface passages, attention was limited to the *steady-state*, spatially periodic, fully developed regime: examples include the works of Patankar et al. [24], Bahaidarah et al. [30], Patankar and Prakash [43], Kelkar and Patankar [47], Ismail et al. [59], and Sebben and Baliga [67]. There have also been many published numerical investigations of *unsteady*, self-sustained, oscillatory, laminar and turbulent flow and heat transfer in the spatially periodic geometries: examples include the works of Acharya et al. [25], Beale and Spalding [28, 29], Hutter et al. [34], Labbé [35], Iacovides et al. [36], Ruck and Arbeiter [37], Amon et al. [48], Zhang et al. [52], Tafti et al. [53], Shah et al. [54], Saidi and Sunden [55], Lamoureux and Baliga [60, 61], Korichi et al. [63], Sui et al. [64], Zheng et al. [65], Ramgadia and Saha [66], Ghaddar et al. [68, 69], Sebben [70], Stone and Vanka [71], and Croce and D'Agaro [72]. In many of the aforementioned numerical investigations, the calculation domain was limited to a single representative geometric module, by imposing temporal and spatial periodicity conditions on specialized formulations of the instantaneous fluid flow and heat transfer in the fully developed region. The validity of this approach to the modeling of *steady* laminar flows in spatially periodic interrupted-surface geometries is well established, for example, in the works of Sparrow et al. [23] and Patankar et al. [24]. Beale and Spalding [28, 29] have investigated temporally periodic *unsteady* laminar flow and heat transfer across in-line and staggered tube banks by considering calculation domains consisting of two complete geometric modules in the mainstream direction. However, the implications (on the numerical solutions) of the number of geometric modules that are included in the calculation domain in the context of temporally periodic *unsteady* laminar flow and heat transfer have not yet been fully established. To do that, it is necessary to study the influence of including different numbers (≥ 1) of adjacent geometric modules in the calculation domain in the fully developed region, in both the mainstream and transverse directions. The number of adjacent modules included in the calculation domain in any particular direction is referred to as the *integer periodicity index* in that direction, following the terminology introduced in the works of Amon et al. [48] and Ghaddar et al. [68, 69].

There is a related issue of whether the spatially fully developed solutions that are obtained by solving the mathematical models of fluid flow and heat transfer from the inlet plane all the way to the fully developed region, and those obtained by solving these models only in the fully developed region (using specialized formulations and imposing instantaneous spatial periodicity conditions on the boundaries of one or multiple adjacent geometric modules), are the same. It has been definitively established that these fully developed solutions are, indeed, the same for *steady* laminar flows in spatially periodic geometries. However, the aforementioned solutions may not be the same for *temporally periodic unsteady* laminar flows in spatially periodic geometries, not just in their instantaneous forms, but also in their time-averaged forms. These issues are addressed in this chapter.

It should be again noted that the issues mentioned in the previous paragraphs have been considered in some earlier published works, but they are still not fully resolved. A few specific examples of such works are provided in this paragraph. A systematic investigation of the influence of the integer periodicity index was undertaken by Sebben [70]. She studied temporally and spatially periodic two-dimensional laminar fluid flow in a straight rectangular duct with spatially periodic, in-line, plate inserts along its longitudinal center plane, using values of the periodicity index in the mainstream direction of one to six. Her results showed that under temporally periodic unsteady conditions, when the value of the integer periodicity index in the mainstream direction is greater than one, multiple (distinct) stable (or metastable) solutions could be obtained. However, Sebben [70] did not study developing flow and heat transfer and did not investigate the influence of the periodicity index in the direction transverse to the mainstream direction, as the geometry she considered did not require or allow it. Sparrow et al. [23] and Bahaidarah et al. [31], for example, have studied developing laminar fluid flow and heat transfer in ducts with spatially periodic interrupted surfaces, but only steady-state cases were considered in these studies. Tafti et al. [53] have studied temporally periodic, developing and fully developed, laminar, two-dimensional fluid flow and heat transfer in an array of louvered fins. However, in their investigation of the fully developed regime, they considered just one geometrical module (periodicity indices of one in both the mainstream and transverse directions). Unsteady, developing, laminar, two-dimensional fluid flow and heat transfer over a bank of flat tubes has also been investigated by Benarji et al. [62]. They used one row of 19 geometrical modules as the calculation domain, which was more than sufficient to ensure the establishment of spatially fully developed conditions before the fluid flow exited; however, they imposed symmetry (rather than spatial periodicity) conditions on the longitudinal boundaries of the aforementioned calculation domain.

4   Problems Considered and Some Related Comments

Temporally periodic, two-dimensional, laminar, constant-property Newtonian fluid flow and forced convection heat transfer in uniform arrays of staggered rectangular plates, in both spatially developing and fully developed regions, are considered in this work. With respect to the notation shown in Figs. 1 and 2, the following four combinations of geometrical parameters were investigated: dimensionless plate thickness $\delta^* = 2\delta/2H = \delta/H = 1/4, 1/8, 1/12$ and $1/16$, each with a dimensionless plate length $L^* = L/2H = 1$. It should be noted here that dimensionless plate thickness values of 1/4 and 1/8 are overly large compared to those commonly encountered in rectangular offset-fin cores of compact heat exchangers [15], but they were used in this study to induce and enhance the unsteady features of the fluid flow phenomena, and thereby facilitate the investigation of the influence of the integer periodicity indices on such flows. The values of Reynolds number (based on the hydraulic diameter of the flow passage and the time-mean average mainstream component of the fluid velocity in the minimum cross-sectional area) ranged from 100 to 1000. At the aforementioned values of Reynolds number, the corresponding fluid flow and heat transfer phenomena are laminar. Furthermore, in the cores of rectangular offset-fin compact heat exchangers, when the values of Reynolds number are less than 800, it has been shown that the fluid flow and heat transfer in the central regions over the fins, away from the sidewalls, can be well approximated as essentially two-dimensional phenomena, as discussed in the works of Suzuki et al. [49] and Ismail et al. [59], for example. The Prandtl number was maintained constant at the value of 0.7 (representative of air).

Depending on the combination of Reynolds number and dimensionless geometric parameters that characterize the aforementioned fluid flows of interest, they could exhibit several different features. With reference to the schematic illustration and notation given in Fig. 1, adjacent to the inlet plane, in a region that could include the first two to ten geometric modules in the mainstream (x) direction, the fluid flow develops (adjusting to the presence of the plates), and it could be steady or unsteady, and the unsteady fluid flow could be temporally periodic or essentially chaotic. Downstream of this developing region, the fluid flow may be either steady and spatially periodic over a geometric module; temporally and spatially periodic over one or more adjacent geometric modules in the mainstream (x) and transverse (y) directions (as shown schematically in Fig. 2); or essentially chaotic. In the aforementioned developing and spatially periodic fully developed regions, the fluid flow could exhibit separated regions adjacent to the leading and trailing edges of the plates, with or without unsteady features that may be temporally periodic and include vortex shedding. These fluid flows may also exhibit asymmetries caused by Coanda-like effects [73, 74]. Mathematical models of these phenomena and their numerical solutions are presented and discussed in this chapter, using results that include values of the time-mean modular friction and Colburn j factors, values of the Strouhal number, and instantaneous streamline plots.

5 Mathematical Models

In the mathematical models presented in this section, it is assumed that the fluid is Newtonian; and the mass density (ρ), dynamic viscosity (μ), specific heat at constant pressure (c_p), and thermal conductivity (k) of the fluid are assumed to be constant. The Prandtl number of the fluid is assigned the constant value of 0.7 (appropriate for air). It is also assumed that viscous dissipation and buoyancy effects are negligible, and attention is restricted to laminar, two-dimensional, fluid flow, and heat transfer.

The calculation domains used in the investigations of the developing and fully developed fluid flow and heat transfer phenomena of interest are akin to those illustrated schematically in Figs. 1 and 2, respectively. Also shown in these figures are some of the boundary conditions and the notation used to describe the geometry of the uniform arrays of staggered plates.

5.1 Developing Fluid Flow and Forced Convection Heat Transfer

The governing equations and boundary conditions for these phenomena, in the context of the aforementioned assumptions, can be cast in the following forms:

Continuity Equation

$$\frac{\partial}{\partial x}(\rho u) + \frac{\partial}{\partial y}(\rho v) = 0 \tag{1}$$

x-Momentum Equation

$$\frac{\partial}{\partial t}(\rho u) + \frac{\partial}{\partial x}(\rho u u) + \frac{\partial}{\partial y}(\rho v u) = -\frac{\partial p}{\partial x} + \frac{\partial}{\partial x}\left(\mu \frac{\partial u}{\partial x}\right) + \frac{\partial}{\partial y}\left(\mu \frac{\partial u}{\partial y}\right) \tag{2}$$

y-Momentum Equation

$$\frac{\partial}{\partial t}(\rho v) + \frac{\partial}{\partial x}(\rho u v) + \frac{\partial}{\partial y}(\rho v v) = -\frac{\partial p}{\partial y} + \frac{\partial}{\partial x}\left(\mu \frac{\partial v}{\partial x}\right) + \frac{\partial}{\partial y}\left(\mu \frac{\partial v}{\partial y}\right) \tag{3}$$

Energy Equation

$$\frac{\partial}{\partial t}(\rho T) + \frac{\partial}{\partial x}(\rho u T) + \frac{\partial}{\partial y}(\rho v T) = \frac{\partial}{\partial x}\left(\frac{k}{c_p} \frac{\partial T}{\partial x}\right) + \frac{\partial}{\partial y}\left(\frac{k}{c_p} \frac{\partial T}{\partial y}\right) \tag{4}$$

In these equations, x and y are Cartesian coordinates in the mainstream and transverse directions, respectively, as shown in Fig. 1; t denotes time; u and v are the instantaneous velocity components in the x- and y-directions, respectively; T is the instantaneous temperature of the fluid; and p is the instantaneous reduced pressure. This reduced pressure is related to the instantaneous static pressure, p_{static}, and the x- and y-direction components of the gravitational acceleration, g_x and g_y, respectively, by the following equation:

$$p = p_{static} - \rho g_x x - \rho g_y y \tag{5}$$

Boundary Conditions
Fluid velocity components and temperature: At open portions of the inlet plane (parts not occupied by the boundaries of the solid plates), u is assumed to be spatially uniform and constant, v is assigned the value of zero, and T is assumed to be spatially

uniform and constant. With the configuration of the inlet plane shown in Fig. 1, the prescribed uniform and constant inlet velocity, u_{in}, is equal to $U_{c\text{-}s\,min}$, the spatial average value of the mainstream component of the fluid velocity in the minimum cross-sectional area of the duct. Thus,

$$u_{\text{open portion of inlet plane}} = u_{in} = U_{c\text{-}s\,min} = \text{constant}$$

$$v_{\text{open portion of inlet plane}} = 0$$

$$T_{\text{open portion of inlet plane}} = T_{in} = \text{constant} \tag{6}$$

In the transverse direction, for both steady and temporally periodic cases with no intrinsic fluid flow asymmetries or maldistributions, the velocity and temperature fields repeat identically in successive *sets* of rows of adjacent geometrical modules. Each such set could consist of n_y (≥ 1) rows of geometrical modules, the integer periodicity index in the transverse (y) direction. With respect to the notation given in Fig. 1, the y-extent of a set of such modules would be $n_y(2H)$. On the open portions of the longitudinal bounding planes of such a set of modules, the spatial periodicity of the instantaneous velocity and temperature fields can be mathematically expressed as follows:

$$u(x, y, t) = u(x, y + n_y(2H), t) = u(x, y + 2n_y(2H), t) = \cdots$$

$$v(x, y, t) = v(x, y + n_y(2H), t) = v(x, y + 2n_y(2H), t) = \cdots$$

$$T(x, y, t) = T(x, y + n_y(2H), t) = T(x, y + 2n_y(2H), t) = \cdots \tag{7}$$

It should be noted that for cases that yield *steady-state* solutions, $n_y = 1$ is adequate for flows with no intrinsic asymmetries or maldistributions. Even for temporally periodic cases, as shown by Lamoureux and Baliga [61], numerical solutions of the aforementioned developing flow problems obtained with $n_y = 1$ and $n_y = 2$ are essentially the same. In this work, therefore, $n_y = 1$ was used, and the calculation domain consisted of a single row of 10 representative geometric modules followed by an open (or plate-less) exit zone in the mainstream direction, and the aforementioned spatially periodicity conditions were imposed on the longitudinal boundaries (aligned with the x-direction), as shown schematically in Fig. 1.

At the outlet plane (see Fig. 1), assuming that the x-extent of the exit zone is long enough to ensure strict outflow conditions (that is, no inflow at the outlet plane), the viscous and conduction transport rates across the outlet plane were assumed to be negligible compared to the corresponding rates of advection transport [9]. This outflow treatment was implemented by imposing the following boundary conditions at the outlet plane:

$$\left(\frac{\partial u}{\partial x} = 0 \;;\; \frac{\partial v}{\partial x} = 0 \;;\; \frac{\partial T}{\partial x} = 0 \right)_{\text{outlet plane}} \tag{8}$$

At the plate–fluid interfaces, the no-slip and impermeability conditions apply. Thus,

$$u_{\text{plate−fluid interface}} = 0 \; ; \; v_{\text{plate−fluid interface}} = 0 \tag{9}$$

In the published investigations of forced convection heat transfer in interrupted-surface geometries, typically, two different thermal boundary conditions at the plate–fluid interface are considered: in one, the plates are maintained at a constant temperature, T_w (this thermal boundary condition is denoted by $[T]$ in this chapter); and in the other, a constant heat flux, q_w'', from the plates to the fluid is imposed at the plate–fluid interfaces (this thermal boundary condition is denoted by $[H]$ in this chapter). Mathematically, these two thermal boundary conditions can be expressed as follows:

$$(T = T_w)_{\text{plate−fluid interface}}^{[T]} \; ; \; \left(-k \partial T / \partial \xi_{\text{surf}} = q_w''\right)_{\text{plate−fluid interface}}^{[H]} \tag{10}$$

In this equation, k is the thermal conductivity of the fluid and ξ_{surf} is the coordinate along the local unit normal to the surface of the plates, pointing into the fluid.

Many of the available experimental data for designing the cores of compact heat exchangers, for example, those given in Kays and London [15], were obtained with thermal boundary conditions that approximate the $[T]$ case. The investigations of developing fluid flow and heat transfer phenomena reported in this chapter were conducted with only the $[T]$ thermal boundary condition, with a long enough extent of the calculation domain in the mainstream (x) direction to ensure spatially periodic conditions are established before (upstream of) the exit zone (see Fig. 1).

Reduced pressure: When the thermophysical properties of the fluid are assumed to remain constant, it is necessary to specify (or fix) the value of the reduced pressure at only one convenient point in the calculation domain, and this value can be assigned any appropriate value [9]. In this investigation, the reduced pressure was fixed at the convenient value of zero at a point on the inlet plane of the calculation domain.

5.2 Temporally and Spatially Periodic Fluid Flow and Forced Convection Heat Transfer

The spatial periodicity conditions, governing equations, and boundary conditions pertaining to the instantaneous velocity and reduced-pressure fields are presented first in this subsection. Following that, these considerations for the temperature field are presented for both the $[H]$ and the $[T]$ boundary conditions, in that order.

Governing Equations and Boundary Conditions for the Velocity and Reduced-Pressure Fields

In the mathematical model of the temporally and spatially periodic fluid flow regime, the instantaneous velocity field is assumed to repeat identically in successive *sets* of adjacent modules. Each such set consists of n_x (≥ 1) and n_y (≥ 1) geometrical modules, the integer periodicity indices in the mainstream (x) and transverse (y) directions, respectively. With respect to the notation given in Fig. 2, the spatial extents of a set of such modules would be $n_x(2L)$ and $n_y(2H)$ in the x- and y-directions, respectively, and the aforementioned spatial periodicity of the instantaneous velocity components, u and v, can be mathematically expressed as follows:

$$u(x, y, t) = u(x + n_x(2L), y, t) = u(x + 2n_x(2L), y, t) = \cdots$$
$$v(x, y, t) = v(x + n_x(2L), y, t) = v(x + 2n_x(2L), y, t) = \cdots \qquad (11)$$

$$u(x, y, t) = u(x, y + n_y(2H), t) = u(x, y + 2n_y(2H), t) = \cdots$$
$$v(x, y, t) = v(x, y + n_y(2H), t) = v(x, y + 2n_y(2H), t) = \cdots \qquad (12)$$

For steady-state conditions, u and v are independent of the time (t), their spatial distributions repeat identically in successive adjacent modules in the spatially periodic fully developed regime, and the calculation domain can be limited to a single geometric module ($n_x = n_y = 1$), as discussed, for example, by Patankar et al. [24]. However, as was mentioned earlier, for temporally and spatially periodic conditions, multiple (distinct) instantaneous solutions may be possible when $n_x > 1$ and/or $n_y > 1$, with significantly different time-averaged results (both local and modular). This possibility was checked out in this work by conducting simulations with different combinations of n_x (≥ 1) and n_y (≥ 1).

In order to have a net mass flow rate in the positive x-direction through the staggered-plate array of interest, there must be a pressure drop in the mainstream (x) direction. Taking this requirement into account, and following the formulation proposed by Patankar et al. [24], in the temporally and spatially periodic regime, the instantaneous reduced pressure, p, as defined in Eq. (5), is expressed as follows:

$$p(x, y, t) = p_{\text{ref}} - \beta x + P(x, y, t) \qquad (13)$$

In this equation, p_{ref} is any convenient reference value of p at any suitable chosen point in the calculation domain. The term $-\beta$ is the modular x-direction gradient of the time-mean reduced pressure that drives the time-mean overall mass flow rate through the plate array:

$$-\beta = \frac{<p(x + n_x(2L), y, t)> - <p(x, y, t)>}{n_x(2L)} \qquad (14)$$

The term $P(x, y, t)$ behaves in a spatially periodic manner akin to the velocity components:

$$P(x, y, t) = P(x + n_x(2L), y, t) = P(x + 2n_x(2L), y, t) = \cdots$$
$$P(x, y, t) = P(x, y + n_y(2H), t) = P(x, y + 2n_y(2H), t) = \cdots \qquad (15)$$

In the context of Eqs. (11)–(15) and the related comments given above, the continuity and momentum equations that govern the fluid flow in the temporally and spatially periodic regime can be expressed as follows:

$$\frac{\partial}{\partial x}(\rho u) + \frac{\partial}{\partial y}(\rho v) = 0 \tag{16}$$

$$\frac{\partial}{\partial t}(\rho u) + \frac{\partial}{\partial x}(\rho u u) + \frac{\partial}{\partial y}(\rho v u) = \beta - \frac{\partial P}{\partial x} + \frac{\partial}{\partial x}\left(\mu \frac{\partial u}{\partial x}\right) + \frac{\partial}{\partial y}\left(\mu \frac{\partial u}{\partial y}\right) \tag{17}$$

$$\frac{\partial}{\partial t}(\rho v) + \frac{\partial}{\partial x}(\rho u v) + \frac{\partial}{\partial y}(\rho v v) = -\frac{\partial P}{\partial y} + \frac{\partial}{\partial x}\left(\mu \frac{\partial v}{\partial x}\right) + \frac{\partial}{\partial y}\left(\mu \frac{\partial v}{\partial y}\right) \tag{18}$$

The boundary conditions on the instantaneous velocity components are the following: $u = v = 0$ on all surfaces of the plates, and the spatial periodicity conditions, namely, Eqs. (11) and (12), apply on the open (fluid) portions of the boundaries of the calculation domain (see Fig. 2) in the x- and y-directions, respectively. The reduced pressure, p, needs to be fixed at only one point in the fluid region: this is achieved by setting both p_{ref} and P to zero (an arbitrary convenient value) at a node located on the inlet plane ($x = 0$) of the calculation domain, and spatial periodicity conditions given in Eq. (15) are applied to P on the x- and y-boundaries of the calculation domain.

With regard to the term β in Eqs. (13), (14), and (17), for any particular computational run, it can be specified as either an adjustable or a prescribed (fixed value) parameter. In the first approach, for each case, the computations are started with a suitable initial value of β, the instantaneous mass flow rate is monitored and time-averaged over a temporal oscillation period, and then, as the computations are continued, the value of β is adjusted iteratively until the desired time-mean overall mass flow rate (or Reynolds number) is obtained to within a desired tolerance. In the second approach, an appropriate value of β is chosen and maintained constant for each case, and the corresponding implied time-mean overall mass flow rate (or Reynolds number) is computed. Unless otherwise specified, the first approach was adopted in this work. In a few cases, the fluid flow was unsteady and essentially chaotic, rather than temporally periodic. In these cases, the aforementioned iterative adjustment of β to achieve a desired value of the Reynolds number was either computationally inefficient (required a very large number of monitored time steps to adjust β and obtain the desired time-mean mass flow rate) or unsuccessful (did not converge to the desired tolerance). In these cases, which are clearly identified in the results section, the second approach (prescribed constant value of β) was used.

Governing Equations and Boundary Conditions for the Temperature Field

The mathematical models for the two thermal boundary conditions considered in this work, [H] and [T], in that order, are presented in this subsection.

Cases with the [H] thermal boundary condition: Expanding on the steady-state formulations put forward in Patankar et al. [24] and Patankar and Prakash [43], the instantaneous temperature in the temporally and spatially periodic region can be expressed as follows:

$$T(x, y, t) - T_{\text{ref}} = \gamma x + \tilde{T}(x, y, t) - \tilde{T}_{\text{ref}} \tag{19}$$

Here, T_{ref} and $\tilde{T}_{\text{ref}}$ are convenient reference values of T and $\tilde{T}$, respectively, at any suitable chosen point in the calculation domain. As the thermophysical properties of the fluid are assumed to be constant, the absolute values of T_{ref} and $\tilde{T}_{\text{ref}}$ are inconsequential. The term γ is given by the following relation:

$$\gamma = \frac{(n_x n_y q_{\text{module}})/\{n_x(2L)\}}{n_y <\dot{m}_{\text{module}}> c_p} = \frac{q_{\text{module}}/(2L)}{<\dot{m}_{\text{module}}> c_p} = \frac{\{2(2L + 4\delta)\}q_w''/(2L)}{<\dot{m}_{\text{module}}> c_p} \tag{20}$$

In this equation, q_{module} is the rate of heat addition per unit depth in each geometrical module, $<\dot{m}_{\text{module}}>$ is the time-mean overall mass flow rate per unit depth in each module, and q_w'' is the prescribed uniform heat flux from the surfaces of the plates to the fluid. The term $\tilde{T}(x, y, t)$ in Eq. (19) behaves in the following spatially periodic manner:

$$\tilde{T}(x, y, t) = \tilde{T}(x + n_x(2L), y, t) = \tilde{T}(x + 2n_x(2L), y, t) = \cdots$$
$$\tilde{T}(x, y, t) = \tilde{T}(x, y + n_y(2H), t) = \tilde{T}(x, y + 2n_y(2H), t) = \cdots \tag{21}$$

With the abovementioned considerations, the energy equation that governs the instantaneous temperature distribution in the fluid can be cast in the following form:

$$\frac{\partial}{\partial t}(\rho\tilde{T}) + \frac{\partial}{\partial x}(\rho u\tilde{T}) + \frac{\partial}{\partial y}(\rho v\tilde{T}) = \frac{\partial}{\partial x}\left(\frac{k}{c_p}\frac{\partial\tilde{T}}{\partial x}\right) + \frac{\partial}{\partial y}\left(\frac{k}{c_p}\frac{\partial\tilde{T}}{\partial y}\right) - \rho u\gamma \tag{22}$$

For cases with the $[H]$ thermal boundary condition, at each point on the plate–fluid interfaces, the following relation applies to the instantaneous temperature field:

$$\left(\partial T/\partial\xi_{\text{surf}} = \gamma\,\partial x/\partial\xi_{\text{surf}} + \partial\tilde{T}/\partial\xi_{\text{surf}} = -q_w''/k\right)_{\text{plate–fluid interface}}^{|H|} \tag{23}$$

Here, ξ_{surf} is the coordinate along the local unit normal to the surface of the plates, pointing into the fluid. Both T_{ref} and $\tilde{T}_{\text{ref}}$ are set to the arbitrary value of zero at a convenient point on the inlet plane ($x = 0$) of the calculation domain. Spatial periodicity conditions, as given by the expressions in Eq. (21), are applied to $\tilde{T}$ on the open (fluid) portions of the x- and y-boundaries of the calculation domain (see Fig. 2).

Cases with the $[T]$ thermal boundary condition: For these cases, the spatially periodic region is characterized using a dimensionless instantaneous temperature field, $\theta^{|T|}$:

$$\theta^{|T|}(x, y, t) = \frac{T_w - T(x, y, t)}{T_w - T_b(x, t)} \tag{24}$$

In this equation, T_b is the instantaneous bulk temperature defined as follows:

$$T_b(x, t) = \frac{\int_{c\text{-}s} \rho c_p u(x, y, t) T(x, y, t) \mathrm{d}y}{\int_{c\text{-}s} \rho c_p u(x, y, t) \mathrm{d}y} \tag{25}$$

This is the classical definition of the bulk temperature. In many published investigations of flows in ducts with spatially periodic interrupted surfaces, for example, Patankar et al. [24], Fullerton and Anand [32, 33], Patankar and Prakash [52], Zhang et al. [52], and Korichi et al. [63], a cross-sectional reference temperature is defined using an equation akin to Eq. (25), but using $|u(x, y, t)|$ in place of $u(x, y, t)$, and a statement saying that the absolute value is necessary to account for negative values of $u(x, y, t)$ in recirculating-flow regions. However, the bulk temperature defined in Eq. (25) is preferred and recommended in this work as the cross-sectional reference temperature for the following reason: it can be directly used in, or deduced from, local and overall energy balances for the cases of interest, even though the fluid flows contain separated regions and recirculation zones, as there are net overall rates of fluid flow and advection of enthalpy in the mainstream direction at all x locations in the calculation domain. For discussions of various proposals for the reference temperature (and the temperature difference) in the definitions of the local and average heat transfer coefficients, and their implications, the reader is referred to the works of Beale and Spalding [28, 29].

The instantaneous dimensionless temperature field, $\theta^{[T]}$, satisfies the following spatial periodicity relations:

$$\theta^{[T]}(x, y, t) = \theta^{[T]}(x + n_x(2L), y, t) = \theta^{[T]}(x + 2n_x(2L), y, t) = \cdots$$
$$\theta^{[T]}(x, y, t) = \theta^{[T]}(x, y + n_y(2H), t) = \theta^{[T]}(x, y + 2n_y(2H), t) = \cdots \tag{26}$$

The formulation of the governing energy equation in terms of $\theta^{[T]}$ involves the x-direction gradient of the instantaneous bulk temperature, which is not known a priori, and has to be obtained at each time by solving an eigenvalue problem. This eigenvalue problem is quite complicated even for steady-state situations, as described, for example, in the works of Patankar et al. [24] and Kelkar and Patankar [47]. For the temporally and spatially periodic cases of interest here, such an eigenvalue formulation gets even more complicated and also computationally expensive to solve. However, this complicated eigenvalue formulation can be circumvented by solving a developing heat transfer problem in the presence of fully developed fluid flow (previously calculated and stored), formulated directly in terms of the instantaneous temperature field, $T(x, y, t)$. This approach was first proposed by Fullerton and Anand [33], who used it for solving steady-state problems. In the work reported in this chapter, the [T] thermal boundary condition was considered only for the temporally periodic, simultaneously developing, fluid flow and heat transfer problem, described earlier in Sect. 5.1.

5.3 *Dimensionless Formulation*

The dimensionless variables and parameters that were used in processing the results of this investigation are presented in this subsection.

Independent Variables and Parameters

$$X = \frac{x}{H} \; ; \; Y = \frac{y}{H} \; ; \; t^* = \frac{t}{H^2/(\mu/\rho)}$$

$$\delta^* = \frac{2\delta}{2H} = \frac{\delta}{H} \; ; \; L^* = \frac{L}{2H} \; ; \; \beta^* = \frac{\beta H}{\rho\{(\mu/\rho)/H\}^2} \tag{27}$$

Dependent Variables

$$u^* = \frac{u}{(\mu/\rho)/H} \; ; \; v^* = \frac{v}{(\mu/\rho)/H} \; ; \; p^* = \frac{(p - p_{\text{ref}})}{\rho\{(\mu/\rho)/H\}^2}$$

$$\theta^{[T]}_{\text{developing}} = \frac{(T - T_{\text{in}})}{(T_w - T_{\text{in}})} \; ; \; \theta^{[H]} = \frac{(T - T_{\text{ref}})}{\gamma H} \; ; \; \theta^{[T]} = \frac{(T_w - T)}{(T_w - T_b)} \tag{28}$$

Fluid Flow Results

The fluid flow results are presented later in this chapter in terms of the dimensionless stream function, time-mean Reynolds number (Re_{D_h}), time-mean modular friction factor (f), and Strouhal number (S), defined as follows:

$$\psi^* = \frac{\psi}{(\mu/\rho)} \; ; \; Re_{D_h} = \frac{\rho <U_{c\text{-}s\,\min}> D_h}{\mu}$$

$$f = \frac{\beta D_h}{\frac{1}{2}\rho(<U_{c\text{-}s\,\min}>)^2} \; ; \; S = \frac{\omega_v(2\delta)}{<U_{c\text{-}s\,\min}>} \tag{29}$$

In Eq. (29), ψ is the stream function ($\partial\psi/\partial y \triangleq u; \; \partial\psi/\partial x \triangleq -v$), D_h is the hydraulic diameter, $<U_{c\text{-}s\,\min}>$ is the time-mean value of the average mainstream (x) component of the fluid velocity in the minimum cross-sectional area, and ω_v is the frequency of the self-sustained oscillations (when they occur) of the cross-stream velocity component, v, at a fixed location along the center line of a plate, at a distance 2δ behind its trailing edge (denoted by the symbol **x** in Fig. 2). In cases where temporally periodic self-sustained oscillations were obtained, the frequency ω_v was calculated as the inverse of the time period of the oscillations. In cases where the self-sustained oscillations were characterized by an effectively chaotic behavior, a discrete Fourier transform algorithm was used with the monitored velocity data to compute its power spectral density, from which the dominant flow oscillation frequency was determined, when it existed. With reference to the notation in Fig. 2, the following definitions apply:

$$D_h = \frac{4A_{c\text{-}s\,\min}}{[A_{\text{wetted}}/(\text{Length})]_{\text{module}}} = 4\frac{(2H - 4\delta)}{2(4\delta + 2L)}(2L)$$

$$A_{c\text{-}s\,\min} = 2H - 4\delta\,;\quad <\dot{U}_{c\text{-}s\,\min}> \,=\, <\dot{m}_{\text{total}}>/\left(\rho n_y A_{c\text{-}s\,\min}\right) \tag{30}$$

Here, $<\dot{m}_{\text{total}}>$ is the time-mean mass flow rate in the mainstream (x) direction and $A_{c\text{-}s\,\min}$ is the minimum cross-sectional area in a geometric module, both per unit depth.

In the developing fluid flow cases considered in this work, $U_{c\text{-}s\,\min}$ was prescribed and maintained constant (invariant with time, thus $<U_{c\text{-}s\,\min}> \,=\, U_{c\text{-}s\,\min}$ and $<\dot{m}_{\text{total}}> \,=\, \dot{m}_{\text{total}}$). Furthermore, in these cases, for each geometric module, β was calculated by monitoring the values of the reduced pressure p at two corresponding spatially periodic points, one on its inlet (upstream) plane and the other on its outlet (downstream) plane, computing the time-mean values of these two monitored reduced pressures, and then using them in Eq. (14), with $n_x = 1$.

Heat Transfer Results for Cases with the [T] Thermal Boundary Condition
These results are presented in terms of a time-mean modular Colburn factor $(j^{[T]})$, defined as follows:

$$j^{[T]} = St^{[T]}Pr^{2/3} = \left(\frac{\langle \bar{h}^{[T]} \rangle}{\rho c_p U_{c\text{-}s\,\min}}\right) Pr^{2/3} \tag{31}$$

In Eq. (31), $St^{[T]}$ is the time-mean modular Stanton number for these cases and Pr is the Prandtl number of the fluid. The time-mean modular heat transfer coefficient, $\langle \bar{h}^{[T]} \rangle$, is defined using the time-mean modular rate of heat transfer per unit depth from the plates to the fluid, $\langle q_{\text{module}} \rangle^{[T]}$, the corresponding modular interfacial area between the plate surfaces and the fluid, A_{wetted}, and a log-mean temperature difference, LMTD:

$$\langle \bar{h}^{[T]} \rangle = \frac{\langle q_{\text{module}} \rangle^{[T]}}{A_{\text{wetted}}\text{LMTD}}$$

$$\langle q_{\text{module}} \rangle^{[T]} = \dot{m}_{\text{module}} c_p (\langle T_{b\,\text{out}} \rangle - \langle T_{b\,\text{in}} \rangle);\quad A_{\text{wetted}} = 2(4\delta + 2L)$$

$$\text{LMTD} = \frac{(T_w - \langle T_{b\,\text{in}} \rangle) - (T_w - \langle T_{b\,\text{out}} \rangle)}{\ln\left(\frac{(T_w - \langle T_{b\,\text{in}} \rangle)}{(T_w - \langle T_{b\,\text{out}} \rangle)}\right)} \tag{32}$$

In Eq. (32), $\langle T_{b\,\text{in}} \rangle$ and $\langle T_{b\,\text{out}} \rangle$ denote the time-mean values of the bulk temperature at the inlet (upstream) and outlet (downstream) planes of a module, respectively, and $\dot{m}_{\text{module}}$ is the mass flow rate per unit depth in each module. For each of the developing fluid flow and heat transfer cases (the $[T]$ thermal boundary condition was used only in these cases), $<U_{c\text{-}s\,\min}> \,=\, U_{c\text{-}s\,\min} = \text{constant}$, thus $<\dot{m}_{\text{module}}> \,=\, \dot{m}_{\text{module}} = \text{constant}$. For these cases, when self-sustained fluid flow oscillations were encountered, the instantaneous values of the bulk temperatures were stored at each time step, and their final time-mean values were computed after the fluid flow and thermal fields became temporally periodic. For each of these cases, $j^{[T]}$ becomes a constant in the temporally and spatially periodic region.

Heat Transfer Results for Cases with the [H] Thermal Boundary Condition
Computer simulations for these cases were done using the formulation described in
Sect. 5.2, only for the spatially periodic fully developed regions. The heat transfer
results for these cases are presented in terms of a time-mean average Colburn factor
($j^{[H]}$), defined for each plate as follows:

$$\bar{h}^{[H]} \triangleq \frac{q''_w}{\{\int_{\text{perimeter plate}} (T_{w,\text{local}} - T_b)\mathrm{d}s\}/(2L + 4\delta)}$$

$$j^{[H]} = St^{[H]} \, Pr^{2/3} = \left[\frac{\langle \bar{h}^{[H]} \rangle}{\rho c_p <U_{c\text{-}s \, \min}>} \right] Pr^{2/3} \qquad (33)$$

In this equation, s is a coordinate along the perimeter of the plate under consideration; $T_{w,\text{local}}$ is the instantaneous local surface temperature of this plate; $\bar{h}^{[H]}$ is the
instantaneous perimeter-averaged value of the local heat transfer coefficient on the
surface of this plate; $<\bar{h}^{[H]}>$ is the time-mean value of $\bar{h}^{[H]}$; and $St^{[H]}$ is the time-
mean perimeter-averaged Stanton number for the plate. For each of these cases, $j^{[H]}$
becomes a constant in the temporally and spatially periodic region. It should also
be noted that the perimeter-average over a single plate in this equation applies when
there are no flow maldistributions caused by Coanda-like effects [73, 74]; otherwise,
the perimeter-average should be done over a suitable number of additional plates.
For discussions of several alternative definitions of local and average heat transfer
coefficients and their implications for the [H] thermal boundary condition, the reader
is referred to the works of Beale and Spalding [28, 29].

5.4 Notes on Initial Conditions

Any set of convenient initial conditions could be used with the problems of interest.
Since attention in the work reported in this chapter was focused mainly on steady and
temporally periodic unsteady fluid flow and heat transfer regimes (and the transient
regimes were not explicitly studied), the initial conditions did not affect the final
results.

6 Synopsis of the Numerical Method

A finite volume method (FVM) based on a co-located primitive-variables formulation
was used to solve the discretized equations. The formulation of this method is based
on those of well-established FVMs described thoroughly in the works of Patankar
[9], Ferziger and Perić [75], and Baliga and Atabaki [76]. Thus, only a very concise
description of this FVM, as applied to the problems of interest, and a note on its
verification are presented in this section.

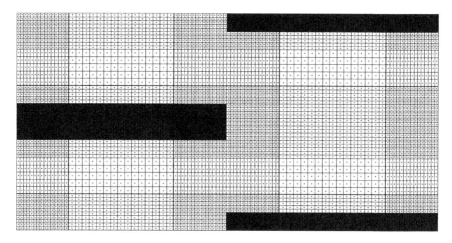

Fig. 3 Schematic representation of the control-volume distribution pattern that was used to discretize each geometric module in the calculation domain

6.1 Domain Discretization

The calculation domains were discretized using the Type-B spatial grid layout of Patankar [9]. Within each geometric module, rectangular control volumes were distributed in a manner akin to the *pattern* illustrated schematically in Fig. 3. Details of this control-volume distribution pattern are given in Sect. 7.1, in the context of grid-independence checks performed prior to the final simulations. As was mentioned earlier, for the cases involving developing fluid flow and heat transfer, the calculation domain consisted of a single row of 10 adjacent geometric modules followed by an open (or plate-less) exit zone in the mainstream direction, as shown schematically in Fig. 1. The total extent of this exit zone in the mainstream (x) direction was set equal to or exceeded ($9L$) in the final computations, as preliminary computations had established that this extent was sufficiently long to ensure that there were no flow reversals at the outlet plane of the calculation domain (so the outflow boundary condition could be imposed there). This exit zone was also discretized using the Type-B grid layout of Patankar [9]. The details of the control-volume distribution pattern in the exit section are given in Sect. 7.3, along with the presentation of results obtained from simulations of developing fluid flow and heat transfer.

6.2 Discretization of the Governing Equations

The governing partial differential equations were first integrated over control volumes associated with each node in the interior of the calculation domain. Algebraic approximations to the resulting integral conservations equations were then derived. These algebraic approximations are referred to as the discretized equations [9]. The algebraic approximations to the advection fluxes in the integral conservation equations were derived using the quadratic upstream interpolation for convective kinematics (QUICK) scheme of Leonard [77]. Piecewise linear interpolation of the dependent variables was used in the derivation of algebraic approximations of the viscous and conduction fluxes. At the boundaries of the calculation domain, however, quadratic interpolation functions were used in order to obtain second-order accuracy in the discretized approximations of the viscous and conductive fluxes there [76]. To avoid pressure checkerboarding in the co-located primitive-variables formulation of this FVM, the momentum interpolation scheme was used to interpolate the velocity components that appear in the mass flux terms [9, 76, 78]. The discretization in time was done using the Crank–Nicolson scheme [9]. The aforementioned procedures ensured second-order accuracy of the FVM.

6.3 Solution of the Discretized Equations

At each time step, a sequential iterative variable adjustment (SIVA) procedure, based on ideas similar to the SIMPLER procedure of Patankar [9], but without the pressure correction step, was used to solve the sets of nonlinear coupled discretized equations [76, 79]. In each overall iteration of this SIVA procedure, linearized and decoupled sets of discretized equations for the velocity components, pressure, and temperature are solved sequentially. An iterative alternating-direction line Gauss–Seidel method, based on either a pentadiagonal matrix algorithm (PDMA) or a circular PDMA [80], was employed for solving the linearized and decoupled sets of discretized equations for the velocity components (u, v) and temperature (T). The discretized equations for the pressure fields $(p$ and $P)$ were solved using a conjugate gradient method with a Jacobi pre-conditioner [81]. At each time step, overall convergence of the SIVA scheme was assumed to be achieved when suitably normalized absolute residues of the discretized equations were all less than 10^{-6}.

6.4 Verification of the Numerical Method

The numerical method used in this work was verified using well-established techniques, such as those discussed, for example, by Patankar [9], Richardson [82], Cotta [83], Roache [84], Cotta and Mikhailov [85], Oberkamf and Roy [86], and Baliga and Lokhmanets [87]. Three of the test problems that were used in these verification tasks, and which are of particular relevance to this investigation, are the following: (1) steady, two-dimensional, spatially periodic fully developed laminar fluid flow and heat transfer in a uniform array of staggered rectangular plates, with checks against the benchmark results of Sebben and Baliga [67]; (2) a stationary square cylinder in a laminar uniform cross flow, with checks against the results of the numerical study of Franke et al. [88] and the experimental results of Davis and Moore [89]; and (3) temporally and spatially periodic fully developed fluid flow and heat transfer in a *single* representative geometric module of a uniform array of staggered rectangular plates, with checks against the corresponding numerical results of Zhang et al. [52]. Additional details of the first two of these three test problems and the results are available in the work of Lamoureux [90], and those of the third test problem are available in the work of Lamoureux and Baliga [60].

With the satisfactory results of the abovementioned test problems at hand, the FVM described in this section was considered verified and suitable for the final computer simulations undertaken in this work.

7 Results and Discussions

The presentations and discussions in this section are organized in the following sequence: (1) a synopsis of grid- and time-step-independence checks, and the procedure that was used for determining the establishment of the temporally periodic, self-sustained, oscillatory regime (when it existed); (2) results obtained from computer simulations of temporally and spatially periodic fluid flow and heat transfer in calculation domains akin to that shown schematically in Fig. 2; and (3) results pertaining to temporally periodic developing fluid flow and heat transfer in calculation domains akin to that shown schematically in Fig. 1.

7.1 Grid- and Time-Step-Independence Checks and Procedure for Determining the Establishment of the Temporally Periodic, Self-sustained, Oscillatory Regime

For the dimensionless plate thickness values of $\delta^* = 1/16$, $1/12$, and $1/8$, the grid-independence checks were done for temporally and spatially periodic fluid flow with two extreme values of Reynolds number (Re_{D_h}), 100 and 1000, and the corresponding heat transfer cases with the $[H]$ thermal boundary condition. For $\delta^* = 1/4$, these grid checks were done only for $Re_{D_h} = 100$, as the solutions were found to be effectively chaotic for $Re_{D_h} = 1000$. Only one representative geometric module ($n_x = 1$ and $n_y = 1$) was considered in all of these grid checks. In each case, the representative geometric module was discretized with rectangular control volumes, distributed according to the *pattern* illustrated schematically in Fig. 3 and the following details: in the x-direction, the control-volume width was set to Δx in the vicinity of the plate edges (in regions of extent $\pm L/4$ about the plate front and rear edges) and $2\Delta x$ elsewhere; in the y-direction, the control-volume height was set to Δy in the vicinity of the plates (in regions of extent $\pm\delta$ about the plate horizontal surfaces) and $2\Delta y$ elsewhere; and $\Delta x = \Delta y$.

An extension of the extrapolation technique proposed by Richardson [82, 87] was used in the grid-independence checks: for a fixed value of the time step, the results obtained with successively finer sets of spatial grids, each having the same pattern (see Fig. 3), were extrapolated (assuming second-order accuracy of the numerical method) to values representative of infinitely fine grids ($\Delta x = \Delta y = 0$), and these extrapolated values were considered essentially grid-independent when they were invariant to within a prescribed tolerance.

The values of the time-mean modular friction and Colburn factors, f and $j^{[H]}$, defined in Eqs. (29) and (33), respectively, were used in these grid checks. The dimensionless time step was set equal to $\Delta t^* = \Delta[t(\mu/\rho)/H^2] = 2.9295 \times 10^{-6}$: preliminary computations had shown that this value of Δt^* ensures a minimum of 1500 time steps per time period, in all cases that resulted in temporally periodic oscillatory fluid flow and heat transfer. The results are summarized in Table 1. For each case, the extrapolated results in this table were obtained using the results yielded by the first (least fine) and third (finest) grids. The absolute relative differences between these extrapolated results and the corresponding extrapolated results yielded by the first and second grids (in each case) were all less than 4.3×10^{-4}. In this sense, the reported extrapolated results reported in Table 1 are considered to be effectively grid-independent.

Based on the abovementioned grid checks, the final chosen spatial discretizations of each geometric module were the following: for $\delta^* = 1/4$, 120×80 control volumes ($\Delta x/H = \Delta y/H = 2.500 \times 10^{-2}$); for $\delta^* = 1/8$, 168×84 control volumes ($\Delta x/H = \Delta y/H = 1.786 \times 10^{-2}$); for $\delta^* = 1/12$, 180×80 control volumes ($\Delta x/H = \Delta y/H = 1.667 \times 10^{-2}$); and for $\delta^* = 1/16$, 192×80 control volumes ($\Delta x/H = \Delta y/H = 1.563 \times 10^{-2}$). The absolute relative differences between the results yielded by these chosen grids and the corresponding extrapolated, effectively grid-independent results reported in Table 1 were all less than 2.2% for f and 0.70% for $j^{[H]}$.

In cases involving developing fluid flow and heat transfer, the calculation domain consisted of a single row of 10 adjacent geometric modules followed by an open (or plate-less) exit zone of extent equal to or exceeding $9L$ in the mainstream direction, as shown schematically in Fig. 1. The final chosen discretizations of each geometric module for these developing fluid flow and heat transfer cases were the same as those mentioned in the previous paragraph. The discretizations of the exit zone of the calculation domains had the following control-volume distribution pattern: in the x-direction, the control-volume width was set equal to $2\Delta x$ immediately adjacent to the interface with the geometric module to the left (see Fig. 1), expanded gradually (using a geometric expansion factor) to $4\Delta x$, and then maintained constant at this value; and in the y-direction, the control-volume height distribution was identical to that used in the geometric modules (discussed above).

The time-step-independence checks were performed with the value of Reynolds number set equal to 1000 (the most challenging of the cases considered in this work). Dimensionless plate thickness values of $\delta^* = 1/8$, $1/12$, and $1/16$ were considered

Table 1 Grid checks for $L^* = 1$ and $\delta^* = 1/4$, 1/8, 1/12, and 1/16

δ^*	Re_{D_h}	Grid	#CVs in 2δ	$\Delta x/H = \Delta y/H$	f	$j^{[H]}$
1/4	100	120×80	20	2.500×10^{-2}	0.93287	0.10036
		300×200	50	1.000×10^{-2}	0.94709	0.10014
		480×320	80	6.250×10^{-3}	0.95104	0.10010
	Extrapolated results				**0.95357**	**0.10007**
1/8	100	168×84	14	1.786×10^{-2}	1.45964	0.14602
		384×192	32	7.813×10^{-3}	1.46998	0.14597
		600×300	50	5.000×10^{-3}	1.47255	0.14595
	Extrapolated results				**1.47433**	**0.14593**
	1000	168×84	14	1.786×10^{-2}	0.25441	0.02609
		240×120	20	1.250×10^{-2}	0.25340	0.02618
		336×168	28	8.929×10^{-3}	0.25292	0.02620
	Extrapolated results				**0.25243**	**0.02622**
1/12	100	180×80	10	1.667×10^{-2}	1.65237	0.16096
		288×128	16	1.042×10^{-2}	1.65896	0.16088
		432×192	24	6.944×10^{-3}	1.66222	0.16084
	Extrapolated results				**1.66482**	**0.16081**
	1000	180×80	10	1.667×10^{-2}	0.26381	0.02707
		504×224	28	5.952×10^{-3}	0.26442	0.02715
		648×288	36	4.630×10^{-3}	0.26460	0.02715
	Extrapolated results				**0.26487**	**0.02717**
1/16	100	192×80	8	1.563×10^{-2}	1.74990	0.16830
		288×120	12	1.042×10^{-2}	1.75540	0.16821
		384×160	16	7.813×10^{-3}	1.75790	0.16818
	Extrapolated results				**1.76111**	**0.16813**
	1000	192×80	8	1.563×10^{-2}	0.26071	0.02745
		288×120	12	1.042×10^{-2}	0.26171	0.02755
		384×160	16	7.813×10^{-3}	0.26217	0.02759
	Extrapolated results				**0.26276**	**0.02764**

in these time-step checks. As was stated earlier, at a Reynolds number of 1000, the fluid flow and heat transfer solutions were effectively chaotic for $\delta^* = 1/4$, so these cases were excluded from the time-step checks. For each of the three cases considered, the spatial grids were the same as those chosen for the final simulations (discussed earlier in this subsection). The values of the time-mean modular friction and Colburn factors, f and $j^{[H]}$, defined in Eqs. (29) and (33), respectively, were used in these time-step checks.

Three different dimensionless time steps were considered in the time-step checks: $\Delta t^* = \Delta[t(\mu/\rho)/H^2] = 2.9295 \times 10^{-6}$, $\Delta t^*/2$, and $\Delta t^*/4$. Again, preliminary

Table 2 Time-step checks: $\delta^* = 1/8$, 1/12, and 1/16; $L^* = 1$; and $\Delta t^* = 2.9295 \times 10^{-6}$

δ^*	Re_{D_h}	Grid	Time step	f	$j^{[H]}$
1/8	1000	168 × 84	Δt^*	0.25441	0.02609
			$\Delta t^*/2$	0.25476	0.02608
			$\Delta t^*/4$	0.25507	0.02608
	Extrapolated results			**0.25511**	**0.02608**
1/12	1000	180 × 80	Δt^*	0.26381	0.02707
			$\Delta t^*/2$	0.26394	0.02707
			$\Delta t^*/4$	0.26406	0.02707
	Extrapolated results			**0.26408**	**0.02707**
1/16	1000	192 × 80	Δt^*	0.26071	0.02745
			$\Delta t^*/2$	0.26077	0.02744
			$\Delta t^*/4$	0.26081	0.02744
	Extrapolated results			**0.26081**	**0.02744**

computations had shown that this value of Δt^* ensures a minimum of 1500 time steps per time period, in all cases that resulted in temporally periodic oscillatory fluid flow and heat transfer.

The results of the time-step checks are summarized in Table 2. For each case, the extrapolated results in this table were obtained using the results yielded by the first and third time steps, Δt^* and $\Delta t^*/4$, respectively. The absolute relative differences between these extrapolated results and the corresponding extrapolated results yielded by the first and second time steps (in each case) were all less than 2.3×10^{-4}; and in this sense, the reported extrapolated results are considered to be essentially time-step-independent. Based on these checks, the time step for the final computations was chosen as $\Delta t^* = \Delta[t(\mu/\rho)/H^2] = 2.9295 \times 10^{-6}$. With this chosen time step, the absolute relative differences between the computed results and the corresponding extrapolated time-step-independent results, as reported in Table 2, were all less than 10^{-4}.

The establishment of the temporally periodic, self-sustained, oscillatory regime was determined by monitoring the cross-stream (y) velocity component (v) at a fixed location along the centerline of a plate at a distance 2δ behind its trailing edge, shown by the mark of "**x**" in Figs. 1 and 2. The temporally periodic regime was assumed to be established when the following conditions were met at the aforementioned location in the calculation domain for each case: (1) the oscillations in the value of v displayed an amplitude that was invariant to within $\pm 0.1\%$ of the mean value between consecutive periodic cycles, (2) the calculated values of f and $j^{[H]}$ varied by less than 0.0001% between two successive cycles, and (3) the difference between the tentative (target) Reynolds number and the time-mean calculated value differed by less than 0.0001%. Since the mass flow rate was maintained at a fixed value for each of the developing fluid flow and heat transfer cases considered in this work, the aforementioned third condition only pertained to simulations undertaken in the spatially periodic fully developed region. For each case, after the establishment of the temporally periodic, self-sustained, oscillatory regime (if it occurred), the simulations were continued for at least 10 additional time periods (cycles), and the time-averaged results were all based on the corresponding instantaneous values calculated at each time step in the aforementioned 10 additional time periods. Even for cases in which no clear temporally periodic behavior was observed, the time-averaged results were recorded when they became essentially invariant over large sampling periods.

7.2   Temporally and Spatially Periodic Fluid Flow and Forced Convection Heat Transfer

The results obtained from computer simulations of fluid flow and forced convection heat transfer in the temporally and spatially periodic regimes are presented and discussed in this subsection. For each of the four combinations of geometric parameters considered ($\delta^* = 1/4$, 1/8, 1/12, and 1/16, each with $L^* = 1$), the following four combinations of integer periodicity indices were investigated: $n_x \times n_y = 1 \times 1, 2 \times 1, 3 \times 1$, and 1×2.

Fluid Flow Simulations with Single and Multiple Geometric Modules
Overview. For each of the four geometries considered and the aforementioned combinations of integer periodicity indices, the fluid flow remained steady (that is, even after significant perturbations, the fluid flow reverted to the steady state and no self-sustained oscillations were established) for Reynolds number, Re_{D_h}, below a critical value. In this work, an attempt was not made to determine these critical values of Re_{D_h} precisely, but it was established that the fluid flow remained steady for the following conditions: for $\delta^* = 1/4$, $Re_{D_h} \leq 100$; for $\delta^* = 1/8$, $Re_{D_h} \leq 400$; for $\delta^* = 1/12$, $Re_{D_h} \leq 400$; and for $\delta^* = 1/16$, $Re_{D_h} \leq 600$. Steady fluid flow was also obtained for $\delta^* = 1/4$, $Re_{D_h} = 200$, and $n_x \times n_y = 2 \times 1$ and 3×1. Self-sustained time-periodic oscillations were obtained for the following conditions: for $\delta^* = 1/4$, $200 \leq Re_{D_h} \leq 400$; for $\delta^* = 1/8$, $600 \leq Re_{D_h} \leq 1000$ ($n_x \times n_y = 1 \times 1$ and 1×2); for $\delta^* = 1/12$, $600 \leq Re_{D_h} \leq 1000$; and for $\delta^* = 1/16$, $800 \leq Re_{D_h} \leq 1000$. Unsteady, non-periodic, effectively chaotic fluid flows, sometimes with Coanda-like effects [73, 74], were obtained for the following conditions: for $\delta^* = 1/4$, $580 \leq Re_{D_h} \leq 1000$; and for $\delta^* = 1/8$, $Re_{D_h} = 974.7$ and $n_x \times n_y = 2 \times 1$, and $Re_{D_h} = 973.9$ and $n_x \times n_y = 3 \times 1$. Some sample results are presented in Figs. 4 and 5.

The results in Fig. 4 reveal that flow recirculation zones form in the vicinity of the leading edges of the plates and cascade down their horizontal surfaces; and behind the plates, vortices are shed and the wakes are unsteady. The plots of the time evolution of the monitored velocity also demonstrate temporal periodicity of the flow.

The results in Fig. 5 are provided as an example of unsteady, non-periodic, essentially chaotic fluid flow. In this case, flow recirculation zones on the horizontal plate

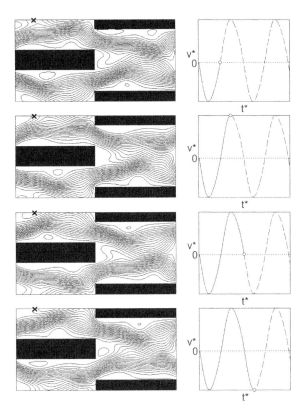

Fig. 4 Instantaneous streamline plots, sampled at different times during a temporally periodic cycle, and plots of the time evolution of the corresponding monitored velocity (at the location indicated by "**x**"), for $\delta^* = 1/4$, $n_x \times n_y = 1 \times 1$, and $Re_{D_h} = 400$

surfaces tend to agglomerate and build up, then get progressively flattened (to some extent), and are finally shed off the trailing edge, in a non-periodic manner.

In some cases with unsteady, non-periodic, essentially chaotic flows, the instantaneous streamline plots revealed relatively large flow recirculation zones that cascaded downstream along the horizontal plate surfaces only in one-half of the calculation domain, while the flow in the other half remained relatively straight. Furthermore, in these cases, the aforementioned flow configuration was found to be a Coanda-like effect [73, 74], in the sense that the different flow streams in the two halves of the calculation domain did not spontaneously switch positions, but an inversion of the flow configuration could be achieved by applying a suitable perturbation. As an example of this type of flow, instantaneous streamline plots sampled at different times, along with plots of the time evolution of the corresponding monitored velocity (v^* vs. t^*), for $\delta^* = 1/4$, $n_x \times n_y = 1 \times 1$, and $Re_{D_h} = 589.3$ are shown in Fig. 6.

Multiple Time-Periodic Solutions. When self-sustained time-periodic oscillations were established with $n_x > 1$ and $n_y = 1$, multiple solutions were obtained for each value of Reynolds number in the abovementioned ranges. In these cases, for each combination of $n_x > 1$ and $n_y = 1$, the possible solutions included one in which the temporally periodic fluid flow field in any one module differed in phase from that

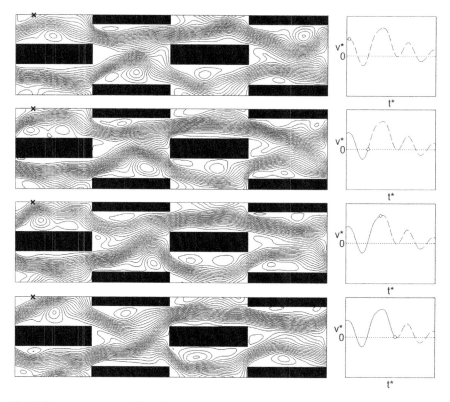

Fig. 5 Instantaneous streamline plots, sampled at different times, and plots of the time evolution of the corresponding monitored velocity (at the location indicated by "**x**"), for $\delta^* = 1/4$, $n_x \times n_y = 2 \times 1$, and $Re_{D_h} = 593.4$

in every other module in the calculation domain: such solutions, which were unique for each combination of $n_x > 1$, $n_y = 1$, and Re_{D_h}, are referred to here as *baseline* results. With $n_x \times n_y = 2 \times 1$ and also $n_x \times n_y = 3 \times 1$, two solutions were obtained for each target Re_{D_h} when time-periodic oscillations were present: one was the baseline result for this case; in the other, the solution corresponding to $n_x \times n_y = 1 \times 1$ was repeated in each of the modules. An example of such multiple time-periodic solutions for $\delta^* = 1/12$, $Re_{D_h} = 1000$, and $n_x \times n_y = 3 \times 1$ is shown in Fig. 7. Instantaneous streamline plots are provided in this figure for both the baseline result and the second, duplicated single-module, solution.

In each case for which multiple temporally and spatially periodic solutions were established, the baseline result was usually obtained by starting the simulation with the fluid at rest and suddenly imposing the modular pressure gradient. The other solution (non-baseline result) was established by starting the simulation with the single-module ($n_x \times n_y = 1 \times 1$) instantaneous solution (at an arbitrary time step) as the initial condition in every module: this single-module solution remained in phase in every module and did not revert back to the baseline solution. In each case, the

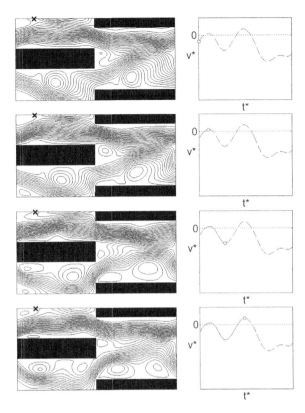

Fig. 6 Instantaneous streamline plots, sampled at different times, and plots of the time evolution of the corresponding monitored velocity (at the location indicated by "**x**"), for $\delta^* = 1/4$, $n_x \times n_y = 1 \times 1$, and $Re_{D_h} = 589.3$

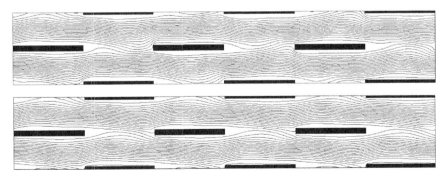

Fig. 7 Instantaneous streamline plots corresponding to two different (multiple) temporally and spatially periodic solutions obtained for $\delta^* = 1/12$, $Re_{D_h} = 1000$, and $n_x \times n_y = 3 \times 1$ (baseline result given in the top plot, and the second, duplicated single-module, solution in the plot below)

baseline and non-baseline solutions remained essentially unchanged (repeated) over at least 20 time periods (in the temporally periodic regime).

In simulations with $n_x \times n_y = 1 \times 2$ for the cases considered, no multiple solutions could be obtained: the time-periodic fluid flow oscillations in each of the two modules (when they were obtained) were the same and identical to those in the corresponding single-module ($n_x \times n_y = 1 \times 1$) solution. Attempts to perturb the flow field and obtain a baseline result were not successful, as the flow field gradually reverted back to the single-module solution.

Effect of Periodicity Indices on Baseline Results. For the cases in which multiple time-periodic solutions were obtained, the single-module and the baseline results were different (unique) for each combination of δ^*, Re_{D_h}, and $n_x \times n_y$. As an example of such results, plots of instantaneous streamlines and variations with time of the monitored dimensionless y-direction velocity components (v^* vs. t^* at the locations indicated by "x") for $\delta^* = 1/8$ and $Re_{D_h} = 800$ and $n_x \times n_y = 1 \times 1, 2 \times 1$, and 3×1 are presented in Fig. 8.

The results presented in Fig. 8 show that for the baseline result obtained with $n_x \times n_y = 2 \times 1$, the monitored values of v^* in the two modules have a phase lag of π radians, and for the baseline result obtained with $n_x \times n_y = 3 \times 1$, the monitored values of v^* in the three successive modules differ in phase by $2\pi/3$ radians. These phase-angle differences are a direct consequence of the imposed spatial periodicity boundary conditions on the instantaneous u, v, and P fields. Furthermore, for this particular case, the time period of the v^* oscillations decreased (thus, the value of the Strouhal number increased) with increasing value of n_x, albeit slightly. In this context, it should be noted that the time period of the v^* oscillations did not necessarily decrease (and the value of the Strouhal number increase) monotonically with increasing value of n_x in all of the various cases studied here.

Fig. 8 Plots of instantaneous streamlines and time evolutions of the corresponding monitored velocities (at the locations indicated by "**x**") for the single-module and two baseline results obtained for $\delta^* = 1/8$ and $Re_{D_h} = 800$ and $n_x \times n_y = 1 \times 1, 2 \times 1$, and 3×1. The scale of t^* is identical in all v^* versus t^* plots

Time-Mean Modular Friction Factor, Modular Colburn Factor, and Strouhal Number

These results for the four values of dimensionless plate thickness, $\delta^* = 1/4$, $1/8$, $1/12$, and $1/16$, each with dimensionless plate length $L^* = 1$, are presented in Tables 3, 4, 5, and 6, respectively.

For cases in which the solutions remained steady, when the simulations were run with multiple modules in the calculation domain, the steady-state flow fields in each of the modules were identical. For these cases, therefore, only the single-module ($n_x \times n_y = 1 \times 1$) results are presented in Tables 3, 4, 5, and 6. For cases in which temporally and spatially periodic oscillatory multiple solutions were obtained, only the single-module and baseline results for multiple modules are presented in Tables 3, 4, 5, and 6, as the other results are identical fits of the single-module solutions into

Table 3 Time-mean single- and multiple-module baseline results for $\delta^* = 1/4$

Re_{D_h}	$n_x \times n_y$	f	$j^{[H]}$	S
100.0	1×1	0.9329	0.1004	Steady
200.0	1×1	0.5604	0.0551	0.2267
200.0	2×1	0.5468	0.0547	Steady
200.0	3×1	0.5468	0.0547	Steady
200.0	1×2	0.5604	0.0551	0.2267
300.0	1×1	0.4892	0.0408	0.2321
300.0	2×1	0.4729	0.0405	0.2684
300.0	3×1	0.4862	0.0407	0.2568
300.0	1×2	0.4892	0.0408	0.2321
400.0	1×1	0.4777	0.0340	0.2409
400.0	2×1	0.4915	0.0342	0.2752
400.0	3×1	0.5083	0.0344	0.2649
400.0	1×2	0.4777	0.0340	0.2409
589.3	1×1	0.4272	0.0262	Chaotic
593.4	2×1	0.4218	0.0239	Chaotic
598.7	3×1	0.4138	0.0262	Chaotic
583.0	1×2	0.4366	0.0259	Chaotic
831.0	1×1	0.4162	0.0222	Chaotic
785.3	2×1	0.4627	0.0253	Chaotic
765.4	3×1	0.4869	0.0255	Chaotic
836.4	1×2	0.4090	0.0219	Chaotic
1010.7	1×1	0.4711	0.0247	Chaotic
998.8	2×1	0.4819	0.0247	Chaotic
993.6	3×1	0.4868	0.0248	Chaotic
987.6	1×2	0.4932	0.0260	Chaotic

Table 4 Time-mean single- and multiple-module baseline results for $\delta^* = 1/8$

Re_{D_h}	$n_x \times n_y$	f	$j^{[H]}$	S
100.0	1×1	1.4597	0.1460	Steady
200.0	1×1	0.8107	0.0819	Steady
300.0	1×1	0.5746	0.0574	Steady
400.0	1×1	0.4504	0.0445	Steady
600.0	1×1	0.3632	0.0360	0.0900
600.0	2×1	0.3644	0.0378	0.1202
600.0	3×1	0.3592	0.0379	0.1292
600.0	1×2	0.3632	0.0360	0.0900
800.0	1×1	0.3141	0.0306	0.0883
800.0	2×1	0.3059	0.0319	0.1188
800.0	3×1	0.2985	0.0318	0.1271
800.0	1×2	0.3141	0.0306	0.0883
1000.0	1×1	0.2544	0.0261	0.1400
974.7	2×1	0.2678	0.0275	Chaotic
973.9	3×1	0.2682	0.0269	Chaotic
1000.0	1×2	0.2544	0.0261	0.1400

Table 5 Time-mean single- and multiple-module baseline results for $\delta^* = 1/12$

Re_{D_h}	$n_x \times n_y$	f	$j^{[H]}$	S
100	1×1	1.6524	0.1610	Steady
200	1×1	0.9109	0.0912	Steady
300	1×1	0.6419	0.0642	Steady
400	1×1	0.5007	0.0499	Steady
600	1×1	0.3655	0.0368	0.0647
600	2×1	0.3681	0.0377	0.0862
600	3×1	0.3712	0.0380	0.0793
600	1×2	0.3655	0.0368	0.0647
800	1×1	0.3048	0.0313	0.0645
800	2×1	0.3047	0.0320	0.0856
800	3×1	0.3077	0.0321	0.0787
800	1×2	0.3048	0.0313	0.0645
1000	1×1	0.2638	0.0271	0.0636
1000	2×1	0.2612	0.0275	0.0845
1000	3×1	0.2650	0.0276	0.0775
1000	1×2	0.2638	0.0271	0.0636

Table 6 Time-mean single- and multiple-module baseline results for $\delta^* = 1/16$

Re_{D_h}	$n_x \times n_y$	f	$j^{[H]}$	S
100	1×1	1.7499	0.1683	Steady
200	1×1	0.9622	0.0958	Steady
300	1×1	0.6765	0.0676	Steady
400	1×1	0.5269	0.0526	Steady
600	1×1	0.3699	0.0368	Steady
800	1×1	0.3046	0.0314	0.0503
800	2×1	0.3034	0.0318	0.0669
800	3×1	0.3068	0.0322	0.0615
800	1×2	0.3046	0.0314	0.0503
1000	1×1	0.2607	0.0275	0.0502
1000	2×1	0.2585	0.0278	0.0667
1000	3×1	0.2615	0.0280	0.0612
1000	1×2	0.2607	0.0275	0.0502

each module of the two- and three-module calculation domains ($n_x \times n_y = 2 \times 1$ and $n_x \times n_y = 3 \times 1$). For such temporally and spatially periodic cases, the time-mean results were calculated by averaging over at least 10 time periods (at least 15,000 time steps), and it was established that these time-mean results were essentially invariant for each of these cases.

For cases where unsteady, non-periodic, essentially chaotic solutions were obtained, the time-mean results were calculated by averaging over at least 100,000 time steps, and it was found that any two of such time-mean results, corresponding to the same case and obtained by averaging over two blocks of over 100,000 time steps, differed by less than 1%. Therefore, these time-mean results are also reported in Tables 3, 4, 5, and 6.

In cases where the solutions are unsteady, non-periodic, and essentially chaotic, the values of β^* were adjusted until the value of Re_{D_h} was reasonably close to the desired value (to within $\pm 5\%$) and then maintained constant. This is the reason why the Re_{D_h} values for such cases are not round numbers, whereas they are for the other cases (in each of which, the value of β^* could be adjusted relatively efficiently to achieve the desired value of Re_{D_h} to within $\pm 0.0001\%$).

The results in Tables 3, 4, 5, and 6 show that for each combination of geometric parameters and Re_{D_h} considered in this work, the single- and multiple-module baseline results for the time-mean values of modular friction factor, Colburn factor, and Strouhal number (for the temporally periodic cases) are relatively similar: the absolute percentage differences between the aforementioned values of f (for $\delta^* = 1/4$) and $j^{[H]}$ (for $\delta^* = 1/8$) are less than 6.4% and 5.1%, respectively, and for the smaller values of δ^*, these differences are lower. Thus, from a first-pass design point of view, it may be considered adequate to undertake only the less expensive one-module ($n_x \times n_y = 1 \times 1$) simulations.

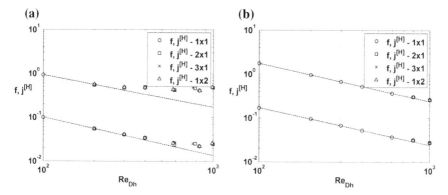

Fig. 9 Variations of f and $j^{[H]}$ with Re_{D_h} obtained by time-averaging single- and multiple-module baseline results for **a** $\delta^* = 1/4$ and $L^* = 1$, and **b** $\delta^* = 1/16$ and $L^* = 1$

The f and $j^{[H]}$ values in Tables 3 and 6 for $\delta^* = 1/4$ and $1/16$ (and $L^* = 1$), respectively, are plotted in Fig. 9a and b. For $\delta^* = 1/4$ and $Re_{D_h} \leq 200$, and $\delta^* = 1/16$ and $Re_{D_h} \leq 600$, f and $j^{[H]}$ vary linearly with Re_{D_h} on the log–log plots shown in these figures, indicating that a power-law correlation could be used to provide an excellent fit to this data. However, for $\delta^* = 1/4$ and $Re_{D_h} \geq 400$, and $\delta^* = 1/16$ and $Re_{D_h} \geq 800$, the fluid flow exhibits unsteady behavior (either self-sustained periodic or non-periodic chaotic oscillations with time), leading to enhanced mixing and higher values of f and $j^{[H]}$ than those that would be obtained if the flow were steady. For these cases, therefore, it is important to solve the mathematical models with the unsteady terms (see Sect. 3) and calculate the values of f and $j^{[H]}$ by suitably time-averaging the unsteady solutions (as was done in this work), rather than by solving steady-state mathematical models (which do not include the unsteady terms, as was done in some of the published works referenced in Sect. 1).

7.3 Temporally Periodic Developing Fluid Flow and Forced Convection Heat Transfer

These results were obtained for each of the four combinations of geometric parameters considered ($\delta^* = \delta/H = 1/4$, $1/8$, $1/12$, and $1/16$, each with $L^* = L/2H = 1$), with the calculation domain consisting of a single row ($n_y = 1$) of 10 consecutive representative geometric modules followed by an open (or plate-less) exit zone in the mainstream direction, and the spatially periodicity conditions given in Eq. (7) imposed at the longitudinal (transverse) boundaries (as shown schematically in Fig. 1). At the open portions of the inlet plane, a uniform velocity was imposed, in accordance with Eq. (6), and the value of u_{in} was chosen to achieve the desired constant value of Re_{D_h} for each case. At the outlet plane, as the x-extent of the exit zone was chosen to be long enough to ensure strict outflow conditions (that is, no inflow at the outlet plane), the standard outflow treatment was applied [9]. In the thermal problem, only the $[T]$ boundary condition was investigated.

Overview. For $\delta^* = 1/4$ and $L^* = 1$, the developing fluid flow exhibited the following behavior: remained steady (even after significant perturbations, the fluid flow reverted to the steady state, and no self-sustained oscillations were established) for $Re_{D_h} \leq 300$; executed self-sustained temporally periodic oscillations for $Re_{D_h} = 400$ and 500; and became unsteady, non-periodic, and essentially chaotic for $600 \leq Re_{D_h} \leq 1000$. For $\delta^* = 1/8$, $1/12$, and $1/16$, the developing fluid flow remained steady for all values of Re_{D_h} considered (100–1000). This behavior of the developing fluid flows is quite different from that of the fluid flows calculated by considering the cases in the temporally and spatially periodic fully developed region (see related discussions presented in the previous subsection). In the developing flow simulations, for all cases in which the fluid flow was steady or exhibited self-sustained temporally periodic oscillations, the time-averaged solutions became essentially identical (spatially periodic fully developed) in successive geometric modules, before the ninth module in the calculation domain. However, for all cases in which the fluid flow was unsteady (exhibited either self-sustained temporally periodic oscillations or essentially chaotic behavior), the instantaneous solutions in successive geometric modules in the mainstream (x) direction always have phase-angle differences. Thus, instantaneous spatial periodicity does not apply over either single or successive multiple modules in the x-direction, even when the time-averaged solutions start exhibiting spatially periodic fully developed behavior. It should also be noted that the fluid flow in the last (10th) module in the mainstream (x-direction) was affected by the exit section and the outflow treatment, but it effectively shielded the upstream modules from such influences.

Time-Mean Modular Friction Factor, Modular Colburn Factor, and Strouhal Number. These results for $L^* = 1$ and $\delta^* = 1/4$, $1/8$, $1/12$, and $1/16$ are presented in Tables 7, 8, 9, and 10, respectively. In all cases, the qualitative behavior of the time-mean modular friction and Colburn factors is similar: the values of these factors are relatively high in the entrance region of the arrays (spatially developing flow), and they progress toward essentially constant values as the flow travels downstream (in the x-direction) and eventually become spatially periodic fully developed (repeat in successive modules). Again, in all cases in which the fluid flow was steady or exhibited self-sustained temporally periodic oscillations, the aforementioned spatially periodic fully developed conditions were established in or before the ninth module in the calculation domain. For these cases, the values of spatially periodic fully developed time-mean modular friction (f) and Colburn ($j^{[T]}$) factors (corresponding to the ninth module) are presented in Table 11.

The values of the fully developed time-mean modular friction factor yielded by simulations of developing and single-module temporally and spatially periodic fluid flow, denoted as $f_{\text{module 9}}^{\text{developing}}$ and $f_{n_x=1, \, n_y=1}^{\text{TSPFD}}$, respectively, are presented together in Table 12, to facilitate a comparative discussion. In this table, the results pertaining to steady fluid flow are presented in *italics* font. Furthermore, as indicated by the

Table 7 Time-mean modular results for developing flow and heat transfer with $\delta^* = 1/4$

Re_{D_h}	Values of f in modules 1–9									S
	1	2	3	4	5	6	7	8	9	
100	1.1133	0.9331	0.9329	0.9329	0.9329	0.9329	0.9329	0.9329	0.9329	Steady
200	0.6846	0.5471	0.5468	0.5468	0.5468	0.5468	0.5468	0.5468	0.5468	Steady
300	0.5261	0.4019	0.4018	0.4018	0.4018	0.4018	0.4018	0.4018	0.4018	Steady
400	0.5407	0.4525	0.4451	0.4462	0.4462	0.4462	0.4462	0.4463	0.4463	0.2204
500	0.5251	0.4229	0.4363	0.4360	0.4388	0.4382	0.4387	0.4387	0.4387	0.2182
600	0.5052	0.4003	0.4489	0.4253	0.4557	0.4280	0.4964	0.5083	0.4938	Chaotic
800	0.5138	0.6077	0.6029	0.5711	0.4991	0.4890	0.4812	0.4655	0.4765	Chaotic
1000	0.5361	0.6152	0.5803	0.5169	0.4873	0.4844	0.4832	0.4755	0.4930	Chaotic

Re_{D_h}	Values of $j^{[T]}$ in modules 1–9									S
	1	2	3	4	5	6	7	8	9	
100	0.1086	0.1014	0.1014	0.1014	0.1014	0.1014	0.1014	0.1014	0.1014	Steady
200	0.0634	0.0553	0.0555	0.0555	0.0555	0.0555	0.0555	0.0555	0.0555	Steady
300	0.0466	0.0386	0.0388	0.0388	0.0388	0.0388	0.0388	0.0388	0.0388	Steady
400	0.0401	0.0367	0.0375	0.0375	0.0376	0.0376	0.0376	0.0376	0.0376	0.2204
500	0.0358	0.0326	0.0331	0.0334	0.0333	0.0334	0.0334	0.0333	0.0334	0.2182
600	0.0324	0.0295	0.0300	0.0299	0.0303	0.0300	0.0303	0.0313	0.0317	Chaotic
800	0.0280	0.0285	0.0294	0.0299	0.0290	0.0284	0.0278	0.0274	0.0274	Chaotic
1000	0.0252	0.0271	0.0265	0.0269	0.0269	0.0266	0.0266	0.0266	0.0264	Chaotic

Table 8 Time-mean modular results for developing flow and heat transfer with $\delta^* = 1/8$

Re_{D_h}	Values of f in modules 1–9									S
	1	2	3	4	5	6	7	8	9	
100	1.8557	1.4597	1.4596	1.4596	1.4596	1.4596	1.4596	1.4596	1.4596	Steady
200	1.1027	0.8111	0.8107	0.8107	0.8107	0.8107	0.8107	0.8107	0.8107	Steady
300	0.8344	0.5769	0.5747	0.5746	0.5746	0.5746	0.5746	0.5746	0.5746	Steady
400	0.6940	0.4566	0.4508	0.4504	0.4504	0.4504	0.4504	0.4504	0.4504	Steady
600	0.5441	0.3357	0.3222	0.3201	0.3198	0.3197	0.3197	0.3197	0.3197	Steady
800	0.4622	0.2754	0.2569	0.2522	0.2511	0.2508	0.2508	0.2508	0.2507	Steady
1000	0.4092	0.2389	0.2181	0.2110	0.2087	0.2080	0.2078	0.2077	0.2077	Steady

Re_{D_h}	Values of $j^{(T)}$ in modules 1–9									S
	1	2	3	4	5	6	7	8	9	
100	0.1505	0.1432	0.1433	0.1433	0.1433	0.1433	0.1433	0.1433	0.1433	Steady
200	0.0894	0.0792	0.0795	0.0795	0.0795	0.0795	0.0795	0.0795	0.0795	Steady
300	0.0661	0.0552	0.0555	0.0556	0.0556	0.0556	0.0556	0.0556	0.0556	Steady
400	0.0537	0.0427	0.0428	0.0429	0.0429	0.0429	0.0429	0.0429	0.0429	Steady
600	0.0404	0.0299	0.0297	0.0297	0.0298	0.0298	0.0298	0.0298	0.0298	Steady
800	0.0333	0.0234	0.0229	0.0229	0.0229	0.0229	0.0229	0.0229	0.0229	Steady
1000	0.0288	0.0195	0.0188	0.0187	0.0187	0.0187	0.0187	0.0187	0.0187	Steady

Table 9 Time-mean modular results for developing flow and heat transfer with $\delta^* = 1/12$

Re_{D_h}	Values of f in modules 1–9									S
	1	2	3	4	5	6	7	8	9	
100	2.1332	1.6524	1.6524	1.6524	1.6524	1.6524	1.6524	1.6524	1.6524	Steady
200	1.2632	0.9119	0.9109	0.9109	0.9109	0.9109	0.9109	0.9109	0.9109	Steady
300	0.9538	0.6466	0.6420	0.6419	0.6419	0.6419	0.6419	0.6419	0.6419	Steady
400	0.7912	0.5115	0.5015	0.5008	0.5007	0.5007	0.5007	0.5007	0.5007	Steady
600	0.6167	0.3765	0.3567	0.3533	0.3528	0.3527	0.3527	0.3527	0.3527	Steady
800	0.5208	0.3090	0.2839	0.2772	0.2754	0.2750	0.2749	0.2749	0.2749	Steady
1000	0.4586	0.2677	0.2408	0.2312	0.2280	0.2270	0.2266	0.2265	0.2265	Steady

Re_{D_h}	Values of $j^{[T]}$ in modules 1–9									S
	1	2	3	4	5	6	7	8	9	
100	0.1637	0.1578	0.1580	0.1580	0.1580	0.1580	0.1580	0.1580	0.1580	Steady
200	0.0982	0.0878	0.0882	0.0883	0.0883	0.0883	0.0883	0.0883	0.0883	Steady
300	0.0730	0.0614	0.0617	0.0618	0.0618	0.0618	0.0618	0.0618	0.0618	Steady
400	0.0594	0.0476	0.0477	0.0478	0.0479	0.0479	0.0479	0.0479	0.0479	Steady
600	0.0449	0.0335	0.0332	0.0332	0.0332	0.0332	0.0332	0.0332	0.0332	Steady
800	0.0371	0.0264	0.0256	0.0256	0.0256	0.0256	0.0256	0.0256	0.0256	Steady
1000	0.0321	0.0220	0.0210	0.0209	0.0208	0.0208	0.0209	0.0209	0.0209	Steady

Table 10 Time-mean modular results for developing flow and heat transfer with $\delta^* = 1/16$

Re_{D_h}	Values of f in modules 1–9									S
	1	2	3	4	5	6	7	8	9	
100	2.2779	1.7499	1.7499	1.7499	1.7499	1.7499	1.7499	1.7499	1.7499	Steady
200	1.3476	0.9636	0.9622	0.9622	0.9622	0.9622	0.9622	0.9622	0.9622	Steady
300	1.0168	0.6828	0.6767	0.6765	0.6765	0.6765	0.6765	0.6765	0.6765	Steady
400	0.8426	0.5404	0.5278	0.5268	0.5268	0.5268	0.5268	0.5268	0.5268	Steady
600	0.6551	0.3987	0.3749	0.3708	0.3701	0.3699	0.3699	0.3699	0.3699	Steady
800	0.5518	0.3276	0.2984	0.2904	0.2883	0.2878	0.2876	0.2876	0.2876	Steady
1000	0.4846	0.2839	0.2531	0.2421	0.2383	0.2371	0.2366	0.2365	0.2364	Steady

Re_{D_h}	Values of $j^{[T]}$ in modules 1–9									S
	1	2	3	4	5	6	7	8	9	
100	0.1699	0.1652	0.1654	0.1654	0.1654	0.1654	0.1654	0.1654	0.1654	Steady
200	0.1024	0.0922	0.0927	0.0928	0.0928	0.0928	0.0928	0.0928	0.0928	Steady
300	0.0764	0.0646	0.0650	0.0651	0.0651	0.0651	0.0651	0.0651	0.0651	Steady
400	0.0623	0.0502	0.0503	0.0504	0.0504	0.0504	0.0504	0.0504	0.0504	Steady
600	0.0472	0.0355	0.0350	0.0350	0.0350	0.0350	0.0351	0.0351	0.0351	Steady
800	0.0390	0.0279	0.0271	0.0270	0.0270	0.0270	0.0270	0.0270	0.0270	Steady
1000	0.0338	0.0234	0.0223	0.0220	0.0220	0.0220	0.0220	0.0220	0.0220	Steady

Table 11 Spatially periodic fully developed time-mean modular friction and Colburn factors yielded by the simulations of developing fluid flow and heat transfer

Re_{D_h}	$\delta^* = 1/4$		$\delta^* = 1/8$		$\delta^* = 1/12$		$\delta^* = 1/16$	
	f	$j^{[T]}$	f	$j^{[T]}$	f	$j^{[T]}$	f	$j^{[T]}$
100	0.9329	0.1014	1.4596	0.1433	1.6524	0.1580	1.7499	0.1654
200	0.5468	0.0555	0.8107	0.0795	0.9109	0.0883	0.9622	0.0928
300	0.4018	0.0388	0.5746	0.05556	0.6419	0.0618	0.6765	0.0651
400	0.4463	0.0376	0.4504	0.04293	0.5007	0.0479	0.5268	0.0504
600	N/A	N/A	0.3197	0.02977	0.3527	0.0332	0.3699	0.0351
800	N/A	N/A	0.2507	0.02292	0.2749	0.0256	0.2876	0.0270
1000	N/A	N/A	0.2077	0.01871	0.2265	0.0209	0.2364	0.0220

notation used, the reported values of $f_{\text{module 9}}^{\text{developing}}$ correspond to the ninth module of the calculation domain in the simulations of developing flow and heat transfer.

The results presented in Table 12 show that in cases where both solutions (developing and single-module temporally and spatially periodic) yield steady fluid flow, the values of $f_{\text{module 9}}^{\text{developing}}$ and $f_{n_x=1,\,n_y=1}^{\text{TSPFD}}$ are identical, as was expected based on related discussions and similar findings in some earlier published works, for example, Acharya et al. [18], Sparrow et al. [23], Patankar et al. [24], Bahaidarah et al. [30], Shah et al. [54], and Stone and Vanka [71]. However, for cases in which both of the aforementioned solutions yield unsteady temporally periodic flows, the values of $f_{n_x=1,\,n_y=1}^{\text{TSPFD}}$ exceed the corresponding values of $f_{\text{module 9}}^{\text{developing}}$ by up to 22%. Furthermore, the $f_{n_x=1,\,n_y=1}^{\text{TSPFD}}$ values pertain to self-sustained temporally and spatially periodic fluid flow for the following conditions: $\delta^* = 1/4$ and $Re_{D_h} = 200$; $\delta^* = 1/8$ and $1/12$, and $600 \leq Re_{D_h} \leq 1000$; and $\delta^* = 1/16$ and $800 \leq Re_{D_h} \leq 1000$. However, for these conditions, the $f_{\text{module 9}}^{\text{developing}}$ values pertain to steady fluid flow. These results illustrate clearly that the assumed spatial periodicity conditions on the instantaneous u, v, and P fields in the simulations of the temporally and spatially periodic flows (using the mathematical models presented in Sect. 5.2) are less restrictive (and thus yield unsteady fluid flow over a wider range of Reynolds number) than the boundary conditions employed in the corresponding simulations of the developing flow (using the mathematical models presented in Sect. 5.1).

8   Conclusion

Mathematical models and numerical predictions of temporally periodic fluid flow and heat transfer in spatially periodic geometries, in both spatially developing and fully developed regions, were presented and discussed in the earlier sections of this chapter. Over the last four decades, such fluid flow and heat transfer phenomena have been the subjects of numerous experimental and numerical investigations that have

Table 12 Values of fully developed time-mean modular friction factor yielded by simulations of developing and single-module temporally and spatially periodic fluid flow. The values presented in *italics* font pertain to steady-state conditions

Re_{D_h}	$\delta^* = 1/4$		$\delta^* = 1/8$		$\delta^* = 1/12$		$\delta^* = 1/16$	
	$f^{\text{developing}}_{\text{module 9}}$	$f^{\text{TSPFD}}_{n_x=1,\,n_y=1}$	$f^{\text{developing}}_{\text{module 9}}$	$f^{\text{TSPFD}}_{n_x=1,\,n_y=1}$	$f^{\text{developing}}_{\text{module 9}}$	$f^{\text{TSPFD}}_{n_x=1,\,n_y=1}$	$f^{\text{developing}}_{\text{module 9}}$	$f^{\text{TSPFD}}_{n_x=1,\,n_y=1}$
100	0.9329	*0.9329*	1.4596	*1.4597*	1.6524	*1.6524*	1.7499	*1.7499*
200	0.5468	*0.5604*	0.8107	*0.8107*	0.9109	*0.9109*	0.9622	*0.9622*
300	0.4018	*0.4892*	0.5746	*0.5746*	0.6419	*0.6419*	0.6765	*0.6765*
400	0.4463	*0.4777*	0.4504	*0.4504*	0.5007	*0.5007*	0.5268	*0.5269*
600	N/A	*N/A*	0.3197	*0.3632*	0.3527	*0.3655*	0.3699	*0.3699*
800	N/A	*N/A*	0.2507	*0.3141*	0.2749	*0.3048*	0.2876	*0.3046*
1000	N/A	*N/A*	0.2077	*0.2544*	0.2265	*0.2638*	0.2364	*0.2607*

appeared in the published literature. Nevertheless, several issues pertaining to their mathematical models and numerical predictions have not been fully resolved in the aforementioned publications. Some of these issues were elaborated and addressed in this chapter in the computationally convenient context of temporally periodic, two-dimensional, laminar, constant-property Newtonian fluid flow and forced convection heat transfer in uniform arrays of staggered rectangular plates.

In the simulations of developing fluid flow and heat transfer, the calculation domains consisted of one row of ten consecutive geometric modules, followed by a plate-free exit zone of suitable length, similar to that shown schematically in Fig. 1. The calculation domains consisted of single and multiple geometric modules, akin to that shown in Fig. 2, in the simulations of the fluid flow and heat transfer in the temporally and spatially periodic regime, with the following combinations of the integer periodicity indices: $n_x \times n_y = 1 \times 1, 2 \times 1, 3 \times 1$, and 2×1. With reference to the notations provided in Figs. 1 and 2, attention in this work was limited to the following set of geometrical parameters: $L^* = 1.0$ and $\delta^* = 1/4, 1/8, 1/12$, and $1/16$. The values of the Reynolds number (Re_{D_h}) ranged from 100 to 1000, and the Prandtl number was fixed at $Pr = 0.7$. The mathematical models of these fluid flow and heat transfer phenomena were described and discussed. A finite volume method (FVM) formulated by extending the ideas contained in some of the seminal contributions of Professor Spalding and his colleagues over the last 50 years [1–12, 76] was used to solve the aforementioned mathematical models. Extensive grid and time-step checks were undertaken, and the results of these checks were used to choose the spatial grids and time steps for the final simulations.

The main findings and conclusions of this investigation are summarized below:

(1) For conditions that lead to *steady* flows, the spatially periodic modular friction and Colburn factors obtained from simulations in the developing region and a single module in the fully developed region are identical, confirming similar conclusions and discussions in earlier published works.

(2) Multiple-module simulations of temporally and spatially periodic fluid flow and heat transfer, using the mathematical models presented in Sect. 5.2, and with $n_x > 1$ and $n_y = 1$, showed the existence of multiple solutions. In these cases, the possible solutions included one in which the temporally periodic fluid flow field in any one module differed in phase from that in every other module in the calculation domain: such solutions were unique for each combination of $n_x > 1$, $n_y = 1$, and Re_{D_h}, and are referred to in this chapter as *baseline* results. With $n_x \times n_y = 2 \times 1$ and also $n_x \times n_y = 3 \times 1$, two solutions were obtained for each Re_{D_h} when time-periodic oscillations were present: one was the baseline result for this case; in the other, the solution corresponding to $n_x \times n_y = 1 \times 1$ was repeated in each of the modules. However, the absolute percentage differences between the corresponding values of f (for $\delta^* = 1/4$) and $j^{[H]}$ (for $\delta^* = 1/8$) were found to be less than 6.4% and 5.1%, respectively, and for the smaller values of δ^*, these differences were even lower.

(3) For cases in which the developing fluid flow and heat transfer phenomena were either steady or temporally periodic, the time-averaged solutions exhibited spatially periodic fully developed behavior in successive geometric modules before the ninth module (in the mainstream direction) of the calculation domain. However, instantaneous spatial periodicity did not apply over either single or successive multiple modules in the mainstream direction, even when the time-averaged developing solutions start exhibiting spatially periodic fully developed behavior.

(4) In cases where the developing and single-module spatially periodic solutions yielded unsteady temporally periodic flows, the values of $f_{n_x=1,\ n_y=1}^{\text{TSPFD}}$ exceeded the corresponding values of $f_{\text{module 9}}^{\text{developing}}$, by up to 22% of the value of $f_{\text{module 9}}^{\text{developing}}$.

(5) The single-module simulations in the spatially periodic region yielded self-sustained temporally periodic oscillations of the fluid flow for the following conditions: $\delta^* = 1/4$ and $Re_{D_h} = 200$; $\delta^* = 1/8$ and $1/12$, and $600 \leq Re_{D_h} \leq 1000$; and $\delta^* = 1/16$ and $800 \leq Re_{D_h} \leq 1000$. However, for the same conditions, the simulations in the developing region yielded steady fluid flow.

(6) The assumed spatial periodicity conditions on the instantaneous u, v, and P fields in the simulations of the temporally and spatially periodic flows are less restrictive than the boundary conditions employed in the corresponding simulations of the developing flow; thus, the former formulation yielded unsteady fluid flow over a wider range of Reynolds number.

On a practical note, for the problems considered in this work, it may be concluded that the time-mean modular results yielded by the computationally economical single-module ($n_x \times n_y = 1 \times 1$) simulations are adequate from a first-pass design point of view, provided differences of up to about 22% from the corresponding results obtained by undertaking developing flow simulations are considered acceptable. Support for this conclusion for the problems investigated in this work, and for problems similar to them, can be obtained by factoring in the following considerations: (1) the dimensional uncertainties that would be incorporated in the geometries of interest (such as the cores of compact heat exchangers, for example) due to inevitable imperfections in the manufacturing processes; and (2) the uncertainties in the mathematical models of cases where turbulent fluid flow and heat transfer phenomena are encountered.

Acknowledgements Financial support of this work by the Natural Sciences and Engineering Research Council of Canada (NSERC) and the Fonds québécois de la recherche sur la nature et les technologies (FQRNT) is gratefully acknowledged by both authors. The second author (on behalf of all his former students, current students, and himself) would also like to express his deep gratitude and admiration for Professor D. B. Spalding for his numerous inspiring, lasting, and peerless contributions to the subject of computational fluid dynamics and heat transfer.

References

1. Patankar, S. V., & Spalding, D. B. (1967). A finite-difference procedure for solving the equations of the two-dimensional boundary layer. *International Journal of Heat and Mass Transfer, 10,* 1389–1411.
2. Patankar, S. V., & Spalding, D. B. (1970). *Heat and mass transfer in boundary layers* (2nd ed.). London, UK: Intertext Books.
3. Runchal, A. K. (1972). Convergence and accuracy of three finite difference schemes for a two-dimensional conduction and convection problem. *International Journal for Numerical Methods in Engineering, 4,* 541–550.
4. Spalding, D. B. (1972). A novel finite difference formulation for differential expressions involving both first and second derivatives. *International Journal for Numerical Methods in Engineering, 4,* 551–559.
5. Patankar, S. V., & Spalding, D. B. (1972). A calculation procedure for heat, mass and momentum transfer in three-dimensional parabolic flows. *International Journal of Heat and Mass Transfer, 15,* 1787–1806.
6. Launder, B. E., & Spalding, D. B. (1974). The numerical computation of turbulent flows. *Computer Methods in Applied Mechanics and Engineering, 3,* 269–289.
7. Pratap, V. S., & Spalding, D. B. (1976). Fluid flow and heat transfer in three-dimensional duct flows. *International Journal of Heat and Mass Transfer, 19,* 1183–1188.
8. Pollard, A., & Spalding, D. B. (1978). The prediction of the three-dimensional turbulent flow field in a flow-splitting tee-junction. *Computer Methods in Applied Mechanics and Engineering, 13,* 293–306.
9. Patankar, S. V. (1980). *Numerical heat transfer and fluid flow.* New York: McGraw-Hill.
10. Patankar, S. V., Pollard, A., Singhal, A. K., & Vanka, S. P. (1983). *Numerical prediction of flow, heat transfer, turbulence, and combustion—Selected works of professor D. Brian Spalding.* New York: Pergamon Press.
11. Artemov, V., Beale, S. B., de Vahl Davis, G., Escudier, M. P., Fueyo, N., Launder, B. E., et al. (2009). A tribute to D.B. Spalding and his contributions in science and engineering. *International Journal of Heat and Mass Transfer, 52,* 3884–3905.
12. Runchal, A. K. (2009). Brian Spalding: CFD and reality—A personal recollection. *International Journal of Heat and Mass Transfer, 52,* 4063–4073.
13. Kakac, S., Bergles, A. E., & Mayinger, F. (1981). *Heat exchangers—Thermal-hydraulic fundamentals and design.* New York: McGraw-Hill.
14. Fraas, A. P. (1989). *Heat exchanger design* (2nd ed.). New York: Wiley.
15. Kays, W. M., & London, A. L. (1984). *Compact heat exchangers* (3rd ed.). Malabar, Florida: Kreiger Publishing Company.
16. Shah, R. K. (2006). Advances in science and technology of compact heat exchangers. *Heat Transfer Engineering, 27,* 3–22.
17. Han, J. C., Glicksman, L. R., & Rohsenow, W. M. (1978). An investigation of heat transfer and friction for rib-roughened surfaces. *International Journal of Heat and Mass Transfer, 21,* 1143–1156.
18. Acharya, S., Myrum, T., Qiu, X., & Sinha, S. (1997). Developing and periodically developed flow, temperature and heat transfer in a ribbed duct. *International Journal of Heat and Mass Transfer, 40,* 461–479.
19. Wang, L., & Sunden, B. (2007). Experimental investigation of local heat transfer in a square duct with various-shaped ribs. *Heat Mass Transfer, 43,* 759–766.
20. Liu, J., Xie, G., & Simon, T. W. (2015). Turbulent flow and heat transfer enhancement in rectangular channels with novel cylindrical grooves. *International Journal of Heat and Mass Transfer, 81,* 563–577.
21. Metwally, H. M., & Manglik, R. M. (2004). Enhanced heat transfer due to curvature-induced lateral vortices in laminar flows in sinusoidal corrugated-plate channels. *International Journal of Heat and Mass Transfer, 47,* 2283–2292.

22. Pedras, M. H. J., & de Lemos, M. J. S. (2001). Simulation of turbulent flow in porous media using a spatially periodic array and a low Re two-equation closure. *Numerical Heat Transfer; Part A, 39,* 35–59.

23. Sparrow, E. M., Baliga, B. R., & Patankar, S. V. (1977). Heat transfer and fluid flow analysis of interrupted-wall channels, with application to heat exchangers. *ASME Journal of Heat Transfer, 99,* 4–11.

24. Patankar, S. V., Liu, C. H., & Sparrow, E. M. (1977). Fully developed flow and heat transfer in ducts having streamwise-periodic variations of cross-sectional area. *ASME Journal of Heat Transfer, 99,* 180–186.

25. Acharya, S., Dutta, S., Myrum, T., & Baker, R. S. (1993). Periodically developed flow and heat transfer in a ribbed duct. *International Journal of Heat and Mass Transfer, 36,* 2069–2082.

26. Wang, G. V., & Vanka, S. (1995). Convective heat transfer in periodic wavy passages. *International Journal of Heat and Mass Transfer, 38,* 3219–3230.

27. Murthy, J. Y., & Mathur, S. (1997). Periodic flow and heat transfer using unstructured meshes. *International Journal for Numerical Methods in Fluids, 25,* 659–677.

28. Beale, S. B., & Spalding, D. B. (1998). Numerical study of fluid flow and heat transfer in tube banks with stream-wise periodic boundary conditions. *Transactions CSME, 22,* 397–416.

29. Beale, S. B., & Spalding, D. B. (1999). A numerical study of unsteady fluid flow in in-line and staggered tube banks. *Journal of Fluids and Structures, 13,* 723–754.

30. Bahaidarah, H. M. S., Anand, N. K., & Chen, H. C. (2005). A numerical study of fluid flow and heat transfer over a bank of flat tubes. *Numerical Heat Transfer, Part A, 48,* 359–385.

31. Bahaidarah, H. M. S., Ijaz, M., & Anand, N. K. (2006). Numerical study of fluid flow and heat transfer over a series of in-line noncircular tubes confined in a parallel-plate channel. *Numerical Heat Transfer; Part B, 50,* 97–119.

32. Fullerton, T. L., & Anand, N. K. (2010). Periodically fully-developed flow and heat transfer over flat and oval tubes using a control-volume finite element method. *Numerical Heat Transfer; Part A, 57,* 642–665.

33. Fullerton, T. L., & Anand, N. K. (2010). An alternative approach to study periodically fully-developed flow and heat transfer problems subject to isothermal heating conditions. *International Journal of Engineering Science, 48,* 1253–1262.

34. Hutter, C., Zenklusen, A., Kuhn, S., & von Rohr, P. R. (2011). Large eddy simulation of flow through a streamwise-periodic structure. *Chemical Engineering Science, 66,* 519–529.

35. Labbé, O. (2013). Large-eddy-simulation of flow and heat transfer in a ribbed duct. *Computers & Fluids, 76,* 23–32.

36. Iacovides, H., Launder, B., & West, A. (2014). A comparison and assessment of approaches for modelling flow over in-line tube banks. *International Journal of Heat and Fluid Flow, 49,* 69–79.

37. Ruck, S., & Arbeiter, F. (2018). Detached eddy simulation of turbulent flow and heat transfer in cooling channels roughened by variously shaped ribs on one wall. *International Journal of Heat and Mass Transfer, 118,* 388–401.

38. Hasis, F. B. A., Krishna, P. M., Aravind, G. P., Deepu, M., & Shine, S. R. (2018). Thermo hydraulic performance analysis of twisted sinusoidal wavy microchannels. *International Journal of Thermal Sciences, 128,* 124–136.

39. Dezan, D. J., Yanagihara, J. I., Jenovencio, G., & Salviano, L. O. (2019). Parametric investigation of heat transfer enhancement and pressure loss in louvered fins with longitudinal vortex generators. *International Journal of Thermal Sciences, 135,* 533–545.

40. London, A. L., & Shah, R. K. (1968). Offset rectangular plate-fin surfaces—Heat transfer and flow friction characteristics. *ASME Journal of Engineering for Gas Turbines and Power, 90,* 218–228.

41. Wieting, A. R. (1975). Empirical correlations for heat transfer and flow friction characteristics of rectangular offset-fin plate-fin heat exchangers. *ASME Journal of Heat Transfer, 97,* 488–490.

42. Cur, N., & Sparrow, E. M. (1979). Measurements of developing and fully developed heat transfer coefficients along a periodically interrupted surface. *ASME Journal of Heat Transfer, 101,* 211–216.

43. Patankar, S. V., & Prakash, C. (1981). An analysis of the effect of plate thickness on laminar flow and heat transfer in interrupted-plate passages. *International Journal of Heat and Mass Transfer, 24,* 1801–1810.

44. Mullisen, R. S., & Loehrke, R. I. (1986). A study of the flow mechanisms responsible for heat transfer enhancements in interrupted-plate heat exchangers. *ASME Journal of Heat Transfer, 108,* 377–385.

45. Joshi, H. M., & Webb, R. L. (1987). Heat transfer and friction in the offset strip-fin heat exchanger. *International Journal of Heat and Mass Transfer, 30,* 69–84.

46. McBrien, R. K., & Baliga, B. R. (1988). Module friction factors and intramodular pressure distributions for periodic fully developed turbulent flow in rectangular interrupted-plate ducts. *ASME Journal of Fluids Engineering, 110,* 147–154.

47. Kelkar, K. M., & Patankar, S. V. (1989). Numerical prediction of heat transfer and fluid flow in rectangular offset-fin arrays. *Numerical Heat Transfer, Part A, 15,* 149–164.

48. Amon, C. H., Majumdar, D., Herman, C. V., Mayinger, F., Mikic, B. B., & Sekulic, D. P. (1992). Numerical and experimental studies of self-sustained oscillatory flows in communicating channels. *International Journal of Heat and Mass Transfer, 35,* 3115–3129.

49. Suzuki, K., Xi, G. N., Inaoka, K., & Hagiwara, Y. (1994). Mechanism of heat transfer enhancement due to self-sustained oscillation for an in-line fin array. *International Journal of Heat and Mass Transfer, 37,* 83–96.

50. Grosse-Gorgemann, A., Weber, D., & Fiebig, M. (1995). Experimental and numerical investigation of self-sustained oscillations in channels with periodic structures. *Experimental Thermal and Fluid Science, 11,* 226–233.

51. Manglik, R. M., & Bergles, A. E. (1995). Heat transfer and pressure drop correlations for the rectangular offset strip fin compact heat exchanger. *Experimental Thermal and Fluid Science, 10,* 171–180.

52. Zhang, L. W., Balachandar, S., Tafti, D. K., & Najjar, F. M. (1997). Heat transfer enhancement mechanisms in inline and staggered parallel-plate fin heat exchangers. *International Journal of Heat and Mass Transfer, 40,* 2307–2325.

53. Tafti, D. K., Zhang, L. W., & Wang, G. (1999). Time-dependent calculation procedure for fully developed and developing flow and heat transfer in louvered fin geometries. *Numerical Heat Transfer; Part A, 35,* 225–249.

54. Shah, R. K., Heikal, M. R., Thonon, B., & Tochon, P. (2001). Progress in numerical analysis of compact heat exchanger surfaces. *Advances in Heat Transfer, 34,* 363–443.

55. Saidi, A., & Sundén, B. (2001). A numerical investigation of heat transfer enhancement in offset strip fin heat exchangers in self-sustained oscillatory flows. *International Journal of Numerical Methods for Heat & Fluid Flow, 11,* 699–717.

56. Muzychka, Y. S., & Yovanovich, M. M. (2001). Modeling the f and j characteristics for transverse flow through an offset strip fin at low Reynolds number. *Enhanced Heat Transfer, 8,* 243–259.

57. Candanedo, J. A., Aboumansour, E., & Baliga, B. R. (2003). Time-mean wall static pressure distributions and module friction factors for fully developed flows in a rectangular duct with spatially periodic interrupted-plate inserts. In *Proceedings of the ASME International Mechanical Engineering Congress and Expo (IMECE 2003),* Washington, D.C., Nov. 16–21 (pp. 1–10).

58. Lamoureux, A., Camargo, L., & Baliga, B. R. (2005). Strouhal numbers and power spectrums for turbulent fully-developed flows in rectangular ducts with spatially-periodic interrupted-plate inserts. In *Proceedings of the Fifth International Conference on Enhanced, Compact and Ultra-Compact Heat Exchangers (CHE 2005),* Whistler, B.C. (pp. 73–80).

59. Ismail, L. S., Ranganayakulu, C., & Shah, R. K. (2009). Numerical study of flow patterns of compact plate-fin heat exchangers and generation of design data for offset and wavy fins. *International Journal of Heat and Mass Transfer, 52,* 3972–3983.

60. Lamoureux, A., & Baliga, B. R. (2007). Temporally- and spatially-periodic laminar flow and heat transfer in staggered-plate arrays. In *Proceedings of the ASME-JSME Thermal Engineering Heat Transfer Summer Conference (HT2007),* Vancouver, B.C., Canada, July 8–12 (pp. 1–10).

61. Lamoureux, A., & Baliga, B. R. (2007). Temporally-periodic developing laminar flow and heat transfer in staggered-plate arrays. In *Proceedings of the Sixth International Conference on Enhanced, Compact and Ultra-Compact Heat Exchangers (CHE 2007)*, Potsdam, Germany (pp. 306–313).

62. Benarji, N., Balaji, C., & Venkateshan, S. P. (2008). Unsteady fluid flow and heat transfer over a bank of flat tubes. *Heat Mass Transfer, 44*, 445–461.

63. Korichi, A., Oufer, L., & Polidori, G. (2009). Heat transfer enhancement in self-sustained oscillatory flow in a grooved channel with oblique plates. *International Journal of Heat and Mass Transfer, 52*, 1138–1148.

64. Sui, Y., Teo, C. J., & Lee, P. S. (2012). Direct numerical simulation of fluid flow and heat transfer in periodic wavy channels with rectangular cross-sections. *International Journal of Heat and Mass Transfer, 55*, 73–88.

65. Zheng, Z., Fletcher, D. F., & Haynes, B. S. (2014). Transient laminar heat transfer simulations in periodic zigzag channels. *International Journal of Heat and Mass Transfer, 71*, 758–768.

66. Ramgadia, A. G., & Saha, A. K. (2016). Numerical study of fully developed unsteady flow and heat transfer in asymmetric wavy channels. *International Journal of Heat and Mass Transfer, 102*, 98–112.

67. Sebben, S., & Baliga, B. R. (1996). A benchmark numerical solution involving steady, spatially-periodic, fully-developed laminar flow and heat transfer. In *Current Topics in Computational Heat Transfer, ASME National Heat Transfer Conference*, Houston, Texas, August 3–6.

68. Ghaddar, N. K., Korczak, K. Z., Mikic, B. B., & Patera, A. T. (1986). Numerical investigation of incompressible flow in grooved channels. Part 1. Stability and self-sustained oscillations. *Journal of Fluid Mechanics, 163*, 99–127.

69. Ghaddar, N. K., Magen, M., Mikic, B. B., & Patera, A. T. (1986). Numerical investigation of incompressible flow in grooved channels. Part 2. Resonance and oscillatory heat-transfer enhancement. *Journal of Fluid Mechanics, 168*, 541–567.

70. Sebben, S. (1996). Temporally and spatially periodic flows in interrupted-plate rectangular ducts. Ph.D. Thesis, Dept. Mech. Eng., McGill University, Montreal, Canada.

71. Stone, K., & Vanka, S. P. (1999). Numerical study of developing flow and heat transfer in a wavy passage. *ASME Journal of Fluids Engineering, 121*, 713–719.

72. Croce, G., & D'Agaro, P. (2006). Two-dimensional and three-dimensional self-sustained flow oscillations in interrupted fin heat exchangers. *Proceedings of the Institution of Mechanical Engineers, Part C: Journal of Mechanical Engineering Science, 220*, 297–307.

73. Schlichting, H. (1979). *Boundary layer theory* (3rd ed.). New York: McGraw-Hill.

74. Tritton, D. J. (1988). *Physical fluid dynamics* (2nd ed.). Oxford: Oxford Univ. Press.

75. Ferziger, J. H., & Perić, M. (2002). *Computational methods for fluid dynamics* (3rd ed.). New York: Springer.

76. Baliga, B. R., & Atabaki, N. (2006). Control-volume-based finite-difference and finite-element methods. In W. J. Minkowycz, E. M. Sparrow, & J. Y. Murthy (Eds.), *Handbook of numerical heat transfer*, Chapter 6 (2nd ed., pp. 191–224). New York: Wiley.

77. Leonard, B. P. (1979). A stable and accurate convective modelling procedure based on quadratic upstream interpolation. *Computer Methods in Applied Mechanics and Engineering, 19*, 59–98.

78. Rhie, C. M., & Chow, W. L. (1983). Numerical study of the turbulent flow past an airfoil with trailing edge separation. *AIAA Journal, 21*, 1525–1532.

79. Saabas, H. J., & Baliga, B. R. (1994). A co-located equal-order control-volume finite element method for multidimensional, incompressible fluid flow, Part I: Formulation. *Numerical Heat Transfer, Part B, 26*, 381–407.

80. Sebben, S., & Baliga, B. R. (1995). Some extensions of tridiagonal and pentadiagonal matrix algorithms. *Numerical Heat Transfer, Part B, 28*, 323–351.

81. Saad, Y. (2003). *Iterative methods for sparse linear systems* (2nd ed.). Philadelphia: Society for Industrial and Applied Mathematics (SIAM).

82. Richardson, L. F. (1910). The approximate arithmetical solution by finite differences of physical problems involving differential equations with application to a masonry dam. *Transactions of the Royal Society of London, Series A, 210*, 307–357.

83. Cotta, R. M. (1994). Benchmark results in computational heat and fluid flow: The integral transform method. *International Journal of Heat and Mass Transfer, 37*, 381–393.
84. Roache, P. J. (2002). Code verification by the method of manufactured solutions. *ASME Journal of Fluids Engineering, 124*, 4–10.
85. Cotta, R. M., & Mikhailov, M. D. (2006). Hybrid methods and symbolic computations. In W. J. Minkowycz, E. M. Sparrow, & J. Y. Murthy (Eds.), *Handbook of numerical heat transfer*, Chapter 16 (2nd ed., pp. 493–522). New York: Wiley.
86. Oberkampf, W. L., & Roy, C. J. (2010). *Verification and validation in scientific computing*. Cambridge, U.K: Cambridge University Press.
87. Baliga, B. R., & Lokhmanets, I. (2016). Generalized Richardson extrapolation procedures for estimating grid-independent numerical solutions. *International Journal of Numerical Methods for Heat & Fluid Flow, 26*, 1121–1144.
88. Franke, R., Rodi, W., & Schönung, B. (1990). Numerical calculation of laminar vortex-shedding flow past cylinders. *Journal of Wind Engineering and Industrial Aerodynamics, 35*, 237–257.
89. Davis, R. W., & Moore, E. F. (1982). A numerical study of vortex shedding from rectangles. *Journal of Fluid Mechanics, 116*, 475–506.
90. Lamoureux, A. (2006). Oscillatory flows in periodically interrupted rectangular passages in heat exchangers. M. Eng. thesis, Dept. Mech. Eng., McGill University, Montreal, Canada.

A Finite Volume Procedure for Thermofluid System Analysis in a Flow Network

Alok Majumdar

Nomenclature

A	Area
C_L	Flow coefficient
$c_{i,k}$	Mass concentration of kth specie at ith node
C_p	Specific heat
D	Diameter
f	Darcy friction factor
g	Gravitational acceleration
g_c	Conversion constant ($=32.174$ lb-ft/lb$_f$-s^2) for English unit ($=1$ for SI unit)
h	Enthalpy
h_{ij}	Heat transfer coefficient
J	Mechanical equivalent of heat (778 ft-lb$_f$/Btu) for English unit ($=1$ for SI unit)
K	Nondimensional head loss factor
K_f	Flow resistance coefficient
K_{rot}	Slip factor of rotating branch
k	Thermal conductivity
L	Length
LH_2	Liquid hydrogen
LN_2	Liquid nitrogen
LO_2	Liquid oxygen
m	Resident mass
$\dot{m}$	Mass flow rate
Nu	Nusselt number
n	Number of branches connected to a node
Pr	Prandtl number

A. Majumdar (✉)
NASA Marshall Space Flight Center, Huntsville, AL 35812, USA
e-mail: alok.k.majumdar@nasa.gov

© This is a U.S. government work and not under copyright protection in the U.S.;
foreign copyright protection may apply 2020
A. Runchal (ed.), *50 Years of CFD in Engineering Sciences*,
https://doi.org/10.1007/978-981-15-2670-1_7

231

P_R Ratio of reservoir pressure and air pressure
p Pressure
Q, q Heat source
Re   Reynolds number (Re $= \rho u D/\mu$)
R Gas constant
r Radius
S Momentum source
s Entropy
T Fluid temperature
T_s Solid temperature
u Velocity
V Volume
x Quality and mass fraction
Y Two-phase factor in Miropolsky correlation
z Compressibility factor

Greek

α Void fraction of air
γ Specific heat ratio
Δ Time step
Δh Head loss
ε Absolute roughness
ε/D Relative roughness
ε_{ij} Emissivity
θ Angle between branch flow velocity vector and gravity vector (deg)
μ Viscosity
ρ Density
σ Stefan-Boltzmann constant
τ Time
ν Kinematic viscosity
ω Angular velocity

Subscripts

a Ambient
c Convection
cr Critical
f Fluid
g Generation
i Node index

ij   Branch index
k   Fluid index
p   Index for neighboring branch
r   Radiation
s   Solid, surface area for shear

1   Introduction

The need for a generalized numerical method for thermofluid analysis in a flow network has been felt for a long time in the aerospace industry. Designers of thermofluid systems often need to know pressures, temperatures, flow rates, concentrations, and heat transfer rates at different parts of a flow circuit for steady-state or transient conditions. Such applications occur in propulsion systems for tank pressurization, internal flow analysis of rocket engine turbopumps, chilldown of cryogenic tanks and transfer lines, and many other applications of gas–liquid systems involving fluid transients and conjugate heat and mass transfer. Computer resource requirements to perform time-dependent, 3-D Navier-Stokes Computational Fluid Dynamics (CFD) analysis of such systems are prohibitive, and therefore, are not practical. A possible recourse is to construct a fluid network consisting of a group of flow branches such as pipes and ducts that are joined together at a number of nodes. They can range from simple systems consisting of a few nodes and branches to very complex networks containing many flow branches simulating valves, orifices, bends, pumps, and turbines. In the analysis of existing or proposed networks, node pressures, temperatures, and concentrations at the system boundaries are usually known. The problem is to determine all internal nodal pressures, temperatures, and concentrations, as well as branch flow rates. Such schemes are known as Network Flow Analysis methods, and they use largely empirical information to model fluid friction and heat transfer.

The oldest method for systematically solving a problem consisting of steady flow in a pipe network is the Hardy Cross method [1] which uses a method of successive approximation to solve for continuity and momentum equation. Flows are assumed for each pipe so that continuity is satisfied at every junction. A correction to the flow in each circuit is then computed from an integral form of momentum equation and applied to bring the circuits into closure balance. The original method was developed for hand calculations, but it has also been widely employed for use in computer-generated solutions. But as computers allowed much larger networks to be analyzed, it became apparent that the convergence of the Hardy Cross method was very slow or even failed to provide a solution in some cases. The other limitation of this method is its inability to extend to unsteady, compressible flow and heat transfer.

SINDA [2], a computer code originally developed at NASA, has been widely used to perform thermal analysis of a solid structure. A solid structure is divided into nodes connected by conductors. Temperatures are calculated at the nodes and heat fluxes are calculated at the conductors by solving the heat conduction equation by a finite

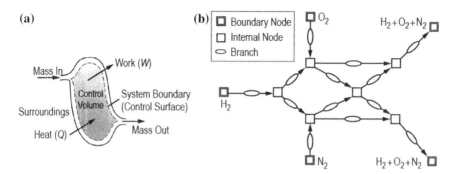

Fig. 1 Extension of control volume analysis to finite volume analysis in fluid network: **a** control volume analysis in classical thermodynamics and **b** finite volume analysis in fluid network

difference method. SINDA/FLUINT [3] and EASY5 [4] are the most widely used commercial network flow analysis codes in Aerospace Industries. SINDA/FLUINT extends SINDA's design philosophy of modeling a solid network to model a flow network. While SINDA conserves energy in the solid node, SINDA/FLUINT also conserves mass in fluid 'lumps' and momentum in fluid 'paths'. EASY5 provides computational modules for various flow elements such as pipe, orifice, and valve. The computational model is capable of solving the continuity and momentum equations for a given boundary condition and geometric parameters. A system model is developed with such elements. The code performs numerical integration of the entire system to ensure there is compatibility between input and output parameters of each element. The details of the numerical scheme of solving the system of equations in a commercial code are often proprietary and not available in the open literature.

The network flow method described in this paper is based on a finite volume procedure of solving mass, momentum, and energy conservation equations. Finite volume procedures are an extension of the control volume analysis performed in classical thermodynamics for mass and energy conservation (Fig. 1). Therefore, a finite volume procedure is a logical choice for solving network flow which is a collection of interconnected control volumes. The finite volume procedure was first developed by Professor Spalding and his students at Imperial College [5] to solve the Navier-Stokes equations in two dimensions. The Navier-Stokes equations were expressed in terms of stream function and vorticity using an upwind scheme [6] to ensure numerical stability for high Reynolds number flows. The governing equations are derived using the principle of conservation of conserved properties. The system of equations was solved by a successive substitution method. This method was successfully applied to solve many recirculating flows which were never solved before. The Navier-Stokes equation in three dimensions was solved in its primitive form by Patankar and Spalding [7]. They used a staggered grid where pressures were calculated at the center of the control volume whereas velocities were located at the boundaries of the control volumes. This finite volume procedure is known as the SIMPLE (Semi-Implicit Pressure-Linked Equation) algorithm. It uses the mass conservation equation to develop pressure corrections using a simplified momentum equation. The pressures and velocities are corrected iteratively until the solution is converged. Turbulence was modeled by defining an effective viscosity which is a function of turbulence properties such as turbulence energy and its dissipation rate and known as Launder and Spalding's [8] k-ε model of turbulence. The turbulence model equations are solved in conjunction with the mass and momentum conservation equations. The SIMPLE algorithm and the two-equation model of turbulence has been implemented in many CFD codes in later years.

Navier-Stokes-based CFD codes, however, are not suitable for thermofluid system analysis. It is not practical to solve for 3-D Navier-Stokes equations in conjunction with turbulence model equations to model a thermofluid system consisting of many fluid components such as pumps, pipes, valves, orifices, and bends. On the other hand, it is possible to solve a 1-D momentum equation with empirical correlations to model frictional effect to determine flow and pressure distribution in a flow network consisting of many such fluid components within reasonable computer time. A modified form of the SIMPLE algorithm was used to compute flow distribution in manifolds [9, 10], where 1-D mass and momentum equations were solved using the Colebrook equation [11] for friction factor to account for the viscous effect. Numerical predictions compared well with experimental data. However, this approach cannot be extended for any arbitrary flow network. A generalized flow network cannot be constructed using a structured coordinate system. In order to develop a numerical method to analyze any arbitrary flow network, the conservation equations for mass, momentum, and energy must be written using an unstructured coordinate system. This paper presents a finite volume procedure for calculating flow, pressure, and temperature distribution in a generalized fluid network for steady-state, transient, compressible, two-phase, and with or without heat transfer. The thermofluid system network is discretized into fluid nodes and branches, solid nodes, and conductors. The fluid nodes are connected with branches, and scalar properties such as pressure, enthalpy, and concentrations are stored in the fluid nodes, and vector properties such as flow rates and velocities are stored in the branches. Solid nodes and fluid nodes are connected by solid to fluid conductors. The conservation equations for mass and energy are solved at fluid nodes and momentum conservation equations are solved at fluid branches in conjunction with the thermodynamic equation of state for real fluids. The energy conservation equation for a solid is solved at the solid nodes. The system of equations is solved by a hybrid numerical method which consists of both the Newton-Raphson and Successive Substitution methods. This procedure has been incorporated into a general-purpose computer program, GFSSP [12–14]. This paper describes several applications of GFSSP that include (1) internal flow in a rocket engine turbopump, (2) compressible flows in ducts and nozzles, (3) pressurization and loading of a cryogenic propellant tank, (4) fluid transient during sudden opening of a valve for priming of a partially evacuated propellant feed line, and (5) a chilldown of cryogenic transfer lines with a phase change and two-phase flows. This paper also describes how to extend the network flow algorithm to perform multidimensional flow calculations.

2   Mathematical Formulation

The mathematical formulation to solve numerically the flow in a network offers a different kind of challenge than solving the Navier-Stokes equations in three dimensions. The Navier-Stokes equations are usually written for the coordinate systems which are topologically Cartesian. In a topologically Cartesian system of coordinates, a control volume can have a maximum of six neighboring control volumes: east, west, north, south, high, and low. The data structure for a 3-D coordinate system can be adapted for deriving the conservation equations for mass, momentum, and energy. On the other hand, a fluid network cannot be fully represented in a 3-D Cartesian coordinate system which has a limitation on the maximum number of neighbors. A fluid network is n-dimensional where n can assume any number. Therefore, its data structure is unique. The network definition and data structure of a flow network will be described in the following section, followed by the description of governing equations, which will include the conservation equations of mass, momentum, energy, and mixture species, as well as auxiliary equations such as the thermodynamic equation of state and empirical equations for friction and heat transfer.

2.1   Network Definitions

A flow network is first discretized into nodes and branches prior to the development of the governing equations. The defining parameters of a network are explained with the help of the example of a counter-flow heat exchanger shown in Fig. 2. In this example, hot fluid in the central tube is cooled by cold fluid in the annulus. The two fluid streams are exchanging energy by heat conduction and convection. This physical system is represented by a network of fluid and solid nodes. The fluid paths in the central tube and annulus are represented by a set of internal and boundary fluid nodes connected by fluid branches. The branch represents a fluid component such as a pipe, orifice, valve, or pump. In this particular case, the pipe and annulus are chosen as branch options. The mass and energy conservation equations are solved at the internal fluid nodes and the momentum equations are solved at the branches. It maybe noted that this concept is similar to the staggered grid concept of the SIMPLE algorithm [7]. The walls, through which heat is transferred from hot fluid to cold fluid, are discretized both axially and radially. Solid to fluid conductors connect solid and fluid nodes and calculate the convective heat transfer rate, and solid to solid conductors connect solid nodes and calculate conduction heat transfer. The energy conservation of a solid is solved at the solid nodes, accounting for heat transfer with neighboring solid and fluid nodes.

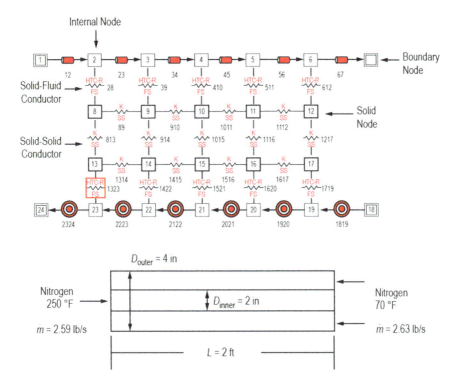

Fig. 2 Flow network representing a counter-flow heat exchanger

2.2 Data Structure

In a flow network, the layout of the nodes cannot be represented by a structured coordinate system (Fig. 1). There is no origin and no preferred coordinate direction to build the network of nodes and branches. In a structured coordinate system, the array of nodes can be constructed in the prespecified coordinate direction. In a 1-D flow network, each node has two neighbors; in a 2-D flow network, each node has four neighbors; and in a 3-D flow network, each node has six neighbors. In a typical flow network, a node can have n number of neighbors. Therefore, a unique data structure needs to be developed to define an unstructured flow network.

Any flow network can be constructed with three elements: (1) Boundary node, (2) internal node, and (3) branch. Each element has properties. Internal nodes and branches, where the conservation equations are solved, have two kinds of properties: geometric and thermofluid. There are two types of geometric properties: relational and quantitative. The data structure of the flow network is shown in Fig. 3. The relational geometric property allows nodes and branches to know their neighbors. Thermofluid properties include pressure, temperature, enthalpy, density, viscosity, etc.

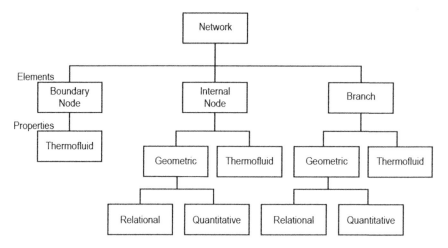

Fig. 3 Data structure for network flow analysis

Each node is designated by an arbitrary number and assigned a pointer to the array where node numbers are stored. The pointers are necessary to access the thermodynamic and thermophysical properties of the node. The relational properties of the node include the number of branches connected to it and the names of those branches. Figure 4 shows an example of these two relational properties of a node in a given network.

Like the nodes, each branch is also designated by an arbitrary number and assigned a pointer to the array where branch numbers are stored. The relational properties of the branch include (a) the names of the upstream and downstream nodes, (b) the number of upstream and downstream branches, and (c) the names of the upstream

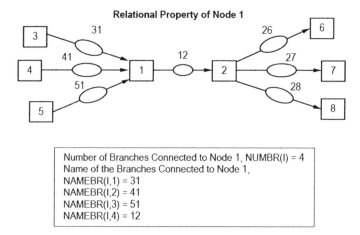

Fig. 4 Example of relational property of a node

Relational Property of Branch 12

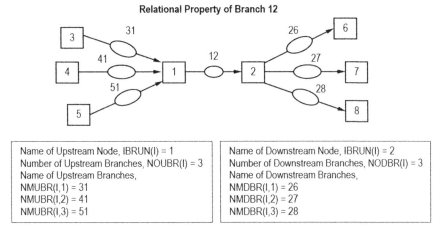

| Name of Upstream Node, IBRUN(I) = 1
 Number of Upstream Branches, NOUBR(I) = 3
 Name of Upstream Branches,
 NMUBR(I,1) = 31
 NMUBR(I,2) = 41
 NMUBR(I,3) = 51 | Name of Downstream Node, IBRUN(I) = 2
 Number of Downstream Branches, NODBR(I) = 3
 Name of Downstream Branches,
 NMDBR(I,1) = 26
 NMDBR(I,2) = 27
 NMDBR(I,3) = 28 |

Fig. 5 Example of relational property of a branch

and downstream branches. Figure 5 shows an example of the relational properties of a branch in a given network.

2.3 Governing Equations

The flow is assumed to be Newtonian, nonreactive, and compressible. It can be steady or unsteady, laminar or turbulent, with or without heat transfer, phase change, mixing or rotation. Figure 6 displays a schematic showing adjacent nodes, their connecting

Fig. 6 Schematic of nodes, branches, and indexing practice

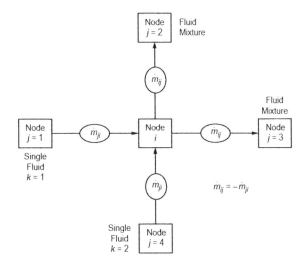

branches, and the indexing system. In order to solve for the unknown variables, mass, energy, and fluid species, conservation equations are written for each internal node and flow rate equations are written for each branch.

Mass Conservation Equation. The following is the mass conservation equation

$$\frac{m_{\tau+\Delta\tau} - m_{\tau}}{\Delta\tau} = -\sum_{j=1}^{j=n} \dot{m}_{ij}. \tag{1}$$

Equation (1) requires that for the unsteady formulation, the net mass flow from a given node must equate to the rate of change of mass in the control volume. In the steady-state formulation, the left side of the equation is zero. This implies that the total mass flow rate into a node is equal to the total mass flow rate out of the node.

Momentum Conservation Equation. The flow rate in a branch is calculated from the momentum conservation equation (Eq. 2) which represents the balance of fluid forces acting on a given branch. A typical branch configuration is shown in Fig. 7. Inertia, pressure, gravity, friction, and centrifugal forces are considered in the conservation equation. In addition to these five forces, a source term, S, has been provided in the equation to input pump characteristics or to input power to a pump in a given branch. If a pump is located in a given branch, all other forces except pressure are zero. The

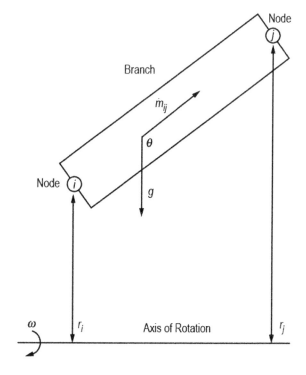

Fig. 7 Schematic of a branch showing gravity and rotation

source term, S, is zero in all branches without a pump or other external momentum source.

$$\frac{(mu)_{\tau+\Delta\tau} - (mu)_{\tau}}{g_c\Delta\tau} + \text{MAX}\left|\dot{m}_{ij},0\right|\left(u_{ij} - u_u\right) - \text{MAX}\left|-\dot{m}_{ij},0\right|\left(u_{ij} - u_u\right)$$

-----Unsteady-----   ----------------Longitudinal Inertia----------------

$$= \left(p_i - p_j\right)A_{ij} + \frac{\rho g V \cos\theta}{g_c} - K_f\dot{m}_{ij}\left|\dot{m}_{ij}\right|A_{ij} + \frac{\rho K_{rot}^2\omega^2 A}{g_c} - \rho A_{norm}u_{norm}u_{ij}/g_c + S .$$

--Pressure--   --Gravity--   --Friction--Centrifugal--Moving Boundary--Source--

$$(2)$$

Unsteady. This term represents the rate of change of momentum with time. For steady-state flow, the time step is set to an arbitrarily large value and this term reduces to zero.

Longitudinal Inertia. This term is important for compressible flows and when there is a significant change in velocity in the longitudinal direction due to change in the area and/or density. An upwind differencing scheme is used to compute the velocity differential.

Pressure. This term represents the pressure gradient in the branch. The pressures are located at the upstream and downstream face of a branch.

Gravity. This term represents the effect of gravity. The gravity vector makes an angle (θ) with the assumed flow direction vector. At $\theta = 180°$ the fluid is flowing against gravity; at $\theta = 90°$ the fluid is flowing horizontally, and gravity has no effect on the flow.

Friction. This term represents the frictional effect. Friction is modeled as a product of K_f, the square of the flow rate, and the area. K_f is a function of the fluid density in the branch and the nature of the flow passage being modeled by the branch. The calculation of K_f for different types of flow passages is described in the later section.

Centrifugal. This term in the momentum equation represents the effect of the centrifugal force. This term will be present only when the branch is rotating as shown in Fig. 7. K_{rot} is the factor representing the fluid rotation. K_{rot} is unity when the fluid and the surrounding solid surface rotate at the same speed. This term also requires knowledge of the distances from the axis of rotation between the upstream and downstream faces of the branch.

Moving Boundary. This term represents the force exerted on the control volume by a moving boundary.

Source. This term represents a generic source term. Any additional force acting on the control volume can be modeled through the source term. In a system level model, a pump can be modeled by this term. A detailed description of modeling a pump by this source term, S, appears in Ref. [13].

In a system level thermofluid model, compressible flow through an orifice is often an option for a branch. Under that circumstance, instead of solving Eq. (2), a simplified form of momentum equation is solved to calculate flow rate through an orifice. If the ratio of downstream to upstream pressure is less than the critical pressure ratio,

$$\frac{p_j}{p_i} < p_{cr},\tag{3a}$$

where

$$p_{cr} = \left(\frac{2}{\gamma+1}\right)^{\frac{\gamma}{\gamma-1}},\tag{3b}$$

then the choked flow rate in the branch is calculated from

$$\dot{m}_{ij} = C_{L_{ij}} A \sqrt{p_i \rho_i g_c \frac{2\gamma}{\gamma-1}(p_{cr})^{2/\gamma}\left[1 - (p_{cr})^{(\gamma-1)/\gamma}\right]}.\tag{3c}$$

If $p_j/p_i > p_{cr}$, the unchoked flow rate in the branch is calculated from

$$\dot{m}_{ij} = C_{L_{ij}} A \sqrt{p_i \rho_i g_c \frac{2\gamma}{\gamma-1}\left(\frac{p_j}{p_i}\right)^{2/\gamma}\left[1 - \left(\frac{p_j}{p_i}\right)^{(\gamma-1)/\gamma}\right]}\tag{3d}$$

Energy Conservation Equations for Fluid and Solid.
Energy Conservation Equation of Fluid. The main purpose of the energy conservation equation in fluid flow calculations is to obtain fluid properties which are primarily functions of pressure and temperature. While pressures are calculated from the mass conservation equation, to obtain temperatures and other properties, the energy equation must be solved. The energy conservation equation can be expressed in terms of enthalpy or entropy. Once pressure and enthalpy or pressure and entropy are known, all thermodynamic and thermophysical properties can be evaluated by using the available computer programs [15–17] that calculate properties of common fluids.

The energy conservation equation in terms of enthalpy for node i, shown in Fig. 6, can be expressed as

$$\frac{m\left(h - \frac{p}{\rho J}\right)_{\tau+\Delta\tau} - m\left(h - \frac{p}{\rho J}\right)_{\tau}}{\Delta\tau} = \sum_{j=1}^{j=n}\left\{\mathrm{MAX}\left[-\dot{m}_{ij}, 0\right]h_j - \mathrm{MAX}\left[\dot{m}_{ij}, 0\right]h_i\right\}$$
$$+ \frac{\mathrm{MAX}\left[-\dot{m}_{ij}, 0\right]}{|\dot{m}_{ij}|}\left[(p_i - p_j) + K_{ij}\dot{m}_{ij}^2\right](v_{ij}A) + Q_i\tag{4}$$

The term $(p_i - p_j)v_{ij}A_{ij}$ represents work input to the fluid due to rotation or having a pump in the upstream branch of the node i. The following term represents viscous work in the upstream branch of the node i where v_{ij} and A_{ij} are velocity and area of the upstream branch.

The energy conservation equation based on entropy is shown in Eq. (5)

$$
\frac{(ms)_{\tau+\Delta\tau} - (ms)_\tau}{\Delta\tau} = \sum_{j=1}^{j=n} \left\{ \mathrm{MAX}\left[-\dot{m}_{ij}, 0\right]s_j - \mathrm{MAX}\left[\dot{m}_{ij}, 0\right]s_i \right\}
$$

$$
+ \sum_{j=1}^{j=n} \left\{ \frac{\mathrm{MAX}\left[-\dot{m}_{ij}, 0\right]}{|\dot{m}_{ij}|} \right\} \dot{S}_{ij,\mathrm{gen}} + \frac{Q_i}{T_i} \tag{5}
$$

The entropy generation rate due to fluid friction in a branch is expressed as

$$
\dot{S}_{ij,\mathrm{gen}} = \frac{\dot{m}_{ij}\Delta p_{ij,\mathrm{viscous}}}{\rho_u T_u J} = \frac{K_f\left(|\dot{m}_{ij}|\right)^3}{\rho_u T_u J}. \tag{5a}
$$

The first term on the right-hand side of the Eq. 5 represents the convective transport of entropy from neighboring nodes. The second term represents the rate of entropy generation in branches connected to the ith node. The third term represents entropy change due to heat transfer.

Energy Conservation Equation of Solid. Typically, a solid node can be connected with other solid nodes, fluid nodes, and ambient nodes. Figure 8 shows a typical arrangement where a solid node is connected with other solid nodes, fluid nodes, and ambient nodes. The energy conservation equation for a solid node i can be expressed

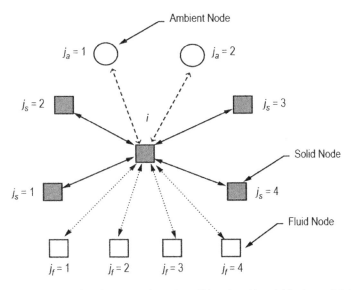

Fig. 8 A schematic showing the connection of a solid node with neighboring solid, fluid, and ambient nodes

as

$$\frac{\partial}{\partial \tau}\left(mC_p T_s^i\right) = \sum_{j_s=1}^{n_{ss}} \dot{q}_{ss} + \sum_{j_f=1}^{n_{sf}} \dot{q}_{sf} + \sum_{j_a=1}^{n_{sa}} \dot{q}_{sa} + \dot{S}_i. \tag{6}$$

The left-hand side of the equation represents the rate of change of temperature of the solid node, i. The right-hand side of the equation represents the heat transfer from the neighboring node and heat source or sink. The heat transfer from neighboring solid, fluid, and ambient nodes can, respectively, be expressed as

$$\dot{q}_{ss} = k_{ij_s} A_{ij_s} / \delta_{ij_s} \left(T_s^{j_s} - T_s^i\right), \tag{6a}$$

$$\dot{q}_{sf} = h_{ij_f} A_{ij_f} \left(T_f^{j_f} - T_s^i\right), \tag{6b}$$

and

$$\dot{q}_{sa} = h_{ij_a} A_{ij_a} \left(T_a^{j_a} - T_s^i\right), \tag{6c}$$

The effective heat transfer coefficients for solid to fluid and solid to ambient nodes are expressed as the sum of the convection and radiation

$$h_{ij_f} = h_{c,ij_f} + h_{r,ij_f}$$
$$h_{ij_a} = h_{c,ij_a} + h_{r,ij_a}$$

$$h_{r,ij_f} = \frac{\sigma\left[\left(T_f^{j_f}\right)^2 + \left(T_s^i\right)^2\right]\left[T_f^{j_f} + T_s^i\right]}{1/\varepsilon_{ij,f} + 1/\varepsilon_{ij,s} - 1}$$

$$h_{r,ij_a} = \frac{\sigma\left[\left(T_a^{j_a}\right)^2 + \left(T_s^i\right)^2\right]\left[T_a^{j_a} + T_s^i\right]}{1/\varepsilon_{ij,a} + 1/\varepsilon_{ij,s} - 1}. \tag{6d}$$

Equation of State and Thermodynamic Properties. The conservation equations for mass, momentum, and energy contain thermodynamic and thermophysical properties of a real fluid. A real fluid can exist in different states as shown in Fig. 9: subcooled liquid (A), saturated liquid (B), a mixture of liquid and vapor (x), saturated vapor (C), and superheated vapor (D). The state of the real fluid in a given node is calculated from its pressure and enthalpy using a thermodynamic property program such as GASP [15] or GASPAK [17]. All these programs use accurate equations of state for thermodynamic properties and correlations for thermophysical properties for common fluids.

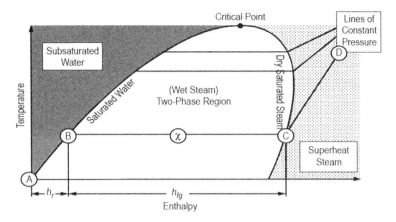

Fig. 9 Thermodynamic state of water

One of the main objectives of using an accurate equation of state is to compute the compressibility factor, z, which is used in the equation of state to compute the resident mass of the node

$$m = \frac{pV}{RTz}. \tag{7}$$

Species Conservation Equation. For a fluid mixture, thermodynamic and thermo-physical properties are also a function of the mass fraction of the fluid species. In order to calculate the properties of the mixture, the concentration of the individual fluid species within the branch must be determined. The concentration for the kth species can be written as

$$\frac{(m_i c_{i,k})_{\tau+\Delta\tau} - (m_i c_{i,k})_\tau}{\Delta\tau} = \sum_{j=1}^{j=n} \{ \text{MAX}[-\dot{m}_{ij}, 0]c_{j,k} - \text{MAX}[\dot{m}_{ij}, 0]c_{i,k} \} + \dot{S}_{i,k}. \tag{8}$$

For transient flow, Eq. (8) states that the rate of increase of the concentration of the kth species in the control volume equals the rate of transport of the kth species into the control volume minus the rate of transport of the kth species out of the control volume plus the generation rate of the kth species in the control volume.

Mixture Properties. A homogeneous mixture of multiple species in a given network can also be modeled provided the properties of the mixture are computed from the properties of the component species.

Temperature. In the absence of phase change, the temperature of the node can be calculated from a modified energy equation which is expressed in terms of specific heat and temperature

$$(T_i)_{\tau+\Delta\tau} = \frac{\sum_{j=1}^{j=n}\sum_{k=1}^{k=n_f} C_{pk,j}x_{k,j}T_j\mathrm{MAX}[-\dot{m}_{ij},0] + (C_{p,i}m_iT_i)_\tau/\Delta\tau + Q_i}{\sum_{j=1}^{j=n}\sum_{k=1}^{k=n_f} C_{pk,j}x_{k,j}\mathrm{MAX}[\dot{m}_{ij},0] + (C_{p,i}m)_{\tau+\Delta\tau}/\Delta\tau}.$$

(9)

Density. For Amagat's model of partial volume, mixture density is expressed as

$$\frac{1}{\rho_{\mathrm{mix}}} = \sum \frac{x_k}{\rho_k}.$$

(10)

ρ_k is evaluated at node pressure, p_i.

For Dalton's model of partial pressures, mixture density is expressed as

$$\rho_{\mathrm{mix}} = \sum \rho_k.$$

(11)

ρ_k is evaluated at partial pressure, p_k which is a product of molar concentration and node pressure, p_i.

Compressibility Factor. The compressibility factor of the mixture, z_i, is expressed as

$$z_i = \sum_{k=1}^{k=n} x_k z_k.$$

(12)

where

$$z_k = \frac{p_i}{\rho_k R_k T_i}.$$

(12a)

Friction Calculation. It was mentioned earlier that the friction term in the momentum equation is expressed as a product of K_f, the square of the flow rate, and the flow area. Empirical information is necessary to estimate K_f. For pipe flow (Fig. 10), length, L, diameter, D, and surface roughness, ε, are needed to compute friction.

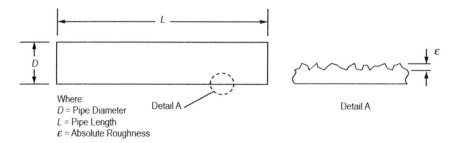

Where:
D = Pipe Diameter
L = Pipe Length
ε = Absolute Roughness

Fig. 10 Pipe parameters to compute friction

K_f can be expressed as

$$K_f = \frac{8fL}{\rho_u \pi^2 D^5 g_c}. \tag{13}$$

The Darcy friction factor, f, is determined from the Colebrook equation [11] which is expressed as

$$\frac{1}{\sqrt{f}} = -2 \log\left[\frac{\varepsilon}{3.7D} + \frac{2.51}{\text{Re}\sqrt{f}}\right]. \tag{13a}$$

To compute friction in a flow-through a restriction with a given flow coefficient, C_L, and area, A, K_f can be expressed as

$$K_f = \frac{1}{2g_c \rho_u C_L^2 A^2}. \tag{14}$$

In classical fluid mechanics, heat loss is expressed in terms of a nondimensional 'K factor'

$$\Delta h = K \frac{u^2}{2g}. \tag{14a}$$

K and C_L are related as:

$$C_L = \frac{1}{\sqrt{K}}. \tag{14b}$$

Reference [13] describes the friction calculations of other fluid components such as valve, bend, and orifice.

Heat Transfer Coefficient. The heat transfer coefficient is determined from empirical correlations.

There are four different options for specifying the heat transfer coefficient:

(1) A constant heat transfer coefficient.
(2) The Dittus-Boelter equation (Eq. 15) for single-phase flow where the Nusselt number is expressed as:

$$\frac{h_c D}{k_f} = 0.023 (\text{Re})^{0.8} (\text{Pr})^{0.33}, \tag{15}$$

where

$$\text{Re} = \frac{\rho u D}{\mu} \text{ and } \text{Pr} = \frac{C_p \mu_f}{k_f}.$$

(3) Miropolsky's correlation [18] for two-phase flow:

$$\text{Nu} = 0.023(\text{Re}_{\text{mix}})^{0.8}(\text{Pr}_v)^{0.4}(Y)$$

$$\text{Re}_{\text{mix}} = \left(\frac{\rho u D}{\mu_v}\right)\left[x + \left(\frac{\rho_v}{\rho_l}\right)(1-x)\right]$$

$$\text{Pr}_v = \left(\frac{C_p \mu_v}{k_v}\right)$$

$$Y = 1 - 0.1\left(\frac{\rho_l}{\rho_v} - 1\right)^{0.4}(1-x)^{0.4}. \tag{16}$$

(4) A new, user-defined correlation can be implemented in the User Subroutine described in Sect. 4.

2.4 Closure

The purpose of the mathematical formulation was to describe the governing equations to solve for the necessary variables of a given thermofluid network. The mathematical closure is shown in Table 1 where each variable and the designated governing equation to solve that variable are listed.

It may be noted that the pressure is calculated from the mass conservation equation although pressure does not explicitly appear in Eq. (1). This is, however, possible in the iterative Newton-Raphson scheme where pressures are corrected to reduce the residual error in the mass conservation equation. This practice was first implemented in the SIMPLE algorithm proposed by Patankar and Spalding [7] and commonly referred to as a 'Pressure Based' algorithm in CFD literature. The momentum conservation equation (Eq. 2), which contains both pressure and flow rate, is solved to calculate the flow rate. The strong coupling of pressure and flow rate requires that the mass and momentum conservation equations be solved simultaneously. In the following section, the numerical method of solving the system of equations listed in Table 1, will be described.

Table 1 Mathematical closure

Variable name	Designated equation to solve the variable
Pressure	Mass conservation (Eq. 1)
Flow rate	Momentum conservation (Eq. 2)
Fluid enthalpy or entropy	Energy conservation of fluid (Eqs. 4 and 5)
Solid temperature	Energy conservation of solid (Eq. 6)
Species concentration	Species conservation (Eq. 8)
Fluid mass	Thermodynamic state (Eq. 7)

3 Numerical Method

A fully implicit iterative numerical method has been used to solve the system of equations described in the previous section. There are two types of numerical methods available to solve a set of nonlinear coupled algebraic equations: (1) The Successive Substitution method and (2) the Newton-Raphson method. In the Successive Substitution method, every conservation equation is expressed explicitly to calculate one variable. The previously calculated variable is then substituted into the other equations to calculate another variable. In one iterative cycle, each equation is visited. The iterative cycle is continued until the difference in the values of the variables in successive iterations becomes negligible. The advantages of the Successive Substitution method are its simplicity to program and its low code overhead. The main limitation, however, is finding the optimum order for visiting each equation in the model. This visiting order, which is called the information flow diagram, is crucial for convergence. Under-relaxation (partial substitution) of variables is often required to obtain numerical stability.

In the Newton-Raphson method, the simultaneous solution of a set of nonlinear equations is achieved through an iterative guess and correction procedure. Instead of solving for the variables directly, correction equations are constructed for all of the variables. The intent of the correction equations is to eliminate the error in each equation. The correction equations are constructed in two steps: (1) The residual errors in all of the equations are estimated and (2) the partial derivatives of all of the equations, with respect to each variable, are calculated. The correction equations are then solved by the Gaussian elimination method. These corrections are then applied to each variable, which completes one iteration cycle. These iterative cycles of calculations are repeated until the residual error in all of the equations is reduced to a specified limit. The Newton-Raphson method does not require an information flow diagram. Therefore, it has improved convergence characteristics. The main limitation to the Newton-Raphson method is its requirement for a large amount of computer memory.

In the present finite volume procedure, a combination of the Successive Substitution method and the Newton-Raphson method is used to solve the set of equations. This method is called SASS (Simultaneous Adjustment with Successive Substitution). In this scheme, the mass and momentum conservation equations are solved by the Newton-Raphson method. The energy and species conservation equations are solved by the Successive Substitution method. The underlying principle for making such a division was that the equations that are more strongly coupled are solved by the Newton-Raphson method. The equations that are not strongly coupled with the other set of equations are solved by the Successive Substitution method. Thus, the computer memory requirement can be significantly reduced while maintaining superior numerical convergence characteristics. Figure 11 shows the flow chart of the numerical scheme.

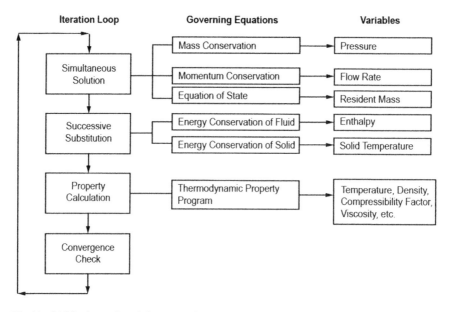

Fig. 11 SASS scheme for solving governing equations

4 Computer Program

This numerical method has been incorporated into a general-purpose computer program, GFSSP [12–14]. There are seven major functions of the computer program:

(1) Development of a flow circuit of fluid and solid nodes with branches and conductors.
(2) Development of an indexing system or data structure to define a network of fluid and solid nodes with branches and conductors.
(3) Generation of the conservation equations of mass, momentum, energy, species concentration, and solid temperatures in respective nodes and branches.
(4) Calculation of the thermodynamic and thermophysical properties of the fluid and solid in nodes.
(5) Numerical solution of the conservation equations.
(6) Input/output.
(7) User-defined modules.

GFSSP consists of three major modules: the Graphical User Interface (GUI) module, the Solver and Property module, and the User Subroutine module. Figure 12 shows the process flow diagram to describe the interaction among the three modules. A flow circuit is created in the GUI and an input data file is created which is read by the Solver and Property module. Specialized input to the model can be applied through a User Subroutine. Such specialized input includes time-dependent

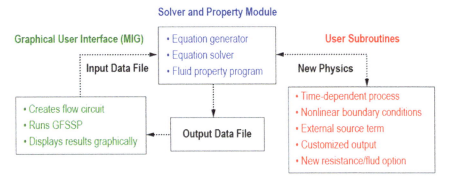

Fig. 12 Process flow diagram showing interaction among three modules

processes; nonlinear boundary conditions; external mass, momentum, and energy sources; customized output; and new resistance and fluid options.

Modeling Interface for GFSSP (MIG) provides the users a platform to build and run their models. Figure 13 shows the main MIG window that consists of menu, toolbar options and a blank canvas. It also allows post-processing of results. MIG allows the user to develop GFSSP models using an interactive 'point and click' paradigm. A

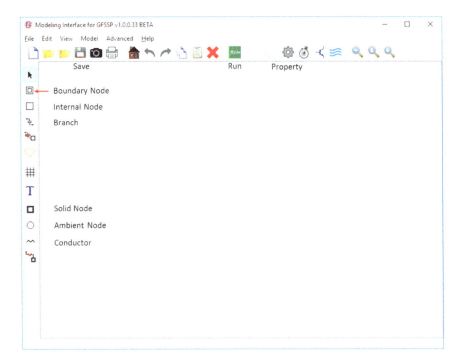

Fig. 13 Modeling interface for GFSSP (MIG)

network flow circuit with conjugate heat transfer is first built using six basic elements: boundary node, internal node, branch, solid node, ambient node, and conductor. Then the properties of the individual elements are assigned. Users are also required to define global options of the model that includes input/output files, fluid specification, and any special options such as rotation, valve operation, etc. During execution of the program, a run manager window opens up and users can monitor the progress of the numerical solution. On the completion of the run, it allows users to open the output file to see the results. It also provides an interface to activate and import data to a plotting program for post-processing. Reference [13] provides a detailed discussion of the data structure, mathematical formulation, computer program, graphical user interface and includes a number of example problems.

5 Applications

The development of this method started in 1994 to develop a computational model of internal flow in a rocket engine turbopump. Since then the described finite volume method for network flow analysis has been successfully applied to simulate a large number of aerospace applications, namely (a) compressible flows in ducts and nozzles, (b) pressurization and loading of a cryogenic propellant tank, (c) fluid transient during sudden opening of a valve for priming of partially evacuated propellant feed line, and (d) chilldown of a cryogenic transfer line with phase change and two-phase flows. Extension of the finite volume-based network flow algorithm has been demonstrated by solving the classical problem of flow inside a driven cavity.

5.1 Flow in a Rocket Engine Turbopump

In this rocket engine turbopump, a turbine, driven by hot gas from a gas generator, drives two pumps for pumping liquid fuel and oxygen before they are ignited in the thrust chamber. Both turbine and pumps are mounted on the same shaft that rotates around 30,000 rpm. There are many design challenges for a successful operation of this complex machine. Network analysis is particularly useful to (1) estimate the axial load on the bearings, (2) ensure appropriate flow through the bearings for cooling, and (3) design the interpropellant seal to prevent any mixing of fuel and oxidizer in the turbopump. References [19] and [20] describe the network flow analysis of internal flow in a rocket engine pump to address the above-mentioned design issues. The numerical predictions of pressure and temperature at various locations in the turbopump compare well with experimental data.

5.2 Compressible Flows in Ducts and Nozzles

The capability to model tank blowdown and flow through a converging–diverging nozzle was demonstrated in Ref. [13] by comparing numerical predictions with analytical solutions. Reference [21] presents a numerical study of the effect of friction, heat transfer, and area change in subsonic compressible flow. The numerical solutions of pressure, temperature, and Mach number have been compared with benchmark solutions for different cases representing the effect of friction, heat transfer, and area change.

5.3 Modeling of a Cryogenic Tank

Modeling of a cryogenic tank is important for the design of liquid propulsion systems. In a liquid propulsion system, cryogenic tanks are subjected to different processes which must be modeled to ensure all fluid properties are within the margin of safe and reliable operation. A robust and accurate network flow analysis method is necessary to simulate processes such as tank loading, boiloff, and tank pressurization.

Tank Loading. One of the very first and longest ground operations before a rocket launch is the loading of cryogenic propellants from the ground storage tanks into the launch vehicle tanks. This process takes several hours because the cryogenic transfer lines and propellant tanks must be chilled down from ambient temperature to liquid propellant temperatures, approximately 20 K for liquid hydrogen (LH$_2$) and 90 K for liquid oxygen (LO$_2$). The primary source of this cooling is the latent heat of vaporization. When cryogenic propellants are introduced into the transfer lines and vehicle tanks, they extract energy from the pipe and tank walls and evaporate. The vaporized propellants are vented from the vehicle tank, either to a flare stack, in the case of hydrogen, or to the atmosphere, in the case of oxygen. A numerical model was developed [22] to model the loading of LH$_2$ and LO$_2$ in the external tank of the Space Shuttle from storage tanks that are a quarter-mile away from the launch site. The model predictions compared well with measured data.

The practice of tank loading in a microgravity environment is quite different from tank loading on the ground. On the ground, under normal gravity, a vent valve on top of the tank can be kept open to vent the vapor generated during the loading process. The tank pressure can be kept close to atmospheric pressure while the tank is chilling down. In a microgravity environment, due to the absence of stratification, such practice may result in dumping large amounts of precious liquid propellant overboard. The intent of the no-vent, chill and fill method is to minimize the loss of propellant during chilldown of a propellant tank in a microgravity environment. The no-vent, chill and fill method consists of a repeated cyclic process of charge, hold, and vent. A numerical model was developed [23] to simulate chilldown of an LH$_2$ tank at the K-site Test Facility at NASA Glenn Research Center and numerical predictions were compared with test data.

Boiloff of Cryogenic Propellants. The cost of loss of propellants due to boiloff in large cryogenic storage tanks is on the order of $1 million per year. One way to reduce this cost is to design a new tank or refurbish existing tanks by using bulk-fill insulation material with improved thermal performance. An accurate numerical model of the boiloff process can help to design a tank with improved boiloff performance. A numerical model of the boiloff in a cryogenic storage tank at NASA Kennedy Space Center was developed [24]. The model developments were carried out in two phases. First, the model was verified with test data from a demonstration tank using liquid nitrogen (LN_2) and LH_2. The verified model was then extended to model the full-scale storage tank and the predictions were compared with field data.

Tank Pressurization. In a liquid propulsion system, cryogenic propellants are stored in an insulated tank. The propellants from the tank are fed to the turbopump by pressurizing the tank by an inert gas such as helium. The tank pressures must be controlled within a certain band for reliable operation of the turbopump. The pressurization of a propellant tank is a complex thermodynamic process with heat and mass transfer in a stratified environment. Numerical prediction of the pressurization process was compared [25] with correlations derived from test data. The agreement between the predictions and correlations was found to be satisfactory. The numerical model developed in Ref. [25] was extended to model the helium pressurization system of a propulsion test article at NASA Stennis Space Center where NASA's Fastrac engine [19] was tested. A detailed numerical model [26] of the tank pressurization system was developed. The model included a helium feed line, control valves, LO_2, and RP-1 (kerosene) tanks, and LO_2 and RP-1 feed lines supplying the propellants to the engine. The control valves of both tanks were modeled to set the pressure within a specified band. The model also accounted for the heat transfer between the helium and propellants and between the helium and the tank wall in the ullage, which is the gaseous space in the tank. The predicted pressure in both tanks compared well with test data.

In long-duration space travel, the cryogenic propellant tanks are self-pressurized due to heat transfer from space to the tank. The ullage pressure is controlled by the thermodynamic vent system (TVS). A TVS typically includes a Joule-Thompson expansion devise, a two-phase heat exchanger, a mixing pump, and a liquid injector to extract thermal energy from the tank without significant loss of liquid propellant. A numerical model of a system level test bed was developed [27] to simulate self-pressurization and pressure control by a TVS. The numerical prediction compared reasonably well with experimental data.

5.4 Fluid Transient Due to Sudden Opening of a Valve

Fluid transient due to sudden opening of a valve is important in propulsion applications when propellant valves are instantaneously opened to feed the thrusters with fuel and oxidizer. The pressure rise could be of the order of 200 atmosphere (20 MPa).

Designers need to have an analytical tool to estimate the maximum pressure and frequency of oscillation to ensure the structural integrity of the propulsion system. A laboratory experiment was performed [28] with water and air to measure the pressure oscillation following a sudden opening of a valve. An 11-m pipe (2.6 cm in diameter) was connected to a water tank at one end and closed at the other end. A valve was placed 6 m from the tank. The valve was initially closed and air at atmospheric pressure was entrapped downstream of the valve. The pressure in the tank was varied from 203 to 710 kPa. This experimental configuration was first modeled [29] assuming a lumped air node with a variable volume. Only the thermodynamics of the air were modeled; the air was considered stagnant. The numerical predictions of pressure oscillation compared well with measurements. Later, a more detailed model of the air–water system was developed and is shown in Fig. 14. In this model, the pipe containing air was also modeled and discretized with several nodes and branches similar to the pipe containing water upstream of the valve. After the opening of the valve, air and water mix, and water penetrates into the air and pushes the air towards the dead end. Boundary node 1 represents the tank, and the restriction in branch 1112 indicates the ball valve. The history of the ball valve opening is shown in Fig. 15.

The comparison between numerical predictions and experimental data is shown in Fig. 16. The frequency of oscillation matches quite well with test data. However, the numerical model predicts a higher peak pressure than test data. The cause of this discrepancy can be attributed to the assumption of a rigid pipe. The experiments were performed in Plexiglas pipe and the elastic deformation of the pipe could be the cause of lower peak pressure in the experiments. More applications and verifications of this procedure are described in Ref. [30].

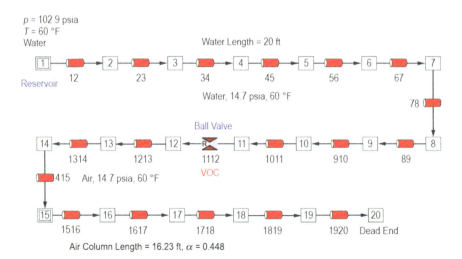

Fig. 14 Computational model of Lee's [28] experimental setup

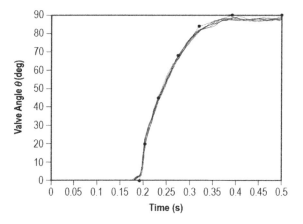

Fig. 15 Ball valve angle change with time [29]

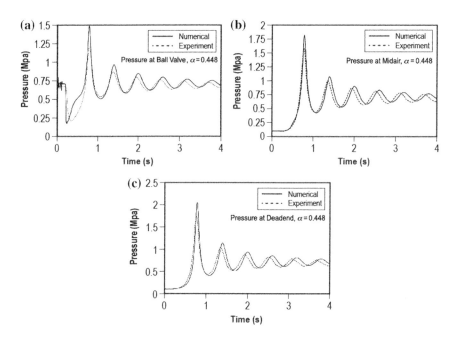

Fig. 16 Pressure comparison between numerical predictions and measured data ($P_R = 7$ and $\alpha = 0.448$) at **a** ball valve, **b** mid-section, and **c** dead end

5.5 Chilldown of a Transfer Line Carrying Cryogenic Fluid

A cryogenic transfer line must be chilled down to cryogenic temperature before steady flow rates can be achieved to engine feed or tank-to-tank propellant transfer.

A numerical model of the chilldown process is useful for optimizing for time to chill down or minimum loss of useful propellants. Cross et al. [31] first applied the present numerical scheme to model chilldown of a cryogenic transfer line. The numerical prediction was compared with an analytical solution to verify the accuracy of the numerical scheme. The verification and validation of the finite volume procedure for the prediction of conjugate heat transfer in a fluid network was performed by comparing the predictions with available experimental results for a long cryogenic transfer line model reported in Ref. [32]. The experimental setup consists of a 200-ft-long, 0.625-in-inside-diameter vacuum-jacketed copper tube supplied by a 300-L tank through a valve and exits to the atmosphere ($\approx$12.05 psia). The tank was filled either with LH_2 or LN_2. At time zero, the valve at the left end of the pipe was opened, allowing liquid from the tank to flow into the ambient pipeline driven by tank pressure.

Figure 17 shows a schematic of the network flow model [32] that was constructed to simulate the cooling of the transfer line. The tube was discretized into 33 fluid nodes (2 boundary nodes and 31 internal nodes), 31 solid nodes, and 32 branches. The upstream boundary node represents the cryogenic tank, while the downstream boundary node represents the ambient where the fluid is discharged. The first branch represents the valve; the next 30 branches represent the transfer lines. Each internal

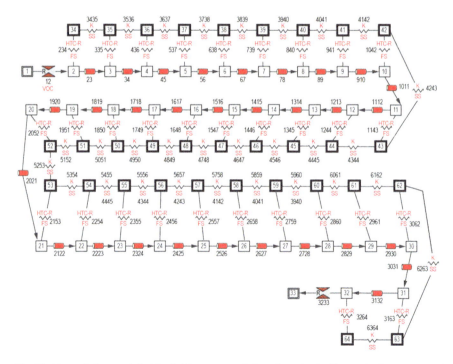

Fig. 17 Network flow model of the fluid system consisting of a tank, pipeline, and valve constructed with boundary nodes, internal nodes, and branches [32]

node was connected to a solid node (nodes 34 through 64) by a solid to fluid conductor. The heat transfer in the wall is modeled using the lumped parameter method, assuming the wall radial temperature gradient is small. The heat transfer coefficient of the energy equation for the solid node was computed from the Miropolskii correlation [18]. The experimental work reported in Ref. [33] did not provide details concerning the flow characteristics for the valve used, nor did they give a history of the valve opening times that they used. An arbitrary 0.05-s transient opening of the valve was used while assuming a linear change in flow area. The measured and predicted chilldown time for LH_2 and LN_2 chilldown at various pressures at saturated and subcooled conditions are shown in Table 2. It may be noted that, at higher pressure, it takes less time to chill down. This is primarily due to increased flow rates at higher inlet pressures. In this experimental program [33], however, flow rates were not measured. The effect of subcooling is not significant for LH_2, but significant for LN_2. Generally, numerical models predicted slightly higher chilldown times than measurements. This discrepancy can be attributed to the inaccuracy of the heat transfer coefficient correlation.

Darr et al. [34] developed correlations for the entire boiling curve based on a large number of chilldown experiments of a short stainless steel tube, 0.6 m long with an

Table 2 Chilldown time for various driving pressures and temperatures for LH_2 and LN_2 [32]

Fluid	Driving pressure (MPa)	Inlet state	Inlet temperature (K)	Experimental chilldown time (s)	Predicted chilldown time (s)
LH_2	0.52	Saturated	27.00	68	70
LH_2	0.60	Saturated	28.11	62	69
LH_2	0.77	Saturated	29.60	42	50
LH_2	1.12	Saturated	31.97	30	33
LH_2	0.25	Subcooled	19.50	148	150
LH_2	0.43	Subcooled	19.50	75	80
LH_2	0.60	Subcooled	19.50	62	60
LH_2	0.77	Subcooled	19.50	41	45
LH_2	0.94	Subcooled	19.50	32	35
LH_2	1.12	Subcooled	19.50	28	30
LN_2	0.43	Saturated	91.98	165	185
LN_2	0.52	Saturated	94.42	150	160
LN_2	0.60	Saturated	96.35	130	140
LN_2	0.25	Subcooled	76.00	222	250
LN_2	0.34	Subcooled	76.00	170	175
LN_2	0.43	Subcooled	76.00	129	140
LN_2	0.52	Subcooled	76.00	100	100
LN_2	0.60	Subcooled	76.00	85	90

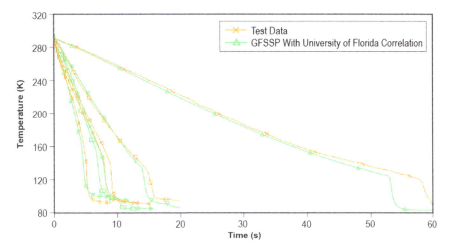

Fig. 18 Downstream wall temperature versus time for vertical upward LN_2 chilldown runs

inner diameter of 1.17 cm, placed inside a vacuum chamber to minimize parasitic heat leak. Flow rates were also measured in addition to temperature history in upstream and downstream locations of the tube. LeClair et al. [35], however, found that the Miropolsky correlation was not adequate for a short tube using LN_2 and used this new correlation. The comparison of numerical predictions with experimental data for five different Reynolds numbers is shown in Fig. 18.

5.6   Extension of Network Algorithm to Model Multidimensional Flow

In thermofluid engineering applications, system level codes are typically used to find flow and pressure distribution in a complex flow network. On the other hand, Navier-Stokes codes are used when detailed knowledge about the flow is needed for design or to investigate a failure scenario. System level codes are often run independently to provide boundary conditions for Navier-Stokes codes. There has not been much success in the integration of these codes to perform any coupled analysis. However, there are situations where integrated analysis brings value to the design. One such example is the propellant feed to a rocket engine from a stratified cryogenic tank. While the bulk of the feed system analysis can be performed by a system level flow network code, the stratification is a multidimensional phenomenon and requires higher fidelity analysis. In order to analyze such problems, an attempt has been made to extend the present network flow algorithm to compute multidimensional flow.

The flow network algorithm described in this paper uses multidimensional conservation equations for scalor properties such as mass (Eq. 1) and energy (Eq. 4). However, the momentum conservation equation (Eq. 2) is one-dimensional. To account for the multidimensional effect, two additional terms need to be introduced in the momentum conservation equation. Equation (2) does not include the transport of longitudinal momentum by shear and transverse inertia. In order to include the multidimensional effect, the shear term must appear in the right-hand side and transverse inertia must appear in the left-hand side of Eq. (2). The shear and transverse inertia can be expressed as follows:

$$\text{Shear Force} = \mu \frac{u_p - u_{ij}}{g_c \delta_{ij,p}} A_s \tag{17}$$

and

$$\text{Transverse Inertia} = \text{MAX}|\dot{m}_{\text{trans},0}|(u_{ij} - u_p) - \text{MAX}|-\dot{m}_{\text{trans},0}|(u_{ij} - u_p). \tag{18}$$

With these two added terms, the momentum equation should be able to model multidimensional flow. It may be noted that the friction term in Eq. (2) will no longer be active because the shear stress term will model the fluid friction. This extended formulation has been tested [36] by computing two-dimensional recirculating flow in a driven cavity. In a square cavity, the flow is induced by shear interaction at the top wall as shown in Fig. 19. The properties and dimension are so chosen that the corresponding Reynolds number is 100 for which Burggraf [6] provided a numerical solution of the Navier-Stokes equation. Figure 20 shows the schematic of the network

Fig. 19 Flow in a shear-driven square cavity

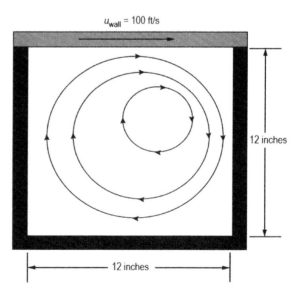

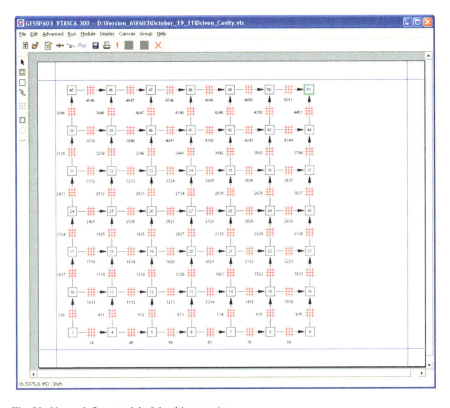

Fig. 20 Network flow model of the driven cavity

flow model of the driven cavity. The comparison between the Burggraf solution and the present prediction of velocity profiles along a vertical plane at the horizontal midpoint is shown in Fig. 21. It may be noted that a 7 × 7 grid network model compares well with the 51 × 51 grid Navier-Stokes solution. The predicted velocity field and pressure contours are shown in Fig. 22. The recirculating flow pattern and stagnation of the flow near the top right corner appear physically realistic.

6 Summary

A finite volume procedure originally developed for solving the Navier-Stokes equation has been implemented in solving the mass, momentum, and energy conservation equations in a flow network consisting of various fluid components. The 1-D momentum equation is solved in fluid components such as pipes, restrictions, pumps, and valves. Fluid friction is calculated using empirical correlations such as friction factor for pipe flows and flow coefficients for orifices and valves. Fluid friction appears

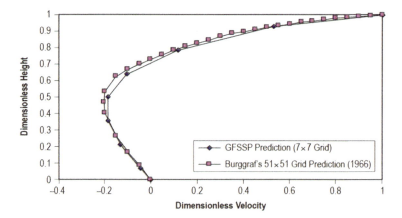

Fig. 21 Shear-driven square cavity centerline velocity distribution

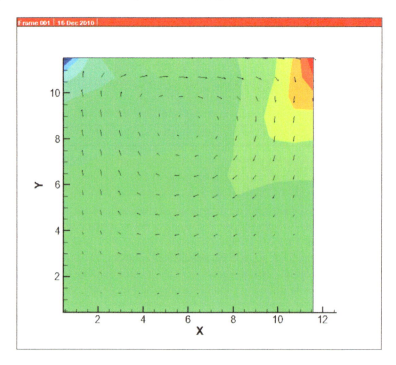

Fig. 22 Predicted velocity field and pressure contours

as a sink term in the momentum equation. Pumps, on the other hand, are modeled as a source term in the momentum equation, which is calculated from pump characteristics or pump horsepower. Mixtures of species and/or phases are assumed homogeneous. Mass or mole averaged properties of the mixture appear in the conservation equations for mass, momentum, and energy. All conservation equations are written in fully implicit form. The mass and momentum conservation equations, as well as the equation of state, are solved simultaneously by the Newton-Raphson method while the energy conservation equations for solid and fluid, and the species conservation equation, are solved by the Successive Substitution method outside the Newton-Raphson loop. The thermodynamic property calculations are also done outside the Newton-Raphson loop. An intuitive and user-friendly graphical user interface helps to build complex models with relative ease. This method has been successfully applied to several aerospace applications, namely, (1) internal flow in a rocket engine turbopump, (2) subsonic compressible flows in ducts and nozzles, (3) pressurization and loading of a cryogenic propellant tank, (4) fluid transient during the sudden opening of a valve for priming of a partially evacuated propellant feed line, and (5) chilldown of a cryogenic transfer line with phase change and two-phase flows. It is possible to perform a coarse grid multidimensional flow calculation within the framework of network flow in some components of a given flow system.

Acknowledgements This paper is dedicated to the memory of Professor D. B. Spalding who inspired and guided the author to develop understanding and expertise in the fascinating field of computational thermofluid dynamics and heat transfer. The author wants to acknowledge NASA Marshall Space Flight Center for the opportunity and resources for the continuous development of GFSSP for the last 25 years. The author also appreciates the contribution and support of Dr. Andre LeClair, Mr. Derek Moody, and other members of the GFSSP development team.

References

1. Streeter, V. L. (1962). *Fluid mechanics* (3rd ed.). McGraw-Hill.
2. Owen, J. W. (Ed.). (1992). *Thermal analysis workbook*. NASA TM-103568, January 1992.
3. SINDA/FLUINT, developed by C&R Technologies. https://crtech.com/sites/default/files/files/Guides-Manuals/Protected/sf60main.pdf.
4. EASY5, developed by MSC Software. http://www.mscsoftware.com/product/easy5.
5. Gosman, A. D., Pun, W. M., Runchal, A. K., Spalding, D. B., & Wolfshtein, M. (1969). *Heat and mass transfer in recirculating flows*. Academic Press.
6. Burggraf, O. R. (1966). Analytical and numerical studies of the structure of steady separated flows. *Journal of Fluid Mechanics, 24*(Part 1), 113–151.
7. Patankar, S. V., & Spalding, D. B. (1972). A calculation procedure for heat, mass and momentum transfer in three dimensional parabolic flows. *International Journal of Heat and Mass Transfer, 15*, 1787–1806.
8. Launder, B. E., & Spalding, D. B. (1974). The numerical computation of turbulent flows. *Computer Methods in Applied Mechanics and Engineering, 3*, 269–289.
9. Majumdar, A. K. (1980). Mathematical modeling of flows in dividing and combining flow manifold. *Applied Mathematical Modelling, 4*, 424–431.
10. Datta, A. B., & Majumdar, A. K. (1980). Flow distribution in parallel and reverse flow manifolds. *International Journal of Heat and Fluid Flow, 2*(4).

11. Colebrook, C. F. (1938–1939). Turbulent flow in pipes, with particular reference to the transition between the smooth and rough pipe laws. *Journal of Civil Engineering and Management* (London), *11*, 133–156.
12. Majumdar, A. K., Bailey, J. W., Schallhorn, P. A., & Steadman, T. E. (2004). *Generalized fluid system simulation program*. U.S. Patent No. 6,748,349, June 8, 2004.
13. Majumdar, A. K., LeClair, A. C., Moore, R., & Schallhorn, P. A. (2013). *Generalized fluid system simulation program, Version 6.0*. NASA/TM—2013–217492, NASA Marshall Space Flight Center, Huntsville, AL, October 2013.
14. Majumdar, A. K. (1999). A second law based unstructured finite volume procedure for generalized flow simulation. In *37th AIAA Aerospace Sciences Meeting Conference and Exhibit*, Reno, NV, January 11–14, 1999, Paper No. AIAA 99-0934.
15. Hendricks, R. C., Baron, A. K., & Peller, I. C. (1975). *GASP—A computer code for calculating the thermodynamic and transport properties for ten fluids: parahydrogen, helium, neon, methane, nitrogen, carbon monoxide, oxygen, fluorine, argon, and carbon dioxide*. NASA TN D-7808, February 1975.
16. Hendricks, R. C., Peller, I. C., & Baron, A. K. (1973). *WASP—A flexible Fortran IV computer code for calculating water and steam properties*. NASA TN D-7391, November 1973.
17. Cryodata Inc., User's Guide to GASPAK, Version 3.20, November 1994.
18. Miropolskii, Z. L. (1963). Heat transfer in film boiling of a steam-water mixture in steam generating tubes. *Teploenergetica, 10*(5), 49–52 (in Russian; translation Atomic Energy Commission, AEC-TR-6252, 1964).
19. Van Hooser, K., Majumdar, A., & Bailey, J. (1999). Numerical prediction of transient axial thrust and internal flows in a rocket engine turbopump. In *35th AIAA/ASME/SAE/ASEE, Joint Propulsion Conference and Exhibit*, Los Angeles, CA, June 21, 1999, Paper No. AIAA 99-2189.
20. Schallhorn, P., Majumdar, A., Van Hooser, K., & Marsh, M. (1998). Flow simulation in secondary flow passages of a rocket engine turbopump. In *34th AIAA/ASME/SAE/ASEE, Joint Propulsion Conference and Exhibit*, Cleveland, OH, July 13–15, 1998, Paper No. AIAA 98-3684.
21. Bandyopadhyay, A., & Majumdar, A. (2007). *Modeling of compressible flow with friction and heat transfer using the generalized fluid system simulation program (GFSSP)*. Paper presented in Thermal Fluid Analysis Workshop, NASA Glenn Research Center, Cleveland, OH, September 10–14, 2007.
22. LeClair, A., & Majumdar, A. (2010). *Computational model of the chilldown and propellant loading of the space shuttle external tank*. Presented at AIAA Joint Propulsion Conference, Nashville, TN, July 2010.
23. Majumdar, A. (2013). *No vent tank fill and transfer line chilldown analysis by GFSSP*. Paper presented at Thermal Fluid Analysis Workshop, NASA Kennedy Space Center, July 29–August 2, 2013.
24. Majumdar, A. K., Steadman, T. E., Maroney, J. L., Sass, J. P., & Fesmire, J. E. (2007). Numerical modeling of propellant boiloff in a cryogenic storage tank. In *Cryogenic Engineering Conference*, Chattanooga, TN, July 16–20, 2007.
25. Majumdar, A., & Steadman, T. (2001). Numerical modeling of pressurization of a propellant tank. *Journal of Propulsion and Power, 17*(2).
26. Steadman, T., Majumdar, A. K., & Holt, K. (1999). Numerical modeling of helium pressurization system of propulsion test article (PTA). In *10th Thermal Fluid Analysis Workshop*, Huntsville, AL, September 13–17, 1999.
27. Majumdar, A., Valenzuela, J., LeClair, A., & Moder, J. (2016). Numerical modeling of self-pressurization and pressure control by a thermodynamic vent system in a cryogenic tank. *Cryogenics, 74*, 113–122.
28. Lee, N. H., & Martin, C. S. (1999). Experimental and analytical investigation of entrapped air in a horizontal pipe. In *Proceedings of the 3rd ASME/JSME Joint Fluids Engneering Conference*, ASME, NY, pp 1–8, 1999.
29. Bandyopadhyay, A., & Majumdar, A. (2014). Network flow simulation of fluid transients in rocket propulsion systems. *Journal of Propulsion and Power, 30*(6), 1646–1653.

30. Bandyopadhyay, A., Majumdar, A., & Holt, K. (2017). Fluid transient analysis during priming of evacuated line. In *AIAA 2017-5004, 53rd AIAA/SAE/ASEE Joint Propulsion Conference*, Atlanta, GA, July 10–12, 2017.
31. Cross, M., Majumdar, A., Bennett, J., & Malla, R. (2002). Modeling of chill down in cryogenic transfer lines. *Journal of Spacecraft and Rockets, 39*(2), 284–289.
32. Majumdar, A., & Ravindran, S. S. (2011). Numerical modeling of conjugate heat transfer in fluid network. *Journal of Propulsion and Power, 27*(3), 620–630.
33. Brennan, J. A., Brentari, E. G., Smith, R. V., & Steward, W. G. (1966). *Cooldown of cryogenic transfer lines—An experimental report.* National Bureau of Standards Report 9264, November 1966.
34. Darr, S. R., Hu, H., Glikin, N. G., Hartwig, J. W., Majumdar, A. K., LeClair, A. C., et al. (2016). An experimental study on terrestrial cryogenic transfer line chilldown I. Effect of mass flux, equilibrium quality, and inlet subcooling. *International Journal of Heat and Mass Transfer, 103*, 1225–1242.
35. LeClair, A. C., Hartwig, J. W., Hauser, D. M., Kassemi, M., Diaz-Hyland, P. G., & Going, T. R. (2018). Modeling cryogenic chilldown of a transfer line with the generalized fluid system simulation program. In *AIAA Joint Propulsion Conference*, Cincinnati, OH, July 9–11, 2018. AIAA Paper No. https://doi.org/10.2514/6.2018-4756.
36. Schallhorn, P., & Majumdar, A. (2012). Implementation of finite based Navier Stokes algorithm within general purpose flow network code. In *50th AIAA Aerospace Sciences Meeting*, Nashville, TN, January 9–12, 2012.

CFD and Turbulence

Turbulent Round Jet Entrainment—A Historical Perspective

Andrew Pollard

1 Introduction

This contribution is dedicated to Brian Spalding, a pioneer in the many realms of turbulence, heat and mass transfer and combustion. Much of this and his history are captured in recent articles by Artemov et al. [3], Runchal [54] and Launder, Patankar and Pollard [32], hereafter referred to as LPP, and many of the contributions in this book highlight certain facets of his sparkling career and life. Irrespective of any theoretical or modelling approaches used to investigate a physical phenomenon, one feature that was central to his work was that experimental confirmation was needed to either confirm or deny the approach taken and approximations that may have been made and indeed to highlight where refinements are necessary. While certainly not the first to have this approach, he did expect all his students to become competent in both experimental and theoretical/computational methods suitable to the problem at hand. I was one who benefitted greatly from this expectation and my career, which spanned eighty percent of the timeline given in the title of this volume. During this time, I was dedicated to the interplay between theoretical, computational and experimental fluid dynamics and its myriad applications and the tussle to reconcile differences between the often divergent information each approach offered. A prime example of this can be found in [31] where theory suggested that the then available experimental data could not be correct and they then demonstrated through very carefully controlled experiments alignment between theory and experiment (and subsequently confirmed through numerical computation, unpublished). I confess that this contribution is not a strict scientific paper in the traditional sense; I follow the long arm of Spalding from one of his elegantly simple but extraordinary insightful ideas embodied in one paper and how it has influenced the evolution of thinking through to the present day.

A. Pollard (✉)
Computational and Experimental Fluid Dynamics Laboratory,
Department of Mechanical and Materials Engineering, Queen's University
at Kingston, Kingston, ON K7L-3N6, Canada
e-mail: pollarda@queensu.ca

© Springer Nature Singapore Pte Ltd. 2020
A. Runchal (ed.), *50 Years of CFD in Engineering Sciences*,
https://doi.org/10.1007/978-981-15-2670-1_8

I also weave into this contribution personal thoughts and interpretations based on myriad discussions with Spalding over our intermittent 40-odd year fraternity.

Throughout his career at Imperial College, and indeed while still at Cambridge pursuing his Ph.D. research, Spalding would consider a problem and, while developing either the theoretical framework to explain the primary mechanisms he would devise elegant experiments to obtain the necessary data to complement the theoretical construct. Even in the very early days of computational fluid dynamics, he focused on the main physical mechanisms, such as parabolic features of boundary layers, which guided the construction of novel computational methods. Note that computer resource limitations also influenced his approach. He also either devised experiments that simply and elegantly enabled data that would support his theory or used existing data from what appeared in the literature. Of course, in the 1950s and 1960s, those data invariably were limited to mean quantities (for example velocity profiles) and what I consider to be first-order measures of flow (for example, skin friction, Nusselt numbers). To place his perspective into context, and Spalding's preference for finding a way to focus on the physical mechanisms, I include below what was reported by Anonymous [1] with reference to papers presented to an AGARD meeting on Combustion and Propulsion:

Four papers were presented in Group IV, the Combustion Session, the one which perpetuated the original purposes of the Panel, starting with Recent Progress in Flame Theory by Mr. D. B. Spalding of Imperial College. Until 1953, when Avery, Longwell, Bragg and van Heerden published separately their three basic ideas, the reasons for the extinction of steady-flow flames at high velocity or low pressure were not clear. It is only possible to study flame phenomena by simplifying the whole flame complex into separate groups of processes—fluid mechanics, heat and mass transfer, chemical reaction–and to treat the gas state and chemical features locally. Even so, combustion theory is too much for busy men to study and so Mr. Spalding offered seven "U.H.T.s"–"useful half-truths": (underline, author) The aerodynamic half-truth: In a given steady-flow combustion chamber, the streamline pattern depends only on overall fuel-air ratio (OFAR); the mixing half-truth: The local fuel/air ratio (LFAR) at any point in a combustion chamber of given shape depends only on the overall fuel/air ratio, and the position of the point; the thermodynamic half-truth: In a given two-stream flow, the state of the local gas mixture is completely specified by the local fuel/air ratio and the local reactedness, r, the relation between the properties being that for adiabatic steady-flow mixing; the chemical half-truth: The local volumetric reaction rate in a flame depends only on the local fuel/air ratio and the local reactedness; the two-stream half-truth: The state of a two-stream steady flame in a given chamber depends only on its dimensionless loading, L2, and the overall fuel/air ratio; the one-stream half-truth: The state of a one-stream steady flame in a given chamber depends only on its dimensionless loading, Li; The superposition half-truth: The reactedness at any point in a steady flame is equal to the sum of the increments of reactedness which would be caused at that point by the volumetric reaction rates prevailing at each increment of flame volume, if that incremental volume alone were reacting. These seven U.H.T.s were explained and elaborated, after which six examples were discussed to make the case that today's problems are no longer restricted to laminar flame propagation and the homogeneous reaction zone. There is hope now that theory can be used to predict the results of a change in chamber geometry and that the chemist can use flame theory much more. There are three outstanding tasks: for the engineer to study models of engine like geometry to see if separate fuel and air induction is better than pre-mixing; for the aerodynamicist to study turbulent flows involving simultaneous jet-mixing, recirculation and chemical reaction; and for the chemist to devise experimental flame arrangements that are amenable to theoretical analysis, are easily set up, controlled and measured, and give widely differing results according to the reaction-kinetic scheme involved.

From these early beginnings, the U.H.T.'s introduced by Spalding guided much of his career, if not explicitly, at least in the spirit of using data to fit his insightful perspective and simplified version of the complex mathematical constructs on turbulence and combustion.

During the last 50 years or so, our community focus has been to increase the granularity of the data to be either predicted or measured, and if I maybe circumspect, only because we have the where-with-all to generate tera-bytes of data, and exa-bytes soon enough! And, to Spalding, this always begged the question, which he often said during our conversations: 'for what purpose?'. This question introduces the two complementary, but often contentious foci: either for engineering application or for the minutiae of the cause–effect for a particular phenomenon? In the case of turbulence alone, as at 2018, for example, we still rely on Kolmogorov's hypothesis for isotropic dissipation, which for many engineering problems is 'about right'; while experimental data to support this hypothesis for complex flows has yet to be confirmed when, for example, multi-dimensional strain rates are at play in regions where small- and large-scale intermittency are evident.

Of the many problems one could choose to focus, here I consider the round, turbulent jet that has played a central role in the careers of both Spalding and the author. This type of jet is ubiquitous both in application and as a canonical flow that enables various theories to be tested. The jet may be heated, it may be composed of a fluid that is different to that which receives it, i.e. it may be either cooler or hotter or of composition, it may be also combusting with concomitant production of thermal radiation, products of combustion, etc.

This contribution is divided into multiple parts. I consider the interaction of theory, experiment and simulation and, as the granularity of data increases, there is a con-comitant need for increased vigilance to ensure that the results from one approach are measured against its counterpart that has the same fidelity. That is, what was once acceptable say 40 years ago where mean and first-order statistics sufficed to characterize a flow, time resolved PIV of today is being developed to enable confirma-tion of direct numerical simulation results. The contribution considers modifications to a round turbulent jet, albeit from a personal perspective and then considers the turbulent–non-turbulent interface which has links to Spalding's ESCIMO theory. I finish by introducing Spalding's population model in the hope that in the future some-one shall pick up this idea and continue to contribute to its development to thereby increase our understanding of the flow physics of turbulent jets and the multi-scale phenomena encompassed therein.

2   The Three-Legged Stool of Science: Experiments, Simulations and Theory

There have been innumerable advances made in the last 60 years in our understanding of some of the basic elements that compose the turbulence and the flow fields of jets. Experimentally, the tools available have expanded from physical probes, such as Pitot probes of various sophistication, to hot wire anemometry that still can not be beaten for temporal resolution. Indeed, given our need to consider finer and finer scales of turbulence, down to the Kolmogorov length scale, nano-scale hot wire probes have been courageously developed, see [4]. Given that a jet at its outer regions experiences large-scale reversed flow, flying hot wires have been employed to ensure there are no biases in the data, see [17, 18]. Laser Doppler anemometry or velocimetry arrived in the 1970s and, as with hot wires, eventually was extended to enable multiple components of velocity to be measured at a single point even though these data were single samples and the statistics of the turbulence assembled through ensemble averages at each point in the flow field. Particle image velocimetry or PIV, another optical method, emerged where whole planes of data could be obtained; however, again, ensemble averages are required to obtain initially two-dimensional velocity vectors which were then with multiple cameras extended to three components. PIV, with the advent of extremely fast repetitive rate lasers (or cameras, now), enable time resolved data to be obtained. A new technique that uses magnetic resonance imaging is emerging as an additional tool in the arsenal of the experimental turbulence researcher, see [21] and more recent advances from Eaton's group at Stanford, see [7, 14]. Even with all these advances, however, limitations and assessment of measurement accuracy remain.

Experimental data are now being generated in volumes and time slices that rival and indeed can easily exceed those generated through simulations for the same problem, see, for example, Pollard et al. [52]. Even so, the limits on the methodology or instruments used are often evolutionary and thus the uncertainty in experimental data may be unknown. A case in point is the use of hot wires to simultaneously measure turbulence velocities. Before embarking on a campaign to perform detailed measurements in a round jet [33], one must be aware of the limitations of data-reduction schemes. Four schemes were implemented and applied to measurements taken in the near field of a free, round jet with single-, cross- and four-wire probes. The results of these three databases were compared with experimental data from the literature and large eddy simulation computations. Comparisons between reduction techniques were also made to examine their respective characteristics and assess how they affected the accuracy of measurements made in a turbulent jet flow. Estimates of the accuracy of each scheme were given as a function of velocity direction and magnitude. It was concluded that four-wire probes, while a useful addition to the experimenter's arsenal, remain limited to low and moderate turbulence intensity flows. The key element here was the use of one facility, one experimenter and the same equipment and calibration methodology and access to LES data from within the same research group so the same initial and boundary conditions could be carefully assessed and implemented and informed decisions made.

Some time ago, I wrote, Pollard [51]:

> I think it is important to note that research performed using either virtual or physical methods usually begins by asking either a question or posing a hypothesis that needs to be tested. Physical methods usually have the tools in place, for example a wind tunnel, while virtual methods have yet to enjoy the same general level of infrastructure, for example, good computing facilities or range of software. Thus, in virtual experiments, the time and effort required to set up the tool may overshadow the reasons for the work. As well, the information available from a simulation is growing quickly so that it is easy to become overwhelmed by the amount generated.

This perspective has changed in the subsequent 20 years: computing infrastructure for the majority of calculations undertaken in both universities and industry now resides on the desks of researchers and designer; computer memory can be easily and cheaply obtained in the terabyte range, etc., and, thanks to Spalding, his consulting company CHAM and its primary product, the comprehensive CFD code PHOENICS and the like; the software is ubiquitous and, in either knowledgeable or well guided hands, relatively straightforward to set up and obtain solutions for the problem considered. Of course, high performance computing environments are critically important to enable the development of new computational methods at exa-scale and to simulate multi-scale, multi-physics problems in complex geometries, see, for example, the LRP for Canada [36], which was instrumental both for and in its generation. Even so, coding bugs are common and bug fixes normally require issuance of software updates which then places in jeopardy the full correctness of anyone previous calculation or simulation. Indeed, many times, Spalding urged me to embrace PHOENICS in my research, but I resisted, for many reasons, but principally because early on in the life of many codes, coding bugs were rife (although that is not to say PHOENICS was suspect). I think the only flow that has been calculated repeatedly where the results are now without much doubt is simple channel flow for which the mean and the time-averaged turbulence characteristics have been consistently reported. All other simulations tend to be one of a kind and as shall be noted below, comparison to experimental data maybe suspect since those data could characterize a problem in which either the initial or boundary conditions are not what is employed in the simulations. A case in point is the round free jet: comparison of the diminution in the mean axial velocity (only; the data set is otherwise remarkably good) using data from [74] would be incorrect since their data were taken in an enclosure that caused the jet not to follow a linear decay rate of x^{-1}.

Indeed, as the granularity of data increases and the questions asked of it become more informed, it is critical that theorists, simulators and experimenters become even more acutely aware of initial and boundary conditions; as well as the limits to those data against comparisons are sought. In some ways, this evolution in our abilities continued to bother Spalding, as noted in his concluding remarks in [65] with minor editorial changes by the author:

It is now nearly half a century since the arrival of the digital computer allowed turbulence models to be used for simulating flow phenomena of practical interest. Two-equation models of the k-epsilon kind soon became popular.

In the 1970s and '80s, success was encouraging and hopes remained high. Gradually however, the limitations of such models became apparent. Researchers developed more elaborate versions, which fitted a few experiments better, and some worse; but all relied on solving equations for time-averaged quantities. Insofar as PDFs were concerned at all, they were guessed rather calculated.

The arrival of bigger computers allowed many researchers to turn with relief to DNS and LES, with the result that research on developing models of the earlier type languished. Consequently, engineers are today still constrained to use the decades-old models which modern researchers scorn; for they have nothing else.

It is therefore time to recognize that

- Neither DNS nor LES will become cheap enough or practical for many decades;
- But there do already exist some cheap-enough-to-use models which can compute PDFs;
- Of these, the multi-fluid model fits most easily into the framework of conventional CFD codes;
- Such models rely on adjustable constants and functions which can be deduced from experimental, DNS or LES studies; and
- That, if this is to proceed as rapidly as it should, modellers, experimenters and DNS/LES practitioners need to talk to one another.

In fact such discussions have taken place, Pollard et al. [52], which asks: Whither Turbulence and Big Data in the twenty-first Century? So perhaps a better question to be asked, for this volume is whither the engineering and science of turbulence? By this I mean to the engineer, interested in say the spread rate of a free jet, what value is there in knowing that the turbulence integral length scales with velocity cubed times the inverse of the dissipation rate, which, by the way, according to George [23] is incorrect anyway; alternately, the researcher interested in the turbulence physics wonders why given the increasing evidence to the contrary the engineer seems content, or worse unaware, to rely on models of turbulence that are inherently unable to capture certain flow phenomena, such as turbulence driven secondary flows that arise from anisotropic turbulence stresses. Indeed, it was the ability to appreciate both perspectives that was a hallmark of Spalding and it became even more important to him in his later years when direct numerical simulation and large eddy simulation focused more and more on the flow physics and ignored, as far as he was concerned, the need to predict those things that an engineer would want to know. Of course, his focus was always on the combustion side of physico-chemical hydrodynamics rather than, say, just turbulence! Yet, fundamental phenomena were always at the core of his thinking and modelling strategies so the introduction of chevrons on the nacelles of new aircraft engines, for example, was developed from intimate knowledge of jet flows including the introduction of tabs and lobes. This database of knowledge was obtained using experiments, simulation (and modelling) and theory of sound generation and propagation.

3   The Turbulent Round Jet [53]

Spalding's interest in jets arose because of their central role in many engineering applications where combustion takes place, see LPP [32]. In these cases, either pre-mixed fuel and air are introduced into a combustion chamber or they are fed separately to mix while combustion takes place. In either case, entrainment of irrotational fluid beyond the limits of the jets causes the jet to spread, which then alters the temperature profiles and subsequently the radiation emitted from the combustion gases. So, even without combustion taking place, it was important to understand and determine the magnitude of the dilution of fuel (or jet source fluid).

Spalding's introduction to free turbulent shear flows seems to have begun with his pioneering efforts to determine the entrainment rates associated with a round jet, which I quote from [53]:

> The mass flow rate m is known to increase with distance x from the nozzle. As a consequence, fluid from the surrounding reservoir is drawn radially inwards towards the jet across its conical surface; this process is known as entrainment. Entrainment is also important in many more practical situations; for example, it controls the flow patterns in combustion chambers and furnaces; it causes 'fire-storms' around large conflagrations; and many mixing devices of the chemical industry rely on entrainment for their effectiveness. If such processes are to be understood and controlled, the quantitative laws which govern the rate of entrainment, i.e. the quantity dm/dx, must be discovered.

While those data obtained remain important and useful as shall be noted below, what remains unique and indicative of his insightfulness is the experimental arrangement implemented to determine dm/dx. In the theoretical derivation for jet entrainment, a basic assumption is that the pressure gradient is everywhere zero, which to first order is acceptable, or perhaps in the vein of a UHT, referred to above. And, it is here that Ricou and Spalding's insight came to the fore: place the jet in a porous walled enclosure and feed the jet sufficient fluid to ensure that all pressure gradients are zero; indeed, they used a micro-manometer that was of Spalding's design [66] for this purpose. From this work, two significant contributions were made in my opinion: the saturation in entrainment rate with Reynolds number, see their Fig. 3, and linear change in entrainment with distance from the jet exit irrespective of the densities of the source and receiver fluid, see their Fig. 4. In the case of saturation, it is notable that beyond about Re = 20,000, the entrainment does not change. In the case of the linear change in entrainment, what are the controlling mechanisms, which subsequently were connected to turbulence structure, for this to be so? Why it required 40 years or so to attempt to come to understand why the first occurred and why it took an additional 10–15 years to definitively conclude the second? Much of the remaining portions of this contribution trace some of the histories behind these questions.

4   The Turbulent Round Jet (1960 Through About 2000)

Between the appearance of the Ricou and Spalding paper and about the year 2000, there were many, papers that focused on characterizing the flow field of round jets, which quantified the mean and turbulence quantities most notably in the far field (beyond, say, $x/D \sim 50$). Focus, too, was on entrainment; however, of the approximately 500 publications that cite this paper in this period, only a handful actually make direct measurements of entrainment.

Early studies on jets focussed on the concepts of self-preservation whereby irrespective of axial location, the profiles of say axial mean velocity can be scaled using one variable, which in the case of the round jet is the jet half-width. Consequently, the Navier–Stokes equations can be reduced to simpler forms and the evolution of the flow can be predicted. These predictions should include both the mean and turbulence quantities. However, self-preservation concepts prior to George [22] assumed that scaling, at least asymptotically, did not depend on the initial conditions of the flow, only its boundary conditions since the flow would adjust to these, lose memory of its origin and only its local properties should be important. An excellent example of this was the paper of Wygnanski and Fiedler, referred to earlier, that demonstrated that their round jet achieved self-preservation in both mean and high order moments of turbulence quantities but the axial velocity did not decay as $1/x$ as expected because the room into which the jet issued was too small and of course the mass of the jet increases due to entrainment, while the momentum integral must be constant. George [22], in what I think is a cornerstone paper, faced with conflicting evidence from a variety of experimental studies, recognized that the fundamental missing piece was to include the initial conditions in whatever analysis was to be undertaken. I quote from his paper:

> It will be argued that the problems presented by the experiments lie not in the experiments themselves nor in the concept of self-preservation, but rather in the restrictive manner in which the self-preservation analyses have been carried out. It will be shown that a more general self-preservation analysis leads to the conclusion that, contrary to previous belief, there exists a multiplicity of self preserving states (for a particular type of flow) and that each state is uniquely determined by its initial conditions.

George and his students considered the free jet, see, for example, Glauser [24], Hussein et al. [28], Citriniti and George [9, 10], and Jung et al. [29], using a variety of experimental techniques, including a 138 hot wire probe arranged in a cylindrical-polar coordinate manner. They focussed mainly at downstream distance of $x/D = 3$, which is well within the potential core region where Kelvin–Helmholtz instabilities are very evident. From those data, they applied the proper orthogonal decomposition of Lumley [37]. The POD attempts to find functions that represent the velocity vector, say $u_i(x, t)$ in an optimal way. The functions are found from the eigenvalue equation that relies on two-point correlations

$$\int_D R_{i,j}(\vec{x}, t, \vec{x}', t')\phi_i(\vec{x}', t')\mathrm{d}(\vec{x}, t) \tag{1}$$

from which the velocity can be reconstructed using

$$u_i^N(\vec{x}.t) = \sum_{n=1}^{N} a_n \phi_i^{(n)}(\vec{x}, t) \tag{2}$$

where the coefficients are obtained from

$$a_n = \int u_i(\vec{x}, t) \phi_i^{(n)*}(\vec{x}, t) \mathrm{d}(\vec{x}, t) \tag{3}$$

and the kinetic energy can be had via the sum of the eigenvalues

$$E = \sum_{n=1}^{\infty} \lambda^{(n)} \tag{4}$$

Thus, the various modes can be calculated to identify and extract the coherent structures as well as to ascertain how the energy modes are distributed. This work clearly separated characterisation of the jet from the more traditional first- and second-order velocity moments. That said, Spalding never thought much of this approach because they detracted from what an engineer needed to know about the flow, especially entrainment and mixing processes, which these efforts, considerable as they were, considered only in passing. But, from a flow physics perspective, this approach and the influential papers from George's group highlighted the complex flow structure in the near field of round jets, identified the energy associated with different length scales within the flow. Furthermore, these are important papers if only because of the care and attention brought to the measurements; the data obtained remain both unique and consequential.

In the near field of a round jet, there exists the centreline region, the shear layer and the outer layer, where attention will be placed later, see Sect. 7. In the centreline region, the flow's maximum axial mean velocity is found on the centreline; here, turbulence evolves with axial position to eventually reach equilibrium and far-field characteristics. In the shear layer, the radial velocity gradient causes vortex cores to form, evolve and pair to create large eddies. These vortex cores are interlinked with axial (or longitudinal) vorticity to create structures referred to as 'braids'. The large eddies break down and form smaller and smaller eddies; that is, the turbulence structures diminish in scale in both time and space. Throughout this process, energy is transferred from the large-scale structures to the smaller. Velocities in the outer layer, typically of order $0.1U_c$, rapidly fall to the free stream value with increasing radial distance from the centreline.

While the axisymmetric and helical modes were known, see [46] and those of George and co-workers, it was the use of the 138 probe array, located in the very near field only, that enabled much of the axial and azimuthal turbulence structure to be captured and dissected into multitude of modes through the application of proper orthogonal decomposition. While George's group contributions focused on the structures and their presumed influence on entrainment and indeed did provide estimates of entrainment using mass and momentum conservation principles (see [28]), direct measurement of this feature was not investigated.

Dahm and Dimotakis [12] considered mixing and entrainment using concentration fields and noted that instantaneous entrainment from the unmarked irrotational fluid external to the jet penetrated deeply into the jet. They considered $1500 \leq Re \leq 20{,}000$. Dimotakis [13] followed this work to classify jets as being above or below what he called the mixing transition. His introduction to the topic is repeated below since it is relevant to the focus of the current contribution:

> A correct description of turbulent mixing is particularly taxing on our understanding of turbulence; such a description relies on an account of the dynamics spanning the full spectrum of scales. Specifically, to describe the entrainment stage that is responsible for the engulfment of large pockets of irrotational fluid species into the turbulent flow region [8], the large-scale flow structures need to be correctly described. Secondly, to describe the subsequent kinematic stirring process responsible for the large interfacial surface generation between the mixing species, the intermediate range of scales must be correctly accounted for. These are below the largest in the flow in size, but above the smallest affected by viscosity and molecular diffusivity. Finally, the dynamics at the smallest scales must be captured to describe the molecular mixing process itself.

Those with a bulk Reynolds number $Re_D \leq 20{,}000$ or Taylor Reynolds number $Re_T = u'\lambda_T/\nu \leq 100$ cannot be considered to be fully turbulent, while jets above these values are indeed fully developed turbulence. Furthermore, from [17]:

> Dimotakis [13] demonstrated that fully developed flow on jet centerline requires a Reynolds number of $Re_D \geq 10000 - 20000$ or a Taylor Reynolds number of $Re_T \geq 100 - 140$, which indicated a transition from coupling of inner and outer scales of turbulence to where this coupling is removed. He concluded that the mixing transition seems to occur at the appearance of the transition from the inertial to dissipation range. Dimotakis explained that the mixing transition may not be traced to the large-scale structure in the flow, which are peculiar to the geometry considered, rather to the physical significance of the various scales of the turbulence and their Reynolds number scaling. That is, there is a decoupling between the large scales and those free from the effects of viscosity. Dimotakis linked the outer scale δ to a variety of other length scales and argued that "viscous decoupling of the outer and inner scales of turbulence is responsible for the transition criterion;" however, the relative magnitude of these length scales does not provide a clear physical reason as to why $Re_D \approx 10000 - 20000$ is the 'magic' criterion that characterizes the start of fully developed turbulent flow.

Throughout this period covered by this section (1961–2000), the paper by Ricou and Spalding was seminal and influenced over 500 papers that dealt with jets and entrainment.

5   The Turbulent Round Jet (2000 and Beyond): Swirl, Tabs and Rings

According to Google, in the first 20 years or so of the twenty-first century, a further 700 papers referenced Ricou and Spalding. Many dealt with combustion, but it remains importantly the basis for research for investigation of entrainment in non-combusting flow situations. In fact, based on perusal of these 700 papers, only about 10% of them refer to the method used by Ricou and Spalding for direct measurement of entrainment. Even so, the general approach inspired its use in variable density and multi-phase flows, see Medrano et al. [42] who considered atomising liquid jets and provide a good list of recent papers on current efforts to characterize entrainment in jets.

Ball et al. [5] noted that at the turn of the century, computational science and engineering (CSE) became a mainstream theme at universities because of the increasing ubiquitous availability of high performance computers. This permitted both large eddy and direct numerical simulation to be entertained by a larger group of researchers. The CSE investments enabled high fidelity calculations to be done that enabled deeper insight into the more fundamental flow physics of jets and of course other flows. For example, Suto et al. [68] used large eddy and direct numerical simulation, to uncover the coherent structures of the developed jet. They used top-hat and 'hyperbolic' inlet mean velocity profiles, but did not note any substantial difference in the structures when well removed from the nozzle exit plane. They proposed a conceptual model of a hairpin-shaped vortex and validated it by two-point correlations and probability density function (PDF) analysis for vortex alignment. They observed pairs of vortices in the jet outer layer and streamwise vortices in the jet central region, the latter being roughly half the jet half-width in size. The hairpin-shaped vortices formed in the background shear of the mean flow 'stand' with 'legs' inclined at roughly 45° to the mean flow direction; these vortices were offered by Suto et al. as being among the universal structures in turbulent shear flows. Suto et al. is a precursor to more recent work on coherent structure eduction in jets.

5.1   Swirl

To gain further insight into the structure of the turbulence in round jets, and their influence on entrainment, we considered LES for round jets including the effects of swirl and tabs, both of which are known to enhance mixing through enhanced entrainment, but the mechanisms remain unclear. McIlwain and Pollard [40] in a time period similar to Suto et al. considered the near fields of turbulent round free jets with swirl numbers of 0.0 and 0.24, which is too weak to enable the formation of recirculating flow within the central core of the jet, and modelled with large eddy simulations that use a dynamic Smagorinsky subgrid scale model. The time-averaged results were found to be in accord with other numerical and experimental data. Instantaneous plots of vorticity were used to examine the evolution of and interaction between coherent structures. Ring structures aligned with the plane normal to the flow form downstream of the jet shear layer and collide with the streamwise braid structures, which interconnect with the ring structures. The resulting interaction causes the rings to break apart into smaller, less organized turbulence structures. The addition of swirl increased the number of streamwise braids, which were found to enhance the breakdown mechanism of the rings. The results suggested that the increased entrainment observed in swirling flows is due to the action of the braids rather than the rings. And, these braids, illustrated in Fig. 1 for the non-swirl configuration, interconnect the Kelvin–Helmholtz primary vortex rings.

5.2 Tabs

Vortex-generating tabs are used in turbulent round jets to increase mixing and flow entrainment, improve combustion efficiency and reduce noise. The tabs are small protrusions that extend into the flow from the nozzle wall, often placed at the nozzle exit. They distort the round shape of the exit velocity profile and manipulate the vorticity in the flow field, thereby controlling the evolution of the large-scale turbulence structures.

A relationship between the presence of large-scale structures of vorticity and the rate of entrainment in a turbulent jet flow was observed in the large eddy simulations of McIlwain and Pollard [40]. The structures form due to flow instabilities downstream from the nozzle. The primary Kelvin–Helmholtz instability mechanism causes the shear layer to roll up into vortex rings. Secondary instabilities also develop that create finger-like pairs of streamwise vortex structures that emerge from the region between two vortex rings and stretch around the jet core [34].

The effect of the shape and tilt (angle of attack) of the tabs, and the combination of several tabs at the nozzle exit, have been previously investigated through experiments. A tab generates a pair of counter rotating streamwise vortices at the nozzle, largely due to the formation of an upstream 'pressure hill' [78]. Enhanced mixing within the flow and an increase in the number of small-scale structures has been observed by Huang and Ho [27]; consequently, an increase in the rate of entrainment of the surrounding fluid is also expected. However, it is not known whether the enhanced mixing in the flow is caused by the increased surface area of the mixing layer that is exposed to the surrounding fluid, or the observed vortex breakdown farther down stream. The effect of a combination of tabs remains controversial: when too many tabs are used, the flow field may settle back into a steadier configuration with a less profound distortion [43, 78]. Zaman [77] concluded:

> Jet spreading for the two tab configurations is found to be the largest among all the cases. This is true in the subsonic regime, as well as in the supersonic regime in spite of the fact that screech is eliminated by the tabs. The dynamics of streamwise vortex pairs produced by the tabs cause the most efficient jet spreading. Thus, a manipulation of the streamwise vortex pairs may hold the key for any further increase in jet spreading.

Although tabs do increase mixing by intensifying the interaction between azimuthal and streamwise turbulence structures, they must be used with a good understanding of their mechanisms to obtain a positive and efficient effect [20]. My students and I embarked on an LES study of round jets with and without tabs [38, 41]. We used single, dual, triple and quadruple tabs arranged symmetrically at the exit

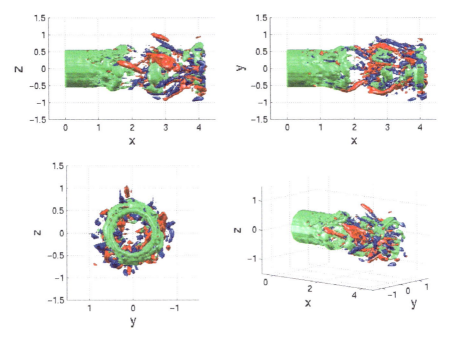

Fig. 1 Instantaneous near field of round jet simulated with large eddy simulation. Green is azimuthal vorticity and red/blue is $+/-$ axial vorticity and are indicative of the 'braids' that link the azimuthal vortex structure that is essentially a Kelvin–Helmholtz instability. $Re_D = 68,000$ which is well beyond the mixing transition value of $Re_D \sim 20,000$

of a round nozzle and considered the flow to a few diameters downstream of the jet exit. Figures 1, 2 and 3 display the azimuthal and axial vorticity distributions for the cases of no, dual and quadruple tabs. They indicate that the number of tabs increases the variegation of the coherent structures with the use of four tabs enhancing this feature most of all. A plot of the flow rate of the jet indicated four tabs was greater than the jet without tabs and better than those with fewer at the downstream extent of the flow domain considered. These simulations provided further evidence of the role of finer scale structural elements to the entrainment process (Fig. 3).

5.3 Rings

Buoyed by the evidence that swirl and tabs increase entrainment, Sadeghi and Pollard [56] embarked on an experimental study of introducing small rings at very small offset distances from the nozzle exit into the shear layer and then into the potential core of a free jet. The rational was to observe the effects of the wakes of these rings on the inner and outer portions of the jet shear layer at various locations in the near to intermediate field of a round jet including the effect of Reynolds numbers. Recall that Ricou

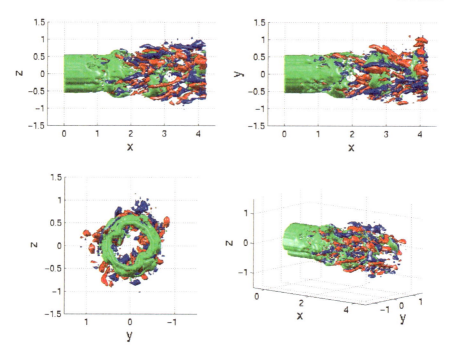

Fig. 2 Instantaneous near field of round jet with 2 tabs, simulated with large eddy simulation. Green is azimuthal vorticity and red/blue is $+/-$ axial vorticity. $Re_D = 68,000$. Thresholds same as used in Fig. 1

and Spalding noted the saturation in mass flux beyond $Re = 20,000$ and Dimotakis explained that the mixing transition may not be traced to the large-scale structure in the flow, which are peculiar to the geometry considered, rather to the physical significance of the various scales of the turbulence and their Reynolds number scaling. That is, there is a decoupling between the large scales and those free from the effects of viscosity. Dimotakis linked the outer scale δ to a variety of other length scales and argued that 'viscous decoupling of the outer and inner scales of turbulence is responsible for the transition criterion;' Furthermore, Fellouah et al. [18] used flying and stationary hot wire measurements to investigate the effect of the Reynolds number that spanned range associated with the mixing transition, on the near to intermediate fields of a round free jet. The specific Reynolds numbers Re_D tested were 6000, 10,000 and 30,000. Their result revealed a close coupling between the mean velocity distribution and the turbulence intensities and the Reynolds shear stress. A significant result was obtained from the velocity spectra, where the inertial sub-range emerged for Reynolds numbers above $Re_D \geq 20,000$. At Reynolds numbers below 20,000, the energy of the large-scale turbulence (low wavenumbers or frequencies) proceeds directly to the dissipation range. Plots of the compensated one-dimensional longitudinal velocity spectra, defined as

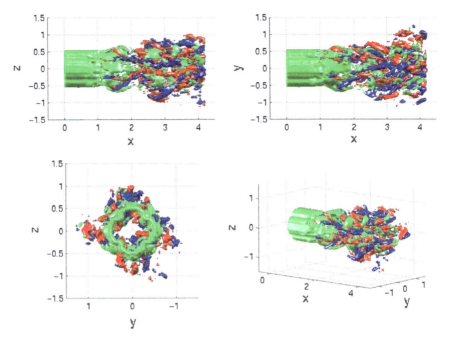

Fig. 3 Instantaneous near field of round jet with 4 tabs, simulated with large eddy simulation. Green is azimuthal vorticity and red/blue is $+/-$ axial vorticity. $Re_D = 68,000$. Thresholds same as used in Fig. 1

$$\epsilon^{-2/3}\kappa_1^{5/3}E_{11}(\kappa_1) \tag{5}$$

were made at various radial locations across the jet and at $x/D = 10, 15$ and 20. When plotted against (product of wavenumber and Kolmogorov length scale) indicated an increased broadening with increasing x/D thereby suggesting a larger community of length scales. They furthermore concluded that the mixing transition occurred at the appearance of the inertial sub-range rather than at the transition from the inertial to dissipation range.

As an aside, with the introduction of Kolmogorov length scale, it maybe of interest to note the connexion between him and Spalding. As readers maybe aware, Spalding was fluent in the Russian language after he vowed to learn it after a visit to Minsk in the early 1960s, see LPP [32]. He obtained Kolmogorov's original paper that described the $k - \omega$ two-equation model of turbulence, and while I recollect, but have no firm proof, he translated this earlier than what appears in [64]. In that paper, Spalding notes:

> The variables which Kolmogorov chose for the characterization of turbulence were the fluctuation energy, b, in his notation...and the frequency, ω, in his notation...The first quantity is actually two thirds of the kinetic energy (usually given by the symbol, k) which appears in the currently popular models; and the second, if multiplied by b, is proportional to the energy dissipation rate (usually given by the symbol ϵ, in contrast to Kolmogorov's use of

the same symbol) which is the second variable for the model invented much later by Harlow and Nakayama [26].

Kolmogorov recognized that the length scale was proportional to $b^{1/2}/\omega$ and so could be deduced from the solution to his equations; and he utilized the quantity b/ω as a measure both of the effective diffusivity of turbulence and of the effective kinematic viscosity. His quantity w is proportional to the square root of W used by Saffman [58] and Spalding [60] as a measure of the fluctuations of vorticity; and his equations can be directly derived from theirs.

Interestingly, irrespective of the modelling challenges, turbulence length scales preoccupied Spalding's thinking from very early in his career.

Subsequently in my career, I became equally inquisitive about turbulent length scales and that, even though many length scales had been suggested, the only one that I believed to be inviolate was Kolmogorov's which states that the smallest length scale must dissipate its energy at the rate of u^3/l, which gives rise to a Reynolds number of unity. Even so, there are integral scales and Taylor micro-scales each of which contributes to our understanding of turbulence.

Sadeghi and Pollard [56] used two different fine rings, with square cross-sections, which were placed very close to the jet exit ($x/D = 0.03$) so as to interfere with the evolution of the naturally occurring flow structures and to further influence the distribution of length scales. The first ring was placed in the middle of the shear layer while, in a separate experiment, the second ring covered a small region of the potential core. Flying and stationary hot wire measurements were carried out to study the near- and intermediate-field development of the jet. Three Reynolds numbers (based on the jet exit mean velocity and the nozzle diameter) were used: 10,000, 30,000 and 50,000. The results showed a considerable reduction in the jet spread rate and turbulence intensity when the passive rings were employed. This is more obvious for the ring placed in the shear layer. The radial profiles of the mean and rms axial velocity and third and fourth moments of fluctuations at several axial positions were considered. The power density spectra revealed the suppression of the initial shear layer instability (shear layer mode) while the jet preferred instability (preferred mode) remained active as the shear layer was modified. In the case when potential core modifications were used, the suppression of the initial shear layer instability was still considerable although there was evidence of another vortical structure behind the ring, which, it was speculated, was due to the vortex shedding behind the ring. In this case too, the jet preferred instability remained active. Therefore, the work confirmed the separation of these two modes. In addition, it provided a comprehensive study of different length scale characteristics in the development region of the jet (e.g., integral scale, Taylor micro scale and Kolmogorov scale) for both the modified and unmodified jets. While it was speculated that entrainment would be affected by the addition of the rings, previous data from [49] and [69] seem to be at odds since they used rings located at different axial locations from the jet exit plane as well as at Reynolds numbers that were below or above the mixing transition value of $\sim$20,000. A good overview of free shear layer control can be found in [19].

6   Higher Order Statistics

The round jet has been investigated experimentally in depth more than computationally (by that I mean LES and DNS), see [5]. While LES has been applied to incompressible round jets, see for example [25] for Re = 4300 and [6] for noise in compressible jet flows at Re = 65,000 and Ma = 0.9, the results are suspect, like any calculation where assumptions are made about the underlying flow physics, which in this case is the subgrid scale model. Direct numerical simulations for jets remain in their infancy, but there is progress albeit slowly. The highest Reynolds number for which DNS has been done, including comparison with experimental data, at the time of writing is about 4700 [72]. Even so, there are significant differences in what is predicted compared to various sets of experimental data, most notably to those of Panchapakesan and Lumley [47] who used a moving hot wire, yet Hussein et al. [28] data seem to be more in support of the DNS results. Most recently, Shin et al. [59] conducted DNS of a round jet at $Re_D = 7290$; however, their focus was on residence times since their main goal was to ultimately consider combusting flows. That said, Shin's group is collaborating with mine using the DNS database to provide further insight to second- and third-order structure functions using experimental data from [57], see [2]. Indeed, it is to the work of Sadeghi et al. that attention is now turned.

Kolmogorov [30] related the mean energy transport to both turbulent advection and molecular diffusion (K41),

$$- \langle (\delta u)^3 \rangle + 6\nu \frac{\mathrm{d}}{\mathrm{d}r} \langle (\delta u)^2 \rangle = \frac{4}{5} \langle \epsilon \rangle r \; , \tag{6}$$

which, for moderate Reynolds numbers encountered under laboratory conditions, is not balanced and assumes global isotropy; where $\delta u \equiv u(x + r) - u(x)$ is the longitudinal velocity increments (for the streamwise velocity component u), r is the distance between two points considered along the streamwise direction, x, and ν is the kinematic viscosity. Here, $\langle \epsilon \rangle$ is the mean dissipation rate of turbulent kinetic energy defined as

$$\langle \epsilon \rangle = \frac{1}{2} \nu \left\langle \left(\frac{\partial u_i}{\partial x_j} + \frac{\partial u_j}{\partial x_i} \right)^2 \right\rangle .$$

The main reason for this imbalance is the inhomogeneous and anisotropic large scales, which are not sufficiently separated from the smallest dissipation scale. Various types of inhomogeneities have been examined in the past for different turbulent flows such as grid turbulence, fully developed channel flow and along the jet centreline. The inhomogeneities associated with large scales were considered by Sadeghi et al. [57] in the shear layer of a round jet since several large-scale phenomena coexist in this region. A new generalized form of Kolmogorov's equation was proposed for the off-centerline (shear layer) of round jet flows:

$$-\left\langle (\delta u)\,(\delta q)^2 \right\rangle + 2\nu\frac{\mathrm{d}}{\mathrm{d}r}\left\langle (\delta q)^2 \right\rangle - \frac{U}{r^2}\int_0^r s^2\frac{\partial}{\partial x}\left\langle (\delta q)^2 \right\rangle \mathrm{d}s$$

$$-2\frac{\partial U}{\partial x}\frac{1}{r^2}\int_0^r s^2\left(\left\langle (\delta u)^2 \right\rangle - \left\langle (\delta v)^2 \right\rangle\right)\mathrm{d}s - 2\frac{\partial U}{\partial y}\frac{1}{r^2}\int_0^r s^2\,\langle (\delta u)\,(\delta v)\rangle\,\mathrm{d}s$$

$$-\frac{1}{r^2}\int_0^r s^2\frac{\partial}{\partial x}\left\langle (u+u^+)\,(\delta q)^2 \right\rangle \mathrm{d}s$$

$$-\frac{1}{r^2}\int_0^r s^2\frac{1}{y}\frac{\partial}{\partial y}\left(y\left\langle (v+v^+)\,(\delta q)^2 \right\rangle\right)\mathrm{d}s - \frac{2}{r^2}\int_0^r s^2\frac{\partial}{\partial x}\,\langle (\delta u)\,(\delta p)\rangle\,\mathrm{d}s$$

$$-\frac{2}{r^2}\int_0^r s^2\frac{1}{y}\frac{\partial}{\partial y}\,(y\,\langle (\delta v)\,(\delta p)\rangle)\,\mathrm{d}s = \frac{4}{3}\,\langle \epsilon \rangle\,r, \qquad (7)$$

where the main sources of the inhomogeneity are the streamwise decay of turbulent energy $\langle q \rangle$, normal stress $\langle u \rangle$, $\langle v \rangle$ production, shear stress $\langle uv \rangle$ production and axial and lateral diffusion. The second- and third-order 'structure' functions as delineated above must equal the dissipation. Even with great care taken to obtain data, the need for the use of Taylor's hypothesis and corrections as suggested by others (see Sadeghi et al.) bias and precision errors were determined to be up to about 10%; and moreover, the last two terms in the above equation include the pressure and Sadeghi et al. noted

> Therefore, similar to previous investigations, the pressure containing terms shall be neglected here, although there is no a priori reason to consider them negligible (and therefore their significance in (7) remains unknown). The contribution from the pressure terms may be interpreted from the deviation of the remaining terms' balance in the scale-by-scale budget Eq. (7). It should be noted that in some previous works (e.g. [35, 48]), the pressure–diffusion terms in the transport equation for the turbulent kinetic energy equation have been "indirectly" interpreted by closing the balance of other terms. The pressure-diffusion terms were found to be negligible on the jet axis, but their significance increased off the centreline.

While we continue to explore the effects of pressure using DNS, it is clear that our thinking has evolved from an engineering approach favoured by Spalding.

The validity of this last equation was investigated using hot wire data obtained for a round turbulent jet at $\mathrm{Re}_D = 50{,}000$. Crucially, we used a flying hot wire to determine that beyond $r/r_{1/2} \sim 1$, stationary hot wire data indicate that the jet spread was broader than those data from the flying wire, which was attributed to the velocity bias from the reverse flow associated with the passing of larger scale structures across the stationary wires in the outer portions of the flow. Indeed, this suggested that determination of entrainment in free shear flows must be considered with great care. And it also further demonstrates that the method of Ricou and Spalding to quantify entrainment, whilst of first order, did not require detailed three-dimensional, time dependent interfacial information, which was implicitly acknowledged by them:

> It is difficult to measure ρu at large values of y where the velocities are small and the flow may be intermittent; the presence of y as a multiplier of u in equation (1) integral of mass flow across any cross section of the jet, author augments the influence of inaccuracies in this region.

Indeed, to better understand the effects of intermittency at both small and large scale, Sadeghi et al. [55] revisited the round jet using both flying and stationary hot wires.

They used standard $5\,\mu m$, $2.5\,\mu m$ diameter wires and the recently developed nano-scale hot wire probe, which was $100\,nm$ by $2\,\mu m$ in cross section length. Crucially, this sensing element enabled data to be taken at sub-Kolmogorov length scales. Moreover, they demonstrated that stationary hot wire data cannot be trusted beyond about the jet half-width because of deleterious effects of the negative velocities encounter there due to large-scale incursions of irrotational fluid, or large-scale intermittency.

The driver for many studies of entrainment is turbulent mixing of heat and species, with the obvious extension to combustion, especially that of the non-premixed variety. Concomitantly, the products of combustion, once an accepted part of the process, are under intense scrutiny as we face the need to minimize them and indeed permitting them to be dumped into the earth's atmosphere. Thus, there is growing interest to more fully understand the dilution that can be achieved through entrainment especially as it relates to minimizing nitrogen oxides, as innovative ideas like that considered by Yimer et al. [76].

7   The Turbulent, Non-Turbulent Interface (TNTI)

The interface between turbulent flow and invariably its irrotational receiver fluid was probably first considered in detail by Corrsin and Kistler [11], wherein they attempted to understand and quantify the intermittent character of signals obtained from stationary hot wires:

> The overall behaviour of the front is described statistically in terms of its wrinkle-amplitude growth and its lateral propagation relative to the fluid as functions of downstream coordinate.
>
> It is proposed and justified that the front actually consists of a very thin fluid layer in which direct viscous forces play the central role of transmitting mean and fluctuating vorticity to previously non-turbulent fluid. Outside this "laminar super-layer" there is presumably a field of irrotational velocity fluctuation (the "non-turbulent" flow) with constant mean velocity. As outlined in the following paragraphs, theoretical analysis based on this general physical picture gives results on front behaviour which are in plausible agreement with experimental result for three turbulent shear flows: rough-wall boundary layer, plane wake, and round jet.
>
> It is shown that the rate of increase of wrinkle amplitude of the front can be roughly explained as a Lagrangian diffusion process, using the statistical properties of the turbulence in the fully turbulent zone.

The fact that they considered three shear flows indicated the broader awareness of the similarities even though the mean velocity profiles and the source of vorticity generation are different.

And, it has only been more recent that some focus has turned to the fine scales along the turbulent, non-turbulent interface (TNTI). A key differentiator that has emerged over the last 60 years is the concept of nibbling at interfaces, which maybe associated with small-scale turbulence effects as opposed to entrainment, which is often related to large-scale effects due primarily to large-scale structures that populate the outer portions of a jet or other free shear layer. In 1960, consideration of the entrainment concept was limited to a handful of publications where closed form solutions to simplified set of equations were obtained that enabled the prediction of mass flow rates beyond what was injected into the quiescent free stream. The flow physics of this phenomenon was limited since the concept of turbulent flow structures had yet to enter the armoury of the fluid mechanics community. At the time of writing, inserting 'turbulent jet entrainment' into Google delivers about 400,000 results, which attests to the importance of this concept in many fields of endeavour.

The turbulent entrainment process and their associated mechanisms control the transport of mass, momentum and scalars from a turbulent region of a fluid to a non-turbulent region. That is, the rotational fluid is separated from its external irrotational counterpart by an interface. After turbulent coherent structures were first identified, see for example [8], entrainment was deemed mainly due to the 'engulfment' of the irrotational non-turbulent region by the large-scale inviscid turbulent motions. Spalding, again at the forefront and continuously aware of advances in non-combusting flows, embraced the idea of turbulence structure and postulated his ESCIMO theory of turbulent combustion [62, 63]. Therein, he described the Engulfment, Stretching, Coherence, Inter-diffusion, Moving Observer and his Eddy Breakup Model ideas. The reader is directed to Feuyo and Malin in this volume for a more detailed review of these.

Later work by Mathew and Basu [39] and Westerweel et al. [73] among others, for example, showed that outward spreading of small-scale vortices ('nibbling') is the dominant process in the entrainment. Those who favour nibbling as the major mechanism of entrainment did not find significant amounts of unmixed fluid within the turbulent fluid; therefore, it has been argued that large-scale engulfment cannot be the major contribution to the entrainment. However, as noted by McIlwain and Pollard [40], swirl and tabs do influence both the large-and small-scale structures thereby increasing entrainment through the breakup of the larger scales, which supports the work of Moser et al. [45] who observed a larger entrainment in a forced-temporal wake compared to the unforced case. Forcing induces large-scale motions in the shear layers, and therefore increases the surface area of the turbulent/non-turbulent interface (TNTI). They did not consider whether this breakup constituted an increase in nibbling although they did speculate about the multi-scale nature as the entrainment increased from a non-manipulated jet. More recent evidence suggests that a full description of the entrainment process will need to account for multi-scale interactions, as proposed by Sreenivasan et al. [67], Philip and Marusic [50] and Reeuwijk and Holzner [71]. The latest instalment of increased granularity in pursuit of what occurs at the TNTI comes from a tri-university effort [44]. They used a round jet at Re = 25,300 and employed different spatial filtering methods applied to two-component PIV velocity data and PLIF scalar concentration data. Importantly, they determined '... at the smallest scales the entrainment velocity is small but is balanced by the presence of a very large surface area, whilst at the largest scales the entrainment velocity is large but is balanced by a smaller (smoother) surface area.' This lends support to the proposition of Townsend [70] and Philip and Marusic [50] who argued that in the entrainment process, viscous nibbling adjusts to the imposed entrainment rate defined by the large-scale parameters of the flow.

8 Multi-fluid Concepts of Spalding and Links to Turbulent Length Scales in Jets

Spalding was well aware of the efforts of computational engineers and scientists to increase the granularity of instantaneous features found in turbulent flows while they remained adherents to the statistical description that can be constructed from time-series analysis of signals. That said, his main focus was always on the more useful aspect of combustion. He began his foray into debunking probability density distributions (PDF, either presumed or not) by considering a laminar gas mixture composed on the inter-mingling of fully burned and fully unburned gases. By analogy, this could also mean the jet fluid and the receiver fluid. He termed the two-fluid as the Eddy Breakup hypothesis or EBU, Spaldung [61]. A little later, he considered multi-fluid models of turbulence, Spalding (2001), where a turbulent mixture can be regarded as a population having an arbitrary number of 'ethnic' components. By this, he maybe suggesting that turbulent 'eddies' can have differences (short, medium or tall in height, say) and they each could have different properties, say velocities, temperature, etc. He referred to some of these as 'promiscuous Mendelian' generation. Further discussion of this topic unfortunately did not occur between us prior to his untimely passing. I refer the reader to the article by Feuyo and Malin in this Volume for further insight into his multi-fluid models of turbulence as well as the large inventory of lecture material (87 entries at http://www.cham.co.uk/search/search.php?zoom_sort=0&zoom_query=population&zoom_per_page=100&zoom_and=0) on his ideas of population models. That said, a key feature of entrainment is the intermittency, and while he and his students worked on certain aspects of this, see Fan [15, 16] and Xi [75] in Feuyo and Malin, he continued to be preoccupied with PDFs since to him they are more useful than traditional statistics or averages. While I believe the population modelling approach may eventually be developed to handle detailed entrainment processes, given the extremely wide range of scales, I think it is unlikely that the details of the nibbling processes, for example, could be fully realized. I admit that because I am seeking higher granularity to understand the full import of the findings in [53], I am sure he would look back at me, with a twinkle in his eye and ask, 'for what purpose?'.

9 Summary

I have provided a survey of turbulent free round jets and entrainment from [53] landmark paper to the present. This paper continues to influence the thinking of those who study flows where a turbulent–non-turbulent interface is encountered. I have taken a decidedly personal perspective and no attempt has been made to cover the full spectrum of contributions to this 'simple' flow. Yet throughout the approximately 60-year span, his simple question 'for what purpose' has guided me and my students

Fig. 4 Photo taken at reception of THMT'15, Sarajevo, September 2015

as we have explored the intricacies of this flow from computational, experimental and theoretical approaches.

Acknowledgements It is with pleasure that I acknowledge the profound influence of Brian Spalding on and in my life and career. He was a brilliant person whether in science or art; his humanity knew few bounds and he always had a smile and a glint of mischievousness in his eye, often accompanied by a glass of some liquid!

References

1. Anonymous. (1958). Report from the combustion and propulsion panel of the NATO AGARD colloquium, Palermo, Flight, 6 June. Further Abstracts of Papers at AGARD Meeting, p. 779, Part 3.
2. Aparece-Scutariu, V., Shin, D., Lavoie, P., Richardson, E. S., & Pollard, A. (2019). Evaluation of scale-by-scale budget equation at off-centreline using a turbulent round jet DNS. Presented in 11th Symposium on Turbulent Shear Flow Phenomena, Southampton, July.
3. Artemov, V., Beale, S., de Vahl Davis, G., Escudier, M., Fueyo, N., Launder, B. E., et al. (2009). A tribute to Spalding, D.B. and his contributions in science and engineering. *International Journal of Heat and Mass Transfer, 52*(17–18), 3884–3905.
4. Bailey, S., Kunkel, G., Hultmark, M., Vallikivi, M., Hill, J., Meyer, K., et al. (2010). Turbulence measurements using a nanoscale thermal anemometry probe. *Journal of Fluid Mechanics, 663*, 160–179.
5. Ball, C. G., Fellouah, H., & Pollard, A. (2012). The flow field of a turbulent round jet. *Progress in Aerospace Sciences, 50*, 1–26.
6. Bogey, C., Bailly, C., & Jove, D. (2003). Noise investigation of a high subsonic, moderate reynolds number jet using a compressible large eddy simulation. *Theoretical Computational Fluid Dynamics, 16*, 273–297. https://doi.org/10.1007/s00162-002-0079-4.

7. Borup, D., Elkins, C., & Eaton, J. (2018). Development and validation of an MRI-based method for 3D particle concentration measurement. *International Journal of Heat and Fluid Flow, 71*, 275–287.

8. Brown, G., & Roshko, A. (1974). On density effects and large structure in turbulent mixing layers. *Journal of Fluid Mechanics, 64*(4), 775–816.

9. Citriniti, J. H., & George, W. K. (1997). The reduction of spatial aliasing by long hot-wire anemometer probes. *Experiments in Fluids, 23*, 217–224.

10. Citriniti, J. H., & George, W. K. (2000). Reconstruction of the global velocity field in the axisymmetric mixing layer utilizing the proper orthogonal decomposition. *Journal of Fluid Mechanics, 418*, 137–166.

11. Corrsin, S., & Kistler, A. L. (1955). Free-stream boundaries of turbulent flows. NACA Technical Report 1244.

12. Dahm, W. J. A., & Dimotakis, P. E. (1987). Measurements of entrainment and mixing in turbulent jets. *AIAA Journal, 25*, 1216–1223.

13. Dimotakis, P. E. (2000). The mixing transition in turbulent flows. *Journal of Fluid Mechanics, 409*, 69–98.

14. Elkins, C. J., Markl, M., Pelc, N., & Eaton, J. K. (2003). 4D Magnetic resonance velocimetry for mean velocity measurements in complex turbulent flows. *Experiments in Fluids, 34*(4), 494–503.

15. Fan, W. C. (1982). ESCIMO applied to the hydrogen-air diffusion flame with refined treatment of biography. Technical Report CFDU/82/9, CFDU, Department of Mechanical Engineering, Imperial College.

16. Fan, W. C. (1988). A two-fluid model of turbulence and its modifications. *Science in China (Series A), 31*(1), 79–86.

17. Fellouah, H., & Pollard, A. (2009). Velocity spectra and turbulence length scale distributions in the near to intermediate region of a round free turbulent jet. *Physics of Fluids, 21*, 115101. https://doi.org/10.1063/1.3258837.

18. Fellouah, H., Ball, C. G., & Pollard, A. (2009). Reynolds number effects within the development region of a turbulent round free jet. *International Journal of Heat and Mass Transfer, 52*(17–18), 3943–3954.

19. Fiedler, H. E. (1998). Control of free turbulent shear flows. In M. Gad-el-Hak, A. Pollard, & J.-P. Bonnet (Eds.), *Flow control fundamentals and practices*. Berlin: Springer.

20. Foss, J. K., & Zaman, K. B. M. Q. (1999). Large-and small-scale vortical motions in a shear layer perturbed by tabs. *Journal of Fluid Mechanics, 382*, 307–329.

21. Fukushima, E. (1999). Nuclear magnetic resonance as a tool to study flow. *Annual Review of Fluid Mechanics, 31*, 95–123.

22. George, W. K. (1989). The self-preservation of turbulent flows and its relation to initial conditions and coherent structures. In W. K. George & R. E. A. Arndt (Eds.), *Advances in turbulence* (pp. 39–73). Hemisphere.

23. George, W. K. (2011). Does turbulence need god? In *Workshop on Models Versus Physical Laws/First Principles, or Why Models Work?* Wolfgang Pauli Institute. https://www.wpi.ac.at/event_view.php?id_activity=127.

24. Glauser, M. N. (1987). Coherent structures in the axisymmetric turbulent jet mixing layer. Ph.D. thesis, State University of New York at Buffalo.

25. Gohil, T. B., Saha, A. K., & Muralidhar, K. (2014). Large eddy simulation of a free circular jet. *ASME Journal of Fluids Engineering, 136*(May), 051205-1–14.

26. Harlow, F. H., & Nakayama, P.I. (1967). Turbulence transport equations. The Physics of Fluids, 10, 2323–2332.

27. Huang, L. S., & Ho, C. M. (1990). Small-scale transition in a plane mixing layer. *Journal of Fluid Mechanics, 210*, 475–500.

28. Hussein, H. J., Capp, S. P., & George, W. K. (1994). Velocity measurements in a high-Reynolds number, momentum-conserving, axisymmetric, turbulent jet. *Journal of Fluid Mechanics, 258*, 31–75.

29. Jung, D., Gamard, S., Woodward, S. H.,& George, W. K. (2002). Downstream evolution of the most energetic pod modes in the mixing layer of a high reynolds number axisymmetric jet. In A. Pollard & S. Candel (Eds.), *Turbulent mixing and combustion. Proceedings of IUTAM Symposium*, Queen's University, Kingston, ON, CA (pp. 23–32). Kluwer.

30. Kolmogorov, A. N. (1941). The local structure of turbulence in incompressible viscous fluids for very large Reynolds numbers. *Doklady Akademii Nauk SSSR, 30*(4), 301–305.

31. Latornell, D., & Pollard, A. (1986). Some observations on the evolution of shear layer instabilities in laminar flow through axisymmetric sudden expansions. *Physics Fluids, 29*(9), 2828–2835.

32. Launder, B. E., Patankar, S.V., & Pollard, A. (2019). Brian Dudley Spalding, biographical memoir fellows of the Royal Society. https://doi.org/10.1098/rsbm.2018.0024.

33. Lavoie, P., & Pollard, A. (2003). Uncertainty analysis of four-sensor hot-wires and their data-reduction schemes used in the near field of a turbulent jet. *Experiments in Fluids, 34*, 358–370.

34. Liepmann, D., & Gharib, M. (1992). The role of streamwise vorticity in the near-field entrainment of round jets. *Journal of Fluid Mechanics, 245*, 643–668.

35. Lipari, G., & Stansby, P. K. (2011). Review of experimental data on incompressible turbulent round jets. *Flow Turbulence and Combustion, 87*, 79–114.

36. LRP—Engines of discovery: The 21st century revolution: The long range PAN for high performance computing in CANADA 2005. https://www.computecanada.ca/wpcontent/uploads/2015/02/LRP.pdf.

37. Lumley, J. L. (1967). The structure of inhomogeneous turbulent flows. In Yaglom & Tatarsky (Eds.), *Atmospheric turbulence and radio wave propagation*. Moscow.

38. Marcouyre, M., Mcilwain, S., & Pollard, A. (2001). Large Eddy simulation of the near field of round jets with vortex generating tabs. In *Proceedings, Turbulent Shear Flow Phenomena*, Stockholm, Sweden, June (Vol. 3, pp. 113–118).

39. Mathew, J., & Basu, A. (2002). Some characteristics of entrainment at a cylindrical turbulence boundary. *Physics of Fluids, 14*(7), 2065–2072.

40. McIlwain, S., & Pollard, A. (2002). Large eddy simulation of the effects of mild swirl on the near-field of a round free jet. *Physics Fluids, 14*(2), 653–661.

41. Mcilwain, S., Holme, T., Waterman, S., & Pollard, A. (2002). Effects of single, dual and quadruple tabs on the near field of round jets. In Pollard & Candel (Eds.), *Proceedings, IUTAM Symposium on Turbulent Mixing and Combustion*. Kluwer Academic Pub., Dortrecht, Netherlands (pp. 377–385).

42. Medrano, F., Fukumoto, Y., Velte, C. M., & HODŹIĆ. (2017). Mass entrainment rate of an ideal momentum turbulent round jet. *Journal of Physical Society of Japan, 86*, 034401 (10 p). https://doi.org/10.7566/JPSJ.86.034401.

43. Mi, J., & Nathan, G. (1999). Effect of small vortex-generators on scalar mixing in the developing region of a turbulent jet. *International Journal of Heat and Mass Transfer,42*, 3919–3926.

44. Mistry, D., Philip, J., Dawson, J. R., & Marusic, I. (2016). Entrainment at multi-scales across the turbulent/non-turbulent interface in an axisymmetric jet. *Journal of Fluid Mechanics, 802*, 690–725.

45. Moser, R., Rogers, M., & Ewing, D. (1998). Self-similarity of time-evolving plane wakes. *Journal of Fluid Mechanics, 367*, 255–298.

46. Mungal, M. G., & Hollingsworth, D. K. (1989). Organized motion in a very high Reynolds number jet. *Physics of Fluids A, 1*, 1615–1623.

47. Panchapakesan, N. R., & Lumley, J. L. (1992). Turbulence measurements in axisymmetric jets of air and helium. Part 1. Air jet. *Journal of Fluid Mechanics, 246*, 197–223.

48. Panchapakesan, N. R., & Lumley, J. L. (1993). Turbulence measurements in axisymmetric jets of air and helium. Part 2. Helium jets. *Journal of Fluid Mechanics, 246*, 225–247.

49. Parker, R., Rajagopalan, S., & Antonia, R. A. (2003). Control of an axisymmetric jet using a passive ring. *Experimental Thermal and Fluid Science,27*, 546.

50. Philip, J., & Marusic, I. (2012). Large-scale eddies and their role in entrainment in turbulent jets and wakes. *Physics of Fluids, 35*, 055108.

51. Pollard, A. (1997). Interaction between CFD and experiments, plenary paper. In *Proceedings, CFD'97*. Victoria, British Columbia: CFD Society of Canada.
52. Pollard, A., Castillo, L., Danaila, L., & Glauser, M. (2017). Whither turbulence and big data in the 21st century? Springer. ISBN 978-3-319-41217-7.
53. Ricou, F., & Spalding, D. B. (1961). Measurements of entrainment by axisymmetrical turbulent jets. *Journal of Fluid Mechanics, 11*(1), 21–32.
54. Runchal, A. K. (2009). Brian Spalding. CFD and reality: A personal recollection. *International Journal of Heat and Mass Transfer, 52*, 4063–4073.
55. Sadeghi, H., Lavoie, P., & Pollard, A. (2018, February). Effects of finite hot-wire spatial resolution on turbulence statistics and velocity spectra in a round turbulent free jet. *Experiments in Fluids, 59*(3). https://doi.org/10.1007/s00348-017-2486-8.
56. Sadeghi, H., & Pollard, A. (2012). Effects of passive control rings positioned in the shear layer and potential core of a turbulent round jet. *Physics of Fluids, 24*, 115103.
57. Sadeghi, H., Lavoie, P., & Pollard, A. (2016). Scale-by-scale budget equation and its self-preservation in the shear-layer of a free round jet. *International Journal of Heat and Fluid Flow, 61*, 85–95.
58. Saffman, P. G. (1970). A Model for Inhomogeneous Turbulent Flow, Proceedings of the Royal Society of London. *Series A: Mathematical and Physical Sciences, 317*(1530), 417–433. https://doi.org/10.1098/rspa.1970.0125
59. Shin, D., Sandberg, R. D., & Richardson, E. S. (2017). Self-similarity of fluid residence time statistics in a turbulent round jet. *Journal of Fluid Mechanics, 823*, 1–25.
60. Spalding, D. B. (1969). The prediction of two-dimensional steady elliptic flows. In *Proceedings of International Seminar on Heat and Mass Transfer in Separated Regions*, Belgrade. http://www.cham.co.uk/phoenics/d_polis/d_enc/turmod/enc_tu.htm.
61. Spalding, D. B. (1971). Mixing and chemical reaction in steady confined turbulent flames. Thirteenth symposium (international) on combustion. *The Combustion Institute, 649*–657.
62. Spalding, D. B. (1976). Mathematical models of turbulent flames: A review. *Combustion Science and Technology, 13*(1–6), 3–25.
63. Spalding, D. B. (1976). *The ESCIMO theory of turbulent combustion*. Technical Report HTS/76/13, Dept. of Mechanical Engineering, Imperial College, University of London.
64. Spalding, D. B. (1991). Kolmogorov's two-equation model of turbulence. In *Proceedings of the Royal Society of London. Series A: Mathematical and Physical Sciences*. https://doi.org/10.1098/rspa.1991.0089.
65. Spalding, D. B. (2010). PTAs and PTBs, facilitating the comparison of experiments and DNS-, LES-, PDF-transport and MFM models of turbulence. In *Proceedings 8th International ERCOFTAC Symposium: Engineering Turbulence Modelling & Measurements*, Marseille, France, June 9–11.
66. Spalding, D. B. (1950). A simple manometer for use in measuring low air velocities. *Journal of Scientific Instruments, 27*(11), 310–312.
67. Sreenivasan, K. R., Ramshankar, R., & Meneveau, C. (1989). Mixing, entrainment and fractal dimensions of surfaces in turbulent flows. *Proceedings of the Royal Society of London, 421*, 79–108.
68. Suto, H., Matsubara, K., Kobayashi, M., & Kaneko, Y. (2004). Coherent Structures in a fully developed stage of a non-isothermal round jet. *Heat Transfer: Asian Research, 33*(5), 342–356.
69. Tong, C., & Warhaft, Z. (1994). Turbulence suppression in a jet by means of a fine ring. *Physical of Fluids, 6*(1), 328–333.
70. Townsend, A. A. (1976). *The structure of turbulent shear flow*. Cambridge University Press.
71. van Reeuwijk, M., & Holzner, M. (2014). The turbulence boundary of a temporal jet. *Journal of Fluid Mechanics, 739*, 254–275.
72. Wang, Z., He, P., Lv, Y., Zhou, J., Fan, J., & Cen, K. (2010). Direct numerical simulation of subsonic round turbulent jet. *Flow Turbulence Combust, 84*, 669–686. https://doi.org/10.1007/s10494-010-9248-5.
73. Westerweel, J., Fukushima, C., Pedersen, J., & Hunt, J. C. R. (2005). Mechanics of the turbulent-nonturbulent interface of a jet. *Physical Review Letters, 95*, 174501.

74. Wygnansky, I., & Fiedler, H. E. (1969). Some measurements in the self-preserving jet. *Journal of Fluid Mechanics, 38*(3), 577–612.
75. Xi, S. T. (1986). Transient turbulent jets of miscible and immiscible fluids. PhD thesis, Imperial College, University of London.
76. Yimer, I., Becker, H. A., & Grandmaison, E. W. (2001). The strong-jet/weak-jet problem: New experiments and CFD. *Combustion and Flame, 124*(3), 481–502.
77. Zaman, K. B. M. Q. (2000). Spreading characteristics of compressible jets from nozzles of various geometries. *Journal of Fluid Mechanics, 383*, 197–228.
78. Zaman, K.B.M.Q., Reeder, M.F., & Samimy, M. (1994). Control of an axisymmetric jet using vortex generators. *Physical of Fluids, A*(6), 778–793.

Eddy-Viscosity Transport Modelling: A Historical Review

K. Hanjalić and B. E. Launder

1 Introduction

The present review provides an account of the evolution of models for the effective stresses in a turbulent shear flow based on the supposition that such stresses are directly proportional to the local mean rate of strain—what is commonly termed the *eddy-viscosity hypothesis*. While such models have many flaws (to which we shall also refer) it must be acknowledged that these weaknesses are often of less importance than the essential reality of enabling CFD software to generate, in a reasonable time and at an acceptable cost, predictions of flow patterns and (more importantly) their impact on interacting solid structures that are sufficiently accurate for many industrial applications. We make no attempt to go into all the details of any model—the published scientific papers by their authors do that. This contribution does, however, try to provide some explanation of why the subject evolved in the way that it has and to give an impression of the roles played by some of the major contributors to the subject.

In many situations, the principal concern in a turbulent flow relates especially to heat- or mass-transport processes that are greatly affected by the turbulent motion. While there is a whole body of literature on the modelling of such processes, time and length constraints meant that here we consider just the modelling of the fluid motion. Where the choice of modelled form has been shaped by heat-transfer considerations, however, that fact is noted; moreover, some of the illustrations chosen relate to such processes. We likewise do not venture into the field of turbulence modelling of combustion, nor of high Mach number or compressible flows. However, we would

K. Hanjalić
Faculty of Applied Sciences, Delft University of Technology, Delft, The Netherlands

B. E. Launder (✉)
Department of Mechanical Aerospace & Civil Engineering, University of Manchester, Manchester, UK
e-mail: brian.launder@manchester.ac.uk

© Springer Nature Singapore Pte Ltd. 2020
A. Runchal (ed.), *50 Years of CFD in Engineering Sciences*,
https://doi.org/10.1007/978-981-15-2670-1_9

draw the reader's attention to the reviews of Jones [40] and Wouters et al. [84] for combustion and that of Barre et al. [1] for high-speed compressible flows, all of which cover a similar time span to that of the present review.

Since this article appears within a volume celebrating the contributions of the late Professor D. B. Spalding to CFD, it is natural that the work by Spalding and other members from his group of co-workers (including the present writers) should be covered more completely than those from elsewhere. Thus, our choice of key proposals and publications is a personal one that may differ in at least some respects from what others attempting a correspondingly brief overview would have selected. Indeed, in choosing illustrative examples, the authors have often turned to those which have appeared in their textbook on closure (Hanjalić and Launder [27], hereafter denoted HL).

The recognition that the motion of a fluid takes on strikingly different appearances depending on flow rate goes back at least to the time of Leonardo da Vinci. The most cogent (and certainly the most lengthy) early written record of the distinctions, however, belongs to the French mathematician and physicist, Boussinesq [3]. His paper (whose seeds had been sown in a lecture to the French Academy of Science 5 years earlier), proposed, by analogy with the kinetic theory of gases, that the strikingly different behaviour could be accounted for by replacing the viscosity coefficient in the Navier (or in contemporary terms, Navier–Stokes) equations by 'a much larger number', later to become known as a 'turbulent' or 'eddy' viscosity

Some years later Osborne Reynolds, Professor of Engineering at Owens College, Manchester, published the results of extensive experiments on the flow of water through pipes [63] bringing out more precisely the hugely important differences between the two flow states: what Reynolds termed 'direct' and 'sinuous' or as nowadays we would say 'laminar' and 'turbulent'. Reynolds also showed that the changeover from the former to the latter occurred for a particular value of a dimensionless group we now call the *Reynolds number*. The paper had been enthusiastically reviewed by two nineteenth-century giants in Fluid Mechanics, Sir George Stokes and Lord Rayleigh. The latter, however, tempered his enthusiasm with the final observation: 'In several passages the Author refers to theoretical investigation whose nature is not sufficiently indicated'.

More than a decade was to elapse before Reynolds felt ready to respond to Rayleigh's gentle criticism. In 1895, he published a paper [64] that, in 2015, was selected as one of sixteen papers from the journal's complete opus, whose impact was to be celebrated in the 350th anniversary publication of *Phil Trans Roy Soc*. The paper contained many innovations (summarized in [47]) but, in the context of the present paper, the most important was his averaging of the equations of motion to produce what is now known as *the Reynolds equations* in which the time-averaged correlations between fluctuating velocities—the kinematic Reynolds stresses, $\overline{u_i u_j}$—appeared as unknowns:

$$\frac{\partial U_i}{\partial t} + \frac{\partial U_i U_j}{\partial x_j} = -\frac{1}{\rho}\frac{\partial P}{\partial x_i} + \frac{\partial}{\partial x_j}\left(\nu\frac{\partial U_i}{\partial x_j} - \overline{u_i u_j}\right) \tag{1}$$

Reynolds, however, did not concern himself with how one might determine these correlations since his goal was to examine the turbulent kinetic energy equation (a step further along in his analysis) and the changing relative magnitudes of the generation and sink terms in that equation with Reynolds number. Later workers, however, recognized that Eq. (1) provided a basis for the computation of turbulent flows. But, evidently, a scheme or 'model' for determining the Reynolds stresses was needed before the Reynolds equations could be used. Thus, the subject now known as *turbulence modelling* was born.

2 Mixing-Length Models of Turbulent Transport

Twenty years were to elapse following Reynolds' direction-setting paper before the first closure model appeared. G. I. Taylor was at the time deeply immersed in both measurement and modelling of the wind velocity variation with height. In Taylor [76] he argued that the eddy viscosity could be expressed 'in the form $1/2\rho\hat{w}d$ where d is the average height through which an eddy moves before mixing with its surroundings and $\hat{w}$ roughly represents the average [turbulent] vertical velocity'. The magnitudes of the quantities $\hat{w}$ and d were inferred from wind data.

A decade later the mixing-length model of Prandtl [61] became the first to provide a simple formulation of the turbulent viscosity in terms of easily definable quantities. Following an analysis similar to Taylor's (of which he was apparently not aware) Prandtl proposed that the relevant mean vertical velocity fluctuation was of similar magnitude to the horizontal velocity fluctuation which, in turn, was approximated by $l|\partial U/\partial y|$, with l being a length scale (analogous to d in Taylor's scheme). Thus, the turbulent shear stress $-\overline{uv}$ could be finally represented by

$$-\overline{uv} \equiv \ell_m^2 |\partial U/\partial y|(\partial U/\partial y) = \nu_t \partial U/\partial y \tag{2}$$

where the kinematic eddy viscosity is given by $\nu_t = \ell_m^2 |\partial U/\partial y|$ and ℓ_m is the eponymous mixing length. Although Prandtl's analysis considered just a simple shear flow, later workers (see, for example, HL, p. 299) have pointed out that the model also emerges quite generally from assuming the shear production and dissipation rates of turbulence energy to be in balance: that leads directly to

$$\nu_t = \ell_m^2 \left(\frac{\partial U_i}{\partial x_j} \left(\frac{\partial U_i}{\partial x_j} + \frac{\partial U_j}{\partial x_i} \right) \right)^{1/2} \tag{3}$$

While the mixing length needed prescribing, in simple flows plausible assumptions led to reasonable variations of velocity. For example, near a wall a linear increase of ℓ_m with wall distance led to reasonable velocity variations or, for a jet or wake, a mixing length proportional to the local flow width. Prandtl's paper triggered reactions and new ideas among the relatively small turbulence research community.

Limitations of his mixing-length concept were soon recognized, focused mainly on the way the mixing length was defined and prompted proposals for its improvement. To circumvent the need for prescribing ℓ_m in terms of wall distance, Von Karman [80] proposed a mixing length defined as the ratio of the first and second mean-velocity derivatives, derived from his so-called 'similarity hypothesis'. While justifiable in some wall-adjacent equilibrium flows, Von Karman's proposal led to ludicrous results wherever there was an inflection point in the mean-velocity profile corresponding to zero second-velocity gradient and, thus, an infinite mixing length. Soon afterwards, referring to Prandtl's mixing-length concept, Taylor [77] sought to improve his earlier ideas. He argued that the theory of 'the dynamics of turbulent motion should be regarded as an effect of diffusion of vorticity rather than as a diffusion of momentum' and duly recovered the mixing-length hypothesis based on a vorticity transport formulation. Apparently, Taylor had already put forward this argument in 1915, a decade before Prandtl's seminal paper, in an essay for which he was awarded the Adams Prize by the University of Cambridge. But the concept, now extended to three-dimensional motion, showed that 'the momentum transport theory of Reynolds and Prandtl agrees with the vorticity transport theory in one case only, namely when the turbulent motion is of a two-dimensional type, being confined to the plane perpendicular to the mean motion'.

Despite these claimed shortcomings, Prandtl's mixing-length hypothesis, unlike its competitors, survived to be used 40 years later in many of the early CFD computations. Its survival may be attributed partly to its appealing simplicity and to the fact that it did, as Eq. (3) indicates, genuinely describe the state of turbulence close to local equilibrium. Indeed, it is still used one-fifth of the way through the twenty-first century for limited situations like, say, the near-wall region of a turbulent boundary layer. In the 1960s, a number of proposals were made (based on a scrutiny of extensive experimental data) for what mixing-length distributions were appropriate for different classes of flow. Here we note inter alia the proposal of Glushko [21]. An interesting conjecture for wall-bounded flows of complex shape was made by Buleev [6] who proposed a 'harmonic mixing length' involving the sum of the reciprocal of the distances from all bounding solid walls, integrated over the duct perimeter.

3   Turbulence Modelling at Entry to the Age of the Digital Computer

On entry to the 1960s, the arrival of digital computers made it feasible to adopt numerical methods to discretize and solve the Reynolds equations of fluid motion and to adopt more ambitious models of turbulence. Indeed, by the end of that decade, a new direction of turbulence modelling research had come into focus.

Although, in terms of strict chronology, the starting point for this 'new direction' should probably be the pioneering work in CFD led by Francis Harlow at the Los Alamos Scientific Laboratories, for the present authors and, indeed many of those connected with engineering CFD, the appearance of the widely used two-dimensional boundary-layer solver developed by Patankar and Spalding [57] represents the natural origin. Like other contemporary researchers developing less ambitious solvers of the Reynolds equations, they had to provide a model of turbulence for the Reynolds stresses. For that purpose, in common with all but one of their competitors, they adopted an algebraic turbulence model, in their case Prandtl's mixing-length hypothesis. Following Escudier [17], for applications to boundary layers, the mixing length, ℓ_m, was prescribed to increase linearly with distance from the wall with slope κ (commonly termed 'the Von Karman constant' and given a value of about 0.41) until it reached a value of about 10% of the boundary-layer thickness and was thereafter held constant. The first major occasion to test the adequacy of the model for two-dimensional attached boundary layers was the 1968 Stanford-AFOSR conference [42]. In such flows, (with x denoting the main flow direction and y the direction normal to the wall) the only element of the turbulent stress tensor affecting the mean flow was the shear stress component, $-\overline{uv}$. Brian Spalding and his student K. H. Ng provided computations for all the test cases, some of which were captured well and some indifferently; boundary layers heading towards separation were rather poorly represented. The other major competitor at this meeting was Peter Bradshaw [4]. Drawing on the extensive experimental work of Townsend [79] and, indeed, of Bradshaw himself which provided several of the test cases (e.g. [5]), he assumed a direct proportionality between turbulent kinetic energy and shear stress with the turbulence energy being computed by way of a transport equation. While, overall, the results from his computations were only slightly better than Spalding's, he attracted great interest among those attending largely because his model was judged to build in better physics. Indeed, it was the case that his computations took account of transport effects on the turbulent shear stress although, like Spalding and the other contributors, the effective turbulent length was prescribed algebraically.

In fact, by the time of the Stanford Conference noted above, Spalding had already shown that the turbulent kinetic energy was a vital contributor in determining heat-transfer rates in separated flows, Spalding [72]. Thus, in the numerical solver for recirculating flows that was then nearing completion in his group [22] he had arranged that a version of Prandtl's later model [62] in which $\nu_t \propto k^{1/2} \ell$ should be embedded in the solver which meant that a transport equation for turbulence energy, k, was solved.

4   Two-Equation Eddy-Viscosity Models

Spalding recognized, however, that to prescribe the turbulent length scale except in very simple flows was impossible: a means must also be provided to compute the length scale, accounting for the effects of convection and diffusion. In fact, but unknown to Spalding at the time, Kolmogorov [45] had already written a short paper proposing just such a scale-determining equation for a 'frequency', ω. This was to be solved along with one for the turbulence energy, to obtain the length scale from which the turbulent viscosity could be computed. However, Kolmogorov's equation contained no input from the mean-strain field, neither directly nor via such mean-strain-affected parameters as the turbulence anisotropy. Thus, it could at best be applicable only to flows without shear. Nevertheless, this was the first proposal for a two-equation model; but, written in the depths of the Second World War (WW2) in a journal off the reading list of Western researchers, it was hardly surprising that it had remained undiscovered. Thus, in the spring preceding the Stanford Conference, Spalding had invited J. C. Rotta to visit Imperial College for several days to deliver group seminars and engage in small-scale discussions. The reason for this interest was that Rotta had earlier published a two-part paper [68] in which he had developed a length-scale equation (or, strictly, the product of length scale and turbulence energy, $k\ell$). Since the papers were in German, Spalding also arranged (for the benefit of his group members) that his new research assistant, Wolfgang Rodi, should provide a written translation in advance of the seminars [66].

Following the visit, Rodi duly developed and applied a Rotta-like model to compute a range of free shear flows [65]. While Rotta had proposed the scheme within the context of second-moment closure (i.e. where transport equations were also solved for the unknown turbulent stresses) Rodi's version employed the turbulent viscosity concept in which transport equations were solved for k and $k\ell$, with the turbulent viscosity being obtained from $\nu_t = c_\mu k^{1/2} \ell$ with c_μ supposedly a constant. The form of the $k\ell$ equation was

$$\frac{Dk\ell}{Dt} = C_{k\ell 1} \ell P_k - C_{k\ell 2} k^{3/2} + \frac{\partial}{\partial y}\left(\frac{\nu_t}{\sigma_{k\ell}} \frac{\partial k\ell}{\partial y} \right) \tag{4}$$

i.e. a positive source involving the shear production rate of turbulence energy (P_k), a sink term including just turbulence parameters, and a gradient-diffusion term with three empirical constant coefficients to be determined from experimental data. The scheme succeeded in predicting a range of mixing layers and plane jets but it returned a much too high spreading rate for the axisymmetric jet in stagnant surroundings and a too low spreading for the weak wake. These weaknesses were ones shared by all RANS closures (not just eddy-viscosity schemes) for the next 20 years. Spalding also set K. H. Ng to test the model's ability to capture the development of two-dimensional boundary layers, many of which had earlier appeared as test cases at the Stanford Conference. Here, however, success was less complete and reasonable agreement with the test flows could only be achieved by the addition of an empirical source term in the $k\ell$ equation dependent upon distance from the wall [55], a feature making it unsuitable for use in wall flows of arbitrary shape.

Independently of these researches, Spalding himself pursued the use of an alternative scale-determining equation for a quantity $W \equiv k/\ell^2$ which initially he used to compute the turbulent concentration fluctuations in a round jet, Spalding [73]. When applied more extensively to boundary layers, however, his efforts [74] were scarcely more successful than when $k\ell$ was adopted as dependent variable. He also encouraged his junior colleague, Brian Launder, to collaborate in the search for the best route for obtaining the length scale. One of Launder's research students, Kemal

Hanjalić, read Russian and that gave him entry to the work of Davydov [14, 15] which, in turn, alerted him to the earlier work of Chou [8]. That paper had proposed the use of the energy dissipation rate, ε, as a transported turbulence property from which the turbulent viscosity was obtained as $\nu_t = c_\mu k^2/\varepsilon$. This idea was attractive as ε was a quantity that directly appeared as a sink in the turbulence energy equation (compared with the rather intangible length scale, ℓ). Moreover, Hanjalić [26] found that with this choice it was possible to match near-wall as well as free flows using just a single source and sink term, analogous to Eq. (4):

$$\frac{D\varepsilon}{Dt} = C_{\varepsilon 1}\frac{\varepsilon}{k}P_k - C_{\varepsilon 2}\frac{\varepsilon^2}{k} + \frac{\partial}{\partial y}\left(\frac{\nu_t}{\sigma_\varepsilon}\frac{\partial \varepsilon}{\partial y}\right) \tag{5}$$

After this initial development at the eddy-viscosity level, however, Hanjalić moved on to develop a further transport equation for $\overline{uv}$, the turbulent shear stress, as his goal was to predict flows where the surfaces of zero shear strain and zero shear stress did not coincide—a phenomenon that clearly could not be accounted for with an eddy-viscosity scheme. As reported below, however, his fellow research student, Bill Jones, carried on further development of the model.

As more results from the alternative two-equation models became available, Spalding announced that the Imperial College group should focus on just a single form and that the $k - \varepsilon$ model was the one to adopt. Shortly thereafter NASA Langley announced a computational challenge comprising 24 diverse free flows as test cases for aspiring CFD predictors, an announcement that attracted 13 groups to the competition. The Imperial College contribution [50], the only entry from outside the USA, provided computations for a range of models for all but one of the test cases and was alone in presenting results where the turbulent length scale was obtained via a transport equation (i.e. the ε equation). In their report [53], the evaluation committee wrote: 'Because of the advent of high-speed digital computers ... new approaches have been developed ... which incorporate more of the physics of turbulence into the describing equations. At this conference the contribution of [the Imperial team] provides an excellent review of several of these new methods'.

While the work at Imperial College may have had the greatest early impact on turbulence modelling, there were several groups in the USA that had taken up the task of turbulence model development (or would shortly do so) though sometimes without rigorous testing. Harlow and Nakayama [28] announced: 'Our goal has been to develop transport equations that enable the macroscopic manifestations of turbulence to be included in high-speed computer calculations ... without introducing such complexity as to render the equations intractable'. They chose a two-equation eddy-viscosity model in which the scale equation was proportional to the Taylor microscale. In their subsequent report [29], however, they switched the subject of their scale equation to (ε/ν).[1]

---

[1] A practice retained by Daly and Harlow [13] but who discarded the eddy-viscosity approach in favour of solving transport equations for the Reynolds stresses.

In an unusual departure from his habitual, deeply fundamental enquiries, Saffman [69] proposed a $k - \omega^2$ eddy-viscosity model[2] (with $v_t = k/\omega$). As a follow-up to his proposals, further model developments with k and ω (rather than ω^2) as dependent variables were made by David Wilcox (in collaboration with various co-workers). His book [81] provides a thorough overview of these developments—what might be called *the Californian school of turbulence modelling*. The culmination of these efforts, documented in Wilcox [81], was a model that was particularly well adapted for boundary layers heading towards separation and thus particularly popular for computing boundary layers on airfoils.

The first papers on the application of two-equation modelling to flows near walls employed what are known as 'wall functions' to cover the thin sublayer right next to the wall where viscous as well as turbulent transport was important. Wall functions are algebraic formulae linking the wall shear stress and heat-transfer rate to the state of turbulence a short distance from it where only turbulent effects are significant. Such schemes saved a large amount of computing time (since a very fine mesh is needed to resolve this extremely thin but highly inhomogeneous sublayer adjacent to the wall) and also avoided the challenge of modelling within it: one simply placed the near-wall grid point outside this viscous sublayer and used the wall function to deliver values for the important wall parameters. But, at entry to the 1970s, such schemes amounted to no more than the 'law of the wall' [9] which delivered reliable values of wall shear stress only for flows in which the near-wall layer could be regarded as in local equilibrium (where, *inter alia*, the total shear stress over the region was uniform). Jones [39], however, specifically sought to predict strongly accelerated flows that underwent partial or complete reversion to laminar where the near-wall turbulence was far from such an equilibrium state. Thus, following on from Hanjalić's doctoral research, he devised what was termed a 'low Reynolds number' version of the model in which various coefficients in the $k - \varepsilon$ model were made functions of the local turbulent Reynolds number, $k^2/v\varepsilon$, Jones and Launder [38], a paper that is sometimes reported as the 'origin' of the $k - \varepsilon$ model.[3] This work was rather successful in accounting for such laminarizing flows, not just in strong accelerations but later in the physically quite different buoyantly-modified flows through vertical pipes, Cotton and Jackson [10], Fig. 1. Here, in up-flow, the rapid radial decrease of shear stress with distance from the wall leads to a thickening of the near-wall sublayer and, as a result, a sudden reduction in Nusselt number; correspondingly, in down-flow, there is a modest augmentation in Nu relative to purely forced convection. The Jones–Launder approach also spawned the development of several competitor schemes, the overall performance of which was assessed by Scheuerer [70] in his doctoral thesis and subsequently by Patel et al. [58] in a journal contribution. The latter concluded from a comparison of ten different formulations that 'the models of Launder and Sharma [48] [just a minor recalibration of the form proposed by Jones

---

[2]The quantity ω^2 differs only by a constant from Spalding's W.

[3]If the form of model applicable to regions excluding the viscosity-affected sublayer is meant (i.e. 'the high Re number k-ε model'), however, a more correct attribution would be to Hanjalić's thesis [26].

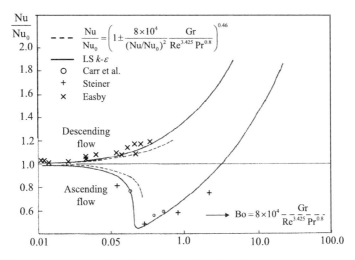

Fig. 1 Effect of buoyancy on Nusselt number in flow through vertical heated pipes. Symbols: Collected experimental data; - - - - - - earlier theoretical estimate; ———— predictions with LS low Re k-ε model. From Cotton and Jackson [10]

and Launder [38]] and, to some extent, Chien [7], Lam and Bremhorst [46] and that of Wilcox and Rubesin [82] yield comparable results and perform considerably better than the others'.

In fact, while it was convenient to think of the near-wall damping of turbulence as due to low Reynolds number effects, it gradually dawned on those working in the field that a quite different mechanism was largely responsible. Launder [49] pointed out that, if one took the turbulent viscosity proportional to $\overline{v^2}k/\varepsilon$ ($\overline{v^2}$ being the normal stress perpendicular to the wall) rather than k^2/ε, the coefficient required to obtain the correct shear stress was effectively constant (rather than one highly dependent upon turbulent Reynolds number). This is strikingly brought out by Fig. 2 which shows that the variation of $\overline{v^2}/k$ across the sublayer (in the figure, x_2 and u_2 denote the wall-normal coordinate and corresponding fluctuating velocity) was essentially the same as that required in a $k - \varepsilon$ model using the turbulent Reynolds number to achieve the required damping, marked '$0.342 f_\mu$'. The question was thus how one should account for the variation of $\overline{v^2}/k$ across the near-wall sublayer. It was Durbin [16] who proposed the concept of *elliptic relaxation* (ER), an eddy-viscosity model in which a separate transport equation was solved for a scalar 'surrogate' of $\overline{v^2}$ in conjunction with an *elliptic relaxation equation* parameter f, in what has become known as the $\overline{v^2} - f$ model. (Although strictly beyond the timescale of this review, it is noted that references [24, 51] independently proposed versions of this model in which a transport equation is solved for the ratio $\zeta \equiv \overline{v^2}/k$, which offers a number of computational advantages.)

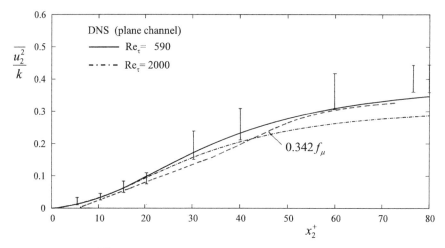

Fig. 2 Variation of $\overline{u_2^2}/k$ (vertical bars denoting uncertainty range of the data) and $0.342f_\mu$ (- - -) across the inner region of turbulent wall flows estimated from experimental data then available [49, 58]. DNS data of $\overline{u_2^2}/k$: — — $Re_\tau = 590$ [54]; – — – — $Re_\tau = 2000$ [30] [*Note u_2 and x_2 correspond to v and y*]

5 The Current Usage of Two-Equation EVMs

While today (in 2020) a large proportion of commercial CFD computations are made with eddy-viscosity models such as those outlined above, several model originators (including the present authors), sensitive to the limitations of that closure level for flows subject to complex strains or force fields, have thrown the weight of their efforts to higher order schemes. In such approaches, closed transport equations (or sometimes truncated algebraic versions thereof) are devised for the second moments (the Reynolds stresses and turbulent heat or mass fluxes) and, sometimes, for higher order moments too. Such schemes lie beyond the scope of the present article, but the books by Wilcox [81], Pope [60], and Hanjalić and Launder [27] provide an account of the development of these types of model and, particularly the last of these, present an array of applications.

But two-equation linear eddy-viscosity models remain the workhorse of the CFD industry and while, to many with a deep knowledge of turbulence and its modelling, their continuing popularity remains a mystery, the reasons are not hard to see. Such models rarely give major convergence problems and, even for complex flow configurations, they can be applied at modest computational cost on an acceptable timescale. Moreover, the accuracy of the predictions they deliver is often good enough to guide industrial design or to trace the cause of a malfunction in some operating plant or machinery. Indeed, it may be said that weaknesses that would seem to be dramatically damaging in a strictly two-dimensional shear flow may turn out to be far less influential in three-dimensional flows—which includes the majority of cases for which commercial CFD is employed.

An excellent illustration of this arises in the development of a turbulent boundary layer on a curved surface. If the flow is *two* dimensional it is well known that on a concave surface the boundary-layer growth rate is significantly higher than on a plane surface while, on a convex surface, it is correspondingly reduced. Solution of the Reynolds stress equations entirely captures this phenomenon,[4] whereas a linear eddy-viscosity model shows virtually no effect of curvature at all. However, for the case of *three*-dimensional flow in a square duct around a 90° bend (measured by Taylor et al. [78]), shown in Fig. 3, the presence of sidewalls creates a really strong secondary flow and in this case the two-equation EVM produces predictions very close to the measured flow pattern. We note that this case was one originally selected for the 1980/81 Stanford Conference [43, 44]. The computations submitted to that meeting by contributing groups produced what many felt to be a rather puzzling outcome; for two of the simplest models (based on mixing-length concepts) gave the best results! Certainly, the curvature was sufficiently strong that, if the flow had been two-dimensional, curvature effects would have been essential to account for. For this highly non-equilibrium flow, however, the main problem was not the use of turbulence

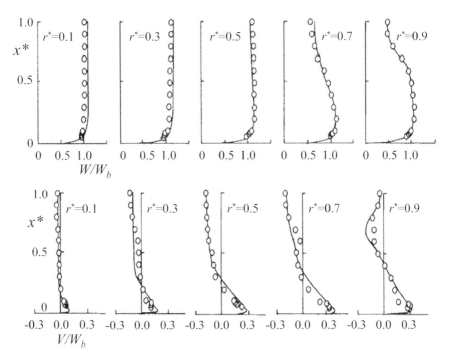

Fig. 3 Velocity profiles $0.25D_h$ downstream of a 90° bend: ○○○: Experiments, Taylor et al. [78]; —— Computations using k-ε EVM with MLH sublayer model. From Iacovides et al. [33]. W and W_b are local and bulk-mean axial velocities; V the negative of the radial mean velocity

---

[4]The small secondary strains associated with the curvature exert strong amplification or damping effects on the turbulent stresses.

models that were insensitive to the streamline curvature but rather the scheme used next to the wall in applying the wall boundary condition. In 1981, only the groups using the simplest models could employ a fine enough grid to resolve the viscosity-affected near-wall sublayer. Other groups employing more elaborate models had to use *wall functions* that assumed a universal log-law variation of velocity with distance from the wall. Now, the radial pressure gradient induced by the bend creates a strong secondary flow down the plane sidewalls carrying the slow-moving near-wall fluid to the inside of the bend. Indeed, the peak secondary velocity occurs right in the viscous region. The implied skewing of the mean-velocity vector within the viscous region was entirely missed by the wall functions then in use. When, a few years later, the sublayer *was* resolved by blending a two-equation EVM used for the main flow with the mixing-length hypothesis across the near-wall sublayer, Iacovides et al. [33], the close agreement with the experimental data shown in Fig. 3 resulted.

It must be acknowledged that not all the weaknesses of assuming the turbulent stresses to be describable by a linear stress–strain law can be ignored if one wants a useful CFD scheme. However, unlike academics in universities who strive to find *universally* applicable models, most industrial users focus on a small subset of flows relating to their particular modelling concerns. In those cases, who is to criticize them if they 'tweak' the supposedly universal set of empirical constants to produce results in harmony with what experiments indicate are the appropriate values for their flows?

In some cases, fundamental turbulence modellers have themselves proposed elaborations with the aim (or at least the hope) of providing a widely valid improvement. The axisymmetric jet has often figured prominently in these attempts; for, while the usual two-equation EVM successfully captures the spreading rate of the plane jet and the plane mixing layer largely independently of the presence of a co-flowing stream, the same set of coefficients in the dissipation rate equation predict the growth of the axisymmetric jet in stagnant surrounds some 30% higher than the experiments from that time indicated. The first remedy for this discrepancy was proposed by Pope [59] who argued that in an axisymmetric flow the mean-flow vortex lines are stretched as the flow develops downstream, whereas no such stretching occurs in a plane two-dimensional flow. He proposed this effect could be captured by including an additional source term in the ε-equation containing the product of strain and vorticity tensors. However, Huang [31] found from computing the axisymmetric jet in a co-flowing stream, measured by Forstall and Shapiro [18], that better agreement was achieved if the Pope correction was omitted. The authors too [23] proposed an additional source term to the ε-equation of the form:

$$S_{\varepsilon}^{HL} \propto k(\partial U_{\ell}/\partial x_m)(\partial U_i/\partial x_j)\varepsilon_{\ell m k}\varepsilon_{ijk} \qquad (6)$$

where ε_{ijk} is the third-rank alternating tensor. This source term likewise removed the anomalous spreading rate for the axisymmetric jet and also improved predictions of the other flows (boundary layers in adverse pressure gradients) considered in their paper. However, over the ensuing decade, more extensive testing revealed that it did not bring universal improvement. A more elaborate modification of the same type has been proposed by Wilcox [81]. His test flows considered both the axisymmetric jet and the plane wake as well as boundary layers approaching separation. His target spreading rate for the half-radius of the axisymmetric jet, $dr_{1/2}/dx$, was taken as 0.086 (from the early recommendation of Rodi [67]). In the 1990s, however, two sets of experiments using flying-hot-wire and LDA instrumentation [32, 56] suggested that a distinctly higher value, 0.095, actually prevailed; so, it would appear that the Wilcox modification also needs some modest re-tuning.

Perhaps the most serious weakness of the linear $k - \varepsilon$ EVM in near-wall flows is the fact that in separated flows or in flows approaching separation the length scale near the wall, if not otherwise constrained, becomes excessively large. A range of approaches have been adopted for removing this weakness. Some have simply adopted a one-equation EVM in a limited near-wall region (where the length scale near the wall is prescribed as for a flat-plate boundary layer), but this is rarely satisfactory as the accuracy of the results is highly sensitive to the thickness chosen for the one-equation region. Other, less arbitrary, forms of damping have been proposed in which empirical sources are added to the ε-equation to augment the near-wall energy dissipation rate (and thus depress the length scale).

Thus, the so-called *Yap correction* [85] adds the term:

$$S_\varepsilon^{\mathrm{YAP}} = \mathrm{Max}\left[0.83 \frac{\varepsilon^2}{k} \left(\frac{l}{c_l x_n} - 1 \right) \left(\frac{l}{c_l x_n} \right)^2, 0 \right] \tag{7}$$

where $l \equiv k^{3/2}/\varepsilon$, $c_l = 2.44$ and x_n denotes normal distance from the wall (rather than a tensor distance). Evidently, if l is greater than the equilibrium value, $c_l\, x_n$, the term acts to increase ε and thus to decrease the length scale. This form has been successfully adopted when treating nearly plane surfaces but the use of the normal distance limits its utility on highly curved or irregular surfaces. Hanjalić [25] proposed replacing explicit reference to wall distance by the gradient of length scale normal to the wall:

$$S_\varepsilon^{H} = \mathrm{Max}\left\{ 0.83 \frac{\varepsilon^2}{k} \left[\left(\frac{1}{c_l} \frac{\partial l}{\partial x_n} \right)^2 - 1 \right] \left(\frac{1}{c_l} \frac{\partial l}{\partial x_n} \right)^2, 0 \right\} \tag{8}$$

Iacovides and Raisee [35] proposed a similar expression in terms of length-scale gradients that directly reflect the Yap formulation,

$$S_\varepsilon^{IR} = \mathrm{Max}\left[0.83 \frac{\varepsilon^2}{k} (F - 1) F^2, 0 \right] \tag{9}$$

where $F \equiv [(\partial l/\partial x_j)(\partial l/\partial x_j)]^{1/2}/c_l$. Again, both forms, (7) and (8), evidently raise the dissipation rate when the gradient of the length scale exceeds that occurring in equilibrium near-wall turbulence, c_l.

Figure 4 shows results from employing two of these correction terms in computing the Nusselt number for flow through a heated straight square duct in which heat

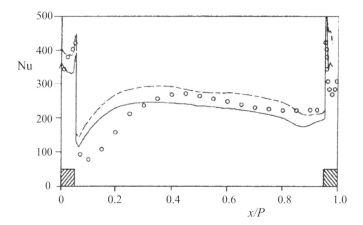

Fig. 4 Application of damping functions for length-scale control. Nusselt number distribution on symmetry line for three-dimensional flow through a square duct with square ribs symmetrically placed on two facing walls. o o o o Experiments, Baughn et al. [2]; - - - - - - Eq. (7); ——— Eq. (9). From Iacovides and Raisee [34]

transfer and mixing are enhanced by ribs affixed to two opposite walls. Unfortunately, results are not shown for the case in which the standard ε equation (without any length-scale damping mechanism) is employed but our experience of other separated flows suggests that in that case the peak Nusselt number would have been at least twice as high as the measured values and probably higher than that! Both length-scale limiters thus provide approximately the correct level of damping.

As a final remark, it should be noted that the latest known version of Wilcox's $k - \omega$ model [83] apparently does not suffer from the above problem, at least in boundary layers in adverse pressure gradients close to separation. It is that feature that makes it especially attractive for predicting flow over wings.

6   The Use of Linear EVMs Within an Unsteady RANS Framework

The previous section has shown that modelling turbulence by way of an EVM closure can, if judiciously applied, lead to predictions of turbulent flows of sufficient accuracy for many industrial applications within an acceptable timescale and cost. The last decade, however, has seen the emergence of large eddy simulation (LES) as a tool ready to be applied to certain industrial flows where the need to achieve high accuracy is essential and the inevitably high costs can be accepted. To conclude this contribution, therefore, we note that if one likewise applies a linear eddy-viscosity model in time-dependent mode to a flow which superficially appears statistically steady, it may well be the case that large-scale structures in the flow are revealed and that the long time-averaged behaviour is different from that obtained with the same model run in steady-flow mode … and in better agreement with experimental observations. The strategy in question is here termed *unsteady RANS*, hereafter denoted URANS.

Essentially, in adopting this strategy, the second-moment turbulence correlations are those arising from an ensemble average rather than a time average of the flow. From this treatment, the resultant computations *may* lead to a steady flow that, if compatible with the boundary conditions, is also two dimensional. As implied above, however, in some circumstances, the computations that emerge are of a continually changing, three-dimensional flow pattern. Thus, one needs to average the flow in time to determine the mean-flow properties and, as a result of that averaging, it becomes evident that, in addition to the turbulent transport caused by the turbulence model, a substantial additional contribution arises from the resolved time-dependent motions. Inevitably, computing such flows in time-dependent mode is considerably more costly than assuming a steady-state behaviour; yet this approach is still at least an order of magnitude less than a conventional LES computation.

Sometimes the time-dependent flow exhibits large-scale regular flow structures reminiscent of some pre-transitional instabilities (even though the finer scale structures, handled by way of the turbulence model, are turbulent). In others, the unsteadiness is irregular and non-repeating though retaining an overall eddy form characteristic of the flow in question—in fact, looking very much like a simulation of the very large turbulent eddies that are present. Indeed, the chosen strategy *is* very like that adopted by large eddy simulations (LES) save that the turbulence model is of conventional RANS type, making no reference to grid dimensions in computing the effective resolved turbulent stresses and indeed generating results independent of the grid (provided a sufficient grid density is used). In comparison with an LES treatment, the modelled contribution with URANS is much larger than the sub-grid-scale part in LES, while the resolved part is, correspondingly, substantially smaller.

The forward time step can usually be much larger than with an LES because the smallest timescales requiring resolution are larger. In consequence, while, for a two-dimensional flow that is steady in the mean, the computational time is obviously substantially greater than if a steady RANS approach were adopted, it will still normally be much less than for a conventional LES treatment. Moreover, a crucial advantage of URANS, where applicable, is its potential to handle flows at very high Reynolds and Rayleigh numbers, which are, at present, still beyond the reach of LES.

The first fully three-dimensional URANS explorations were those of Tatsumi et al. [75] and Kenjereš and Hanjalić [41]. In fact, in both these studies, the turbulence models employed were of a higher order than the linear EVMs considered in this paper and are thus not included here. In the case of the Kenjereš–Hanjalić work, this was in order to capture turbulent mixing in Rayleigh–Bénard convection where mean velocities are zero. Nevertheless, in those cases, as in the examples below, the computations were made with precisely the model for the unresolved turbulent transport originally proposed for steady-flow applications. The generic cases presented here are those that arise within the narrow, enclosed cylindrical space between confined, coaxial circular discs a short distance apart, one or both of which rotate. The flow configuration is an idealization of that which occurs in turbine-disc cavities or a stack of rotating computer discs. Conventionally, these flows have been examined assuming the flow to be steady and axisymmetric. The three-dimensional time-dependent results shown below are drawn from the computations of Zacharos [86]. The effects of the unresolved turbulence have been accounted for with the standard high Reynolds number k-ε EVM using the Yap correction, Eq. (7). Within the low Re sublayer, analytical wall functions (AWF) have been used [11] which enable an economical yet reasonably comprehensive treatment of the sublayer region where the mean-velocity vector undergoes strong skewing.

The first example is of co-rotating discs at a spin Reynolds number of 1.46×10^5. The shroud is stationary and this induces a radial motion outwards near the two rotating discs (the Ekman layer) with a return flow in the central region. Figure 5 (left) provides a pictorial view for the case where the disc spacing is half the radius; it shows the flow surface where the circumferential velocity is half the disc speed. In this case, the time-dependent examination has revealed four large eddies in both the upper and lower halves of the cavity. While the flow is globally symmetric about the midplane, the four 'eddies' in the upper half are rotated by $45°$ relative to those in the lower half. The axial vorticity on the geometric symmetry plane, Fig. 5 (right), thus exhibits eight trailing 'tails'.

Quite different large-scale structures appear in the case where one of the discs is stationary (the so-called rotor–stator cavity). The number of separate lobes, in this case, is known to depend on the disc spacing and the Reynolds number, Czarny et al. [12], though the most commonly arising structure is a double lobe as seen in the flow visualization photograph in Fig. 6. The three-dimensional, time-dependent calculation by Zacharos [86] also predicts such a two-eddy form for the same values of H/R and Re_Ω. In fact, the computational contours shown are for different levels

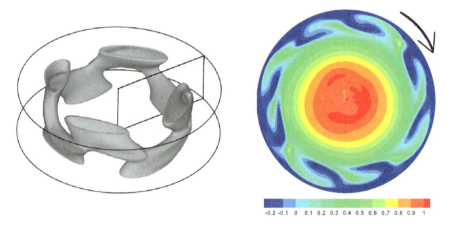

-0.2 -0.1 0 0.1 0.2 0.3 0.4 0.5 0.6 0.7 0.8 0.9 1

Fig. 5 Flow in a co-rotating disc cavity, $H/R = 0.5$, $\mathrm{Re}_\Omega = 1.46 \times 10^5$. Left: Surface of circumferential velocity equal to half disc spin velocity (from [86]). Right: Normalized axial vorticity contours on geometric symmetry plane (from [36])

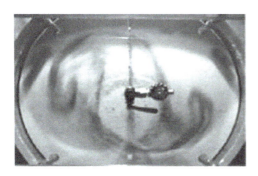

Fig. 6 Flow within a rotor–stator cavity. $H/R = 0.195$, $Re_\Omega = 0.9 \times 10^6$. Left: Streak-line flow visualization, Czarny et al. [12]; Right: Computations showing contours of $k^{1/2}/\Omega R$

of $k^{1/2}/\Omega R$. The comparison is relevant, however, since the regions (in the left-hand figure) where the dye remains concentrated indicate areas where turbulent agitation is minimal and thus where the modelled k-levels are very low (which appear as the dark blue zones in the right-hand figure).

As a third example on rotating flows, Fig. 7 considers the case of flow between *contra*-rotating discs, a case which is further complicated by the fact that one disc rotates at twice the speed of the other. Iacovides et al. [37] had earlier examined this flow using a variety of turbulence models ranging from two-equation eddy-viscosity schemes up to second-moment closure but assuming the flow to be axisymmetric and steady. Their results achieved at best indifferent agreement with the experimental data of Gan et al. [19], the greatest discrepancy occurring at $r = 0.85R$ close to the slower disc where the computed radial velocity was towards the disc centre whereas the measurements showed an outward flow. Moreover, near the faster disc the boundary-layer thickness was more than twice as thick as measured. The flow predicted by the unsteady three-dimensional computations, Fig. 7a, shows major improvements in resolving the Ekman layers near each disc and, overall, a significantly closer agreement with the experiment. Figure 7b provides a snapshot of the spiral flow structure present in this case near the slower moving disc (the dark and light surfaces show contours of two different values of axial velocity towards and away from the upper surface).

The foregoing examples have perhaps provided a flavour of the potential of URANS in tackling really challenging CFD problems. The approach is rapidly evolving, however, with a variety of new strategies whose permanence at present is difficult to judge. We note, however, the PITM (partially integrated turbulent transport) method of Schiestel and Dejoan [71] or PANS (partially averaged Navier–Stokes) of Girimaji [20]. Another approach is the SAS (scale-adaptive simulations) method of Menter and Egorov [52], which uses no grid size information, but instead introduces the von Karman length scale (containing the second derivative of the instantaneous velocity), making the model sensitive to inherent flow instabilities. One should note,

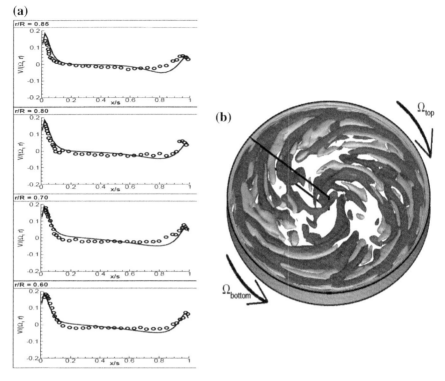

Fig. 7 Flow in a counter-rotating disc cavity with the faster disc (at $x/s = 0$) rotating at twice the speed of the slower disc. **a** Profiles of radial velocity at different radial locations. Symbols: Experiments, Gan et al. [19]; —— URANS k-ε EVM with analytical wall functions, Iacovides et al. [36]; **b** Visualization of coherent vortex structure via iso-contours of axial velocity towards and away from the slower disc with velocity $\pm 0.005 \Omega_{\text{fast}} R$, Zacharos [86]

however, that all these and other strategies are still under development and that their scope and limitations have not yet been fully exemplified.

References

1. Barre, S., Bonnet, J.-P., Gatski, T. B., & Sandham, N. D. (2002). Ch. 19: Compressible high speed flows. In B. Launder & N. Sandham (Eds.), *Closure strategies for turbulent and transitional flows*. CUP.
2. Baughn, J. W., Hechanova, A. E., & Yan, X. (1992). An experimental study of entrainment effects on the heat transfer from a flat surface to a heated circular impinging jet. *ASME Journal of Heat Transfer, 113*, 1023–1028.
3. Boussinesq, J. (1877). Essai sur la théorie des eaux courantes. *Mémoires présentés par divers savants à l'Académie des Sciences, 23*(1), 1–680.
4. Bradshaw, P., Ferriss, D. H., & Atwell, N. P. (1967). Calculation of boundary layer development using the turbulent energy equation. *Journal of Fluid Mechanics, 28*, 593–616.

5. Bradshaw, P. (1966). The turbulence structure of equilibrium boundary layers. NPL. Rep. 1184, National Physical Laboratory, Teddington, UK.

6. Buleev, N. J. (1962). *Theoretical model of the mechanism of turbulent exchange in fluid flows* (p. 957). Moscow; AERE Trans: Teploperedacha (Heat Transfer), USSR Academy of Sciences.

7. Chien, K.-Y. (1982). Prediction of channel and boundary layer flows with a low-Reynolds-number turbulence model. *AIAA Journal, 20*, 33–38.

8. Chou, P. Y. (1945). On the velocity correlations and the solution of the equations of turbulent fluctuation. *Quarterly of Applied Mathematics, 3*, 38–54.

9. Clauser, F. H. (1954). Turbulent boundary layers in adverse pressure gradients. *Journal of Aerosol Science, 21*, 91–108. (See also: The turbulent boundary layer. *Advances in Applied Mechanics, 4*, 1–51, Academic Press, New York, 1956).

10. Cotton, M. A., & Jackson, J. D. (1990). Vertical tube air flows in the turbulent mixed-convection regime using a low-Reynolds-number k-ε model. *International Journal of Heat and Fluid Flow, 33*, 275–286.

11. Craft, T. J., Gerasimov, A. V., Iacovides, H., & Launder, B. E. (2002). Progress in the generalization of wall-function treatments. *International Journal of Heat and Fluid Flow, 23*, 148–160.

12. Czarny, O., Iacovides, H., & Launder, B. E. (2002). Precessing vortex structures within rotor-stator disc cavities, *Flow, Turbulence and Combustion, 69*, 51–61.

13. Daly, B. J., & Harlow, F. H. (1970). Transport equations in turbulence. *Physics of Fluids, 13*, 2634–2649.

14. Davydov, B. I. (1959). On the statistical dynamics of an incompressible turbulent fluid. *Doklady Akademii Nauk SSSR, 127*, 768–771.

15. Davydov, B. I. (1961). On the statistical dynamics of an incompressible turbulent fluid. *Doklady Akademii Nauk SSSR, 136*, 47–50.

16. Durbin, P. A. (1991). Near-wall turbulence closure modelling without damping functions. *Theoretical and Computational Fluid Dynamics, 3*, 1–13.

17. Escudier, M. P. (1966), The distribution of mixing length in turbulent flows near walls. Imperial College, Heat Transfer Section Rep. TWF/TN/1.

18. Forstall, W., & Shapiro, A. H. (1952). Momentum and mass transfer in coaxial gas jets, ASME. *Journal of Applied Mechanics, 17*, 399–408.

19. Gan, X., Kilic, M., & Owen, J. M. (1995). Flow between contra-rotating discs. *Journal of Turbomachinery, 117*, 298–305.

20. Girimaji, S. S. (2006). Partially-averaged Navier-Stokes method for turbulence: A Reynolds-averaged Navier-Stokes to direct numerical simulation bridging method. *ASME Journal of Applied Mechanics, 73*, 413–421.

21. Glushko, G. S. (1965). Turbulent boundary layer on a flat plate in an incompressible fluid. *Izv. Akad. Nauk SSSR, Mekh. 4*, 13–23.

22. Gosman, A. D., Pun, W. M., Runchal, A. K., Spalding, D. B., & Wolfshtein, M. (1969). *Heat and mass transfer in recirculating flows*. London: Academic Press.

23. Hanjalić, K., & Launder, B. E. (1980). Sensitizing the dissipation equation to secondary strains. *ASME Journal of Fluids Engineering, 102*, 34–40.

24. Hanjalić, K., Popovac, M., & Hadžiabdić, M. (2004). A robust near-wall elliptic-relaxation eddy-viscosity turbulence model for CFD. *International Journal of Heat and Fluid Flow, 25*, 1047–1051.

25. Hanjalić, K. (1996). some resolved and unresolved issues in modelling non-equilibrium and unsteady turbulent flows. In W. Rodi & G. Bergeles (Eds.), *Engineering turbulence modelling and experiments-3* (pp. 3–18). Amsterdam: Elsevier.

26. Hanjalić, K. (1970). *Two-dimensional asymmetric flow in ducts*. Ph.D. thesis, Faculty of Engineering, University of London, UK.

27. Hanjalić, K., & Launder, B. E. (2011). *Modelling turbulence in engineering and the environment: Second-moment routes to closure*. CUP.

28. Harlow, F. H., & Nakayama, P. I. (1967). Turbulence transport equations. *Physics of Fluids, 10*, 2323.

29. Harlow, F. H., & Nakayama, P. I. (1968). Transport of turbulence energy decay rate. Rep LA-3854, Los Alamos Scientific Laboratory, Los Alamos.

30. Hoyas, S., & Jimenez, J. (2006). Scaling of velocity fluctuations in turbulent channels up to $Re_\tau = 2003$. *Physics of Fluids, 18,* 0011702.

31. Huang, P. G. (1986). *The computation of elliptic turbulent flows with second-moment-closure models.* Ph.D. thesis, Faculty of Technology, University of Manchester.

32. Hussein, J. H., Capp, S., & George, W. K. (1994). Velocity measurements in a high-Reynolds-number momentum-conserving axisymmetric jet. *Journal of Fluid Mechanics, 258,* 31–75.

33. Iacovides, H., Launder, B. E., & Loizou, P. A. (1987). Numerical computation of flow through a 90-degree bend. *International Journal of Heat and Fluid Flow, 8,* 320–325.

34. Iacovides, H., & Raisee, M. (1999). Recent progress in the computation of flow and heat transfer in internal cooling passages of turbine blades. *International Journal of Heat and Fluid Flow, 20,* 320–328.

35. Iacovides, H., & Raisee, M. (1997). Computation of flow and heat transfer in 2D rib-roughened passages. In K. Hanjalić & T. Peeters (Eds.), *Turbulence heat & mass transfer 2—supplement.* The Netherlands: Delft University Press.

36. Iacovides, H., Launder, B. E., & Zacharos, A. (2009). The economical computation of unsteady flow structures in rotating cavities. In *Proceedings of the 6th International Symposium on Turbulence & Shear Flow Phenomena* (pp. 797–802).

37. Iacovides, H., Nikas, K., & Te Braak, M. (1996). *Turbulent flow computations in rotating cavities using low Reynolds number models* (Paper 96-GT-59). Birmingham, UK: ASME Gas Turbine Congress.

38. Jones, W. P., & Launder, B. E. (1972). The prediction of laminarization with a two-equation model of turbulence. *International Journal of Heat and Fluid Flow, 15,* 301–314.

39. Jones, W. P. (1971). *Laminarization in strongly accelerated boundary layers.* Ph.D. thesis, Faculty of Engineering, University of London.

40. Jones, W. P. (2002). Ch. 20: The joint scalar probability density function method. In: B. Launder & N. Sandham (Eds.), *Closure strategies for turbulent and transitional flows.* CUP.

41. Kenjereš, S., & Hanjalić, K. (1999). Transient analysis of Rayleigh-Bénard convection with a RANS model. *International Journal of Heat and Fluid Flow, 20,* 329–340.

42. Kline, S. J., Morkovin, M. V., Sovran, G., & Cockrell, D. J. (Eds.). (1968). Computation of turbulent boundary layers-Vol 2. In *AFOSR-IFP-Stanford Conference*, August 18–25.

43. Kline, S. J., Cantwell, B. J., & Lilley, G. M. (1981). *Proceedings of the AFOSR-HTTM-Stanford Conference on Complex Turbulent Flows: Comparison of Computations and Experiments, Vol. 1, Objectives, Evaluation of Data, Specification of Test Cases.* Thermosciences Division, Stanford University, Stanford, CA.

44. Kline, S. J., Cantwell, B. J., & Lilley, G. M. (1982). *Proceedings of the AFOSR-HTTM-Stanford Conference on Complex Turbulent Flows: Comparison of Computations and Experiments, Vol. 2, Taxonomies, Reporters' Summaries, Evaluation and Conclusions.* Thermosciences Division, Stanford University, Stanford, CA.

45. Kolmogorov, A. N. (1942). Equations of turbulent motion in an incompressible fluid. *Izvestiya Akademii Nauk SSR*, Serie *Physics, 6*(1–2), 56–58.

46. Lam, C. K., & Bremhorst, K. (1981). Modified form of the k-ε model for predicting wall turbulence. *Journal of Fluids Engineering, 103,* 456–460.

47. Launder, B. E. (2015). First steps in modelling turbulence and its origins: A commentary on Reynolds (1895) 'On the dynamical theory of incompressible viscous fluids and the determination of the criterion'. *Philosophical Transactions of the Royal Society A, 20140231,* 373.

48. Launder, B. E., & Sharma, B. I. (1974). Application of the energy-dissipation model of turbulence to the calculation of flow near a spinning disc. *Letters in Heat and Mass Transfer, 1,* 131–138.

49. Launder, B. E. (1986). Low Reynolds number turbulence near walls. Rep TFD/86/4, Mechanical Engineering Dept, UMIST, Manchesster, UK.

50. Launder, B. E., Morse, A. P., Rodi, W., & Spalding, D. B. (1973). Prediction of free shear flows: a comparison of the performance of six turbulence models. In *Free turbulent shear flows* (Vol. 1, pp. 361–426) (Conference held at Langley Research Center, July 20–21, 1972) NASA Rep SP-321.

51. Laurence, D. R., Uribe, J. C., & Utyuzhnikov, S. V. (2004). A robust formulation of the v2–f model. *Flow, Turbulence and Combustion, 75*, 169–185.

52. Menter, F. R., & Egorov, Y. (2010). Scale-adaptive simulation method for unsteady turbulent flows prediction, Pt 1: Theory and model description. *Flow, Turbulence Combustion, 85*, 113–138.

53. Morkovin, M. V., Bushnell, D. M., Edelman, R. B., & Libby, P. A. (1973). Report of the conference evaluation committee. In *Free turbulent shear flows* (Vol. 1, pp. 673–695). (Conference held at Langley Research Center, July 20–21, 1972) NASA Rep SP-321.

54. Moser, R. D., Kim, J., & Mansour, N. N. (1999). DNS of turbulent channel flow up to $Re_\tau = 590$. *Physics of Fluids, 11*, 943.

55. Ng, K. H., & Spalding, D. B. (1972). Turbulence model for boundary layers near walls. *Physics of Fluids, 15*, 20–30.

56. Panchapakesan, N. R., & Lumley, J. L. (1993). Turbulence measurements in axisymmetric jets of air and helium; Part 1: Air jets. *Journal of Fluid Mechanics, 246*, 197–223.

57. Patankar, S. V., & Spalding, D. B. (1967). *Heat and mass transfer in boundary layers*. London: Morgan-Grampian Press.

58. Patel, V. C., Rodi, W., & Scheuerer, G. (1985). Turbulence models for near-wall and low-Reynolds-number flows. *AIAA Journal, 23*, 1308–1319.

59. Pope, S. B. (1978). An explanation of the round-jet/plane-jet anomaly. *AIAA Journal, 16*, 3311–3340.

60. Pope, S. B. (2000). *Turbulent flows*. Cambridge, UK: Cambridge University Press.

61. Prandtl, L. (1925). Bericht über Untersuchungen zur ausgebildeten Turbulenz. *ZAMM, 5*, 136.

62. Prandtl, L. (1945). Über ein neues Formelsystem für die ausgebildeten Turbulenz. Nachrichten von der Akad. der Wissenschaft in Göttingen.

63. Reynolds, O. (1883). An experimental investigation of the circumstances which determine whether the motion of water should be direct or sinuous and the law of resistance in parallel channels. *Philosophical Transactions of the Royal Society, 174*, 935–982.

64. Reynolds, O. (1895). On the dynamical theory of incompressible viscous flows and the determination of the criterion. *Philosophical Transactions of the Royal Society, 186A*, 123–164.

65. Rodi, W., & Spalding, D. B. (1970). A two-parameter model of turbulence and its application to free jets. *Wärme und Stoffübertragung, 3*, 85–95.

66. Rodi, W. (1968). Statistical theory of non-homogeneous turbulence: Translation of J.C. Rotta's papers into English. Imperial College Mech. Eng. Dept. Rep. TWF/TN/38 and TWF/TN/39.

67. Rodi, W. (1975). A review of experimental data of uniform-density, free turbulent boundary layers. In B. E. Launder (Ed.), *Studies in convection 1* (pp. 79–165). London: Academic Press.

68. Rotta, J. C. (1951). Statistische Theorie nichthomogener Turbulenz: Part 1: *Zeitschrift für Physik, 129*, 547–573; Part 2: *Zeitschrift für Physik, 131*, 51–77.

69. Saffman, P. G. (1970). A model for inhomogeneous turbulent flow. *Proceedings of the Royal Society, A317*, 417–433.

70. Scheuerer, G. (1983). Entwicklung eines Verfahrens zur Berechnung zweidimensionaler Grenzschichten an Gasturbinenschauffeln. Dr Ing Thesis, University of Karlsruhe, Germany.

71. Schiestel, R., & Dejoan, A. (2005). Towards a new partially integrated transport model for coarse grid and unsteady turbulent flow simulations. *Theoretical and Computational Fluid Dynamics, 18*, 443–468.

72. Spalding, D. B. (1967). Heat transfer from turbulent separated flows. *Journal of Fluid Mechanics, 27*, 97–109.

73. Spalding, D. B. (1971). Concentration fluctuations in a round turbulent jet. *Chemical Engineering Science, 26*, 95–107.

74. Spalding, D. B. (1971b). The k-W model of turbulence. Mech. Eng. Dept. Rep TW/TN/A/16, Imperial College, London.

75. Tatsumi, K., Iwai, H., Neo, E. C., Inaoka, K., & Suzuki, K. (1999). Prediction of time-mean characteristics and periodical fluctuation of velocity and thermal fields of a backward-facing step, 1167–1172. In S. Banerjee & J. K. Eaton (Eds.), *Turbulence & shear-flow phenomena—1*. New York: Begell House.
76. Taylor, G. I. (1915). Eddy motion in the atmosphere. *Philosophical Transactions of the Royal Society, A215,* 1–26.
77. Taylor, G. I. (1932). The transport of vorticity and heat through fluids in turbulent motion. *Proceedings of the Royal Society, 135*(828), 685–702.
78. Taylor, A. M. K., Whitelaw, J. H., & Yianneskis, M. J. (1982). Curved duct with strong secondary motion—Velocity measurements of developing laminar and turbulent flow. *Journal of Fluids Engineering, 104,* 350–359.
79. Townsend, A. A. (1956). *The structure of turbulent shear flows*. Cambridge University Press.
80. Von Kármán, Th. (1930). Mechanische Ähnlichkeit und Turbulenz. In *Proceedings of the 3rd International Congress on Applied Mechanics*, Stockholm, Pt. 1, 85.
81. Wilcox, D. C. (1998). *Turbulence modeling for CFD* (2nd ed.). La Cañada, California: DCW Industries.
82. Wilcox, D. C., & Rubesin, M. W. (1980). Progress in turbulence modelling for complex flow fields including effects of compressibility. NASA Tech. Rep: TP 1517, NASA AMES Research Center.
83. Wilcox, D. (2006). *Turbulence modeling for CFD* (3rd ed.). La Cañada, California: DCW Industries.
84. Wouters, H. A., Peeters, T. W., & Roekaerts, D. (2002). Joint velocity-scalar PDF methods. In B. Launder & N. Sandham (Eds.), *Closure strategies for turbulent and transitional flows*. CUP.
85. Yap, C. R. (1987). *Turbulent heat and momentum transfer in recirculating and impinging flows*. Ph.D. thesis, Faculty of Technology, University of Manchester.
86. Zacharos, A. (2010). *The use of unsteady RANS in the computation of 3-dimensional flow in rotating cavities*. Ph.D. thesis, Faculty of Engineering & Physical Sciences, University of Manchester, UK.

Studies on Mixed Convection and Its Transition to Turbulence—A Review

Somenath Gorai and **Sarit K. Das**

Nomenclature

Dimensionless Numbers

Re	Reynolds number ($\rho V D / \mu$)
Ri	Richardson number (Gr/Re^2)
Pr	Prandtl number ($\mu C_p / k$)
Ra	Rayleigh number ($Gr.Pr$)
Nu	Nusselt number (hD/k)
Kn	Knudsen number (λ/L); 'λ' is mean free path
Ha	Hartman number
$\overline{Nu}$	Average Nusselt number
Nu_T	Nusselt number for forced turbulent convection
Gr	Grashof number
Gr_q	Grashof number based on heat flux ($g\beta D^4 \dot{q}/v^2 k$)
Gr_D	Grashof number based on diameter ($g\beta D^3 \Delta T/v^2$)
Gr_L	Grashof number based on constant wall temperature ($g\beta L^3 \Delta T/v^2$)
Gr_T	Thermal Grashof number ($g\beta D^3 \Delta T/v^2$)
Gr_M	Solutal Grashof number ($g\beta_M D^3 \Delta\omega/v^2$)

The article is dedicated to the memory of Prof. Brian Spalding with whom the corresponding author spent an exciting week at St. Petersburg, Russia during a conference organized by Prof. D. Leontiev. Prof. Spalding's (along with Prof. Launder) work in turbulent mixed convection is the inspiration behind this review.

S. Gorai · S. K. Das (✉)
Department of Mechanical Engineering, Indian Institute of Technology Ropar, Rupnagar, Punjab 140001, India
e-mail: skdas@iitrpr.ac.in

S. Gorai
e-mail: somenathgorai22@gmail.com

$\overline{Ra}$	Average Rayleigh number
$\overline{Re}$	Average Reynolds number
$\widehat{Nu}$	Weighted average Nusselt number
Gz	Graetz number $\left(\frac{D}{L}Re.Pr\right)$
Re_{cr}	Critical Reynolds number
Re_{θ}	Transition momentum thickness Reynolds number
Re_{qt}	Reynolds value at the start of quasi-turbulent flow regime
Re_x	Local Reynolds number
Bo	Buoyancy parameter $(Gr_q/Re^m Pr^n)$
Bo_2	Buoyancy parameter $(Gr_q/Re^{2.5} Pr)$
K	Buoyancy parameter $(Gr/Re^{2.5})$
ER	Expansion ratio (outlet to inlet height ratio)
E	Enhancement ratio (ratio of heat input with porosity and without porosity)

Abbreviations

ACFD	Asymptotic Computational Fluid Dynamics
AWF	Analytical Wall Functions
CFD	Computational Fluid Dynamics
CMM	Compound Matrix Method
CMSIP	Coupled Modified Strongly Implicit Procedure
DNS	Direct Numerical Simulation
FCD	Forced Convection Developing
FD	Fully Developed
LBM	Lattice Boltzmann Method
LES	Large Eddy Simulation
MCD	Mixed Convection Developing
OpenFOAM	Open source Field Operation And Manipulation
PIV	Particle Image Velocimetry
RANS	Reynolds Averaged Navier-Stokes
RNG	Renormalization Group Method
SST	Shear Stress Transport
UHF	Uniform Heat Flux
UWT	Uniform Wall Temperature

Greek Letters

ε	Rate of dissipation of turbulence energy
ω	Specific rate of dissipation

φ_v	Viscosity ratio (μ_b/μ_w)
μ_b	Dynamic viscosity at bulk temperature (Pa s)
μ_w	Dynamic viscosity at the wall temperature (Pa s)
λ	Buoyancy parameter (Gr/Re^2)
$\emptyset$	Volume fraction
φ	Angle (degree or radian)
θ	Dimensionless temperature
ρ	Density (kg/m^3)
γ	Intermittency in $\gamma - Re_\theta$ transition model
β	Thermal expansion coefficient (K^{-1})

Other Symbols

k	Thermal conductivity (W/m K)
k	Turbulence kinetic energy (J)
Sp. Gr.	Specific gravity
g	Acceleration due to gravity (m/s^2)
a, b, c	Constants in Eqs. 8, 12
L	Length (m)
D	Diameter (m)
D_h	Hydraulic diameter (m)
d_e	Equivalent diameter of a channel (m) ($2hb/(h+b)$; h, b are channel height and width)
A	Aspect ratio (−)
Q	Heat rate (W)
$\dot{q}$	Heat flux (W/m^2)
q_1	Cold wall heat flux (W/m^2)
q_2	Hot wall heat flux (W/m^2)
x, y	Axial and transverse coordinates (m)
u, v	Axial and transverse velocities (m/s)
X, Y	Dimensionless axial and transverse coordinates
U, V	Dimensionless axial and transverse velocity
x	Distance from the initial point of heating (m)
C_p	Specific heat at constant pressure (J/kg K)
C_f	Skin friction coefficient (−)
~	Approximately
©	Copyright

Subscripts

lam, l	Laminar
b	Properties of fluid at bulk temperature
q	Based on heat flux
M	Mixed convection in Eq. 7
trans	Transition
c, cr	Critical value
lc	Laminar flow at critical Reynolds number
cr_2	Critical value at second wall
x	Local value

Superscripts

t	Turbulent

1   Introduction

Mixed convection is the combined form of natural and forced convection heat transfer, which has been a subject matter of study of researchers in the field of nuclear reactors, solar energy systems, heat exchangers, supercritical boilers, cooling of electronic devices, extraction of geothermal energy and many more areas because of its varying nature heat transfer phenomena. At the interface of natural and forced convection, this mode of heat transfer can be categorized into either opposing (buoyancy force and fluid flow are in the opposite direction) or aiding (buoyancy and fluid flow are in the same direction), which decides whether the enhancement or impairment of heat transfer will take place.

In the early understanding of convective heat transfer, forced and free convective heat transfer were studied separately and any intermixing of these two were ignored. When the attention was given on such possibilities it was limited to laminar and transitional flow studies initially and later extended to turbulent flows. Published articles on mixed convection indicate that most of the early studies concentrated on the correlations for heat transfer and fluid flow behavior in mixed convection regimes. The simultaneous effect of buoyancy forces and externally applied inertia forces is responsible for this mixed regime. Detailed reviews on mixed convection till date can be found in review articles by Jackson et al. [61], Jackson [59], Galanis and Behzadmehr [44], Meyer [80] and Dawood et al. [35].

In this review, attention is focused on mixed convection in vertical and horizontal plates, pipes, channels, ducts, cavities, etc., comprising of analytical, computational, and experimental work concentrating on the transitional regime of flow. The studies can be categorized as follows.

2 Overview of Laminar Mixed Convection

It can be safely said that when both Reynolds number, Re (for forced convection) and Rayleigh number, Ra (for natural convection) are in the laminar regime, the mixed convection is also laminar.

2.1 Theoretical Studies on Laminar Mixed Convection

The distortion of the velocity field and the pattern of the convection in the fluid decides the heat transfer enhancement or impairment in case of aiding (upward direction passed a heated surface) and opposing (upward direction passed a cooled surface) mixed convection.

In laminar flow variations of density and viscosity with temperature cause distortions in the flow field and affect the rate of heat transfer to fluids. From the mid-twentieth century, Martinelli and Boelter [78], Hallman [54], Hanratty et al. [56], Morton [85] and many others started analyzing fully developed flow model. They were only able to describe the variation of density and viscosity for a fully developed model during heating and cooling in a vertical tube with isothermal wall. Thereafter with the emergence of computers attempts were made to obtain solutions for the developing flow. Rosen and Hanratty [103] used the boundary layer integral method and power series to obtain the temperature and velocity profiles following earlier work by Pigford [90]. A comprehensive study was presented by Chen et al. [27] on laminar mixed convection for isothermal vertical, inclined, and horizontal flat plates. Simple correlations for local and average Nusselt numbers were developed and found to be similar to that of numerically predicted as well as experimental results. Articles suggested the buoyancy effect and forced flow enhances the heat transfer by ~20% in case of pure forced and pure free convection, respectively. Boulama and Galanis [22] also presented exact analytical solutions for upward fully developed flow, steady-state laminar mixed convection between two vertical parallel plates. The results show that in uniform wall temperature (UWT) case it depends on a single parameter called combined buoyancy parameter: $(Gr_T + Gr_M)/Re$ and in uniform heat flux (UHF) case it depends on three parameters: Gr_T/Re, Gr_M/Re and q_1/q_2. Solutions also revealed that the UHF case is valid when the net heating effect is positive. The various plots depict that there is a significant improvement in heat transfer rate near the walls due to the buoyancy effect. The flow reversal phenomena and heat transfer characteristics of the fully developed laminar flow mixed convection in vertical heated channels were analyzed analytically by Cheng et al. [29]. The velocity distribution, temperature distribution as well as Nusselt number (Nu) variation exhibit a flow reversal near the colder wall within the channel below the threshold value of Re/Gr.

Gavara et al. [46] studied analytically a laminar boundary layer in mixed convection from non-isothermal vertical plates using a perturbation technique. The heat transfer rate and fluid velocity inside the boundary layer were analyzed with the variation of governing parameters. Universal perturbation functions were obtained once and for all which can be used to estimate heat transfer for any type of wall temperature variation without solving each time. Analytical solutions for mixed convection in a vertical micro-channel along with numerical results using Lattice Boltzmann Method (LBM) were obtained by Avramenko et al. [7]. The velocity profiles, temperature profiles, and Nusselt number variations showed that the effect of Knudsen number (Kn) was more pronounced near the wall whereas near the centerline Rayleigh number (Ra) effect was stronger. This is because higher Rayleigh number increases the velocity gradient at the wall.

Thus, the above studies indicate that enhancement or impairment of heat transfer may happen depending on whether the natural convection is aiding or opposing the flow direction. Distortion in velocity field due to density and viscosity variation decides the enhancement or deterioration in heat transfer of a laminar mixed convection. Quite a few studies indicate that the buoyancy effect in pure forced convection and forced flow in pure free convection enhances ~20% of heat transfer. There may be flow reversal beyond a threshold value of Re/Gr and mostly happens near the colder wall. Few studies implemented perturbation techniques to estimate heat transfer rate and fluid velocity inside the boundary layer. Universal perturbation functions were obtained for any type of vertical wall temperature variation. Due to complexity in analytical solutions largely numerical techniques were used for analysis, which is discussed next.

2.2 Numerical Studies on Laminar Mixed Convection

Numerical solutions of heat transfer for upward flow of air by taking variable physical properties with uniform wall temperature were obtained by Bradley and Entwistle [23]. The numerical models for developing mixed convection flow taking account of variations of density and viscosity were also developed by Lawrence and Chato [72] and by Marner and McMillan [75] for the same boundary conditions. An interesting result is presented in Figs. 1 and 2, which shows the local Nusselt value increases near the point of maximum velocity distortion with the thermal entry length and then decreases further downstream as the fluid and wall temperature difference minimizes. Zeldin and Schmidt [126] used an iterative method and solved the full elliptic equations to avoid the use of marching procedures. They inferred that the velocity profile is different from the forced convection and the maximum velocity may not occur at the centerline. In buoyancy-aiding flow, the velocity near the wall increases due to the

Fig. 1 Typical developing dimensionless axial velocity profiles [reproduced with permission from Marner and McMillan [75] © ASME]

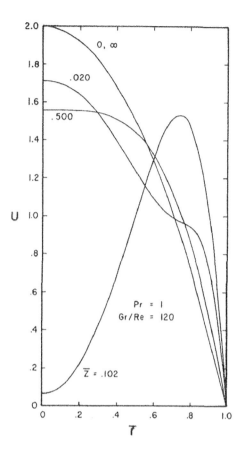

difference in wall and fluid temperature and decreases at the center, hence the heat transfer coefficient increases, and velocity profile becomes unstable. In buoyancy-opposing flow, the velocity reduces near the wall and hence heat transfer decreases. The velocity gradient at the wall approaches zero causing the instability. Tanaka et al. [115] predicted a fully developed upward flow by $k - \varepsilon$ model of turbulence in a heated vertical pipe and compared with experimental results taking nitrogen gas as a test fluid. A numerically predicted regime for mixed convection was plotted between Reynolds number (Re) and Grashof number (Gr) as shown in Fig. 3. The upper left part shows forced convection regime and lower right part endorse natural convection. Velusamy and Garg [117] described the velocity and temperature distribution in fully developed case for vertical elliptic duct. They found that the ratio of Nusselt number to friction factor is higher in the case of the elliptical duct as compared to a circular duct that leads to a substantial rate of heat transfer. Aydin [8] described the aiding and opposing mechanism of mixed convection in shear and buoyancy-driven cavity. Velocity and temperature distributions shown in Fig. 4 were carried out by the finite difference scheme. They obtained a mixed convection range

Fig. 2 Typical developing
dimensionless temperature
profiles [reproduced with
permission from Marner and
McMillan [75] © ASME]

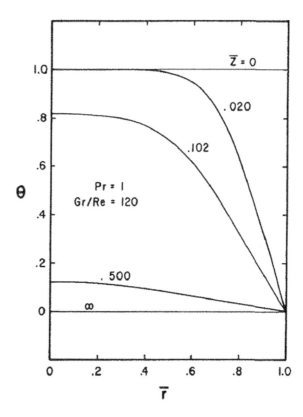

$(0.01 \leq Gr/Re^2 \leq 100)$ for a ratio of Gr and Re^2, named it as mixed convection
parameter (Gr/Re^2).

Various theoretical investigations demonstrate that pressure drop is significantly
affected by heat transfer and vice versa. Semi-empirical correlations given by Eqs. 1
and 2 for pressure drop were developed by Joye [64]. The equations are valid for
laminar, constant wall temperature boundary conditions in vertical, internal aiding
flows. The measurement of pressure drop is quite difficult in vertical mixed convection particularly for liquids. Hence a correction factor needs to be applied due to
the density difference of fluid present in the manometer and the conduit. In Fig. 5,
pressure drop versus volumetric flow rate for forced convection with and without
the influence of buoyancy shows substantial pressure drop for laminar and turbulent
flow up to Reynolds number of ~7000.

$$\Delta P_{\mathrm{lam}} = \left(128\mu L\, Q/\pi D^4\right)(\rho g)\left(Sp.Gr/\varphi_v^{0.38}\right). \tag{1}$$

$$\Delta P/\Delta P_{\mathrm{lam}} = 1 + 1565 Gr_L^{3/4} Pr^{1/2}/(0.952 + Pr)^{3/4} Re^2\, (L/D)^2\big|_b. \tag{2}$$

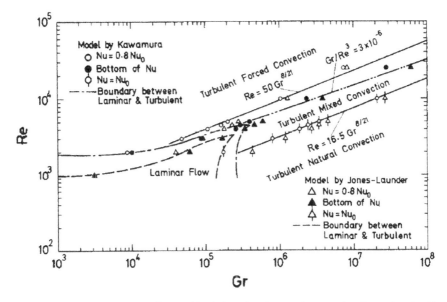

Fig. 3 Predicted regime map for combined forced and natural convection [reproduced with permission from Tanaka et al. [115] © Elsevier]

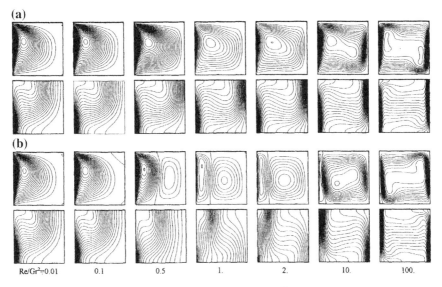

Fig. 4 Streamlines and isotherms at different values of Gr/Re^2 for **a** aiding **b** opposing buoyancy [reproduced with permission from Aydin [8] © Elsevier]

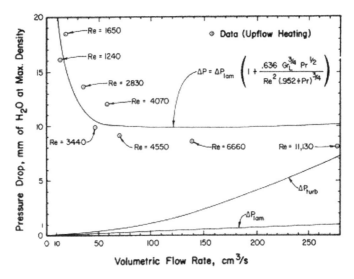

Fig. 5 Variation of pressure drop with volumetric flow rate [reproduced with permission from Joye [64] © Elsevier]

The average Nusselt number was calculated numerically in a vertical channel flow discharging over the horizontal isothermal surface by Nobari and Beshkani [86] having varying area of vertical channels and analysis showed that a diverging channel was more efficient than a convergent one with respect to heat transfer. The influence of linearly heated side walls or/and cooled right wall on mixed convection was carried out numerically by Basak et al. [14] in a Lid-driven square cavity. Another similar investigation on mixed convection flows within a square cavity with uniform and non-uniform heating of bottom wall was done by Basak et al. [15]. The vertical walls were kept isothermally cooled and the top wall was well insulated moving from left to right. In both studies the flow pattern signifies that as the value of Re increases from 1 to 100, a transition from natural convection to forced convection occurs depending on the value of Gr irrespective of Prandtl number (Pr). The effect of natural convection is dominant up to $Re = 10$ for higher value of Gr and thereafter with the increase of Re, forced convection becomes dominant. The stability of mixed convection flow past vertical flat plate using Compound Matrix Method (CMM) was studied by Venkatasubbaiah and Sengupta [118] to solve linearized equations and using Direct Numerical Simulations (DNS) with Boussinesq approximation. For the first time, existence of critical buoyancy parameter (K_{cr}) was found theoretically by Sengupta and Subbaiah [105] which was also observed experimentally by Wang [121]. New similarity solution of steady laminar mixed convection boundary layer flow over a permeable vertical flat surface for convective boundary condition was dealt by Subhashini et al. [111]. It was observed that the buoyancy assisting force produces significant velocity rise near the wall within the boundary layer for low Prandtl number which increases with the buoyancy parameter (λ). The reason mentioned was that the effect of buoyancy force is more for low Prandtl number fluid due to its low viscosity. This low viscosity increases the velocity within the boundary layer as the buoyancy force aids favorable pressure gradient to it. A numerical investigation by Rana and Bhargava [101] was carried out for steady, laminar incompressible flow of nanofluid along the vertical plate with temperature-dependent heat source or sink. The objective was to analyze the development of steady laminar boundary layer flow and heat transfer for different types of nanoparticles like Ag, Cu, CuO, Al_2O_3, and TiO_2. Results showed that there was a decrease in the average Nusselt number for all of them. Due to enhanced heat transfer capabilities of nanofluids, recent studies have inclined toward using it as working fluid in many engineering applications. Singh et al. [108] investigated computationally mixed convection heat transfer phenomena of non-Newtonian nanofluids past upward through vertical circular cylinder using a commercial Computational Fluid Dynamics (CFD) solver FLUENT. Analyzing the effect of various parameters it was found that the average Nu also increases with the increase in Re and volume fraction (ϕ) whereas it has very less effect of Richardson number (Ri). Also numerical investigation of pure mixed convection for different heater configurations in a ferrofluid-filled Lid-driven cavity was carried out by Rabbi et al. [99]. Analysis was carried out for two different geometric heater configurations (triangular and semi-circular) under externally applied magnetic field considering ferrofluid (Fe_3O_4–water) as working fluid and modeled as single-phase fluid. The analysis was carried out for a wide range of $Ri = 0.1-10$, $Re = 100-500$, Hartmann number ($Ha = 0-100$) and volume fraction ($\phi = 0-0.15$) of ferrofluid. The overall performance in heat transfer for both the configurations was quantitatively investigated by the average Nusselt number. Finally, it was concluded that the triangular notched cavity seems to produce less value of average Nu compared to semi-circular notched cavity. It was also observed that higher Ri increases the heat transfer rate even though higher Ha decreases it. Extended from it, they suggested to carry out experiments to get the actual scenario.

Compound effects of shear and buoyancy for mixed convection in a 2D enclosure was numerically studied by Onyejekwe [87]. It was found that the core flow is not only sensitive to the boundary conditions but also to the movement of the walls. Under the effect of aiding buoyancy mixed convection flow and heat transfer around a long cylinder of square cross section were investigated in the vertical open configuration by Sharma et al. [106]. The local and average Nu were calculated to make a thorough study of heat transfer and it was observed that it increases with the increase in Ri and/or Re. The mixed convection basically distorts the streamline, isotherms and increases the drag coefficient hence increases the rate of heat transfer. The aim of the article by Ismael et al. [58] was to study numerically and to measure the effects of Ri, slip parameter and the direction of the sliding walls on flow and heat transfer inside a Lid-driven square cavity filled with water. It was interesting to see that there was no effect of Ri on isotherms in absence of partial slip, but for non-zero partial slip, the average Nu increases with the increase of Ri. Al-asadi et al. [6] illustrated the characteristics of heat transfer and fluid flow in an inclined circular pipe for nanofluids. They found that the wall shear stress and velocity increases as the Re increases, while the surface temperature decreases. Furthermore, they inferred that the surface temperature increases as the inclination angle increases and the heat transfer is enhanced in assisting flow compared to that of opposing flow. Babajani et al. [9] also studied solar cells cooling using nanofluid. An increase in temperature gradient and heat transfer near the solar cells was observed due to the presence of nanoparticles. It was also observed that there was not much impact of inclination angle on average Nu at low Ri. However, at high Ri value the maximum heat transfer in solar cells occurs with minimum inclination angle of the channel.

Many problems which were very difficult to solve analytically became simple with the advent of powerful computers and various software packages like ANSYS-Fluent, COMSOL Multiphysics, MATLAB code, OpenFOAM, etc. Among various research works on laminar mixed convection, one was by Balaji et al. [11]. They obtained correlations for average Nusselt number by a general methodology called Asymptotic Computational Fluid Dynamics (ACFD) to disturb the limiting solutions of forced and natural convection. This methodology was first developed by Gersten and Herwig [47]. Because of its asymptotic correctness, this approach worked well and was also physically consistent.

In coaxial double duct heat exchangers experimental and numerical study of mixed convection with flow reversal was carried out by Mare et al. [74]. Velocity vectors of water in a vertical parallel ascending flow in a heat exchanger was determined by Particle Image Velocimetry (PIV) technique experimentally and were in very good agreement with numerical results. Both numerical and experimental observations showed that the flow reversal occurs simultaneously in the inner tube and the annulus for heating as well as for cooling of the flow which must be avoided for affecting the stability and heat transfer. Fu et al. [43] also numerically investigated the reversal of mixed flow in three-dimensional vertical rectangular channel. The compressibility of fluid was taken into consideration which means Boussinesq approximation is no longer valid. They found that at high Ri the natural convection governs the flow and thermal field, ultimately directs the mechanisms of the flow reversal.

Rahman et al. [100] performed a computational work to analyze the heat transfer, temperature and flow profile in a channel heated from different sides. The location of heater plays an important role in the distribution of heat within a cavity. They simulated a two-dimensional cavity with different locations of the heater and observed the highest heat transfer when the heater was in the right vertical wall and the flow was from left toward right. Reichl et al. [102] compared the heat transfer mechanisms and velocity field in a compound parabolic concentrator by CFD simulations with experimentally obtained local temperature measurements. The simulations were performed in ANSYS whereas experiments were carried out with PIV. Fontana et al. [42] analyzed numerical mixed convection in partially open cavities heated from below using a transient three-dimensional model. The results confirm the existence of two different processes controlling the formation of recirculation inside the cavity. One of them associated with the shear, called hydrodynamic recirculation and another one associated with the buoyancy forces, called thermal recirculation. Li et al. [73] carried out a three-dimensional numerical analysis on fully developed laminar mixed convection in parabolic trough solar receiver tube. The aim was to analyze the effect of the buoyancy force induced by the non-uniform heat flux on the laminar flow and heat transfer characteristics in the solar receiver tube. Results showed that natural convection increases the heat transfer when Gr is up to a threshold value and heat transfer deterioration takes place when Ri is greater than 12.8.

Burgos et al. [24] did a numerical study of laminar mixed convection for both steady and unsteady flows in a square open cavity by Lattice Boltzmann Method (LBM). Observations indicated that the effect of the buoyancy force was negligible for $Ri < 0.1$. Although for $Ri = 1$ and 10, buoyancy effects were significant, and responsible for the development of the upstream secondary vortex. The observations shed a new light on the characteristics of the oscillatory instability and the role of the Reynolds and Richardson numbers. Mixed convection heat transfer characteristics in Lid-driven cavity containing triangular block with constant heat flux was carried out numerically by Gangawane [45] and analyzed the effect of Re, Pr and Gr. It was observed that increase in Re increases the Nu only up to a critical value of Re ($180 \leq Re \leq 200$), thereafter heat transfer rate decreases. These studies reveal that a blockage will also control the convection flow and heat transfer in a cavity. Recently mixed convection heat transfer through metal foams partly filled in a vertical channel has been investigated by Kotresha and Gnanasekaran [69] using ANSYS-Fluent. The main objective was to quantify the effect of metal foam thickness on thermal performance and fluid flow characteristics. Results indicated that the Nu increases with the increase of partial filling of metal foams.

With all the above studies we can say that various numerical models may be used for complex problems. The commercial softwares available are quite efficient and reduce the effort of mathematical modeling. Authors started using iterative methods in case of strong buoyancy effects to avoid the use of marching procedures. They inferred that the maximum velocity may not be at the centerline and may be at the wall depending on the difference in wall and fluid temperature. An existence of critical buoyancy parameter for instability was obtained numerically and validated experimentally. Investigations also concluded that heat transfer is significantly affected using nanofluids which are much better in heat conduction. The local and average Nu are found to be increasing with the increase of Re and Ri, but not having much of impact of the inclination angle on it. Another important conclusion has been that natural convection which is responsible for flow reversal affects both the stability and heat transfer. In the next section, we will discuss some experimental observations on laminar mixed convection.

2.3 Experimental Studies on Laminar Mixed Convection

Any experimental work in natural convection is difficult to perform due to low range of velocities that too in mixed convection is more complex. Since earlier studies didn't measure all the temperatures which cause a problem in determining the Nusselt number accurately. Hallman [55] carried out an experimental study on laminar mixed convection in a vertical tube of test section approximately ~36 inches in length with 58 thermocouples. The aim was to examine the transition, limiting case for pure laminar forced convection and the thermal entrance effects. Concluding remarks reveal that the solutions are applicable at far from the entrance for positive values of Ra and may be for small negative Ra value. However not much could be obtained from experiments regarding transition to turbulent flow for large negative Rayleigh number and the unsteady case.

Joye [63] compared the existing correlations with his experimental investigation for opposing flow in mixed convection in a vertical tube for various Gr and a range of Re (700–25,000). The mixed convection region exists between Re values of 4,000 and 10,000. Correlations for Nu presented by Jackson et al. [61] for opposing flow, predicted quite similar results except for the region $Re < 4000$. It was observed that Nu reduces as the Gr reduces in this range of Re. Kamath et al. [65] investigated experimentally on aiding mixed convection in a vertical channel carrying aluminum metal foam. Experiments were conducted for a large range of Richardson number ($0.005 \leq Ri \leq 1032$). Correlations (Eqs. 3 and 4) for Nu were developed with Ri, Re and porosity (Ø), valid for a specific range of these parameters tabulated in Table 1. It was observed that porosity increases the pressure drop but it also enhances the heat transfer.

$$Nu_{\text{porous}} = 0.223 Re^{0.723} \emptyset^{6.81} \left(1 + Ri^{0.46} Re^{-0.73}\right). \tag{3}$$

$$E = 1 + 2.188 Ri^{0.06} Re^{0.016} \emptyset^{11.9}. \tag{4}$$

To study the local and average heat transfer for hydrodynamically fully developed, thermally developing and fully developed laminar air flow inside a horizontal circular cylinder, an experiment was conducted by Mohammed and Salman [83]. They found that the Nu increases as the heat flux increases. It was also concluded that free convection reduces the heat transfer at low Re and enhances for higher Re. A correlation (Eq. 5) was also developed for average Nu in terms of average values of Ra and Re, which was validated with Mori et al. [84].

$$\overline{Nu} = 3.19 \left(\overline{Ra}/\overline{Re}\right)^{0.26}. \tag{5}$$

In a vertical rectangular channel, the effect of film evaporation on mixed convective heat and mass transfer was investigated experimentally by Cherif et al. [30] and was compared with the numerical study of a laminar flow. Tian et al. [116] experimentally studied mixed convection in an asymmetrically heated, narrow, inclined rectangular channel to reduce the rise of fuel and cladding temperature of plate-type fuel element in reactors. The aim was to observe the secondary flow. It was found that higher inclination angle can enhance the thermal instability while a higher value of Re delays the thermal instability.

All the above experimental literature clarifies that difficulties in measuring temperatures leads to an inaccurate measurement of Nusselt number. Results showed that Nu decreases as the Gr reduces. Correlation was obtained for Nu in terms of Ra and Re. It was concluded that free convection reduces the heat transfer at low

Table 1 Various correlations and its range of parameters for different flow conditions

Eq. no.	Correlation	Authors	Range of parameters	Flow condition
(1), (2)	$\Delta P_{\text{lam}} = (128\mu L Q/\pi D^4)(\rho g)(Sp.Gr/\varphi_v^{0.38})$ $\Delta P/\Delta P_{\text{lam}} = 1 + 1565 Gr_L^{3/4} Pr^{1/2}/(0.952 + Pr)^{3/4} Re^2 (L/D)^2 \big|_b$	Joye [64]	$Gr_L \leq 1.22 * 10^{13}$ $Re \leq 11,000$ $Pr \leq 4.53$	Laminar mixed convection in vertical internal aiding flows with constant wall temperatures
(3)	$Nu_{\text{porous}} = 0.223 Re^{0.723}\varphi^{6.81}\left(1 + Ri^{0.46} Re^{-0.73}\right)$	Kamath et al. [65]	$0.005 \leq Ri \leq 1032$ $24 \leq Re \leq 3730$ $0.9 \leq \phi \leq 0.95$	Assisted mixed convection in a porous (Al metal foam) vertical channel
(4)	$E = 1 + 2.188 Ri^{0.06} Re^{0.016}\varphi^{11.9}$	Kamath et al. [65]	$0.005 \leq Ri \leq 254$ $48 \leq Re \leq 3730$ $0.9 \leq \phi \leq 0.95$	Assisted mixed convection in a porous (Al metal foam) vertical channel
(5)	$\overline{Nu} = 3.19\left(\overline{Ra/Re}\right)^{0.26}$	Mohammed and Salman [83]	$400 \leq Re \leq 1600$ $14768 \leq Ra \leq 12,950,400$ taking $Pr = 0.71$ for air	Mixed convection in a horizontal circular cylinder
(6)	$\dfrac{Nu}{Nu_T} = 1.9 Bo_2^{0.18}\varphi^{0.17}$	Poskas and Poskas [92]	$0.076 \leq Bo_2 \leq 0.7$ $20° \leq \varphi \leq 90°$	Turbulent opposing mixed convection in inclined flat channels
(7)	$Nu_M = 4.36\left(1 + \dfrac{Gr^{0.468}}{750+0.24Re}\right)$	Behzadmehr et al. [18]	$Gr \leq 5 * 10^7$ $1000 \leq Re \leq 1500$	Fully developed upward mixed convection in vertical tubes with constant heat flux

(continued)

Table 1 (continued)

Eq. no.	Correlation	Authors	Range of parameters	Flow condition
(8)	$Nu_{\text{trans}} = Nu_l + \{\exp[(a - Re)/b] + Nu_t^c\}^c$	Tam and Ghajar [113]	Follow Eqs. 9 and 10	Forced and mixed convection in plain horizontal tubes with uniform wall heat flux
(9)	$Nu_l = 1.24\left[\left(\frac{RePrD}{x}\right) + 0.025(GrPr)^{0.75}\right]^{1/3} \left(\frac{\mu_b}{\mu_w}\right)^{0.14}$	Tam and Ghajar [113]	$3 \leq x/D \leq 192$ $280 \leq Re \leq 3800$ $40 \leq Pr \leq 60$ $1000 \leq Gr \leq 28,000$ $1.2 \leq \mu_b/\mu_w \leq 3.8$	Forced and mixed convection in plain horizontal tubes with uniform wall heat flux
(10)	$Nu_t = 0.023Re^{0.8}Pr^{0.385}\left(\frac{X}{D}\right)^{-0.0054}\left(\frac{\mu_b}{\mu_w}\right)^{0.14}$	Tam and Ghajar [113]	$3 \leq x/D \leq 192$ $7000 \leq Re \leq 49,000$ $4 \leq Pr \leq 34$ $1.1 \leq \mu_b/\mu_w \leq 1.7$	Forced and mixed convection in plain horizontal tubes with uniform wall heat flux
(11)	$Nu^{10} = Nu_l^{10} + \left[\frac{\exp[(2200-Re)/365]}{Nu_{lc}^2} + \frac{1}{Nu_t^2}\right]^{-5}$	Tam and Ghajar [113], Churchill [31]	Re, Pr, Nu_l, Nu_{lc} and Nu_t depends on flow conditions	Laminar, transition and turbulent regime for fully developed near-isothermal flow in smooth tubes
(12)	$Nu = \widehat{Nu}(aGr^{0.25} + bRe^{0.5})$	Balaji et al. [11]	All values of Gr and Re for laminar flow and a = 0.319, b = 0.24	2D, steady, laminar mixed convection in square lid-driven cavity
(13)	$\widehat{Nu} = 1 - 0.54Ri^{-0.25}\left[1 + \left(\frac{0.559}{Ri}\right)^{0.8}\right]^{-0.0625}$	Balaji et al. [11]	$0 \leq Ri \leq \infty$	2D, steady, laminar mixed convection in square lid-driven cavity
(14)	$Nu = \frac{d\theta}{dY} - Pr.Re.V\theta$	Cheng [28]	Not mentioned in Cheng [28]	2D square lid-driven cavity

(continued)

Table 1 (continued)

Eq. no.	Correlation	Authors	Range of parameters	Flow condition
(15)	$Nu = 7.96A^{1.1}Re^{0.18}Ri^{-0.02}$	Cheng [28]	$0.1 \leq Ri \leq 1000$ $10^7 \leq Gr \leq 5*10^9$ $2200 \leq Re \leq 12.000$	2D square lid-driven cavity
(16)	$Re_{cr_2} = 8.46Gr_q^{0.33}$	Poskas et al. [95]	$2.3*10^7 \leq Gr_q \leq 9.4*10^9$	Opposing mixed convection in vertical flat channel in laminar-turbulent transition region
(17)	$Nu = 0.02\left(Gr_q Re\right)^{0.244}$	Poskas et al. [95]	$1.5*10^3 \leq Re \leq 6.6*10^4$ $2*10^{11} \leq Gr_q Re \leq 1.6*10^{14}$	Opposing mixed convection in vertical flat channel in laminar-turbulent transition region
(18)	$Re_{cr} = 2524 - 0.82(192 - \frac{x}{D})$	Everts and Meyer [38]	$\frac{x}{D} = 3–192$	Developing and fully developed flow in smooth horizontal pipes in transitional flow regime
(19)	$Re_{qt} = 8791 - 7.69(192 - \frac{x}{D})$	Everts and Meyer [38]	$\frac{x}{D} = 3–192$	Developing and fully developed flow in smooth horizontal pipes in transitional flow regime

Re and enhances at higher *Re*. Correlations for mixed convection in porous media for *Nu* with *Ri*, *Re* and porosity (Ø) shows that porosity increases the pressure drop as well as the heat transfer.

3 Overview of Turbulent Mixed Convection

Contrary to laminar mixed convection when the Rayleigh number for free convection and Reynolds number for forced convection, both are in the turbulent regime, it is said to be turbulent mixed convection. The effects of buoyancy can be speculated very easily in the laminar case whereas it is quite different in turbulent flows. Detailed reviews were presented by Jackson et al. [61] and Jackson [59] on turbulence and heat transfer characteristics in vertical passages.

3.1 Theoretical Studies on Turbulent Mixed Convection

Polyakov [91] discussed analytically the growth of secondary free convection currents in forced turbulent flows in horizontal pipes with weak thermo-gravitational effects. The results were compared with experimental data obtained for water flow and air flow, where substantial thermo-gravitational influence was present on turbulent flow and the heat exchange. The overall resistance to the momentum and heat transport in turbulent forced and mixed convective flows was treated by employing "wall functions" to the thin near-wall viscosity affected sub-layer by Craft et al. [32]. Wall functions are of different types, its purpose is to solve the differential equations across the sub-layer with some algebraic formulae or low-cost routes. Their aim was to accurately model the flow in viscosity-affected-sublayer region in a suitable form for use in CFD. Farrugia and Micallef [40] did analytical study upon turbulent plumes evolving in a vertical upward flow in the mixed convection regime and compared with the experimental results as well as the simulations obtained using CFD. However the results were applicable for a particular range of parameters but it was in good agreement with experimental as well as numerical data obtained. Suga et al. [112] improved the performance of analytical wall-functions (AWF) developed by Craft et al. [33] while predicting the turbulent heat transfer for recirculating and impinging flows. To account for the variations of parameters they introduced a functional behavior into the coefficient of eddy viscosity of AWF which was also validated for different flows. The results confirm that for all flows and heat transfer tested the modification improves the performance.

Theoretical studies related to turbulent flows however started later but the equations are quite complex to solve in case of a turbulent flow. Many authors like T. J. Craft, put efforts to develop some analytical wall functions to overcome the problem of solving differential equations and tried to solve by some algebraic formulae. This approach requires only a fraction of the computational effort as that of a turbulence model. Thus, this approach can be recommended for industrial calculations, even though the performance is often poor due to its inherent limitations. Thus, the entire exercise boils down to the accuracy of the wall function chosen. The search for a universal wall function remains elusive till date and the prospect in this direction doesn't appear to be very promising. It is obvious that computational ease is only possible with various numerical models having different ways of treating turbulence.

3.2 Numerical Studies on Turbulent Mixed Convection

Supercritical boilers in power plants and supercritical water in cooling nuclear reactors makes turbulent mixed convection essential to study. Studies on supercritical pressure found huge loss of heat transfer for upward flow near critical point in heated tubes. The effect was named "pseudo-boiling" and thought it to be like the film-boiling. Through experiments by Shitsman [107], Ackerman [3] and Jackson et al. [62], it became clear that the effect is due to buoyancy and not due to film-boiling. The fact of localized heat impairment is not only true for fluids at high pressure but also occurs for liquids and gases at normal pressure alike the experimental results by Hall and Price [53], Steiner [110], Kenning et al. [68] and Fewster [41] etc.

An early study by Hsu and Smith [57] showed unpredicted patterns and mentioned that the heat transfer coefficient in turbulent mixed convection is less as compared to forced convection alone. In fact, for downward flow in heated tubes buoyancy force enhances the turbulent properties causing higher heat transfer coefficient than the forced flow alone. In conclusion as the buoyancy becomes more and more dominant, heat transfer for upward flow also increases with the same heat transfer coefficients for these two cases. Studies on turbulent mixed convection with water and air, there is less indication about the influence of buoyancy on heat transfer. Later considerable work mentioned by Hall et al. [52] and Hall and Jackson [51] shows such effects near the critical points. Using a variety of computational formulations and turbulence models, attempts were made to simulate buoyancy-induced turbulent convective heat transfer in vertical tubes. One of them is by Tanaka et al. [114]. They used modified Reichardt's eddy diffusivity model. This approach was not suitable for buoyancy-induced flow because the turbulent viscosity was described as a function of distance from the surface lacking reference to the local properties of the flow. Walklate [120] used both $k - \varepsilon$ models and mixing length models to simulate the experiments of Carr et al. [25]. He found that the low $Re \ k - \varepsilon$ models performed better than both the mixing length model and standard $k - \varepsilon$ models.

Launder and Spalding [71] numerically predicted a turbulent flow and recommended that turbulence models are best served as per computational economy, range of applicability and physical reality. Skiadaressis and Spalding [109] predicted the flow and heat transfer characteristics for the turbulent flow of air in the developing and fully developed region of a circular horizontal pipe. They compared the numerical results with experimental data reported by Petukhov and Strigin [89] and Petukhov [88] in fully developed flow and found them to be fairly good in agreement. Other similar numerical work was also carried out by Abdelmeguid and Spalding [1]. Jackson et al. [61] examined low $Re\ k - \varepsilon$ turbulence model of Launder and Sharma [70] for turbulent mixed convection of a developing air flow in vertical tubes. It was in good agreement with the experimental heat transfer data and flow profile measurements. It was also suggested that the low Re two-equation models make the simplest formulation for turbulent mixed convection flows. Further, Mikielewicz [82] carried out a relative study of the performance of various turbulence models and found that the low $Re\ k - \varepsilon$ turbulence model as the most suitable model.

After 1980s, for better results several upgraded turbulence models have been used. Direct Numerical Simulations (DNS) is one of those methods. Kasagi and Nishimura [66] conducted one of the earliest studies of mixed convection in vertical tubes with DNS. Simulations obtained with fixed Re and varying Gr provided detailed information with visualization than could be obtained with experiments. Later You et al. [125] conducted DNS study in a vertical heated tube with uniform physical properties and Boussinesq approximation. Bae et al. [10] also followed DNS study of mixed convection at supercritical pressure where the effects of notable non-uniformity of fluid properties had been taken into consideration. The past study by Sakurai et al. [104] provided radiation effects for different turbulence statistics over a horizontal channel using DNS in an optically thin medium. Simulations provided a reliable database, however the information gathered using DNS regarding the interaction between turbulence-radiation was unsatisfactory.

More recently Reynolds-Averaged-Navier–Stokes (RANS) equations were solved by Marocco et al. [77] for investigation of a turbulent mixed convection of liquid metal with $Pr = 0.021$ for a concentric heated annulus. Results indicate that there is a great influence of Reynolds number for low-high Prandtl numbers on the onset, laminarization and recovery of heat transfer. They have also seen that in a pipe flow liquid metals with $Pr = 0.25$ behave like air or water whereas in concentric annular flow they behave differently. The reason is due to the substantial contribution of molecular heat transfer in liquid metals. Also high thermal conductivity of liquid metals makes them more attractive in applications where the effective heat transfer is the prime concern. Bieder et al. [20] carried out CFD simulation and experimental validation of steady-state mixed convection sodium flow using reference code TrioCFD and FLUENT employing RANS equations. Turbulence modeling was also incorporated to study the interaction of sodium flow and thermal stratification that had been analyzed experimentally. Marocco [76] numerically investigated a fully developed turbulent forced and mixed convection heat transfer to a liquid metal flowing upwards in a concentric annulus by using LES. The LES results were validated with DNS results. The simulations were performed with the open source code OpenFOAM. It was illustrated that for mixed convection, when turbulence is reduced, better results were obtained taking a much coarser mesh in circumferential direction. For low-Pr number fluids, the Nu becomes function of only the buoyancy number (Bo) at higher Re. In general, conclusions that can be made on the behavior of liquid metals in mixed convection is that it is mandatory to perform detailed simulations.

Studies on turbulent mixed convection confirm that the influence of buoyancy increases the turbulence causing higher heat transfer coefficient. It was evident from experiments that there is heat loss at critical points and this is true not only for fluids at high pressure but also for liquids and gases at normal pressure. Among various turbulence models low $Re\ k - \varepsilon$ model was inferred as the simplest model. However, for better accuracy, 1980 onwards RANS, LES, and DNS have overtaken other turbulence models. The only limitation is that they require a high computing power especially for complex geometries in the presence of turbulence. More recently RANS equations were solved for the investigation of turbulent mixed convection of liquid metals. Due to the substantial molecular heat transfer and higher thermal conductivity, liquid metals become more effective in heat transfer. The complexity as well as order of accuracy increases as we move from RANS and LES to DNS, but the accuracy is restricted by the numerical scheme used as described by Biswas and Eswaran [21]. The present results serve only as guidelines for future simulations as well as for the experiments.

3.3 Experimental Studies on Turbulent Mixed Convection

Among various experimental works of Prof. Robertas Poskas in mixed convection, one was by Vilemas et al. [119]. They performed experimental examination of local heat transfer in a vertical gas-cooled tube for turbulent mixed convection. Correlations were obtained to calculate local heat transfer along the tube for weak and strong buoyancy effects but these are unable to provide much information in the intermediate region. Furthermore, Poskas and Poskas [92] investigated experimentally the local turbulent opposing mixed convection heat transfer in an inclined flat channel under stably stratified flow conditions. Three characteristic regions were identified: region without buoyancy ($Bo_2 \leq 0.7$), transition ($0.7 \leq Bo_2 \leq 1.7$) and region with buoyancy ($Bo_2 \geq 1.7$). A correlation Eq. 6 was also developed to calculate heat transfer for the region without buoyancy. A similar experiment was also performed by Poskas et al. [94] for unstably stratified flow conditions with a slightly varying range of parameters. The only difference was, in this study the bottom wall is heated whereas in the previous study the upper wall was heated. They also obtained similar three regions as mentioned in his previous article with a minor variation: region without buoyancy ($Bo_2 \leq 0.7$), transition ($0.7 \leq Bo_2 \leq 2$) and region with buoyancy ($Bo_2 \geq 2$).

$$\frac{Nu}{Nu_T} = 1.9 Bo_2^{0.18} \varphi^{0.17}. \tag{6}$$

Aicher and Martin [4] summarizes experimental results for both aiding and opposing flow conditions and furnishes own experimental results influencing length-to-diameter ratio on heat transfer in vertical tubes. Finally, they provided a new empirical correlation that provides better results than all available correlations. Wang et al. [123] reported an experimental study of buoyancy influenced convective heat transfer upward and downward flow through a vertical plane passage under turbulent condition. It was observed that the results were similar to that of circular tubes explained in preceding papers "Part 1" and "Part 2", except the buoyancy was slightly weaker than circular tubes and the onset of buoyancy-induced heat transfer impairment was delayed.

Jackson [60] performed a few experiments on fluid flow and convective heat transfer to fluids at supercritical pressure between two horizontal planes. The influence of buoyancy and deterioration of heat transfer were analyzed. Yang et al. [124] assessed turbulent mixed convection heat transfer numerically and experimentally in a vertical open cavity. The velocity and temperature profiles were obtained by hot-wire anemometry and PIV for the range of parameters: $4.6 * 10^4 \leq Re \leq 5.4 * 10^4$ and $Gr \leq 1.8 * 10^{13}$. A secondary upward flow was observed near the heated walls due to buoyancy. Among various turbulence models low Re $k - \varepsilon$ turbulence model provided the most accurate results in the presence of strong buoyancy forces.

It is noticeable from all available experimental results that there are heat transfer results available for strong and weak buoyancy effects, but these are unable to provide much information in the intermediate region. From experiments, a buoyancy parameter (Bo) was obtained which differentiates this transition region from the regions with and without buoyancy. Few correlations were developed for aiding and opposing flow conditions, which signifies the effect of length-to-diameter ratio on heat transfer. Nusselt number plots showed that the nature of heat transfer is the same for aiding and opposing cases. Only a few authors have discussed this intermediate regime (between weak and strong natural convection effect). Experiments suggested that the flow features and vortex characteristics can be well visualized through advanced techniques such as interferometry or PIV.

4 Transition Behavior and Criteria from Laminar to Turbulent Mixed Convection

It can be said that the fluid flow and heat transfer characteristics in laminar and turbulent mixed convection are reasonably well understood theoretically, numerically and experimentally. However, literature available for laminar to turbulent transition in mixed convection are scarce.

4.1 Theoretical Studies on Transition

Since transition is a phenomenon that can be observed just beyond or close to a certain critical parameter, and no such analytical formulation can be made. Hence the studies concentrated on results of simulations and experiments only.

4.2 Numerical Studies on Transition

Very few studies delineate the transition between laminar and turbulent mixed convection. Numerical study by Behzadmehr et al. [17, 18] for upward mixed convection of air flow in a long vertical tube were conducted for two values of $Re = 1000$ and 1500, and a range of Grashof number ($Gr \leq 10^8$) by using low $Re \; k - \varepsilon$ turbulence model under Boussinesq approximations. Corresponding to laminar–turbulent transition and relaminarization of the flow, two critical Gr were identified for each Re. The critical value observed from laminar to turbulent condition was at $Gr = 8 * 10^6$ for $Re = 1000$ and at $Gr = 2 * 10^6$ for $Re = 2000$, results were validated with Metais and Eckert [79]. Finally, for $Gr = 7 * 10^7$ the fully developed flow field became turbulent for $Re = 1500$ and laminar for $Re = 1000$. This transition from turbulent to laminar is called relaminarization and was due to the laminarization of buoyancy-induced acceleration. A correlation given in Eq. 7 had been developed for fully developed upward mixed convection in vertical tubes with uniform heat flux which was valid for both laminar and turbulent conditions within a range: $1000 \leq Re \leq 1500$.

$$Nu_M = 4.36\left(1 + \frac{Gr^{0.468}}{750 + 0.24Re}\right). \tag{7}$$

Tam and Ghajar [113] examined many experimental works and collected data points to know in detail the heat transfer behavior in the transition region under a uniform wall heat flux boundary condition for plain horizontal tubes. Also, many correlations were recommended for different regimes to predict the heat transfer. Among them, Tam and Ghajar [113] provided the most accurate correlation as indicated below in Eqs. 8–10, following a popular correlation obtained by Churchill [31] mentioned in Eq. 11. Finally, a flow regime map was presented to determine the boundary between forced and mixed convection in horizontal tubes having different inlet configurations. Another correlation for Nusselt number proposed by Balaji et al. [11] mentioned in Eqs. 12 and 13, valid for the whole range of parameters as long as the flow is laminar. Cheng [28] in his article discussed the characteristics of heat transfer in a lid-driven square cavity with respect to Reynolds number ($10 \leq Re \leq 2200$), Grashof number ($10^2 \leq Gr \leq 4.84 * 10^6$), Richardson number ($0.01 \leq Ri \leq 100$) and Prandtl number ($0.01 \leq Pr \leq 50$). In this article he answered a question whether heat transfer would be continuously increased with the increase of Gr and Re while keeping Pr and Ri constant and said that it is true only for $Ri = 0.01$. It was also mentioned that heat transfer does not increase as expected with the increase of both Reynolds and Grashof number which is very surprising in mixed convection flows. The reason may be due to the change of flow structure, the kinetic energy and the heat transfer mechanism. To elucidate the cause, the streamlines and isotherms were compared for $Ri = 1.0$ and varying $Re = 300, 375, 376$ and 900. When the Re and Gr both continuously increased, isotherms get clustered at the bottom wall and steep temperature gradients with thin boundary layers formed in vertical directions. This steep temperature gradient and weekend mixing of hot and cold fluids reduces the conduction and convection mode of heat transfer respectively. Equation 14 was used to understand the heat transport process between $Re = 375$ and 376, a plot was presented for the conduction $(d\theta/dY)$ and convection $(-Pr.Re.V\theta)$ heat distributions for $Ri = 1$. It was seen that the conductive heat transfer was negligible in the interior region and significant near the horizontal wall whereas the convective mode is comparatively small in the upper half. Hence combined effect stipulates drop of heat transfer rate, Nu from $Re = 375$ to $Re = 376$. Similar results were obtained when analysis was done at $Re = 713, 376, 248, 129$ and 61 for $Ri = 0.5, 1, 2, 10$ and 100 respectively. Also, the decrease of total kinetic energy reduces the convective heat transfer in the cavity. Surrounded by many correlations, Prasad and Koseff [98] developed a correlation given in Eq. 15 between aspect ratio (A), Re and Ri which was mentioned invalid by Cheng [28] for laminar forced and mixed convection flows. However, Cheng [28] has compared his correlation with Balaji et al. [11] and found in excellent agreement for $Ri = 0.01$ when $a = 0.2$ and $b = 0.21$ and suitable for a compact range of Re and Gr when $Ri = 1.0$.

$$Nu_{\text{trans}} = Nu_l + \left\{\exp[(a - Re)/b] + Nu_t^c\right\}^c. \tag{8}$$

$$Nu_l = 1.24\left[\left(\frac{RePrD}{x}\right) + 0.025(GrPr)^{0.75}\right]^{1/3}\left(\frac{\mu_b}{\mu_w}\right)^{0.14}. \tag{9}$$

$$Nu_t = 0.023Re^{0.8}Pr^{0.385}\left(\frac{X}{D}\right)^{-0.0054}\left(\frac{\mu_b}{\mu_w}\right)^{0.14}. \tag{10}$$

$$Nu^{10} = Nu_l^{10} + \left[\frac{\exp[(2200 - Re)/365]}{Nu_{lc}^2} + \frac{1}{Nu_t^2}\right]^{-5}. \tag{11}$$

$$Nu = \widehat{Nu}\left(aGr^{0.25} + bRe^{0.5}\right). \tag{12}$$

$$\widehat{Nu} = 1 - 0.54Ri^{-0.25}\left[1 + \left(\frac{0.559}{Ri}\right)^{0.8}\right]^{-0.0625}. \tag{13}$$

$$Nu = \frac{d\theta}{dY} - Pr.Re.V\theta. \tag{14}$$

$$Nu = 7.96A^{1.1}Re^{0.18}Ri^{-0.02}. \tag{15}$$

Poskas et al. [93] used turbulence transition models with ANSYS-Fluent code and investigated the opposing mixed convection heat transfer in an inclined flat channel. The numerical modeling results have been compared with the experimental heat transfer results performed with the same boundary conditions and found well in agreement. It was shown that the vortices which exist in the channel makes the velocity profiles unsymmetrical. The parameters responsible for this was not only the buoyancy parameter alone but also the length of the channel. As the influence of buoyancy parameter becomes stronger, the instability increases, and the flow becomes turbulent. The author had performed modeling using different laminar and transitional turbulence models: $k - k_l - \omega$, $Shear\ Stress\ Transport\ (SST)$ and $Reynolds\ stress - \omega$ model. After the analysis it was concluded that in case of laminar model for low and high Re number the vortices are formed at the beginning of the heated part and diminished along the channel as shown in Fig. 6. The $Reynolds\ stress - \omega$ model gives almost similar results for low Re number. But other two transition models $k - k_l - \omega$ and SST showed clear vortical flow at the very beginning of the heated part of the channel. Laminar and transitional mixed convection flow over a backward-facing step in a horizontal channel was carried out numerically by Barrios-Pina et al. [13]. DNS were performed in a two-dimensional horizontal channel of expansion ratio, $ER = 2$ at step level and observed that the transition from steady to disordered flow appears by period-doubling bifurcations. RANS turbulence models were numerically studied by Abdollahzadeh et al. [2] for laminar–turbulent transition in convection heat transfer. They investigated different turbulence models used for simulation of transition heat transfer in forced convection over a flat plate and for natural/mixed convection between two flat plates. In forced convection over a flat plate Stanton number (St) and the skin friction coefficient (C_f) with local Reynolds number (Re_x) were studied while in natural/mixed convection the inlet velocity, inclination angle, heated wall temperature and Richardson number were studied in detail. Results are shown here in Figs. 7, 8 and 9 for different turbulence models. It was found that the increase in inlet velocity initially promotes the transition and further increase of it delays the transition. Also increase in inclination angle delays the transition point whereas increase of wall temperature and Richardson number accelerates it. When inclination increases, it reduces the magnitude of buoyancy force which causes a delay in transition.

Akyuzlu [5] performed an unsteady numerical study using Coupled Modified Strongly Implicit Procedure (CMSIP) and Standard $k - \varepsilon$ turbulence model to simulate the heat transfer features and circulation patterns of compressible flows in a square cavity during transition from laminar to turbulent mixed convection. The author used a modified version of implicit procedure which was proposed by Chen [26]. The study was to analyze the discrete heating of densified liquid propellants used in cryogenic storage tanks. There appears to be lack of stability in the core flow region, however, they diminish as the flow becomes stronger. The steady and unsteady results are presented here in Figs. 10, 11, 12, 13 and 14. The main difference between the results of compressible flow and incompressible flow has been observed in the location of the center of the circulation. In the case of compressible flow, the circulation center moves toward the bottom of the enclosure with varying Rayleigh number $(10^3 \leq Ra \leq 10^5)$ and is at the geometric center in case of incompressible flow for all the said Ra value. Validation of the proposed mathematical model was done from Ghia et al. [49] and Davis [34], results were found in good match.

Literature available where the transition from laminar to turbulent mixed convection occurred are very less in number. However, with the effort of few researchers various numerical models and correlations have been developed to know about the heat

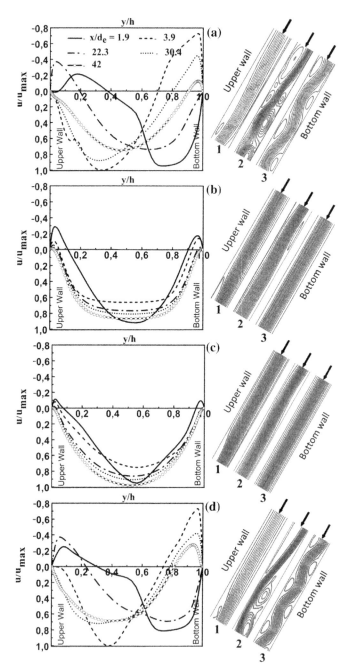

Fig. 6 Dynamics of velocity profiles and flow structure for $Re_{in} = 2.3 * 10^3$, $Gr_{qin} = 3.9 * 10^8$ using different models: **a** laminar, **b** $k - k_l - \omega$, **c** SST, **d** Re stress $- \omega$. 1—$x/D_h = -2.2$ to 0 (the end of hydrodynamic stabilization region and beginning of heating); 2—$x/D_h = 1.9$–3.9; 3—$x/D_h = 39.2$–42 [reproduced with permission from Poskas et al. [93] © IOP Science]

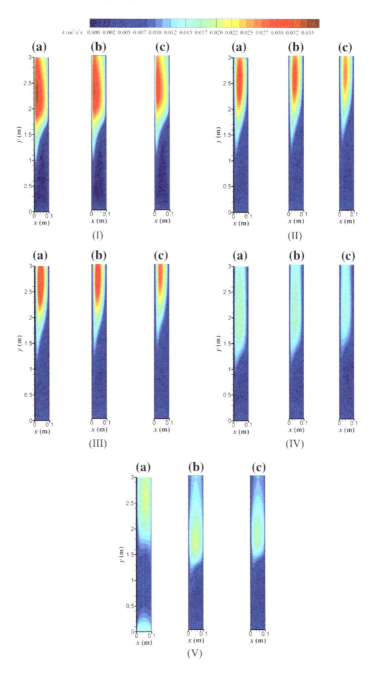

Fig. 7 Turbulent kinetic energy contour of (I) $k - k_l - \omega$ (II) Low-Re $k - \omega$ SST (III) $\gamma - Re_\theta$ (IV) RNG $k - \varepsilon$ (V) Low-Re $k - \varepsilon$ models for $\theta = $ **a** $0°$, **b** $15°$ and **c** $30°$ with constant wall temperature of $T_p = 396.15$ K and $V_{in} = 0.5$ m/s [reproduced with permission from Abdollahzadeh et al. [2] © Elsevier]

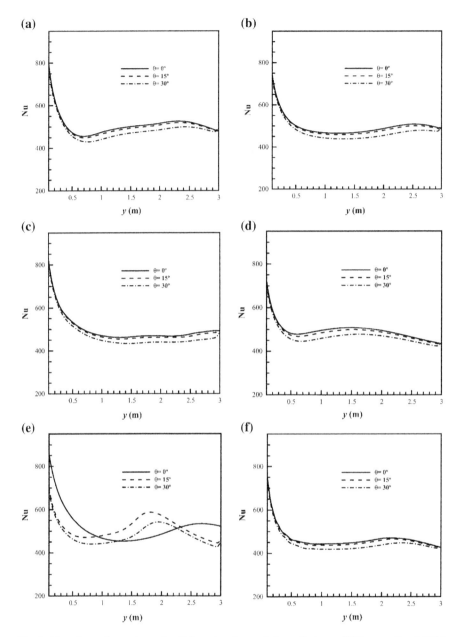

Fig. 8 Nusselt number on heated wall for **a** $k - k_l - \omega$ **b** Low-$Re\,k - \omega$ SST **c** $\gamma - Re_\theta$ **d** RNG $k - \varepsilon$ **e** Low-$Re\,k - \varepsilon$ and **f** V-SA models for different channel inclination angles [reproduced with permission from Abdollahzadeh et al. [2] © Elsevier]

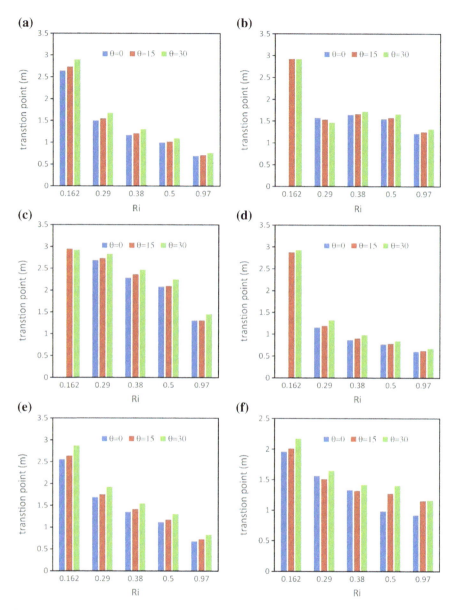

Fig. 9 Start point of transition for **a** $k - k_l - \omega$ **b** Low-$Re\ k - \omega$ SST **c** $\gamma - Re_\theta$ **d** RNG $k - \varepsilon$ **e** Low-$Re\ k - \varepsilon$ and **f** V-SA models for different channel inclination angles [reproduced with permission from Abdollahzadeh et al. [2] © Elsevier]

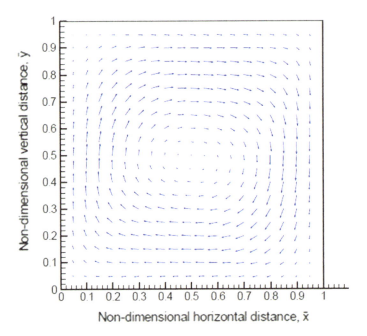

Fig. 10 Steady state velocity vectors at $Ra = 10^4$ [reproduced with permission from Akyuzlu [5] © ASME]

transfer behavior and its characteristics in this regime. Numerical studies obtained critical Gr for different Re corresponding to laminar–turbulent transition and also for relaminarization of the flow. Prasad and Koseff [98], Balaji et al. [11] and others developed correlations for mixed convection. The correlation developed by Prasad and Koseff [98] was found contradictory by Cheng [28] for laminar forced mixed convection. Cheng [28] validated his correlation with Balaji et al. [11] and found in excellent agreement for $Ri = 0.01$, and suitable for a compact range of Re and Gr when $Ri = 1.0$. There are many numerical works on transition. Few were by Barrios-Pina et al. [13], Poskas et al. [93], Akyuzlu [5] and Abdollahzadeh et al. [2]. They discussed about laminar and turbulent transitional models. Among them, Akyuzlu [5] used some modified strongly implicit procedure called CMSIP (Coupled Modified Strongly Implicit Procedure) for simulation of compressible flows. This model was well-validated with other published simulation results. Further, Abdollahzadeh et al. [2] found that the increase of inlet velocity, inclination angle delays the laminar–turbulent transition point whereas the decrease of wall temperature and Ri accelerates the transition. These numerical results will be quite effective for experimentation as discussed in the next section.

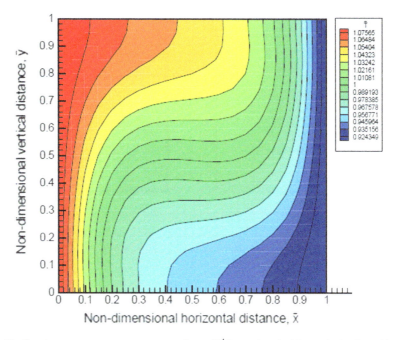

Fig. 11 Steady state temperature contours at $Ra = 10^4$ [reproduced with permission from Akyuzlu [5] © ASME]

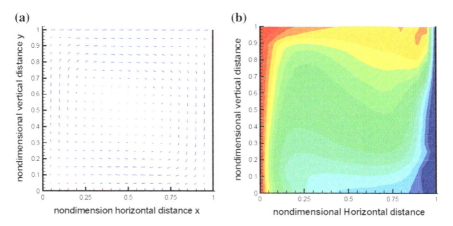

Fig. 12 **a** Velocity vector field at $\bar{t} = 11$ for $Ri = 0.01543$, $Re = 3000$ **b** temperature contours at $\bar{t} = 11$ for $Ri = 0.01543$, $Re = 3000$ [reproduced with permission from Akyuzlu [5] © ASME]

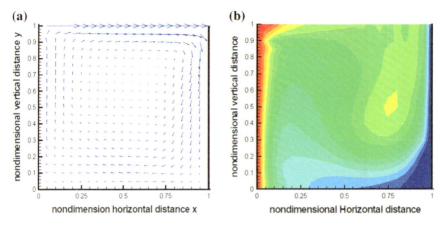

Fig. 13 **a** Velocity vector field at $\bar{t} = 13$ for $Ri = 0.01543$, $Re = 3000$ **b** temperature contours at $\bar{t} = 13$ for $Ri = 0.01543$, $Re = 3000$ [reproduced with permission from Akyuzlu [5] © ASME]

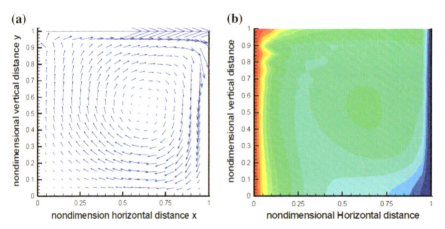

Fig. 14 **a** Velocity vector field at $\bar{t} = 20$ for $Ri = 0.01543$, $Re = 3000$ **b** temperature contours at $\bar{t} = 20$ for $Ri = 0.01543$, $Re = 3000$ [reproduced with permission from Akyuzlu [5] © ASME]

4.3 Experimental Studies on Transition

Kemeny and Somers [67] concluded from experimental data that the transition normally either by a reversal in wall temperature or a small rise in temperature than expected as the distance from the entry increases. Metais and Eckert [79] studied available experimental data and tentatively indicated a transition laminar–turbulent regime which was plotted in between Re and $Gr * Pr$ as shown in Fig. 15. Barozzi et al. [12] reported experimentally and numerically the sharp entry and transition effects for laminar combined convection in vertical tubes. They also investigated a possible criterion of transition based on the axial location. The experimental data were

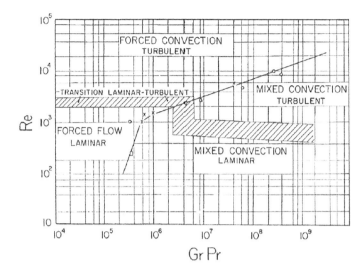

Fig. 15 Regimes of free, forced and mixed convection for flow through horizontal tubes along with transition laminar-turbulent regime [reproduced with permission from Metais and Eckert [79] © ASME]

evident of transition from laminar flow. Grassi and Testi [50] performed experiments for developing upward flow in a circular duct under transitional mixed convection. Heat transfer loss was observed due to laminarization of the turbulent flow, which was characterized by two non-dimensional numbers, Graetz number (Gz) and Grashof number. The heat transfer along the tube showed a non-monotonic, transitional behavior with a minimum at laminarized zone. Behzadmehr et al. [19] expressed experimentally the onset of laminar–turbulent mixed convection transition phenomena in a vertical heated tube. This article dealt with the nature of temperature and velocity variations using experimental data obtained at $Re = 1000, 1300$ and 1600 for a wide range of Gr. Observing the average temperature at the center of the tube, a point of deviation was detected for a value of $Gr/Re > 1500$ indicating the periodic thermal instability. In a vertical flat channel, the local opposing mixed convection heat transfer was experimentally investigated by Poskas et al. [95] with symmetrical heating in a laminar–turbulent transition region. Results revealed that there was a notable increase in heat transfer with the air pressure (can be said Gr) and the heat transfer was much more intensive in vortex flow region than it was in case of turbulent forced convection. Due to this vortex flow under the influence of buoyancy effect the transition period from laminar to turbulent was delayed which is clearly dependent on the air pressure. The critical Reynolds number (Re_{cr_2}) and Nusselt number correlation were also developed as mentioned in Eqs. 16 and 17.

$$Re_{cr_2} = 8.46 Gr_q^{0.33}. \tag{16}$$

$$Nu = 0.02 \left(Gr_q Re \right)^{0.244}. \tag{17}$$

Wang et al. [122] carried out an experimental investigation of breakdown type transition from laminar to turbulent flow in a vertical narrow channel. Findings indicated that the wall heating leads to a delay in transition from laminar to turbulent. At the point of laminar flow breakdown, the critical Re value increases with the increase of fluid and wall temperature difference. The local Nu at that point also increases with the increase of inlet Re. The flow and heat transfer behavior in the transition regime signifies that the change of viscosity was not the dominant factor for delay in transition. Poskas et al. [97] investigated experimentally opposing mixed convection heat transfer at different air pressure (0.1–0.4 MPa) in a vertical flat channel in the transition region. Their results revealed a considerable increase in heat transfer rate upon increase in the buoyancy effect for a certain value of x/d_e, which is related to the formation or disappearance of vortices. Thus, the results obtained changed the concept of transition from laminar to turbulent flow under a consequential buoyancy effect, and therefore the transition to turbulent flow occurs at higher Re. Poskas et al. [96] performed another similar experimental work of opposing mixed convection for higher pressure (0.7–1.0 MPa). They also found that the Reynolds number increases with the increase in buoyancy. Same correlation (Eq. 16) was suggested to obtain the critical Reynolds number (Re_{cr2}) for a stabilized flow.

According to literature by Meyer [80], it has been inferred that not much work has been done in the laminar–turbulent transitional flow regime for tubes. The challenges with experimental work and methodologies have been discussed along with the state of art in terms of different types of inlets, boundary conditions, tube geometries, tube surface roughness, and different working fluids. Everts and Meyer [36] investigated the heat transfer characteristics of developing flow in the transitional regime. They found that the start and end of transition were delayed as the flow developed. Bashir and Meyer [16] studied the heat transfer in the laminar and transitional regime of a vertical tube in an upward direction. It was found that for fully developed vertical flow the transition was delayed as compared to the horizontal flow, where the secondary flow caused earlier transition. Buoyancy effects strongly influenced the start of transition and became negligible at the end of the transition. Everts and Meyer [38] studied heat transfer behavior in developing and fully developed flow in smooth horizontal tubes within the transitional regime which was rarely done in previous articles if any. It was observed that the Re at which the transition started was independent of the axial distance and transition takes place at a time throughout the whole tube. However, the end transition was dependent on the axial position and appeared as the flow approached fully developed. Apart from it, another finding was that the free convection effect aids faster transition from laminar to turbulent. Correlations presented as Eqs. 18 and 19 were originally developed by Ghajar and Tam [48] for the start and end of transitional flow regime in developing and fully developed flow. It was said that heat transfer characteristics observed were much different in developing flow as that of fully developed flow. Another work by Everts and Meyer [39] compared the pressure drop and heat transfer characteristics of developing and fully developed flow in horizontal tubes for all the four regimes identified by Everts and Meyer [37]. The effect of free convection on the development of local heat transfer characteristics in horizontal circular tubes, heated with constant heat flux were experimentally investigated by Meyer and Everts [81]. Developing and fully developed flow in the laminar and transitional flow regimes were studied with water as the test fluid. Three different regions FCD (Forced Convection Developing), MCD (Mixed Convection Developing), and FD (Fully Developed) were defined qualitatively as well as quantitatively for developing laminar flow. The transitional flow regime along the tube length was divided into four regions: laminar, transitional, quasi-turbulent and turbulent flow. It was found that the transition was faster with the increase in free convection effect and Reynolds number.

$$Re_{cr} = 2524 - 0.82\left(192 - \frac{x}{D}\right). \tag{18}$$

$$Re_{qt} = 8791 - 7.69\left(192 - \frac{x}{D}\right). \tag{19}$$

The above research reveals that the transition from laminar to turbulent flow is usually by a reversal in wall temperature or a small rise in temperature than expected as the distance from the entry increases. Some authors found that the onset of laminar–turbulent transition is clearly dependent on the vortex flow caused by the buoyancy effect. Few correlations were also developed indicating critical parameters. Wang et al. [122] were probably the first to study the breakdown type transition instead of buoyancy-induced type transition. They confirmed that the wall heating delays the transition and change of viscosity is not the dominant cause for the delay. Further, experimental work by Poskas et al. [96, 97] also emphasized that transition depends on vortical flows caused due to buoyancy effect. It was also observed that in the case of vortical flow heat transfer rate was remarkably higher than that of turbulent flow and the transition to turbulent flow occurred at higher Reynolds numbers. This changed the concept of transition occurred under a buoyancy effect only. Published results showed that: a. heat transfer behavior is quite different in developing and fully developed flows, b. Reynolds number at which the transition from laminar to turbulent started is independent of axial distance, and c. buoyancy force or free convection plays an essential role for transition.

5   Summary and Future Direction

Heat transfer by combined forced and natural convection, commonly known as mixed convection occurs both in natural and engineered systems. Due to its complex nature and limited applicability the studies in mixed convection are few and far between in literature. The present review provides a comprehensive overview of the compelling research in the field of mixed convection over half a century because of the enhanced heat transfer ability of combined free and forced convection as compared to that of free and forced convection alone. This review covers theoretical, numerical, and experimental studies associated with laminar and turbulent mixed convection. Apart from that, studies related to laminar–turbulent transition are also discussed. The first observation is that, there are ample amount of theoretical as well as experimental work present associated to laminar and turbulent mixed convection. However, an acceptable and clear criterion corresponding to the laminar–turbulent transition is still elusive. Detailed reviews in mixed convection were presented by Jackson et al. [61], Jackson [59], Galanis and Behzadmehr [44], Meyer [80] and Dawood et al. [35].

From these studies, conclusion can be made regarding the enhancement or deterioration of heat transfer in mixed convection. It is obvious that aiding flow will enhance and opposing flow will impair the heat transfer. From the mid-twentieth century Martinelli and Boelter [78], Hallman [54], Hanratty et al. [56], Morton [85] and many others started analyzing the flow field and heat transfer in mixed convection. Their studies were restricted to analytical and experimental work. The studies confirm that the buoyancy effect in forced convection and forced flow in free convection enhance the heat transfer by ~20%. Recent studies prefer using nanofluids as working fluid in many engineering applications due to its enhanced heat transfer capability. Some of the studies favor ferrofluids and analysis was carried out for a large range of Ri, Re, Ha and volume fraction (ϕ). It was observed that higher Ri enhances the heat transfer rate. Various flow profiles depict that hydrodynamic recirculation and thermal recirculation are responsible for the flow reversal in laminar mixed convection. Experiments were conducted for wide ranges of Ri, Re and Gr and correlations were developed for various parameters. A mixed convection regime was proposed by Everts and Meyer [37] when $Ri = 0.1 - 10$.

Contrary to laminar mixed convection the effect of buoyancy is very difficult to predict in case of turbulent flows. Detailed reviews were presented by Jackson et al. [61] and Jackson [59] on turbulence and heat transfer in mixed convection. Theoretical studies related to turbulent flows however started later and the equations are quite complex to solve. Few authors put efforts to develop some analytical wall functions to overcome the problem of solving differential equations. However, the performance is often poor due to its inherent limitations. It is obvious that computational ease is only possible with various numerical models having different ways of treating turbulence. Hence the search for a numerical wall function remains elusive yet and the prospect in this direction doesn't appear to be very encouraging. In the early studies on turbulent mixed convection with water and air, there was less indication about the influence of buoyancy on heat transfer. Later, a number of works show such effects near the critical points as mentioned by Hall et al. [52] and Hall and Jackson [51]. Earlier works found a huge loss of heat transfer for upward flow near critical points. The effect was named "pseudoboiling" and thought it to be like the film-boiling. Through experiments, it became clear that the effect is due to buoyancy and not due to film-boiling. With the advent of computers, it became possible to solve various two-equation turbulence models. From 1980 onwards, advanced models like RANS, LES, and DNS have overtaken conventional turbulence models. The only limitation is that they require a high computing power especially for complex geometries in the presence of turbulence. Some of the experimental studies proposed the buoyancy parameter (Bo) that differentiates the region with and without buoyancy as a characteristics number. Few correlations were also developed using the buoyancy parameter, but these are unable to provide much information in the intermediate regime.

Literature available for the transition from laminar to turbulent mixed convection are very less in number. Also, analytical formulation can't be made for laminar–turbulent transition, hence studies were focused on computational and experimental observations of this phenomenon. Research reveals that the transition from laminar to turbulent flow is usually by a reversal in wall temperature or an unforeseen rise in temperature from the entry point. It was established that Gr and Re has individual significance in transition and relaminarization. Three things: a. change of flow structure, b. heat transfer mechanism and, c. the kinetic energy may be the reason for less heat transfer with the increase of both Gr and Re. Some of the authors found the onset of this transition is clearly dependent on the vortex flow caused by the buoyancy effect. Few correlations were also developed indicating critical parameters. Various turbulence models considering inlet velocity, inclination angle, heated wall temperature, and Richardson number were used to analyze the transition. It was concluded that the increase of inlet velocity and inclination angle delays the transition whereas the increase in wall temperature and Ri accelerates it. Experiments done by Professor Robertas Poskas confirm that not only the buoyancy parameter alone but also length of the channel is responsible for the asymmetrical velocity profile. Finally, studies on mixed convection show that: a. heat transfer behavior is quite different in developing and fully developed flows, b. Reynolds number at which the transition from laminar to turbulent started is independent of axial distance, and c. buoyancy force or free convection plays an essential role in transition.

Thus, it can be said that the physics of fluid flow and heat transfer in laminar mixed convection is reasonably well understood. The flow features, wall temperature distribution and flow reversal in this regime are well explained through a relatively larger number of studies both theoretically and experimentally. However, in comparison, the studies on turbulent mixed convection appear to be less conclusive. The search for an appropriate wall function remained the dream of all researchers in this field. It appears that there does not exist any universal wall function since the two regimes of forced and natural convection are physically quite different processes and their interaction does not have any general feature. In our opinion, this elusive search of a standard wall function would not lead anywhere and research in this direction appears to be a futile exercise. On the other hand, using brute force computing such as LES and DNS is also not bringing any better understanding of turbulent mixed convection. Under the circumstances, experiments appear to be the way out where the flow features and vortex characteristics can be revealed through advanced experimental techniques such as interferometry or PIV. The experiments performed by Professor Josua P. Meyer and his co-workers at the heat transfer laboratory of the Department of Mechanical and Aeronautical Engineering, University of Pretoria may be used as the benchmark for future studies. A clear and acceptable criterion behind laminar–turbulent conversion in mixed convection is still evasive. The experimental studies are often in patches with limited ranges of Rayleigh, Richardson, and Reynolds numbers. There is a need to carry out comprehensive experiments with a wide range of these parameters for both aiding and opposing flows to indicate any absolute criteria for transition. Such experimental data will also serve as a database for validation of future computational work either by using turbulence models or by using brute force computing such as DNS or LBM.

References

1. Abdelmeguid, A. M., & Spalding, D. B. (1979). Turbulent flow and heat transfer in pipes with buoyancy effects. *Journal of Fluid Mechanics, 94*(2), 383–400.
2. Abdollahzadeh, M., Esmaeilpour, M., Vizinho, R., Younesi, A., & Pàscoa, J. C. (2017). Assessment of RANS turbulence models for numerical study of laminar-turbulent transition in convection heat transfer. *International Journal of Heat and Mass Transfer, 115,* 1288–1308. https://doi.org/10.1016/j.ijheatmasstransfer.2017.08.114.
3. Ackerman, J. W. (1970). Pseudoboiling heat transfer to supercritical pressure water in smooth and ribbed tubes. *Journal of Heat Transfer, 92*(3), 490. https://doi.org/10.1115/1.3449698.
4. Aicher, T., & Martin, H. (1997). New correlations for mixed turbulent natural and forced convection heat transfer in vertical tubes. *International Journal of Heat and Mass Transfer, 40*(15), 3617–3626.
5. Akyuzlu, K. M. (2013). A numerical study of unsteady mixed convection in a square cavity: Transition from laminar to turbulent flow. In *ASME 2013 International Mechanical Engineering Congress and Exposition,* V08AT09A047–V08AT09A047.
6. Al-asadi, M. T., Mohammed, H. A., Kherbeet, A. S., & Al-aswadi, A. A. (2017). Numerical study of assisting and opposing mixed convective nano fluid flows in an inclined circular pipe. *International Communications in Heat and Mass Transfer, 85*(May), 81–91.
7. Avramenko, A. A., Tyrinov, A. I., Shevchuk, I. V., Dmitrenko, N. P., Kravchuk, A. V., & Shevchuk, V. I. (2017). Mixed convection in a vertical flat microchannel. *International Journal of Heat and Mass Transfer, 106,* 1164–1173.
8. Aydin, O. (1999). Aiding and opposing mechanisms of mixed convection in a shear-and buoyancy-driven cavity. *International Communications in Heat and Mass Transfer, 26*(7).
9. Babajani, M., Ghasemi, B., & Raisi, A. (2017). Numerical study on mixed convection cooling of solar cells with nanofluid. *Alexandria Engineering Journal, 56*(1), 93–103. https://doi.org/10.1016/j.aej.2016.09.008.
10. Bae, J. H., Yoo, J. Y., & McEligot, D. M. (2008). Direct numerical simulation of heated CO_2 flows at supercritical pressure in a vertical annulus at Re = 8900. *Physics of Fluids, 20*(5), 055108. https://doi.org/10.1063/1.2927488.
11. Balaji, C., Hölling, M., & Herwig, H. (2007). A general methodology for treating mixed convection problems using asymptotic computational fluid dynamics (ACFD). *International Communications in Heat and Mass Transfer, 34*(6), 682–691. https://doi.org/10.1016/j.icheatmasstransfer.2007.03.006.
12. Barozzi, G. S., Dumas, A., & Collins, M. W. (1984). *Sharp entry and transition effects for laminar combined convection of water in vertical tubes.* 235–241.
13. Barrios-Pina, H., Viazzo, S., & Rey, C. (2012). A numerical study of laminar and transitional mixed convection flow over a backward-facing step. *Computers & Fluids, 56,* 77–91. https://doi.org/10.1016/j.compfluid.2011.11.016.
14. Basak, T., Roy, S., Sharma, P. K., & Pop, I. (2009). Analysis of mixed convection flows within a square cavity with linearly heated side wall(s). *International Journal of Heat and Mass Transfer, 52*(9–10), 2224–2242.
15. Basak, T., Roy, S., Sharma, P. K., & Pop, I. (2009). Analysis of mixed convection flows within a square cavity with uniform and non-uniform heating of bottom wall. *International Journal of Thermal Sciences, 48*(5), 891–912. https://doi.org/10.1016/j.ijthermalsci.2008.08.003.

16. Bashir, A. I., & Meyer, J. P. (2017). Heat transfer in the laminar and transitional flow regimes of smooth vertical tube for upflow direction. In *13th International Conference on Heat Transfer, Fluid Mechanics and Thermodynamics*.

17. Behzadmehr, A., Galanis, N., & Laneville, A. (2002). Laminar-turbulent transition for low Reynolds number mixed convection in a uniformly heated vertical tube. *International Journal of Numerical Methods for Heat and Fluid Flow, 12*(7), 839–854. https://doi.org/10.1108/09615530210443052.

18. Behzadmehr, A., Galanis, N., & Laneville, A. (2003). Low Reynolds number mixed convection in vertical tubes with uniform wall heat flux. *International Journal of Heat and Mass Transfer, 46*(25), 4823–4833. https://doi.org/10.1016/S0017-9310(03)00323-5.

19. Behzadmehr, A., Laneville, A., & Galanis, N. (2008). Experimental study of onset of laminar-turbulent transition in mixed convection in a vertical heated tube. *International Journal of Heat and Mass Transfer, 51*(25–26), 5895–5905. https://doi.org/10.1016/j.ijheatmasstransfer.2008.04.005.

20. Bieder, U., Ziskind, G., & Rashkovan, A. (2018). CFD analysis and experimental validation of steady state mixed convection sodium flow. *Nuclear Engineering and Design, 326*, 333–343. https://doi.org/10.1016/j.nucengdes.2017.11.028.

21. Biswas, G., & Eswaran, V. (2002). *Turbulent flows: Fundamentals, experiments and modeling*. Retrieved from https://books.google.co.in/books?id=2MdXPgAACAAJ.

22. Boulama, K., & Galanis, N. (2004). Analytical solution for fully developed mixed convection between parallel vertical plates with heat and mass transfer. *Journal of Heat Transfer, 126*(3), 381–388. https://doi.org/10.1115/1.1737774.

23. Bradley, D., & Entwistle, A. G. (1965). Developed laminar flow heat transfer from air for variable physical properties. *International Journal of Heat and Mass Transfer, 8*(4), 621–638. https://doi.org/10.1016/0017-9310(65)90049-9.

24. Burgos, J., Cuesta, I., & Salueña, C. (2016). Numerical study of laminar mixed convection in a square open cavity. *International Journal of Heat and Mass Transfer, 99*, 599–612.

25. Carr, A. D., Connor, M. A., & Buhr, H. O. (1973). Velocity, temperature, and turbulence measurements in air for pipe flow with combined free and forced convection. *Journal of Heat Transfer, 95*(4), 445. https://doi.org/10.1115/1.3450087.

26. Chen, K.-H. (1990). A primitive variable, strongly implicit calculation procedure for two and three-dimensional unsteady viscous flows: Applications to compressible and incompressible flows including flows with free surfaces. Iowa State University Capstones, Theses and Dissertations.

27. Chen, T. S., Armaly, B. F., & Ramachandran, N. (1986). Correlations for laminar mixed convection flows on vertical, inclined, and horizontal flat plates. *Journal of Heat Transfer, 108*(4), 835–840.

28. Cheng, T. S. (2011). Characteristics of mixed convection heat transfer in a lid-driven square cavity with various Richardson and Prandtl numbers. *International Journal of Thermal Sciences, 50*(2), 197–205. https://doi.org/10.1016/j.ijthermalsci.2010.09.012.

29. Cheng, C., Kou, H., & Huangt, W. (1990). Flow reversal and heat transfer of fully developed mixed convection in vertical channels. *Journal of Thermophysics and Heat Transfer, 4*(3), 375–383.

30. Cherif, A. S., Kassim, M. A., Benhamou, B., Harmand, S., Corriou, J. P., & Ben Jabrallah, S. (2011). Experimental and numerical study of mixed convection heat and mass transfer in a vertical channel with film evaporation. *International Journal of Thermal Sciences, 50*(6), 942–953. https://doi.org/10.1016/j.ijthermalsci.2011.01.002.

31. Churchill, S. W. (1977). Comprehensive correlating equations for heat, mass and momentum transfer in fully developed flow in smooth tubes. *Industrial and Engineering Chemistry Fundamentals, 16*(1), 109–116. https://doi.org/10.1021/i160061a021.

32. Craft, T. J., Gant, S. E., Gerasimov, A. V., Iacovides, H., & Launder, B. E. (2006). Development and application of wall-function treatments for turbulent forced and mixed convection flows. *Fluid Dynamics Research, 38*, 127–144. https://doi.org/10.1016/j.fluiddyn.2004.11.002.

33. Craft, T. J., Gerasimov, A. V., Iacovides, H., & Launder, B. E. (2002). Progress in the generalization of wall-function treatments. *International Journal of Heat and Fluid Flow, 23*, 148–160.
34. Davis, D. V. (1983). Natural convection of air in a square cavity. A bench mark numerical solution. *International Journal for Numerical Methods in Fluids, 3*, 249–264.
35. Dawood, H. K., Mohammed, H. A., Sidik, N. A. C., Munisamy, K. M., & Wahid, M. A. (2015). Forced, natural and mixed-convection heat transfer and fluid flow in annulus: A review. *International Communications in Heat and Mass Transfer, 62*, 45–57. https://doi.org/10.1016/j.icheatmasstransfer.2015.01.006.
36. Everts, M., & Meyer, J. P. (2015). Heat transfer of developing flow in the transitional flow regime. In *Proceeding of First Thermal and Fluids Engineering Summer Conference* (pp. 1051–1063). https://doi.org/10.1615/TFESC1.fnd.012660.
37. Everts, M., & Meyer, J. P. (2018). Flow regime maps for smooth horizontal tubes at a constant heat flux. *International Journal of Heat and Mass Transfer, 117*, 1274–1290. https://doi.org/10.1016/j.ijheatmasstransfer.2017.10.073.
38. Everts, M., & Meyer, J. P. (2018). Heat transfer of developing and fully developed flow in smooth horizontal tubes in the transitional flow regime. *International Journal of Heat and Mass Transfer, 117*, 1331–1351. https://doi.org/10.1016/j.ijheatmasstransfer.2017.10.071.
39. Everts, M., & Meyer, J. P. (2018). Relationship between pressure drop and heat transfer of developing and fully developed flow in smooth horizontal circular tubes in the laminar, transitional, quasi-turbulent and turbulent flow regimes. *International Journal of Heat and Mass Transfer, 117*, 1231–1250. https://doi.org/10.1016/j.ijheatmasstransfer.2017.10.072.
40. Farrugia, P. S., & Micallef, A. (2012). Turbulent plumes evolving in a vertical flow under a mixed convection regime. *International Journal of Heat and Mass Transfer, 55*(7–8), 1931–1940.
41. Fewster, J. (1976). *Mixed forced and free convective heat transfer to supercritical pressure fluids flowing in vertical pipes*. The University of Manchester.
42. Fontana, É., Capeletto, C. A., da Silva, A., & Mariani, V. C. (2015). Numerical analysis of mixed convection in partially open cavities heated from below. *International Journal of Heat and Mass Transfer, 81*, 829–845.
43. Fu, W.-S., Lai, Y.-C., Huang, Y., & Liu, K.-L. (2013). An investigation of flow reversal of mixed convection in a three dimensional rectangular channel with a finite length. *International Journal of Heat and Mass Transfer, 64*, 636–646.
44. Galanis, N., & Behzadmehr, A. (2008). Mixed convection in vertical ducts. In *Proceedings of 6th IASME/WSEAS International Conference on Fluid Mechanics and Aerodynamics* (pp. 35–43).
45. Gangawane, K. M. (2017). Computational analysis of mixed convection heat transfer characteristics in lid-driven cavity containing triangular block with constant heat flux: Effect of Prandtl and Grashof numbers. *International Journal of Heat and Mass Transfer, 105*, 34–57.
46. Gavara, M. R., Dutta, P., & Seetharamu, K. N. (2012). Mixed convection adjacent to non-isothermal vertical surfaces. *International Journal of Heat and Mass Transfer, 55*(17–18), 4580–4587.
47. Gersten, K., & Herwig, H. (1992). *Strömungsmechanik, Grundlagen der Impuls-, Wärme- und Stoff-Übertragung aus Asymptotischer Sicht*. Braunschweig/Wiesbaden: Vieweg-Verlag. Google Scholar.
48. Ghajar, A. J., & Tam, L. M. (1995). Flow regime map for a horizontal pipe with uniform wall heat flux and three inlet configurations. *Experimental Thermal and Fluid Science, 10*(3), 287–297. https://doi.org/10.1016/0894-1777(94)00107-J.
49. Ghia, U., Ghia, K. N., & Shin, C. T. (1982). High-Re solutions for incompressible flow using the Navier Stokes equations and a multigrid method. *Journal of Computational Physics, 48*, 387–411. https://doi.org/10.1016/0021-9991(82)90058-4.
50. Grassi, W., & Testi, D. (2006). Developing upward flow in a uniformly heated circular duct under transitional mixed convection. *International Journal of Thermal Sciences, 45*(9), 932–937. https://doi.org/10.1016/j.ijthermalsci.2005.11.007.

51. Hall, W. B., & Jackson, J. D. (1971). Heat transfer near the critical point. *Advances in Heat Transfer, 7*(1), 86.
52. Hall, W. B., Jackson, J. D., & Watson, A. (1967). Paper 3: A review of forced convection heat transfer to fluids at supercritical pressures. *Proceedings of the Institution of Mechanical Engineers, Conference Proceedings, 182*(9), 10–22. https://doi.org/10.1243/PIME_CONF_1967_182_262_02.
53. Hall, W. B., & Price, P. H. (1970). Mixed forced and free convection from a vertical heated plate to air. *International Heat Transfer Conference, 4*, 19.
54. Hallman, T. M. (1956). Combined forced and free-laminar heat transfer in vertical tubes with uniform internal heat generation. *Transaction ASME, 78*(8), 1841–1851.
55. Hallman, T. M. (1961). *Experimental study of combined forced and free laminar convection in a vertical tube, N.A.S.A.T.N. D-1104.*
56. Hanratty, T. J., Rosen, E. M., & Kabel, R. L. (1958). Effect of heat transfer on flow field at low Reynolds numbers in vertical tubes. *Industrial and Engineering Chemistry, 50*(5), 815–820.
57. Hsu, Y.-Y., & Smith, J. M. (1961). The effect of density variation on heat transfer in the critical region. *Journal of Heat Transfer, 83*, 176. https://doi.org/10.1115/1.3680510.
58. Ismael, M. A., Pop, I., & Chamkha, A. J. (2014). Mixed convection in a lid-driven square cavity with partial slip. *International Journal of Thermal Sciences, 82*(1), 47–61. https://doi.org/10.1016/j.ijthermalsci.2014.03.007.
59. Jackson, J. D. (2006). Studies of buoyancy-influenced turbulent flow and heat transfer in vertical passages. *International Heat Transfer Conference 13, KN-24.* https://doi.org/10.1615/IHTC13.p30.240.
60. Jackson, J. D. (2013). Fluid flow and convective heat transfer to fluids at supercritical pressure. *Nuclear Engineering and Design, 264*, 24–40. http://dx.doi.org/10.1016/j.nucengdes.2012.09.040.
61. Jackson, J. D., Cotton, M. A., & Axcell, B. P. (1989). Studies of mixed convection in vertical tubes. *International Journal of Heat and Fluid Flow, 10*(1), 2–15.
62. Jackson, J. D., Lutterodt, K. E., & Weinberg, R. (2003). Experimental studies of buoyancy-influenced convective heat transfer in heated vertical tubes at pressures just above and just below the thermodynamic critical value. In *Proceedings of the International Conference on Global Environment and Advanced Nuclear Power Plants (GENES4/ANP2003), 36*(4), Paper No. 1177.
63. Joye, D. D. (1996). Comparison of correlations and experiment in opposing flow, mixed convection heat transfer in a vertical tube with Grashof number variation. *International Journal of Heat and Mass Transfer, 39*(5), 1033–1038.
64. Joye, D. D. (2003). Pressure drop correlation for laminar, mixed convection, aiding flow heat transfer in a vertical tube. *International Journal of Heat and Fluid Flow, 24*(2), 260–266. https://doi.org/10.1016/S0142-727X(02)00238-2.
65. Kamath, P. M., Balaji, C., & Venkateshan, S. P. (2011). Experimental investigation of flow assisted mixed convection in high porosity foams in vertical channels. *International Journal of Heat and Mass Transfer, 54*(25–26), 5231–5241. https://doi.org/10.1016/j.ijheatmasstransfer.2011.08.020.
66. Kasagi, N., & Nishimura, M. (1997). Direct numerical simulation of combined forced and natural turbulent convection in a vertical plane channel. *International Journal of Heat and Fluid Flow, 18*(96), 88–99.
67. Kemeny, G. A., & Somers, E. V. (1962). Combined free and forced-convective flow in vertical circular tubes-experiments with water and oil. *Journal of Heat Transfer, 84*(4), 339–345. https://doi.org/10.1115/1.3684389.
68. Kenning, D. B. R., Poon, J. Y., & Shock, R. A. W. (1973). *Local reductions in heat transfer due to buoyancy effects in upward turbulent flow.* Atomic Energy Research Establishment.
69. Kotresha, B., & Gnanasekaran, N. (2018). Investigation of mixed convection heat transfer through metal foams partially filled in a vertical channel by using computational fluid dynamics. *Journal of Heat Transfer, 140*(11), 112501–112511. https://doi.org/10.1115/1.4040614.

70. Launder, B. E., & Sharma, B. I. (1974). Application of the energy-dissipation model of turbulence to the calculation of flow near a spinning disc. *Letters in Heat and Mass Transfer, 1*(2), 131–137.

71. Launder, B. E., & Spalding, D. B. (1974). The numerical computation of turbulent flows. *Computer Methods in Applied Mechanics and Engineering, 3*, 269–289. https://doi.org/10.1016/B978-0-08-030937-8.50016-7.

72. Lawrence, W. T., & Chato, J. C. (1966). Heat-transfer effects on the developing laminar flow inside vertical tubes. *Journal of Heat Transfer, 88*(2), 214–222. https://doi.org/10.1115/1.3691518.

73. Li, Z. Y., Huang, Z., & Tao, W. Q. (2016). Three-dimensional numerical study on fully-developed mixed laminar convection in parabolic trough solar receiver tube. *Energy , 113*, 1288–1303. https://doi.org/10.1016/j.energy.2016.07.148.

74. Mare, T., Galanis, N., Voicu, I., Miriel, J., & Sow, O. (2008). Experimental and numerical study of mixed convection with flow reversal in coaxial double-duct heat exchangers. *Experimental Thermal and Fluid Science, 32*(5), 1096–1104. https://doi.org/10.1115/1.3449722.

75. Marner, W. J., & McMillan, H. K. (1970). Combined free and forced laminar convection in a vertical tube with constant wall temperature. *Journal of Heat Transfer, 92*(3), 559–562.

76. Marocco, L. (2018). Hybrid LES/DNS of turbulent forced and aided mixed convection to a liquid metal flowing in a vertical concentric annulus. *International Journal of Heat and Mass Transfer, 121*, 488–502. https://doi.org/10.1016/j.ijheatmasstransfer.2018.01.006.

77. Marocco, L., Alberti di Valmontana, A., & Wetzel, T. (2017). Numerical investigation of turbulent aided mixed convection of liquid metal flow through a concentric annulus. *International Journal of Heat and Mass Transfer, 105*, 479–494. https://doi.org/10.1016/j.ijheatmasstransfer.2016.09.107.

78. Martinelli, R. C., & Boelter, L. M. K. (1942). Analytical prediction of superimposed free and forced convection in a vertical pipe, 5. University of California. *Publications in Engineering*, 23–58.

79. Metais, B., & Eckert, E. R. G. (1964). Forced, mixed, and free convection regimes. *Journal of Heat Transfer, 86*(2), 295. https://doi.org/10.1115/1.3687128.

80. Meyer, J. P. (2014). Heat transfer in tubes in the transitional flow regime. In *Proceedings of the 15th International Heat Transfer Conference, Kyoto, Paper KN03* (pp. 10–15).

81. Meyer, J. P., & Everts, M. (2018). Single-phase mixed convection of developing and fully developed flow in smooth horizontal circular tubes in the laminar and transitional flow regimes. *International Journal of Heat and Mass Transfer, 117*, 1251–1273.

82. Mikielewicz, D. P. (1994). *Comparative studies of turbulence models under conditions of mixed convection with variable properties in heated vertical tubes*. The University of Manchester.

83. Mohammed, H. A., & Salman, Y. K. (2007). Experimental investigation of mixed convection heat transfer for thermally developing flow in a horizontal circular cylinder. *Applied Thermal Engineering, 27*(8–9), 1522–1533. https://doi.org/10.1016/j.applthermaleng.2006.09.023.

84. Mori, Y., et al. (1966). Forced convective heat transfer in uniformly heated horizontal tubes 1st report—Experimental study on the effect of buoyancy. *International Journal of Heat and Mass Transfer, 9*, 453–463.

85. Morton, B. R. (1960). Laminar convection in uniformly heated vertical pipes. *Journal of Fluid Mechanics, 8*(2), 227–240. https://doi.org/10.1017/S0022112060000566.

86. Nobari, M. R. H., & Beshkani, A. (2007). A numerical study of mixed convection in a vertical channel flow impinging on a horizontal surface. *International Journal of Thermal Sciences, 46*(10), 989–997. https://doi.org/10.1016/j.ijthermalsci.2006.11.012.

87. Onyejekwe, O. O. (2012). Combined effects of shear and buoyancy for mixed convection in an enclosure. *Advances in Engineering Software, 47*(1), 188–193. https://doi.org/10.1016/j.advengsoft.2011.11.002.

88. Petukhov, B. S. (1976). Turbulent flow and heat transfer in pipes under considerable effect of thermogravitational forces. In *Heat Transfer and Turbulent Buoyant Convection. Seminar of International Centre for Heat and Mass Transfer, Dubrovnik, Yugoslavia, Hemisphere Publishing Corporation, Washington D.C.* (Vol. 2, pp. 701–717).

89. Petukhov, B. S., & Strigin, B. K. (1968). Experimental investigation of heat transfer with viscous-inertial-gravitational flow of a liquid in vertical tubes. *Teplofizika Vysokikh Temperatur, 6,* 933–937.
90. Pigford, R. L. (1955). Non-isothermal flow and heat transfer inside vertical tubes. *Chemical Engineering Progress Symposium Series, 51,* 79–92.
91. Polyakov, A. F. (1974). Development of secondary free-convection currents in forced turbulent flow in horizontal tubes. *Journal of Applied Mechanics and Technical Physics, 15*(5), 632–637.
92. Poskas, P., & Poskas, R. (2003). Local turbulent opposing mixed convection heat transfer in inclined flat channel for stably stratified airflow. *International Journal of Heat and Mass Transfer, 46*(21), 4023–4032.
93. Poskas, P., Poskas, R., & Gediminskas, A. (2012). Numerical investigation of the opposing mixed convection in an inclined flat channel using turbulence transition models. *Journal of Physics: Conference Series, 395*(1), 012098. https://doi.org/10.1088/1742-6596/395/1/012098.
94. Poskas, R., Poskas, P., & Sabanskis, D. (2005). Local turbulent opposing mixed convection heat transfer in inclined flat channel for unstably stratified airflow. *International Journal of Heat and Mass Transfer, 48*(5), 956–964. https://doi.org/10.1016/j.ijheatmasstransfer.2004.09.025.
95. Poskas, P., Poskas, R., Sirvydas, A., & Smaizys, A. (2011). Experimental investigation of opposing mixed convection hear transfer in the vertical flat channel in a laminar-turbulent transition region. *International Journal of Heat and Mass Transfer, 54*(1–3), 662–668. https://doi.org/10.1016/j.ijheatmasstransfer.2010.09.004.
96. Poskas, R., Sirvydas, A., & Bartkus, G. (2016). Experimental investigation of opposing mixed convection heat transfer in a vertical flat channel in the transition region. 2. Analysis of local heat transfer in the case of the prevailing effect of buoyancy and generalization of data. *Heat Transfer Research, 47*(8), 745–751. https://doi.org/10.1615/HeatTransRes.2016012394.
97. Poskas, R., Sirvydas, A., Kolesnikovas, J., & Kilda, R. (2013). Experimental investigation of opposing mixed convection heat transfer in a vertical flat channel in the transition region. 1. Analysis of local heat transfer. *Heat Transfer Research, 44*(7), 589–602. https://doi.org/10.1615/HeatTransRes.v44.i7.10.
98. Prasad, A. K., & Koseff, J. R. (1996). Combined forced and natural convection heat transfer in a deep lid-driven cavity flow. *International Journal of Heat and Fluid Flow, 17*(5), 460–467.
99. Rabbi, K. M., Saha, S., Mojumder, S., Rahman, M. M., Saidur, R., & Ibrahim, T. A. (2016). Numerical investigation of pure mixed convection in a ferrofluid-filled lid-driven cavity for different heater configurations. *Alexandria Engineering Journal, 55*(1), 127–139. https://doi.org/10.1016/j.aej.2015.12.021.
100. Rahman, M. M., Öztop, H. F., Rahim, N. A., Saidur, R., Al-Salem, K., Amin, N., et al. (2012). Computational analysis of mixed convection in a channel with a cavity heated from different sides. *International Communications in Heat and Mass Transfer, 39*(1), 78–84. https://doi.org/10.1016/j.icheatmasstransfer.2011.09.006.
101. Rana, P., & Bhargava, R. (2011). Numerical study of heat transfer enhancement in mixed convection flow along a vertical plate with heat source/sink utilizing nanofluids. *Communications in Nonlinear Science and Numerical Simulation, 16*(11), 4318–4334. https://doi.org/10.1016/j.cnsns.2011.03.014.
102. Reichl, C., Hengstberger, F., & Zauner, C. (2013). Heat transfer mechanisms in a compound parabolic concentrator: Comparison of computational fluid dynamics simulations to particle image velocimetry and local temperature measurements. *Solar Energy, 97,* 436–446. https://doi.org/10.1016/j.solener.2013.09.003.
103. Rosen, E. M., & Hanratty, T. J. (1961). Use of boundary-layer theory to predict the effect of heat transfer on the laminar-flow field in a vertical tube with a constant-temperature wall. *AIChE Journal, 7*(1), 112–123. https://doi.org/10.1002/aic.690070126.
104. Sakurai, A., Matsubara, K., Takakuwa, K., & Kanbayashi, R. (2012). Radiation effects on mixed turbulent natural and forced convection in a horizontal channel using direct numerical simulation. *International Journal of Heat and Mass Transfer, 55*(9–10), 2539–2548. https://doi.org/10.1016/j.ijheatmasstransfer.2012.01.006.

105. Sengupta, T. K., & Subbaiah, K. V. (2006). Spatial stability for mixed convection boundary layer over a heated horizontal plate. *Studies in Applied Mathematics, 117*(3), 265–298. https://doi.org/10.1111/j.1467-9590.2006.00355.x.

106. Sharma, N., Dhiman, A. K., & Kumar, S. (2012). Mixed convection flow and heat transfer across a square cylinder under the influence of aiding buoyancy at low Reynolds numbers. *International Journal of Heat and Mass Transfer, 55*(9–10), 2601–2614. https://doi.org/10.1016/j.ijheatmasstransfer.2011.12.034.

107. Shitsman, M. E. (2006). Natural convection effect on heat transfer to a turbulent water flow in intensively heated tubes at supercritical pressures. *ARCHIVE: Proceedings of the Institution of Mechanical Engineers, Conference Proceedings 1964–1970 (Vols. 178–184), Various Titles Labelled Volumes A to S, 182*(39), 36–41. https://doi.org/10.1243/PIME_CONF_1967_182_265_02.

108. Singh, A. K., Harinadha, G., Kishore, N., Barua, P., Jain, T., & Joshi, P. (2015). Mixed convective heat transfer phenomena of circular cylinders to non-newtonian nanofluids flowing upward. *Procedia Engineering, 127,* 118–125. https://doi.org/10.1016/j.proeng.2015.11.434.

109. Skiadaressis, D., & Spalding, D. B. (1977). Prediction of combined free and forced convection in turbulent flow through horizontal pipes. *Letters in Heat and Mass Transfer, 4*(1), 35–39.

110. Steiner, A. (1971). On the reverse transition of a turbulent flow under the action of buoyancy forces. *Journal of Fluid Mechanics, 47*(3), 503–512.

111. Subhashini, S. V., Samuel, N., & Pop, I. (2011). Effects of buoyancy assisting and opposing flows on mixed convection boundary layer flow over a permeable vertical surface. *International Communications in Heat and Mass Transfer, 38*(4), 499–503. https://doi.org/10.1016/j.icheatmasstransfer.2010.12.041.

112. Suga, K., Ishibashi, Y., & Kuwata, Y. (2013). An analytical wall-function for recirculating and impinging turbulent heat transfer. *International Journal of Heat and Fluid Flow, 41,* 45–54.

113. Tam, L. M., & Ghajar, A. J. (2006). Transitional heat transfer in plain horizontal tubes. *Heat Transfer Engineering, 27*(5), 23–38. https://doi.org/10.1080/01457630600559538.

114. Tanaka, H., Ayao, T., Masaru, H., & Nuchi, N. (1973). Effects of buoyancy and of acceleration owing to thermal expansion on forced turbulent convection in vertical circular tubes—criteria of the effects, velocity and temperature profiles, and reverse transition from turbulent to laminar flow. *International Journal of Heat and Mass Transfer, 16*(6), 1267–1288.

115. Tanaka, H., Shigeo, M., & Shunichi, H. (1987). Combined forced and natural convection heat transfer for upward flow in a uniformly heated, vertical pipe. *International Journal of Heat and Mass Transfer, 30*(1), 165–174.

116. Tian, C., Wang, J., Cao, X., Yan, C., & Ala, A. A. (2018). Experimental study on mixed convection in an asymmetrically heated, inclined, narrow, rectangular channel. *International Journal of Heat and Mass Transfer, 116,* 1074–1084. https://doi.org/10.1016/j.ijheatmasstransfer.2017.09.099.

117. Velusamy, K., & Garg, V. K. (1996). Laminar mixed convection in vertical elliptic ducts. *International Journal of Heat and Mass Transfer, 39*(4), 745–752.

118. Venkatasubbaiah, K., & Sengupta, T. K. (2009). Mixed convection flow past a vertical plate: Stability analysis and its direct simulation. *International Journal of Thermal Sciences, 48*(3), 461–474. https://doi.org/10.1016/j.ijthermalsci.2008.03.019.

119. Vilemas, J. V., Poškas, P. S., & Kaupas, V. E. (1992). Local heat transfer in a vertical gas-cooled tube with turbulent mixed convection and different heat fluxes. *International Journal of Heat and Mass Transfer, 35*(10), 2421–2428.

120. Walklate, P. J. (1976). *A comparative study of theoretical models of turbulence for the numerical prediction of boundary-layer flows.* University of Manchester Institute of Science and Technology.

121. Wang, X. A. (1982). An experimental study of mixed, forced, and free convection heat transfer from a horizontal flat plate to air. *Journal of Heat Transfer, 107*(3), 738. https://doi.org/10.1115/1.3247494.

122. Wang, C., Gao, P., Wang, Z., & Tan, S. (2012). Experimental study of transition from laminar to turbulent flow in vertical narrow channel. *Annals of Nuclear Energy, 47,* 85–90. https://doi.org/10.1016/j.anucene.2012.04.018.

123. Wang, J., Li, J., & Jackson, J. D. (2002). Mixed convection heat transfer to air flowing upwards through a vertical plane passage: Part 3. *Chemical Engineering Research and Design, 80*(3), 252–260.

124. Yang, G., Huang, Y., Wu, J., Zhang, L., Chen, G., Lv, R., et al. (2017). Experimental study and numerical models assessment of turbulent mixed convection heat transfer in a vertical open cavity. *Building and Environment, 115,* 91–103. https://doi.org/10.1016/j.buildenv.2017.01.016.

125. You, J., Yoo, J. Y., & Choi, H. (2003). Direct numerical simulation of heated vertical air flows in fully developed turbulent mixed convection. *International Journal of Heat and Mass Transfer, 46*(9), 1613–1627. https://doi.org/10.1016/S0017-9310(02)00442-8.

126. Zeldin, B., & Schmidt, F. W. (1972). Developing flow with combined forced-free convection in an isothermal vertical tube. *Journal of Heat Transfer, 94,* 211–223. https://doi.org/10.1115/1.3449899.

Large Eddy Simulation of Flows of Engineering Interest: A Review

S. Sarkar

1 Introduction

Transition from laminar to turbulent flows subjected to a wide variety of possible disturbances is quite common in engineering applications. Flows are often complex involving separation and reattachment with turbulent structures. A wide scale of eddies are observed ranging from as large as the size of the flow domain with low-frequency fluctuations to as small as the Kolmogorov microscale of high-frequency fluctuations at which dissipation takes place. Further, the shear layer or the boundary layer instability is occasionally seen with vortex stretching mechanism and breakdown, making the flow unsteady, three-dimensional, and very irregular. This turbulent motion, therefore, affects greatly the distribution of velocity, temperature, and concentration of flow species, and subsequently dictates the flow mixing, heat and mass transfer. Thus, successful simulation of transitional and turbulent flows plays an important role in computational fluid dynamics (CFD) research.

The complexities associated with the turbulent motion can be resolved by the solution of unsteady three-dimensional (3D) Navier–Stokes (N-S) equations without introducing any model and the method is called direct numerical simulation (DNS). Since philosophically the method should resolve all the scales of motion, demanding large numbers of grid points, it can be shown that the grid points required and the cost of computation increase roughly with Re^3. In DNS, most of the computational effort is used up for the smallest dissipative motions, whereas the energy and anisotropy are predominantly contained in the larger scales of motion. This restricts the application of DNS only with flows of relatively low Reynolds number (Re) and simple geometry to understand the physics of the flow. In contrast to DNS, one can employ the Reynolds averaged Navier–Stokes (RANS) equations, where all turbulent fluctuations are averaged and a turbulence model is used to mimic the effect

S. Sarkar (✉)
Department of Mechanical Engineering, Indian Institute
of Technology Kanpur, Kanpur 208016, India
e-mail: subra@iitk.ac.in

© Springer Nature Singapore Pte Ltd. 2020
A. Runchal (ed.), *50 Years of CFD in Engineering Sciences*,
https://doi.org/10.1007/978-981-15-2670-1_11

of turbulence. This approach requires much less computational effort, however, the turbulence scales and their interactions are not resolved and hence, it cannot be used to understand the physics of turbulence.

In LES, only the larger-scale motions are resolved on a given grid system by solving 3D time-dependent N-S equations and the smaller-scale motions being universal in nature are accounted for by a subgrid-scale (SGS) model. Thus, a substantial computational cost of explicitly representing the small-scale motions is avoided as compared to the DNS. Further, LES resolves the dynamics of vortices, their nonlinear interactions, and flow instability and it is, therefore, often a preferred method of making certain advances of complex transitional and turbulent flows. LES can be expected to be more accurate and reliable than Reynolds stress models for flows, where larger scales are significant, such as the flow past a bluff body involving unsteady separation and vortex shedding. In brief, LES can be an alternative with the benefit of being a considerably lower computational effort to study the effect of turbulence.

The LES formulation in physical space and its application to the Navier–Stokes equations have been discussed. The review starts with the features of turbulence, LES, and Smagorinsky model, but the focus is on the dynamic subgrid model and on evaluating how well these models reproduce the true impact of small scales on large-scale physics and how they perform in numerical simulations. The hybrid LES-RANS is then discussed in brief to model the effect of near-wall turbulent structures in view of the compromise in accuracy vis-à-vis the gain in efficiency and computational cost. A few examples of LES are presented here, where the majority is performed by the author and his research group. The test cases involve the flow environments often encountered in engineering applications, such as boundary layer transitions, shear layer instability, and flow separation followed by breakdown and reattachment. The paper concludes with an appraisal of LES, but the author would like to state that the review is not meant for an exhaustive cataloging of the work published in the area. The material presented here reflect the author's view focusing on LES of transitional flows, which would benefit the beginners in the field.

2 Features of Turbulence

Turbulent flows are unsteady, three-dimensional, highly vortical, random, and irregular with evidence of a wide range of *scales of motion* [32]. The flow reaches a chaotic non-repeating form describable only in statistical terms. Further, vortex stretching, twirling, and breakdown are essential features of turbulence that leads to energy exchange between different scales. It exhibits a nonlinear process due to advection, a non-Gaussian process in which the triple correlations and high-order moments are not zero. In fact, these correlations are responsible for energy exchange between different scales of motion. A vital property of turbulence that has received much attention in past decades is *scale invariance*, which means that certain features of the flow remain invariant in different scales of motion. Such an observation leads to a simple relationship between small- and large-scales of motion in a turbulent flow and is of interest to LES. The larger scales are anisotropic and influenced by the boundary conditions, whereas, small-scales motion are universal and isotropic. These small scales are not explicitly resolved in an LES; they must be modeled.

The idea of scale invariance in turbulence was first introduced by Richardson [52] in qualitative terms. It states that the large eddies are unstable and thus break down transferring energy (*by an inviscid process*) to somewhat smaller eddies. These smaller eddies transfer energy to yet smaller eddies by a similar breakup process. This is referred to as *energy cascade*, in which energy is transferred to successively smaller and smaller eddies until the Reynolds number based on local micro-scale $Re(l) = u(l)l/v$ becomes sufficiently small that the molecular viscosity is effective in dissipating the kinetic energy (l, $u(l)$ being isotropic length and velocity scale, v is kinematic viscosity). Thus, dissipation is end of the sequence of processes and rate of dissipation ε is determined by the first process in the sequence, which is the transfer of energy from the largest eddies. Following experimental observations, ε scales as u_0^3/l_0, independent of kinematic viscosity v, where l_0, u_0 being the characteristic scales of larger eddies.

Kolmogorov [30] addressed another fundamental question regarding the size of the smallest eddies that are responsible for dissipating the energy in the form of three hypotheses. A chaotic scale-reduction process kills the directional biases of large-scale eddies, where energy is transferred successively to smaller and smaller eddies. As a consequence, the statistics of the small-scale motions are *isotropic* and *universal*. This leads to the Kolmogorov's first similarity hypothesis as, "*the small-scale turbulent motions have a universal form that is uniquely determined by v and ε*". This defines the smallest possible dissipation eddies, known as *Kolmogorov's scales*. Given the two parameters v and ε, there exist unique length, velocity, and time scales as

$$\eta \equiv (v^3/\varepsilon)^{1/4}, \ u_\eta \equiv (\varepsilon v)^{1/4} \text{ and } \tau_\eta \equiv (v/\varepsilon)^{1/2}.$$

Further, the Reynolds number based on the Kolmogorov scales becomes unity, i.e., $\eta u_\eta/v = 1$. This is consistent with the concept of energy cascade that the break down of eddies proceeds until the Reynolds number $u(l) \cdot l/v$ becomes small enough for dissipation to be effective. The ratios of the smallest to largest scales can be determined from the definitions of the Kolmogorov scales and from the scaling $\varepsilon \sim u_0^3/l_0$. These result in

$$\eta/l_0 \sim (Re)^{-3/4}, u_\eta/u_0 \sim (Re)^{-1/4} \text{ and } \tau_\eta/\tau_0 \sim (Re)^{-1/2},$$

where l_0, u_0, and τ_0 being length, velocity, and time scale of the largest eddies. Thus, at high Re, scales of the smallest eddies are very small as compared to those of the largest eddies. As a consequence, there exists a range of scales l that are sufficiently small as compared to l_0, but yet very large as compared to η. The Re based on these scales [$u(l) \cdot l/v$] is, therefore, high enough that the motion is little affected by viscosity. This leads to the Kolmogorov's second hypothesis as, "*at a sufficiently high Reynolds number, the statistics of the motions of scale l in range $l_0 \gg l \gg \eta$ have a universal form that is uniquely determined by ε and independent of ν*".

The above discussion indicates the existence of two subranges: the *dissipation range* (say, $l < l_{DI} = 60\eta$) and *the inertial subrange* ($l_0 > l > l_{DI}$). The second hypothesis is valid in the inertial subrange, where the viscous effects can be negligible and the motions are determined by the inertial effects. On the contrary, viscous effects are significant in the dissipation range, attributing to the dissipation of energy. Further, the central importance of energy cascade lies in the assessment of the rate of energy $E(l)$ at which it is transferred from larger to smaller scales, satisfying the identity, $E(l) = u(l)^2 / \tau(l) = \varepsilon$. Pope [51] has made a detailed discussion on the scales of motions. Thus in the inertial subrange, the effects of viscosity, boundary conditions, and large-scale structures are not important in quantitative sense and the dimensional analysis [15, 30] leads to the following well-known universal power-law spectrum,

$$E(k) = c_k \varepsilon^{2/3} k^{-5/3}. \tag{2.1}$$

Here, c_k is the Kolmogorov constant, k the magnitude of wave number, and ε the dissipation rate of kinetic energy by molecular viscosity ν. The equation is well supported by a number of experiments and implies scale invariance in the sense that the quantity $\gamma^{5/3} E(\gamma k)$ remains invariant in the inertial subrange under the scale transformation $k \to \gamma k$. Several other features of turbulence exhibiting scale invariance have been well documented in the literature [72, 73]. Figure 1 shows the schematic of energy spectra for a generic turbulent flow, illustrating different regimes, i.e., energy-containing range, inertial subrange, and dissipation range in the spectral space. Further, the production of turbulent kinetic energy occurs in the larger scale, low-frequency domain and then energy is transferred through inertial subrange without any appreciable viscous loss. Finally, it is dissipated by molecular viscosity at the low scale and high frequency domain.

Fig. 1 Schematic of energy spectra for a turbulent flow

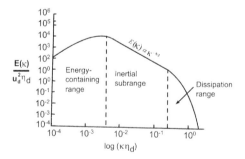

3   Introduction and Hierarchy of Modeling

The deeper insights into relationships between large and small scales are the root of the emergence of large eddy simulation. The approach with LES is to resolve most of the turbulent kinetic energy (TKE) of flow, while modeling most of the dissipation. The possibility of scale separation arises here because of the fact that TKE is determined by large-scales motion, whereas, ε by the small scales [65]. In LES, we separate the motion into small and large scales by spatially filtering the velocity field with a convolution kernel, $G(r, x)$ [33, 34]. The kernel eliminates scales smaller than the filter width Δ. Thus, a filtering operation is defined to decompose the velocity $u(x, t)$ into the sum of a filtered (or resolved) component $\bar{u}(x, t)$ and a residual (or subgrid-scale, SGS) component $u'(x, t)$. The filtered velocity field $\bar{u}(x, t)$ is time-dependent and three-dimensional, representing the motion of the large eddies. The Navier–Stokes equations are used to derive the evolution of the filtered velocity field, containing the *residual-stress tensor* (or SGS stress tensor). The residual-stress tensor is modeled by the *eddy-viscosity hypothesis* to ensure the closure. The filtered N-S equations with modeling of SGS terms are solved by a suitable numerical algorithm.

Turbulent flows are characterized by a wide range of spatial and temporal scales. Consequently, DNS with the highest fidelity can be performed by solving the N-S equations, resolving all the turbulence scales down to the Kolmogorov scale η. However, it is prohibitively expensive for flows of engineering applications, particularly at high Re. On the contrary, the entire range of turbulent scales is modeled for a RANS simulation with the lowest fidelity. Thus, RANS simulations are computationally affordable, where, the turbulence scales and their interactions are not resolved and hence, they cannot be used to understand the physics of turbulence. This hierarchy of modeling is illustrated in Fig. 2, where the top is represented by DNS, being the most flow physics resolving and computationally expensive approach and the bottom by RANS, being the most empirical and computationally affordable.

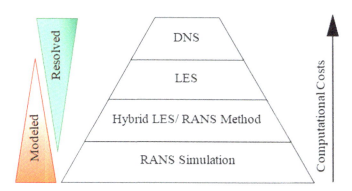

Fig. 2 A schematic representation of the hierarchy, computational costs, and the amounts of resolved versus modeled physics following Sagaut et al. [56]

A compromise between DNS and RANS simulations at two ends of the spectrum is the LES, in which only the larger scale, being most energetic, are resolved, while scales below a cutoff threshold are filtered out and reflected as SGS stress tensor. In LES, the filter and hence grid are sufficiently fine to resolve the bulk (say 80%) of TKE everywhere in the flow field [51]. In some cases, the grid and filter are too large to resolve the energy-containing motions and a substantial fraction of the energy resides in the residual motions. Such a simulation is referred to as very large eddy simulation (VLES). Thus, VLES is less expensive and the simulation is strongly dependent on the modeling of the residual stress. LES has significantly reduced computational cost as compared to DNS while simulating the shear flows that are not influenced by the wall boundaries. Unfortunately, LES also becomes prohibitively expensive, calling for large numbers of grids, for wall-bounded flows at high Re due to the small yet energetic scales dominating the dynamics in the near-wall regions [69]. This challenge has led to the development of methods combining with the LES for free shear regions and RANS or other simplified models (e.g., boundary layer equation or the law of the wall) for the under-resolved near-wall regions. Such approaches are referred to as hybrid LES-RANS method [7, 17] or LES with near-wall modeling (NWM) [6, 49]. In case of LES with NWM, grids are thus refined at a level to resolve 80% of the energy in the far field region away from the wall, while the near-wall motions are not resolved and their influences are modeled.

4   Filtered Navier–Stokes Equations

As already stated, large-scale eddies are directly resolved in LES, while small-scale motions are modeled. Thus, a low-pass spatial filter is often applied to a variable, resulting in the filtered velocity field $\bar{u}(x, t)$, where scales smaller than filter width Δ are removed. The filtered counterpart $\bar{u}(x, t)$ can now be resolved on a relatively coarser grid spacing h. The filter width is considered proportional to the grid spacing, or simply by $\Delta = h$.

A generic filtering operation, which was introduced by Leonard [33], can be defined mathematically in the physical space as a convolution product as

$$\bar{u}(x, t) = \int G(r, x)u(x - r, t)\mathrm{d}r \qquad (4.1)$$

where integration is performed over the entire flow domain and the filter function G satisfies the following normalization condition,

$$\int G(r, x)\mathrm{d}r = 1. \qquad (4.2)$$

Now, the velocity field can be decomposed as

$$u(x, t) = \bar{u}(x, t) + u'(x, t) \tag{4.3}$$

where $u'(x, t)$ is the residual field. Although this appears analogous to the Reynolds decomposition, they are different. Here, the filtered velocity $\bar{u}(x, t)$ is a random field and the filtered of residual is not zero, i.e., $\overline{u'}(x, t) \neq 0$.

The most commonly used filters for performing spatial scale separation are the *Box or top-hat filter*, *Gaussian filter*, and *Spectral or Sharp cutoff filter*. In the box filter, $\bar{u}(x)$ is simply the average of $u(x')$ in the interval $x - \Delta/2 < x' < x + \Delta/2$, whereas a Gaussian distribution with mean zero and variance $\sigma^2 = \Delta^2/12$ is used in the Gaussian filter function. The value of variance is so chosen to match the second moments of Gaussian and box filters. The sharp cutoff filter eliminates all the Fourier modes of wave number greater than the cutoff wave number, $\kappa_c = \pi/\Delta$. Figure 3 illustrates the difference between instantaneous velocity $u(x)$ and the corresponding filtered field $\bar{u}(x)$ using the Gaussian filter. It is evident that $\bar{u}(x)$ follows the general trends of $u(x)$, but the small-scale fluctuations have been removed. Similar features also appear in the residual field $u'(x)$.

Conservation of mass, momentum, and energy lead to the governing equations, which are referred to as the Navier–Stokes equations. There is no need to impose a filter to the governing equations explicitly if a finite volume method is employed to solve the equations numerically integrating over control volumes. This is equivalent to convolution with a top-hat filter and therefore, this is referred to as implicit filtering. However, there exists a shortcoming with implicit filtering, i.e., truly mesh-independent results can never be achieved. With the refinement of mesh, small-scale motions are resolved shifting the cutoff filter to higher wave numbers. Hence, in the ultimate limit $\Delta \rightarrow 0$, the effect of the SGS model vanishes and the simulation turns to a DNS. Thus, it is almost impossible to distinguish between numerical and modeling errors if implicit filtering is employed.

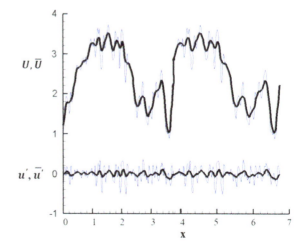

Fig. 3 Difference between the velocity field $u(x)$ and the corresponding filtered field $\bar{u}(x)$, denoted by blue and black line, respectively. Lower curves represent the residual field $u'(x)$ (blue line) and the filtered residual field $\overline{u'}(x, t)$ (black line) [51]

The governing equations of LES are obtained by applying a homogeneous filter to the flow field. Thus, the filtered mass and momentum equations for Newtonian incompressible flow in conservative form can be expressed as

$$\partial_i \bar{u}_i = 0, \tag{4.4}$$

$$\partial_t \bar{u}_i + \partial_j (\bar{u}_i \bar{u}_j) = -\partial_i \bar{p}/\rho + 2\partial_j (\nu \bar{S}_{ij}) - \partial_j \tau_{ij}, \tag{4.5}$$

where $\bar{S}_{ij} = 1/2(\partial_j \bar{u}_i + \partial_i \partial \bar{u}_j)$ and $\tau_{ij} = \overline{u_i u_j} - \bar{u}_i \bar{u}_j$.

Here, $\bar{u}_i$ is the filtered velocity, $\bar{p}_i$ the filtered pressure, ν kinematic viscosity, and ρ is the density of fluid. $\bar{S}_{ij}$ is filtered rate of strain, τ_{ij} is subgrid (SGS) stress tensor, which represents the effect of subgrid motion on the resolved field of LES and needs to be modeled to close the equations. Subscript i indicates the tensor notation. For details, it is recommended to refer to Sagaut [55].

The filtered momentum equation brings out the nonlinear convective term $\overline{u_i u_j}$, which is decomposed [33] in terms of the filtered quantity $\bar{u}_i$ and its fluctuations u'_i, where $u'_i = u_i - \bar{u}_i$. This decomposition leads to

$$\tau_{ij} = L_{ij} + C_{ij} + R_{ij} = \overline{u_i u_j} - \bar{u}_i \bar{u}_j,$$

where $L_{ij} = \overline{\bar{u}_i \bar{u}_j} - \bar{u}_i \bar{u}_j$, $C_{ij} = \overline{\bar{u}_i u'_j} + \overline{\bar{u}_j u'_i}$, and $R_{ij} = \overline{u'_i u'_j}$.

The term L_{ij}, known as Leonard tensor, represents the interaction between large scales. The cross-stress-tensor C_{ij} signifies the interactions between large and small scales, while the Reynolds subgrid tensor R_{ij} reflects the interactions between subgrid scales.

5   The Smagorinsky Model

The Smagorinsky model [35, 68], which is based on eddy-viscosity concept, is the simplest model for the deviatoric part of residual stress and can be expressed as

$$\tau_{ij}^r = \tau_{ij} - 1/3 \, \delta_{ij} \tau_{kk} = -2\nu_r \bar{S}_{ij}, \tag{4.6}$$

where ν_r is eddy viscosity of the residual motion.

Now, Eq. 4.5 becomes

$$\partial_t \bar{u}_i + \partial_j (\bar{u}_i \bar{u}_j) = -\partial_i \bar{P}/\rho + 2\partial_j [(\nu + \nu_r) \bar{S}_{ij}]. \tag{4.7}$$

The eddy viscosity ν_r is analogous to the mixing-length hypothesis and is modeled as the product of a length scale (Δ) and a velocity scale ($\sim\Delta.\bar{S}$), where $\bar{S} = (2\bar{S}_{ij}\bar{S}_{ij})^{1/2}$. Thus,

$$v_r = (C_s \Delta)^2 \bar{S}. \tag{4.8}$$

In the above equation, C_s is Smagorinsky constant. A value of 0.23 is theoretically valid in homogeneous isotropic turbulent flow simulation. However, a reduced value C_s in the range 0.1–0.17 is often used based on the flow environment. The length scale Δ is dependent on mesh, and is taken as $\Delta = (\Delta x \cdot \Delta y \cdot \Delta z)^{1/3}$.

The Smagorinsky model is absolutely dissipative; it is incapable of predicting the reverse cascade regions and overestimates the subgrid-scale dissipation. Part of the problem is due to the fact that the model is only based on local large-scale quantities and predicts nonzero residual stresses even in the laminar region. Part of the limitations has been overcome by using the low-Re model of Voke [82], which is derived following the dissipation spectrum and is expected to simulate the transitional flow. The work of Sarkar and Voke [63] has indicated that simple corrections to the Smagorinsky model are worth pursuing.

In the computation of turbulent flows, the length scale is often multiplied by a damping function to account for the fact that the growth of turbulence structures is inhibited by the presence of the wall. The damping function, suggested by Van Driest [81], is often used modifying the length scale $(l = C_s \Delta)$ in the Smagorinsky model following the exponential relation as

$$l = C_s \left[1.0 - \exp\left(-y^+ / A^+ \right) \right] \Delta, \tag{4.9}$$

where $A^+ = 25$ and $y^+ = y u_\tau / \nu$, and y being the distance to the nearest wall. There are also some other variants of damping function such as the one suggested by Piomelli et al. [50] as

$$l = C_s \left[1.0 - \exp\left(-y^{+3} / A^{+3} \right) \right]^{1/2} \Delta. \tag{4.10}$$

It is believed that these functions stated above would ensure the proper behavior of the SGS stress near the wall.

6 Dynamic Subgrid-Scale Model

The dynamic subgrid-scale model exercises a methodology for determining an appropriate local value of the Smagorinsky coefficient using the resolved scales of motion. Thus, the problem of the Smagorinsky model, i.e., the value of C_s remains the same throughout the domain or even being nonzero in the laminar regime of flow is eliminated. The model was proposed by Germano et al. [20] and modified by Lilly [36]. Here, the assumption of scale invariance is utilized by applying the coefficient estimated from the resolved scales to the SGS range. Formally, the dynamic procedure is based on the Germano identity [19], which states that the subgrid tensors corresponding to two different levels of filtering can be related by an exact relation as

$$L_{ij} = T_{ij} - \tilde{\tau}_{ij} \tag{4.11}$$

where $L_{ij} = \overline{\tilde{\bar{u}_i \bar{u}_j}} - \tilde{\bar{u}}_i \tilde{\bar{u}}_j$.

Here, τ_{ij} and T_{ij} are the subgrid tensors corresponding to the grid and test filter level. The tilde symbol designates the test filtering at a characteristics scale $\tilde{\Delta}$, with $\tilde{\Delta} > \bar{\Delta}$. The value of the test filter is often taken twice the value of grid filter. The significance of this identity is that L_{ij} being the resolved stress is known in terms of $\bar{u}$, whereas T_{ij} and τ_{ij} are unknown.

It may be assumed that these two subgrid stress tensors, τ and T, can be modeled by the same constant C_s for both the grid and test filtering level. This is expressed analytically as

$$\tau_{ij} - \frac{1}{3}\tau_{kk}\delta_{ij} = -2C_s\Delta^2 \bar{S}\,\bar{S}_{ij}, \tag{4.12}$$

$$T_{ij} - \frac{1}{3}T_{kk}\delta_{ij} = -2C_s\tilde{\Delta}^2 \tilde{S}\,\tilde{S}_{ij}. \tag{4.13}$$

Introducing the above two relations into Eq. (4.11), the following equation is obtained

$$L_{ij} - \frac{1}{3}L_{kk}\delta_{ij} \equiv L_{ij}^d = C_s M_{ij}, \tag{4.14}$$

where $M_{ij} = -2\Delta^2 \overline{\bar{S}\,\bar{S}_{ij}} - 2\tilde{\Delta}^2 \tilde{S}\,\tilde{S}_{ij}$.

In LES, both L_{ij} and M_{ij} are known in terms of $\bar{u}$ and thus, Eq. (4.14) of course may be used to estimate C_s. However, a residual E_{ij} is defined to minimize the error while estimating C_s. The expression is given by

$$E_{ij} = L_{ij} - \frac{1}{3}L_{kk}\delta_{ij} - C_s M_{ij}. \tag{4.15}$$

The above definition of error function would lead to six different values of C_s. To elevate this problem, Germano et al. [20] proposed a contracting relationship with the resolved strain tensor to conserve to a single relation, and thus to obtain a single value of C_s. This lead to a solution of the partial differential equation as

$$\frac{\partial E_{ij} \bar{S}_{ij}}{\partial C_s} = 0. \tag{4.16}$$

This method works well except the situation when the stress tensor $\bar{S}_{ij}$ cancels out. As a remedy to this problem, Lilly [36] proposed a least square method to calculate the constant C_s as

$$C_s = \frac{M_{ij} L_{ij}^d}{M_{kl} M_{kl}}. \tag{4.17}$$

This is the foundation of dynamic Smagorinsky model, but still it demands some attention. Now, the problem with the evaluation of may arise from the fact, although the denominator neither can be zero nor negative, that the numerator becomes frequently negative over a large volume fraction of real turbulent flow simulation. This gives a negative eddy viscosity, and very likely a negative total viscosity. This is not a flaw of the model as negative value of the constant indicates the backscatter from smaller to larger scales or backward energy cascade, which is physically possible. But, if negative total viscosity is present constantly for a period of time in one region of flow, the simulation will be unstable. As a remedy, the denominator and numerator of the above equation are averaged in the homogeneous direction, evaluating C_s as

$$C_s = \frac{\left\langle M_{ij} L_{ij}^d \right\rangle}{\langle M_{kl} M_{kl} \rangle}. \tag{4.18}$$

The symbol $\langle \ \rangle$ denotes the averaging in the homogeneous direction.

The dynamic model has generally been applied to many flows of engineering interest of varying degrees of complexity. In many flow situations, such as rotating channel flow, mixing layers, flow over bluff bodies, and in the turbulent boundary layer over a bump, the dynamic model predicts the distributions of eddy viscosity better than the traditional Smagorinsky model leading to better flow predictions [54, 87]. It has also been reported that the dynamic model is successful in resolving transitional flows, shear layer instability, vortex stretching, and breaks down, developing three-dimensional motions [60, 62].

7 Similarity Model

In the similarity model, the assumption of scale invariance is used in a literal sense. Here the velocity field at scales below Δ is postulated to be similar to that at scales above Δ. In other words, an explicit incorporation of Leonard stress is translated into the model. Thus, it is suggested that τ_{ij} must also be similar to a stress tensor constructed from the resolved velocity field as

$$\tau_{ij} = C \left(\overline{\tilde{u}_i \tilde{u}_j} - \overline{\tilde{u}}_i \overline{\tilde{u}}_j \right) \tag{4.19}$$

where C is the model constant and an overbar represents a second filter at some scale $\alpha \Delta$ with $\alpha \geq 1$. There are many different forms of the similarity model, differing in detail, i.e., in the value of C and α. Bardina et al. [4], who was the first to introduce the similarity model, used $\alpha = 1$, Liu et al. [40] proposed $\alpha = 2$, whereas Akhavan et al. [1] used $\alpha = 4/3$. The value of C is close to unity and it depends on the filter used.

Further, Bardina et al. [4] depicted high correlations between real and modeled stresses, typically as high as 80% with Gaussian or box filters, illustrating the successful proposition of the similarity model. Another desirable feature of the similarity model is that it resolves the backscatter of energy. However, the similarity model alone, while implemented in simulations, fails to dissipate enough energy such that it often tends to yield inaccurate results. To elevate this deficiency, Bardina et al. [4] proposed to add the Smagorinsky term, which is dissipative. The resulting model is known as the mixed model, which can be expressed as

$$\tau_{ij} = C\left(\overline{\overline{u}_i \overline{u}_j} - \tilde{\overline{u}}_i \tilde{\overline{u}}_j\right) - 2(C_s \tilde{\Delta})^2 \tilde{S} \, \tilde{S}_{ij} \qquad (4.20)$$

This mixed model combines the strengths of both the similarity and the Smagorinsky models. The dynamic procedure may also be used in evaluating the value of C_s while implementing the mixed model. For example, LES of Zang et al. [90] with a dynamic mixed model predicts well the recirculating flows. Similarly, Wu and Squires [87] have successfully applied the dynamic mixed model to predict the three-dimensional boundary layers.

8 Numerical Method

The philosophy of LES is to resolve the large-scale, energy dominating structures while modeling only the fine-scale eddies that are usually homogeneous and isotropic. In incompressible flow, the governing equation is often discretized following the finite volume method on a staggered mesh. This means that an implicit filtering approach is adopted, where the velocity components at a grid point are interpreted as the volume average. Any scales of motion, that are smaller than the mesh or control volume, are accounted for a subgrid-scale stress.

Most of the engineering solutions with RANS solvers employ second-order upwind schemes to predict the flow features. However, it is a topic of debate whether the upwind methods would be appropriate for LES. The desirable features of a numerical scheme suitable for LES are to be non-dissipative and conservative, not only for mass and momentum but also for kinetic energy. This is the reason why the second-order central differencing, being non-dissipative and conservative, is often used in LES for spatial discretization. These requirements rule out the use of low-order upwind methods, being highly dissipative and would likely contaminate the turbulent structures, although there exist numbers of higher-order upwind schemes, revealing quality LES data. The pressure-velocity coupling has been achieved either by SIMPLE algorithm [48], MAC [9], or projection method [24] and the viscous fluxes are approximated by the central differences.

The higher-order central schemes are always preferable, but their use in complex configurations is rather difficult. Further, it is difficult, at least for incompressible flows, to construct a high-order energy-conserving scheme [46]. This is one of the reasons why the second-order central difference scheme on a staggered mesh is still a popular choice because it conserves global kinetic energy on the uniform Cartesian meshes. This energy conservation property has not yet been proven for the non-uniform meshes. The second-order central differencing is likely the option with the increasing use of LES on body-fitted curvilinear grids for applications to flows of engineering interest. Numerical schemes with unstructured mesh are very appealing for LES of complex engineering flows following the pioneering work by Jansen [27], but it has not gained its momentum till now. Further, LES on unstructured mesh appears to be too memory and/or CPU intensive with relatively large numbers of grid points for simulations of turbulence.

With respect to time advancement, implicit schemes allowing larger time steps are often used. However, they are in general more expensive because at each time step nonlinear equations have to be solved. Furthermore, large time steps are unlikely to be used in LES in order to resolve certain time scales for accurate simulations of turbulence. Hence, explicit schemes seem to be more preferred for LES than implicit schemes. The Adams–Bashforth scheme is one of the popular choices among the majority, working with LES using explicit schemes. Further, it is not essential to use much higher-order schemes either since the time steps are usually small in LES.

For compressible flow, there are several studies, where the central second-order schemes in the cell-centered finite volume formulation have been used for LES because of their low intrinsic dissipation. A higher-order, non-dissipative (or low intrinsic dissipation) central scheme is a preferred one for LES. However, these central schemes have disadvantages of leading to odd/even pressure and velocity oscillations in regions of large cell Reynolds number. A possible approach to correct this behavior without destroying the global low-dissipation character of the method is to locally add an upwind correction, where such spurious oscillations are encountered. More details on these schemes can be found in the article by Mary and Sagaut [41], where good LES results were obtained for the flow past an airfoil in a near stall configuration. Another approach, which is also widely used for compressible LES is the upwind biased with a third-order MUSCL interpolation scheme of the AUSM+(P) family [37–39]. Time integration is generally carried out by means of a third-order compact Runge–Kutta scheme [47].

8.1 Boundary Conditions

The imposition of boundary conditions, particularly at inlet, is very important for a realistic simulation. Unlike the RANS, where the inflow boundary conditions can be simply specified according to experimental data, LES demands the specification of three components of instantaneous velocity at the inflow planes at each time step. Nevertheless, the specified instantaneous velocity at each time step should possess all the characteristics of stochastically varying turbulent structures with scales down to the Kolmogorov scale (i.e., compatible turbulence intensities, length scales, and spectrum). These are almost impossible to be obtained from any experimental data.

The inflow boundary conditions are usually specified in LES and can be classified into two categories. The simplest one is to specify the mean flow with synthetically generated random perturbations. Although it is the easiest method and computationally inexpensive, the random disturbances are nothing like the real turbulence as they have no correlation either in space or in time and have usually a flat spectrum similar to that of the white noise. Therefore, they decay rapidly and it takes a certain distance downstream from the inlet plane for a desired realistic turbulence to develop. Moreover in some cases, the use of random noise at the inlet does not develop turbulence at all. Some improved versions have been developed to generate the instantaneous velocity components at the inlet in such a way to correlate with certain properties of a turbulence signal, moments, or power spectrum. No method are available till date to generate inflow turbulence with all the desired characteristics such as turbulence intensity, shear stresses, length scales, power spectrum, and proper turbulent structures as mentioned earlier. Specification of an appropriate inflow boundary condition is thus a bottleneck of LES.

The final one is the so-called precursor simulation technique, where another simulation (precursor simulation) is basically performed and the data are stored to provide the input for the required simulation. Precursor methods can generate the most realistic turbulence information at the inflow boundary, but the penalty is that it is too expensive. Thus, the synthetic method is a compromise to generate inflow conditions having at least a few characteristics of turbulence with much cheaper efforts. Details on inlet boundary conditions for LES can be found in a review by Tabor and Baba-Ahmadi [77]. At the outlet, a convective boundary condition is usually employed.

9   Bottlenecks of LES

Even after intensive research and development in the field of LES over the last three decades, there still exist many challenges and issues regarding its successful implementations. Development of accurate SGS models, near-wall resolutions, inflow boundary conditions, LES for compressible flows, turbulent combustions, and aeroacoustics are a few important issues.

Although several SGS models have been developed, the simple Smagorinsky model and its variants are still being widely used. Considering the present status of SGS models, particularly the dynamic versions, resolution of fully turbulent flow can be simulated with good confidence. However in many situations, it is difficult to implement mesh to resolve more than 80% of the turbulent kinetic energy even with an advanced SGS model. An improvement of SGS models is definitely needed to accurately simulate flow features such as transition, relaminarization, and turbulent combustion. In most practical engineering flows, the Reynolds number being very large, it is very expensive to perform a wall-resolved LES satisfying the logarithmic law of the wall. Moreover, it has been reported that the near-wall grid requirement is further stringent with $\Delta x^+ \simeq 20$, $\Delta y^+ \simeq 1$, and $\Delta z^+ \simeq 10$ to resolve transitional flow, where the requirement is almost similar to the DNS [57–60]. Till date, limited studies have been made on compressible LES as compared to incompressible flows. There will be extra terms in subgrid stress and turbulent heat flux due to compressibility effect. On the other hand, while treating combustion, the turbulent mixing of air–fuel and different chemical reactions can occur over a broad spectrum of length and time scales, making accurate SGS modeling very challenging.

10   LES of Flows of Engineering Interest

A few examples of LES of different flows of engineering applications, performed by the author and his research group, are presented here. The first example illustrates the evolution of boundary layer on the suction surface of a low-pressure turbine (LPT) blade subjected to periodic passing wakes. The LES data elucidate the wake-induced transition, apart from the wake kinematics and its turbulence within the blade passage. Next is about the capability of LES in resolving the transition of a laminar separation bubble, which is commonly evidenced in low Re flows. The separated shear layer undergoes transition via Λ-vortex-induced breakdown and reattaches as a turbulent layer downstream. By means of these examples, the paper shows that the transitional and turbulent flows of engineering relevance can be well documented by LES as an alternative to DNS with a moderate computational cost.

10.1   *Flow Past an LPT Blade with Incoming Wakes*

Transition, in general, and its applications to turbomachinery flows in particular, have been the subject of much research. In a modern jet engine, the LPT supplies power to the fan and the first stages of the compressor. As the LPT is relatively heavy with a large number of stages, small improvements in its efficiency have a significant effect on the overall efficiency of an engine. Efforts toward the decrease of weight of LPT and thus the fuel consumption by reducing the number of blades come up against the problem of boundary layer separation on the suction surface of blades. Wakes shed from the upstream vane rows interact with this separated boundary layer on the downstream blades resulting in a highly unsteady flow. Under such an environment, the flow features depend on the receptivity to external disturbances and the internal growth mechanisms of the shear layer [23]. A combination of bypass transition and separation-induced transition may be observed depending on the wake frequency, turbulence level (*Tu*), and *Re*. Considerable progress has been made over the last three decades illustrating excitation of boundary layer subjected to periodic passing wakes [60, 63, 64, 74, 75, 85, 86].

The LES of Sarkar [58, 59] illustrated that small-scale fluctuations of advecting wake controlled the transition of highly diffusive boundary layer on the suction surface, apart from the wake kinematics within the blade passage. The cascade geometry considered was the high-lift LPT blade (T106), which has been extensively tested in different laboratories [74, 75]. The effects of wakes from upstream blade row have been traditionally studied through experiments by sweeping a row of wake-generating bars upstream of a cascade. The moving row of wake-generating bars have been replaced by a precursor LES of the flow past a thin cylinder, diameter (D) being 2% of the blade chord. The wake data with $Tu = 10\%$ were interpolated at the inlet plane of cascade considering the kinematics of flow. For the cascade, Re based on the chord and the isentropic exit velocity was 1.6×10^5. The cascade pitch was kept the same as the bar pitch, yielding a reduced blade passing frequency, $f_r = fC/V_{2is}$ of 0.68, f being the frequency.

The LES for the cascade was performed with 7.0 million grid points having the near-wall mesh resolution as $5 < \Delta x^+ < 80$, $5 < \Delta z^+ < 20$, and $1 < \Delta y^+ < 3.5$. On the contrary, DNS of Wu and Durbin [85] used 57 million points to elucidate wake-induced transition on the LPT blade. The results of LES [58, 59] have been validated against the experiment [74, 75]. It was reported that LES data could be qualitatively used to describe the evolution of the boundary layer perturbed by passing wakes on the suction surface. Further without wake passing, the boundary layer on the suction surface of the blade remains laminar till 60% of surface length before separation. It then undergoes transition, becomes turbulent, and reattaches near the trailing edge. The apparent suppression of boundary layer in the time-averaged sense is attributed to wake passing, temporal data support, and the existence of separation bubble over the rear half of the blade. LES of the author [57–60] illustrates the transition mechanism resulting from the interactions of passing wakes and the separation bubble on the suction surface of LPT. The computation required 0.3 micros/iteration/grid based on 32-processors of an IBM iDataPlex dx360 platform.

The filtered, time-dependent, incompressible mass and momentum equations in fully covariant form were solved on a staggered grid using the finite volume method. The explicit second-order accurate Adams–Bashforth scheme was used for the momentum advancement. The pressure equation was discrete Fourier transformed in one dimension (in which periodicity of the flow and so the uniformity of the geometry were imposed) and solved iteratively using multi-grid acceleration in the other two dimensions. The spatial discretization was second-order accurate using a symmetry-preserving central difference scheme, which was widely used in LES owing to its non-dissipative and conservative property. The Smagorinsky model with a reduced model constant of 0.125 was used for the computation economy. The low-Reynolds number model of Voke [82], which was derived following the dissipation spectrum, has been also used to simulate the transitional flow.

The distortion of a wake segment (characterized by bowing, reorientation, elongation, and stretching) as it convects within the LPT passage is depicted in Fig. 4. After being segmented at the leading edge, wakes are quickly broken down into random small-scale structures forming a bow in the blade passage. The pressure-side wake suffers from severe stretching and thinning with the decay of turbulence. On the contrary, the enhancement of turbulence occurs with growing random small-scale structures in the apex region. Thus, this highly strained wake fluid and small-scale eddies interact periodically with the separated boundary layer on the suction surface,

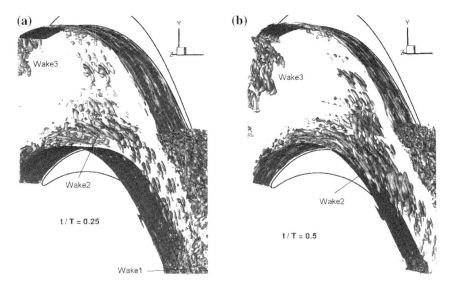

Fig. 4 Instantaneous isosurface of vorticity |ω| at two time intervals during the wake passing cycle (*T* being wake passing time)

attributing to the unsteady boundary layer. The growth mechanisms of the shear layer due to receptivity to the external disturbances along the suction side of the LPT will be described now.

A migrating wake can be modeled as a negative jet, which impinges on the blade and splits into two streams; one pointing downstream that accelerates the flow downstream of the approaching wake and the other pointing upstream that retards the flow after the wake has passed. Figure 5 depicts the snapshots of isocontours of spanwise vorticity along the mid-span over the rear half of the suction surface. Prior to t/T at about 0.5, when the wake center is ahead of the steady-flow separation, the shear layer depicts the development of a cat's eye pattern, typical to the Kelvin–Helmholtz (K-H) instability (Fig. 5a). The location of the separation that shifts with time is marked by an arrow. The shear layer thus becomes unstable by inviscid mechanism, attributing to the enhanced receptivity of perturbations and finally breaks down to turbulence near the trailing edge, when the wake is ahead of the separation bubble. As the wake approaches the region of steady-flow separation ($t/T > 0.5$), the negative jet effect of the wake initiates rollups of the shear layer evolving large-scale vortices that convect downstream at a speed of about 37% of the local free stream (Fig. 5b). Finally, a train of apparently three prominent coherent vortices appears as the wake moves past the separated region and these vortices retain their presence far downstream. The large amplitude of pressure fluctuations is observed on the rear half of the suction surface because of these convective vortices [60, 63]. Once the influence of the passing wake subsides, the boundary layer again begins to separate.

To visualize the three-dimensional flow structures, isosurface of instantaneous vorticity is presented in Fig. 6. When the wake is ahead of separation, longitudinal

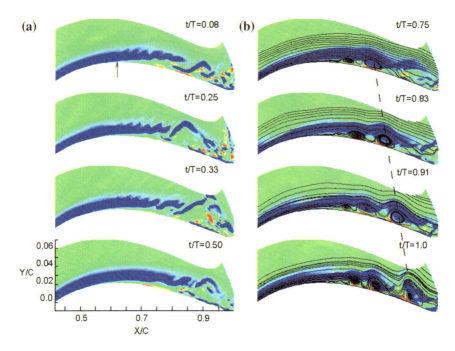

Fig. 5 Snapshots of **a** instantaneous spanwise vorticity: wake is ahead of separation, **b** spanwise vorticity superimposed with streamlines: wake is crossing the separation over the rear half of the suction surface

streaks appear, breaking down into small and irregular structures downstream of $S/So = 0.9$ (Fig. 6a). The very similar flow structures were observed by Alam and Sandham [2] while studying the transition of a laminar separation bubble through DNS. Thus, perturbations are amplified because of the separated boundary layer and then the nonlinear interactions via the vortex stretching process evolve these streaks, which are features of transition. Figure 6b elucidates the formation of 3D vortex loops via excitation of the separated layer when the wake is crossing over the rear half of the suction surface. Once formed, they convect downstream and grow in strength before breaking down to small structures. These 3D vortex loops retain their identity far downstream and have an influence on transition. Further, the phase-averaged contours of turbulent kinetic energy (TKE) and production (not presented because of space) on the rear half of the suction surface elucidated that the transition during the wake-induced path is governed by a mechanism that involves the formation and convection of these vortices followed by the production of turbulent kinetic energy.

The unsteady boundary layer on a turbomachinery blade develops along two separate paths: one is the wake-induced path that lies approximately under the wake trajectory and the other is a transitional path between wakes. In the present case, LES illustrates multi-modal transition, i.e., different modes of traction may coexist in time and space. Further, it is noteworthy that the wake-induced path here differs from the traditional wake-induced transition because of the formation of the coherent

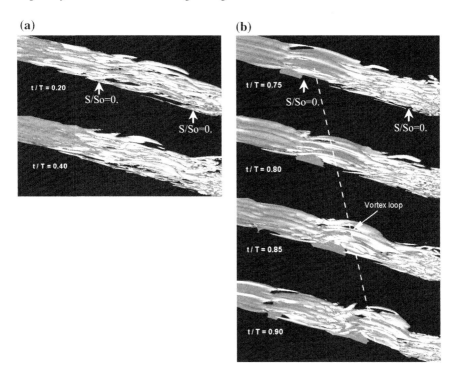

Fig. 6 Instantaneous isosurface of spanwise component of vorticity: **a** illustrating transition of separated boundary layer when the wake is ahead of separation, **b** illustrating the formation of 3D vortex loop via excitation of separated boundary layer as the wake crosses the rear half of the suction surface

vortices associated with the separation bubble as explained earlier. Figure 7 depicts the computed phase-averaged shape factor $\langle H \rangle$ as a function of space and time in the S-T diagram. It may qualitatively describe the states of the unsteady boundary layer on the suction surface during the wake passing cycle. To aid visualization, the data are reproduced to form three wake passing cycles. Two trajectory lines A and B denote the local maximum and minimum velocity owing to the kinematics of the passing wakes, i.e., the negative jet effect. Different states of the boundary layer are marked in Fig. 7, where the wake-induced path and the path between wakes are represented by arrows.

The wake path lies in between the trajectories A and B, where the level of shape factor is increased attributing to the appearance of coherent vortices (marked by C and D) that thicken the shear layer. A wake-induced transition influenced by coherent vortices occurs along the trajectory B. Drop in the shape factor along the trajectory B that lags the wake passing somewhat is observed. The turbulence created by the breakdown of coherent vortices also merges with the strip. This results in a wide turbulent strip downstream of $S/So = 0.85$ as depicted in Fig. 7. The line E, drawn at 37% of the free-stream speed, becomes parallel to the trailing edge of the turbulent

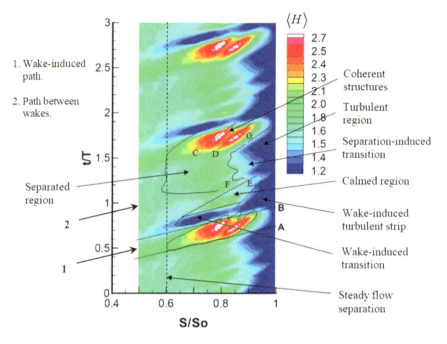

Fig. 7 Space–Time (S-T) diagram depicting phase-averaged shape factor

strip. Thus, the end of the strip is controlled by the convection speed of the coherent vortices.

After the wake has passed, the stimulus for early transition is removed and the turbulent boundary layer relaxes to its pre-transitional state. This relaxation process results in a calmed region, which is reflected by a slow increase of shape factor. Further, the boundary layer on the rear half of the suction surface becomes inflexional and begins to separate under the action of the adverse pressure gradient (the separated region is approximately marked). Between the wakes, the separation-induced transition becomes prominent and the flow remains turbulent downstream of $S/So =$ 0.9 (within the trajectories E and G). Thus, the transition between wakes occurs via the excitation of the shear layer over the separated region. When compared with the experiment [75], some discrepancies in the state of the boundary layer are observed particularly between the wakes. The calmed region following the wake-induced transitional strip has not been well predicted. However, the time-dependent nature and transitional flow owing to the separated boundary layer are qualitatively resolved. The discrepancies with the experiments may occur for several reasons. The difference in wake turbulence intensity and length scale might have a remarkable effect on transition. This may also explain the limitations of LES and uncertainties in subgrid-scale models. Thus, this example illustrates that LES can be successfully used to describe the complex flow field and turbulence statistics within a real gas turbine blade. It further explains phenomena such as wake-induced transition on an LPT blade.

In industrial scenario, URANS has been the primary choice for simulations of unsteady flow in turbomachinery till date. Michelassi et al. [45] investigated the unsteady flow within the same LPT blade (T106) subjected to periodic passing wakes and computed the capabilities of URANS against LES, DNS [84], and experiment [74]. Several URANS computations were used with or without a transition model [44] and a realizability constraint [12] on the turbulence time scale to prevent excessive production in the stagnation region. The details of grids, models, and the reported computational costs are provided in Table 1. The dynamics of wakes within the passage illustrates a fair qualitative agreement between the phase-averaged TKE predicted by turbulence model with the DNS and LES, Fig. 8. Further, the figure clearly shows how turbulent diffusion increases from DNS to URANS. The LES plot reports only the resolved stresses, where the peak level is slightly smaller than the DNS. The blade–wake interaction can be quantified by the kinetic energy loss coefficient defined as $\zeta = (P_{0,\,\text{inlet}} - P_{0,\,\text{exit}})/(P_0,\,\text{inlet} - P_{\text{inlet}})$, P_0 and P being the total and static pressure, respectively. Figure 9 shows the distribution of ζ at 40% downstream of axial chord along the pitch. It can be inferred that LES and DNS

Table 1 The details of grids, models, and computational [45]

Simulation	Grid	Model	Transition model	Realizability	CPU time in hours (processors)
DNS	$1014 \times 260 \times 64$		NA	NA	500 (64)
LES	$448 \times 144 \times 32$	Dynamic	NA	NA	190 (32)
URANS-01	384×144	$k - \omega$	No	Yes	12 (1)
URANS-02			No	No	
URANS-03			Yes	Yes	
URANS-04			Yes	No	
URANS-05			Yes	Except separation	

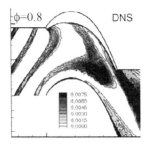 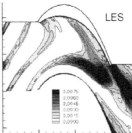 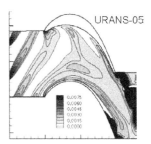

Fig. 8 Turbulent kinetic energy at a phase $\phi = 0.8$, DNS, LES, and URANS

Fig. 9 Variation of kinetic
energy loss at 40%
downstream of the axial
chord by Michelassi et al.
[45]

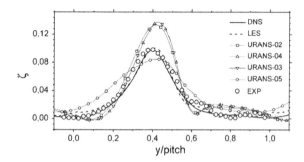

are far superior to URANS for unsteady flows. It is worthwhile to note that LES is
remarkably close to DNS with much lesser computational cost, Table 1.

In the context of heat transfer in HPT vanes, where the correct prediction of
blade temperature is critically important, RANS models fail to provide reasonable
accuracy. A comparative study between RANS and LES to estimate the surface heat
transfer distribution on an HPT guide vane was presented by Fransen et al. [14]. It can
be noted in Fig. 10 that RANS utterly fails in predicting the heat transfer distribution
for transitional flows with high *fst* as compared to LES. In reality, RANS initiates an
early onset of transition with almost instant augmentation of heat transfer distribution
depicting the turbulent region. Thus, RANS/URANS is not suitable for complex
flow phenomena such as separation, stall, and transition prevalent in turbomachinery
flows, although being numerically inexpensive. In such situations, DNS and LES are
far better choices to resolve all the unsteady flow features. Further, LES and DNS

Fig. 10 Heat transfer
distribution over an HPT
blade surface by Fransen
et al. [14]

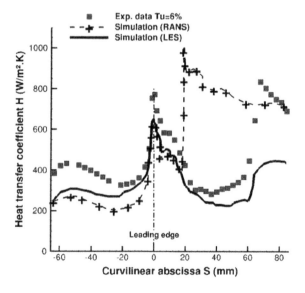

results can be utilized to tune the RANS models for better prediction of mean flow quantities.

10.2 LES of a Separation Bubble

The second example manifests the transition mechanism of a laminar separation bubble (LSB) through LES. The LSB is commonly found in low Reynolds number flow, which is encountered in many engineering applications such as flow over LPT blades, or in airfoils. In LPT blades, high levels of diffusion accompanied by the relatively low Reynolds number flows lead the laminar boundary layer to separation on the suction surface. A rapid transition then occurs within the shear layer owing to enhanced receptivity of free-stream disturbances. Nonlinear breakdown of spatially growing traveling waves in the shear layer is believed to be the cause of transition. The resultant turbulent-like layer re-energizes the flow and reattaches to form an LSB. Another very important situation, where LSB appears, is flow over the leading edge of an airfoil at a low Reynolds number. The separation bubble dictates the evolution of the downstream boundary layer adversely affecting the stall characteristics of an aerofoil. The existence of an LSB was first recognized by Jones [28]. A remarkable realization in the bubble structures was attained with the work of Gaster [18] for a wide range of Re and varying pressure gradient. The flow structure of a time-averaged LSB is reproduced in Fig. 11, following Horton [26].

A number of studies [8, 22, 53, 57, 71, 83] have been conducted illustrating the transition of a separated boundary layer that shares the characteristics of an attached boundary layer and a free shear layers. Most of the studies elucidate that the transition of a separated boundary layer occurs in a similar fashion of free shear layer with a high receptivity to perturbations. Amplification of certain selective frequencies

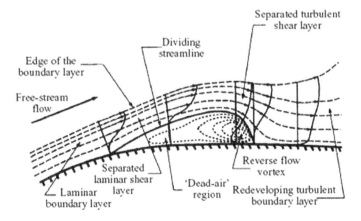

Fig. 11 Classical structure of a short laminar separation bubble

of background disturbances is observed leading to the formation of K-H rolls and shedding of large-scale vortices in the vicinity of reattachment. There are evidences that the transition of a laminar separated boundary layer occurred via oblique modes and vortex-induced breakdown resulting in turbulent reattachment [2]. A secondary instability was seen in the second half of the bubble with considerable growth of three-dimensional motions and the evolution of hairpins attributing to breakdown [89]. Recent studies illustrate that the shear layer was unstable through K-H instability for low fst, whereas the primary K-H instability was bypassed for high fst [3].

In spite of several studies made, there are uncertainties in understanding the transition of an LSB under different flow conditions. Although the majority of studies support that the separated boundary layer is inviscidly unstable, there are evidences of viscous effect in initiating transition [2, 61, 83]. In reality, the transition mechanism of an LSB includes the following features, i.e., receptivity, linear growth of disturbance, and then nonlinear effects followed by the breakdown. The principal objective of this illustration is to explore the ability of LES in resolving the physics of transition of an LSB, its flow structures, and turbulent statistics near and after the reattachment. Results of LES have been compared with DNS data available in the literature.

An LSB on a flat plate was created by a suction profile following the Gaussian distribution of the wall-normal velocity component at the upper boundary that produced an adverse pressure gradient. The 3D unsteady N-S equations for incompressible flow in the Cartesian coordinate were normalized with respect to inlet displacement thickness (δ_{in}^*), and the inlet free-stream velocity (U_∞). The model proposed by Germano et al. [20] and Lilly [36] was used here to include the effect of subgrid motions, where the model coefficient is dynamically calculated instead of input a priori. A streamwise length of 200, flow-normal length of 10, and spanwise length of 30 were taken with a mesh of $200 \times 64 \times 64$ after a grid resolution study. The near-wall resolution at $x = 170$ (where an almost conical layer appeared) was $\Delta x^+ = 20$, $\Delta y^+ = 1.0$, and $\Delta z^+ = 10$. At the inlet, a Blasius velocity profile was specified for the streamwise velocity component corresponding to $\mathrm{Re}_{\delta_{in}^*} = 500$, the wall-normal and the spanwise velocity components were set to zero. At the outlet, a convective boundary condition was imposed. On the lower boundary, a no-slip condition was applied, whereas a Dirichlet boundary condition was used to the streamwise velocity component ($u = 1.0$) and the other two components were set to zero at the upper boundary. In the simulation, a disturbance strip was imposed at the wall and upstream of separation to trigger the transition. The disturbance was specified to the normal velocity following the work of Alam and Sandham [2].

Figure 12 shows the streamline superimposed with the streamwise velocity (u) and u_{rms} profile. This illustrates the time-averaged shape of the bubble and the growth of three-dimensional motion over the LSB. The growth of u_{rms} occurs in the second half of the bubble, which becomes maximum near the reattachment. After reattachment, it takes several bubble lengths downstream for the near-wall turbulent characteristics to develop. The dead-air region and the reverse flow vortex are also indicated. The pictorial views of the bubbles from the LES and that of DNS are almost the same. A few important parameters, which are used to describe the features of an LSB, are

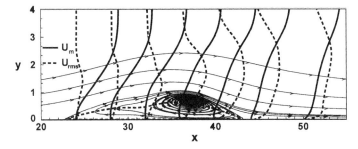

Fig. 12 Time-averaged streamlines superimposed with streamwise velocity and $U_{\mathrm{r.m.s}}$ profile from LES

momentum thickness at separation (θ_S), length of transition (lt), that is the distance from separation point to the point of minimum skin friction (C_f), and the length of separation bubble ($1_b/\delta_{\mathrm{in}}^*$). Apart from these, Reynolds numbers based θ_S and lt are other two vital parameters that characterize the flow. The values of these parameters are specified in Table 2 and compared with the corresponding values obtained from the DNS [2]. The θ_S and Re_{θ_s} predicted by LES compare well with those of DNS, but the bubble length ($1_b/\delta_{\mathrm{in}}^*$) and Re_{lt} are overpredicted.

The variation of mean skin friction coefficient, which is normalized by local free-stream velocity, is presented in Fig. 13. It is noted here that the flow is locally forced to leave the upper boundary by imposing suction and hence the potential flow is

Table 2 Data related to mean bubble shape

Case	Re_{θ_s}	Separation point (x/δ_{in}^*)	Reattachment point (x/δ_{in}^*)	$1_b/\delta_{\mathrm{in}}^*$	$\theta_s/\delta_{\mathrm{in}}^*$	Re_{lt}
DNS	246	24.2	40.6	16.4	0.49	6667
LES	210	22	43	21.3	0.42	8500

Fig. 13 Variation skin friction from the LES and that of DNS

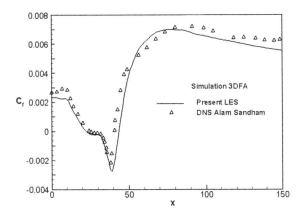

distorted in the vicinity of the bubble. This does not allow having a unique value of free-stream velocity. As a remedy to this problem, Spalart and Strelets [71] defined a pseudo-velocity by integrating the spanwise vorticity, which is used here as the local free-stream velocity. When compared with the corresponding DNS results [2], the LES shows a good agreement. The separation and reattachment points, and hence the bubble length are evaluated from C_f. The plateau in C_f distribution after the separation corresponds to the dead-air region of the bubble, whereas the reverse flow vortex region is associated with larger negative values of C_f.

The instantaneous flow field is very revealing and can be used to explain the transition of a laminar separation bubble, development of three-dimensional motions and breakdown near reattachment. Figure 14a shows contours of streamwise velocity in the x-y plane (side view) for $z = 30.0$. The darkest grayscale represents the separation region. It illustrates the thickening of the shear layer over the bubble, its roll-up forming large-scale vortices, which is typical of the K-H instability. These large-scale vortices retain their structures far downstream leading to predominant outer layer activities and slow relaxation to a canonical layer. Figure 14b shows the top view (x-z plane) of streamwise velocity contours at a wall-normal location ($y =$

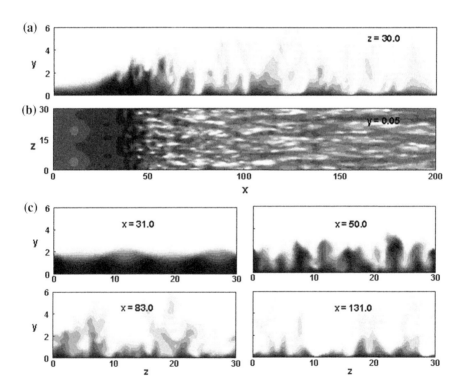

Fig. 14 Instantaneous contours of streamwise velocity due to the LSB with turbulent reattachment. Maximum level is 0.98, minimum level is -0.13, and the darkest color shows reversed flow. **a** (x, y)-plane at $z = 30.0$, **b** (x, z)-plane at $y = 0.05$, and **c** different (y, z)-planes

0.1). The top view illustrates that the initial flow field is two-dimensional and the boundary layer separates as laminar. The imposed perturbations seem to be growing with the appearance of a sinuous spanwise undulation and the separated layer remains laminar up to $x = 30$. Three-dimensionality appears downstream of $x = 35$ and breakdown occurs near $x = 43$ with the appearance of small-scale eddies. The development of the low-speed longitudinal streaks, which are the characteristics of near-wall turbulent flow, appears downstream of reattachment and visible far downstream. Figure 14c shows the side views (y-z plane) of streamwise velocity contours for different streamwise sections at the same time. The contours at $x = 31.0$ exhibit the two-dimensionality, although the spanwise symmetry of $0.38\ L_z$ is slightly distorted due to the transitional shear layer. The two-dimensionality is further disturbed downstream with the appearance of large-scale spanwise vortices, which constantly eject fluid from the inner layer and promote mixing. The initial symmetry is completely destroyed downstream of $x = 83$ with the development of the near-wall fine-scale structures.

Isosurface of the spanwise component of instantaneous vorticity is presented in Fig. 15, which is very helpful to visualize the three-dimensional flow structure. The separated shear layer, which is two-dimensional initially, is distorted by nonlinear interactions and Λ-vortices appear in the transition region due to the vortex stretching mechanism. Breakdown to small-scale and random structures with complete loss of orientation occurs just downstream of reattachment. The presence of longitudinal streaks is also evident after reattachment, which is the characteristic of the turbulent layer. The LES results are very consistent with the DNS of Alam and Sandham [2] in resolving the flow structure. However, the appearance of Λ-vortices was not observed in DNS of an LSB by Spalart and Strelets [71]. Further, the growth of disturbances

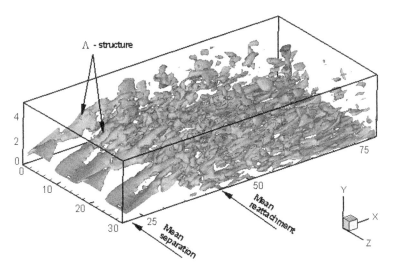

Fig. 15 Isosurface of spanwise vorticity depicting Λ-structures and breakdown

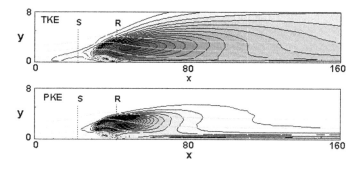

Fig. 16 Contours of turbulent kinetic energy (TKE) and production over LSB

and the development of three-dimensional motions can be illustrated by the evolution of velocity fluctuations (r.m.s. values). Almost no growth of fluctuations is observed in the first 25% of bubble length. The fluctuations increase very rapidly in the second half of the bubble, reaching a maximum value near the mean reattachment point and then tend to drop slowly illustrating the end of transition. Thus, augmentation of three-dimensional motions leading to breakdown to turbulence occurs in the second half of the bubble (last 75% of the bubble length). The shear layer is highly anisotropic with very high values of streamwise velocity fluctuations.

The contours of TKE and production are depicted in Fig. 16, where both TKE and production increase rapidly over the rear half of the bubble. The peaks shift away from the wall illustrating that turbulence is dominant in the outer part of the shear layer ($35 < y^+ < 100$). Production attains its maximum just after the reattachment ($x = 46$) and the peak value progressively decreases downstream. It is apt to mention that production occurs in the region of high turbulence stress and high spatial velocity gradient, aligned in the same direction. Thus, high production near the reattachment is attributed to the concentration of relatively large-scale vortices and turbulence because of breakdown. Further, the boundary layer downstream of reattachment appears very different from an equilibrium turbulent boundary layer illustrating a slow relaxation. Far downstream, a new peak in production emerges near the wall ($y^+ \simeq 7$) because of the wall effect. TKE profiles also reflect a similar trend. This figure also indicates that it takes several bubble lengths to relax the boundary layer to the canonical layer. This test case describes the capability of LES in successfully resolving the transition mechanism of a laminar separation bubble, which is commonly evidenced in low Reynolds number flows.

11 Hybrid LES-RANS

The construction of LES is such that it provides unsteady data, resolving most of the turbulent kinetic energy K of the flow while modeling most of the dissipation ε. As already stated, this separation is possible due to the fact that K is determined by the large scale of motion and ε by the small scales. The concept, although clear and simple, works well for high Reynolds number flows remote from boundaries even with the simple models. Similarly, flows past a bluff body, being dominated by large-scales motion, can be well resolved by LES without employing a fine mesh and an accurate result can thus be achieved at an affordable cost [31, 54]. But, it is a challenging task to make accurate predictions of wall-bounded flows with LES at high Re and LES becomes, unfortunately, more costly than RANS computations even by a factor of 100. Indeed, the requirement of near-wall grid resolution is the main reason why LES is too expensive for engineering flows, which was one of the lessons learned in the LESFOIL project [11, 42]. On the contrary, the mean quantities of the flow variables can be predicted with RANS models for a wide range of flows at a moderate cost, while averaging all the fluctuating quantities. Thus, the RANS model cannot resolve turbulence but only can mimic its effect by providing good surface characteristics.

The objective of hybrid LES-RANS [10, 29, 78, 80, 88] is to eliminate the stringent requirement of near-wall resolution for the wall-dominated turbulence such as the flow past an aerofoil. A low Re number turbulence model with RANS is used for the near-wall region, whereas LES is employed in the outer region resolving large-scale structures. Thus, the effects of the near-wall turbulent structures are mimicked by the RANS turbulence model, rather than being resolved. Although the concept of mingling LES with RANS is not theoretically valid, it appears appealing considering the fact that the hybrid LES-RANS approach yields good results with an affordable resolution near the wall, provided the treatment of interface between URANS and LES region has been successfully incorporated. Another point is to note that the hybrid LES-RANS method leads to a substantial loss of fidelity vis-à-vis considerably lower computation cost.

11.1 Principal Approaches to Couple LES with RANS

The structural similarity between the filtered equations of LES and time-averaged equations of RANS suggests the concept of unified modeling. The obvious similarity is further enhanced by the usage of the eddy viscosity concept for most SGS terms and the fact that the employed models are often derived from RANS counterparts. As a consequence, not only the governing equations but also many of the turbulence models exhibit a structural similarity. Thus, transport equations coupling LES and RANS approach can be expressed as

$$\partial_t \bar{u}_i + \partial_j (\bar{u}_i \bar{u}_j) = -\partial_i \bar{p}/\rho + 2\partial_j \left(\nu \bar{S}_{ij}\right) - \partial_j \tau_{ij}^{\text{model}}. \tag{4.21}$$

If an eddy viscosity concept is employed then,

$$\partial_t \bar{u}_i + \partial_j (\bar{u}_i \bar{u}_j) = -\partial_i \bar{p}/\rho + 2\partial_j \left[(\nu + \nu_t) \bar{S}_{ij}\right]. \tag{4.22}$$

A transition from LES to RANS can be achieved in several ways. One way of blending is a weighted sum of RANS and LES model as

$$\tau_{ij}^{\text{model}} = f^{\text{RANS}} \tau_{ij}^{\text{RANS}} + f^{\text{LES}} \tau_{ij}^{\text{LES}}. \qquad (4.23)$$

In this equation, f^{RANS} and f^{LES} are local blending coefficients determined by the local value of a given criterion.

Another strategy is to use a pure LES model in one part of the domain and a pure RANS model in the remaining part, where a boundary between the RANS and the LES zone is specified. Here, the governing equation for the velocity, however, is the same in both zones with no particular adjustment other than switching the model term at the interface. Thus, the resolved velocity is continuous. Furthermore, if the interface remains constant in time, it is called a hard interface. If it changes in time depending on the computed solution, it is termed as a soft interface. A detailed discussion has been made in the literature [17, 79].

11.2 Assessment of Hybrid LES-RANS

Noticeable progress has been made in hybrid LES-RANS models to explore at high *Re* and even for separated flows. There exist two kinds of LES-RANS hybrid methodology: one is a wall-modeled large eddy simulation (WMLES) and the other being detached-eddy simulation (DES). In WMLES, RANS is applied only to a thin layer near the wall, where the layer thickness may be much smaller than the boundary layer thickness [13, 67]. This method captures most of the turbulence inside the boundary layer by LES without resolving the smallest scales above the viscous sublayer. In DES, a major part of the attached boundary layer is treated by RANS, while LES is only applied in the separated flow regions [66, 76]. In delayed DES (DDES), the shifting from RANS to LES within the boundary layer is prevented by using a generic formulation which depends on the eddy-viscosity, the local velocity gradient, and the wall distance [21, 70]. Here, the problem of grid-induced separation, which is often observed due to shifting from RANS to LES in traditional DES, is elevated by DDES. If the model coefficients are dynamically evaluated, then it is termed as dynamic DDES. As an example, He et al. [25] developed a dynamic DDES based on the k-ω SST model of Menter et al. [43] and the well-established dynamic k-equation subgrid-scale model, where two model coefficients were computed dynamically at the grid and test filter levels.

A schematic indicates the effect of dynamic DDES ensuring that at least a large portion of the boundary layer is simulated using the RANS model, while it is smoothly turned to LES at the edge of the boundary layer, Fig. 17. A significant decrease of the shielding function was also noticed with dynamic DDES resulting in the twisted hybrid interface. Thus, the undesired growth of eddy viscosity in the far-wall region is alleviated with the resolved eddies, as sketched in the figure. This would

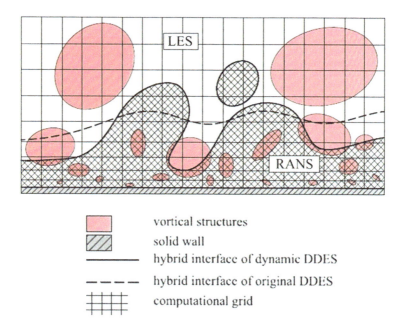

vortical structures
solid wall
——————— hybrid interface of dynamic DDES
– – – – hybrid interface of original DDES
computational grid

Fig. 17 Schematic illustrating the effect of dynamic DDES following He et al. [25]

definitely improve the quality of simulation with a resolution of relatively more refined finer scales. The simulation of developed channel flow with the dynamic DDES model showed that the problem of mismatch in the log-layer region was significantly improved [25, 67]. In a review by Fröhlich and Terzi [17], a large number of hybrid methods depending on the various interfaces in time and space, such as hard, soft, and on a unified formulation of blending of LES/RANS, were discussed. It has been shown with numbers of examples that if judiciously applied hybrid LES-RANS formulations have merits to reproduce solutions of the same order of LES with relatively limited grids. However, there exist disadvantages and one such issue is the numerical discretization scheme. In general, numerical methods, which are optimized for the RANS, are poor choices for DNS/LES and vice versa. Low dissipation schemes are preferred for LES/DNS, whereas the high numerical stability resulting in an upwind scheme is choice for RANS and trade-offs are considerably difficult to make.

A typical example of separated flow past a periodic hill was illustrated by He et al. [25]. The streamwise and spanwise directions were $9H$ and $4.5H$, respectively, where H was the hill height. The Re based on the hill height and the bulk velocity was 10,595. The mesh used was $218 \times 80 \times 90$ nodes in the streamwise, spanwise, and wall-normal directions. The original DDES model showed an excellent agreement of mean streamwise velocity and its *rms* fluctuations with the experiment [5]. The skin friction distribution also compared well with the LES results [16], except near the strong recirculation region. The overall agreement improved with the dynamic

Fig. 18 Contours of shielding function in-flow over periodic hills for different models following He et al. [25]

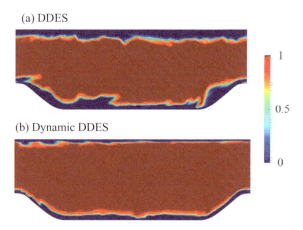

DDES. The contours of the shielding function calculated by the original DDES model and that of the dynamic DDES model are presented in Fig. 18. The original DDES model is a quite conservative and comparatively large portion of the RANS region which occurs in the flow domain, killing a large portion of the turbulence or small-scale motions near the wall. However, the dynamic DDES model is definitely better off in constraining the damping mechanism by a much thinner RANS region near the wall. The benefit of dynamic DDES is appreciated by contours Q-criterion illustrating that it can resolve much smaller vortices in the flow field with the same level of mesh as compared to the original DDES model, Fig. 19. Further, small-scale structures near the upper wall are also successfully captured by the dynamic DDES.

Fig. 19 Isosurface of vorticity in flow over periodic hills computed by different models following He et al. [25]

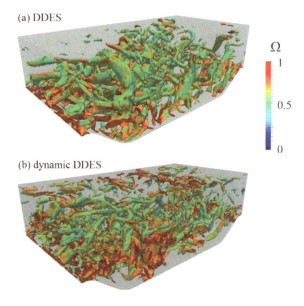

Another example of a hybrid LES-RANS method by He et al. [25] was the simulation of the impinging jet with heat transfer, where the RANS was used for the flow in pipe and in a thin layer close to the wall near the jet impingement. RANS was switched to the LES immediately at the exit of the pipe. It was reported that the original DDES model substantially overestimates the time-averaged Nusselt number in the stagnation region, while the agreement with the experiment is relatively better with the results from the dynamic LES and the dynamic DDES. The flow structures are depicted in Fig. 20 by isosurface of Q-criterion. The ring-like vortical structures near the wall are attenuated due to the high dissipation of the original DDES model. These flow structures, originally generated in the shear layer of the jet, do not break down after impingement on the wall and propagate downstream. In contrast, abundant small-scale eddies are resolved by the dynamic LES or the dynamic DDES model. A relatively large number of ring-like vortical structures appear in the shear

Fig. 20 Isosurface of Q-criterion depicting flow structures of an impinging jet simulated by different models

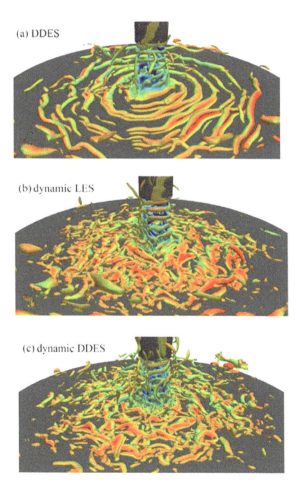

layer of the jet and then break down to smaller eddies immediately after the impingement. This elucidates that the dynamic DDES model can estimate the steady internal
pipe flow by RANS and then switches rapidly to the dynamic LES at the exit of the
pipe by attenuating the modeled turbulent eddy viscosity. Near the jet impingement,
the WMLES is activated in the vicinity of the wall. This example illustrates that
the dynamic DDES model is comparable with the dynamic LES while resolving the
complex flow field of engineering interests.

I must mention here that the present status of CFD, particularly toward numerics, turbulence modeling, and thus LES, would have been incomplete without the
contributions of Prof. Spalding. We the present generation, practicing CFD either in
industries or academics, must remain indebted to him. Thus, a review on LES in a
commemorative issue as a tribute to Prof. D. Brian Spalding is very pertinent.

12 Summary

The LES formulation in physical space and its application to the Navier–Stokes
equations have been discussed apart from features of turbulence in brief. Several
SGS models have been reviewed including major challenges as near-wall modeling,
inflow boundary conditions, and numerical methods. Two examples of LES of flows
of engineering interest, such as excitation of the boundary layer on an LPT blade by
periodic passing wakes and transition of a laminar separation bubble, performed by
the author and his research group are presented. These test cases involve boundary
layer transitions, shear layer instability, separation followed by breakdown, and reattachment of flow. LES with proper resolution of flow is definitely superior to RANS
for situations, where unsteadiness and large-scale structures dominate the flow. LES
is worth pursuing in making certain advances in the fundamental understanding of
complex turbulent flows, where the eddy motions and their interactions are the key
to comprehend the flow. The LES calculations are certainly affordable than DNS, but
they are fairly expensive for wall-dominated turbulence and particularly in resolving
transitional flow. Although good progress has been made in the hybrid LES-RANS
method, illustrating its contribution to modeling the turbulent flows of engineering
applications, it is still an active field of research and yet to be matured. Continued
research in this direction would definitely enhance its range of applicability. However,
the hybrid LES-RANS methods would lead to a considerable loss of fidelity.

Acknowledgements I would like to thank Ms. Sonalika Srivastava and Mr. Souvik Naskar for their
help in editing the manuscript.

References

1. Akhavan, R., Ansari, A., Kang, S., & Mangiavacchi, N. (2000). Subgrid-scale interactions in a numerically simulated planar turbulent jet and implications for modeling. *Journal of Fluid Mechanics, 408,* 83–120.
2. Alam, M., & Sandham, N. D. (2000). Direct numerical simulation of short laminar separation bubbles with turbulent reattachment. *Journal of Fluid Mechanics, 410,* 1–28.
3. Anand, K., & Subrata, S. (2016). Features of a laminar separated boundary layer near the leading-edge of a model aerofoil for different angles of attack: An experimental study. *ASME Journal of Fluids Engineering, 139,* 021201–021214.
4. Bardina, J., Ferziger, J. H., & Reynolds, W. C. (1980). Improved subgrid scale models for large eddy simulation. *American Institute of Aeronautics and Astronautics,* 80–1357.
5. Breuer, M., Peller, N., Rapp, C., & Manhart, M. (2009). Flow over periodic hills-numerical and experimental study in a wide range of Reynolds numbers. *Computers & Fluids, 38,* 433–457.
6. Cabot, W., & Moin, P. (1999). Approximate wall boundary conditions in the large-eddy simulation of high Reynolds number flow. *Flow Turbulence Combustion, 63,* 269–291.
7. Chaouat, B. (2017). The state of the art of hybrid RANS/LES modeling for the simulation of turbulent flows. *Flow, Turbulence and Combustion, 99*(2), 279–327.
8. Cherry, N. J., Hiller, R., & Latour, M. E. M. P. (1984). Unsteady measurements in a separated and reattaching flow. *Journal of Fluid Mechanics, 144,* 13–46.
9. Chorin, A. J. (1968). Numerical solution of the Navier-Stokes equations. *Mathematics of Computation, 22,* 745–762.
10. Davidson, L., & Billson, M. (2006). Hybrid LES-RANS using synthesized turbulent fluctuations for forcing in the interface region. *International Journal of Heat and Fluid Flow, 27,* 1028–1042.
11. Davidson, L., Cokljat, D., Frohlich, J., Leschziner, M., Mellen, C., & Rodi, W. (Eds.). (2003). LESFOIL: Large eddy simulation of flow around a high lift airfoil. In *Notes on numerical fluid mechanics* (Vol. 83). Springer.
12. Durbin, P. A. (1996). On the k–ε stagnation point anomaly. *International Journal of Heat and Fluid Flow, 17,* 89–90.
13. Foroutan, H., & Yavuzkurt, S. (2014). A partially-averaged Navier-Stokes model for the simulation of turbulent swirling flow with vortex breakdown. *International Journal of Heat and Fluid Flow, 50,* 402–416.
14. Fransen, R., Morata, E. C., Duchaine, F., Gourdain, N., Gicquel, L. Y. M., & Vial, L. (2012). *Comparison of RANS and LES in high pressure turbines* (pp. 12–25). Toulouse: Safran Turbomeca, CERFACS.
15. Frisch, U. (1995). *Turbulence, the legacy of A. N. Kolmogorov.* Cambridge: Cambridge Univ. Press.
16. Fröhlich, J., Mellen, C. P., Rodi, W., Temmerman, L., & Leschziner, M. A. (2005). Highly resolved large-eddy simulation of separated flow in a channel with streamwise periodic constrictions. *Journal of Fluid Mechanics, 526,* 19–66.
17. Fröhlich, J., & von Terzi, D. (2008). Hybrid LES/RANS methods for the simulation of turbulent flows. *Progress in Aerospace Sciences, 44*(5), 349–377.
18. Gaster, M. (1968). Growth of disturbances in both space and time. *Physics of Fluids, 11*(4), 723–727.
19. Germano, M. (1992). Turbulence: The filtering approach. *Journal of Fluid Mechanics, 238,* 325–336.
20. Germano, M., Piomelli, U., Moin, P., & Cabot, W. H. (1991). A dynamic subgrid-scale eddy viscosity model. *Physics of Fluids A, 3,* 1760–1765.
21. Gritskevich, M. S., Garbaruk, A. V., Schütze, J., & Menter, F. R. (2012). Development of DDES and IDDES formulations for the k-ω shear stress transport model. *Flow, Turbulence and Combustion, 88,* 431–449.
22. Hain, R., Kahler, C. J., & Radespiel, R. (2009). Dynamics of laminar separation bubbles at low-Reynolds-number aerofoils. *Journal of Fluid Mechanics, 630,* 129–153.

23. Halstead, D. E., Wisler, D. C., Okiishi, T. H., Walker, G. J., Hodson, H. P., & Shin, H.-W. (1997). Boundary layer development in axial compressors and turbines—Part 1 of 4: Composite picture; Part 2 of 4: Compressors; Part 3 of 4: LP turbines; Part 4 of 4: Computations and analyses. *ASME Journal of Turbomachinery, 119,* 114–127, 426–444, 225–237, 128–139.
24. Harlow, F. H., & Welch, J. E. (1965). Numerical calculation of three-dependent viscous incompressible flow of fluid with free surfaces. *Physics of Fluids, 8,* 2182–2188.
25. He, C., Liu, Y., & Yavuzkurt, S. (2017). A dynamic delayed detached-eddy simulation model for turbulent flows. *Computers & Fluids, 146,* 174–189.
26. Horton, H. (1969). *A semi-empirical theory for the growth and bursting of laminar separation bubbles.* HM Stationery Office.
27. Jansen, K. (1996). Large-eddy simulation of flow around a NACA 4412 airfoil using unstructured grids. In *Annual research briefs* (pp. 225–232). Center for Turbulence Research, Stanford University and NASA Ames.
28. Jones, B. M. (1934). Stalling. *Journal of Royal Aeronautical Society, 38,* 753–770.
29. Kenjereš, S., & Hanjalić, K. (2005). LES, T-RANS and hybrid simulations of thermal convection at high Ra numbers. *International Journal of Heat and Fluid Flow, 27,* 800–810.
30. Kolmogorov, A. N. (1941). The local structure of turbulence in incompressible viscous fluid for very large Reynolds number. *Proceedings of the USSR Academy of Sciences, 30,* 301.
31. Krajnovic, S., & Davidson, L. (2005). Flow around a simplified car, Part II: Understanding the flow. *Journal of Fluids Engineering, 127*(5), 919–928.
32. Launder, B. E. (1991). An introduction to the modeling of turbulence. In *VKI lecture series 1991–02,* March 18–21, 1991. Von Karman Institute of Fluid Dynamics.
33. Leonard, A. (1974). Energy cascade in large eddy simulations of turbulent fluid flows. *Advances in Geophysics, 18,* 237.
34. Lesieur, M., & Métais, O. (1996). New trends in large-eddy simulations of turbulence. *Annual Review of Fluid Mechanics, 28,* 45–82.
35. Lilly, D. K. (1967). The representation of small scale turbulence in numerical simulation experiments. In *Proceedings of the IBM Scientific Computing Symposium on Environmental Sciences* (Vol. 195).
36. Lilly, D. K. (1992). A proposed modification of the Germano subgrid-scale closure method. *Physics of Fluids A, 4,* 633–635.
37. Liou, M.-S. (1996). A sequel to ausm: Ausm+. *Journal of Computational Physics, 129,* 364–382.
38. Liou, M.-S. (2006). A sequel to AUSM, Part II: AUSM+-up for all speeds. *Journal of Computational Physics, 214,* 137–170.
39. Liou, M.-S., & Steffen, C. J. (1993). A new flux splitting scheme. *Journal of computational Physics, 107,* 23–39.
40. Liu, S., Meneveau, C., Katz, J. (1994). On the properties of similarity subgrid-scale models as deduced from measurements in a turbulent jet. *Journal of Fluid Mechanics, 275,* 83–119.
41. Mary, I., & Sagaut, P. (2002). LES of a flow around an airfoil near stall. *AIAA Journal, 40,* 1139.
42. Mellen, C., Frohlich, J., & Rodi, W. (2003). Lessons from LESFOIL project on large eddy simulation of flow around an airfoil. *AIAA Journal, 41*(4), 573–581.
43. Menter, F., Kuntz, M., & Langtry, R. (2003). Ten years of industrial experience with the SST turbulence model. *Turbulence, Heat and Mass Transfer, 4.*
44. Michelassi, V. (1997). Shock-boundary layer interaction and transition modelling in turbomachinery flows. *IMechE Part A Journal of Power and Energy, 211,* 225–234.
45. Michelassi, V., Wissink, J. G., & Rodi, W. (2003). Direct Numerical simulation, large eddy simulation and unsteady Reynolds averaged Navier-Stokes simulations of periodic unsteady flow in a low-pressure turbine cascade: A comparison. *IMechE Part A Journal of Power and Energy, 217,* 403–411.
46. Morinishi, Y. (1995). Conservative properties of finite difference schemes for incompressible flow. In *Annual research brief* (pp. 121–132). Center for Turbulence Research, NASA Ames/Stanford University.

47. Nair, K. M., & Sarkar, S. (2017). Large eddy simulation of self-sustained cavity oscillation for subsonic and supersonic flows. *Journal of Fluids Engineering, 139,* 011102.
48. Patankar, S. V., & Spalding, D. B. (1972). A calculation procedure for the heat, mass and momentum transfer in three-dimensional parabolic flows. *International Journal of Heat and Mass Transfer, 15,* 1787–1805.
49. Piomelli, U., & Balaras, E. (2002). Wall-layer models for large-eddy simulation. *Annual Review of Fluid Mechanics, 34,* 349–374.
50. Piomelli, U., Ferziger, J., Moin, P., & Kim, J. (1989). New approximate boundary conditions for large eddy simulations of wall-bounded flows. *Physics of Fluids A, 1,* 1061–1068.
51. Pope, S. B. (2000). *Turbulent flows.* Cambridge: Cambridge University Press.
52. Richardson, L. F. (1922). *Weather prediction by numerical process.* Cambridge: Cambridge University Press.
53. Robert, S. K., & Yaras, M. I. (2005). Boundary layer transition affected by surface roughness and free-stream turbulence. *ASME Journal of Fluids Engineering, 127*(3), 449–457.
54. Rodi, W., Ferziger, J. H., Breuer, M., & Pourquié, M. (1997). Status of large eddy simulation: Results of a workshop. *Journal of Fluids Engineering, 119,* 248–262.
55. Sagaut, P. (2000). *Large eddy simulation for incompressible flows: An introduction.* Springer.
56. Sagaut, P. (2013). *Multiscale and multiresolution approaches in turbulence: LES, DES and hybrid RANS/LES methods: Applications and guidelines* (2nd ed.). Imperial College Press.
57. Samson, A., & Sarkar, S. (2016). Effects of free-stream turbulence on transition of a separated boundary layer over the leading-edge of a constant thickness airfoil. *ASME Journal of Fluids Engineering, 138*(2), 021202(19).
58. Sarkar, S. (2007). Effects of passing wakes on a separating boundary layer along a low-pressure turbine blade through large-eddy simulation. *Proceedings of the Institution of Mechanical Engineers, Part A: Journal of Power and Energy, 221,* 551–563.
59. Sarkar, S. (2008). Identification of flow structures on a LP Turbine due to periodic passing wakes. *ASME Journal of Fluids Engineering, 130,* 061103.
60. Sarkar, S. (2009). Influence of wake structure on unsteady flow in a low pressure turbine blade passage. *ASME Journal of Turbomachinery, 131*(041016), 1–14.
61. Sarkar, S., Babu, H., & Sadique, J. (2016). Interactions of separation bubble with oncoming wakes by LES. *ASME Journal of Heat Transfer, 138*(2), 021703–021712.
62. Sarkar, S., & Sarkar, Sudipto. (2010). Vortex dynamics of a cylinder wake in proximity to a wall. *Journal of Fluids and Structures, 26,* 19–40.
63. Sarkar, S., & Voke, P. R. (2006). Large-eddy simulation of unsteady surface pressure over a LP turbine due to interactions of passing wakes and inflexional boundary layer. *ASME Journal of Turbomachinery, 128,* 221–231.
64. Schulte, V., & Hodson, H. P. (1998). Unsteady wake-induced boundary layer transition in high lift LP turbine. *ASME Journal of Turbomachinery, 120,* 28–35.
65. Schumann, U. (1995). Stochastic backscatter of turbulence energy and scalar variance by random sub-grid scale fluxes. *Proceedings of the Royal Society of London. Series A, 451,* 293–318.
66. Shur, M., Spalart, P., Strelets, M., & Travin, A. (1999). Detached-eddy simulation of an airfoil at high angle of attack. *Engineering Turbulence Modeling Experiments, 4,* 669–678.
67. Shur, M., Spalart, P. R., Strelets, M. K., & Travin, A. K. (2008). A hybrid RANS-LES approach with delayed-DES and wall-modeled LES capabilities. *International Journal of Heat and Fluid Flow, 29*(6), 1638–1649.
68. Smagorinsky, J. (1963). General circulation experiments with the primitive equations. I. The basic experiment. *Monthly Weather Review, 91,* 99.
69. Spalart, P. R. (2009). Detached-eddy simulation. *Annual Review of Fluid Mechanics, 41,* 181–202.
70. Spalart, P. R., Deck, S., Shur, M., Squires, K., Strelets, M. K., & Travin, A. (2006). A new version of detached-eddy simulation, resistant to ambiguous grid densities. *Theoretical and Computational Fluid Dynamics, 20,* 181–195.

71. Spalart, P. R., & Strelets, M. K. (2000). Mechanisms of transition and heat transfer in a separation bubble. *Journal of Fluid Mechanics, 403,* 329–349.
72. Sreenivasan, K. R. (1991). Fractals and multifractals in turbulence. *Annual Review of Fluid Mechanics, 23,* 539–600.
73. Sreenivasan, K. R., & Antonia, R. A. (1997). The phenomenology of small-scale turbulence. *Annual Review of Fluid Mechanics, 29,* 435–472.
74. Stadtmüller, P. (2001). Investigation of wake-induced transition on the LP turbine cascade T106 A-EIZ, DFG-VerbundprojectFo 136/11, version 1.0, University of the Armed Forces Munich, Germany.
75. Stieger, R., Hollis, D., & Hodson, R. (2003). Unsteady surface pressures due to wake induced transition in laminar separation bubble on a LP turbine cascade. ASME Papar No. GT2003-38303.
76. Strelets, M. (2001). Detached eddy simulation of massively separated flows. In *39th AIAA, Aerospace Sciences Meeting and Exhibit,* Reno, NV.
77. Tabor, G. R., & Baba-Ahmadi, M. H. (2010). Inlet conditions for large eddy simulation: A review. *Computers & Fluids, 39*(4), 553–567.
78. Temmerman, L., Hadžiadbic, M., Leschziner, M., & Hanjalić, K. (2005). A hybrid two-layer URANS-LES approach for large eddy simulation at high Reynolds numbers. *International Journal of Heat and Fluid Flow, 26,* 173–190.
79. Tucker, P. G. (2011). Computation of unsteady turbomachinery flows: Part 2-LES and hybrids. *Progress in Aerospace Sciences, 47*(7), 546–569.
80. Tucker, P., & Davidson, L. (2004). Zonal *k-l* based large eddy simulation. *Computers & Fluids, 33*(2), 267–287.
81. Van Driest, E. R. (1956). On turbulent flow near a wall. *Journal of the Aeronautical Sciences, 23,* 1007.
82. Voke, P. R. (1996). Subgrid-scale modeling at low mesh Reynolds number. *Theoretical and Computational Fluid Dynamics, 8*(2), 131–143.
83. Watmuff, J. H. (1999). Evolution of a wave packet into vortex loops in a laminar separation bubble. *Journal of Fluid Mechanics, 397,* 119–169.
84. Wissink, J. G. (2002). DNS of separating, low Reynolds number flow in a turbine cascade with incoming wakes. In *5th International Symposium on Engineering Turbulence Modelling and Experiments* (pp. 731–740). Mallorca, Spain: Elsevier.
85. Wu, X., & Durbin, P. A. (2001). Evidence of longitudinal vortices evolved from distorted wakes in turbine passage. *Journal of Fluid Mechanics, 446,* 199–228.
86. Wu, X., Jacobs, R. G., Hunt, J. R. C., & Durbin, P. A. (1999). Simulation of boundary layer transition induced by periodically passing wakes. *Journal of Fluid Mechanics, 398,* 109–153.
87. Wu, X., & Squires, K. D. (1998). Numerical investigation of the turbulent boundary layer over a bump. *Journal of Fluid Mechanics, 362,* 229–271.
88. Xiao, X., Edwards, J., & Hassan, H. (2003). Inflow boundary conditions for LES/RANS simulations with applications to shock wave boundary layer interactions, Reno, NV, AIAA paper 2003-0079.
89. Yang, Z. Y., & Voke, P. R. (2001). Large-eddy simulation of boundary layer separation and transition at a change of surface curvature. *Journal of Fluid Mechanics, 439,* 305–333.
90. Zang, Y., Street, R. L., & Koseff, J. (1993). A dynamic mixed subgrid-scale model and its application to turbulent recirculating flows. *Physics of Fluids A, 5,* 3186–3196.

CFD and Reactive Flows

Brian Spalding and Turbulent Combustion

Norberto Fueyo and Michael R. Malin

1 Introduction

Brian Spalding regarded the simulation of combusting flows as one of the main driving forces behind the emergence of CFD as another tool in the engineer's toolkit:

> It could be reasonably argued that it was the needs of the combustion engineers in the aerospace industry which brought the CFD-software business into existence, the reason being that the complexity of the combustion process left expensive experimentation as the only alternative. [107]

Brian Spalding is well known for his contributions to Engineering and Science in numerous fields, including numerical methods, thermodynamics, turbulence, heat and mass transfer, multiphase and free-surface flows. Combustion featured prominently among the subject matters of Spalding's research writings. A quick survey of his scientific production shows that about 32% of his contributions in his 67-year academic career were directly related to combustion (Fig. 1).

This chapter presents an overview of Spalding's contribution to the modelling and simulation of combustion. In doing so, we attempt to follow his thinking as he progressed from the integral models of the early years to the multi-fluid models of the latter ones. The focus is therefore in finding and conveying his line of thought, rather than in the mathematics or the numerics of the result; the latter can be found in the references cited, and in many others that are not included for reasons of space. We do not attempt, for the same reasons of space, to be encyclopaedic in our account.

N. Fueyo (✉)
Universidad de Zaragoza, María de Luna 3, 50018 Zaragoza, Spain
e-mail: Norberto.Fueyo@unizar.es

M. R. Malin
Concentration Heat and Momentum Limited (CHAM) Bakery House,
40 High Street Wimbledon Village, London SW19 5AU, UK
e-mail: mrm@cham.co.uk

© Springer Nature Singapore Pte Ltd. 2020
A. Runchal (ed.), *50 Years of CFD in Engineering Sciences*,
https://doi.org/10.1007/978-981-15-2670-1_12

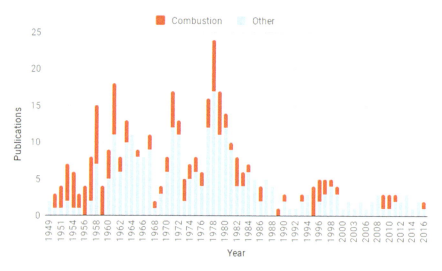

Fig. 1 Spalding's research writings on combustion

Spalding's masterful command of the written language was for many as engaging as the scientific ideas it conveyed. His writings are full of powerful images, literary references, witticisms and flares of plain humour. In this recollection of his contributions to combustion science, we often cite his own words; we signify this by using double quotation marks "⊙", while we reserve single quotation marks '⊙' for all other, non-citation uses.

We reproduce a few of the figures and diagrams that also define Spalding's production, from hand-made graphical depictions of his ideas in the early years to his use of 'ASCII Art' in the latter ones.

The chapter is structured in six sections. After this introduction, in the second section we cover the 'early' years, from his Ph.D. thesis to his first uses of computers. The third one is dedicated to the popular Eddy-Breakup Model, and its (perhaps less popular) sequel the ESCIMO Model. The fourth and fifth sections cover Spalding's most recent thinking on turbulent combustion: the Two- and Multi-Fluid Models of turbulence and combustion. We close the chapter with some conclusions.

2 The Early Years

2.1 The Pre-computer Years

Spalding's scientific interest in the intricate field of combustion dates back to his research leading to his Ph.D. thesis at University of Cambridge, which he concluded in 1951 [81]. The thesis combined Spalding's masterful command of the theoretical grounds with a set of simple but ingenious experiments. The research addressed the prediction of the rate of combustion from a liquid-fuel surface (a sphere) into a surrounding gas, and presented a rigorous and elegant treatment of the associated, complex heat and mass-transfer phenomena between the liquid phase and the surrounding gas. Spalding studied both "envelope" and "wake" flames, and the "breakdown" of the latter (or their extinction, as it would be more commonly called nowadays). The then-called *Transfer Number* was for the first time presented, and a whole thesis chapter was dedicated to its calculation for several configurations (Chap. 3); this Transfer Number was later called, and is still known as, *Spalding's Number*, B. In his preface to a re-typed, 1987 edition version of the thesis [103], he humorously declared to be "pleased to find" that his re-reading it "did not cause me much embarrassment", despite "[some] misplacings of the word 'only' and the omissions of hyphens and commas". (Later in his career, he would try to spare his own students from such predicament by publishing a style booklet entitled 'The Writing of Technical Reports' [96].)

In this pre-computer age, Spalding recognised two important limitations of his theory: first, the need to use constant values for the properties; second, that the gas flow-field had to be analytically prescribed.

It is perhaps this latter one that prompted Spalding to initially explore the use of the then-called analogue *computers*. These were devices used to mimic the behaviour of physical processes, such as heat transfer. Spalding, now at Imperial College, contrived several of these devices to investigate combustion and heat-transfer problems [72]. One of such devices was created for the physical simulation of bluff-body-stabilised flames [82] using heated air (Fig. 2). The 'combustion' chamber consisted in a Cartesian array of heating rods downstream of a baffle. Each rod was equipped with a thermocouple, and the heating element in each rod was turned on or off (manually!) as a function of the local temperature, thus representing the effect of the activation energy on the reaction rate.

The device was successful in predicting 'flame' extinction [82], and the predictions compared reasonably well, given the model limitations, with real-flame experiments. In the subsequent discussion of this paper [82], Owen Saunders (Fellow of the Royal Society 1965) that "an extremely ingenious tool had been devised which brought together the chemistry and physics of combustion in a new way".

Forty years later, reflecting on this piece of research, Spalding recognised [107] that the Cartesian arrangement of the rods was perhaps not unlike the discretisation techniques he would later pioneer; and that he was unaware at the time "of the turbulence-chemistry interaction problem", to which he would dedicate much of his time in later years (see below).

2.2 Spalding 'Goes Digital'

By the early 50s, even before Spalding developed his analogue combustion chamber, the applications of digital computers to combustion science were starting to

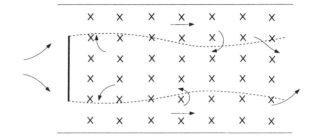

Fig. 2 Schematic of Spalding's 'analogue combustion chamber'. The crosses represent the heating rods; the dashed line represents "the region where the air temperature appreciably exceeds that at inlet" (re-created from [82])

emerge [23]. Spalding's first published use of computers to a combustion problem is perhaps the Adler and Spalding paper where they presented numerical solutions of one-dimensional, laminar, premixed flames with simplified chemistry [1].

Unlike most of the premixed flames studied until then, the focus in this paper was on flames subject to enthalpy gradients; such situations are often encountered in practice when the flame products transfer heat, through conduction or radiation, with the reactor wall or with the surroundings. They concluded that, compared to adiabatic flames, positive enthalpy gradients increased the burning rate, while negative ones decreased it.

The computer used was a Ferranti Mark I*, similar to the one shown in Fig. 3.

2.3   The First CFD Combustion Calculations

The next large stride towards the numerical simulation of combustion was the development of a calculation method for parabolic flows as part of Suhas Patankar's 1967 Ph.D. thesis [59]. The method was, in 1969, used in the GENMIX code [92] to simulate laminar and turbulent jet flames for the first time.

For further progress to be made, a method was needed to remove the parabolic restriction from the problem formulation. Akshai Runchal and Micha Wolfshtein joined Brian Spalding's team of students soon after Patankar to work on elliptic and turbulent flows. As a consequence, the first method for solving elliptic flows was created around 1968 [20]; it was the 'vorticity-stream-function' method, and it would be extensively used for a few years for calculating mainly inert, recirculating flows.

The method used, also for the first time, the concept of 'upwind differencing', the creation of which Spalding famously attributed to "[his] childhood experience of having lived near a pig-sty, and therefore known well how the direction of the wind influenced the strength of the influence of near neighbours" [107].

Soon after the creation of the 'vorticity-stream-function' method, Spalding (together with WM (Sam) Pun) started its application to chemically reacting flows, resulting in 1967 in perhaps the first simulation of a recirculating, reacting flow using

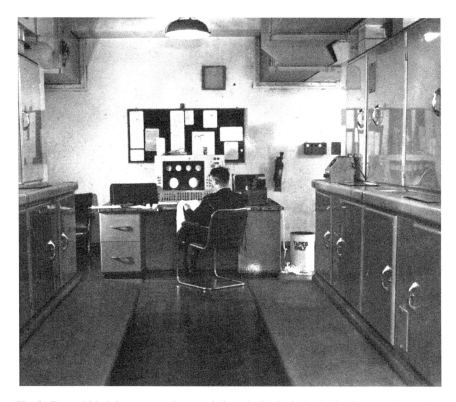

Fig. 3 Ferranti Mark I computer: the console is at the back; the logic circuits are in the cabinets along the sides of the aisle (Courtesy of the University of Manchester, with additional thanks to Prof Jim Miles, School of Computer Science; James Peters, National Archive for the History of Computing; and Fujitsu Services Limited.)

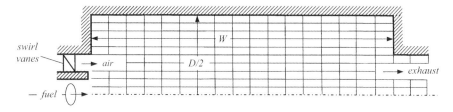

Fig. 4 First elliptic flame calculation: domain and mesh (re-created from [68])

CFD [68]. This work considered the simulation of a non-premixed, swirling flame in a combustion chamber, re-created in Fig. 4 from the original figure in [68].

Turbulence was represented with a simple effective-viscosity model, and combustion with a mixed-is-burned hypothesis based on the Burke–Schumann Simple Chemically Reacting System (SCRS) [87]. The aim of the paper was to demonstrate the feasibility of performing calculations of recirculating, reacting flows. The calculations were performed on an IBM 7090 computer, at a cost of between 2 and 6 minutes per case for a 161-cell mesh.

It was around this time that Spalding [84] began to consider the problem of how to predict a finite thickness for the reaction zone of a turbulent diffusion flame when using an SCRS combustion model based on fast chemistry. The SCRS assumes that fuel and oxidiser cannot coexist at the same location, even at different times; and this leads to a very thin reaction zone, with peaky flame profiles, rather than a thicker flame brush. For more realistic predictions, Spalding [84, 86, 111, 112] recognised the need to account for the fluctuation in mixture strength with time; and hence for the fluctuations in species mass fractions, in temperature and in fluid density, which are often large compared to their mean values.

For this purpose, Spalding devised a modified model in which the reactants cannot exist at the same location at the same time, but can occur at the same location at different times. This was done by assuming a rectangular wave variation of the mixture fraction f with time, and the magnitudes of the fluctuations in f were determined by solving an additional transport equation for its variance. This approach amounts to using an assumed probability density function (PDF) for the mixture fraction in the form of a double-delta function, and it was used later by Gosman and Lockwood [19], Khalil et al. [37] and Serag-Eldin [73]. Other workers soon adopted the method, but to obtain improved results the double-delta function was replaced by more realistic PDF forms, such as the clipped Gaussian or the beta distributions (see Jones and Whitelaw [35]).

2.4 The Gateway to 3D Combustion Calculations

In the fast-paced late 60s and early 70s, the Imperial College group led by Spalding soon came up with what would become the mainstream method for three-dimensional recirculating flows. It was based on the solution of the so-called primitive variables: velocity and pressure. This was an extension of the SIMPLE procedure [62], so far used only for 3D boundary layers, to recirculating flows.

Their landmark 1974 publication [61] about the simulation of a gas-turbine combustion chamber contained all the ingredients needed for three-dimensional CFD modelling of combustion. (However, Spalding credits Zuber [131] as the creator of the first 3D chamber model in 1972.) Patankar and Spalding's model used SIMPLE for solving the velocity–pressure coupling; the $k - \epsilon$ model for turbulence; a six-flux radiation model [60]; and a single-step, equilibrium model for the chemical reaction.

This pioneering method would be soon afterwards exploited by Spalding's student Amr Serag-Eldin. His 1977 Ph.D. thesis [73] would highlight, through the comparison of calculations and experiments, the many deficiencies of the physical models available at the time for combustion CFD modelling. Among these, the need was clearly seen for modelling the interaction between turbulence and chemistry, and the effect of chemical kinetics. Spalding would dedicate a significant fraction of his academic career to redress such deficiencies.

3   EBU and ESCIMO

3.1   The Eddy-Breakup Model

Spalding's most recognisable contribution to combustion is arguably his Eddy-Breakup (EBU) model. (There is ample documented evidence that Spalding thought this model too long-lived, as we will discuss below.)

In its original form, the EBU model was first published in 1971 [112]. Spalding was researching the reasons why the spread angle in confined, premixed, turbulent flames was nearly independent of the operating conditions, including the unburned mixture velocity, its composition, its temperature or its level of turbulence. The hypothesis that the fresh mixture burned immediately following entrainment into the burnt gases had already been made, and Spalding noted that he had tested it numerically [83], and found that, although it "could fairly well explain the observed phenomena", his validation against experiments showed that "the assumption [was] far too crude" and that unburned fuel could be found even well inside the flame.

While Spalding's first approach to the problem in 1967 [83] had used an integral formulation, he now had at his disposal powerful new tools: computational methods to solve, via discretisation, partial differential equations in parabolic flows; and increasingly sophisticated models for the turbulence energy and its length scale.

The original EBU model was formulated in terms of the reactedness variable τ:

$$\tau = \frac{m_f - m_{f,unburned}}{m_{f,burned} - m_{f,unburned}} \quad , \tag{1}$$

where m_f is the local value of the mass fraction of fuel.

From its definition τ can be regarded as a non-dimensional fuel mass fraction.

Central to the EBU model, and to Spalding's view of turbulent combustion, is the supposition that "the mixture is mainly composed of alternating fragments of unburned gas and almost-fully burned gas" (and thus largely non reactive), and that it is "at the interfaces between the hot and cold lumps", with intermediate compositions, where combustion takes place, with a maximum reaction rate (per unit volume) of, say, $\dot{m}_{max}$. Spalding then further argued that the breaking down of these parcels into increasingly small ones (down to the Kolmogorov scale) is the cause of turbulence decay; and that the decay rate, for turbulence in equilibrium, is proportional to the density and velocity gradient, i.e. equal to $0.35\rho|\partial\bar{u}/\partial y|$, where $\bar{u}$ is the average velocity.

From this, he assumed that the breakdown of parcels into sufficiently small ones for heat conduction and chemical reaction to be significant proceeds at the same rate, and thus is (in terms of mass per unit volume per unit time):

$$\dot{m}_{mix} = C(1-\tau)\rho\left|\frac{\partial\bar{u}}{\partial y}\right| \quad , \tag{2}$$

which is the essence of the Eddy-Breakup model.

It is interesting to note that in this first version Spalding included the influence of the kinetic rate, crudely represented by a conversion rate (mass per unit volume per unit time):

$$\dot{m}_{kin} = \tau \dot{m}_{max} \quad , \tag{3}$$

so that the effective reaction rate is a harmonic blend of both the kinetic and the mixing rates, thus ensuring that the overall rate is dictated by the slower process:

$$\dot{m}_\tau = \left[\frac{1}{\dot{m}_{kin}} + \frac{1}{\dot{m}_{mix}} \right] \quad . \tag{4}$$

At the time, more sophisticated turbulence closure models were being developed that allowed for the transport of turbulence statistics, thus dispensing with the hypothesis of local equilibrium. Spalding was at the time exploring one of these models, namely, the $k - W$ model [85], where the transported variables are the turbulence kinetic energy and the square of the local frequency of the turbulent motion. Spalding immediately recognised the opportunity to use it in his EBU model, and proposed the use of W to replace the velocity gradient $|\partial u / \partial y|$ in Eq. 2. Of course, in the end it would be the kinetic energy dissipation rate ϵ that would be used, and ϵ / k would be the mixing rate used in the EBU model [57].

Spalding also stressed other aspects susceptible to refinement; among them, the fact that the mixing fluid fragments may not be just in burned and unburned states, but perhaps closer in compositional space. This is an idea that he would develop 24 years later as his Multi-Fluid Models of turbulent combustion (see below); at the time, however, he would propose the writing and solution of equations for the root mean square of the concentration fluctuation, which would become in time the second ingredient (together with the mixing frequency ϵ / k) of his EBU model.

Refined derivatives of the EBU model would appear later, such as the popular Eddy-Dissipation model (EDM) [50] and the Eddy-Dissipation Concept (EDC) [48, 49]. The EDM differs from the EBU in that it replaces the square root of the concentration fluctuations [57] in the reaction rate with the mean concentration of the deficient species (fuel for lean, or oxidiser for rich mixtures). Further, the EDM computes the reaction rate from the minimum of three rates based on the mean oxygen mass fraction, mean fuel mass fraction and the mean product mass fraction. Although the original EBU approach was developed for premixed combustion, the EDM can be used for both premixed and non-premixed combustion [50]. The EDC is an extension of the EDM that considers detailed reaction kinetics on small scales, which are modelled as a perfectly stirred constant-pressure reactor with initial conditions taken from the prevailing cell composition and temperature. The reaction rates are determined by Arrhenius expressions, and proceed over a Kolmogorov residence time.

Despite these advances from his highly influential EBU concept, Spalding regarded his Multi-Fluid Models (see below) as the heir apparent to the EBU.

3.2   The ESCIMO Model

Spalding's quest for physical fidelity in his combustion models was clearly reflected in the genesis of the EBU model as portraying the interaction of fluid parcels with different degrees of reactedness. It may be argued that, for the EBU model, the translation of such interaction into the model equations was limited by the mathematical and computational tools available at the time. Spalding would in fact often use the EBU model imagery (the colliding fluid fragments) as the basis for his successor models, while disapproving of its mathematical embodiment as inadequate.

The first significant step in the sophistication of the mathematical description was the ESCIMO model. Around 1976, Spalding contrived a means for modelling the creation of scalar isosurface by the turbulent flow. His idea was first published in a review of mathematical models of turbulent flames [89] as "the stretch-cut-superpose model of turbulent scale reduction", and is illustrated in Fig. 5.

As it was often the case in Spalding's scientific career, the spark of intuition soon developed into a new theory. He now turned his attention to the "*coherent* bodies of gas which are squeezed and stretched during their travel through the flame" [91] (the emphasis is ours). The essential ingredients of the new model were

- An equation for the time evolution of the thickness λ of a two-part layer (of the kind portrayed in Fig. 5) in a turbulent flow:

$$\frac{D\lambda}{Dt} = -\lambda R \quad , \tag{5}$$

where R is a stretch rate, such as $R = |\partial \overline{u}/\partial y|$ in quasi-unidirectional flow.
- A probability density function for the ages $a = \exp(Rt) - 1$ of the gas parcels present at any point.
- A functional relationship, derived from the above assumptions, between the average reactedness $\overline{\tau}$ (obtained from a transport equation) and the average reaction rate $\overline{S}_\tau$.

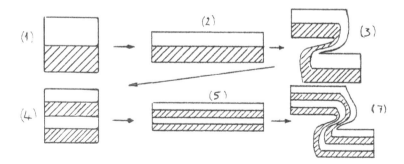

Fig. 5   The stretch-cut-superpose model of turbulent scale reduction (reproduced from [88])

FIG.1: THE ROLLING-UP OF THE INTERFACE (VORTEX SHEET)
BETWEEN TWO STREAMS TO FORM "BILLOWS".

Fig. 6 Engulfment or rolling up of the iso-surface at the edge of a mixing-region (reproduced from [90])

The above ideas were soon developed into the ESCIMO model of turbulent combustion [88, 93, 110]. ESCIMO stands for Engulfment, Stretching, Coherence, Interdiffusion, Moving Observer:

- Engulfment is the formation of layers of fluids at the mixing-region edges (Fig. 6).
- Stretching is the reduction in length scale by turbulence, for instance as mathematically described by Eq. 5.
- Coherence is the preservation of this layered structure as it moves with the flow.
- Inter-diffusion is the all-important role of molecular diffusion, enhanced by the stretching and coherence processes which increase the scalar iso-surface density.
- Moving Observer is the use of Lagrangian, parcel-attached, coordinates under which the interplay between diffusion and chemical reaction can be regarded as a one-dimensional, unsteady process.

The ESCIMO mathematical theory appears to have been formulated largely on intuitive grounds and it divides the analysis into two parts: (1) a biographic (Lagrangian) part, in which the details of reaction and molecular diffusion within folds are treated as essentially one-dimensional; and (2) a demographic (Eulerian) part, which involves the specification and description of the fold distribution. The biographic part considers what happens within the fold between its birth by engulfment, and death by re-engulfment or escape from the entire region of interest. It is here that the one-dimensional transient problem is solved including the chemical kinetics, which are assumed to obey laminar laws. The dynamics of turbulent flow are addressed in the demographic part, which "concerns the statistics of the population of coherent fluid parcels: how many of each kind are born, where they travel to, to what environmental conditions they are subjected to, and how long they live." [93].

Spalding recognised that "the components of ESCIMO are [...] familiar ones; but they [were] linked together in a special way to form a predictive method for turbulent flames" [88]; and likened this process to the creation of new paintings from the arrangement of a finite number of colours on the canvas.

It is interesting to note that the ESCIMO model shares many features with the Linear Eddy Model (LEM) Model developed almost a decade later by Kerstein [36]. As with Spalding's ESCIMO, Kerstein's LEM also solves a differential equation for the one-dimensional diffusion-reaction process at the microscale. While in Spalding's ESCIMO the structure is convected in a Lagrangian frame, Kerstein's LEM used a cell-based, Eulerian strategy whereby fragments of the one-dimensional eddy are moved from between neighbouring cells, the size of the exchanged fragment being proportional to the convection and diffusion contributions in the discretised transport-equation coefficients. In ESCIMO, scalar gradients are increased through 'squashing' as the fluid parcel ages; in LEM, the gradients are increased by a stochastic process of artificially compressing and replicating fragments of the one-dimensional field, a process termed a 'triplet map'.

The ESCIMO model was applied to several problems, including baffle-stabilised premixed flames [46, 58], premixed well-stirred reactors [120], natural-gas diffusion flames [119] and hydrogen-air diffusion flames [14, 47, 119].

4 Two-Fluid Models

The ESCIMO model was relatively short-lived. Spalding reflected that the necessary parameters could be adjusted "to make predictions fit a limited set of experimental data", but that "time-averaged concentration profiles [...] proved hard to reproduce via ESCIMO" [97]. He attributed these difficulties to the fact that the then-prevailing turbulence model, the $k - \epsilon$ one, did not account for intermittency at the outer regions of the jet. Thus ESCIMO "emphasises the small-scale non-uniformities while leaving the large-scale ones to be handled (poorly until now) by the hydrodynamic turbulence model" [98]. He then moved on to remedy this difficulty with a new class of turbulence models: the two-phase turbulence models.

In the mid-70s, Spalding contrived a means for solving two-phase flows, with interphase slip, in an Eulerian–Eulerian framework. The so-called Inter-Phase Slip Algorithm, or IPSA [95, 113], was implemented in a number of codes. It is still nowadays one of the main algorithms for solving the Eulerian equations in multiphase flows.

In a canonic example of Spalding's ability to think unconventionally in the search for solutions, soon after developing IPSA he saw an opportunity to use the new technique to evolve a new class of turbulence models. Spalding's motivation was to address some shortcomings of conventional turbulence models, such as their inability to represent the intermittency of turbulent flow or of counter-gradient diffusion (although the latter is within the capabilities of some second-moment closures).

Conventional turbulence models employ a statistical approach [40, 71] where the unknown statistically averaged values of the turbulent fluxes of mass, heat and momentum are closed either by solving transport equations for these quantities directly, or by using a turbulence model based on the Boussinesq [6] eddy-viscosity relationship. Most commonly used are eddy-viscosity models employing transport equations for the turbulence kinetic energy and a length scale determining variable, usually the turbulence dissipation rate [22] or turbulence frequency [40]. During the 1960s Spalding [85] pioneered the development of such models.

The new turbulence model exploited the analogy between turbulence and two-phase flow, and was called the two-fluid model of turbulence [99–101, 104, 114].

The two-fluid model is based on the notion that a turbulent fluid can be represented as a mixture of two fluids, which can be distinguished from each other in several ways, such as for example by defining the two fluids as hotter and colder, turbulent and non-turbulent, lighter and heavier, faster and slower, upward-flowing and downward-flowing, etc. Each fluid has its own temperature, composition variables, velocity components, volume fractions etc, and the two fluids interact with each other through the sharing of space and the exchange of mass, heat and momentum by the physical processes of tearing, folding, inter-diffusion and separation. The volume fractions can be regarded as 'probabilities of presence', and the interspersed fluid fragments can be characterised by one (or more) local fragment size distributions. Spalding was quick to point out that in the past "several authors have advocated thinking of a turbulent flow as a mixture of two intermingling fluids. The idea formed a part of the thinking of Reynolds [70] and Prandtl [67] as they considered how mass, momentum and energy were transported in turbulent fluids." Spalding light-heartedly likened Reynolds' intermingling-fragments concept as "stew", and Boussinesq's enlarged viscosity as "thick soup".

During the subsequent development of the two-fluid model at Imperial College it was discovered that ideas similar to those embodied in the two-fluid concept had already been proposed by several workers [11, 13, 39, 42, 117]. All of these studies were concerned with developing methods to predict intermittency within the framework of conditional zone-averaged conservation equations for the turbulent and non-turbulent zones.

4.1 Mathematical Formulation

The two-fluid turbulence model requires the simultaneous solution of two sets of differential equations which are coupled through laws defining the exchange of mass, momentum, energy and chemical species between the two fluids, which are assumed to be arranged in fragments. An additional constraint is that the volume fractions of the two fluids must sum to unity. Spalding [97, 100, 114] proposed expressions for determining the rates of exchange between the intermingling fluid fragments. These interaction terms involve the specific area of the interface between the two fluids, which Spalding proposed should be computed via a differential equation for the fragment size, with mechanisms that control its growth or diminution. In less advanced work, the fragment size is taken as proportional to some characteristic dimension of the flow.

Another important term appearing in the differential equations is the 'phase-diffusion' term, which originates from time-averaging the convection terms in the phase-continuity equations. Consequently, a related term also appears in the other differential equations where it augments the within phase diffusion term. The term is intended to represent the random-motion flux associated with the movement of the interface between the two fluids, i.e. the to-and-fro motion which occurs in addition to the mean motion of the phase in question. Spalding [99] proposed modelling this type of motion in terms of an effective diffusion coefficient and the local gradient of the volume fraction, with the former being taken as proportional to the mean rate of strain and fragment size.

4.2 Applications to Inert Turbulent Flows

Intermittent and other turbulent flows. Initial developments of the two-fluid model at Imperial College were concerned with predicting the intermittency of free turbulent shear layers. Both Malin [51–54, 116] and Xi [123], working under Spalding's direction, developed and applied two-fluid models for predicting the intermittency factor and flow variables in the turbulent and non-turbulent zones of free turbulent shear layers. These two models differed from Spalding's [99] original proposal mainly in their use of either Prandtl's mixing length [51, 54, 123] or a two-equation $k - \epsilon$ model [52–54, 116] for determining the fragment size, rather than compute it from a differential equation. Xi used a constant eddy-viscosity for modelling turbulent diffusion, whereas Malin used the eddy-viscosity computed from the turbulence model within the turbulent phase. When the $k - \epsilon$ model was used, it was modified to account for the additional production of turbulence at the interface.

Xi applied his model to steady and transient jets, both with and without buoyancy, and reported good agreement with measurements of mean and conditioned flow variables, jet-tip penetration rates, and intermittency profiles. For heated jets and wakes, Malin's model also produced results that were in reasonable agreement with both the conditioned and unconditioned data, but the intermittency factor was overpredicted for jets. This was attributed mainly to the interphase mass-transfer model not allowing for turbulent fluid to enter the non-turbulent category and so account for the physical mechanism of decay and dissipation. This deficiency was later addressed by Fan [15], who obtained much improved intermittency profiles for the round jet by adding an interphase mass sink term proportional to the mean rate of strain.

These early studies prompted Spalding [101, 104, 114, 116] to abandon the idea of classifying the two fluids as 'turbulent' and 'non-turbulent' in favour of a 'faster' and 'slower' split where turbulent fragments are presumed to be present in both fluids. The thinking was that this new approach would prove more general, because the earlier classification would be inadequate for internal flows, where downstream of the initial entrance region, there can be no non-turbulent fluid. Spalding [104] established a connection with Prandtl's mixing length theory to make three major modifications to his original two-fluid model [99]. The first was the use of a symmetric mass-transfer law between the two fluids to allow mass transfer to take place in both directions. The second was the introduction of a shear-related source in the cross-stream momentum equations, the purpose of which was to express "the tendency of a shear layer to breakup into a succession of vortices". The third modification concerned within-phase diffusion, for which the eddy-viscosity was replaced with a diffusion coefficient proportional to the fragment size and the slip between the two fluids. This was based on the notion that the turbulent-transport properties arise from "the mixing in the wakes of the fragments in relative motion".

The revised two-fluid model was applied first by Spalding [104] to a Couette flow with heat transfer, in which the shear-induced source of cross-stream momentum drives one of the fluids upwards while the other moves downwards. Further applications were undertaken by Ilegbusi [26–28] at Imperial College under Spalding's supervision, and these culminated in joint publications [30–32] which reported satisfactory agreement with experimental data on flat-plate boundary layers and Couette flow [31], and pipe and channel flows and plane and round jets with and without heat transfer [30, 32].

All of the foregoing studies computed the fragment size from Prandtl's mixing length, but another of Spalding's students, Fueyo [18], developed the model further by computing the fragment size from a differential equation which accounted for fragment growth by entrainment and agglomeration, and fragment breakup by shear. The model was applied to the round jet and the predictions were found to be in satisfactory agreement with the experimental data. An interesting outcome of this study was that after a certain jet-development region, the centreline values of the fragment size were found to be proportional to the Prandtl mixing length.

Rayleigh–Taylor Instability. Also at Imperial College, Andrews [2] developed a two-fluid model of different density fluids for the simulation of flows involving Rayleigh–Taylor instability (RTI). Such flows are of interest in astrophysics, geophysics and nuclear fusion; and they occur whenever a heavy fluid is placed above a light one in a force field, usually gravity. The interface between the two fluids is unstable to any disturbances and becomes increasingly distorted before degenerating finally into a turbulent mixing process. Andrews [2, 4] performed a RTI experiment by overturning a stably stratified tank of two fluids, and then applied a two-fluid model of the turbulence to simulate the mixing process. This model classified the fluids as light and heavy, and a differential equation was solved for the mixing length scale with account taken of the growth and reduction in size of the fluid fragments. The results were reported to be in good agreement with the experiments.

Andrews [3] also considered the de-mixing heated-saline experiment of Stafford [118] where the RTI mixing of cold water over hot saline is followed by a de-mixing phenomenon, which occurs when there is a reversal in the pressure gradient. This is an example of a double-diffusive process where the salinity and temperature make opposing contributions to the vertical density gradient. Simulations with the two-fluid model predicted successfully both the mixing and de-mixing phenomena. Conventional turbulence models cannot address 'counter-gradient' de-mixing because the gradient-diffusion hypothesis means that once mixing has taken place, de-mixing cannot happen.

Two-fluid descriptions of RTI have also been used by Youngs [124–127], Chen et al. [9], Llor [44, 45] and more recently by Kokkinakis et al. [38], who compared the results of a two-fluid model with those of a modified two-equation turbulence model with reference to high-resolution implicit large-eddy simulations (ILES) of compressible Rayleigh–Taylor mixing. Both the single and two-fluid models were reported to produce good agreement with ILES with respect to the self-similar mixing width; peak turbulent kinetic energy growth rate, as well as volume fraction and turbulent kinetic energy profiles. Despite being computationally more demanding, the two-fluid model was preferred because of its ability to represent the degree of molecular mixing directly by transferring mass between the two phases, and also because of its potential to model de-mixing when the acceleration reverses sign.

Metallurgical Applications. Ilegbusi [29, 33] applied the two-fluid model of turbulence to compute the flow distribution in a continuous-casting tundish, which is a broad, open bath containing molten metal with one or more holes in the bottom. These feed into an ingot mould during the casting process. Tundishes are characterised by highly turbulent flow regions near the inlet and outlet, and essentially quiescent (laminar) regions elsewhere in the bath. Consequently, the fluids were defined as turbulent and non-turbulent, and it was argued that the two-fluid concept was ideal for representing the coexisting zones of turbulent and non-turbulent flow. For both water models and steel systems, the two-fluid model produced better agreement with the measured residence times than the single-fluid $k - \epsilon$ model. The longer residence times obtained with the $k - \epsilon$ model were believed to be due to its tendency to overpredict mixing between the highly turbulent and largely quiescent regions.

Later, Sheng and Jonsson [76] and Anestis [5] employed similar two-fluid turbulence models to simulate transient flow and heat transfer in a tundish. The original liquid in the bath was defined as one fluid, and the inlet stream as the other fluid. Both studies reported that the two-fluid model showed better agreement with the measurements than the single-fluid $k - \epsilon$ model, and especially in transition and mixed-convection regions, where the two-fluid model predicted less turbulent transport. These studies confirmed Ilegbusi's [29, 33] earlier result that the two-fluid model captures the physics better when representing systems with localised highly turbulent regions and largely quiescent regions elsewhere.

Huang [24, 25], Ilegbusi [25, 34], and Pfender and Chang [65] all reported on two-fluid modelling of turbulent plasma jets, which are of interest in materials processing, including spray deposition. In all of these studies, the fluids were defined as hot (the plasma gas-argon) and cold (the ambient fluid). The results showed that the model can predict the observed unmixing and intermittency that escapes the more conventional turbulence models, which only account for gradient-diffusion mixing.

Other Applications. Shen et al. [74, 75] employed a two-fluid model based on the work of Fan [15] and Spalding [99] to simulate the turbulent stratified flow of lighter density fluid above a denser stream in a channel. This model classified the fluids as light and heavy and the $k - \epsilon$ turbulence model was used within each phase to determine the eddy-viscosity and fragment size. The model produced good agreement with measured velocity and relative density profiles, and the collapse of the turbulence under the influence of strong stable stratification was predicted correctly.

Yu et al. [128] applied two-fluid turbulence modelling to simulate the flow and heat transfer characteristics of air curtains in an open refrigerated display cabinet. The asymmetric and symmetric forms of the mass-transfer rate were investigated, as well as a weighted mean of the two relations. The fluids were classified as turbulent (the air curtains) and non-turbulent (the ambient air outside the cabinet), and the $k - \epsilon$ model was used within the turbulent phase. The two-fluid model was found to give better agreement with the measurements than the single-fluid $k - \epsilon$ model, and so a better prediction of the air curtain, the thermal field outside the case, and the cold-air overspill from the case into the store. This same two-fluid model was used later by Cao et al. [8] in combination with machine-learning methods to produce a strategy for optimising the design of air curtains for open-display cabinets.

Liu et al. [43] applied the revised two-fluid model of Spalding for predicting the turbulent flow in a closed-conduit polychromatic UV disinfection reactor. These types of reactors are widely used in treating both drinking water and wastewater. In addition to the turbulence modelling, this challenging application also included a fluence-rate model for the UV light-intensity distribution, and a microbial inactivation kinetic model to represent the fluence response of target microorganisms. In the two-fluid model, the fragment size was computed from the differential equation of Fueyo [18]. Overall, the results compared reasonably well with the measurements, but no better than those produced by four other single-fluid two-equation turbulence models and a second-moment closure.

More recently, Zhang et al. [129] used Spalding's two-fluid concept to develop a turbulence model based on the EMMS (energy-minimization multi-scale) method [41] for fluid-solid systems. This EMMS-based turbulence model regards single-phase flow as a mixture of turbulent and non-turbulent fluids; and a multi-scale analysis divides the energy of the turbulence into three separate scales: a molecular scale, an eddy scale and a macro scale. A turbulence stability condition plus several constraint equations are then used to close a set of turbulent dynamic equations, which are then solved to produce optimised values of the volume fraction and diameter (fragment size) of the turbulent eddies. For any given flow system, using as input the superficial velocity of the turbulent eddies and some other parameters, the volume fraction and fragment size are computed in advance of the CFD simulation. The EMMS-based turbulence model then uses a table in the CFD simulation to modify the local values of the eddy-viscosity produced by a conventional turbulence model. This novel approach was applied to simulate lid-driven cavity flow and turbulent forced convection in an empty room. The EMMS-based models showed improved performance over the unmodified turbulence models by capturing the detailed secondary and tertiary vortices in the corners of both the cavity and the room. It was argued that the standard models were less successful because they regard the fluid to be in a turbulent state everywhere, but near walls and especially in the corners, viscosity dominates rather than inertia, and so the fluid should be close to laminar instead of fully turbulent. In contrast, the EMMS-based turbulence model treats the flow everywhere as a mixture of laminar and turbulent fluids.

4.3 Applications to Combustion

Spalding's inspiration for the development of two-fluid models of turbulence was perhaps the fragmentariness exhibited by many classes of turbulent flow. Such fragmentariness is seldom more apparent than in a turbulent flame, where burned and unburned fragments intermingle at the flame front. To describe graphically the model concept, Spalding used the image of a rugged coastline, "with many islands and land-locked lagoons" [100] as shown in Fig. 7. (For historical interest, the figure also shows one Spalding's hallmark lecture panels. In a private conversation in 2008 with one of the present authors, NF, Spalding humorously shared his belief that such format was possibly his greatest contribution to Science.)

Internal combustion engines were perhaps the inspiration for Spalding's Two-Fluid Models, possibly because they exhibit, like no other flow, "patchiness": sharp changes in fluid properties (temperature, composition, perhaps even velocity) over small distances [98].

Combustion Two-Fluid Models are described by the averaged multiphase Eulerian equations. For a generic variable ϕ_i in phase i (such as momentum per unit mass,

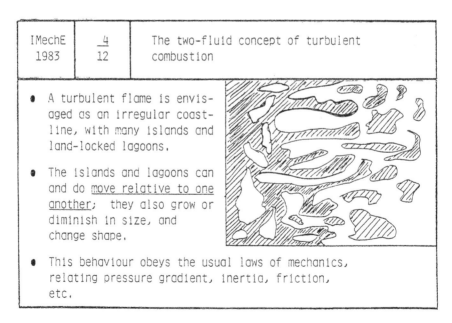

Fig. 7 Spalding's lecture panel, portraying a flame front as an irregular coastline, with islands and land-locked lagoons. Reproduced from [100]

mass fraction, enthalpy per unit mass), its transport equation is (the overbar denoting averaging being dropped for simplicity)

$$\frac{\partial (\rho_i r_i \phi_i)}{\partial t} + \nabla \cdot (\rho_i r_i \phi \mathbf{v}_i) - \nabla \cdot (\Gamma_r \phi_i \nabla r_i) - \nabla \cdot \left(\Gamma_\phi r_i \nabla \phi_i\right)$$
$$= S_{\phi_i} + f_{ji} \left(\phi_j - \phi_i\right) + [[\dot{m}_{ji}]]\phi_j - [[-\dot{m}_{ji}]]\phi_i \quad , \tag{6}$$

where

- r_i is the volume fraction of phase i.
- Γ_ϕ is a diffusion coefficient, encompassing turbulent diffusion if an eddy-viscosity model is used.
- S_{ϕ_i} is a source term, such as a component of the pressure gradient in the momentum equation.
- f_{ji} is an interphase diffusion coefficient (such as drag, or heat transfer between the phases by combustion).
- $\dot{m}_{ji}$ is the interphase mass-transfer rate from phase j into phase i.
- $[[\odot]]$ indicates the maximum of 0 and $\odot$.

The equation for the volume fraction r_i of fluid i is

$$\frac{\partial (\rho_i r_i)}{\partial t} + \nabla \cdot (\rho_i r_i \mathbf{v}_i) - \nabla \cdot (\Gamma_r \nabla r_i) = \dot{m}_{ji} \quad . \tag{7}$$

The mass-transfer rate between the fluids $\dot{m}_{ji}$ is a crucial parameter in Two-Fluid Models. It accounts for the entrainment of one fluid into the other. Several formulations were proposed at the time, usually involving a relative phase velocity, $|\mathbf{v}_i - \mathbf{v}_j|$ and a fragment size l (or equivalently the interface surface area per unit volume l^{-1}), for which a model needs to be provided.

Spalding [102] discusses the potential of the two-fluid model for simulating a wide range of combustion problems, but practical applications appear to have been limited to transient one-dimensional premixed flames [55, 102, 122] and unbounded fire plumes [56].

Spalding regarded [106] his earlier Eddy Breakup Model as a precursor to his Two-Fluid Model, in which the two fluids were the fresh reactants and the burned products. He also viewed the Eddy-Dissipation Concept of Magnussen [49], flamelet models of combustion [63], and the Bray–Libby–Moss model [7] as two-fluid models embodied in single-phase frameworks.

4.4 Concluding Remarks

Spalding's Two-Fluid Model was a novel approach for modelling turbulence that provided an alternative formulation capable of predicting intermittency and counter-gradient diffusion. Also, unlike any single-fluid turbulence model, the model can portray adequately the interactions of pressure gradients and density fluctuations which are major sources of generation of turbulent motion in certain applications. Despite these advantages the model failed to gain widespread popularity because it offered no superiority over conventional models for most engineering and environmental applications. For general usage, the two-fluid concept was hard to grasp because of uncertainty over the best characteristic to use for distinguishing the two fluids. Another deterrent was the increased computer time and convergence demands associated with the need to solve twice as many differential equations, as compared to conventional models.

Spalding later concluded that in one respect, the Two-Fluid Model was over-elaborate on the grounds that the relative velocities of the fluids are often small enough to be neglected, or computed by way of an algebraic-slip approximation, which permits extension to the treatment of the relative motion of more than two fluids, which is what is needed for greater realism. This led Spalding to turn his attention to multi-fluid turbulence models.

5   Population-Type Multi-fluid Models

In the mid-90s, Spalding concluded that two fluids did not suffice to represent the complexity of the composition field in combustion situations, and adopted the idea of Multi-Fluid Models of turbulent combustion. Unlike Two-Fluid Models, which were multiphase in nature, Spalding's Multi-Fluid Models are single-phase, and based on the idea of discretising one or several relevant fluid properties. The notion of discretising properties was not new to Spalding. The six-flux model of radiation discretised the radiation fluxes in positive and negative directions [94], and non-slip clouds of particles had been discretised into particle sizes.

At the time, Spalding was aware of the particle-based Monte Carlo methods for solving the multi-dimensional composition Probability Density Function (PDF) in turbulent combustion [66], and often saw the newly born Multi-Fluid models as a more economical way to compute the PDF [79]. Striving as always to find the best word to describe an idea, he often referred to these Multi-Fluid Models as "Population Models", in that the fluid is represented by an ensemble of fluids, each fluid in the ensemble being a population member.

5.1   A Four-Fluid Model

The inspiration for Multi-Fluid Models appears to have presented itself to Spalding around 1995 [115] as he reflected on the limitations of his popular Eddy-Breakup Model, and its widely adopted successor the Eddy-Dissipation Model of Magnussen.

The Eddy-Breakup Model, under the Multi-Fluid lens, consisted of just two "fluids": one is the fresh mixture of fuel and oxidant, and other is the completely burned products of their combustion. Of course, intermediate states of partially burned mixture do exist in the flame, but they were assumed to be confined to very thin layers in the interface between both, and hence were neglected in the overall thermodynamics of the mixture. Flamelet models [64] place special importance on this interface region. Spalding would at some point draw a similarity between flamelet models and his Multi-Fluid ones [78], the nexus being the notion of an 'encounter' between fluid parcels whereby they "approach and make contact; remain in contact for a short time, say T_{cont}; and then separate". This process, and the contact time, is the essence of his Multi-Fluid Models.

The first stepping-stone in the development of a Multi-Fluid Model was a Four-Fluid Model. The model was originally contrived as a means of obtaining greater realism than the original Eddy-Breakup Model could afford. While the original Eddy-Breakup Model consisted only of fresh, unburned reactants (R) and fully burned products (P), the new Four-Fluid Model allowed for two mixtures of these: a non-reacting mixture (N) of reactants and products (with given, fixed proportions) that is too cold to burn, and a mixture (K) of reactants and products (with given, fixed proportions) that can burn (their burning rate being controlled by chemical kinetics). Therefore, unlike the single pathway from R to P in the original Eddy-Breakup Model, the new model allowed for three parallel routes:

$$R + P \rightarrow K \rightarrow P \tag{8}$$

$$R + P \rightarrow N; N + P \rightarrow K \rightarrow P \tag{9}$$

$$R + K \rightarrow N; N + P \rightarrow K \rightarrow P \tag{10}$$

Thus, while the original Eddy-Breakup Model consisted only for states R and P, the Four-Fluid Model allows for intermediate states N and K; and, crucially, both mixing and chemical reaction have an influence on the overall rate of conversion, as will be discussed next.

The last reaction in all the pathways is the creation of newly burned products P from flammable mixture K; such creation was controlled by chemical kinetics, thus allowing "chemical-kinetic effects to be introduced rationally" in the model:

$$\dot{m}_P = m_K R_{chem} \quad , \tag{11}$$

where R_{chem} is a kinetic rate (e.g., given by an Arrhenius law).

The remaining steps are mixing-controlled, and are modelled, similar to the Eddy-Breakup Model, using the turbulence kinetic energy k and its rate of dissipation ϵ; for instance, for the first reaction, the rate of creation of mass of K fluid would be

$$\dot{m}_K = C m_R m_P R_{mix} \quad , \tag{12}$$

and the corresponding sinks for R and P. Here, C is a constant that may vary for each reaction step.

The mixing rate R_{mix} may be given by a mechanical mixing rate, such as (in the original Eddy-Breakup model):

$$R_{\text{mix}} = \frac{\epsilon}{k} \qquad . \tag{13}$$

The Four-Fluid Model was demonstrated by Spalding in simple geometries. The first one was, incidentally, "the same problem as that for which the EBU was invented, namely, that of steady turbulent flame spread in a plane-walled duct". He reported [80] that his new Four-Fluid Model exhibited a similar qualitative behaviour as in the experiments of [121] *viz* that the rate of spread of the flame very little depends on the fuel-air ratio, the velocity or the temperature of the incoming mixture.

A transient version of the Four-Fluid Model was also used to predict flame propagation in a duct with baffles [16], and an explosion in an off-shore platform [17].

More recently, Hampp and Lindstedt [21] have used the multi-fluid concept of Spalding to identify, experimentally, burning regimes at low Damköhler numbers. Hampp and Lindstedt employ simultaneous Mie scattering, PIV and OH-PLIF to identify, in the flame, fluid fragments in one of the following categories: fresh reactant fluid, mixing fluid, mildly reacting fluid, strongly reacting fluid and product. This classification is very similar to that employed by Spalding in the four-fluid model just described; Hampp and Lindstedt, however, use two categories to represent the reacting fluid.

5.2   Population Models

Spalding soon concluded that the main benefit of the foregoing Four-Fluid Model would be educational [106] (because of its simplicity and economy), and that more realism would require more fluids. The generalisation was in the form of population models, where the flow is made up of a number of 'fluids'. The key ideas in Multi-Fluid Models are

1. Fluids are distinguished from one another through one or several properties (such as mixture fraction, or the concentration of a chemical species, or temperature) called Population-Distinguishing Attributes.
2. The value of such a property in a fluid is constant, and different from that in other fluids, so that all the fluids combined *discretise* the property.
3. Fluid parcels coalesce and mix, exchanging mass and properties and creating intermediate, or "offspring", fluids.
4. Further, the increase in the value of the property in the flow is represented by mass transfer from fluids with low values to fluids with high values (and conversely for the decrease); this process is referred to below as *transport in population space*.
5. The fluids also have non-discretised properties, called Continuously Varying Attributes. These are also exchanged as the fluids coalesce and mix.

5.2.1 The Population-Distinguishing Attributes

In Multi-Fluid Models, the properties that are used to distinguish each fluid are called Population-Distinguishing Attributes, or PDAs. In a non-premixed flame, for instance, mixture fraction is a prime candidate for a PDA: the mixture-fraction space will be discretised, and different fluids will have different (but fixed) values for the mixture fraction; this is a one-dimensional fluid population. Additional properties may be needed for other problem classes. For instance, a reaction-progress variable, such as reactedness, may be additionally required, resulting in a two-dimensional fluid population.

Figure 8 represents a one-dimensional fluid population where the population-distinguishing attribute is a generic variable ξ. This is discretised into 'bins', each with a constant value of the variable ξ; for fluid γ, this value is ξ_γ, and m_γ is the mass fraction of the overall fluid with this value of ξ.

For the mass fraction of fluid γ, m_γ, a conservation equation is solved

$$\frac{\partial(\rho m_\gamma)}{\partial t} + \nabla \cdot \left(\rho \mathbf{v} m_\gamma\right) - \nabla \cdot (\Gamma_\gamma \nabla m_\gamma) = \dot{m}_\gamma \quad . \tag{14}$$

To be clear, this is not the conservation equation for the mass fraction of a species γ, as in a conventional turbulence model; it may be, in a Multi-Fluid Model, the mass fraction of fluid γ with a given fixed value, for the mass fraction of a chemical species. The terms on the left-hand side of the equation are the usual transient, convection, and diffusion terms in the Eulerian conservation equation, representing accumulation and transport in physical space. The local velocity $\mathbf{v}$ is often presumed to be the same for all the fluids, albeit Spalding envisaged ways to remove this limitation, for instance, if the density of the several fluids differ so much as to make their different response to body forces relevant; also, the diffusion coefficient Γ_γ is taken as being the same for all fluids.

The source term in Eq. 14 accounts for two distinct processes in the context of Multi-Fluid Models: the exchange of matter between fluids as they collide and mix, $\dot{m}_\gamma^{\text{mixing}}$, and transport in population space (for instance, in *compositional* space due

Fig. 8 A one-dimensional fluid population using a generic distinguishing attribute ξ, arbitrarily varying between 0 and 1

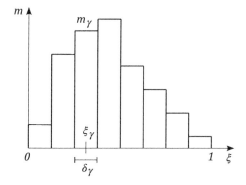

to chemical reaction if the attribute γ is composition-related), $\dot{m}_\gamma^{\text{convection}}$:

$$\dot{m}_\gamma = \dot{m}_\gamma^{\text{mixing}} + \dot{m}_\gamma^{\text{convection}} \quad . \tag{15}$$

These are explained in the next subsections.

5.2.2 Fluid-Fragment Collision and Mass Exchange

The notion of fluid-fragment collision and mass exchange is central to Spalding's Multi-Fluid Models; it is often referred to by Spalding as "coupling and splitting", and is akin to the concept of micro-mixing, or molecular mixing, used in transported-PDF combustion models [69]. The physical image of 'Spalding's micro-mixing' is that fragments of two fluids briefly collide, partially mix, and then separate leaving some "offspring". In his own words [108]:

1. "Two fragments of fluid are brought into temporary contact by the random turbulent motion [...];
2. Molecular and smaller scale turbulent mixing processes cause intermingling to occur [between] the coupling fragments;
3. Before the intermingling is complete, however, the larger scale random motions cause the fragments to be plucked away again, with the result that the amounts of the material having the compositions of the parent fluids [...] are diminished, while some fluid material of intermediate composition has been created [...]."

The rate of micro-mixing can be computed in several ways. A general model proposed by Spalding is

$$\dot{m}_\gamma^{\text{mixing}} = \sum_{\alpha,\beta} \rho F_{\alpha\beta}^\gamma m_\alpha m_\beta T_{\alpha\beta} \quad . \tag{16}$$

In this equation, $F_{\alpha\beta}^\gamma$ is the fraction of fluid from the encounter between fluids α and β that enters fluid γ; and $T_{\alpha\beta}$ is a turbulence frequency (with dimensions s^{-1}).

Spalding stated [108] that "the crux of multi-fluid modelling lies in the formulae chosen for [these] functions", and that "physical intuition, mathematical analysis, guess-work and computational parsimony all play a part in their choices".

He nevertheless suggested a simple, intuition-based equation for $F_{\alpha\beta}^\gamma$:

$$F_{\alpha\beta}^\gamma = \begin{cases} -0.5 & \text{if } (\gamma = \alpha \text{ or} \gamma = \beta) \text{ and } (\beta > \alpha + 1) \\ 0 & \text{if } \gamma < \alpha \text{ or } \gamma > \beta \text{ or } \beta = \alpha + 1 \\ \frac{1}{\beta - \alpha - 1} & \text{otherwise} \end{cases} \quad . \tag{17}$$

For the mixing frequency $T_{\alpha\beta}$, he suggested that it should be independent of the parent fluids. His most often-used proposal, connecting with the origins of the Eddy-

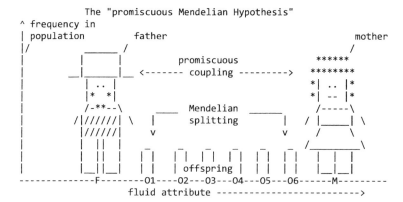

Fig. 9 The "promiscuous-Mendelian" mixing model as illustrated by Spalding [108] using ASCII art

Breakup Model and its successors, is to use a kinetic-energy decay-frequency, such as ϵ/k if the $k - \epsilon$ turbulence model is used.

Which parent fluids α and β collide and the nature of the offspring γ are key aspects of Spalding's micro-mixing models. He termed the above-described model as "promiscuous-Mendelian". The model was often illustrated by Spalding using the ASCII art rendition shown in Fig. 9 [108]. The promiscuity attribute is given because any two parent fluids α and β may collide; the Mendelian aspect is due to the offspring fluids γ possessing, in varying proportions, the attributes of either parent.

5.2.3 Transport in Population Space

When a flow property (such as temperature, or mixture fraction) is a PDA, the property is discretised and given a *constant* value from this discretisation in each of the fluids making up the population (see Fig. 8). For instance, if a PDA is temperature then the fluid is regarded as made up of N component fluids having different temperature levels $T_1 < T_2 < \cdots < T_N$.

This bears the question of how the flow property, e.g. temperature, increases or decreases its value at a point in the domain as a consequence of property sources and sinks (for instance, changes in temperature in the flow through heating). Fluid-fragment collision (or "coupling and splitting", or micro-mixing), discussed above, changes the amount mass in the involved fluids, but not the value of the property as a whole in the flow, since there is no net gain or loss of property in the exchange brought about by the collision.

Therefore, in Multi-Fluid Models, sources or sinks of a property are implemented by shifting the fluid population distribution towards higher or lower values of the property, via fluid-to-fluid mass transfer.

Let us assume that, for a fluid γ, a PDA ξ has a value ξ_γ (constant); that the mass fraction of fluid with such value for ξ is m_γ; and that the property ξ is changing locally in the flow at a rate $\dot\xi$ (see Fig. 8); then, Spalding [108] indicates that the mass source of fluid γ is

$$\dot m_\gamma^{\text{convection}} = \begin{cases} \dot\xi \rho \frac{m_{\gamma-1}}{\delta_{\gamma-1}} & \text{(from fluid } \gamma - 1 \text{ to fluid } \gamma) \text{ if } \dot\xi > 0 \\ \dot\xi \rho \frac{m_{\gamma+1}}{\delta_{\gamma+1}} & \text{(from fluid } \gamma + 1 \text{ to fluid } \gamma) \text{ if } \dot\xi < 0 \end{cases} \tag{18}$$

Here, δ_γ is the width of the γ interval in PDA space, see Fig. 8. Opposite sinks must be provided for the donor fluid, so that mass is preserved.

5.2.4 Continuously Varying Attributes

PDA's have discrete values, each represented by a fluid. In addition, the fluids may have non-discretised properties, which Spalding termed Continuously Varying Attributes, or CVA's. For instance, in a turbulent combusting flow one may choose the mixture fraction as a PDA, and hence have a number of fluids each with a set value for the mixture fraction; then, each fluid may have a continuously varying value of (for instance) nitrogen-oxide mass fraction.

The transport equation for a continuously varying property ϕ_γ in fluid γ (the local mass fraction of which is m_γ) is

$$\frac{\partial(\rho m_\gamma \phi_\gamma)}{\partial t} + \nabla \cdot (\rho \mathbf{v} m_\gamma \phi_\gamma) - \nabla \cdot (\Gamma_\gamma \nabla m_\gamma \phi_\gamma) = S_{\phi_\gamma} \quad . \tag{19}$$

The last term on the left-hand side represents diffusion processes, and the term on the right-hand side is the source of ϕ_γ. Such source will usually encompass three distinct contributions.

The first one is the within-fluid source, due, for instance, to chemical reaction (if the CVA ϕ_γ is chemical species) or to heat transfer by radiation (if the CVA ϕ_γ is temperature or enthalpy).

The second contribution to the source term, say $S_{\phi_\gamma}^{\text{mixing}}$, is the contribution from the coupling and splitting process. In the simplest model, when fragments of (say) fluid α and β collide to create fluid γ, the source of ϕ_γ due to mixing is taken as the average of the parents' values:

$$S_{\phi_\gamma}^{\text{mixing}} = \dot m_\gamma^{\alpha\beta} \frac{\phi_\alpha + \phi_\beta}{2} \quad . \tag{20}$$

(Readers familiar with transported-PDF methods will recognise the similarity with the Linear Mean Square Estimation, or LMSE, model of Dopazo [10].)

The third contribution is an 'inter-fluid' diffusion of the CVA; for instance, if the CVA is enthalpy, the heat exchange due to fragment-to-fragment heat transfer between two fluids.

For both PDA's and CVA's, the local mean of the property can be computed from the fluid population distribution; for instance, the local average of CVA ϕ is

$$\overline{\phi} = \sum_{\gamma} m_{\gamma} \phi_{\gamma} \quad . \tag{21}$$

Other moments can be computed similarly.

5.3 Towards Population CFD

In the Four-Fluid Model presented in Sect. 5.1, the fluids were distinguished by their reactedness. Spalding's next step towards 'population CFD' was to increase not only the number of fluids, but also the number of attributes used to distinguish the fluid population.

In this Fourteen-Fluid Model [105], the compositional space is discretised along the reacted fuel fraction *and* the mixture-fraction dimensions, as indicated in Fig. 10. (The reacted fuel fraction measures reaction progress, and is akin to reactedness.) Thus fluids 1 and 14 are the original fluid streams (say air and fuel, respectively); fluids 13–16 are those with stoichiometric composition, and a varying progress of reaction. Dashed fluids in Fig. 10 are not compositionally accessible (because the reacted fuel fraction cannot exceed the mixture fraction), and thus even if the underlying discretisation of the compositional space is 5 × 4, only fourteen fluids are used.

Thus,

- "Fluids 6, 11, 16 and 20 are supposed to contain no unburned fuel; they thus represent completely burned gases, of various fuel-air ratios, which can react no further.

Fig. 10 Fluid population in the Fourteen-Fluid Model

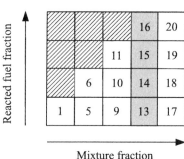

Mixture fraction

- Fluids 5, 9, 13, 14, 17 and 18 contain finite amounts of free fuel; but they are regarded as being too cold to burn, like fluid [N] above.
- Fluids 10, 15 and 19 both contain fuel and are hot enough to burn; it is therefore they which, like fluid [K] in the four-fluid model, carry out the chemical reaction process. Fluid 10 thus becomes transformed into 11, fluid 15 into 16, and fluid 19 into 20."

This Fourteen-Fluid Model was applied by Spalding to the simulation of a Bunsen burner-type flame [105].

Spalding also exemplified the use of models with large numbers of fluids [108], albeit for 0-dimensional (in space) problems. Thus, he used a 100-fluid model to solve a "well-stirred reactor" (a partially stirred reactor is a more common name), where unreacted and fully reacted mixtures entered, and a partially reacted mixture exited). Since there is a single mixture fraction, a one-dimensional population suffices, with reaction progress (or reactedness) as the distinguishing attribute.

He also demonstrated a similar calculation of a non-premixed reactor, where fuel and oxidant enter separately, mix and burn. In this case, a two-dimensional population of fluids is required, with mixture fraction and reactedness as population-distinguishing attributes. He used increasingly large 'population grids' of 3×3, 5×5, 7×7 and 11×11 fluids, showing that the last two where accurate enough.

Practical applications of Multi-Fluid Models to turbulent diffusion flames have been reported by Zhubrin [130].

Around 2010, Spalding started using the concepts of "populational modelling" and even "populational CFD" [109] to describe this new type of model that, he thought, would revolutionise CFD. Indeed, Spalding last ideas on the subject were published posthumously with the appellation "The Discretised Population Model of Turbulence" [77].

6 Conclusions

Spalding's nearly 70-year scientific career is heaving with ground-breaking contributions; a large fraction of them were in the field of combustion.

His underlying vision of turbulent combustion was that of fragments of fluids with different compositions and temperatures colliding and mixing, thus creating opportunities for chemical reaction to take place.

This was indeed the *leit-motiv* of his Eddy-Breakup Model. The limited tools available at the time resulted in its embodiment in a simple mathematical expression that, although widely used due to its simplicity and much refined in subsequent evolutions, was clearly seen by Spalding as constraining.

A first attempt at removing the constraints was the ESCIMO model. Although it did enjoy some success, it was short-lived. The ESCIMO model is similar in concept to Kerstein's Linear Eddy Model.

From the early 1980s, and thanks to the emergence and widespread use of two-phase Eulerian–Eulerian models, Spalding turned his attention to their use for the modelling of turbulent flows in general, and of turbulent combustion in particular. For these, these multiphase models had the advantage of allowing for locally high density differences, and thus for accounting for the effect of body forces in turbulent combustion, such as counter-gradient diffusion.

Multiphase two-fluid models quickly evolved in the 1990s into single-phase, multi-fluid models. Spalding viewed these models as an alternative to transported-PDF models, and used the term "populational CFD" to refer to extensions of his theory to other fields, including turbulent non-reacting flows. He would dedicate to his multi-fluid models, or populational CFD, a great fraction of the latter part of his career; they often featured in his talks and presentations as he continued to travel the world to share, with undiminished stamina, his vision of combustion, CFD and Science.

Annex: A Rough Chronology

See Table 1.

Table 1 A rough timeline of Spalding's career in Combustion. Dates are approximate

DBS	Others	Milestone
1951		Ph.D. Thesis, Pembroke College, University of Cambridge
	1953	First numerical computations of combustion [23]
1954		First numerical computations "by graphical means"
1955		Textbook 'Some Fundamentals of Combustion', Butterworth Scientific
1955		Analogue combustion model
1961		First numerical calculations
1968		First CFD calculation of recirculating flows with combustion
1969		GENMIX
1970		Presumed-PDF modelling of turbulent diffusion flames
1971		Numerical prediction of laminar flame propagation
1971		Eddy Breakup Model
	1972	First 3D simulation of a combustion chamber
1973		First 3D combustion simulation
	1974	Transport equation for the PDF [12]
1976		ESCIMO
	1977	Eddy Dissipation Model
	1977	BML model
1980		IPSA: Eulerian, multiphase algorithm
	1981	First Monte Carlo method for solving the transport eq. for the PDF [66]
	1981	Eddy Dissipation Concept
1982		Two-Fluid Models of turbulence
	1984	Computational implementation of flamelet models by Peters [63]
1995		Multi-Fluid Models of combustion

References

1. Adler, J., Spalding, D. B., & Jones, H. (1961). One-dimensional laminar flame propagation with an enthalpy gradient. *Proceedings of the Royal Society of London. Series A. Mathematical and Physical Sciences, 261*(1304), 53–78.
2. Andrews, M. J. (1986). *Turbulent mixing by Rayleigh–Taylor instability*. Ph.D. thesis, Imperial College, University of London.
3. Andrews, M. J. (1986). A two-fluid model of de-mixing in the saline experiment. CFDU/86/11, Imperial College, London.
4. Andrews, M. J., & Spalding, D. B. (1990). A simple experiment to investigate two-dimensional mixing by Rayleigh-Taylor instability. *Physics of Fluids A: Fluid Dynamics, 2*(6), 922–927.
5. Anestis, S. (2015). Tundish operation of a two fluid model using two scale k-epsilon turbulence model in a real power law fluid. *American Journal of Science and Technology, 2*(2), 55–73.
6. Boussinesq, J. V. (1877). *Théorie de l'écoulement tourbillonnant*. Paris: Mémoire présente par la division savante.
7. Bray, K. N. C., Libby, P. A., & Moss, J. B. (1985). Unified modeling approach for premixed turbulent combustion—Part I: General formulation. *Combustion and Flame, 61*(1), 87–102.

8. Cao, Z., Gu, B., Han, H., & Mills, G. (2010). Application of an effective strategy for optimizing the design of air curtains for open vertical refrigerated display cases. *International Journal of Thermal Sciences, 49*(6), 976–983.

9. Chen, Y., Glimm, J., Sharp, D. H., & Zhang, Q. (1996). A two-phase flow model of the Rayleigh-Taylor mixing zone. *Physics of Fluids, 8*(3), 816–825.

10. Dopazo, C. (1973). *Non-isothermal turbulent reactive flows: Stochastic approaches.* Ph.D. thesis, State University of New York, Stony Brook.

11. Dopazo, C. (1977). On conditioned averages for intermittent turbulent flows. *Journal of Fluid Mechanics, 81*, 433–438.

12. Dopazo, C., & O'Brien, E. E. (1974). An approach to the autoignition of a turbulent mixture. *Acta Astronautica, 1*, 1239–1266.

13. Duhamel, P. (1981). A detailed derivation of conditioned equations for intermittent turbulent flows. *Letters Heat Mass Transfer, 8*, 491–502. Dec.

14. Fan, W. C. (1982). ESCIMO applied to the hydrogen-air diffusion flame with refined treatment of biography. Technical Report CFDU/82/9, CFDU, Dept. Mechanical Engineering, Imperial College.

15. Fan, W. C. (1988). A two-fluid model of turbulence and its modifications. *Science in China (Series A), 31*(1), 79–86.

16. Freeman, D. J., & Spalding, D. B. (1995). The multi-fluid turbulent combustion and its application to simulation of gas explosions. *The PHOENICS Journal of Computational Fluid Dynamics and its Applications, 8*(3), 225–295.

17. Freeman, D. J., & Spalding, D. B. (1997). Structural forces induced by gas explosions in offshore oil platforms. In *Application of fluid dynamics in the safe design of topsides and superstructures.* The Institute of Marine Engineers.

18. Fueyo, N. (1990). *Two-fluid models of turbulence for axisymmetrical jets and sprays.* Ph.D. thesis, Imperial College.

19. Gosman, A. D., & Lockwood, F. C. (1973). Prediction of influence of turbulent fluctuations of the flow and heat transfer in furnaces. Technical Report HTS/73/53, Dept. of Mechanical Engineering, Imperial College, University of London.

20. Gosman, A. D., Pun, W. M., Runchal, A. K., Spalding, D. B., & Wolfshtein, M. (1969). *Heat and mass transfer in recirculating flows.* New York: Academic Press.

21. Hampp, F., & Lindstedt, R. P. (2017). Quantification of combustion regime transitions in premixed turbulent DME flames. *Combustion and Flame, 182*, 248–268.

22. Harlow, F. H., & Nakayama, P. I. (1968). Transport of turbulence energy decay rate. Report LA-3854, Los Alamos Scientific Laboratory of the University of California, Los Alamos, New Mexico, January 1968.

23. Hirschfelder, J. O., Curtiss, C. F., & Campbell, D. E. (1953). The theory of flame propagation IV. *The Journal of Physical Chemistry, 57*(4), 403–414.

24. Huang, P. (1993). *A turbulent swirling arc model and a two-fluid turbulence model for thermal plasma sprays.* Ph.D. thesis, University of Minnesota.

25. Huang, P. C., Hebeylein, J., & Pfender, E. (1995). A two-fluid model of turbulence for a thermal plasma jet. *Plasma Chemistry and Plasma Processing, 15*(1), 25–46.

26. Ilegbusi, O. J. (1984). Progress in the development of a two-fluid model of turbulence for near-wall flows—I. Technical Report CFDU/84/7, Imperial College, University of London.

27. Ilegbusi, O. J. (1984). Progress in the development of a two-fluid model of turbulence for near-wall flows—II. Heat transfer characteristics of a boundary layer. Technical Report CFDU/84/8, Imperial College, University of London.

28. Ilegbusi, O. J. (1984). Progress in the development of a two-fluid model of turbulence for near-wall flows—III. Preliminary investigation of a stably-stratified boundary layer. Technical Report CFDU/84/9, Imperial College, University of London.

29. Ilegbusi, O. J. (1994). Application of the two-fluid model of turbulence to tundish problems. *ISIJ International, 34*(9), 732–738.

30. Ilegbusi, O. J., & Spalding, D. B. (1987). Application of a two-fluid model of turbulence to turbulent flows in conduits and free shear-layers. *Physicochemical Hydrodynamics, 9*, 161–181.

31. Ilegbusi, O. J., & Spalding, D. B. (1987). A two-fluid model of turbulence and its application to near-wall flows. *PhysicoChemical Hydrodynamics, 9*, 127–160.
32. Ilegbusi, O. J., & Spalding, D. B. (1989). Prediction of fluid flow and heat transfer characteristics of turbulent shear flows with a two-fluid model of turbulence. *International Journal of Heat and Mass Transfer, 32*(4), 767–774.
33. Ilegbusi, O. J. (1994). The two-fluid model of turbulence and its application in metals processing. *Journal of Materials Processing and Manufacturing Science, 3*, 143.
34. Ilegbusi, O. J. (1995). Mathematical modelling of mixing and unmixedness in plasma jets. In V. R. Vollerand, S.P. Marsh, & N. El Kaddah (Eds.), *Materials processing in the computer age–II*.
35. Jones, W. P., & Whitelaw, J. H. (1982). Calculation methods for reacting turbulent flows: A review. *Combustion and Flame, 48*, 1–26.
36. Kerstein, A. R. (1988). A linear-eddy model of turbulent scalar transport and mixing. *Combustion Science and Technology, 60*(4–6), 391–421.
37. Khalil, E. E., Spalding, D. B., & Whitelaw, J. H. (1975). The calculation of local flow properties in two-dimensional furnaces. *International Journal of Heat and Mass Transfer, 18*(6), 775–791.
38. Kokkinakis, I. W., Drikakis, D., Youngs, D. L., & Williams, R. J. R. (2015). Two-equation and multi-fluid turbulence models for Rayleigh-Taylor mixing. *International Journal of Heat and Fluid Flow, 56*, 233–250.
39. Kollmann, W. (1984). Prediction of the intermittency factor for turbulent shear flows. *AIAA Journal, 22*, 486–492.
40. Kolmogorov, A. N. (1942). Equations of motion of an incompressible turbulent fluid. *Izvestiya Akademii Nauk SSR, Seriya Fizicheskaya, 6*, 56–58.
41. Li, J., Tung, Y., & Kwauk, M. (1988). Method of energy minimization in multi-scale modeling of particle-fluid two-phase flow. In *Circulating fluidized bed technology* (pp. 89–103). Elsevier.
42. Libby, P. A. (1975). On the prediction of intermittent turbulent flows. *Journal of Fluid Mechanics, 68*, 273–295.
43. Liu, D., Wu, C., Linden, K., & Ducoste, J. (2007). Numerical simulation of UV disinfection reactors: Evaluation of alternative turbulence models. *Applied Mathematical Modelling, 31*(9), 1753–1769.
44. Llor, A. (2005). *Statistical hydrodynamic models for developed mixing instability flows* (Vol. 681). Springer Science & Business Media.
45. Llor, A., & Bailly, P. (2003). A new turbulent two-field concept for modeling Rayleigh-Taylor, Richtmyer-Meshkov, and Kelvin-Helmholtz mixing layers. *Laser and Particle Beams, 21*(3), 311–315.
46. Ma, A., Noseir, M., & Spalding, D. B. (1980). An application of the ESCIMO theory of turbulent combustion. In *18th Aerospace Sciences Meeting*, volume AIAA Paper No. 80-0014. AIAA.
47. Ma, A. S. C., Spalding, D. B., & Sun, R. L. T. (1982). Application of "ESCIMO" to the turbulent hydrogen-air diffusion flame. In *Nineteenth Symposium (International) on Combustion* (Vol. 19, No. 1, pp. 393–402).
48. Magnussen, B. F. (1981). *On the structure of turbulence and a generalized eddy dissipation concept for chemical reaction in turbulent flow*. American Institute of Aeronautics and Astronautics.
49. Magnussen, B. F. (1989). Modeling of NOx and soot formation by the eddy dissipation concept. In *International Flame Research Foundation First Topic Oriented Technical Meeting* (pp. 17–19).
50. Magnussen, B. F., & Hjertager, B. H. (1977). On mathematical modeling of turbulent combustion with special emphasis on soot formation and combustion. In *Symposium (International) on Combustion* (Vol. 16, No. 1, pp. 719–729).
51. Malin, M. R. (1983). Progress in the development of an intermittency model of turbulence. Technical Report CFDU/83/7, Imperial College, University of London.

52. Malin, M. R. (1986). *Turbulence modelling for flow and heat transfer in jets, wakes and plumes.* Ph.D. thesis, Imperial College, University of London.

53. Malin, M. R. (1983). Calculations of intermittency in self-preserving free turbulent jets and wakes. Technical Report CFDU/83/10, Imperial College, University of London.

54. Malin, M. R. (1984). Prediction of the temperature characteristics in intermittent free turbulent shear layers. Technical Report CFDU/84/1, Imperial College, University of London.

55. Markatos, N. C., & Kotsifaki, C. A. (1994). One-dimensional, two-fluid modelling of turbulent premixed flames. *Applied Mathematical Modelling, 18*(12), 646–657.

56. Markatos, N. C., Pericleous, K., & Cox, G. (1986). A novel approach to the field modelling of fires. *Physico Chemical Hydrodynamics, 7,* 01.

57. Mason, H. B., & Spalding, D. B. (1973). Prediction of reaction rates in turbulent pre-mixed boundary-layer flows. In *Combustion Institute European Symposium* (pp. 601–606).

58. Noseir, M. (1980). *Application of the ESCIMO theory of turbulent combustion.* Ph.D. thesis, Imperial College, University of London.

59. Patankar, S. V. (1967). *Heat and mass transfer in turbulent boundary layers.* Ph.D. thesis, Imperial College, University of London.

60. Patankar, S. V., & Spalding, D. B. (1973). Mathematical models of fluid flow and heat transfer in furnaces: A review. *Journal of the Institute of Fuel, 46*(388), 279–283.

61. Patankar, S. V., & Spalding, D. B. (1974). *Heat Transfer in Flames, chapter Simultaneous predictions of flow patterns and radiation for three-dimensional flames.* Wiley.

62. Patankar, S. V., & Spalding, D. B. (1972). A calculation procedure for heat, mass and momentum transfer in three-dimensional parabolic flows. *International Journal of Heat and Mass Transfer, 15*(10), 1787–1806.

63. Peters, N. (1984). Laminar diffusion flamelet models in non-premixed turbulent combustion. *Progress in Energy and Combustion Science, 10*(3), 319–339.

64. Peters, N. (1986). Laminar flamelet concepts in turbulent combustion. In *21st Symposium on Combustion* (pp. 1231–1250). The Combustion Institute.

65. Pfender, E., & Chang, C. H. (1998). Plasma spray jets and plasma-particulate interaction: Modeling and experiments. *Proceedings of the International Thermal Spray Conference, 1,* 315–327.

66. Pope, S. B. (1981). A Monte Carlo method for the PDF equations of turbulent reactive flow. *Combustion Science and Technology, 25*(5–6), 159–174.

67. Prandtl, L. (1925). Bericht uber untersuchungen zur ausgebildeten turbulenz. *Zeitschrift fur angewandte Mathematik und Physik, 5,* 136–139.

68. Pun, W. M., & Spalding, D. B. (1967). A procedure for predicting the velocity and temperature distributions in a confined, steady, turbulent, gaseous diffusion flame. In *Proceedings of 18th International Aeronautical Congress,* Belgrade. London: Pergamon Press.

69. Ren, Z., & Pope, S. B. (2004). An investigation of the performance of turbulent mixing models. *Combustion and Flame, 136*(1), 208–216.

70. Reynolds, O. (1874). On the extent and action of the heating surface for steam boilers. *Manchester Literary and Philosophical Society, 14,* 7–12.

71. Reynolds, O. (1895). On the dynamical theory of incompressible viscous fluids and the determination of the criterion. *Philosophical Transactions of the Royal Society of London A, 186,* 123–164.

72. Robertson, A. F. (1961). Visit to European Fire Research Laboratories. Report 7373, National Bureau of Standards.

73. Serag-Eldin, M. A. (1977). *The numerical prediction of the flow and combustion processes in a three-dimensional combustion chamber.* Ph.D. thesis, Imperial College.

74. Shen, Y. M., Ng, C.-O., & Chwang, A. T. (2003). A two-fluid model of turbulent two-phase flow for simulating turbulent stratified flows. *Ocean Engineering, 30*(2), 153–161.

75. Shen, Y. M., Wang, Y. L., & Liang, C. (2000). Numerical simulation of stratified two-phase flow in an aquatic environment. *Journal of Environmental Sciences, 12*(2), 149–153.

76. Sheng, D. Y., & Jonsson, L. (2000). Two-fluid simulation on the mixed convection flow pattern in a nonisothermal water model of continuous casting tundish. *Metallurgical and Materials Transactions B, 31*(4), 867–875.

77. Spalding, B. (2018). The discretised population model of turbulence. *International Journal of Heat and Fluid Flow, 70*, A3–A7.
78. Spalding, D. B. Connexions between the multi-fluid and flamelet models of turbulent combustion. http://www.cham.co.uk/phoenics/d_polis/d_lecs/mfm/flamelet.htm.
79. Spalding, D. B. MFM: The economical route to PDFs. http://www.cham.co.uk/phoenics/d_polis/d_lecs/mfm/mfm_pdf.htm.
80. Spalding, D. B. Turbulence models in PHOENICS. http://www.cham.co.uk/phoenics/d_polis/d_enc/turmod/enc_tu64.htm.
81. Spalding, D. B. (1951). *The combustion of liquid fuels*. Ph.D. thesis, University of Cambridge.
82. Spalding, D. B. (1957). Analogue for high-intensity steady-flow combustion phenomena. *Proceedings of the Institution of Mechanical Engineers, 171*(1), 383–411.
83. Spalding, D. B. (1967). The spread of turbulent flames confined in ducts. In *Symposium (International) on Combustion* (Vol. 11, No. 1, pp. 807–815).
84. Spalding, D. B. (1969). Predicting the performance of diesel engine combustion chambers. *Proceedings of the Institution of Mechanical Engineers, 184*(10), 241–299.
85. Spalding, D. B. (1969). The prediction of two-dimensional steady, turbulent elliptic flow. In *Heat and mass transfer in flows with separated regions and measurement techniques, Herceg-Novi* (pp. 1–13). International Centre for Heat and Mass Transfer.
86. Spalding, D. B. (1970). Mathematische modelle turbulenter flammen. *VDI-Berichte, 149*, 25–30.
87. Spalding, D. B. (1971). The calculation of combustion processes. Technical Report RF/TN/A/2, Dept. of Mechanical Engineering, Imperial College, University of London.
88. Spalding, D. B. (1976). The ESCIMO theory of turbulent combustion. Technical Report HTS/76/13, Dept. of Mechanical Engineering, Imperial College, University of London.
89. Spalding, D. B. (1976). Mathematical models of turbulent flames: A review. *Combustion Science and Technology, 13*(1–6), 3–25.
90. Spalding, D. B. (1977). Chemical reaction in turbulent fluids. Technical Report HTS/77/11, Dept. Mechanical Engineering, Imperial College, University of London.
91. Spalding, D. B. (1977). Development of the eddy-break-up model of turbulent combustion. In *Symposium (International) on Combustion* (Vol. 16, No. 1, pp. 1657–1663).
92. Spalding, D. B. (1977). *GENMIX: A general computer program for two-dimensional parabolic phenomena*. Oxford and New York: Pergamon Press.
93. Spalding, D. B. (1978). A general theory of turbulent combustion. *Journal of Energy, 2*(1), 16–23.
94. Spalding, D. B. (1980). Mathematical modelling of fluid-mechanics, heat-transfer and chemical-reaction processes. Technical Report HTS/80/1, Dept. of Mechanical Engineering, Imperial College, University of London.
95. Spalding, D. B. (1980). Numerical computation of multi-phase fluid flow and heat transfer. In C. Taylor & K. Morgan (Eds.), *Recent advances in numerical methods in fluids* (pp. 139–167). Pineridge Press Ltd.
96. Spalding, D. B. (1980). The writing of technical reports. Technical Report CFD/80/3, Dept. of Mechanical Engineering, Imperial College, University of London.
97. Spalding, D. B. (1982). Chemical reaction in turbulent fluids. *Physicochemical Hydrodynamics, 4*, 323–336.
98. Spalding, D. B. (1982). Representations of combustion in computer models of spark ignition engines. Technical Report CFD/82/18, Dept. of Mechanical Engineering, Imperial College, University of London.
99. Spalding, D. B. (1983). Chemical reaction in turbulent fluids. *Physicochemical Hydrodynamics, 4*(4), 323–336.
100. Spalding, D. B. (1983). Towards a two-fluid model of turbulent combustion in gases with special reference to the spark ignition engine. In *Conference on Combustion in Engineering*, number Paper No C53/83 (pp. 135–142). IMechE.
101. Spalding, D. B. (1984). Progress in the development of a two-fluid model of turbulence. In *5th Conference on Physico-Chemical Hydrodynamics*, Tel Aviv, Israel.

102. Spalding, D. B. (1986). The two-fluid model of turbulence applied to combustion phenomena. *AIAA Journal, 24*(6), 876–884.
103. Spalding, D. B. (1987). The combustion of liquid fuels. Technical Report CFD/87/8, Dept. of Mechanical Engineering, Imperial College, University of London.
104. Spalding, D. B. (1987). A turbulence model for buoyant and combusting flows. *International Journal for Numerical Methods in Engineering, 24,* 1–23.
105. Spalding, D. B. (1995). Multi-fluid models of turbulent combustion. In *CTAC95 Conference, Melbourne, Australia.*
106. Spalding, D. B. (1995). Multifluid models of turbulence. http://www.cham.co.uk/phoenics/d_polis/d_lecs/mfm/mfmlec95.htm#3.2b.
107. Spalding, D. B. (1999). CFD as applied to combustion: Past, present, and future. http://www.cham.co.uk/phoenics/d_polis/d_lecs/cmbstr4/cmbstr4.htm.
108. Spalding, D. B. (2010). Turbulent mixing and chemical reaction; the multi-fluid approach (1995–2010). http://www.cham.co.uk/phoenics/d_polis/d_lecs/mfm/mfm00.htm, 2010.
109. Spalding, D. B. (2011). Mapping turbulent combustion. In *Ninth Australian Heat and Mass Transfer Conference, Melbourne, Australia.*
110. Spalding, D. B. (1979). The influences of laminar transport and chemical kinetics on the time-mean reaction rate in a turbulent flame. In *Seventeenth Symposium (International) on Combustion* (Vol. 17, No. 1, pp. 431–440).
111. Spalding, D. B. (1971). Concentration fluctuations in a round turbulent free jet. *Chemical Engineering Science, 26*(1), 95–107.
112. Spalding, D. B. (1971). Mixing and chemical reaction in steady, confined turbulent flames. In The Combustion Institute (Ed.), *13th Symposium (International) on Combustion*, Pittsburgh (pp. 649–657).
113. Spalding, D. B. (1981). IPSA 1981: New developments and computed results. Technical Report HTS/81/2, Dept. of Mechanical Engineering, Imperial College, University of London.
114. Spalding, D. B. (1984). Two-fluid models of turbulence. In *NASA Langley Workshop on Theoretical Approaches to Turbulence*, Hampton, Virginia.
115. Spalding, D. B. (2009). Population models of heat and mass transfer. In *6th International Symposium on Turbulence, Heat and Mass Transfer*, Rome, Italy.
116. Spalding, D. B., & Malin, M. R. (1984). A two-fluid model of turbulence and its application to heated plane jets and wakes. *Physicochemical Hydrodynamics, 5,* 339–361.
117. Spiegel, E. A. (1972). Qualitative model for turbulent intermittency. *The Physics of Fluids, 15*(8), 1372–1376.
118. Stafford, L. G. (1982). An experimental investigation of turbulent mixing due to buoyancy forces in stably stratified media. Technical Report CFDU/82/01, Dept. of Mechanical Engineering, Imperial College, University of London.
119. Sun, R. L. (1982). *Application of the ESCIMO theory to turbulent diffusion flames.* Ph.D. thesis, Imperial College, University of London.
120. Tam, L. T. (1981). *The theory of turbulent flow with complex chemical kinetics.* Ph.D. thesis, Imperial College, University of London.
121. Williams, G. C., Hottel, H. C., & Scurlock, A. C. (1948). Flame stabilization and propagation in high velocity gas streams. In *Symposium on Combustion, Flame, and Explosion Phenomena* (Vol. 3, pp. 21–40). Baltimore: Williams and Wilkins.
122. Wu, Z. Y. (1986). *Studies in unsteady flame propagation.* Ph.D. thesis, Imperial College, University of London.
123. Xi, S. T. (1986). *Transient turbulent jets of miscible and immiscible fluids.* Ph.D. thesis, Imperial College, University of London.
124. Youngs, D. L. (1984). Numerical simulation of turbulent mixing by Rayleigh-Taylor instability. *Physica D: Nonlinear Phenomena, 12*(1–3), 32–44.
125. Youngs, D. L. (1989). Modelling turbulent mixing by Rayleigh-Taylor instability. *Physica D: Nonlinear Phenomena, 37*(1–3), 270–287.
126. Youngs, D. L. (1994). Numerical simulation of mixing by Rayleigh-Taylor and Richtmyer-Meshkov instabilities. *Laser and Particle Beams, 12*(4), 725–750.

127. Youngs, D. L. (1995). Representation of the molecular mixing process in a two-phase flow turbulent mixing model. In R. Young, J. Glimm, & B. Boston (Eds.), *Proceedings of the 5th International Workshop on Compressible Turbulent Mixing* (pp. 83–88). Singapore: World Scientific.

128. Yu, K. Z., Ding, G. L., & Chen, T. J. (2007). Simulation of air curtains for vertical display cases with a two-fluid model. *Applied Thermal Engineering, 27*(14–15), 2583–2591.

129. Zhang, L., Qiu, X., Wang, L., & Li, J. (2014). A stability condition for turbulence model: From EMMS model to EMMS-based turbulence model. *Particuology, 16*, 142–154.

130. Zhubrin, S. V. (2002). Multi-fluid modelling of turbulent combustion. In *IX International PHOENICS Users Conference*, Moscow.

131. Zuber, I. (1972). Ein Mathematiches Modell des Brennraums (A mathematical model of the combustion chamber). Monographs and Memoranda 2, Staatliches Forschungsinstitut fuer Maschinenbau, Bechovice, Czechoslovakia.

Hypotheses-Driven Combustion Technology and Design Development Approach Pursued Since Early 1970s

Hukam C. Mongia, Kumud Ajmani and Chih-Jen Sung

1 Introduction

It is appropriate to start this chapter by reminiscing the state of the gas turbine combustion design process in 1972, the year Mongia joined AiResearch Manufacturing Company in Phoenix, a division of Garrett Corporation. He had to rely on simple design tools for combustion system design and development process including (1) Airflow and pressure distributions around the combustor and hot-side gas temperatures; (2) Liner wall temperature levels and gradients that his on the job teacher Jack Haasis told him that he can guess better than the state-of-the-art one-dimensional heat transfer analysis; (3) Highly empirical correlations for Sauter mean diameter (SMD) of pressure atomizing sprays, combustion efficiency, lean blowout, and ignition fuel/air ratio; and (4) Very limited poor quality design database. But thanks to the CFD revolution that occurred in the late 1960s and early 1970s for which a lot of credit should be given to Professor Spalding, his team, and the three key Garrett management team members (Dr. Monte Steel, Carl Paul and Dr. John Mason) who were willing to provide resources that made possible the first successful demonstration of the effectiveness of an empirical/analytical design methodology as summarized in [1, 2].

Both the NASA Lewis Research Center under its Pollution Reduction Technology Program (PRTP) and the US Army AVRADCOM of Fort Eustis under its Combustor Design Criteria (ACDC) Program as summarized in [3–5] facilitated the first two

H. C. Mongia (✉)
CSTI Associates, LLC, Yardley, PA, USA
e-mail: hmongia43@hotmail.com

K. Ajmani
CFD Nexus, LLC, Cleveland, OH, USA

C.-J. Sung
University of Connecticut, Storrs, CT, USA

© Springer Nature Singapore Pte Ltd. 2020
A. Runchal (ed.), *50 Years of CFD in Engineering Sciences*,
https://doi.org/10.1007/978-981-15-2670-1_13

platforms for formulating and validation of Empirical Analytical Design Methodology (EADM) as expressed by the following key sentences duplicated from the abstract sections of [3, 4]:

CFD "**when used in the proper context in conjunction with empirical methodology** will reduce the design and development time and cost associated with gas turbine combustion systems."

"The baseline, Concept I, configuration met all design objectives with no hardware modifications. Concept II met the design objectives after only one major hardware modification. — The combustor development (process) was significantly reduced, meeting the main objective of the (ACDC) program."

It is perhaps an appropriate time to reflect on the gems of "empirical/analytical design methodology" pursued by Mongia and his coworkers during his 37-year period in industry (1972–2009), as summarized in Sect. 2. Here, we summarize that the success of EADM depended strongly on the process of (1) Defining and pursuing verification of key hypotheses germane to advancing technology consistent with program milestones; (2) Calibration of semi-analytical and CFD combustion models with engineering data on element test rigs that simulate "real" combustors' operation (in Sect. 3); and (3) The art of design supported by models qualitatively (in Sect. 4). The resulting empirical/analytical design methodology was used for advancing combustion technology by relying on CFD simulations to provide qualitative guidance continued through the end of 1995.

The anchored design methodology [6] formulated in 1994 with models' refinements pursued through 1998 [7, 8], which allowed computational control volumes approaching one million, was considered mostly applicable for scaling and advancing rich-dome combustion products. It offered significant advantage over semi-analytical models even when they both have comparable accuracy levels, as summarized in Sect. 5. We expected that use of the anchored design tools will reduce dependence on hypotheses and the art of design, but that was not the case for the development of the GE90-110/115 engine combustion system.

The first-generation lean-dome combustion products did not fully measure up to expectations leading to intensive lean technology development effort starting in 1995. Realizing the limitations of the in-house codes for handling comprehensive combustion system modeling approach, we switched to a commercial code in 2001 without fully realizing the level of effort required to get useful Phase I computational tool which finally happened in June 2007, as summarized in Sect. 6 in regard to its accuracy levels. But even with comprehensive combustion system analysis capable of handling 20 million mesh size or more, the dependence on hypotheses and art of design continues.

In spite of all the effort we have put in since the early 1970s for improving the accuracy of models' predictions in regard to critical design requirements (but excluding pattern factor and radial profile factor), the model accuracy is approximately three times the experimental standard deviation (σ_{exp}) in regard to gaseous emissions, operability, and liner hot spots. Consequently, the spirit of empirical/analytical modeling concept continues to play key role in making critical design-related decisions which is consistent with reality, namely, CFD can complement and shorten significantly the gas turbine combustion design and substantiation process but it will continue to depend on the ingenuity of the designers in regard to hypotheses motivated by lessons learned from their empirical design database.

As required contractually by sponsored technology programs, gas turbine original equipment manufacturers OEM's publish technical papers honestly while simultaneously minimizing details considered critical for maintaining competitive edge. However, after his retirement from GE in February 2009, Mongia got involved in fundamentals of technology development, application of CFD and diagnostics in association with his coauthors leading to "share everything openly" including application of the NASA National Combustion Code (NCC) in the lean-direct injection (LDI) technology development. Therefore, we summarize most recent modeling and diagnostics activities for gaseous emissions and operability relevant to both rich- and lean-dome combustors in Sect. 7 followed by summary in Sect. 8.

2 Hypotheses-Driven Combustion Technology and Design Development Process

The Army Combustor Design Criteria (ACDC) Program's use of empirical/analytical design methodology [2] needs to be further explained as illustrated in Fig. 1. Fundamental and applied combustion research as well as OEM's combustion technology and product development activities went through transformational changes in the early 1970s perhaps caused primarily by the Clean Air Act. However, the intensity with which the boxes 5 and 6 were allowed to interact with the so-called conventional approach, namely, boxes 1 through 4 shown in Fig. 1, varied enormously in the early 1970s through 2009, the year Mongia retired. After successful demonstration of the empirical/analytical design methodology in ACDC, NASA T1 Concept 3 and several other technology programs described briefly in this chapter, application of CFD in combustion technology and product development programs was claimed by

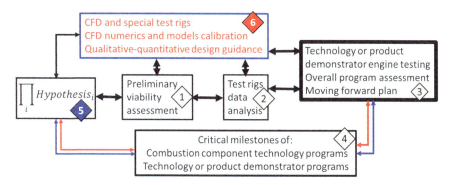

Fig. 1 Conventional combustion technology or product development process comprised of boxes 1 through 4 has been complemented by boxes 5 and 6 starting early 1970s

almost all the engine OEM's. However, the extent of success or failures of CFD in accomplishing the critical design metrics cannot be quantified accurately because of the complexity of the circumstances as evident from Fig. 1.

Mongia considers himself to be blessed for having interacted personally with dozens of combustion fundamental research scientists including Professor Spalding; they were or are awesome that he used to call some of them as "gods of combustion". But nobody knew this list for obvious reasons. Similarly, Mongia has interacted with equally "street smart" dozens of senior and peer combustion technologists and designers and promising next-generation combustion engineers that have contributed to advancing combustion technologies and products since the early 1970s. For example, the swirl cup, a term coined by GE, in various shapes, forms and designs started appearing in GE gas turbine engines in the late 1960s and became an essential entity in all its modern engines starting early 1970s. Mongia came to know about the use of swirl cup in GE's Dual-Annular Combustion (DAC) technology program in 1975, see [9]. Mongia was a test engineer working for his first-team leader Mike Wood who also designed his version of dual radial swirler design in 1973. But his approach called NASA T1 Concept 3 [10–12] for staged combustion technology program was distinctly different from DAC for obvious reasons.

Mongia calls combustion engineers as "mortals" in contrast to "gods of combustion" because the former struggle throughout their careers for meeting conflicting design requirements, see [13]. They try to make best use of combustion fundamentals, work hard to meet technology and product design requirements, and continue to discover new applicable combustion science in "real" hardware. Let us illustrate this point by sharing the development of TAPS technology and products described later in this chapter. The first 3 years of the TAPS technology development effort costing approximately $10 million entailed conceptual design and single-cup testing of the various configurations illustrated through Fig. 2 reproduced from US Patent 6,367,262, April 9, 2002.

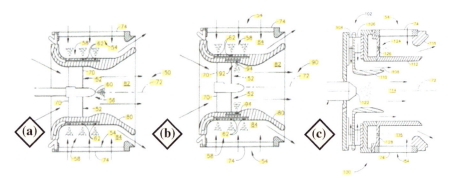

Fig. 2 TAPS1 Mod1 conceptual configurations involving different arrangements of swirlers and fuel insertion devices, axial, radial, and combination as claimed in US Patent 6,367,262, April 9, 2002

It followed through several inventions in the ensuing years that can be broadly divided into the following categories:

TAPS1 Mod2: US Patents 6,381,964, May 7, 2002; and 6,389,815, May 21, 2002. Improving Low-Power Emissions and/or fuel nozzle: US Patents 6,405,523, June 18, 2002; 6,865,889, March 15, 2005; 7,010,923, March 14, 2006; 7,878,000, February 1, 2011; 8,171,735, May 8, 2012; and 8,555,645, October 15, 2013. Improving Operability: US Patents 6,453,660, Sep 24, 2002; 8,001,761, August 23, 2011; and 8,607,575, December 17, 2013. TAPS2 and TAPS3 for 50% and 75% further reduction in high-power NO_x compared to TAPS1: US patents 6,418,726, June 16, 2002; 6,484,489, November 26, 2002; 7,464,553, December 16, 2008; 7,565,803, July 28, 2009; 7,581,396, September 1, 2009; and 7,762,073, July 27, 2010. Multi-Injection Concept, an alternate to TAPS: US patents 6,363,726, April 2, 2002; 6,474,071, November 5, 2002; and 6,609,377, August 26, 2003.

For the benefit of readers not familiar with the US Patent Office (USPO) process for awarding patents, both OEM and USPO conduct meticulous investigation for application and award of patents, especially when it comes to their use in commercial products because of its financial ramifications. OEM's are very careful in regard to patent infringement violations; see for example, the New York Times article by Don Clark and Daisuke Wakabayashi dated April 16, 2019 on Apple and Qualcomm patent violation dispute settlement.

It is a strange coincidence that we included in this chapter that fuel nozzle and wall carboning and resulting hot-section distress were serious issues in the engine products of the 1970s to come across nozzle carbon buildup issue encountered in one of the LEAP-1B engines. This carboning issue could be responsible for creating hot-spot in the high-pressure turbine leading to blade failure as reported in the Aviation Daily article of April 18, 2019 by Sean Broderick and Guy Norris.

As to the extent of level of effort required for the development of CFD tools and combustion technology, especially to educate young readers involved in fundamental and applied combustion research, we share a few examples from Mongia's experience since early 1970s. There is no hesitation in stating that Mongia would not have ventured into including boxes 5 and 6 in his design process shown schematically in Fig. 1 without direct and indirect help he received from the dozens of fundamental combustion researchers since the early 1970s through the end of 2008. It obviously includes Professor Spalding and his team and many more; there are too many people to mention (and they know who they are), and manuscript page limitation does not allow us to do full justice in describing qualitative contributions of their important publications in all the hypotheses Mongia used since early 1970s. A generic description of these hypotheses (rich-quick quench-lean combustion for rich-dome, and equivalent extended list for lean-dome combustion) leaves out "all of the trade secrets" from the general-purpose publications. Mongia had high regards for all of his competitors; they were equally competent if not more. He did not consider "wise" to second-guess one's competitor and try to copy their works, especially if one wants to become a leader in one's chosen field of expertise.

As to application of CFD in the design process, Mongia's views expressed openly in February 1979 have not changed as of now. Namely, assign at least two Ph.D.'s with expertise in CFD and combustion modeling as part of the technology team; and hope, (1) Code will be "ready" to provide the guidance; (2) Provide "qualitative" guidance; and (3) Set quantitative guidance goal as long-term prospects. In all the programs that Mongia was involved personally as team leader, first or second line manager, CFD was never "fully ready" for providing "qualitative" guidance. Mongia kept on increasing CFD and combustion modeling manpower over the years in his department in addition to seeking more collaboration with the research community in order to achieve the abovementioned objectives (1) and (2). These were the "best" CFD and combustion modeling people and they worked hard. Mongia had more than 4 CFD and combustion modeling people by end of 1983, 10 by end of 1993, and 30 by end of 2007. In all these programs, Mongia thought he got the design guidance qualitatively, but with a reservation expressed by one of his peers at the end of the four concepts selected by Mongia for full-annular pizza pie rig testing. He said, "I see that you used CFD in selecting these concepts but I don't understand how you made the decision." Mongia shrugged his shoulders; got hardware fabricated and test results met his expectations.

Looking back, Mongia feels that he was always using composite of boxes 5 and 6 (in Fig. 1) coupled with continuously evolving conventional combustion technology and product design development process. He was fortunate enough to have help from a highly competent dedicated professionals throughout his 37-year industry career. When Mongia asked one of his right-hand persons how many people have been involved in the development of TAPS technology during 1998 and 2003, he counted more than 100. Of course, we used applicable combustion fundamentals; sought help from more than four universities together with very esteemed professors who have provided us with highly valuable insight which along with "street smart" combustion engineers have made possible to make progress from lean-dome DAC's summarized in Fig. 3a to GEnx shown in Fig. 3b. But the "same" TAPS technology when applied to LEAP-X products produced mixed results, namely, comparable takeoff NO_xEI at 33.4 takeoff pressure ratio (PR), but worse than DAC's for the higher pressure ratios. In spite of the best comprehensive CFD modeling capability developed since 2001, more than 10 years of fundamental research on TAPS, the best combustion team that Mongia was proud of pulling together, his own experience in developing TAPS technology and its scale-up to GEnx proven through TG6, he can only second-guess the circumstances that led to the results summarized in Fig. 3b. It should be mentioned that Mongia thought he knew the reasons why DAC did not get lower NO_x at nominal 40 pressure ratio conditions before marching on TAPS technology development path. He would not dare to conclude that something went wrong with LEAP-X product for lack of on-hand experience. It is for this reason, Mongia never felt comfortable in assessing other OEM's technology and/or modeling capabilities.

Therefore, the main objective of this chapter is to summarize examples of hypotheses-based combustion technology programs that Mongia was personally involved, see [1, 2, 4, 10–12, 14–27], models formulation, calibration/validation, and application of CFD. The second objective is to show typical comparison between

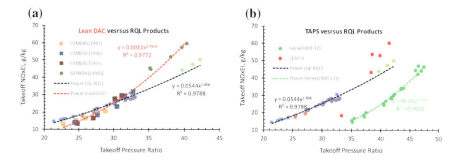

Fig. 3 Evolution of three combustion technology products, namely, GE Rich-Quench-Lean (RQL), lean-domes' DAC and TAPS technologies. **a** Lean Dual-Annular Combustion products (CFM DAC's tested in 1995, 1996 and 1997) and GE90DAC tested in 1995 compared with GE RQL products, CFM56-Tech Insertion, CF34-10, GE90-110/115 and CF6-80C2. RQL's takeoff $NO_x EI = 0.0544PR^{1.806}$ whereas that of the lean DAC's $NO_x EI = 0.00111PR^{2.9502}$. **b** GE Twin Annular Premixing Swirl stabilized (TAPS) products GEnx (tested during 2009 and 2012) and LEAP-X models compared with GE RQL; GEnx takeoff $NO_x EI = 3.52 \times 10^{-5}PR^{3.6535}$

predictions (for consistently formulated approach applicable during design process) and data from "real" combustors, as summarized in Sects. 5 and 6; references for the most relevant publications have been provided for the readers interested in details. There are thousands of outstanding benchmark quality diagnostics and model validation publications reported by "gods" of combustion which we chose not to include in our list of publications because of page limitations.

3 Combustors Internal Flow Field Data and CFD Models' Calibration

The ACDC Program and Concept III of the NASA Pollution Reduction Technology Program (PRTP) were the first two programs that conducted several engineering benchmark quality experiments in order to calibrate the first-generation of CFD-based combustion models that could afford only very coarse mesh simulations as summarized in this section. As summarized in [3], a systematic set of engineering experiments were undertaken to calibrate six analytical models, namely, (1) quasi-one-dimensional model for correlating with pressure and airflow distributions around the combustor, (2) fuel insertion model, (3) liner cooling model, (4) emissions model, (5) transition mixing model, and (6) three-dimensional combustor performance model, as described in [5]. We will illustrate the extent of these experiments by picking two examples, starting with jet mixing investigation that included the following dilution jet orifice configurations and operating conditions:

Single row of two, four and six orifices, and two rows of six orifices, inline and staggered arrangement were investigated in order to provide thermocouple-based jet mixing characteristics that were used for calibrating three-dimensional turbulent mixing model. The jet air supply pressure was varied in order to investigate mixing of dilution jet air (at ~300 K) in a 12.6 cm diameter cylindrical test section placed downstream of a can combustor that provided hot-side nominal air temperature, pressure and velocity of 1200 K, 10 atm and 40 m/s, respectively. The orifice pressure drop was varied between 1 and 10% with the corresponding jet velocity variation between 40 and 150 m/s and resulting jet momentum ratios of $J = 4-60$, and jet Reynolds numbers of $Re_j = 1.4 \times 10^5 - 6.8 \times 10^5$.

Since relative circumferential location of the primary and dilution jet orifices is an important design variable for controlling mixing in the dilution zone with attendant impact on combustor exit temperature quality, we have summarized only one experimental data expressed in terms of isothermal temperature contours in Figs. 4, 5, 6, and 7 for both inline and staggered orifice configurations. For the staggered configuration: hot-side axial velocity, density, jet velocity, and density were $U_{hot} = 39$ m/s, $\rho_{hot} = 2.93$ kg/m^3, $V_j = 72$ m/s, and $\rho_j = 11.73$ kg/m^3 giving $J = 13.5$ and $Re_j = 2.65 \times 10^5$. For the inline configuration: $U_{hot} = 39$ m/s, $\rho_{hot} = 2.95$ kg/m^3, $V_j = 66.7$ m/s, and $\rho_j = 11.75$ kg/m^3 giving $J = 11.6$ and $Re_j = 2.48 \times 10^5$. The center of the first-row jets is at $x/D_j = 0.0$ whereas that of the second row is $x/D_j = 5.9$ implying that the downstream edge of the first-row jet is at $x/D_j = 0.5$ whereas upstream edge of the second row jet is at $x/D_j = 5.4$.

Figure 4 shows measured contours at $x/D_j = 0.71$ (namely, $0.21D_j$ downstream of the edge of the orifice) and $x/D_j = 4.95$ and 5.3, respectively, for the staggered and inline configurations, with the corresponding axial locations being $0.45D_j$ and $0.1D_j$ upstream of the edge of the second row orifice. As evident from the data slightly upstream of the second row orifice, the effect of staggered versus inline dilution orifice configuration continues downstream as summarized in Fig. 5 with further corroboration by Fig. 6. Therefore, what has been known empirically for controlling hot-streaks in gas turbine combustors by both inline and staggered two rows of primary/dilution orifices can be explained by this simple experiment. What was interesting to observe that three-dimensional jet mixing CFD models could reproduce this qualitatively with less than 40,000 nodes providing impetus for using three-dimensional combustor CFD modeling techniques for optimizing combustor exit temperature quality; now a standard practice among all OEM's albeit with several million mesh sizes.

The second example is on reacting flow in a "real" can combustor shown in Fig. 7: A 5-in. (12.7 cm) can combustor with six equispaced primary orifices (1.12 cm diameter) at $x = 9.09$ cm, followed by inline six dilution orifices (1.42 cm diameter) at $x = 17.21$ cm cooled by "English" louver cooling slots with the corresponding metering cooling holes placed at $x = 4.37, 11.52, 19.90$, and 28.98 cm; here, x denotes axial distance downstream of the face of the fuel nozzle.

Two types of fuel nozzles (natural gas nozzle and air-assisted airblast nozzle) were used for making gaseous emissions measurements within the combustor covering up to eight axial stations (ranging between $x = 6$ and 26.2 cm) and five circumferential stations for each of the preselected ten radial stations, namely, $y = 0, 0.60, 1.07, 1.80, 2.42, 3.02, 3.63, 4.23, 4.84$, and 5.44 cm. A total of 13 data sets were taken covering combustor overall pressure drop between 2 and 6%, inlet pressure P3 between 2

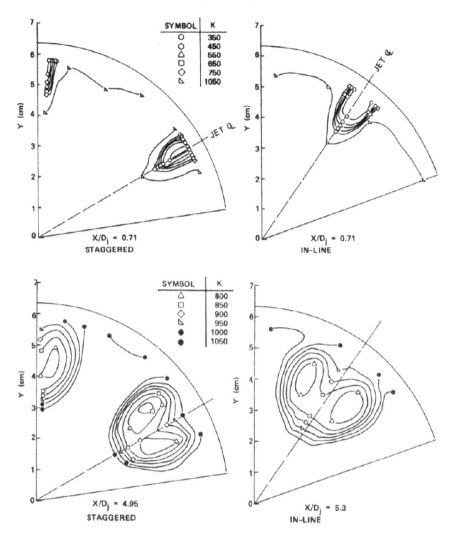

Fig. 4 Mixing characteristics of inline and staggered dilution jets configurations in a nonreacting can combustor test section upstream of the second row of orifices. *Source* [3]

and 10 atm., inlet temperatures T3 between 366 and 623 K, overall fuel/air ratio FAR between 0.006 and 0.015 and resulting combustor exit adiabatic temperature T4 between 643 and 1170 K. The nominal test conditions and detailed emissions (unburned fuel, CO, NO_x, and CO_2) data are reported in [3] including mesh for three-dimensional model simulations shown in Fig. 8 and comparison between model predictions and data for all of the 13 data sets. Here, we have summarized key findings for the two cases, both for Jet-A fuel; the simulated idle in somewhat more details shown in Figs. 7, 9 and 10, and the simulated maximum power shown in Fig. 11.

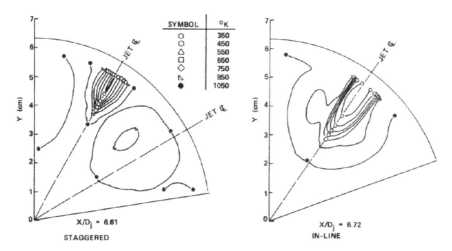

Fig. 5 Mixing characteristics immediately downstream of the second row located at $x/D_j = 5.9$; dilution jet orifice diameter $D_j = 7.19$ mm can radius $= 8.8D_j$. *Source* [3]

No one would dare to show now a comparison between data and simulations for $32 \times 16 \times 14$ mesh size; but that is all we could afford to do in the 1970s. It is perhaps appropriate here to remind the readers that Professor Spalding's publication in 1979, see [28], used $33 \times 22 \times 12$ nodes for three-dimensional CFD simulation of a research combustor. The only difference is that we were dealing with a "real" can combustor. While a modified eddy break-up model for a two-step kinetic scheme along with five-discrete size spray deterministic Lagrangian formulation was used for a practical can combustor, that is where we started from in the application of CFD in the development of combustion technology and product in the late 1970s. Readers may want to refer to [1, 2, 4, 14–20, 29] for details. We started RANS simulations of "real" combustors with ~7,000 nodes in the middle 1970s and increased continually to ~75,000 by early 1990s. Starting 1995, the mesh size increased from 200,000 to 600,000 until 2001 when we decided to increase steadily from one million to 10–20 million mesh size. It is a true statement to make that we made use of "applicable CFD simulations along with hypotheses" (see boxes 5 and 6 in Fig. 1) in all the combustion technology and product development activities starting with the Army ACDC and NASA PRTP in the middle 1970s through 2007 in which Mongia was involved personally as team leader or hands-on manager, a long period of more than 30 years.

No one will disagree with the statement: multielement emissions probe inserted in practical combustors (can or annular) cannot provide profiles of fuel/air ratio, NO_x, CO, and unburned fuel and resulting combustion efficiency accurate enough for calibrating or validating turbulent combustion models. But what was the alternative for young combustion engineers in the middle 1970s or is in 2019 who want to apply CFD in the development of combustion technology or more challenging combustion products? Combustion technology and product development was our primary

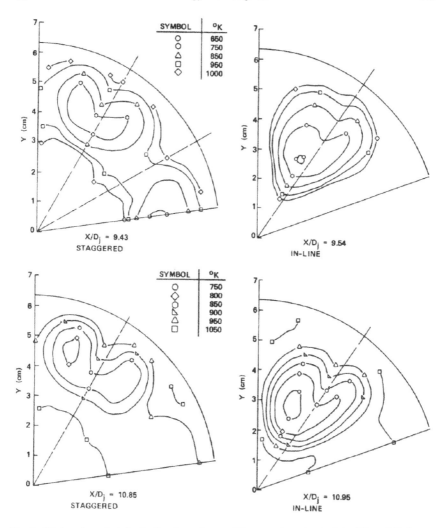

Fig. 6 Mixing characteristics further downstream of the second row located at $x/D_j = 5.9$. *Source* [3]

job responsibility. All industry and sponsored technology and product development activities are schedule and resource driven. Formulation, calibration, and validation of CFD-based combustion models tried to follow both short-term and long-term goals, namely, use hypotheses-based CFD models to support ongoing hardware programs while keeping in mind the ultimate objective: a hardware consistent modeling methodology that reproduces aerothermal performance, emissions, and operability of product combustors with accuracy comparable with data expressed in terms of standard deviation σ_{exp} as summarized in [13]. After putting in considerable resources in this area since 1975, the models have not reached this goal in spite of increasing mesh

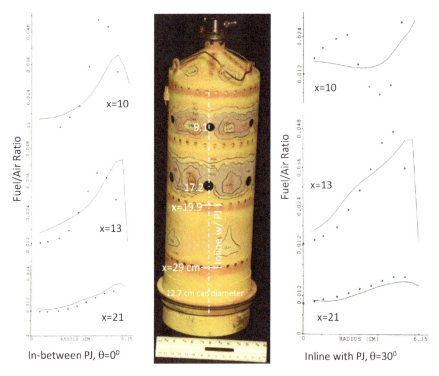

Fig. 7 Comparison between data (symbols) and predicted profiles of fuel/air ratio inline ($\theta = 30°$) and in-between ($\theta = 0°$) primary jets of a can combustor at simulated idle condition, $P3 = 2.02$ atm., $T3 = 371$ K, FAR $= 0.0104$, $\Delta P = 3.1\%$. *Source* [3]

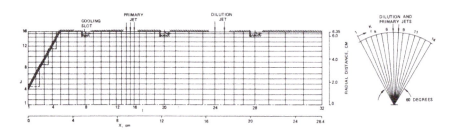

Fig. 8 $32 \times 16 \times 14$ grids (5040 control volumes) were used for the simulations presented in Figs. 7 and 9, 10 and 11. *Source* [3]

size to 20 million or more, use of unstructured grids, URANS, LES and its equivalent time-filtered Navier–Stokes (TFNS) simulations perhaps due to our inability to reproduce accurately spray characteristics and resulting fuel/air mixing and turbulent combustion with attendant impact on accurately predicting aerothermal design, performance, operability, and emissions characteristics relevant to meeting design requirements. Tens of thousands of research articles have been published in the last

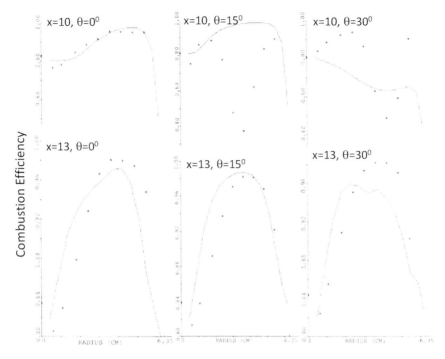

Fig. 9 Comparison between data (symbols) and predicted profiles of combustion efficiency along the three planes $\theta = 0°$, 15° and 30° at two axial stations in the intermediate combustion zone. Primary jets are along $\theta = 30°$ and in-between primary jets along $\theta = 0°$. Refer to Fig. 7 for operating conditions and Fig. 8 for mesh description. *Source* [3]

50 years covering these topics with attendant advances in turbulent spray combustion sciences relevant to gas turbine combustors, and CFD has become an integral part of the design process. We will share examples from the formative years and how they have helped us address technology and product development challenges as explained in Sect. 4.

Due to page limitation, we will make summary statements for the figures covered in Sect. 3. Although we have not presented CFD simulations for the jet mixing data summarized in Figs. 4, 5, 6 and 7, we can say that predicted temperature contours were in qualitative agreement with the data.

It is hard to make good quality emissions measurements 0.9 cm downstream of the primary jet (at $x = 10$ cm, Fig. 7); but CFD gave us qualitatively good agreement with fuel/air ratio profiles data in addition to combustion efficiency data at idle summarized in Figs. 9 and 10. Similar summary statement can be made for the maximum power condition summarized in Fig. 11.

Similar level of comprehensive internal emissions mapping was conducted, albeit at one atmosphere only but for different T3 and FAR conditions, for two product annular combustors and a staged combustion annular combustor [1]. Schematic

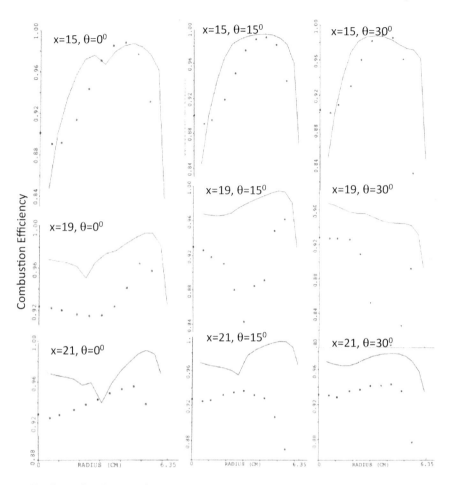

Fig. 10 Comparison between data (symbols) and predicted profiles of combustion efficiency along the three planes $\theta = 0°$, $15°$ and $30°$ at three axial stations, ~2 cm upstream and downstream of dilution jets located at $x = 17.2$ cm, and further downstream by additional 2 cm. Refer to Fig. 7 for operating conditions. *Source* [3]

description of two of these combustors (Fig. 12) along with typical data and predictions (namely, radial profiles of fuel/air ratio in Fig. 13 and axial variations of total fuel, unburned fuel and CO mass fractions in Figs. 14 and 15) gave us confidence in making qualitative use of CFD in the combustion technology and development process. It is perhaps CFD "**when used in the proper context in conjunction with empirical methodology**" gave us the insight for expeditiously developing combustion technologies summarized in next section. Details on the models' calibrations are given in [1, 3, 6–8, 11–13, 18, 29–45].

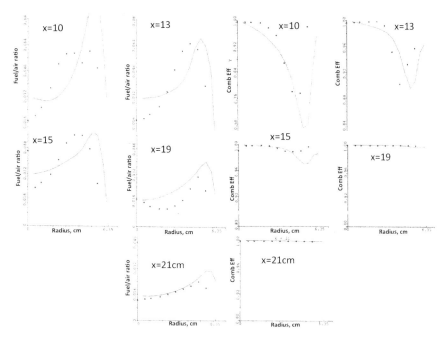

Fig. 11 Comparison between data (symbols) and predicted profiles of fuel/air ratio and combustion efficiency for the θ-plane in-between the primary jets (namely $\theta = 0°$) in a can combustor burning Jet-A fuel at simulated max power operating condition, $P3 = 9.99$ atm., $T3 = 623$ K, FAR $= 0.0148$, $\Delta P = 3.2\%$. Refer to hardware and the corresponding mesh in Figs. 7 and 8, respectively. *Source* [3]

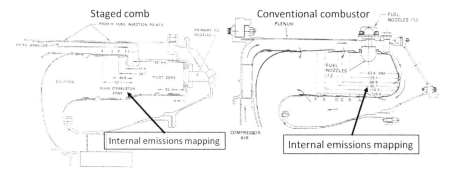

Fig. 12 Internal emissions mapping of a staged Concept III annular combustor, left part from [1] and its counterpart conventional diffusion flame annular combustor shown on right

Fig. 13 Comparison
between measured and
predicted radial profile of
fuel/air ratio at axial station
#4 of the staged combustor's
two planes lying inline and
in-between
premix/prevaporating jets.
Source [1]

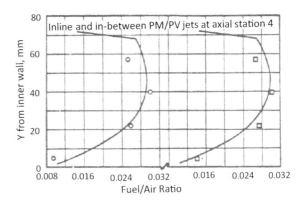

Fig. 14 Comparison
between measured and
predicted circumferentially
and radially averaged axial
profiles of total fuel,
unburned fuel and CO (all in
mass fractions) for the staged
combustor shown in Fig. 12

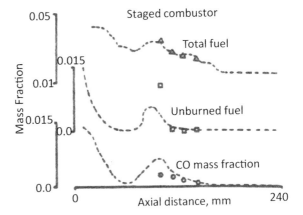

4 Addressing Technology Challenges During 1970–1990s

Empirical/analytical design methodology [31, 32] helped develop several combustion technologies as described in [1, 2, 4, 14–20, 29, 30]. The first application included two ACDC combustion concepts for small engines (in 1978) and subsequently increasing its temperature rise ΔT from 1635 to 2100 °F. We will give an overview of how CFD was used to provide insight for these combustors. In parallel was the development of NASA PRTP Concept 3 (in 1977), two high ΔT (2400 and 2900 °F) combustors, the two first product combustors (1986), two near-stoichiometric temperature high-performance combustors (1993), and entitlements for ultralow NO_x premix/pre-vaporized mixer (1993). To be covered in next three sections are RQL combustor for the largest turbofan engine (1996), the second-generation lean-dome combustion technology TAPS demonstration (2003) and its product introduction in GEnx (2009), entitlement ultralow NO_x partially premixed mixer, and NASA LDI-2 and LDI-3 technology demonstrations in 2014 and 2018, respectively.

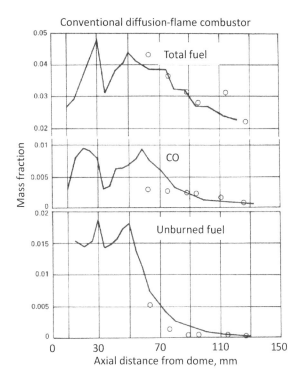

Fig. 15 Comparison between measured and predicted radially and circumferentially averaged axial profiles of total fuel, unburned fuel and CO (all in mass fractions) for the conventional diffusion flame combustor shown in Fig. 12

Internal flow mapping (for example, jet mixing summarized in Figs. 4, 5 and 6) of can combustor emissions and three-dimensional CFD simulations (Figs. 7, 8, 9, 10 and 11), three annular combustors (two of these shown in Fig. 12 and overview of comparison between data and CFD simulations summarized in Figs. 13, 14 and 15) led to a comfortable conclusion: CFD can be used to reliably predict qualitatively internal flow field characteristics of technology and product gas turbine combustors leading to expeditious development of combustion technologies for addressing the challenges commonly encountered in the 1975–1993 period. These included product combustors with high NO_x, CO, HC and smoke emissions, pattern factor, local liner hot spots, liner and fuel nozzle carboning with attendant frequent interruptions caused by unscheduled repair and overhaul shop visits. Qualitative applications of CFD-based combustion models helped us rapidly develop low-emissions, high-performance combustion technologies as summarized in several publications; [31, 32] provide good summary of these CFD-based design tools which do not need to be discussed here. However, we provide some examples in this section to illustrate how CFD predictions were used to facilitate technology development during the early period of empirical/analytical design methodology evolution starting with the Army ACDC Concepts I and II leading to the first high-ΔT (2200 °F) small annular combustor compared to 1635 °F ΔT Concept I without adversely impacting other design requirements.

The left part of Fig. 16 shows schematic layout of the ACDC Concept I and Concept II Mod 1, indicating that both combustors' outer liner wall diameter is 9.65 in. (24.5 cm) and channel height is 1.73 in. (4.39 cm). They also used the same set of ten air-assisted airblast nozzles inserted through the outer liner wall with the identical axial and circumferential location. However, since these combustors' lengths are different (being 3.8 in. and 4.27 in., respectively), the x-location of the spray origin from the respective domes is different for these combustors, as will be described shortly. Consistent with the practice of the 1970s, Concept I has three rows of orifices identified as 2, 4, and 6 for the outer wall but only two rows for the inner wall identified as 11 and 13 in Fig. 16. The spray origin is located at $x = 1.9$ cm, $y = 2.9$ cm from the inner wall at $\theta = 9°$ compared to the center of the dome swirler being $x = 0$, $y = 2.2$ cm and $\theta = 18°$. The spray with 90° included angle is injected towards the dome with specified back and down angles. Both combustors' sea-level design point is identical, namely, combustor airflow rate Wa3.1 = 2.7 lb/s, inlet pressure $P3 = 147$ psi, inlet air temperature $T3 = 665$ °F, fuel/air ratio $F/A = 0.0268$, and the corresponding combustor exit temperature $T4 = 2300$ °F. When a technology program for higher temperature rise demonstration was called for, this small combustor with corrected airflow rate $W_C = 0.397$ lb/s was able to increase its exit temperature $T4$ to 2850 °F by simply adding a row of primary orifices through the inner liner resulting in three rows of orifices for the inner wall also, as illustrated in the right part of Fig. 16. Amazingly, this 2200 °F ΔT combustor had pattern factor (PF) less than 0.2 along with its sea-level idle lean blowout fuel/air ratio (LBO F/A) less than 0.005. It also had relatively uniform liner wall temperature characteristics and attendant acceptable liner hot spots.

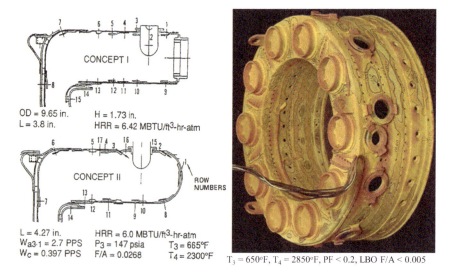

Fig. 16 Army Combustor Design Criteria (ACDC) Concepts I and II met all design objectives; later followed by Concept I potential higher temperature rise capability demonstration [4]

Concept I was designed by using an empirical/analytical design methodology as described in [2, 4]. It included analytical investigations of ten modifications by making use of the three-dimensional Combustor Performance Model that had been calibrated by using an extensive engineering database described in [3] with some examples shown previously in Figs. 7, 8, 9, 10 and 11. We show Figs. 17 and 18 as an illustration for augmenting empirical design know-how of a typical experienced combustor designer's perspective.

One nozzle sector of 36° of Concept I with dome swirler center located at $\theta = 18°$ was simulated by 13 θ-nodes, combustor length of 12.6 cm divided into 30 x-nodes, channel height of 4.4 cm covered by 19 y-nodes. Recall, Serag-Eldin and Spalding [28] used $33 \times 22 \times 12$ nodes for three-dimensional CFD simulation of a research combustor, not much different from $30 \times 19 \times 13$ used here. The main difference is that we used it for developing an advanced technology reverse-flow annular combustor shown in Fig. 17. The radius of the inner wall is 7.85 cm. The two outer wall

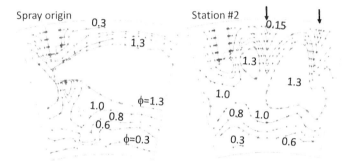

Fig. 17 Predicted fuel/air contours (at two axial stations in the primary zone) expressed as equivalence ratio ϕ of the Army Combustor Design Criteria Concept I at its design point; see Fig. 16 for design conditions and legends [4]

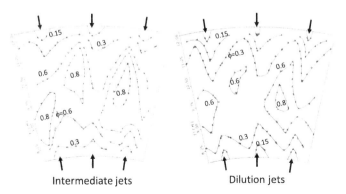

Fig. 18 Predicted ϕ contours at the intermediate and dilution orifices axial planes #4–11 and #6–13; see Fig. 16 for design conditions and legends [4]

(OD) primary jets are located at $x = 3.2$ cm and $\theta = 18°$ and $30°$. The six impinging intermediate jets (three each through OD and inner wall ID, respectively) are located at $x = 5.1$ cm, and six impinging dilution jets are located at $x = 7.6$ cm with circumferential locations of $\theta = 6°$, $18°$, and $30°$, respectively. The three OD cooling slot lips are located at $x = 1.4$, 4.3 and 6.6 cm, respectively; and the four ID cooling slot lips are located at $x = 1.4$, 4.3, 6.6 and 9.8 cm, respectively. The centerline of the nozzle shroud is located at $x = 2.7$ cm whereas its spray exits at $x = 1.9$ cm, $y = 2.92$ cm from the inner wall and $\theta = 9°$ which may be compared with the center of the dome swirler being $x = 0$, $y = 2.2$ cm and $\theta = 18°$. All these combustor details were considered adequately represented by $30 \times 19 \times 13$ nodes resulting in 5236 control volumes; no one will accept this statement in 2019.

The predicted fuel/air distribution expressed as equivalence ratio ϕ is shown in Figs. 17 and 18 for the selected four axial stations, namely the spray origin located at $x = 1.9$ cm, primary orifices located at $x = 3.2$ cm, intermediate orifices at $x = 5.1$ cm, and dilution orifices at $x = 7.6$ cm. These locations have been identified as stations 2, 4, 6, 11, and 13, respectively in Fig. 16. It should be made clear that here as well through the early 1990s our objective was to make qualitative assessment of the predicted flow field characteristics including mean velocity components, turbulence kinetic energy k, turbulent mixing times k/ε (here ε is dissipation rate of k), spray transport and evaporation and attendant fuel/air distribution, concentration of unburned fuel and CO (as determined by two-step kinetic scheme and modified eddy break-up time approach calibrated with the extensive data sets described in [3]), temperature distribution, combustor exit temperature quality and adiabatic wall temperature levels and gradients.

Due to page limitations, we have not shown predicted contours of velocity components, unburned fuel, CO and temperature here. Suffices to say that from these predictions along with empirical correlating parameters including combustor reference velocity (V_R), heat release rate (HRR), loading parameter (LP), cold and hot residence times τ_C and τ_H, summarized in Table 1, we expected to have low PF for both Concepts I and II, lower values of LBO FAR, NO_x, smoke and idle HCEI for Concept I compared to Concept II, also summarized in Table 1.

It is hard to make direct comparison of the Concept I and II (CI and CII) combustors' emissions with the modern low-emissions rich-dome combustion products developed during 1995–2009 timeframe. We can, however, summarize these results by comparing takeoff NO_xEI with idle COEI and the latter with idle HCEI by selecting one of the smallest product engine combustors identified as 2.8 in Fig. 19. Product combustor 2.8 belongs to a group of low-emissions combustors (LEC) of modern propulsion engines covering takeoff pressure ratio between 20 and 40 with the corresponding takeoff thrust ranging between 60 and 500 kN. We got both CI and CII combustors showing better emissions characteristics than the best LEC technology products developed between 1995 and 2009; the credit goes to the proper use of the empirical/analytical methodology or "Good Karma".

Let us illustrate one more example on how CFD was used in the Army program in regard to eliminating liner and dome wall carboning problems that were very common in combustors during the 1970s; a typical case is illustrated in Fig. 20a. This product

Table 1 Design, performance and emissions characteristics of Army Combustor Design Criteria ACDC Concept I and Concept II Mod 1

Design and performance parameters	ACDC Concept I	ACDC CII Mod 1	Performance and emissions	ACDC Concept I	ACDC CII Mod 1
V_R (m/s)	7.80	7.88	LBO FAR	0.0031	0.0048
HRR (J/m³ s)	6.63E+07	6.16E+07	SAE Smoke number	0	8.5
LP (kg/s m³)	1.283	1.192	NOx, kg/kW-hr.-cycle	0.034	0.041
τ_C (ms)	9.77	10.51	Max Power NOxEI	14.2	15.0
τ_H (ms)	3.98	4.28	Idle HCEI	0.9	1.65
$\Delta P/P$ (%)	2.5	1.7	Idle COEI	16.7	11.18
Pattern factor	0.211	0.219	Idle comb efficiency	99.85	99.57
Radial profile factor	0.014	0.044	Max Power Eff.	99.99	99.99
T^0_{Wmax} (F)	1475	1600			

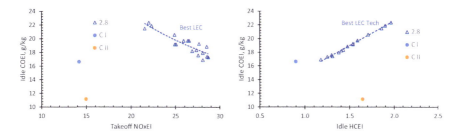

Fig. 19 Emissions characteristics of Concept I and Concept II Mod 1 compared with the best modern low-emissions combustion product

combustor exhibited excessive amount of dome and liner wall carboning, which when built in chunks big enough will lead to breaking of the carbon clinkers and passing through turbine hot-section leading to blades' erosion and attendant loss in power up to 10%. Based on experience, we related its occurrence to the extent of fuel rich pockets and separating flow regions near the wall. We tried indirect verification of this hypothesis in Concept I design process.

Figure 21 shows predicted contours of fuel/air ratio (expressed as equivalence ratio ϕ) at the Concept I dome as affected by the spray cone angle α and spray Sauter mean diameter D_{32}. When a spray with $\alpha = 90°$ and $D_{32} = 50$ μm was replaced with $\alpha = 60°$ and $D_{32} = 50$ μm, we see increased degree of fuel rich pockets. On the other hand, decreasing D_{32} from 50 to 15 μm, we do not see any fuel rich pockets as shown in Fig. 21. The main inference from these CFD simulations is that when we

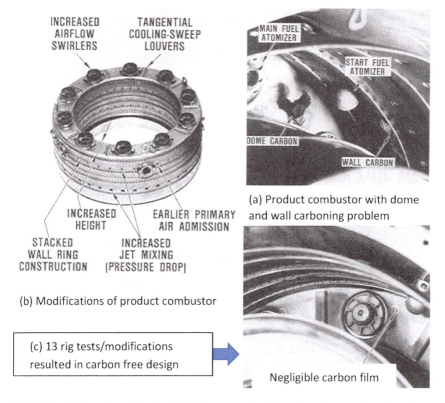

Fig. 20 Application of empirical/analytical design methodology for reducing carbon formation in a product combustor

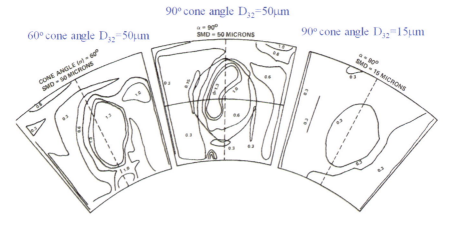

Fig. 21 Three-dimensional CFD predicted profiles of unburned fuel (expressed as equivalence ratio) on the Concept I showing effect of spray angle (α) and Sauter mean diameter (D_{32}) on unburned fuel distribution at the dome [4]

(a)   **(b)**

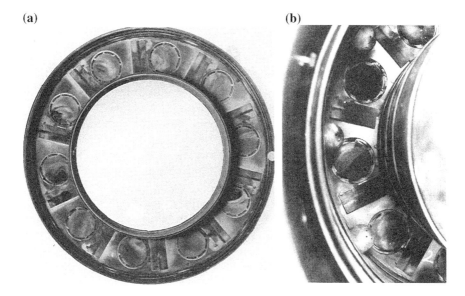

Fig. 22 Indirect qualitative substantiation of three-dimensional CFD predicted contours of unburned fuel and flow field characteristics near the ACDC Concept I at design point by comparing sooting after one rig testing with 90-deg spray, Part (**a**), compared to slightly more sooting observed after 600-4 min simulated takeoff-idle cycles shown in Part (**b**). *Source* [4]

use the injector with "right" combination of spray cone angle and D_{32}, we should not see any dome carboning.

Figure 22a shows typical hardware condition of the ACDC Concept I after a rig shutdown, indicating very small sooting which is generally consistent in combustors having no carbon deposition problems. This combustor also went through 600 simulated takeoff-idle power thermal cycles; each cycle comprised of 2-min steady-state operation at takeoff conditions followed by snapped reduction (in less than 1 s) to a reduced level of fuel flow rate in order to simulate idle fuel/air ratio, followed by 2-min steady-state operation at idle FAR; and then snap back to takeoff fuel flow rate. Figure 22b shows typical dome condition after completing 600 thermal cycles, indicating an acceptable level of sooting and no carboning. It is interesting to note that PF of this combustor at the start of thermal cycling was 0.219 compared to 0.178 measured after completing 600 thermal cycles.

When a proper set of hypotheses as summarized in Fig. 20b was used along with the empirical/analytical design methodology, we were able to get rid of dome and liner wall carboning problem of the product combustor (Fig. 20a) as illustrated in Fig. 20c. This was not the last time we encountered carboning problem. We encountered one in the late 1980s and another one in the middle 1990s. All of these problems got eventually resolved with conventional or hypotheses-based design approach. It is, therefore, not surprising to discover nozzle carboning problem in the LEAP-1B engines where 1% of the engines in field operation are under close observation for avoiding turbine blade failure.

5   On Predictability of Rich-Dome Combustors

None of the CFD simulations conducted before 1994 exceeded 100,000 computational control volumes, and the complex combustors' configurations were simulated by stairstep Cartesian/cylindrical coordinate system. Therefore, CFD simulations were used to provide qualitative guidance which along with combustor design experience know-how facilitated significant advances in both technology and combustor products. Mongia and his team continued to meet or exceed management expectations and enjoyed 20-year streak of continuing success of empirical/analytical design methodology for which credit should duly be given to the insight provided by CFD simulations.

However, starting 1994, our focus shifted to higher resolution simulations but still limited to less than one million control volumes that used curvilinear orthogonal coordinate system which allowed better representation of product combustors. With improved spray and turbulent combustion modeling capabilities as described in [6], we wanted to set our CFD simulation accuracy goals comparable with state-of-the-art semi-empirical gaseous emissions models as described in [46]. Here, the emphasis was on developing low-emissions rich-dome combustion products; although limitations were established on how much this approach can be extended for application in lean-domes' development, as summarized in [7, 8]. This methodology, known popularly as "anchored CFD", became the workhorse for developing rich-dome low-emissions combustion products as described in [21, 22]. Until the comprehensive combustion system analysis approach was developed [13, 33–36], the anchored CFD was used cautiously for providing guidance in the design and development activities of lean-domes described in [23–25]. Readers may want to go through these publications, namely [6–8, 13, 21–25, 33–36, 46], in order to get details on the modeling and resulting technology and combustion products. In Figs. 23, 24, 25, 26, 27 and 28, we will illustrate capability of the anchored models (Figs. 23, 25 and 27) compared to their counterparts' semi-empirical models (Figs. 24, 26 and 28) for NO_x, CO and HC emissions for several product combustors operating from idle to maximum power.

Fig. 23 NO_xEI data from five product combustors compared with anchored CFD predictions; adopted from [6]

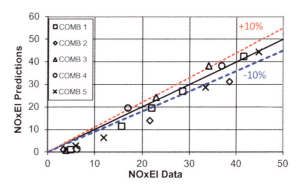

Fig. 24 NO$_x$EI data from
four rich-dome product
combustors compared with
semi-empirical correlation;
adopted from [46]

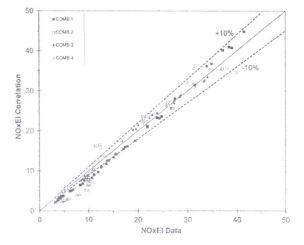

Fig. 25 COEI data from five
product combustors
compared with anchored
CFD predictions; adopted
from [6]

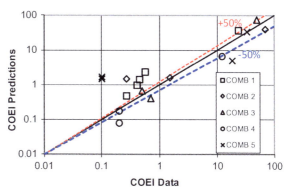

Fig. 26 COEI data from
four rich-dome product
combustors compared with
semi-empirical correlation;
adopted from [46]

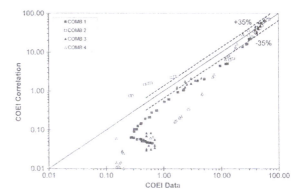

Fig. 27 HCEI data from five product combustors compared with anchored CFD predictions; adopted from [6]

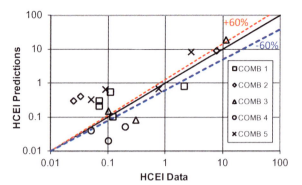

Fig. 28 HCEI data from four rich-dome product combustors compared with semi-empirical correlation; adopted from [46]

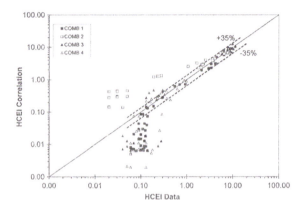

Since majority of NO_x emissions contribution comes from high-power operation, accuracy of the model predictions is generally considered relevant for high-power. On the other hand, model accuracy for CO and HC is generally based on idle emissions. For the selected five product combustors summarized in Figs. 23, 25 and 27, CFD predictions for NO_x are within $\pm 10\%$ whereas those of COEI and HCEI are approximately $\pm 50\%$ and $\pm 60\%$, respectively. The corresponding semi-empirical emissions prediction accuracy levels are $\pm 10\%$, $\pm 35\%$ and $\pm 35\%$, respectively. Some from the old school of combustion designers may come to wrong conclusion, namely, why do we need CFD simulations when they are not more accurate than the semi-empirical correlations? For obvious reasons, middle of the road, empirical/analytical design methodology is the right approach which has been practiced by Mongia and his coworkers since 1975.

6 On Predictability of Lean-Dome Combustors

A comprehensive combustion system analysis methodology was kicked off in 2001 while simultaneously successfully adopting partially premixed laminar flamelet modeling considered critical for providing radical concentrations required to solve turbulent transport equations to compute "accurately" HC, CO, NO_x emissions for the four types of combustors, namely, dry low NO_x combustors [37], modern rich-dome aviation combustors [38], industrial diffusion flame combustors with water injection [39], and second-generation lean-dome propulsion engine combustors [24, 25]. These emissions-related publications based on composite laminar flamelet modeling approach did not use comprehensive combustion system analysis as described in [13]. Only two publications, [35] and [36], have shared comprehensive analysis based composite laminar flamelet predictions; the former for the second-generation lean-domes and the latter for rich-domes known popularly as RQL. In [36], only NO_xEI emissions results for a product combustor were published, getting high-power NO_x prediction capability comparable with the anchored CFD shown in Fig. 23.

   While the overview level details of the comprehensive system analysis were provided in [13], suffices to say here that this approach does not require user to provide pressure and airflow distribution around the combustor, inlet conditions for the various holes and swirlers including profiles of the mean and turbulence quantity of the dependent variables. The user can conduct simulations for any engine operating conditions of interest, namely, compressor discharge pressure $P3$ and temperature $T3$, airflow rate and combustor fuel flow rate. The combustor diffuser inlet profiles of the total and static pressure distributions are provided by compression system design tools. Validated injector comprehensive models are used to compute spray conditions required for combustion system analysis (CSA) simulation. Combustion system hardware is simulated to the accuracy consistent with manufacturing tolerances by using unstructured mesh requiring 15–25 million size as described in [13].

   Even though most of the simulations have been reported by using RANS, the advantage of LES is well established as described in [40–42]; the latter produces better agreement with data on average exit temperature (θ_{avg}) profiles without having to use turbulent Schmidt number (Sc_t) as a variable in RANS simulations; a practice commonly used by most of the OEM's for both θ_{avg} and maximum exit temperature (θ_{max}) profiles. Figure 29 shows a typical comparison for a selected combustor. In spite of the obvious advantage of LES over RANS, the former is used by OEM's only as a last resort to sort through critical design challenges because it requires at least an order of magnitude bigger computer resources and a need for thoroughly anchoring the LES modeling strategy before it can be used as part of combustor design toolkit. A recent example described in next section is the use of a time-filtered Navier–Stokes (TFNS) for the lean-direct injection (LDI) technology development effort.

   There were many reasons that motivated us to develop and validate comprehensive CSA approach; perhaps the most important one being technology and product development of the second-generation lean-domes [24, 25]. Here, we wanted more

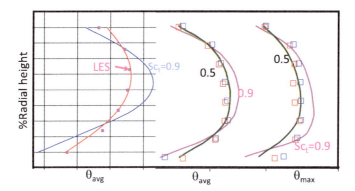

Fig. 29 Typical comparison between data and simulations with combustor exit plane normalized average and maximum temperature, θ_{avg} and θ_{max}, radial profiles showing effect of turbulent Schmidt number Sc_t. All calculations except one are with RANS simulations

accurate description of the combustor internal flow field and temperature contours (see Figs. 30 and 31) in order to relate these more reliably in regard to NO_x and CO emissions (see Figs. 32 and 33) and tradeoff for combustor exit temperature profiles shown in Fig. 34 for one of the two combustors covered in these figures. Even though Figs. 32 and 33 appear to show better agreement with data compared to anchored CFD (see Figs. 23 and 25), there is a lot more to be done before the job is completed; the least is to get equally good agreement with other second-generation lean-domes' engine data which takes us to introductory part of the next section.

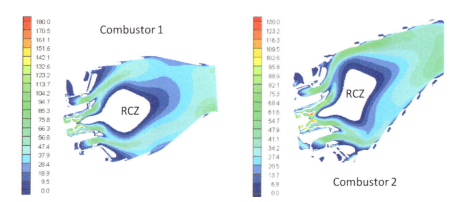

Fig. 30 Reacting flow mean axial-velocity contours (m/s) predicted by comprehensive combustion system analytical modeling approach [35]

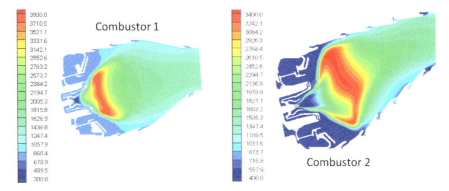

Fig. 31 Mean gas temperature contours (°F) predicted by comprehensive combustion system analytical modeling approach [35]

Fig. 32 NO$_x$ EI data from two lean-dome combustors compared with predictions of comprehensive combustion system analytical modeling approach [34]

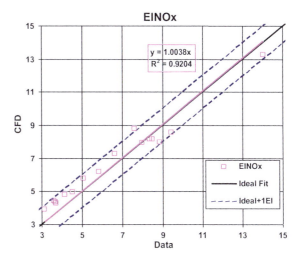

7 Recent Lean and Rich-Dome Combustors' Fundamental Investigations

While the second-generation lean-domes show the potential of further reducing emissions [24], it also raises concerns about challenges in regard to operability, autoignition, and flashback. It is remarkable to see these lean-domes operating reliably at takeoff pressure ratio of 47.5 in the GEnx-1B engine while simultaneously achieving 27% reduction in LTO NO$_x$ compared to its counterpart rich-dome Trent1000; see Table 2. However, at lower takeoff pressure ratio of 37.5, it demonstrated 43% NO$_x$ reduction, a 16 point competitive loss as it went to its highest operating pressure. Can CFD help in recovering this loss? Yes, it can as evident qualitatively for the predicted contours of two lean-dome Combustors 1 and 2 shown previously in Figs. 30

Fig. 33 COEI data from two lean-dome combustors compared with predictions of comprehensive combustion system analytical modeling approach [35]

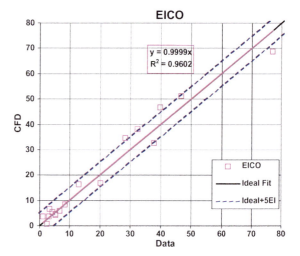

Fig. 34 Typical comparison between data and predicted average combustor exit temperature profile [35]

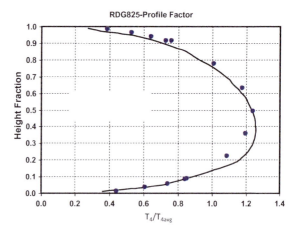

and 31. The TALON X rich-dome technology-based combustion products emissions certified in 2017 have achieved another significant step in reducing NO_x compared to its counterpart RQL combustors. While its competing lean-dome LEAP has 13% lower NO_x compared to TALON X at 33 pressure ratio, its NO_x at 41.5 pressure ratio does not show any advantage over RQL; another opportunity for CFD tools to help.

The next generation of very large engines GE90-X is expected to operate above 50 pressure ratio which provides impetus for looking into more robust lean-dome technologies as described in [47]. One of these approaches called swirl venturi lean-direct injection first-generation technology (LDI-1) was described in [48]. Its second-generation technology, LDI-2, described in [26] relied on high-resolution CFD simulations reported in several publications [e.g., 27, 49–52]. Both RANS and TFNS simulations have been used; here, we will give an overview of back to back

Table 2 Design and technology evolution of propulsion engines during last 40 years in regard to rich or lean-dome technology (RD/LD), takeoff fan bypass ratio (BPR) and pressure ratio (PR), rated thrust and landing-takeoff NO_x emissions of the selected 8 engines emissions certified during 1976 and 2017

Tech	Engine model	BPR	PR	kN	NOx	Year	Tech	Engine model	BPR	PR	kN	NOx	Year
RD	TFE731-2	2.6	13.9	15.6	40.5	1976	LD	LEAP-1B	8.6	41.5	130	57.8	2016
RD	HTF7500	4.2	23.6	34	45.9	2010	RD	CF6-6D	5.9	24.7	175	66.4	1979
RD	CMFM56-2	6.0	25.2	107	42.8	1983	RD	Trent1000	8.9	48.7	358	69.9	2017
RD	TALON X	11.6	38.1	147	31.3	2017	LD	GEnx-1B	8.6	47.5	349	49.4	2012

comparison between RANS and TFNS results reported in [52] where 21.34 million all-tetrahedral elements were used to resolve details of a three-cup third-generation LDI-3 flametube configuration shown schematically in Fig. 35.

CFD computations with OpenNCC were reported for a generic NASA N+3 ICAO engine cycle for single-aisle, mid-range aircraft of 25,000 lb. Thrust class at an operating takeoff pressure ratio of 40. OpenNCC CFD results reported include flow field predictions and emissions comparisons for the idle and approach conditions. While the details of the hardware, models, mesh generation strategy, numerical scheme, and convergence criteria were reported in [27, 49–52], here we will focus on the key results as summarized in Table 3 and Figs. 36, 37, 38 and 39.

An initial evaluation of OpenNCC was performed for the central, five-injector cup of the three-cup, nineteen-element geometry. Figure 36 shows axial-velocity contours of the OpenNCC results at various cross sections for this five-injector cup.

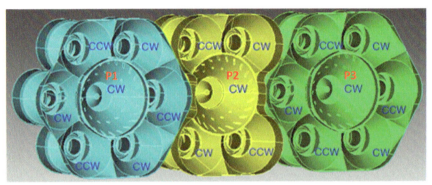

Three pilots (P1, P2 and P3) with clockwise (CW) swirler vanes arrangement
(6+4+6) main mixers with CW and counterclockwise (CCW) swirlers arrangement

Fig. 35 Dome-layout with main- and pilot-injector elements for a 3-cup LDI-3 flametube simulated by 21.34 million all-tetrahedral elements. *Source* [52]

Table 3 Comparison between data and predictions; adopted from [52]		Experimental data	OpenNCC predictions, value (% error)
	Effective area, in^2	2.672	2.738 (2.5)
	Idle NO$_x$EI, g NO$_2$/kg fuel	4.7	4.4 (6.5)
	Idle COEI, g CO/kg fuel	51	1.6 (97)
	Approach NO$_x$EI, g NO$_2$/kg fuel	5.7	3.7 (35)
	Approach COEI, g CO/kg fuel	50	0.5 (99)

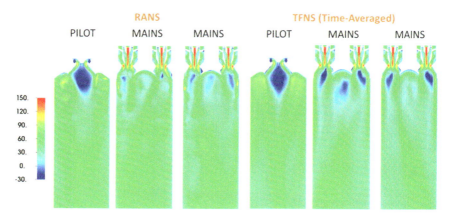

Fig. 36 Axial-velocity (m/s) contours in three axial cross sections through centerline of two main injectors and the pilot injector. Left: RANS solution. Right: TFNS solution from [52]

A strong, prominent central recirculation zone (CTRZ) behind the pilot was predicted by both RANS and TFNS methods. One major difference was that TFNS predicts a relatively symmetric and uniform flow pattern behind each of the four main injectors. The RANS predictions showed relatively large variations in the flow patterns behind the four main injectors. The corresponding reacting flow solutions for temperature and NO mass fraction, as predicted by RANS and TFNS, are shown and compared in Fig. 37a and b, respectively. Based on the single-cup geometry evaluation with OpenNCC, the TFNS methodology was chosen for further study of the full three-cup, nineteen-element geometry.

Table 3 summarizes the OpenNCC TFNS results, as compared with measured experimental data [52]. The predicted effective area, as computed from nonreacting computations, compares very well with experimental data (2.5% error). Figures 38 and 39 show the contours of predicted NO and CO, for the idle and approach conditions, respectively. For the idle case, only the pilot injector of each of the three cups is fueled. For the approach case, all five injectors of the central cup and the pilot injectors of each seven-element module are fueled.

The predicted $NO_x EI$ for the idle case provides the best comparison with data (6.5% error), followed by $NO_x EI$ prediction of the approach case (35% error). However, the prediction of emissions for COEI for both cases is in large error when compared to experimental data. The current results suggest that considerably more effort needs to be invested in improving modeling capabilities (chemical kinetic models, spray modeling, turbulence-chemistry interaction, etc.), particularly for CO predictions at low-power cycle conditions (idle, approach).

In order to meet continuing stringent regulatory requirements, the rich-dome combustion products have recently encountered smoke emissions challenges as summarized in Fig. 40 for 13 propulsion engine models ranging in takeoff thrust from 30 to 514 kN whereas the corresponding takeoff pressure ratio varies between 15.8 and 49.4. Small and medium-size engines with takeoff pressure ratio less than 35 fall

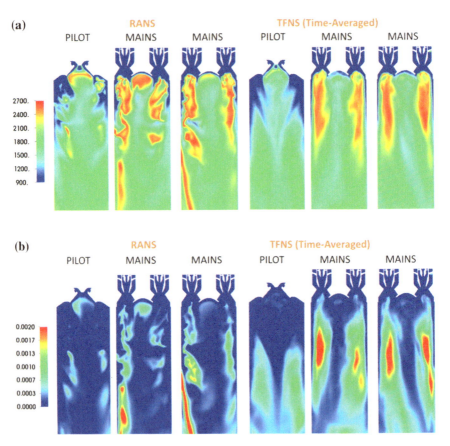

Fig. 37 **a** Temperature (K) contours in three axial cross sections through centerline of two main injectors and the pilot injector. Left: RANS solution. Right: TFNS solution from [52]. **b** NO mass fraction contours in three axial cross sections through centerline of two main injectors and the pilot injector. Left: RANS solution. Right: TFNS solution from [52]

along Group 1 engines for both landing-takeoff (LTO) NO_x and smoke emissions expressed as % of regulatory standards, with the former as %CAEP8. OEMs prefer to have 20–30% margin from the regulatory standards. All large engines used to meet LTO NO_x standards, namely % of CAEP4 or CAEP6. In regard to smoke they fell in Group 2 during 1995–2011 timeframe. However, after 2011 Group 2 NO_x group led to Group 3 and Group 4 for smoke emissions which are considered uncomfortably too close to the regulatory levels. TALON X technology created further reduction in NO_x leading to Group 3 NO_x engines; but one of these engine models belongs to Group 4 smoke engines. This clearly makes tradeoff between NO_x and smoke emissions more intense while simultaneously ensuring high-altitude ignition capability more difficult to maintain.

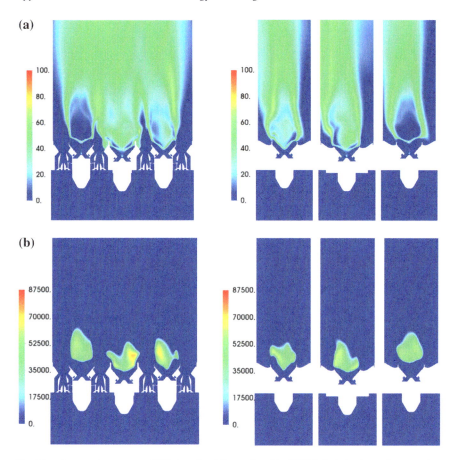

Fig. 38 **a** Isometric contours of NO (ppm) at idle predicted by TFNS. Left: horizontal center plane including Mains and Pilots. Right: vertical plane with composite of three pilot injectors. **b** Isometric contours of CO (ppm) at idle predicted by TFNS. Left: horizontal center plane including Mains and Pilots. Right: vertical plane with composite of three pilot injectors

One OEM has decided to go with lean-dome propulsion engines and has already in revenue service two engines (namely, LEAP and GEnx), and plans to introduce GE90-X. These three engines cover takeoff thrust and operating pressure ratio ranges of 120–500 kN and 33 to 50+, respectively. While lean-domes produce considerably lower NO_x and smoke emissions compared to rich-domes, they are accompanied by more challenging operability issues. Therefore, long-term accuracy objectives of the phenomenological combustion models and CFD simulations need to be comparable with data measurement capabilities expressed in terms of standard deviation σ summarized in Table 4. From here, we can set our long-term CFD accuracy goals comparable with experimental error, namely, 5% for LTO NOx, and 20% for LTO HC, LTO CO, and maximum SAE smoke number.

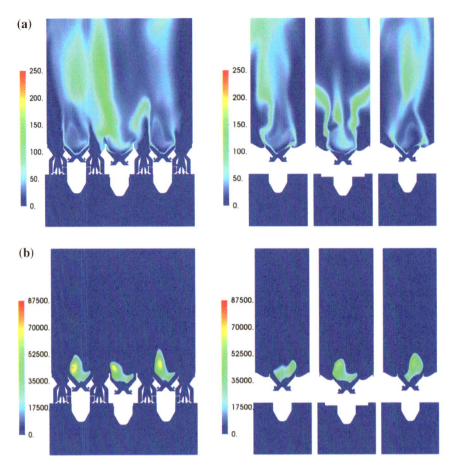

Fig. 39 a Isometric contours of NO (ppm) at approach predicted by TFNS. Left: horizontal center plane including Mains and Pilots. Right: vertical plane with composite of three pilot injectors. **b** Isometric contours of CO (ppm) at approach predicted by TFNS. Left: horizontal center plane including Mains and Pilots. Right: vertical plane with composite of three pilot injectors

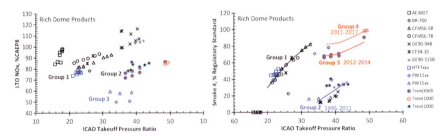

Fig. 40 NO$_x$ and smoke emissions tradeoffs encountered in recently certified combustion products

Table 4 Current status of engine emissions measurement standard deviation σ for HC, CO, NO_x and smoke number (SN)

LTO HC	Range of σ	%	LTO CO	Range of σ	%	LTO NO_x	Range of σ	%	SN	Range of σ	%
<0.24	25	100	<5	25	75				<5	29%	57%
<1	17%	25	<10	8.2%	25				<10	7%	29%
>1	8%	17%	>10	5.1%	8.2%	>30	1.7%	5.9%	>10	7%	14%

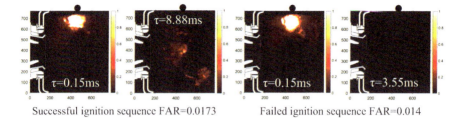

Successful ignition sequence FAR=0.0173   Failed ignition sequence FAR=0.014

Fig. 41 An illustration of high-speed flow visualization showing successful ignition at 0.0173 FAR and ignition failure at 0.014 FAR; from [53]

Accurate high-resolution CFD simulations for applying fundamentals learned in regard to operability experiments conducted under controlled laboratory environment for both rich and lean-dome concepts need to be extended for providing guidance during the engine development process. For example, as reported in [53], rich-dome combustor rig experiments were conducted parametrically including ignition/lean blowout testing at ambient pressure/temperature, high-altitude (simulated combustor pressures of 35, 55, 80 kPa, and pressure drop of 1–5%) ignition testing, and ignition process visualization with a high-speed camera providing 6750 frames/s, shutter speed of 1/51,000 s with resolution of 768 pixels × 768 pixels. The high-speed camera was used for studying sequence of ignition kernel (shown in Fig. 41 to appear at time $\tau = 0.15$ ms) followed by its growth and propagation and eventually sustained combustion in the primary zone at fuel/air ratio (FAR) above ignition (here 0.024) at $\tau = 8.88$ ms. On the other hand, when below ignition FAR, here 0.021 FAR, the ignition kernel does not grow and we found no flame in the primary zone at $\tau = 3.55$ ms. CFD simulations are planned to be undertaken to model the entire process starting with ignition kernel, its growth and propagation to achieving self-sustaining combustion or extinction. Validated models will be used for studying tradeoff between NO_x, smoke and operability for a typical gas turbine combustor.

A very extensive operability related fundamental investigation is currently ongoing as described in [54–57] with several more planned to be published including CFD simulations relevant to swirl venturi LDI mixers, as well as the same set of configurations without flare that resembles modern airblast injectors. Advanced diagnostics used include two-dimensional time-resolved particle image velocimetry (PIV) system, OH*, CH*, and NO_2* chemiluminescence signals with an intensified charge-coupled device camera with species-specific bandpass filters for each radical, namely 310 ± 2 nm with 10 nm full width half maximum (FWHM) for OH*, 430 ± 2 nm with 10 nm FWHM for CH*, and 750 ± 10 nm with 70 nm FWHM for NO_2*.

We want to summarize key findings on the weak and strong swirling flow field established by LDI and airblast configurations with 45° and 60° (also called baseline) outer swirler vane angles; inner swirler vane angle is 60° for all the four configurations summarized in Figs. 42, 43, 44 and 45. Direct flame images overlaid with velocity vector maps for the baseline LDI configurations with and without the venturi flare section are shown in Fig. 42 for a pressure drop of 3% and $\phi = 0.65$. Without

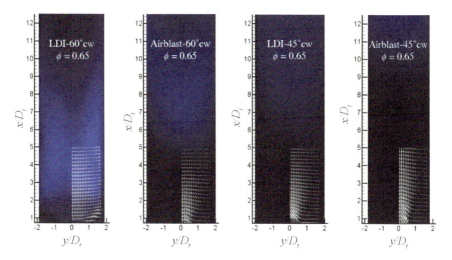

Fig. 42 Flame structure variations at 3% pressure drop and $\phi = 0.65$ for the LDI-60°cw, Airblast-60°cw, LDI-45°cw, and Airblast-45°cw configurations, represented by mean direct flame images overlaid with mean velocity vectors

the venturi flare, both the Airblast-60°cw and Airblast-45°cw configurations exhibit lifted flames, although the Airblast-60°cw appears to have a stronger flame than Airblast-45°cw, with the primary flame zone of Airblast-60°cw located several throat diameter (D_t) lower than for the Airblast-45°cw case. Comparing the LDI-45°cw with its airblast counterpart, while overall flame structure is similar, the flame sits closer to the dump plane for the LDI configuration. However, for the LDI-60°cw configuration with and without the venturi flare, the flame is either anchored to the dump plane or lifted solely due to the flare.

The source of this change in flame behavior is apparent from the velocity field results, as shown in Fig. 43. The minor differences in flame structure between the two 45°-vane angle cases can be explained by the swirling jet structure; whereas for the LDI-45°cw case the jet is radially larger and axially shorter, removal of the venturi flare results in a thinner and dramatically longer jet. This behavior suggests that for a given flame propagation, the flame will not be able to stabilize until further downstream due to slower expansion of the swirling jet. More interesting, however, are the 60°cw cases. Removal of the venturi flare from the LDI-60°cw configuration results in a flow field much more akin to that found in the LDI-45°cw configuration, namely, a swirling jet flow in lieu of a strong center recirculation zone. Clearly, the removal of the flare section significantly weakens overall swirl strength to the point that vortex breakdown no longer occurs. Considering the above in terms of swirl number, the results are somewhat intuitive. For a given vane configuration, tangential velocities are essentially fixed; the addition or removal of a flare section will alter this motion very little. However, due to the Coanda effect—that fluid streamlines tend to follow solid boundaries—flow passing through the venturi flare will tend to expand, in the process reducing axial velocity. This in turn effectively increases the

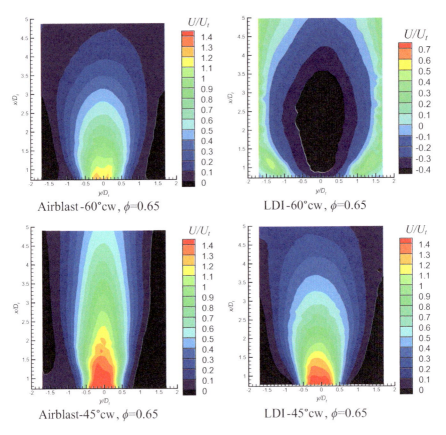

Fig. 43 Mean reacting axial-velocity fields at $\phi = 0.65$ for the Airblast-60°cw, LDI-60°cw, Airblast-45°cw, and LDI-45°cw configurations

Fig. 44 Lean blowout limits of the four selected configurations

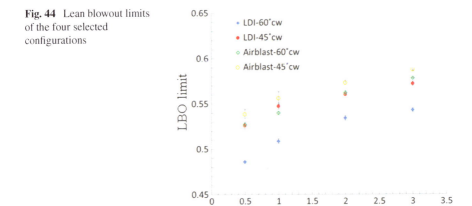

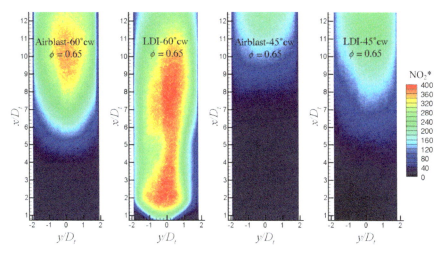

Fig. 45 NO$_2$* chemiluminescence contours at $\phi = 0.65$ for the Airblast-60°cw, LDI-60°cw, Airblast-45°cw, and LDI-45°cw configurations

swirl number by decreasing the flux of axial momentum. One interesting implication is that the Coanda effect is typically influenced by the immersion depth of the mixer relative to the dome; a future study will explore the extent to which apparent swirl strength is influenced by swirl assembly insertion depth.

In terms of operability, the aforementioned alterations to the flow field greatly impact LBO limit. As shown in Fig. 44, across the pressure drop range tested, the LBO limits for the three configurations not exhibiting a center recirculation zone are roughly similar, while the LDI-60°cw configuration exhibits LBO limits consistently 5–10% lower. Also shown in Fig. 45, NO$_2$* emissions are strongly tied to the presence of the center recirculation zone. For each of the configurations resulting in a lifted flame, including Airblast-45°cw, Airblast-60°cw, and LDI-45°cw, NO$_2$* emissions are relatively low. Some correlation with swirl strength exists; the Airblast-45°cw configuration exhibits the lowest emissions, followed by the LDI-45°cw, and then the Airblast-60°cw case. However, when a center recirculation zone is present, as for the LDI-60°cw case, high NO$_2$* signal is observed from just after the dump plane to 11–12 D_t downstream. This provides further evidence that for a single swirl injector, a design tradeoff exists between operability, stability, and NO$_x$ emissions.

Even though we would like to have modern CFD tools' capabilities better than what is summarized in Table 3 for LDI or their predecessors for TAPS summarized in Figs. 32, 33 and 34, it is important to recognize that these tools have been used successfully for developing lean-dome technologies (LDI-3 and TAPS3) and products (GEnx) and will play important role for creating next-generation TAP3 based potential products identified as TAPS3PP in Fig. 46. It is interesting to notice that the first-generation dual-annular combustion know-how led to GE90 DACI product for large 40 pressure ratio engines. As is generally the case, its NO$_x$ technology as measured in terms of takeoff NO$_x$EI versus takeoff pressure ratio was not better than

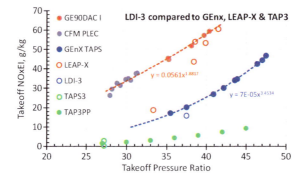

Fig. 46 Takeoff NO$_x$ emissions index versus takeoff pressure ratio of several propulsion engine combustion products including rich-dome pre-low-emissions CFM56 products certified in early 1990s (CFM PLEC), the first dual-annular 40-pressure ratio lean-dome combustor (GE90DAC I) which provided impetus to develop second-generation lean-dome GEnx based on TAPS1 technology

low-pressure ratio rich-dome technology, the CFM56 identified as CFM PLEC. As expected, the GE90 DAC I emissions results provided impetus for developing the second-generation lean-dome technology and products (e.g., GEnx), known popularly as TAPS as described in [24, 25]. Compared to GE90 DAC I at 40.4 pressure ratio, GEnx achieved 68% reduction in takeoff NO$_x$EI. As leading combustion designers have experienced several times before, the GEnx scale-up for the CFM56 application (known as LEAP-X) did not turn out good. Its takeoff NO$_x$EI at 41.5 pressure ratio turned out to be 2.1 times GEnx value for the same pressure ratio. On the other hand, for the lower pressure ratio of 33.4, its NO$_x$EI was only 1.37 times GEnx value. When and if the competitive marketing situation arises, we are sure that the TAPS-based LEAP-X will be followed by further reduction in NO$_x$ emissions comparable with GEnx or beyond by applying TAPS3 technology and producing potential TAPS3 based product, TAP3PP.

It is equally interesting to note that LDI-3 technology (see Fig. 35) was developed by using CFD showing it slightly lower than GEnx takeoff NO$_x$EI, namely, 22% lower at 37.5 pressure ratio, even though we need to further improve its prediction capability, as summarized in Table 3. What we are trying to convey is CFD when used properly will always provide good guidance even when its accuracy levels are significantly off compared to measurement uncertainties.

8 Summary

The empirical/analytical design methodology (EADM) reported by Mongia and his coworkers in 1977–1978 [1, 2] perhaps provided impetus for other gas turbine OEM's to pursue variants of theirs consistent with their vision. Figure 1 gives a schematic representation of EADM which comprises of mutually interacting five elements. Can we include elements 1 through 5 as part of continuously evolving conventional design process that is uniquely different for each major OEM's? How closely coupled are elements 5 and 6? It is, therefore, fair assumption to make that EADM's of all major OEM's are uniquely different; moreover, they include trade secrets critical for maintaining competitive edge. Mongia did not feel qualified enough to make assessment calls for EADM's pursued by others. This chapter's focus is only on EADM used in programs where Mongia was personally involved as team leader. We used CFD simulations in conjunction with applicable hypotheses supported by subcomponent engineering tests or not in order to seek qualitative guidance during continuously evolving design process even for a given project. This approach has been used successfully in several programs as recognized by customers as described in [1, 2, 4, 6, 9–11, 14–25, 29]. These programs included NASA staged combustion Concept 3 (in 1977), two Army combustor concepts for small engines (1978), 2100, 2400 and 2900 °F temperature rise combustors (1981–1983), the two first product combustors (1986), two near-stoichiometric temperature high-performance combustors (1993), entitlements for ultralow NO_x premix/pre-vaporized and partially premixed mixers (1993, 2008), an RQL combustor for the largest turbofan engine (1996), the second-generation lean-dome combustion technology TAPS demonstration (2003) and its product introduction in GEnx (2009), NASA LDI-2 and LDI-3 technology demonstrations in 2014 and 2018, respectively. Mongia agrees with the statement made by one of his peers in 2005, "I see that you use CFD in selecting designs. But I don't understand your decision process."

The key to success of EADM includes (1) Proper use of the methodology (Fig. 1), customers' trust, freedom to define and pursue evolving technical approach; (2) Combustion hypotheses driven by test results, see Fig. 20b as an example for addressing dome and liner wall carboning; (3) Element test configurations tested under relevant initial and boundary conditions, cf. Sect. 3; (4) Establishing limits of models' capabilities, cf. Sects. 5 and 6; (5) Optimum use of resources, team members, time, test facilities, design experience driven by test results backed by semi-analytical and CFD design tools; and (6) Art of design and application of tools. CFD has become an integral part of combustion design practice; but its usefulness will be influenced greatly by how well these six elements are integrated. Section 7 summarizes vision of the authors on how CFD can be helpful in developing technology; and ongoing applicable fundamental research effort for both rich- and lean-dome combustion concepts. Even though the scope of the effort required for the abovementioned (3) and (4) has exceeded the original estimate by a factor of up to 5; and we do not see clearly how models' accuracy levels can become comparable with experimental data scatter, "genuine" CFD design tools development will be pursued by enlightened combustion designers.

References

1. Reynolds, R. S., Kuhn, T. W., & Mongia, H. C. (1977). Advanced combustor analytical design procedure and its application in the design and development testing of a premix/prevaporized combustion system. In *1977 Spring Meeting of the Central States Section of the Combustion Institute*.
2. Mongia, H. C., & Smith, K. F. (1978). An empirical/analytical design methodology for gas turbine combustor. In *AIAA1978-998*.
3. Bruce, T. W., Mongia, H. C., & Reynolds, R. S. (1979). Combustor design criteria validation Volume I—Element tests and model validation. In *USARTL-TR-55A (AD-A0-67657)*, March 1979.
4. Mongia, H., Reynolds, R., Coleman, E., & Bruce, T. (1979). Combustor design criteria validation Volume II—Development testing of two full-scale annular gas turbine combustors. In *USARTL-TR-55B (AD-A0-67689)*, March 1979.
5. Mongia, H. C., & Reynolds, R. S. (1979). Combustor design criteria validation Volume III—User's Manual. In *USARTL-TR-55C (AD-AO-66793)*.
6. Danis, A. M., Burrus, D. L., & Mongia, H. C. (1996). Anchored CCD for gas turbine combustor design and data correlation. In *ASME1996-GT-143*.
7. Hura, H. S., Joshi, N. D., & Mongia, H. C. (1998). Dry low emissions premixer CCD modeling and validation. In *ASME1998-GT-444*.
8. Hura, H. S., & Mongia, H. C. (1998). Prediction of NO emissions from a lean dome gas turbine combustor. In *AIAA1998-3375*.
9. Bahr, D. W., & Gleason, C. C. (1975). Experimental clean combustor program Phase I—Final report. NASA CR-134737.
10. Bruce, T. W., Davis, F. G., Kuhn, T. E., & Mongia, H. C. (1977). Pollution reduction technology program small jet aircraft engines phase 1 final report. NASA CR-135214.
11. Bruce, T. W., Davis, F. G., Kuhn, T. E., & Mongia, H. C. (1978). Pollution reduction technology program small jet aircraft engines phase 2 final report. NASA CR-159415.
12. Bruce, T. W., Davis, F. G., Kuhn, T. E., & Mongia, H. C. (1981). Pollution reduction technology program small jet aircraft engines phase 3 final report. NASA CR-165386.
13. Mongia, H. C. (2008). Recent progress in comprehensive modeling of gas turbine combustion. In *AIAA2008-1445*.
14. Mongia, H. C., Coleman, E. B., & Bruce, T. W. (1981). Design and testing of two variable geometry combustors for high altitude propulsion engines. In *AIAA1981-1389*.
15. Kuhn, T. E., Mongia, H. C., Bruce, T. W., & Buchanan, E. (1982). Small gas turbine augmenter design methodology. In *AIAA1982-1179*.
16. Sanborn, J. W., Mongia, H. C., & Kidwell, J. R. (1983). Design of a low-emission combustor for an automotive gas turbine. In *AIAA1983-0338*.
17. Roesler, T. C., Mongia, H. C., & Stocker, H. L. (1992). Analytical design and demonstration of a low-cost expendable turbine engine combustor. In *AIAA1992-3754*.
18. Mongia, H. C. (1993). Application of CFD in combustor design technology. In *AGARD CP-536*, pp. 12-1/12-18.
19. Mongia, H. C. (2011). Engineering aspects of complex gas turbine combustion mixers Part I: High ΔT. In *AIAA2011-0107*.
20. Mongia, H. C. (2011). Engineering aspects of complex gas turbine combustion mixers Part II: High T3. In *AIAA2011-0106*.
21. Mongia, H. C. (2011). Engineering aspects of complex gas turbine combustion mixers Part III: 30 OPR. In *AIAA2011-5525*.
22. Mongia, H. C. (2011). Engineering aspects of complex gas turbine combustion mixers Part IV: Swirl cup. In *AIAA2011-5526*.
23. Joshi, N. D., Mongia, H. C., Leonard, G., Stegmaier, J. W., & Vickers, E. C. (1998). Dry low emissions combustor development. In *ASME98-GT-310*.
24. Mongia, H. C. (2003). TAPS—A fourth generation propulsion combustor technology for low emissions. In *AIAA2003-2657*.

25. Mongia, H. C. (2011). Engineering aspects of complex gas turbine combustion mixers Part V: 40 OPR. In *AIAA2011-5527*.
26. Tacina, K. M., Podboy, D. P., He, Z. J., Lee, P., Dam, B., & Mongia, H. C. (2016). A comparison of three second-generation swirl-venturi lean direct injection combustor concepts. In *AIAA2016-4891*.
27. Tacina, K. M., Chang, C. T., He, Z. J., Lee, P., Dam, B., & Mongia, H. C. (2014). A second-generation swirl-venturi lean direct injection combustion concept. In *AIAA2014-3434*.
28. Serag-Eldin, M. A., & Spalding, D. B. (1979). Computations of three-dimensional gas-turbine combustion chamber flows. *Journal of Engineering for Gas Turbines and Power, 101*, 326–336.
29. Mongia, H. C., Reynolds, R. S., & Srinivasan, R. (1986). Multidimensional gas turbine combustion modeling: Applications and limitations. *AIAA Journal, 24*(6), 890–904.
30. Mongia, H. C. (1988). A status report on gas turbine combustion modeling. In *AGARD CP-422*, pp. 26-1/26-14.
31. Mongia, H. C. (2001). A synopsis of gas turbine combustor design methodology evolution of last 25 years. In *15th International Symposium on Air Breathing Engines, ISABE-2001-1086*.
32. Mongia, H. C. (2001). Gas turbine combustor liner wall temperature calculation methodology. In *AIAA2001-3267*.
33. Held, T. J., & Mongia, H. C. (1998). Application of a partially premixed laminar flamelet model to a low emissions gas turbine combustor. In *ASME1998-GT-217*.
34. Mongia, H. C. (2004). Perspective of gas turbine combustion modeling. In *AIAA2004-0156*.
35. Mongia, H., Krishnaswami, S., & Sreedhar, P. S. V. S. (2007). Comprehensive gas turbine combustion modeling methodology. In *Fluent's International Aerospace CFD Conference*, June 18, 2007, Paris.
36. Sripathi, M., Krishnaswami, S., Danis, A. M., & Hsieh, S. Y. (2014). Laminar flamelet based NOx predictions for gas turbine combustors. In *GT2014-27258*.
37. Held, T. J., Mueller, M. A., Li, S. C., & Mongia, H. C. (2001). A data-driven model for NO_x, CO and UHC emissions for a dry low emissions gas turbine combustor. In *AIAA2001-3425*.
38. Stevens, E. J., Held, T. J., & Mongia, H. C. (2003). Swirl cup modeling Part VII: Partially-premixed laminar flamelet model validation and simulation of a single-cup combustor with gaseous n-heptane. In *AIAA 2003-0488*.
39. Giridharan, M. G., Held, T. J., & Mongia, H. C. (2003). A wet emissions CFD model based on laminar flamelet approach. In *ISABE2003-1087*.
40. Kim, W. W., Menon, S., & Mongia, H. C. (1999). Large eddy simulations of reacting flow in a dump combustor. *Composites Science and Technology, 143*, 25–62.
41. Grinstein, F. F., Young, G., Gutmark, E. J., Hsiao, G., & Mongia, H. (2002). Flow dynamics in a swirl combustor. *Journal of Turbulence, 3*(30), 1–19.
42. Wang, S., Yang, V., Hsiao, G., Hsieh, S. Y., & Mongia, H. C. (2007). Large eddy simulations of gas-turbine swirl injector flow dynamics. *Journal of Fluid Mechanics, 583*, 99–122.
43. Srinivasan, R., Reynolds, R., Ball, I., Berry, R., Johnson, K., & Mongia, H. C. (1983). Aerothermal modeling program Phase I final report. NASA CR-168243.
44. Nikjooy, M., Mongia, H. C., Sullivan, J. P., & Murthy, S. N. B. (1993). Flow interaction experiment, aerothermal modeling Phase II final report, Volumes I and II. NASA CR-189192.
45. Nikjooy, M., Mongia, H. C., McDonell, V. G., & Samuelsen, G. S. (1993). Fuel injector-air swirl characterization, aerothermal modeling Phase II final report, Volumes I and II. NASA CR-189193.
46. Danis, A. M., Pritchard, B. A., & Mongia, H. C. (1996). Empirical and semi-empirical correlation of emissions data from modern turbo propulsion gas turbine engines. In *ASME1996-GT-86*.
47. Ren, X., Sung, C. J., & Mongia, H. C. (2018). On lean direct injection research. In *Energy for propulsion. Green energy and technology* (pp. 3–26). Springer, Singapore.
48. Tacina, R., Lee, P., & Wey, C. (2005). A lean-direct-injection combustor using a 9-point swirl-venturi fuel injector. In *ISABE2005-1106*.
49. Ajmani, K., Mongia, H. C., & Lee, P. (2013). Evaluation of CFD best practices for combustor design: Part I—Nonreacting flows. In *AIAA2013-1144*.

50. Ajmani, K., Mongia, H. C., & Lee, P. (2013). Evaluation of CFD best practices for combustor design: Part II—Reacting flows. In *AIAA2013-1143*.

51. Ajmani, K., Mongia, H. C., Lee, P., & Tacina, K. M. (2018). CFD led designs of prefilming injectors for gas turbine combustors. In *GT2018-75329*.

52. Ajmani, K., Mongia, H. C., Lee, P., & Tacina, K. M. (2018). CFD predictions of N + 3 cycle emissions for a three-cup gas-turbine combustor. In *AIAA2018-4957*.

53. Mi, X., Zhang, C., Hui, X., Lin, Y., Mongia, H. C., Twarog, K., et al. (2019). Ignition failure modes during ignition kernel propagation in swirl spray flames. In *Second Asia and Middle East Forum of the Global Power and Propulsion Society*, September 16–18, 2019.

54. Ren, X., Xue, X., Sung, C. J., Brady, K. B., Mongia, H. C., & Lee, P. (2016). The impact of venturi geometry on reacting flows in a swirl-venturi lean direct injection airblast injector. In *AIAA2016-4650*.

55. Ren, X., Xue, X., Sung, C. J., Brady, K. B., & Mongia, H. C. (2018). Fundamental investigations for lowering emissions and improving operability. *Propulsion and Power Research, 7*(3), 197–204.

56. Ren, X., Xue, X., Brady, K. B., Sung, C. J., & Mongia, H. C. (2019). The impact of swirling flow strength on lean-dome LDI pilot mixers' operability and emissions. In *Experimental thermal and fluid science* (in press).

57. Ren, X., Xue, X., Brady, K. B., Sung, C. J., & Mongia, H. C. (2019). Lean-dome pilot mixers' operability fundamentals. In *Innovations in sustainable energy and cleaner environment*. Springer, Singapore (in press).

Heat and Mass Transfer in Fuel Cells and Stacks

S. B. Beale⑩**, S. Zhang**⑩**, M. Andersson**⑩**, R. T. Nishida**⑩**, J. G. Pharoah and W. Lehnert**⑩

Nomenclature

A	Cell area, m^2
b	Blowing parameter

Submitted to: 50 Years of CFD in Engineering Sciences. A Commemorative Volume in Honour of D. Brian Spalding. Springer.

S. B. Beale (✉) · S. Zhang · M. Andersson · W. Lehnert
Forschungszentrum Jülich GmbH, Institute of Energy and Climate Research, IEK-3, 52425 Jülich, Germany
e-mail: s.beale@fz-juelich.de

S. Zhang
e-mail: s.zhang@fz-juelich.de

W. Lehnert
e-mail: w.lehnert@fz-juelich.de

S. B. Beale · J. G. Pharoah
Mechanical and Materials Engineering, Queen's University, Kingston, ON K7L 3N6, Canada
e-mail: pharoah@queensu.ca

M. Andersson
Department of Energy Sciences, Lund University, 22100 Lund, Sweden
e-mail: martin.andersson@energy.lth.se

R. T. Nishida
Department of Engineering, University of Cambridge, Trumpington Street, Cambridge CB2 1PZ, UK
e-mail: rn359@cam.ac.uk

W. Lehnert
Modeling in Electrochemical Process Engineering, RWTH Aachen University, 52425 Aachen, Germany

JARA-HPC, 52425 Jülich, Germany

© Springer Nature Singapore Pte Ltd. 2020
A. Runchal (ed.), *50 Years of CFD in Engineering Sciences*,
https://doi.org/10.1007/978-981-15-2670-1_14

B	Spalding number
C	Source term coefficient, [units of solved variable]
c	Specific heat, J/(kg K)
D_h	Hydraulic diameter, m
E	Nernst potential, V
E_{act}	Activation energy, J/mol
F	Faraday's constant, 96,485 C/mol
F	Distributed resistance, kg/(m^2 s)
f	Friction factor
g	Mass transfer coefficient, kg/(m^2 s)
$g*$	Zero-flux mass transfer coefficient, kg/(m^2 s)
H	Height, m
H_{fg}	Heat of evaporation, J/mol
h	Heat transfer coefficient, W/(m^2 K)
i''	Current density, A/m^2
i_0''	Exchange current density, A/m^2
j''	Diffusion flux, kg/(m^2 s)
k_R	Reaction rate, m/s
L	Length, m
M	Molecular weight, kg/mol
$\dot{m}''$	Mass flux, kg/(m^2 s)
P	Péclet number
p	Pressure, N/m^2
Q	Reaction quotient, [arbitrary]
R	Universal gas constant, 8.31446 J/(mol K)
R	Resistance, Ohm m^2
Re	Reynolds number
r	Volume fraction
$\dot{S}$	Source term, [units of solved variable] $\times$ kg/s
$\dot{S}''$	Source term, [units of solved variable] $\times$ kg/(m^2 s)
s	Saturation
Sh	Sherwood number
T	Temperature, K
t	Time, s
U	Superficial velocity, m/s
$\boldsymbol{u}$	Interstitial velocity, m/s
u	Velocity, m/s
V	Cell voltage, V, source term value, kg/s
x	Displacement, m, mole fraction
y	Mass fraction
z	Charge number

Greek Letters

α	Charge transfer coefficient, volumetric heat transfer coefficient W/(m^3 K)
Γ	Exchange coefficient, kg/(m s)
γ	Reaction order
δ	Cell half-width, m
η	Overpotential, V
κ	Permeability, m^2
μ	Dynamic viscosity, kg/(m s)
ν	Kinematic viscosity, m^2/s
ρ	Density, kg/m^3
τ	Tortuosity
Φ	Polarisation

Subscripts

b	Bulk
e	Electrode
eff	Effective
g	Gas
H_2	Hydrogen
H_2O	Water
int	Interface
l	Liquid
O_2	Oxygen
NB	Neighbour value
t	Transformed substance state
P	Nodal value
w	Wall
0	Ambient, external

1 Introduction

Prof. D. B. Spalding made many important contributions to engineering science. Amongst the best-known are, perhaps, (i) the reacting flow equations in laminar flames [1, 2]; (ii) the rationalised mass transfer formulation [3]; (iii) together with S. V. Patankar, the development of the semi-implicit method for pressure-linked equations (SIMPLE) [4], which has been, and continues to be, applied in original and modified forms in numerous powerful and widely used computational fluid dynamics (CFD) codes, such as PHOENICS, Fluent, STAR-CD, CFX and OpenFOAM, to name but a few; (iv) with B. R. Launder and others, significant contributions to turbulence modelling [5]. (v) An important and pioneering analysis, once again with S. V. Patankar, was related to the calculation of heat exchanger performance by means of a 'distributed resistance analogy' [6], a subject which Spalding continued to pursue personally throughout his later career [7]; and (vi) a generalisation of the SIMPLE algorithm to multi-phase flow problems by means of the inter-phase slip (IPSA) algorithm [8].

The present authors have been engaged in the development and application of CFD models for fuel cells for a number of years. Mathematical models of both solid oxide fuel cells (SOFCs) [9–16] and polymer electrolyte membrane fuel cells (PEFCs) [17–19] of varying complexity have been proposed. Early models of fuel cells were implemented in simple 2-D FORTRAN codes, for example, the early finite-difference scheme of Achenbach [20], wherein the continuity and energy equations were solved, but momentum was not considered and perfect mixing was assumed throughout. Such models rapidly gave way to detailed 3-D formulations of the physicochem-ical hydrodynamics and the chemistry and electrochemistry of complex reactions. Amongst the first papers on the modelling of fuel cells using CFD were those by Ding et al. [21] and He and Chen [22], who simulated molten carbonate fuel cells using PHOENICS. This work describes how the present authors, their colleagues and co-workers employed Spalding's mass transfer formulation and Patankar and Spald-ing's distributed resistance analogy (DRA) concept in the heart of the development of generic models of fuel cell stacks.

1.1 Mass Transfer

Spalding's generalisation of the convective mass transfer problem [3] was an excep-tional achievement. In a single unified formulation, it brought together a diverse body of knowledge that had previously consisted of a multiplicity of different approaches, depending on the area of application or trade, and spanning the fields of aerospace, process engineering, hydraulics, etc. Spalding's remarkable book on the subject [23] consisted of only two chapters (one very short and one very long). According to the author, it was to be followed by a second volume, which never materialised. Never-theless, the approach became the standard for mechanical engineering courses at the postgraduate level; see, for example, Mills and Coimbra [24] and Kays et al. [25].

Figure 1 shows an illustrative example from Spalding [23], namely a hydrogen-peroxide steam generator in which an unconsolidated bed of catalyst-impregnated stones are held in a wire basket. As is noted in the text [23]:

> A special complication is that, as in boiling heat transfer, the result of the process impedes its further existence; for the gas formed by the decomposition at the catalyst layer tends to prevent access of further liquid.
>
> D. B. Spalding

Fig. 1 Catalytic vessel for the decomposition of hydrogen peroxide. Reproduced from Spalding [23], with permission

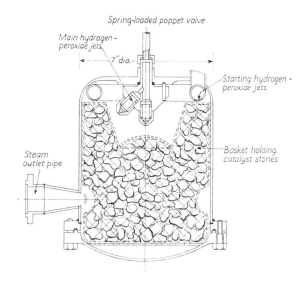

A similar problem is found with PEFCs, where the liquid water formed by the reaction prevents access of the counter-flowing gaseous oxygen to the catalyst sites. Indeed, late in his career, Spalding was to participate in a study of the subject of two-phase counter-flow in PEFCs, see [26].

Spalding's 'Ohm's law' of mass transfer, hereinafter referred to as *Spalding's law of mass transfer*, states that the rate of mass flux, $\dot{m}''$, is given by

$$\dot{m}'' = g \times B$$
$$\text{Mass flux} = \text{Conductance} \times \text{Driving force} \tag{1}$$

B is a mass transfer driving force, generally referred to nowadays as a *Spalding number*,

$$B = \frac{y_b - y_w}{y_w - y_t} \tag{2}$$

and g is a conductance, or mass transfer coefficient, which remains positive and finite as $\dot{m}''$, $B \to 0$. The subscripts refer to mass fraction values in the bulk, y_b, at the wall, y_w, and the so-called 'transferred substance state' (T-state for short):

$$y_t = \frac{\dot{m}_i''}{\dot{m}''} \tag{3}$$

For the reader unfamiliar with the notation, an explanation is given in the book by Spalding [23].[1] When a single pure substance is transferred, for example, across an osmotic membrane [27], $y_t = 1$ for the transferred substance and $y_t = 0$ for the considered substance, assuming binary mass transfer. Conversely, if a series of substances are transferred simultaneously, then

$$\dot{m}'' = \sum_{i=1}^{n} \dot{m}''_i \tag{4}$$

in Eq. (1) and, in the absence of chemical reactions, $0 \leq y_t \leq 1$ for each species. By convention, $\dot{m}''$ is positive for injection and negative for suction. The reader should note that for problems involving heterogeneous chemical reactions, $-\infty \leq y_t \leq +\infty$, with the limits occurring for catalysis, when $\dot{m}'' = 0$, but g is finite.

The derivation of Spalding's law consists of two simple steps: (i) the substitution of the rate term for the diffusion flux (Fick's law), $j'' = -\rho D \partial y / \partial n = g(y_w - y_b)$; and (ii) the combination of the sum of the total (convective and diffusive) fluxes in the bulk of the flow and the convective flux (only) beyond the wall, $\dot{m}'' y_t = \dot{m}'' y_w - j''$ (control-volume analysis). The beauty in the definition of the Spalding number, B, in Eq. (2) is that, not only is it a non-dimensional number, but also that it is a non-dimensional number of enormous generality and utility unlike, for example, the 'concentration polarisation' of Sherwood et al. [28], defined as $\Phi = y_w / y_b - 1$. In hindsight, a more general definition might have been $\Phi = (y_w - y_b) / (y_b - y_t)$. Nonetheless, the expression of mass transfer problems in terms of polarisation is popular in the field of chemical engineering, see, for example, Bird et al. [29].

In electrochemistry monographs [30, 31], it is not usual to refer to the effect of concentration/mass fraction in terms of a mechanical driving force, B, nor in terms of Φ (although it is seen on occasion), but rather as a so-called 'concentration over-potential', η, sometimes called a 'concentration loss'. In other words, the reduction in electric potential due to the chemical activity of charged species at a wall or electrode being lower than that within the bulk of the fluid. This results from the consumption of reactants and/or production of products by the electrochemical reaction, affecting concentration gradients within the fluid near the electrode(s). This

---

[1]

It is perhaps a pity that the first symbol to be introduced should be so elaborate as $\dot{m}''$; at any rate, an immediate explanation of the symbol is called for. The dot and the dashes have the significances given to them by Jakob (1949), who introduced them into heat transfer work: the superscribed dot signifies "per unit time"; one dash signifies "per unit length"; two dashes signify "per unit area"; and three dashes signify "per unit volume". In accordance with this convention, we elsewhere employ the symbol $\dot{m}'$ for mass flow rate per unit width of a fluid film, and the symbol $\dot{m}'''_j$ for the mass rate of generation of chemical substance j per unit volume due to a chemical transformation; similarly $\dot{q}''$ will be used for the rate of heat transfer per unit area.

D. B. Spalding

approach obfuscates to some extent the simple intuitive relationship described by Spalding's law, Eq. (1). However, in electrochemistry, it is considered quite natural to write most finite-rate processes: mass transfer, kinetics (charge transfer), Ohmic resistance and so forth, in terms of overpotentials. Paradoxically, heat transfer is not generally expressed in terms of an overpotential.

The conductance, g, which is often non-dimensionalised as a Sherwood number, may be obtained (i) from physical experiments [32]; (ii) by analytical or numerical solutions to the Sturm–Liouville system of equations for simple geometries, e.g., Sherwood et al. [28]; or (iii) by performing detailed CFD calculations [33]. At high $\dot{m}''$, g is not constant. The 1-D solution of the convection–diffusion equation may be written as follows:

$$B = \exp(b) - 1 \tag{5}$$

where $b = \dot{m}'' / g^*$ is a non-dimensional blowing parameter and $g^* = g$, $\lim \dot{m}'' \to 0$. Beale showed that this simple solution conformed well to numerical solutions for plane and rectangular ducts [33]. This leads to the following expression for $g(\dot{m}'')$:

$$\frac{g}{g^*} = \frac{b}{\exp(b) - 1} = \frac{\ln(1 + B)}{B} \tag{6}$$

All of the above arguments also apply in equal measure to heat transfer problems involving mass transfer at a boundary, with temperature T (or enthalpy) replacing mass fraction, y, and a heat transfer coefficient h, with associated h^*, supplanting g^*. This is also true for the momentum equation.

Numerous authors have made the erroneous presumption that Eq. (1) may be supplanted by the following expression:

$$\dot{m}'' = g(y_w - y_b) \tag{7}$$

replacing the diffusion flux with the convection flux on the left side of Eq. (7). As is carefully explained in Spalding's book [23], Eq. (7) is only valid for very specific situations. Beale demonstrated that Eq. (7) is generally not valid for oxygen transport in SOFCs [34], but may be appropriate in modified form in the cathodes of PEFCs [35], where the concentration of dissolved oxygen is extremely small.

1.2 CFD Boundary Conditions

The T-state value provides a simple and compact way to prescribe mass transfer boundary conditions in CFD [33], namely,

$$j_w'' = -\Gamma \left.\frac{\partial y}{\partial n}\right|_w = \dot{m}''(y_t - y_w) \tag{8}$$

This Robin formulation converges more readily than the conventional practice of prescribing individual convective fluxes and mass fractions [36] as Von Neumann conditions; see Eq. (24), below. The reader will note that y_w is the boundary value of y, i.e., on the surface $\partial\Omega$, not the value, y_p, in the nearest computational cell within the finite volume, Ω. Starting from the 1-D convection–diffusion equation, it may be readily shown that

$$j_w'' = m'' \exp(-P) \times (y_t - y_P) \tag{9}$$

$$y_w = \frac{1}{\exp(P)} y_P + \frac{\exp(P) - 1}{\exp(P)} y_t \tag{10}$$

where y_p is the interior nodal value of y within Ω, adjacent to $\partial\Omega$, and

$$P = \rho\, u_w\, \delta / \Gamma \tag{11}$$

is a cell Péclet number [37]. Generally speaking, for mass transfer problems, P is small and $\exp(P)$ can be expanded in a Taylor series, neglecting terms of $O(P^2)$:

$$j_w'' = \frac{\rho\, u_w}{1 + P} (y_t - y_P) = \rho\, u_w (y_t - y_P) - \frac{\Gamma}{\delta} (y_t - y_P) \tag{12}$$

$$y_w = \frac{1}{1 + P} y_P + \frac{P}{1 + P} y_T \tag{13}$$

Many finite-volume codes that employ source term linearization, corresponding to a Robin (linear) boundary condition, $S = j''A$; In Spalding's notation, the source term is composed of a coefficient, C, and a value, V [38].[2]

$$S = C(V - y_P) \tag{14}$$

Patankar [37] employs the altogether similar

$$S = S_C + S_P\, y_P \tag{15}$$

Equation (12) then yields [27]:

$$C = \left(\frac{P}{1 + P} \right) \frac{\Gamma A}{\delta} \tag{16}$$

$$V = y_T \tag{17}$$

---

[2]Sometimes a geometric factor G pre-multiplies C, $S = GC(V - y_p)$, depending on the patch type (volume, area, etc.). For convenience, it is excluded here.

$$S_C = \left(\frac{P}{1+P}\right)\frac{\Gamma A}{\delta} y_T \qquad (18)$$

$$S_P = -\left(\frac{P}{1+P}\right)\frac{\Gamma A}{\delta} \qquad (19)$$

where A is the area of the cell boundary, $\partial\Omega$, and δ is the distance from the boundary to the cell node. Equations (16)–(19) represent a central-difference formulation. Conventional wisdom is that C should be positive in Eq. (16) and S_p negative in Eq. (19), otherwise divergence may occur [37]. This is clearly the case for $P > 0$ (inflow). For $P < 0$, the source term should therefore be coded as a fixed flux/source (Von Neumann), in other words, not linearised. In fact, for values of $|P| \ll 1$, typically encountered, it was found that the linear form of Eqs. (16)–(17) will still result in convergence even if S_P is positive. This may be because y_w is not necessarily bounded by y_P and y_T, depending on the value of y_t and P. A discussion of boundendess is considered beyond the scope of the present chapter. Nevertheless, the reader should appreciate that for a mass transfer outflow boundary condition, the flux/gradient is generally non-zero. A similar treatment for Eq. (9) yields

$$C = \frac{P}{\exp(P)} \times \frac{\Gamma A}{\delta} \qquad (20)$$

$$V = y_T \qquad (21)$$

$$S_c = \frac{P}{\exp(P)} \times \frac{\Gamma A}{\delta} y_t \qquad (22)$$

$$S_p = -\frac{P}{\exp(P)} \times \frac{\Gamma A}{\delta} \qquad (23)$$

for an exponential scheme [37]. It is left as an exercise to the reader to construct source terms based on schemes aside from central-differences and exponential formulations. Equations (16)–(23) represent the generalised convection–diffusion boundary condition with the T-state values representing the inlet/outlet value. In the limit $P \to 0$, $S \to \Gamma A(y_t - y_P)/\delta$, whereas for $P \gg 1$, $S \to \rho A u_w$. It is worthy of note that modern CFD codes, such as described in Ref. [39], may not necessarily prescribe boundary conditions as a linearized source terms, S, but rather a boundary mesh is constructed. This is then populated with values based directly on the wall value Eq. (13) directly or by extrapolation from the flux/gradient Eq. (12).

The alternative convention is to code the species sources and sinks manually [29]:

$$j_i'' = \dot{m}_i''(1 - y_i) - \sum_{j \neq i} \dot{m}_j'' y_i \qquad (24)$$

The physical interpretation of Eq. (24), is as follows: The first term is the increase in y_i due to injection at rate $\dot{m}_i''$ of species i. The second term is the decrease in y_i due to injection at rate $\dot{m}_j''$ of species j, $j \neq i$. Equation (24) has several disadvantages compared to Eq. (9). (i) y_j appears in the equation for y_i, an unnecessary and potentially unstable formulation; (ii) for binary diffusion, there are two such equations with two terms for ternary diffusion, three equations with three terms each, and so forth. (iii) The terms in Eq. (24) must be coded as Von Neumann boundary conditions, whereas Eq. (9) is a Robin-type condition; (iv) the $\dot{m}_i''$ appearing in Eq. (24) are the individual species fluxes, whereas in Eq. (9) $\dot{m}'' = \rho\, u_w$ is the overall convective mass flux, Eq. (4). Thus, it is seen that Spalding's early mass transfer formulation directly impacts upon modern CFD best practices for problems involving scalar transport with injection/suction at walls.

1.3 Distributed Resistance Analogy for Heat Exchanger Design

The original formulation of Patankar and Spalding [6] is, for a somewhat idealised shell-and-tube heat exchanger design, illustrated in Fig. 2. At a time when computer memory was small and clock-speeds were slow by today's standards, the possibility of performing flow calculations around numerous cylinders was not feasible, and therefore volume averaging was employed. The method has also been referred to in the literature as a 'porous media approach', as in the work of Nield and Bejan [40], who failed to cite the earlier work [6]. Heat exchangers are, generally speaking, ordered rather than random structures, and the term 'porous' is therefore somewhat misleading. The adjective 'homogenised' is also sometimes applied to such a class of models. The basic methodology proposed [6] was

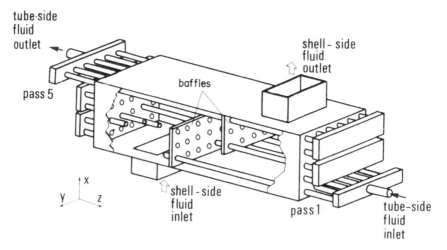

Fig. 2 Heat exchanger design reproduced by Patankar and Spalding [6], with permission

to regard the space within the shell as uniformly filled with fluid, through which however is distributed, on a fine scale, a resistance to fluid motion

S. V. Patankar and D. B. Spalding

and to substitute rate terms in place of diffusion terms,

$$\rho_i \frac{\partial \boldsymbol{u}_i}{\partial t} + \rho_i \boldsymbol{u}_i \text{ grad } \boldsymbol{u}_i = -\text{grad } p_i + F_i \boldsymbol{u}_i \tag{25}$$

The densities were partial densities, $\rho_i = r_i \rho$, so that the shell-side volume fractions, r_i, were included. The distributed resistances, F_i, may be computed from analytical, numerical or experimental results, and may be anisotropic and/or non-homogeneous. Patankar and Spalding only considered continuity and momentum on the shell-side, the tube-side flow was considered to be fully developed internal flow, for which analytical solutions are readily available, and thereby not requiring a numerical solution. The distributed resistance is obtained as

$$F = \tfrac{1}{2} f \rho |\boldsymbol{u}| / D_h \tag{26}$$

where f is a friction factor, often correlated with laminar flow as $f = a/\text{Re}^m$. Heat exchangers frequently operate in a transitional regime in which the boundary layers are laminar, but there is substantial unsteadiness or 'turbulence' in the free stream. Beale [41] proposed the following simple correlation:

$$f = a(c + 1/\text{Re})^m \tag{27}$$

where a, c, m are fitted parameters. The energy equation(s) are assumed to take the form:

$$\rho_i c_v \frac{\partial T_i}{\partial t} + \rho_i c_p \boldsymbol{u}_i \cdot \text{grad} T_i = \sum_{j \neq i} \alpha_{ij} (T_j - T_i) \tag{28}$$

These are solved, not only in the shell-and-tube fluid spaces, but also in the metallic tube wall, where the convective term is, of course, absent. The volumetric heat transfer coefficients, α_{ij}, are obtained from the more usual area-specific h-values. A partial elimination algorithm removed the T_j terms from the solution for T_i in Eq. (28). Details may be found in Spalding [8]. The volumetric heat transfer coefficient may be written as

$$\alpha = \tfrac{1}{2} St \rho |\boldsymbol{u}| / D_h \tag{29}$$

Although the distributed resistance analogy was an exciting and innovative approach, the conservative heat exchanger community proved stubbornly resistant to exploring the idea for many years, and traditional 'presumed flow' methods, such as the so-called 'Bell–Delaware method' [42], typically incorporated into proprietary computer codes, prevailed as the gold standard for many years. Slowly, however, the Patankar and Spalding approach was taken up by a number of groups [43–48] and today enjoys a measure of popularity among the heat exchanger community. The reader will note that in the original DRA [6], only continuity, momentum and energy were considered, not species mass fractions or chemical reactions. The substitution of rate terms for diffusion terms based on Spalding's mass transfer analysis, above, is a natural extension to the DRA, as shown below, and is a requirement in fuel cell stack models where the mass fractions at the electrode walls must be computed from the bulk gas values in order to prescribe the cell potential (in other words, the concentration overpotential).

2 Fuel Cell Model

A fuel cell is an electrochemical device that converts chemical energy into electricity and heat. The fuel may be hydrogen or a hydrogen-rich hydrocarbon. Two common types of fuel cells are the SOFC with operating temperatures of 600–800 °C, and PEFC. For the PEFC, the low-temperature PEFC (LT-PEFC) has an operating temperature of 60–80 °C and the high-temperature PEFC (HT-PEFC) around 160 °C. Figure 3a, b is a schematic of an idealised SOFC; the cell consists of a current-conducting ceramic or steel interconnect within which gas channels for air and fuel are machined. The solid 'ribs' may be replaced with wire mesh, resulting in the mixing of the gas. The channels are separated by two electrochemically active electrodes, and an ion-conducting electrolyte. In planar SOFCs, the gas channels are

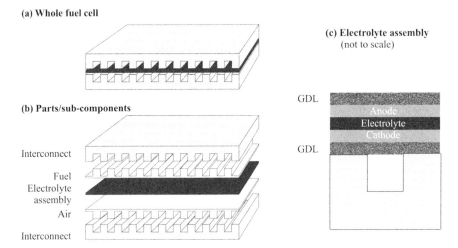

Fig. 3 Schematic of an idealised solid oxide fuel cell for co-flow/counter-flow (left) and electrolyte assembly of polymer electrolyte fuel cell (right)

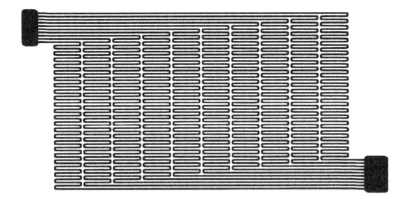

Fig. 4 Flow field configuration for Jülich high-temperature polymer electrolyte fuel cell

typically straight, with the flow of fuel and air being co-flow, counter-flow or cross-flow (not shown in diagram). In an SOFC, oxygen O^{2-} anions are transferred across the electrolyte, whereas in a PEFC, cations in the form of H^+ protons are transported across a polymer electrolyte membrane which is covered by electrochemically active electrodes on both sides. The exterior of this electrode–electrolyte assembly typically consists of non-active porous transport layers to distribute the gases on the electrode surfaces. The misnomer of 'gas diffusion layer', or GDL, is frequently employed for these, see Fig. 3c. For PEFCs, the interconnectors are generally referred to as bipolar plates (BPPs), which can be made of various materials, for example, carbon. The gas passages may be of complex multi-pass serpentine design, in order to maximise the dwell time within the active region of the cell. Figure 4 shows the flow-field configuration for a Jülich 400 cm^2 HT-PEFC design. Fuel cells are generally operated in stacks in order to increase the voltage, V. A fuel cell stack is hydraulically in parallel, but electrically in series. Figure 5 illustrates a 5-cell HT-PEFC stack.

Along with the usual equations of concentration, heat and momentum, some additional equations are required, namely,

$$V = E - \eta_{an} - \eta_{cat} - i''R \tag{30}$$

$$E = E^0 + \frac{RT}{z\mathrm{F}} \ln Q \tag{31}$$

$$\dot{m}'' = \left(\prod_{i=1}^{n} x_i^{\gamma_i} \right) k_R \tag{32}$$

where V is the cell potential, E is the ideal or Nernst potential and R is the local cell resistance $R = R(T)$. k_R is the reaction rate, which is a function of the exchange current density; see Eqs. (38)–(39), below, x_i is the mole fraction and γ_i is the reaction order (for species i). The Nernst potential is a function of Q, the reaction quotient,

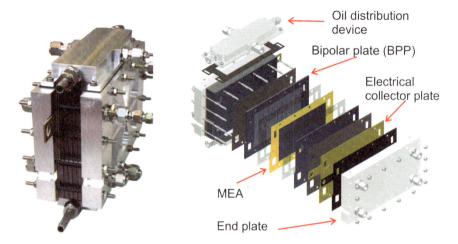

Fig. 5 Five-cell polymer electrolyte fuel cell stack, from [49] with permission

which is in turn a function of the species mass/mole fractions at the electrodes. For example, for a hydrogen–oxygen reaction, with $z = 2$,

$$Q = \frac{p_{H_2} p_{O_2}^{1/2}}{p_{H_2O}} \tag{33}$$

By redefining E^0, Eq. (33) may be rewritten in terms of mole fractions $Q = x_{H_2} x_{O_2}^{1/2} \big/ x_{H_2O}$, or normalised pressures p_i/p_0, which has the advantage that the quotient is non-dimensional. Equation (33) is based on the ideal gas assumption, which is appropriate for high-temperature SOFCs. For a PEFC, the pressures are generally replaced by the activities, a_i. The normalised partial pressures, p_i/p_0, or mole fractions, x_i, of the reactants/products are obtained directly from the mass fractions, y_i. Spalding's mass transfer analysis allows the electrode/wall value to be computed from the bulk value as

$$y_w = \frac{y_b + y_t B}{1 + B} \tag{34}$$

Faraday's law relates the individual species mass fluxes to the current density:

$$\dot{m}_i'' = \pm \frac{M_i i''}{z_i F} \tag{35}$$

where Eq. (4) yields the overall mass flux at the anode and cathode. Substituting Eq. (35) into Eq. (7) and setting $y_w = 0$ as corresponding to the maximum mass flux possible and associated limiting current density,

$$i''_{\text{lim}} = \mp \frac{Fz}{M} g y_b \tag{36}$$

The concept of the limiting current density is used to define a 'concentration overpotential':

$$\eta_{\text{conc}} = \frac{RT}{zF} \left(1 - \frac{i''}{i''_{\text{lim}}}\right) \tag{37}$$

that is, the amount by which the cell voltage, V, Eq. (30), is reduced when bulk values, rather than wall values of mass/mole fraction, are employed in the computation of the reaction coefficient, Q, Eq. (33). This type of formulation is frequently inaccurate, e.g., for concentrated systems and higher rates of mass transfer, but nonetheless is widely adopted throughout the electrochemical community. Of course, when a detailed CFD simulation tool is employed, such lumped-parameter estimates are unnecessary.

The system of equations, Eqs. (30)–(31), is closed by means of one or two Tafel or Butler–Volmer equations at the anode and/or the cathode. The latter is given by

$$i'' = i''_0 \left(\exp(\alpha_1 \eta F / RT) - \exp(-\alpha_2 \eta F / RT)\right) \tag{38}$$

For a Tafel equation [31], only the first term (the forward reaction) on the right side of Eq. (38) is considered. The exchange current density, i''_0, is usually, though not necessarily always, written in terms of an Arrhenius-type expression:

$$i''_0 = i''_{0,\text{pre}} \prod_{i=1}^{n} \left(\frac{p_i}{p_{i,\text{ref}}}\right)^{\gamma_i} \exp(-E_{\text{act}} / RT) \tag{39}$$

where $i''_{0,\text{pre}}$ is a pre-exponential factor, γ_i is the reaction order and E_{act} is activation energy. Under the circumstances the reaction rate is just,

$$k_R = \frac{M_i i''_0 \left(\exp(\alpha_1 \eta F / RT) - \exp(-\alpha_2 \eta F / RT)\right) \exp(E_{\text{act}} / RT)}{z_i F} \tag{40}$$

assuming $x_i = p_i / p_{i,\text{ref}}$.

2.1 Detailed Numerical Model

The governing equations for a fuel cell mathematical model are the detailed continuity, momentum and energy equations:

$$\text{div}(\rho \boldsymbol{U}) = 0 \tag{41}$$

$$\mathrm{div}(\rho \boldsymbol{U} y_i) = \mathrm{div}\left(\rho D^{\mathrm{eff}} \mathrm{grad}\ y_i\right)_i \tag{42}$$

$$\mathrm{div}(\rho \boldsymbol{U} \boldsymbol{U}) = -\mathrm{grad}\ p + \mathrm{div}(\mu \mathrm{grad} \boldsymbol{U}) - \frac{\mu}{\kappa} \boldsymbol{U} \tag{43}$$

$$\mathrm{div}\left(\rho \boldsymbol{U} c_p T\right) = \mathrm{div}\left(k^{\mathrm{eff}} \mathrm{grad}\ T\right) + \dot{q}''' \tag{44}$$

For porous regions, $\boldsymbol{U}$ is a superficial velocity (superficial and interstitial velocities are identical in the gas channels). It is noted that Eqs. (41)–(44) are for the case of stationary operation. Dynamic models would require additional (transient) terms. The mass source/sinks at the electrodes do not appear as source terms in the overall continuity, Eq. (41), and individual species equations, Eq. (42), as these are prescribed as boundary conditions, according to Eq. (35). The overall mass flux is computed from Eq. (4) and thence the individual T-state values, Eq. (3). The reader will note that in the porous layers, only the pressure gradient and Darcy terms (first and third terms on the right hand side) of Eq. (43) are active, whereas in the gas channels, the latter term is absent, so Eq. (43) reduces to the incompressible Navier–Stokes equation. The matter is discussed further in Beale [41]. Transport properties are not constant, as mass is being removed at the cathode and added at the anode, or the opposite depending on the fuel cell type, in addition to changes in temperature and pressure. Details as to how these are enumerated may be found in Beale et al. [16].

2.2 Distributed Resistance Analogy for Fuel Cell Stacks

Planar SOFCs are generally made with straight passages. Conversely, PEFCs frequently contain serpentine passages with bends as well as straight regions. Figures 4 and 5 show an example of one such geometry. Therefore, one may write the distributed resistance analogy in the following form [50]:

$$\mathrm{div}(\rho \boldsymbol{U}) = r \dot{m}''' \tag{45}$$

$$\mathrm{div}\left(\rho \frac{\boldsymbol{U} \boldsymbol{U}}{r}\right) = -\frac{r}{\tau^2} \mathrm{grad}\ p + \mathrm{div}\left(\frac{1}{\tau^2} \mu\ \mathrm{grad}\ \boldsymbol{U}\right) - \frac{r}{\tau^2} \boldsymbol{F_D} \cdot \boldsymbol{U} \tag{46}$$

$$\mathrm{div}(\rho \boldsymbol{U} y_i) = \mathrm{div}\left(\frac{r}{\tau^2} \Gamma \mathrm{grad}\ y_i\right) + r \dot{m}_i''' \tag{47}$$

$$\mathrm{div}\left(\rho \boldsymbol{U} c_p T\right) = \mathrm{div}\left(\frac{r}{\tau^2} k \mathrm{grad}\ T\right) + r \sum_{NB} \alpha(T_{NB} - T) + r \dot{q}''' \tag{48}$$

where r is the volume fraction[3] of each solid or fluid space, $\sum_{\text{Spaces}} r = 1$. In the fluid spaces, τ is the geometric tortuosity, obtained by integration along the serpentine path-length:

$$\tau = \frac{1}{L} \int_0^s \mathrm{d}s \qquad (49)$$

where L is the straight-line distance between 0 and s. For an SOFC with straight passages, $\tau = 1$. For the PEFC shown in Fig. 4, $\tau = 1$ in the upper and lower entrance/exit (triangular) regions, whereas in the central parallelogramical zone, $\tau = 9$.

Regardless of whether a direct numerical model (DNM) or a DRA is employed, either the mean cell current density (galvanostatic condition) or the cell voltage (potentiostatic) must be prescribed. The latter is straightforward and may readily be applied to single cells. The former is generally required for stacks, in which, the individual cell voltages are adjusted until

$$V = V^* + N_R\left(\bar{i}''^* - \bar{i}''\right) \qquad (50)$$

where $\bar{i}''$ is the desired current density and V^* and $\bar{i}''^*$ are the set voltage and computed mean current density at the previous iteration and N_R is a relaxation parameter.

3   Illustrative Examples

Beale and Zhubrin [51] considered a planar SOFC geometry operating under conditions of cross-flow using the commercial software PHOENICS [52]. Four 'spaces', corresponding to fuel, air, interconnect and electrolyte (including electrodes) were tessellated using a structured single-block rectilinear mesh, as shown in Fig. 6. The use of 'halo' cells to isolate the four spaces allowed for multiple pressures to be solved (for the fuel and air spaces). Multiple pressure fields present problems for some CFD codes. (Recall Patankar and Spalding [6] solved only a single-set of pressure-corrected momentum equations on the shell-side space). In the so-called multiply shared spaces methods (MUSES) method, Zhubrin and Spalding [53] solve the problem. In addition, air and fuel manifolds were constructed as risers and down-comers. Both a single cell and a 10-cell stack were employed. Figure 7 shows a plan view of a comparison for a single cell between the DRA and DNM approaches, in terms of local current density and local Nernst potential, data which are impossible to measure, in detail, at the present time. Agreement between the two approaches is excellent. Figure 8 is an elevation view of a 10-cell stack. It can be seen that the

---

[3]In the interests of readability indices for the volume fractions, $r = r_k$, and state-variables of each 'space' have been excluded.

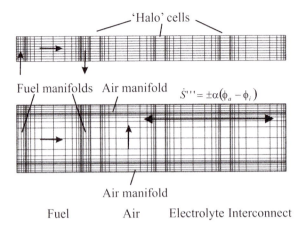

Fig. 6 Computational mesh employed in Beale SB and Zhubrin, from [51] with permission

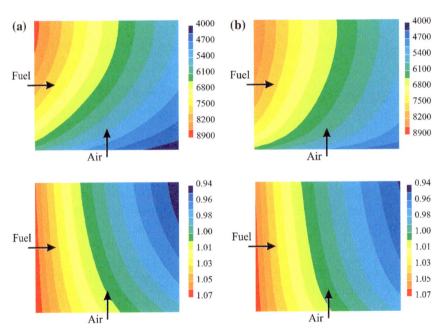

Fig. 7 Comparison of current density (A/m^2) and Nernst potential (V) for a single cell: **a** DRA **b** DNM, from [51] with permission

stack temperatures vary with height and that agreement between the DRA and DNM are excellent. The local variations between the 'spaces' (air, fuel, interconnect, electrolyte) are smeared out in the DRA. However, the bulk values may be computed from the inter-phase heat transfer coefficients.

Fig. 8 Comparison of temperature (°C) for 10-cell stack **a** DRA **b** DNM, from [51] with permission

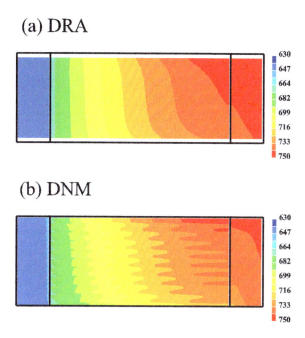

(a) DRA

(b) DNM

Sometime later, Kvesić et al. [49–54] independently developed a methodology similar to that of Beale and Zhubrin. Specific details of the model closures varied, but the general ideas were comparable. The commercial software, Fluent, was employed. The application was an HT-PEFC similar to the design shown in Fig. 5, which is further discussed below. Subsequently, Nishida et al. [55–57] developed a DRA-based SOFC model in the open-source programming environment, OpenFOAM [58]. Once again, four 'spaces' were employed. Because more modern unstructured meshes were employed in the work and, moreover, with the ability to instantiate new class members within the object-oriented paradigm, the requirement for multiple meshes disconnected via 'halo' cells is eliminated. Figure 9a shows stream tubes in the manifolds and active area for an 18-cell SOFC stack, based on a Jülich Mark-F geometry. The stream tubes are coloured by oxygen mass fraction. Figure 9a, b shows local temperature distributions in the centre of the stack under (b) conditions when the stack is mounted in a furnace with a radiation boundary condition at the sides and conduction boundary conditions at the top and bottom, and (c) perfectly insulated (adiabatic) walls. Figure 10 shows a quantitative comparison of DRA temperatures with those measured near the surface in an experimental stack, operating in a furnace from [55]. It can be seen that there is good quantitative and qualitative agreement. The DRA model allows for temperatures in the centre of the stack to be predicted, something which is not possible with existing experimental capabilities.

Zhang et al. [50] modified the model presented by Nishida et al. [55–57] for an oil-cooled (with 5 'spaces') HT-PEFC DRA model. Fuel cells and stacks are generally characterised in terms of a polarisation, or current density–voltage, diagram.

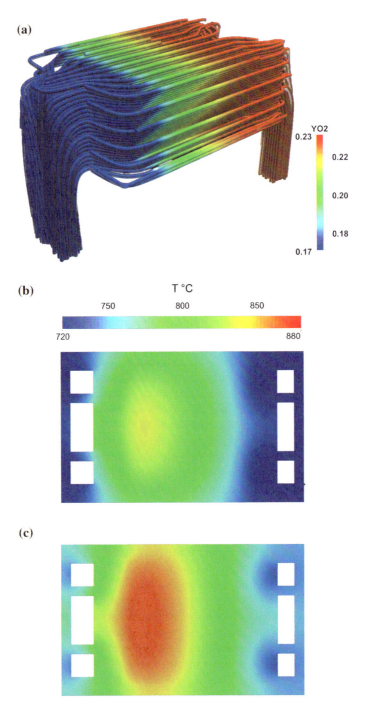

Fig. 9 Detailed results for the Jülich Mark-F 18-cell SOFC stack. **a** Stream tubes for the air-side, coloured by oxygen mass fraction; **b** temperature distribution at the centre of the stack under steady-state operation for radiation; and **c** insulated boundary conditions. The air flows from right to left in counter-flow with the fuel, from Nishida et al. [55] with permission

Fig. 10 Comparison of DRA temperatures with experimental measurements for SOFC operated in a furnace, from Nishida et al. [55] with permission

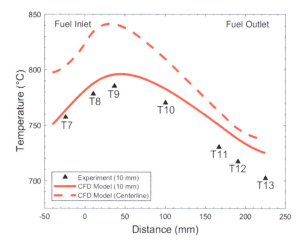

Fig. 11 Polarisation diagram for an HT-PEFC, from Zhang et al. [50] with permission

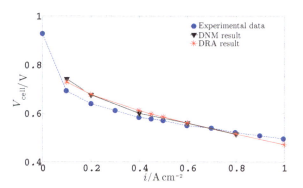

Figure 11 is one such representation from the work of Zhang et al. [50]. It can be seen that as the current density, i'', increases from zero, the voltage, V, initially at the open circuit potential, E, decreases. Typically, losses at low-i'' are primarily due to activation or kinetic losses, η_{an}, η_{cat}, in Eq. (30), see also Eq. (38). These are generally larger in PEFCs than in SOFCs due to the higher operating temperatures of the latter. At intermediate current densities, the Ohmic, $i''R$-term, Eq. (30), contributes to the linear reduction in Fig. 11. At high current densities, mass transfer issues may become a problem. This is a function of the air and fuel utilisation, namely the ratio of reactant to total fuel/oxidant available for an SOFC, or stoichiometric values (inverse of utilisation) in PEFCs. However, Beale et al. [36] point out that regardless of stoichiometry, according to Spalding's law, Eq. (1), or the simplified form, Eq. (7), the minimum mass fraction, at the electrode mid-way between the gas channels, is a function of current density, i''. Should y_{O_2} or y_{H_2} (assuming hydrogen is the fuel) approach zero, oxidant/fuel starvation will occur locally, a highly undesirable situation that would invalidate the Nernst equation in the form of Eqs. (31) and (33). In

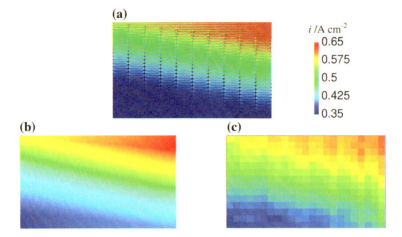

Fig. 12 Local current density distribution in a high-temperature polymer electrolyte membrane fuel cell stack. Mean current density 0.45 A/cm². **a** Detailed numerical model, **b** distributed resistance analogy, **c** experimental results from [49] with permission

any event, such a situation is highly undesirable, in that it would lead to degradation of the cell.

Figure 12 shows local current density, i'', in terms of a comparison for a high-temperature PEFC for a DRA model. It can be seen that agreement between the DNM, DRA and the experimental results are excellent, and that the highest degree of resolution is provided by the DNM. Both the DNM and DRA provide substantially more details than are obtained from experimental results.

4 Present and Future Work

The research described in this chapter is a continuing activity. Current ongoing developments by the present authors include extending the models described to consider electrolysers, that is, fuel cells operated in reverse (i.e., converting electricity to chemical energy). Both solid oxide electrolyser cells (SOECs) and polymer electrolyte cells (PEECs) are studied. Low-temperature PEFCs, operating at or below 100 °C, present new problems in that both liquid and gaseous water may be present in the channels and porous GDLs. In low-temperature PEECs, the same problem exists as is shown for the hydrogen-peroxide/oxygen catalytic converter, Fig. 1, which was the device that introduced Spalding to the subject of mass transfer, around 1945. The product gas (oxygen) formed at the catalyst layer impedes access by the reactant liquid (water or hydrogen peroxide). In low-temperature PEFCs, the opposite problem is present at the air electrode, namely the produced liquid water impedes the transport of the oxygen to the reaction sites. Here, Spalding and others came up with a solution for the computing the rate of mass transfer from a heat balance, for problems involving evaporation or condensation. Starting from the Clausius–Clapeyron relationship,

$$p_v(T) = C \exp(-H_{fg}/RT) \tag{51}$$

If the water vapour partial pressure, p_v, (or the mass fraction, y_v) at the liquid–gas interface is known, then the interfacial temperature, T, is fixed according to Eq. (51), or equivalent. The mass flux may therefore be written as

$$\dot{m}'' = \frac{h_g(T_g - T_{int}) + h_l(T_l - T_{int})}{H_{fg}} \tag{52}$$

where h_g and h_l are heat transfer coefficients for the gas and liquid phases, respectively. Equation (52) provides the required mass flux from heat flux. The reader will note that incorporating the prescribed interfacial temperature according to Eq. (51) is a Dirichlet condition for the energy equation. This is numerically more stable than prescribing the heat flux as a Neumann condition, as has been implemented in some fuel cell CFD codes (heat from mass transfer problem).

In addition to two-phase DNMs based on a Eulerian–Eulerian formulation, where two 'phases' and several 'species' are solved for with combined heat and mass transfer, the present authors have also developed a two-phase DRA where there are (i) multiple 'spaces' (air, fuel, oil, electrolyte, current collectors); (ii) multiple 'phases' (solid, liquid, gas); and (iii) multiple 'species' (hydrogen, oxygen, water, etc.). Therefore, both inter-component and inter-phase terms are present in the governing equations, which substantially complicates the DRA algorithm, but at the same time, extends its range of use.

5   Conclusions

Spalding's methods were [59],

> practical, imaginative, useful, and packed with maximum possible rigour.
> D. Bradley

His ideas have endured over more than half a century and remain very relevant today. One of the fascinating facts surrounding the enduring success of CFD is the repeated application of existing techniques to new, previously unvisited areas. It is in this context that concepts such as the mass transfer driving force and distributed resistance analogy can readily be applied to novel electrochemical applications, such as fuel cells, electrolysers and batteries. Computational resources have increased by orders of magnitude, to the point where it is now possible to perform detailed performance calculations involving the numerical solution of Poisson equations for the ionic and electronic electric fields, and the Stefan–Maxwell equations for mass transfer on small stacks. Nevertheless, it is still true today, as it was 40 years ago, that such calculations cannot be performed for large stacks of 10–100 cells, and certainly not with the kinds of computer resources available to the average engineering company. The field of CFD is still expanding, and the application of fuel cell and electrolyser technology is very likely to play a significant role in the future of energy conversion. CFD has proved to be a powerful tool in numerous energy conversion applications, such as the design of the blades of wind turbines and the siting of their locations in the environment. Electrochemical processes and products represent another exciting area for which mathematical modelling may provide important solutions leading to disruptive technological breakthroughs.

References

1. Spalding, D. B. (1957). I. Predicting the laminar flame speed in gases with temperature-explicit reaction rates. *Combustion and Flame, 1,* 287–295. https://doi.org/10.1016/0010-2180(57)90015-9.
2. Spalding, D. B. (1957). II. One-dimensional laminar flame theory for temperature-explicit reaction rates. *Combustion and Flame, 1,* 296–307. https://doi.org/10.1016/0010-2180(57)90016-0.
3. Spalding, D. B. (1960). A standard formulation of the steady convective mass transfer problem. *International Journal of Heat and Mass Transfer, 1,* 192–207.
4. Patankar, S. V., & Spalding, D. B. (1972). A calculation procedure for heat, mass, and momentum transfer in three-dimensional parabolic flows. *International Journal of Heat and Mass Transfer, 15,* 1787–1806.
5. Launder, B. E., & Spalding, D. B. (1979). *Lectures in mathematical models of turbulence.* New York: Academic Press.
6. Patankar, S. V., & Spalding, D. B. (1974). A calculation procedure for the transient and steady-state behavior of shell-and-tube heat exchangers. In N. Afgan & E. U. Schlünder (Eds.), *Heat exchangers: Design and theory sourcebook* (pp. 155–176). Washington, D.C.: Scripta Book Company.
7. Spalding, D. B. (2005). A simplified CFD method for the design of heat exchangers. In *ASME Summer Heat Transfer Conference*, San Fransisco (Vol. HT2005-72760). New York: ASME International.
8. Spalding, D. B. (1980). Numerical computation of multi phase flow and heat transfer. In C. Taylor (Ed.), *Recent advances in numerical methods in fluids* (pp. 139–167). Swansea: Pineridge Press.
9. Beale, S. B., Ginolin, A., Jerome, R., Perry, M., & Ghosh, D. (2000). Towards a virtual reality prototype for fuel cells. *Phoenics Journal of Computational Fluid Dynamics and Its Applications, 13,* 287–295.
10. Beale, S. B., & Dong, W. (2002). Advanced modelling of fuel cells: Phase II heat and mass transfer with electrochemistry. Technical Report No. PET-1515-02S, National Research Council of Canada, Ottawa.
11. Beale, S. B., Lin, Y., Zhubrin, S. V., & Dong, W. (2003). Computer methods for performance prediction in fuel cells. *Journal of Power Sources, 118,* 79–85. https://doi.org/10.1016/S0378-7753(03)00065-X.
12. Ackmann, T., de Haart, L. G. J., Lehnert, W., & Stolten, D. (2003). Modeling of mass and heat transport in planar substrate type sofcs. *Journal of the Electrochemical Society, 150,* A783–A789. https://doi.org/10.1149/1.1574029.

13. Lin, Y., & Beale, S. B. (2006). Performance predictions in solid oxide fuel cells. *Applied Mathematical Modelling, 30,* 1485–1496. https://doi.org/10.1016/j.apm.2006.03.009.

14. Andersson, M., Yuan, J. L., & Sunden, B. (2012). SOFC modeling considering electrochemical reactions at the active three phase boundaries. *International Journal of Heat and Mass Transfer, 55,* 773–788. https://doi.org/10.1016/j.ijheatmasstransfer.2011.10.032.

15. Le, A. D., Beale, S. B., & Pharoah, J. G. (2015). Validation of a solid oxide fuel cell model on the international energy agency benchmark case with hydrogen fuel. *Fuel Cells, 15,* 27–41. https://doi.org/10.1002/fuce.201300269.

16. Beale, S. B., Choi, H. W., Pharoah, J. G., Roth, H. K., Jasak, H., & Jeon, D. H. (2016). Open-source computational model of a solid oxide fuel cell. *Computer Physics Communications, 200,* 15–26. https://doi.org/10.1016/j.cpc.2015.10.007.

17. Lin, Y., & Beale, S. B. (2005). Numerical predictions of transport phenomena in a proton exchange membrane fuel cell. *Journal of Fuel Cell Science and Technology, 2,* 213–218. https://doi.org/10.1115/1.2039949.

18. Scholta, J., Escher, G., Zhang, W., Kuppers, L., Jorissen, L., & Lehnert, W. (2006). Investigation on the influence of channel geometries on PEMFC performance. *Journal of Power Sources, 155,* 66–71. https://doi.org/10.1016/j.jpowsour.2005.05.099.

19. Schwarz, D. H., & Beale, S. B. (2009). Calculations of transport phenomena and reaction distribution in a polymer electrolyte membrane fuel cell. *International Journal of Heat and Mass Transfer, 52,* 4074–4081. https://doi.org/10.1016/j.ijheatmasstransfer.2009.03.043.

20. Achenbach, E. (1994). Three-dimensional modelling and time-dependent simulation of a planar solid oxide fuel cell stack. *Journal of Power Sources, 73,* 333–348. https://doi.org/10.1016/0378-7753(93)01833-4.

21. Ding, J., Patel, P., Farooque, M., & Maru, H. C. (1997). A computer model for direct carbonate fuel cells. In *Proceedings 4th International Symposium on Carbonate Fuel Cell Technology* (Vol. 97-4, pp. 17–138).

22. He, W., & Chen, Q. (1998). Three-dimensional simulation of a molten carbonate fuel cell stack under transient conditions. *Journal of Power Sources, 73,* 182–192. https://doi.org/10.1016/S0378-7753(97)02800-0.

23. Spalding, D. B. (1963). *Convective mass transfer; an introduction*. London: Edward Arnold.

24. Mills, A. F., & Coimbra, C. F. (2016). *Mass transfer* (3rd ed.). Upper Saddle River, New Jersey: Prentice Hall.

25. Kays, W. M., Crawford, M. E., & Weigand, B. (2005). *Convective heat and mass transfer* (4th ed.). New York: McGraw-Hill.

26. Beale, S. B., Schwarz, D. H., Malin, M. R., & Spalding, D. B. (2009). Two-phase flow and mass transfer within the diffusion layer of a polymer electrolyte membrane fuel cell. *Computational Thermal Sciences, 1,* 105–120. https://doi.org/10.1615/ComputThermalScien.v1.i2.10.

27. Beale, S. B., Pharoah, J. G., & Kumar, A. (2013). Numerical study of laminar flow and mass transfer for in-line spacer-filled passages. *Journal of Heat Transfer, 135,* 011004. https://doi.org/10.1115/1.4007651.

28. Sherwood, T. K., Brian, P. L. T., Fisher, R. E., & Dresner, L. (1965). Salt concentration at phase boundaries in desalination by reverse osmosis. *Industrial and Engineering Chemistry Fundamentals, 4,* 113–118. https://doi.org/10.1021/i160014a001.

29. Bird, R. B., Stewart, W. E., & Lightfoot, E. N. (2002). *Transport phenomena* (2nd ed.). New York: Wiley.

30. Bard, A. J., & Faulkner, L. R. (2001). *Electrochemical methods* (2nd ed.). New York: Wiley.

31. Gileadi, E. (1993). *Electrode kinetics for chemists, chemical engineers, and material scientists*. New York: Wiley-VCH.

32. Goldstein, R. J., & Cho, H. H. (1995). A review of mass-transfer measurements using naphthalene sublimation. *Experimental Thermal and Fluid Science, 10,* 416–434. https://doi.org/10.1016/0894-1777(94)00071-F.

33. Beale, S. B. (2005). Mass transfer in plane and square ducts. *International Journal of Heat and Mass Transfer, 48,* 3256–3260. https://doi.org/10.1016/j.ijheatmasstransfer.2005.02.037.

34. Beale, S. B. (2004). Calculation procedure for mass transfer in fuel cells. *Journal of Power Sources, 128*, 185–192. https://doi.org/10.1016/j.jpowsour.2003.09.053.
35. Beale, S. B. (2015). Mass transfer formulation for polymer electrolyte membrane fuel cell cathode. *International Journal of Hydrogen Energy, 40*, 11641–11650. https://doi.org/10.1016/j.ijhydene.2015.05.074.
36. Beale, S. B., Reimer, U., Froning, D., Jasak, H., Andersson, M., Pharoah, J. G., et al. (2018). Stability issues of fuel cell models in the activation and concentration regimes. *Journal of Electrochemical Energy Conversion, 15*, 041008. https://doi.org/10.1115/1.4039858.
37. Patankar, S. V. (1980). *Numerical heat transfer and fluid flow*. New York: Hemisphere.
38. Retrieved June 25, 2018, from http://www.Cham.Co.Uk/phoenics/d_polis/d_enc/coval.Htm.
39. Weller, H. G., Tabor, G., Jasak, H., & Fureby, C. (1998). A tensorial approach to computational continuum mechanics using object-oriented techniques. *Journal of Computational Physics, 12*, 620–631. https://doi.org/10.1063/1.168744.
40. Nield, D. A., & Bejan, A. (1999). *Convection in porous media* (2nd ed.). New York: Springer.
41. Beale, S. B. (2012). A simple, effective viscosity formulation for turbulent flow and heat transfer in compact heat exchangers. *Heat Transfer Engineering, 33*, 4–11. https://doi.org/10.1080/01457632.2011.584807.
42. Bell, K. J. (2004). Heat exchanger design for the process industries. *Journal of Heat Transfer, 126*, 877–885. https://doi.org/10.1115/1.1833366.
43. Butterworth, D. (1977). The development of a model for three-dimensional flow in tube bundles. *International Journal of Heat and Mass Transfer, 21*, 253–256. https://doi.org/10.1016/0017-9310(78)90231-4.
44. Rhodes, D. B., & Carlucci, L. N. (1983). Predicted and measured velocity distributions in a model heat exchanger. In *Proceedings CNS/ANS International Conference on Numerical Methods in Nuclear Engineering*, Montreal.
45. Theodossiou, V. M., & Sousa, A. C. M. (1988). Flow field predictions in a model heat exchanger. *Computational Mechanics, 3*, 419–428. https://doi.org/10.1007/BF00301142.
46. Prithiviraj, M., & Andrews, M. J. (1998). Three dimensional numerical simulation of shell-and-tube heat exchangers. Part I: Foundation and fluid mechanics. *Numerical Heat Transfer Part A, 33*, 799–816. https://doi.org/10.1080/10407789808913967.
47. Prithiviraj, M., & Andrews, M. J. (1998). Three-dimensional numerical simulation of shell-and-tube heat exchangers. Part II: Heat transfer. *Numerical Heat Transfer Part A, 33*, 817–828. https://doi.org/10.1080/10407789808913968.
48. Sundén, B. (2007). Computational fluid dynamics in research and design of heat exchangers. *Heat Transfer Engineering, 28*, 898–910. https://doi.org/10.1080/01457630701421679.
49. Kvesić, M., Reimer, U., Froning, D., Lüke, L., Lehnert, W., & Stolten, D. (2012). 3D modeling of a 200 cm^2 HT-PEFC short stack. *International Journal of Hydrogen Energy, 37*, 2430–2439. https://doi.org/10.1016/j.ijhydene.2011.10.055.
50. Zhang, S., Beale, S. B., Reimer, U., Nishida, R. T., Andersson, M., Pharoah, J. G., et al. (2018). Simple and complex polymer electrolyte fuel cell stack models: A comparison. *ECS Transactions, 86*, 287–300. https://doi.org/10.1149/08613.0287ecst.
51. Beale, S. B., & Zhubrin, S. V. (2005). A distributed resistance analogy for solid oxide fuel cells. *Numerical Heat Transfer Part B-Fundamentals, 47*, 573–591. https://doi.org/10.1080/10407790590907930.
52. Retrieved June 27, 2019 from http://www.Cham.Co.Uk/.
53. Retrieved June 25, 2019 from http://www.Cham.Co.Uk/phoenics/d_polis/d_lecs/plant/hex/hex.Htm.
54. Kvević, M. (2013). *Modellierung und simulation von hochtemperatur-polymerelektrolyt-brennstoffzellen* (Vol. 158). Jülich: Forschungszentrum Jülich GmbH. ISBN 978-3-89336-853-8.
55. Nishida, R., Beale, S., Pharoah, J., de Haart, L., & Blum, L. (2018). Three-dimensional computational fluid dynamics modelling and experimental validation of the jülich mark-f solid oxide fuel cell stack. *Journal of Power Sources, 373*, 203–210.
56. Nishida, R. T. (2013). Computational fluid dynamics modelling of solid oxide fuel cell stacks. M.Sc., Queen's University.

57. Nishida, R. T., Beale, S. B., & Pharoah, J. G. (2016). Comprehensive computational fluid dynamics model of solid oxide fuel cell stacks. *International Journal of Hydrogen Energy, 41,* 20592–20605. https://doi.org/10.1016/j.ijhydene.2016.05.103.

58. Retrieved June 27, 2019 from https://openfoam.Org/.

59. Bradley, D. (2018). Personal communication.

Modeling Proton Exchange Membrane Fuel Cells—A Review

Ayodeji Demuren⓪ and Russell L. Edwards

1 Introduction

This paper is written in commemoration of D. Brian Spalding (DBS), who made significant contributions to the theory and practice of combustion. He wrote his Ph.D. thesis on the combustion of liquid fuels in 1951 [1], followed by several papers in this field [2, 3] and culminating in a couple of textbooks. A general theory on turbulent combustion [4] and combustion and mass transfer [5]. Combustion relies on thermochemical processes to derive energy from burning fuels. This invariably occurs at high temperatures, and the thermodynamic efficiency is limited by the Carnot cycle. On the other hand, fuel cells are electrochemical devices which convert energy from simple fuel and oxidant chemical reactions, at the anode and the cathode, directly into electrical energy. These can occur over a wide range of temperatures, and the thermodynamic efficiency is not limited by the Carnot cycle, and so are expected to deliver energy from the fuels more efficiently, due to the direct nature of the process. An excellent treatise on fuel cells by Dicks and Rand [6] shows that this is mostly true, especially at lower temperatures. In effect, fuel cells are like continuous batteries, which can operate endlessly as long as the electrodes are supplied with fuel and oxidant, respectively, and the conditions for the chemical reactions are maintained. In contrast to secondary (rechargeable) batteries, neither the electrodes nor the electrolyte is consumed, so no recharging is necessary. In the early days of the automobile [7], battery-powered electric vehicles were in competition with those powered by internal combustion engines (ICE): a competition that was won easily by the latter, because of their reliability and range. Fuel cells have the potential to increase the range of electric vehicles, thereby making them viable again. But recent improvements in the technology of secondary batteries have greatly expanded options for battery-powered electric vehicles. There are many different types of fuel

A. Demuren (✉) · R. L. Edwards
Department of Mechanical and Aerospace Engineering,
Old Dominion University, Norfolk, VA, USA
e-mail: ademuren@odu.edu

© Springer Nature Singapore Pte Ltd. 2020
A. Runchal (ed.), *50 Years of CFD in Engineering Sciences*,
https://doi.org/10.1007/978-981-15-2670-1_15

cells and technology implementations, which are reviewed in the excellent treatise [6]. The most common of these is the polymer electrolyte membrane or proton exchange membrane fuel cells (PEMFC), which use thin layers of polymer-based solid electrolyte, along with catalyst infused conductive electrodes. They are the basis of the so-called Hydrogen economy.

PEMFC are electrochemical energy conversion devices that produce electrical energy from the chemical energy present in hydrogen fuel when it reacts with oxygen in a cell. Water and fractional waste heat are the byproducts. Because of the low operational temperature, typically 25–80 °C, PEMFC has been used in automotive, marine, and portable electronics applications where they produce useful electrical energy at high efficiency, without creating the pollutants associated with combustion of fossil fuels. They exhibit high efficiency in the range of 60% when used for electrical energy conversion and 80% when there is cogeneration of electrical and thermal energy. The main hindrances to wider adoption have been high cost, low lifespan, and wider availability of hydrogen fuel. The suitability of PEM fuel cells for particular applications is determined in terms of power density, cost per kilowatt, and durability (a lifetime of operation in hours). Each cell can only produce a fraction of a volt, during operation, so systems have to be built with stacks of cells, in series, to obtain useable voltage. Depending on application, systems have been created with power of only a few Watts to hundreds of Kilowatts. At the low end are devices for powering mobile electronics, and at the high end are powertrains for public transportation such as electric busses. Durability goals also vary based on application, which could be a few thousand operating hours in transportation use, to hundreds of thousands of hours in stationary power generation use. Material degradation of key components typically limits durability. Performance degrades gradually over time. The choice of cell design and operating conditions will typically influence some vectors of performance degradation, such as membrane degradation, or the reduction of effective catalyst surface area with agglomeration or aging.

Water and thermal management are fundamentally intertwined problems which will influence the durability, performance, and efficiency of a PEM fuel cell. The reactant gases are commonly highly humidified even though the device produces water in operation. The humidification is done to ensure high protonic conductivity and efficiency from the water-absorbing (PEM) membrane at the heart of the fuel cell.

Complexity of fuel cell design, along with the thinness of several of the membranes involved make experimental measurements extremely difficult. Hence, computational fluid dynamics (CFD) approach is widely used for PEMFC design and simulation [8]. These mostly follow the computational techniques developed at Imperial College under the supervision of DBS. Early pioneering computational modeling effort was isothermal and only one-dimensional. It was helpful in establishing the fundamental models and relationships, but is generally thought to be inadequate in analyzing three-dimensional flow effects that interact with electrochemical reactions of the PEMFC. Later models account for three-dimensional effects that many analytical solutions will neglect or not consider. This approach has been implemented successfully in many commercial CFD codes, however, computational costs remain quite high. The PEMFC is a multiscale problem. Costs are driven by the requirements to mesh/discretize the flow channels (~mm thickness), simultaneously with the extremely thin catalyst and electrolyte layers, the membrane electrode assembly (MEA), that are two to three orders of magnitude thinner.

Another approach is to omit the thin MEA from the computational domain, treating the MEA as an interface, between the anode and cathode flow domains. The MEA can be treated as a reacting wall, with consumption/production source terms on either side to mimic its operation. The interface model [9, 10] then matches these effects as boundary conditions to both domains. Models considering similar state, dimensionality, and effects (physics) will then differ, primarily, in how they treat the MEA [11]. The physics of the MEA include water transport, heat transport, gaseous diffusion, and the kinetics of the reactions occurring in the anode and cathode. A multitude of subsequent studies have improved the understanding of each area from the time when CFD models were first being developed and refined [12] Some widely used modeling assumptions have been called into question by direct experimental evidence [13]. Many analytical approaches to model catalyst layer operation have also been published.

2 PEM Fuel Cell Fundamentals

The PEMFC is an electrochemical energy conversion device that converts the internal chemical energy of the reactants (the hydrogen fuel and the oxygen oxidizer), into electrical energy. The basic components of the PEMFC are shown in a cutaway diagram in Fig. 1. The left side is the negative, or anode terminal, and the right side the positive, or cathode terminal. Electrical connection to an external circuit is made via the electrically conductive current collector plates. Both current collector plates typically have gas flow channels that direct the flow of the hydrogen fuel to the anode side, and the oxygen or air oxidizer to the cathode side. The fuel and oxidizer, collectively referred to as reactants, are typically supplied in carefully metered amounts as pressurized, humidified gas streams.

At the anode, hydrogen is oxidized in the hydrogen oxidation reaction (HOR) into its constituent protons (H^+) and electrons (e^-). The HOR occurs at reaction sites (catalyst particles) distributed through the thickness of the anode catalyst layer (ACL) where diatomic hydrogen is split into two protons and two electrons:

$$H_2 \rightarrow 2H^+ + 2e^- \tag{1}$$

The protons produced by this reaction move through the thickness of the solid electrolyte membrane to reach the cathode catalyst layer (CCL) under the influence of the electric field. Electrons reach the CCL through the external circuit, and oxygen gas reaches the CCL by diffusing through the diffusion media. At reaction sites, distributed through the thickness of the CCL, the oxygen reduction reaction (ORR) combines oxygen, protons, and electrons, creating water as the reaction product:

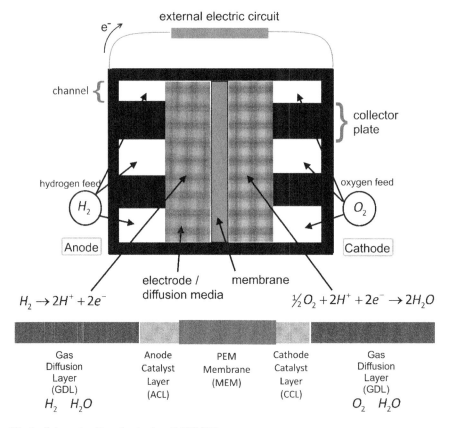

Fig. 1 Schematic slice of a single cell PEMFC

$$O_2 + 4H^+ + 4e^- \rightarrow 2H_2O \tag{2}$$

The product water hydrates the membrane, and will leave the cell through the reactant gas streams. Water movement within the membrane depends on operating conditions. Both anode and cathode gas streams exchange water with the MEA, and remove product water from the device.

Collector plates serve both as the electric terminals of the cell, in which role they collect electrons generated at the catalyst layers, as well as flow channels to distribute the humidified gases over the surface of the anode and cathode. Collector plates have been made from graphite and various metals to have the desired electrical conductivity and resistance to corrosion.

Gas diffusion layers (GDL) contact both sides of the MEA. These provide electronic and heat conduction between the collector plates and the MEA. They are highly porous in order to allow the gases in the flow channels to diffuse through to the MEA. The diffusion media typically is made of carbon cloth or carbon paper materials, with thicknesses ~150–400 μm typical.

Catalyst layers are attached to both the anode (ACL) and cathode (CCL) sides of the central polymer electrolyte membrane (PEM). These are porous structures consisting of three phases. Fine catalyst particles of platinum (Pt) or an alloy thereof provide the reaction sites distributed throughout the CL thickness. These are supported on complex structures of carbon particles that provide electronic conduction within the catalyst layers. This carbon-supported platinum catalyst (often denoted as PT/C) technology creates a large number of active catalyst sites and a catalytic surface area several orders of magnitude greater than the planar electrode surface area, so that the reaction can proceed at a feasible rate. A dispersed electrolyte phase serves, likewise, to allow protonic conduction to the reaction sites. The remainder void volume consists of pores to allow for gas permeation. Typical thicknesses are 5–15 μm.

The electrolyte membrane serves to separate the two respective gas flows. The membrane must, according to its function, be highly conductive of the protons while simultaneously nonconductive for electrons in order to avoid short-circuiting the external flow of electricity. Nafion has been the common electrolyte material. It absorbs water, possibly 40% or more by weight, when in contact with humidified gases, and when so hydrated, becomes proton-conductive. Protonic conductivity rises significantly with water content, allowing the device to operate efficiently. Typical electrolyte membrane thicknesses range from (25 μm) to (175 μm). Thinner membranes have been more popular for some time.

Coupled chemical reactions occur on opposite sides of the planar catalyzed membrane. Convective and diffusive transport of reactant gases occur within separated flow channels on both the anode and cathode sides. Each gas permeates through a gas diffusion layer (GDL) and reaches the respective anode and cathode catalyst layers where corresponding electrochemical reactions occur. Continuous operation requires conduction of protons through the central membrane MEA, as well as conduction of electrons through the GDL, graphite flow field plates, and an external electrical circuit.

The overall chemical reaction can be written, combining Eqs. (1) and (2), as

$$H_2 + \frac{1}{2}O_2 \rightarrow H_2O \tag{3}$$

The performance, or device efficiency, of a PEM fuel cell is typically assessed through a voltage measurement conducted over a range of current densities, known as a polarization curve or characteristic curve. The curve is typically analyzed by computing an ideal, thermodynamic voltage, and then subtracting estimates of various polarizations (losses) from it. Under ideal conditions, the maximum thermodynamic work that can be done in the reversible reaction, at constant pressure and temperature, is the change in Gibb's free energy of the products and reactants. For the reaction of Eq. (3), 1 mol of hydrogen reacts with half a mol of oxygen to produce 1 mol of water. In the process two mols of electrons are exchanged through the external circuit. The corresponding change in Gibb's free energy is

$$\Delta g_f = \left\{ (\Delta g_f)_{H_2O} - [(\Delta g_f)_{H_2} + \frac{1}{2}(\Delta g_f)_{O_2}] \right\} \tag{4}$$

For liquid water product at 80 °C, $\Delta g_f = -228.2 \frac{KJ}{mol}$

The corresponding charge is $-2F$, where F, the Faraday number, is the charge of 1 mol of electrons.

Hence, the electrical work done, (current × voltage) is $-2FV$.

Therefore, if the process is reversible, with no losses, the reversible, or maximum open-circuit voltage (OCV) is

$$V_r = \frac{\Delta g_f}{-2F} = \frac{-228,200}{-2 * 96,485} = 1.18 \text{ V} \tag{5}$$

Once current is drawn from the cell, there will be losses and the process is no longer reversible, and the cell voltage will be reduced. The polarization curve, (Fig. 2) shows how the cell voltage decreases with current density [14]. These losses are associated with decreased efficiency in the conversion of the chemical energy to electrical energy in the PEMFC. The maximum efficiency, based on the higher heating value (HHR), $\Delta h_f = -285.84 \frac{KJ}{mol}$, in which the reaction product is liquid, is 80%, compared to typical operational efficiency of 60%.

It is noted that the maximum efficiency decreases slightly with temperature, whereas that of a heat engine, limited by the Carnot cycle increases with temperature, with a crossover point around 1000 K [14]. Therefore, fuel cells which operate at high temperatures, such as the Solid Oxide Fuel Cell (SOFC) may have lower nominal efficiency than a comparable heat engine.

The *Nernst equation* can be used to calculate changes in OCV, based on concentration or partial pressures of reactants and products, as

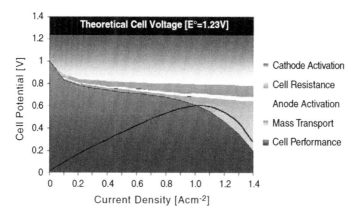

Fig. 2 Polarization curve showing sources of performance losses in a PEMFC operating at 25 °C. Power curve is also shown [14]

$$V_r = V_r^o + \frac{RT}{2F} \ln\left(\frac{P_{H_2} \cdot P_{O_2}^{1/2}}{P_{H_2O}}\right) \tag{6}$$

where V_r^o is the OCV at standard pressure, R is the universal gas constant, T is the absolute temperature, and the P's are the respective partial pressures. Therefore, higher stoichiometric values of reactants would produce higher OCV. Also the use of pure oxygen at the cathode instead of air.

Performance losses within the PEMFC can be categorized based on different mechanisms. The major categories of losses within the cell are activation (or kinetic), ohmic, and mass transport (or concentration) polarizations.

Activation losses describe the voltage drop required to drive both the anode and cathode chemical reactions at rates greater than equilibrium. Also known as activation polarization, kinetic losses tend to dominate the voltage drop occurring in the PEMFC at low-current densities. As shown in Fig. 2, activation losses occurring from the ORR at the CCL tend to be more significant than those in the ACL. These losses are described by models of electrochemical kinetics, and are greatly influenced by catalyst site temperature, gas composition, and the effective catalyst site surface area, which is itself dependent upon the complex catalyst layer roughness and composition. The *Tafel equation* can be used to calculate the activation losses in terms of exchange current density as

$$\Delta V_{act} = \frac{RT}{2\alpha F} \ln\left(\frac{i}{i_o}\right) \tag{7}$$

where i_o is the exchange current density, and α is the charge transfer coefficient with a typical value between 0.1 and 0.5. The higher the value of i_o, the more active is the electrode and the smaller the activation voltage loss. This equation represents the limiting behavior of the Butler–Volmer equations, to be discussed in a later section.

Ohmic losses within the PEMFC are 'iR' type losses. These tend to dominate the voltage drop occurring in the PEMFC at mid-range current densities. Thus the measured voltage curve decreases almost linearly with current density. Relatively minor ohmic losses come from electronic conduction losses within PEMFC components and contact resistances. Primary measurable ohmic losses come from the ionic resistance of the membrane, which is dependent upon its state of hydration and temperature. The use of thin membranes helps to minimize ohmic losses.

Mass transport, or concentration, losses occur under high current density operation. These arise from the inability to replenish the supply of reactants at the reaction sites fast enough. When reactant consumption levels grow large enough, gas transport limitations develop in the diffusion media and catalyst layers. These arise through the normal effective diffusivity of the GDL media and catalyst layers, and are further exacerbated if liquid water accumulates and blocks the pores of the GDL used for gas diffusion. Mass transport losses which arise from reduced reactant concentration, or partial pressures, at the reaction sites within the catalyst layers can be estimated from the Nernst equation (6). With the inability to supply reactants at required rates, the voltage curve turns down precipitously, which is indicative of large overpotentials leading to a limiting current density value.

2.1 Fuel Cell Stacks

For any single cell, the operational voltage has been found to be in the range of 0.6 to 0.7 V, therefore, cells have to be stacked together in series to build a working engine. The flow channels are within bipolar plates which supply the anode on one side and the cathode on the other. Figure 3 shows a picture of a 5-cell stack. Practical stacks use many more cells. For example in automobile applications [15], stacks with 200

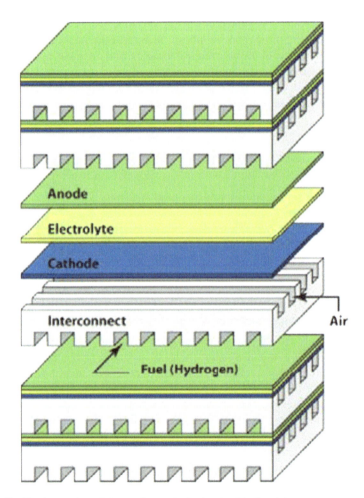

Fig. 3 5-cell polymer electrolyte membrane stack, showing bipolar plates

cells are not uncommon. With an active cell area of 500 cm^2, total power of 60 kW could be delivered, assuming 1 A/cm^2 current density. In these cases, with tightly packed cells, thermal and water management are of critical concern, requiring in some cases, extra cooling channels. Design and flow patterns in bipolar plates are also critical to ensure adequate supply of reactants to the GDL and reaction sites [16].

3 Mathematical Model

The governing equations typically used to model a PEM fuel cell with computational fluid dynamics are presented. They are equations of flow, heat transfer, and current conduction. The flows are laminar and involve variable properties, such as density. In the porous diffusion media, the Brinkman equations solve for fluid velocity. The solid and fluid phases are assumed to be in thermodynamic equilibrium

The various computational subdomains or regions of the PEM fuel cell are described in Fig. 4 with a cutaway view representing the cross section of a single channel on each side. The divisions mirror the major components of the device.

Gas flows freely through the anode and cathode gas channels and also in the porous regions of the gas diffusion layers (GDL). The GDL has a porosity ε, which is the fraction of a control volume occupied by gas, between zero and 1. Porosity is zero in pure solid regions (the collector plates) and unity in the flow channels. Physical properties of the fluid are averages taken over the volume of the pores. The density and viscosity of the fluid in the porous regions are properties that can be

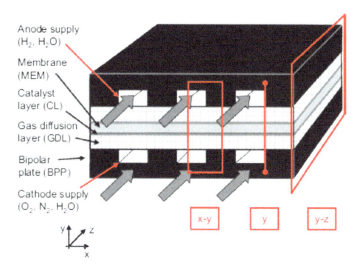

Fig. 4 PEMFC geometry: schematic illustration of different computational domains (1-D—y-direction; 2-D—x-y or y-z directions; 3-D—x-y-z directions) [17]

measured experimentally in a free flow region and so all properties are continuous with those in the adjacent free flow. The flow velocity in the porous regions is defined as a superficial volume average, or Darcy velocity. This average over a unit volume (entire volume comprising both solid matrix and pore) defines velocity as the volume flow rate per area of the porous medium. When defined this way, the fluid velocity is continuous at gas channel-GDL boundaries.

The continuity and momentum equations (8) and (9), taken together, are known as the *Brinkman equations*, which must be solved to yield the velocity field.

$$\nabla \cdot (\rho \vec{u}) = 0 \tag{8}$$

$$\frac{\rho}{\varepsilon^2}(\vec{u} \cdot \nabla \vec{u}) = -\nabla p + \nabla \cdot \left[\frac{\mu}{\varepsilon} \left\{ \nabla \vec{u} + (\nabla \vec{u})^T - \frac{2}{3}(\nabla \cdot \vec{u})\bar{\bar{I}} \right\} \right] - \bar{\bar{K}}^{-1} \mu \vec{u} \tag{9}$$

where μ (kg m^{-1} s^{-1}) is the dynamic viscosity of the mixture, $\vec{u}$ is the mass-average velocity vector (m s^{-1}) in the Cartesian coordinate system, $\bar{\bar{K}}$ is the permeability tensor of the porous medium.

The conservation of species equations are applied to the same domains and represent mass flux from convection and diffusion, as well as sources which arise from phase change and/or electrochemical reactions. The cathode gas has three species (oxygen = 1, water = 2, and nitrogen = 3) and the anode uses two species (hydrogen = 1, water = 2). The mass fractions ω_i of species i at the cathode are given by Eq. (10) where $\tilde{D}_{ik}$ ($i, k = 1, 2, 3$) are the multicomponent Fick Diffusivities, the components of the multicomponent Fick diffusivity matrix, which are needed to solve the problem. These are symmetric, i.e., $\tilde{D}_{ik} = \tilde{D}_{ki}$. The multicomponent Fick Diffusivities are determined from the multicomponent Maxwell–Stefan diffusivities D_{ik}.

$$\begin{aligned}
\nabla \cdot (\rho \vec{u} \omega_1) &= \nabla \cdot \left[\rho \omega_1 \sum_k \tilde{D}_{1k} \left(\nabla x_k + \frac{\nabla p}{p}(x_k - \omega_k) \right) \right] \\
\nabla \cdot (\rho \vec{u} \omega_2) &= \nabla \cdot \left[\rho \omega_2 \sum_k \tilde{D}_{2k} \left(\nabla x_k + \frac{\nabla p}{p}(x_k - \omega_k) \right) \right] \\
\omega_3 &= 1 - \omega_1 - \omega_2
\end{aligned} \tag{10}$$

where x_k represents the corresponding mole fraction.

The heat transfer is governed by Eq. (11)

$$\nabla \cdot \left(\rho c_p \vec{u} T - k_{\text{eff}} \nabla T \right) = Q \tag{11}$$

where

$$k_{\text{eff}} = (1 - \varepsilon)k_{\text{GDL}} + \varepsilon k \tag{12}$$

The thermal conductivity of the GDL material is strongly anisotropic, with in-plane ($k_{\text{GDL},=}$) and thru-plane ($k_{\text{GDL},\perp}$) values. It is represented by a tensor

$$k_{GDL} = \begin{bmatrix} k_{GDL,=} & 0 & 0 \\ 0 & k_{GDL,=} & 0 \\ 0 & 0 & k_{GDL,\perp} \end{bmatrix} \tag{13}$$

Significant thermal contact resistances have been recognized as occurring at the current collector plate and the GDL. At other component interfaces, contact resistances are commonly ignored [18]. Thermal contact resistance between multiple layers of GDL was found to be negligible [19], and the contact resistance with solid plates (R_{ct}) was found to be about $1\text{--}1.5 \times 10^{-4}$ m^2 K/W at typical cell compaction pressures.

The conservation of electron charge equation must be solved in the electronically conductive domains. Voltage losses arising from electronic conduction have been ignored in many models; thought to be insignificant due to the high electrical conductivity of the respective materials. Current flux $\overrightarrow{J_e}$ in Eq. (14) (A m^{-2}) is represented as the flow of positive charges in the direction of reduced electrical potential, in the presence of an electric field $\overrightarrow{E}$ (V m^{-1}), and current must be conserved as Eq. (14) where Φ_e is the scalar electric potential and σ_e (S m^{-1}) the electrical conductivity. The electrical conductivity of the current collector plate (CCL) is isotropic, but that of the GDL is a tensor: it has separate in-plane ($\sigma_{e,=}$) and thru-plane ($\sigma_{e,\perp}$) values as in Eq. (15)

$$\nabla \cdot \overrightarrow{J_e} = \nabla \cdot \left(\sigma_e \overrightarrow{E} \right) = \nabla \cdot (\sigma_e(-\nabla \Phi_e)) = 0 \tag{14}$$

$$\sigma_e = \begin{bmatrix} \sigma_{e,=} & 0 & 0 \\ 0 & \sigma_{e,=} & 0 \\ 0 & 0 & \sigma_{e,\perp} \end{bmatrix} \tag{15}$$

Thru-plane resistivity values of 0.08 (Ω cm) and in-plane resistivity values of 0.006 (Ω cm) have been reported as typical [20]. In the catalyst layers, current production is governed by the Butler–Volmer relations [21]. Thus the charge equations for the electron and proton transport in the solid and electrolyte media, respectively are

$$\nabla \cdot (\sigma_s(-\nabla \Phi_s)) = S_s \tag{16}$$

$$\nabla \cdot (\sigma_{el}(-\nabla \Phi_{el})) = S_{el} \tag{17}$$

where the source terms in the Anode Catalyst Layer are

$$S_s = -j_a; \; S_{el} = j_a; \tag{18}$$

and the source terms in the Cathode Catalyst Layer are

$$S_s = -j_c; \; S_{el} = j_c; \tag{19}$$

j_a, j_c are the local current density at the anode and cathode sides, respectively, given by the *Butler–Volmer relations*

$$j_a = \left(ai_o^{\text{ref}}\right)_a \left(\frac{P_{H_2}}{P_{H_2}^{\text{ref}}}\right)^{\frac{1}{2}} \left[\exp\left(\frac{\alpha_a^a F}{RT}\right)\eta_a - \exp\left(\frac{-\alpha_c^a F}{RT}\right)\eta_a\right] \qquad (20)$$

$$j_c = \left(ai_o^{\text{ref}}\right)_c \left(\frac{P_{O_2}}{P_{O_2}^{\text{ref}}}\right) \left[\exp\left(\frac{\alpha_a^c F}{RT}\right)\eta_c - \exp\left(\frac{-\alpha_c^c F}{RT}\right)\eta_c\right] \qquad (21)$$

where a is the electrocatalytic surface area per unit volume and η_a and η_c are overpotentials and i_o is the corresponding exchange current density, at the anode and cathode, respectively.

Resistive heating gives a minor volumetric heat source Q (W m^{-3}) from the dissipation of electrical energy within the conductive GDL and CCL domains described by Eq. (22)

$$Q = \vec{J}_e \cdot \vec{E} \qquad (22)$$

In summary, the total number of conservation equations are 8 (anode) or 9 (cathode) scalar equations for the same number of unknown variables to be solved for. The continuity and momentum equations give the pressure p and the three components of $\vec{u}$. The conservation of species equations give the 2 scalar mass fractions in the anode domain, with variables $\omega_1 = \omega_{H_2}$ and $\omega_2 = \omega_{H_2O}$, or the 3 scalar mass fractions, with variables $\omega_1 = \omega_{O_2}$ and $\omega_2 = \omega_{H_2O}$, and $\omega_3 = \omega_{N_2}$, on the cathode side. The conservation of energy equation gives the scalar variable T (temperature) and the remaining conservation of electronic current equation gives the scalar potential Φ_e. This system of equations is solved by standard CFD methods [11, 17], many of which use the SIMPLE algorithm [22], or one of its many derivatives (SIMPLER, SIMPLEC, PISO, etc.).

4 Methodology Review

4.1 *Empirically Based 0-D Models*

The earliest experimental studies of the PEM fuel cell typically involved determination of the polarization curve, the measure of its performance. This is done under specific test conditions such as cell temperature, and reactant gas flow rates and humidity levels. As voltage levels are synonymous with efficiency, such comparisons have direct validity. In a recent paper, Wilberforce et al. [23] used measurements to analyze the performance of a 5-cell stack, under varying conditions of temperature and oxygen supply pressure. Figure 5 shows variations in OCV, stack current,

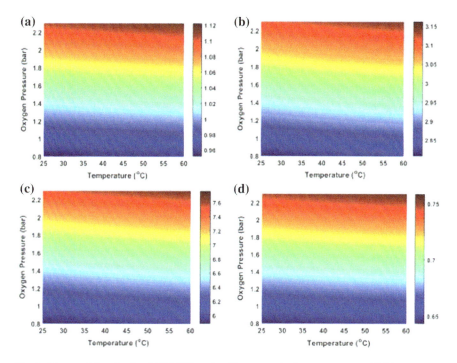

Fig. 5 Measured data for 5-cell PEMFC stack with constant hydrogen pressure of 2.5 bar; **a** open-circuit voltage (V), **b** stack current (A), **c** maximum power (W), **d** fuel cell efficiency [23]

maximum power and efficiency, derived from the data. Performance increases with oxygen gas pressure, as expected from the Nernst equation (6). Furthermore, at any given gas pressure levels, performance improved with increase in temperature. This can be explained with the increased activity level, and corresponding decrease in activation losses, with temperature rise in the cells.

In these zero-dimensional (0-D) models, the measured curve was compared against the thermodynamic voltage, and several mechanisms of voltage losses assessed from the measured data. The polarization curve can usually be fitted with a number of empirical functions that estimate the open-circuit voltage, along with average values of the different loss effects [6, 24], such as gas crossover, kinetic losses, ohmic losses, and mass transport effects. One example of this approach illustrated how the shape of the curve indicates different fundamental loss mechanisms, where one loss mechanism was clearly dominant in various current density ranges [25]. 0-D models are not true predictive models but serve as empirical fits of the experimental observation, indicative of expected performance. Laurencelle et al. [26] found a good correlation fit for the experimental data of the polarization curve. Cooper et al. [27] concluded that the status of these "characterization" equations is settled, and that the analysis and fitting procedures are unlikely to experience significant further development.

Experimental data are of utmost importance in calibrating and validating CFD models. Distributed diagnostic measurements have been developed to provide benchmark data for fuel cell model validation. The term distributed diagnostics refers to the efforts to measure local spatial variations in temperature, current, cell resistance or conductivity, species concentration, etc. These tools are ideally used to validate multidimensional models. Local polarization curves have been produced [28] from a small portion of a segmented cell. Current mapping is commonly used as a term to describe distributed current density measurement. Distributed current density measurements were identified as a key to understanding fuel cell systems [29], where very small cells were found to have nearly uniform current distribution, and larger area cells would have variations in current density, oxygen concentration, humidity, water flooding, and/or temperature. The measurement techniques began with magnetic loop arrays that are embedded in the current collector plates [30, 31]. This technique has the advantage of being done independently of cell operation, but can only be used with a single cell: it is not compatible with fuel cell stacks. Other approaches involved using segmented cells: individual cells that were electrically isolated from each other. These segmented cells could provide spatial resolutions about the size of a single gas channel [32]. Some of these approaches have been criticized in that a segmented MEA will disrupt the current distribution reaching the current collectors, thus giving a measurement that is not indicative of the full-sized MEA. On the other hand, Mench and Wang [29] used an un-altered MEA with an electrically segregated (segmented) flow field.

Distributed temperature and species measurements have been used for evaluating model agreement. Attempts to measure the thru-membrane temperature profile have involved embedding thermocouples in the diffusion media of a PEMFC [33], as well as embedding micro-thermocouples between two layers of membrane sheets [34, 35]. Temperature measurements were plotted at discrete points along the length of the serpentine flow path. These works showed that commonly used isothermal assumptions are not well justified at high current densities, as temperatures varied as much as 10 °C within the electrolyte. Species distribution was determined with a gas chromatograph which measured mole fractions of the various species present in the gas channels of an operating PEMFC [36]. This allowed for an experimental verification of the water vapor mole fraction in the PEMFC flow-fields at steady state. Measurements could only be taken about every 2 min, but this technique was not workable in the presence of liquid water. Later the same group combined in situ data measurement of species distribution, current distribution, and high-frequency resistance distribution with improved transient response to nearly reach real time (1 s per point) [37]. Edwards and Demuren [38] measured transient response of a PEM fuel cell to step changes in load current under various conditions of temperature and humidity. They found good correlation with a triple-term exponential curve fit. The first and fastest time scale is associated with hydration effect on protonic resistance in the catalyst layers. The second, which is one order of magnitude slower is associated with hydration effect on protonic resistance in the electrolyte. And the third time scale, which is two orders of magnitude slower is consistent with effects of heat generation in the MEA on activation losses.

4.2 Electrochemical and Transport Model Calibration Studies

Reactant crossover measurement is relevant to PEMFC modeling because the reactant crossover, or leakage, reduces the open-circuit cell voltage (OCV) by a noticeable amount from the reversible voltage or "Nernst" value. There are several methods to determine gaseous reactant (hydrogen) crossover through the proton exchange membrane. Accounting for crossover can be done by adding a crossover current to the model for kinetic losses or with an arbitrary reduction in the OCV of a fuel cell. The crossover is usually unresolved; i.e., it is assumed to be spatially averaged over the area of the MEA. In situ electrochemical methods for measuring crossover include cyclic voltammetry (CV) [39] and linear sweep voltammetry (LSV) [27]. Crossover can be measured directly, and levels of 1–2 mA/cm^2 are thought to be typical. LSV can also detect the presence of an internal short-circuit through the presence of a positive slope of the current/potential curve.

Active Surface Area Measurement typically uses CV in fuel cell studies to measure the electrochemically active surface area (ECSA) of catalyst layers. ECSA testing is used to evaluate the performance of catalysts, modifications to catalyst layer microstructure(s) and/or catalyst loading, and the presence of adsorbed surface poison. CFD models use either ECSA or exchange current density estimates to calculate kinetic losses of both the anode and cathode [40, 41]. They are often not known precisely and have been treated as parameters which are adjusted to align model predictions with experimental data. Typically, area values are reported with respect to the catalyst loading. Values of around 50 m^2/g Pt [24], with ordinary platinum catalysts, are reported in well-designed PEMFC cathodes. ECSA values have been found, experimentally, to drop with decreasing humidity and/or ionomer volume fraction [42], and most CFD models have failed to account for this.

Ohmic Losses in the PEMFC are comprised of ionic resistance losses (mostly from the electrolyte), electronic contact resistances between components, and bulk electronic resistance. Test equipment is now commonly available to make in situ ohmic resistance measurements on an operating PEMFC. The most common in situ methods are the current-interrupt and high-frequency resistance (HFR) methods. These produce the sum of (i) electronic resistances, and (ii) ohmic resistances arising in the membrane of a PEMFC [43]. The two methods are nearly equivalent and have been frequently used for kinetics research and as a validation aid.

It is typically desired to separate the ohmic resistance components. Electronic resistances might be implicitly assumed to be negligible or may be measured separately (ex-situ) within a "dummy" cell (one where the MEA has been removed). This latter procedure was used in kinetics research [40, 41, 44]. The previously mentioned in situ methods do not address, or include, catalyst layer ionic and electronic resistances. Further research has been focused on experimentally assessing ionic resistance losses in catalyst layers [45]. Ionic conduction losses have been shown to be significant, while electronic conduction losses within PEMFC catalyst layers have frequently been considered negligible [41, 46].

Diffusional Losses in Catalyst Layers have been investigated. In an operating PEMFC, reactant gases must move through some portion of the thickness of each catalyst layer to reach the available reaction sites. This process was thought to be dominated by diffusion instead of pressure-driven flow [24], and so significant experimental and modeling research has been invested to develop accurate gas diffusion transport models, applicable to the catalyst layer composition. As hydrogen has a greater diffusional coefficient, and anode losses are not typically as significant, most effort has focused on oxygen diffusion through CCL media. Key early works investigated the microstructure of catalyst layers, finding that there existed two distinct pore size distributions within typical CL structures, and that pore sizes were between nanometer and sub-micrometer [47, 48]. The larger pores tended to be partially filled with ionomer, with the effect that oxygen diffusion coefficient decreased as ionomer loading increased [49] and thus the performance of the MEA suffered. On the other hand, computer models used much higher oxygen diffusion coefficients in PEMFC catalyst layers. Estimates of effective values were in the range $0.1–0.3$ cm^2 s^{-1} at 80 °C, which justified ignoring these diffusional losses [50, 51]. Experimental measurements of effective diffusion coefficients were taken in 2005 to be in the range $0.006–0.007$ cm^2 s^{-1} at 80 °C [52], from evaluation of a typical CL structure. In a 2010 study, measured effective diffusion coefficients dropped from 0.024 to 0.001 cm^2 s^{-1} at 80 °C as ionomer loadings and humidity were increased [53]. Later measurements, in 2011, found a value of 0.002 cm^2 s^{-1} at 80 °C [54]. The Bruggeman correction, which was commonly used for estimating effective diffusion coefficients of porous catalyst layers, was found to underestimate catalyst layer tortuosity [53, 54].

Pore size distributions within the CL structures were found to be similar to the mean free path of oxygen molecules. Inaccessible pores were present, which reduced the porosity by 20–60% from the value which was expected, based purely on density calculations. Oxygen diffusion was found to take place within the "transition region" where both Knudsen and molecular diffusion must be taken into account [54]. By 2012, experimental measurements at 0% RH were shown to be consistent with such a model of diffusion, when porosity and tortuosity effects were properly assessed [55]. The presence of inaccessible pores in the CL structures served to reduce porosity and the effective diffusion coefficient. Intensive, pore-scale models have been developed based on these findings. These models are intended to serve as a valid analytical basis for macro-homogeneous models, providing effective transport properties. They computationally reconstruct catalyst layer microstructures and then devise effective transport properties. The choice of pore-scale models (referred to as a reconstruction algorithm) did not substantially affect effective transport properties that these models were to predict [56]. It was more important to account for the reduced porosity [57] resulting from Nafion loading.

Electrode Kinetic parameter estimation research has progressed significantly. Both anode and cathode electrodes are expected to continue utilizing the current carbon-supported platinum (Pt/C) technology [58]. CFD models require kinetic data to estimate voltage or efficiency losses occurring in these reactions. Each follows Butler–Volmer kinetics [20, 24], where the reaction rate is described by two parameters: (i) exchange coefficients and (ii) exchange current density. Many CFD models presented kinetic values without clear justification, treating these as free parameters which are adjusted to produce agreement between model and experimental data. Kinetic model parameters, such as exchange current density, are sometimes discussed in terms of the electrochemically active surface area (ECSA), which is then related to catalyst loading. The practice in several studies is to use kinetic model parameters devised with consideration of ACL/CCL composition.

The hydrogen oxidation reaction (HOR) occurring at the anode is quite facile (rapid). Under these conditions, the Butler–Volmer relations, simplify to linear kinetics [24]. Voltage losses are typically very small, often too small to measure. Experimental results suggest the possibility to lower the anode platinum catalyst loading, thereby saving cost, without significantly degrading performance [59]. Estimates of the HOR kinetics parameters available in 2007 were compared and found to show variations of orders of magnitude. Neyerlin et al. [44] presented HOR kinetics parameters from a H_2 pump test, with the suggestion that the prior rotating disk electrode (RDE) experiments were compromised by unacknowledged mass transport limitations. Butler–Volmer model parameters could only be determined from experimental data to within a factor of 2. Durst et al. [58] offered more precise kinetic data estimates and developed temperature-dependence as well.

The oxygen reduction reaction (ORR) occurring in the cathode is more complicated and much slower, and voltage losses from it are more substantial. Under these conditions, the Butler–Volmer relation, Eq. (21), reduces to Tafel kinetics, Eq. (7) [20, 24], where a Tafel slope, reaction order, and exchange current density are required. Substantial work has gone into cathode catalyst layer development and the characterization of the cathode ORR reaction. The cathode reaction follows Tafel kinetics with reaction order (with respect to oxygen concentration) of about 0.5 [60]. There has been a substantial disagreement in modeling the cathode reaction between much older and more recent works. Experimental work combined with intensive microscale modeling was able to explain how previous experiments produced kinetic data with double the theoretically correct Tafel slope due to undiagnosed ionic conduction losses in the cathode catalyst layer and unrealized gas diffusion limitations [61]. The suggestion of these and later workers is that older rotating disk electrode (RDE) kinetic experiments and the derived data are in some doubt. Wang [51] presented calculations in 2007 which showed how the oxygen concentration levels, temperature, and ionic conductivity within typical cathode catalyst layers could be considered uniform. Ionic conductivity, however, was still highly hydration dependent. Neyerlin and Gu [40] improved the ohmic loss compensation techniques to better measure in situ ORR kinetic data. Later works extended the technique to account for incomplete catalyst layer utilization [41], which occurs under high current density and/or low humidity operation when effective catalyst layer resistances become significantly large. Wang and Feng [50] expanded the analytical solutions used by Neyerlin et al. [40]. By 2009, Liu et al. [45] verified the correctness of these solutions by testing a range of electrode compositions with different humidity levels. The ionomer volume fraction of about 0.13 ($I/C > 0.6$), and relative humidity of 30% [62], were consistent with the analytical solution for cathode catalyst layer performance. These values mirror most typical catalyst layer compositions.

With recognition/compensation of appropriate catalyst layer resistance, a model of ORR kinetics could be devised that is consistent across different operating conditions. Experimentally determined ORR kinetics were found to be almost independent of humidity and electrode composition and Pt catalyst loading. The remaining loss, attributed to humidity variation in typical operational conditions, can be explained by the experimentally observed shift in ECSA from humidity alone [42]. Thus, a model of ORR kinetics was presented that is consistent with and viable across different operating conditions and catalyst layer compositions. CFD models that do not account for cathode catalyst layer resistances can still match experimental polarization data fairly well, and even while operating at low humidity conditions where catalyst layer losses should be large enough to introduce significant error. These models can match experimental *V-I* data by doubling the Tafel slope of the ORR [63] and then treating the exchange current density as a free parameter of the problem, i.e., a calibration choice could be made to match particular experiments, even with neglect of the underlying physics of the CCL catalyst loading.

Water Transport in PFSA Membranes has been an area of active investigation. Water Transport models mostly follow two very distinct modeling approaches: the hydraulic model of Bernardi and Verbrugge [64] and the "diffusive" approach of Springer and Zawodzinski of Los Alamos National Laboratory [65]. The latter approach is used by most CFD models. By 2008, the diffusive approach was the de facto model in commercial software [66]. Many works reported membrane water transport property values devised mostly ex-situ. These have had the goal of measuring water sorption, membrane conductivity, water diffusion coefficients, and electroosmotic (EO) drag coefficients following the "diffusive" model framework. The intention was that diffusive models used in CFD simulations would be more useful and accurate with appropriate transport parameters (property values).

The "diffusive" approach begins with determination of an equilibrium water uptake value. Zawodzinski et al. [67] measured equilibrium water uptake curves from membranes exposed to humidified gas and liquid water. They developed an empirical fit for water content as a function of water activity (relative humidity) based upon weight gain data for selected membranes. This was supported with data from both liquid water uptake and vapor phase uptake [68]. Hinatsu et al. [69] refined the apparatus for these measurements and reported water uptake data by Nafion with liquid through a range of temperatures, as well as the vapor-equilibrated uptake curve through 80 °C. Improved measurement data were reported by Jalani et al. [70]. Onishi et al. [71] examined water uptake in more detail, raising the possibility that prior water uptake data were dependent upon "*thermal history,*" and that many water uptake studies might have been influenced by these effects. Kusoglu et al. [72] investigated the influences of geometrical constraint and compression upon water uptake. The previous water uptake curves were mostly empirical in nature, with expressions made to fit the experimental data.

Previous investigations of membrane ionic conductivity had shown that a strong dependence on water content could be expected. Zawodzinski et al. [68] were able to measure conductivity levels which exhibited linear dependence on water content at 30 °C. Halim et al. [73] also conducted conductivity experiments, using AC impedance measurements of liquid-equilibrated membranes. They found similar activation energies for multiple membranes and that the variation of conductivity with temperature followed the Arrhenius law. Though these early relationships still appear to be in use in CFD models; later works have proposed relationships in which membrane ionic conductivity takes somewhat different forms [71, 74].

The membrane water transport model of Springer and coworkers [65, 67, 68] requires a self-diffusion coefficient of water, and an electroosmotic drag coefficient (representing the number of water molecules carried with each proton that crosses the membrane). Both are temperature and water content dependent. Attempts to measure these parameters have generated much disagreement between various groups and authors. The earliest work, from 1991 where Zawodzinski et al. [67] used pulsed-field gradient spin-echo 'H NMR measurements of 'H intradiffusion coefficients in Nafion 117 membranes (in the absence of a water concentration gradient). Measures of electroosmotic drag were also reported as being 2.5 (liquid-equilibrated membrane) and 0.9 (vapor-equilibrated membrane). Motupally et al. [75] reported permeation experiments to measure water diffusion. Their empirical correlation is used in many CFD models for the diffusion coefficient. Note that this work assumes the validity of the equilibrium assumption: water uptake data from equilibrated membranes is used to assign water content values on opposite sides of a membrane during permeation experiments. The experiment limited itself to one particular membrane thickness. A multitude of approaches has been used to estimate the electroosmotic drag, such as a hydrogen pumping cell by Ye and Wang [76]. The transport parameters, however, were subject to a great deal of disagreement based on the method used and results obtained, as indicated by Ge et al. [77]. Nevertheless, there was general agreement that the resulting water content profile in thin (50 μm) membranes was nearly linear and anode dryout was not expected in the thin membranes [76].

Membrane Water Content Imaging techniques have been refined over the years. Some of these works attempted to validate the membrane water transport models in situ. Buchi and Scherer [78] made membrane resistance measurements in an operating fuel cell with a known, thick composite membrane. Membrane resistance was found to increase significantly with current density, which was attributed to the drying out of the membrane on the anode side. A later work, by the same authors [79] used distributed local resistance measurements from evenly spaced points between the anode and the cathode. Drying out effects could clearly be seen where the membrane resistances increased at the anode side as current density increased. The effect was more pronounced with thicker membranes and less noticeable in the thinner ones. The characteristic drying out of the membrane on the anode side was observed. This was caused by the nonlinearity of the water transport in the membrane, and the variation of the diffusion coefficient with water content. However, only a qualitative agreement of diffusive models with the experimental data was achieved [80].

Further sorption/desorption (wetting and drying) experiments, performed on Nafion membranes, were used to investigate water transport. Nafion membranes had very flat water content profiles as they gained or lost water; the rate of water redistribution within the membrane was found to be much faster than the water loss.

It was also seen that the uptake of water is an order of magnitude faster in the presence of liquid water, compared to water vapor. In both liquid/vapor cases the sorption kinetics are controlled by the transfer process at the membrane surface [81, 82]. Water content equilibrium values were seen to match previously determined sorption curves via gravimetric tests [83]. Imaging was done during permeation and operational fuel cell experiments, which utilized micro-Raman spectroscopy found the presence of significant interfacial resistance, influenced by temperature and humidity. The local water content at the edges of the membrane was not in equilibrium with the water activity in the gas phase during these permeation experiments. In all cases [83, 84], equilibrium water content was slowly reestablished when water flux ceased, and linear water content profiles were observed in steady operation with membranes less than 100 μm in thickness.

Examinations of water content imaging, during permeation experiments, with X-ray microtomography by Hwang et al. [85] suggested the presence of an interfacial resistance to water transport. Operational PEMFC water content imaging measurements with the MRI technique by Tsushima et al. [86] captured the anode side depletion of water as current density increased. Later, in 2010, Tsushima et al. [87] were able to perform MRI measurements in PEMFC at higher realistic operating temperatures, and the technique (environmental MRI or EMRI) was extended beyond most of the earlier works which were limited to room-temperature measurements. This work showed linear water profiles developing as the membrane hydration levels increased. They inferred that the diffusion coefficient of water in Nafion was maximum at dimensionless water content level as low as 3–5. These results call into question the assumption of equilibrium of water content (between the gaseous phase and the membrane water content) employed with most membrane water transport models. That equilibrium assumption has been widely employed in many CFD models.

Nonequilibrium Water Transport in PFSA Membranes was introduced around 2004 by Chen et al. [88] and Berg et al. [89]. They modeled nonequilibrium membrane water uptake rates. The concept was to build a model of water transport in the membrane that was not based on equilibrium absorption from the adjoining water vapor. Instead, the rate of absorption/desorption of water into/from the membrane was driven by a gap or jump in the water content level between the actual water content and that determined by equilibrium sorption values. In 2007 Majsztrik et al. conducted a more thorough series of sorption, desorption, and permeation experiments, with the observation that diffusion was the only one of three resistances to water sorption (interfacial resistance, diffusion, and membrane swelling) to accommodate water uptake [90]. Which resistance was the controlling one was said to vary with humidity, sample thickness, and temperature. Monroe et al. [91] did another group of water permeation experiments and claimed interfacial resistance limits the overall water transfer rates when membrane thickness is less than a critical value of about 300 μm. State-of-the-art membranes are about 30–100 μm thick.

Ge et al. [92] developed mass transfer coefficients which were not constant, but rather were dependent on the local volume fraction of water at the membrane edges. The diffusion coefficient was also modeled as being dependent on the water volume fraction in the membrane (which is water content dependent). Ge et al. [77] later determined the electroosmotic drag coefficient using measurements in an operational fuel cell. They compared measured and calculated resistance from the membrane to validate the model. The agreement was good over a range of thicknesses and current densities. The membrane property values (correlations) produced were reasonably consistent with prior works. This work neglected catalyst layer effects and losses, and still required meshing and solving conservation equations throughout the membrane. This work also relied heavily upon the water uptake data of Hinatsu et al. [69], which was later called into question by Onishi et al. [71].

Adachi [93] reported experimental investigations of interfacial water transport resistance. Adachi et al. [94, 95] noted significant interfacial resistance in membranes less than 200 μm in thickness. The presence of catalyst layers, on both sides of the membrane, was found to not change water permeation significantly, and so it was suggested that membrane water permeation experiments could be correlated to those of an operating PEMFC.

Transport in diffusion layers (GDL) has been studied extensively. Gas transport in porous materials depends on their porosity, tortuosity, and permeability. The macro-homogeneous models commonly treat the diffusion layers as isotropic [11]. Two nearly equivalent approaches exist for modeling gas transport in these porous materials [9, 52]. Computational works might not use the permeability values supplied by manufacturers, however, because the properties are expected to change under GDL compression. Mench [24] also described how catalyst and microporous layers (MPL) needed to take into account both molecular and Knudsen diffusion.

Electronic conduction in the GDL is often neglected in earlier published CFD models because the commercially available carbon papers have good electrical conductivity. Meng and Wang [96, 97] observed that electron conduction could have significant effects in altering the current density distribution in the second and third dimensions of a PEMFC model. Properties of the electronically conducting phase will be needed, such as the electrical conductivity of that solid phase. It is commonly recognized that these GDL materials do not have an isotropic electronic conductivity: higher values occur in the in-plane than in the through-plane direction [11, 20].

The thermal conductivity of the diffusion media might be similarly anisotropic, though most early works tended to treat the GDL as isotropic. Attempts to measure thermal conductivity of GDL materials intensified, post 2003, when more effort was put into non-isothermal fuel cell modeling. Mench et al. [34, 35] developed an estimate of thermal conductivity of the diffusion media from in situ micro-thermocouple measurements. Soon afterward, several research groups began to use ex-situ direct measurement methods to determine the thermal conductivity of fuel cell materials, particularly the diffusion media. Measurements included thermal conductivity of the GDL media itself, along with its temperature and compression dependence; and also a contact resistance which was itself compression dependent. Khandelwal and Mench [98] found the thru-plane thermal conductivity for a common carbon paper between 1 and 2 W/m K, which decreases with temperature due to the presence of a carbonized thermosetting resin used as a binder. They found thermal conductivity affected by the presence of added PTFE in the GDL as well. They also noted the presence of a

significant thermal contact resistance which asymptotically reached its lowest limiting value with compression pressure of about 2 MPa. Hysteresis is observed in the
contact resistance due to the compression and deformation and breakage of the fibers
in the media. Models describing the thermal conductivity of porous materials have
been compared with this experimental data [99]. Another work examined thru-plane
thermal conductivity and contact resistance of GDL materials with varying compression and levels of liquid water [19]. Each changed conductivity significantly. Sadeghi
et al. [100] made additional measurements showing an air-pressure dependence. At
low gas pressures, the surface contact resistance became greater than internal conduction losses. Finally, Teertstra et al. [101] showed the thermal anisotropy of the GDL
materials, with in-plane thermal conductivity levels 10 times those in the thru-plane
direction.

4.3 Computational Fluid Dynamics Models

These works might model a single flow channel of a PEMFC, a complete fuel cell,
or even a complete stack, consisting of several cells. CFD models are typically
characterized, by time or space, as transient or steady state; combined with their scale
and dimensionality; as isothermal/non-isothermal and single-phase/multiphase in the
treatment of the presence of liquid water. 1-D models consider the direction through
the MEA exclusively. These make up the majority of early modeling works. 2-D
models additionally consider the direction downstream, or down the flow channel,
and 3-D models further consider effects in the cross-flow direction. Some works
have varying dimensionality by region. Figure 4 shows a typical orientation of the
coordinate systems. 1-D models are in the y-direction. 2-D models follow either the
x-y (in-plane or sandwich) or y-z (along-the-channel) orientations. 3-D models use
the full x-y-z coordinate directions.

Most CFD models employ the continuum assumption, so that exact details of
material microstructure are neglected. Important sub-regions of the fuel cell such
as diffusion layers, membrane, and catalyst layers are handled this way. Diffusion
layers are commonly considered as randomly oriented porous structures that are
defined by a porosity and permeability. In examining catalyst layers, a porosity and
surface area per unit volume might be considered. CFD codes typically use the finite-
volume, finite-difference or finite-element method for solving flow and heat transfer
equations in both single- and two-phase flow. Electrochemical effects are introduced
via add-on modules available for many commercial software packages.

Isothermal models are those that do not consider the presence of temperature
gradients within the fuel cell, and thus, do not account for heat transfer. Isothermal
models can consider temperature effects, but simply treat temperature as an input to
the problem. This assumption is commonly used in early CFD models. Discrepancies
in the treatment of liquid water are also common. The simplest approach is the single-
phase one, where the total water amount is considered without regard to whether it
is in vapor or liquid form. The gas and liquid phases share the same velocity as they

are in the same fluid mixture. Two-phase models consider the treatment of liquid water presence by predicting a liquid saturation. Saturation values greater than zero characterize flooding and reduce the diffusion coefficients in the gas phase. This has been the focus of a tremendous amount of effort in using CFD to model the PEMFC and is considered critical [102]. However, single-phase models are generally considered valid as long as the saturation value of liquid within the GDL is kept small. Thus they are assumed to be appropriate for low-current density and low humidity operations where liquid water production rate is low and is not expected to accumulate [11].

1-D CFD models were developed for PEMFC simulation in the early 1990s by researchers at Los Alamos National Laboratory [65, 68] and at General Motors [64, 103]. They established the methodology for subsequent modeling efforts, in particular for isothermal conditions. With the adjustment of a couple of parameters, they succeeded in predicting polarization curves in agreement with experimental data. They also predicted low levels of catalyst utilization in the CL [103]. Subsequent models [74, 104–107] expanded on the water treatment and thermal management, accounting for two-phase flow and non-isothermal treatment with heat generation and transport. These models gave insight into the average performance of a fuel cell, but are lacking in simulating distributed properties as observed in real fuel cells. Such variations within a cell are important for the optimization of performance of practical fuel cells with large surface areas. On the other hand, 1-D models require manageable number of grid points (~1000), and can produce comparable results of overall performance of PEMFC as 3-D models [108], which require millions of grid points for resolution. Figure 6 shows very similar predictions of polarization curves between the 1-D and 3-D single-phase models. Improved prediction is achieved, not by increased dimensionality resolution, but solely by going to a two-phase simulation model. The degree of calibration and the number of empirical parameters used varied widely from model to model.

2-D CFD models of PEMFC were developed in the late 1990s. They represent the best compromise between 1-D and 3-D models, in terms of computational details and cost [12]. Number of grid points required for resolution increases from the order of thousands to tens to hundreds of thousands. Gurau et al. [109] presented an along-the-channel (y-z), 2-D model of the flow channels and the MEA, which allowed for computation of variation in reactants, hydrogen and oxygen, in the flow channels as the reaction progresses. The model uses the SIMPLE algorithm and a staged solution process. Similar models were developed by Yi and Nguyen [110] and Um et al. [111]. Siegel et al. [112] presented a 2-D finite-element, single-phase model which studied effectiveness of the catalyst layer. It was subsequently extended to study two-phase flow to account for flooding from water production at the cathode [113]. Computational efficiency of 2-D models was utilized for PEMFC optimization studies [12, 114]. These studies combined the CFD capabilities of COMSOL software FEMLAB with the optimization utilities of MATLAB/Simulink.

2-D sandwich (x-y) orientation models were presented in [115–118], which used a variety of commercial CFD software. Sui and Djilali [115] used the CFD-ACE+ software to investigate the coupling between electronic and mass transport in the

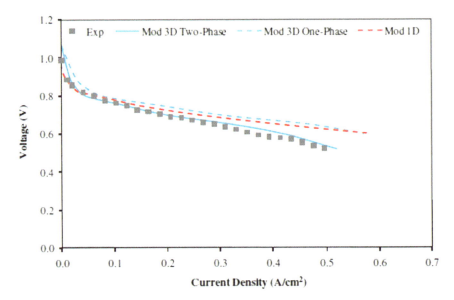

Fig. 6 Comparison of predicted polarization curves for 1-D and 3-D models to experimental data [108]

MEA. They found that either process could be dominant, depending on operating conditions. Losses due to gas crossover effects were investigated by Seddiq et al. [116]. Lin et al. [117] used an existing solver along with an optimization technique, based on the simplified conjugate gradient method to study how the performance of PEMFC is influenced by design parameters such as MEA porosity and flow channel geometry. Meng [118] used FLUENT to solve the transient, non-isothermal, two-phase problem in a PEMFC undergoing a step change in cell voltage, and found that the heat transfer significantly increased the transient response time.

Beale et al. [119] used the PHOENICS [120] code to calculate two-phase flow and mass transfer within the cathode GDL of a PEMFC. A modified version of the interphase slip algorithm (IPSA) [121] was utilized. In this formulation, the liquid and gas flows have different pressures which are determined by solving corresponding Darcy's laws within the GDL.

3-D CFD models of the PEMFC are very computationally intensive, and so are mostly based on commercial codes, with electrochemistry modules built as add-ons. Gridpoints required for adequate resolution of a single cell, in the x-y-z directions shown in Fig. 4, are in the range of hundreds of thousands to several million [17], especially when different arrangements of flow channels have to be simulated. Pourmahmoud et al. [122] detailed the mesh requirements for the single-domain approach with a commercial CFD code. This approach meshes all regions including the MEA (thin membrane and catalyst layers), requiring a large number of cells with high-skew ratio. Altogether, about 60 cells minimum must be used in the MEA through-plane direction, with about 20 grid points to cover the channel width and

~100 points in the flow direction. There are about 20–60 individual channels in the complete flow field, yielding about 6 million points to model even a single small fuel cell [123]. The MEA cells have a very high-aspect ratio, which negatively impacts the solution convergence [66]. Wu et al. investigated MEA meshing requirements as well [124]. An overly fine computational mesh (used in the through-plane direction) caused the solution to become unstable. The number of nodes used in the in-plane directions had much less effect on stability. Figure 7 shows a picture of the typical regions that need to be meshed in a complete model of a single cell [125].

Zhou and Liu expanded the earlier two-dimensional work of Gurau et al. [109] to three dimensions [126], which was later extended with a multiphase mixture formulation by You and Liu [127, 128]. Shimpalee and coworkers, presented 3-D interface models [9, 129], based on the commercial CFD package FLUENT, with added user-coded modules. A later work added 2-phase water treatment [130]. Membrane and catalyst layers, the MEA, was not meshed in these works but was treated as an interface, with zero thickness, separating the anode and cathode flow-fields. Water transport and ohmic potential drop across the MEA were treated with simplifying linear approximations. Calculations then established the water flux and ohmic drop between the anode and cathode and created source terms on both sides of the interface to produce (i) reactant consumption/product creation, (ii) water flux, and (iii) ohmic loss. Sui and Djilali [131] compared computed results of water flux from a single-domain model to those from the interface model of Shimpalee and coworkers. They found significant discrepancies in the presence of large gradients in water content. Errors were said to arise from the diffusion term, and were greater in the thinner membranes than in thicker ones; (diffusion is stronger relative to electro-osmotic drag in the thinner membranes). Mazumder [132] showed that a linear water content profile was appropriate for diffusion dominated drag effects in the membrane; i.e., for thin membranes. Edwards [10] has formulated an improved interface model which uses corrected formulations of the HOR and ORR kinetics and water transport within the MEA.

Um and Wang [133] presented a single-domain approach that utilizes a single set of governing equations in all sub-regions of the fuel cell, avoiding the use of an interface model. The initial model was isothermal and single-phase. Later, Um and Wang [134] added a detailed MEA submodel where the water content distribution within the membrane was resolved. Spatial variations in ionic resistance and reaction rate within the catalyst layer were observed. Wang and Wang [135] expanded the model with the addition of a variable flow model, with mass source terms in the continuity equation and variable gas density in the momentum equations. Mass consumption terms impacted the anode flow field by reducing the pressure drop in a serpentine flow channel.

Computational models need to be properly validated, but experimental data is mostly lacking in details. Recently, distributed current density data, measured in a 10×10 grid in a PEMFC as was used to validate the 3-D model of Wang and coworkers [136]. Results show significant variation in current density between gas inlet and outlet regions. Deviations between data and simulation were generally within $\pm 20\%$. Figure 8 shows comparison of cell polarization curves and deviations

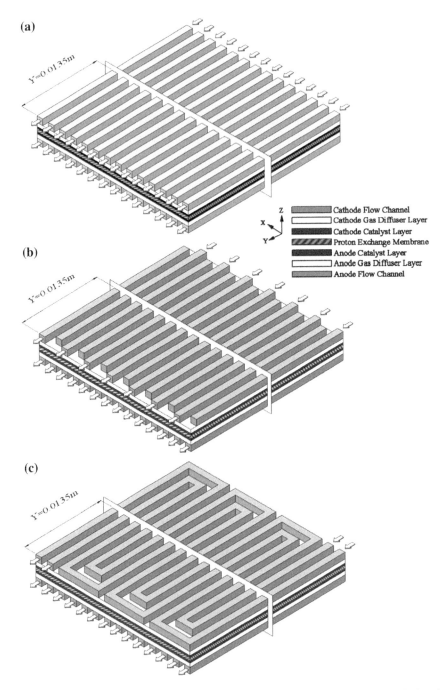

Fig. 7 Single fuel cell with different flow channel arrangements **a** parallel, **b** interdigitated, **c** serpentine [125]

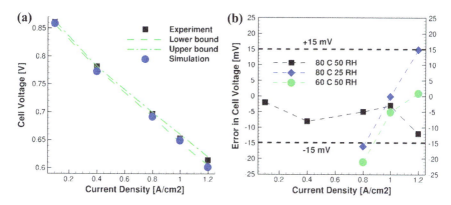

Fig. 8 Comparison of distributed (10×10) cell test data with simulation; **a** polarization, **b** error in cell voltage [136]

in cell voltages, three different test conditions, varying temperature, and relative humidity.

Figure 9 shows the distribution of the relative errors in local values of current density, between simulation and measured data, at two average current densities. Largest errors are found near the gas inlet (top), and the gas outlet (bottom). Other 3-D models will benefit from this type of validation effort.

Non-Isothermal models consider heat transport and solve an additional conservation of energy equation. Thermal management is intertwined with water management through temperature effects, whereby the membrane water content depends on the relative humidity of gases, and the saturation pressure of the gases is an exponential function of temperature. Bvumbe et al. [137] considered different modes of heat generation in the MEA. Berning et al. studied thermal management and how different heating terms affected various loss mechanisms [138]. The catalyst layers contain

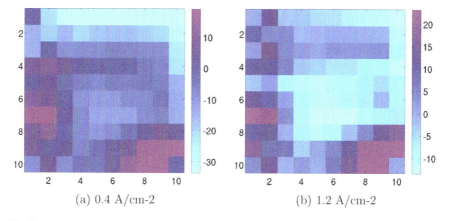

Fig. 9 Relative errors in local values of current density at 80 °C and 50% RH [136]

reversible and irreversible heat terms, and joule heating (ohmic heating) which must be considered. Temperature distributions will depend on the thermal boundary conditions of the model, as well as the thermal properties (i.e. thermal conductivities) of various materials such as the diffusion layer and bipolar plates. The temperature distribution affects water management. Higher temperatures encourage dryout, especially at the anode, with attendant increase in protonic resistance. But at high current densities, water production at the cathode may lead to flooding, thereby impeding adequate supply of oxygen to the cathode. The model of Mazumder and Cole [139, 140] omitted diffusion or electro-osmotic drag in the membrane phase, and also water transport. Models which assume a constant membrane hydration [141] essentially ignore the transport of water and thus predict constant ohmic loss in the PEMFC membrane. They fail to reproduce experimentally observed reduction in membrane ohmic resistance (water gain) at low humidity and high current operation. Likewise, the increase in membrane ohmic resistance with high current density, under high humidity conditions. Some CFD models such as Schwarz and Beale [142] focus on multiphase effects by modeling diffusion in microporous layers, without the difficulties of modeling in the MEA layer. Others have proposed a multiscale approach in which first a CFD model is used for the largest scales of the problem, the macroscopic scales, where problems of heat and mass transfer are solved. But microstructural effects such as pore-filling are used to model water transport throughout the fine, small-scale structure of catalyst layers [143]. The effects of very fine catalyst layer structure (of too small a scale to be resolved by the macroscopic CFD model) on PEMFC performance are handled by mesoscale models.

Commercial CFD software have been used to solve the PEMFC problem in one-two- or three-dimensions. FLUENT appears to be the most popular, with its available fuel cell module or subroutines. It has been used for both 2-D and 3-D simulations. COMSOL Multiphysics (FEMLAB) has also been widely used for 1-D, 2-D, and 3-D simulations. STAR-CD has been widely used mostly for 3-D simulations. To a lesser extent, other commercial software, CFD-ACE+, CFX and PHOENICS, have also been used. In most cases, the SIMPLE (or SIMPLER, SIMPLEC) algorithm is deployed, though in some cases the PISO algorithm is used for its superior coupling quality. One way or the other, the guiding hand of D. Brian Spalding is seen in the development of CFD modeling of PEM fuel cells [144].

5   Summary and Conclusions

PEMFC Modeling approaches have evolved from the early works of Springer and colleagues at Los Alamos National Labs. These treat water transport with the single-phase approach, where water flux is governed by electroosmotic drag and diffusion. Modeling the behavior of the PEMFC with computational fluid dynamics requires understanding of the coupled physics of flow, heat, water, and charge transport in the MEA and diffusion layers. Different approaches have been used to analyze thermal, water, and charge transport, however, there is some convergence. Several commercially available software packages have incorporated the appropriate physics in add-on modules or subroutines, with flexibility for user modifications.

New developments in in-situ fuel cell experiments, and advanced imaging techniques with in-situ and ex-situ experiments, have enhanced the calibration and validation of the MEA water transport models used in many CFD models. Water management is critical to effective operation of PEMFC. The MEA needs to be hydrated for protonic conductivity, but excessive water produces flooding which blocks supply of reactant gases to reaction sites in the catalyst layers.

There are two main methodological approaches; one- or two-domain. The former meshes and solves the governing differential equations in all parts of the fuel cell, including the electrolyte membrane and catalyst layers, the MEA. Because the catalyst layers are typically 1–2 orders of magnitude thinner than the gas channels, this approach entails adding large numbers of thin, high-aspect ratio cells to the problem, with negative consequences for solution stability and convergence. The other method is the interface approach which does not mesh the MEA, but treats it as an interface which separates the anode and cathode regions. Properties must be prescribed within the interphase, based on empirical relations and analytical solutions, to be used as boundary conditions for the regions. The reliance on experimental data suggests that interface model may need continuous updating as better data become available.

CFD models are now capable of fairly accurate simulation of a single PEMFC in steady or transient operation. Stacks of a few cells can also be simulated with the aid of parallel computing. However, large stacks which contain tens to hundreds of cells are still beyond the resolution ability of most CFD software. These have to be solved in segments, with some clever way devised to patch them together.

References

1. Spalding, D. B. (1951). The combustion of liquid fuels. In *Engineering*. Cambridge, U.K.: Cambridge University Press.
2. Spalding, D. B. (1953). The combustion of liquid fuels. *Symposium (International) on Combustion, 4*(1), 847–864.
3. Spalding, D. B. (1955). Some fundamentals of combustion. In *Gas turbine series* (Vol. 2, 250 pp.). Butterworth's Scientific Publications.
4. Spalding, D. B. (1978). A general theory of turbulent combustion. *Journal of Energy, 2*(1), 16–23.
5. Spalding, D. B. (1979). *Combustion and mass transfer: A textbook with multiple-choice exercises for engineering students* (1st ed., p. vii, 409 pp.). Oxford; New York: Pergamon International Library of Science, Technology, Engineering, and Social Studies, Pergamon Press.
6. Dicks, A., & Rand, D. (2018). *Fuel cell systems explained* (3rd ed.). West Sussex: Wiley.
7. Stone, R. (2003). Competing technologies for transportation. In G. Hoogers (Ed.), *Fuel cell technology handbook*. Boca Raton, FL: CRC Press.
8. Wang, Y., et al. (2011). A review of polymer electrolyte membrane fuel cells: Technology, applications, and needs on fundamental research. *Applied Energy, 88*(4), 981–1007.

9. Dutta, S., Shimpalee, S., & Van Zee, J. W. (2001). Numerical prediction of mass-exchange between cathode and anode channels in a PEM fuel cell. *International Journal of Heat and Mass Transfer, 44,* 2029–2042.

10. Edwards, R. L., & Demuren, A. (2018). Interface model of PEM fuel cell membrane steady-state behavior. *International Journal of Energy and Environmental Engineering.*

11. Wang, C.-Y. (2004). Fundamental models for fuel cell engineering. *Chemical Reviews, 104*(10), 4727–4766.

12. Guvelioglu, G. H., & Stenger, H. G. (2005). Computational fluid dynamics modeling of polymer electrolyte membrane fuel cells. *Journal of Power Sources, 147*(1–2), 95–106.

13. Bednarek, T., & Tsotridis, G. (2017). Issues associated with modelling of proton exchange membrane fuel cell by computational fluid dynamics. *Journal of Power Sources, 343,* 550–563.

14. Hoogers, G. (2003). Fuel cell components and their impact on performance. In G. Hoogers (Ed.), *Fuel cell technology handbook.* Boca Raton, FL: CRC Press.

15. Hoogers, G. (2003). Automotive applications. In G. Hoogers (Ed.), *Fuel cell technology handbook.* Boca Raton, FL: CRC Press.

16. Macedo-Valencia, J., et al. (2016). 3D CFD modeling of a PEM fuel cell stack. *International Journal of Hydrogen Energy, 41.*

17. Siegel, C. (2008). Review of computational heat and mass transfer modeling in polymer-electrolyte-membrane (PEM) fuel cells. *Energy, 33,* 1331–1352.

18. Pharoah, J. G., & Burheim, O. S. (2010). On the temperature distribution in polymer electrolyte fuel cells. *Journal of Power Sources, 195*(16), 5235–5245.

19. Burheim, O., et al. (2010). Ex situ measurements of through-plane thermal conductivities in a polymer electrolyte fuel cell. *Journal of Power Sources, 195*(1), 249–256.

20. Barbir, F. (2013). *PEM fuel cells: Theory and practice.* Academic Press.

21. Wu, H.-W. (2016). A review of recent development: Transport and performance modeling of PEM fuel cells. *Applied Energy, 165,* 81–106.

22. Patankar, S. V., & Spalding, D. B. (1972). A calculation procedure for heat, mass and momentum transfer in three-dimensional parabolic flows. *International Journal of Heat and Mass Transfer, 15,* 1787–1806.

23. Wilberforce, T., et al. (2017). Modelling and simulation of proton exchange membrane fuel cell with serpentine bipolar plate using MATLAB. *International Journal of Hydrogen Energy.*

24. Mench, M. (2008). *Fuel cell engines.* Wiley.

25. Wood, D. L., Yi, J. S., & Nguyen, T. V. (1998). Effect of direct liquid water injection and interdigitated flow field on the performance of proton exchange membrane fuel cells. *Electrochimica Acta, 43*(24), 3795–3809.

26. Laurencelle, F., et al. (2001). Characterization of a Ballard MK5-E proton exchange membrane fuel cell stack. *Fuel Cells, 1*(1), 66–71.

27. Cooper, K. R., et al. (2007). *Experimental methods and data analyses for polymer electrolyte fuel cells.* Scribner Associates, Inc.

28. Ju, H., & Wang, C.-Y. (2004). Experimental validation of a PEM fuel cell model by current distribution data. *Journal of the Electrochemical Society, 151*(11), A1954–A1960.

29. Mench, M. M., Wang, C. Y., & Ishikawa, M. (2003). In situ current distribution measurements in polymer electrolyte fuel cells. *Journal of the Electrochemical Society, 150*(8), A1052–A1059.

30. Wieser, C., Helmbold, A., & Gülzow, E. (2000). A new technique for two-dimensional current distribution measurements in electrochemical cells. *Journal of Applied Electrochemistry, 30*(7), 803–807.

31. Geiger, A. B., et al. (2004). An approach to measuring locally resolved currents in polymer electrolyte fuel cells. *Journal of the Electrochemical Society, 151*(3), A394–A398.

32. Natarajan, D., & Van Nguyen, T. (2005). Current distribution in PEM fuel cells. Part 1: Oxygen and fuel flow rate effects. *AIChE Journal, 51*(9), 2587–2598.

33. Vie, P. J. S., & Kjelstrup, S. (2004). Thermal conductivities from temperature profiles in the polymer electrolyte fuel cell. *Electrochimica Acta, 49*(7), 1069–1077.

34. Mench, M., Burford, D. J., & Davis, T. W. (2003). In situ temperature distribution measurements in an operating polymer electrolyte fuel cell, Paper No. 42393. In *Proceedings of the 2003 International Mechanical Engineering conference and Exposition (IMECE)*. Washington, DC: ASME.

35. Burford, D. J., Davis, T. W., & Mench, M. M. (2004). Heat transport and temperature distribution in PEFCs, IMECEC 2004-59497. In *Proceedings of the 2004 International Mechanical Engineering conference and Exposition (IMECE)*, Anaheim, CA.

36. Mench, M. M., Dong, Q. L., & Wang, C. Y. (2003). In situ water distribution measurements in a polymer electrolyte fuel cell. *Journal of Power Sources, 124*(1), 90–98.

37. Dong, Q., Kull, J., & Mench, M. M. (2005). Real-time water distribution in a polymer electrolyte fuel cell. *Journal of Power Sources, 139*(1–2), 106–114.

38. Edwards, R. L., & Demuren, A. (2016). Regression analysis of PEM fuel cell transient response. *International Journal of Energy and Environmental Engineering, 7*(3), 329–341.

39. Stevens, D. A., & Dahn, J. R. (2003). Electrochemical characterization of the active surface in carbon-supported platinum electrocatalysts for PEM fuel cells. *Journal of the Electrochemical Society, 150*(6), A770–A775.

40. Neyerlin, K. C., et al. (2006). Determination of catalyst unique parameters for the oxygen reduction reaction in a PEMFC. *Journal of the Electrochemical Society, 153*(10), A1955–A1963.

41. Neyerlin, K. C., et al. (2007). Cathode catalyst utilization for the ORR in a PEMFC. *Journal of the Electrochemical Society, 154*(2), B279–B287.

42. Soboleva, T., et al. (2011). PEMFC catalyst layers: The role of micropores and mesopores on water sorption and fuel cell activity. *ACS Applied Materials & Interfaces, 3*(6), 1827–1837.

43. Cooper, K. R., & Smith, M. (2006). Electrical test methods for on-line fuel cell ohmic resistance measurement. *Journal of Power Sources, 160*(2), 1088–1095.

44. Neyerlin, K. C., et al. (2007). Study of the exchange current density for the hydrogen oxidation and evolution reactions. *Journal of the Electrochemical Society, 154*(7), B631–B635.

45. Liu, Y., et al. (2009). Proton conduction and oxygen reduction kinetics in PEM fuel cell cathodes: Effects of ionomer-to-carbon ratio and relative humidity. *Journal of the Electrochemical Society, 156*(8), B970–B980.

46. Kulikovsky, A. A. (2010). Introduction. In A. A. Kulikovsky (Ed.), *Analytical modelling of fuel cells* (pp. xiii–xv). Amsterdam: Elsevier.

47. Uchida, M., et al. (1995). Investigation of the microstructure in the catalyst layer and effects of both perfluorosulfonate ionomer and PTFE-loaded carbon on the catalyst layer of polymer electrolyte fuel cells. *Journal of the Electrochemical Society, 142*(12), 4143–4149.

48. Uchida, M., et al. (1996). Effects of microstructure of carbon support in the catalyst layer on the performance of polymer-electrolyte fuel cells. *Journal of the Electrochemical Society, 143*(7), 2245–2252.

49. Xie, J., et al. (2004). Ionomer segregation in composite MEAs and its effect on polymer electrolyte fuel cell performance. *Journal of the Electrochemical Society, 151*(7), A1084–A1093.

50. Wang, Y., & Feng, X. (2008). Analysis of reaction rates in the cathode electrode of polymer electrolyte fuel cell I. Single-layer electrodes. *Journal of the Electrochemical Society, 155*(12), B1289–B1295.

51. Wang, Y. (2007). Analysis of the key parameters in the cold start of polymer electrolyte fuel cells. *Journal of the Electrochemical Society, 154*(10), B1041–B1048.

52. Stumper, J., Haas, H., & Granados, A. (2005). In situ determination of MEA resistance and electrode diffusivity of a fuel cell. *Journal of the Electrochemical Society, 152*(4), A837–A844.

53. Yu, Z., & Carter, R. N. (2010). Measurement of effective oxygen diffusivity in electrodes for proton exchange membrane fuel cells. *Journal of Power Sources, 195*(4), 1079–1084.

54. Shen, J., et al. (2011). Measurement of effective gas diffusion coefficients of catalyst layers of PEM fuel cells with a Loschmidt diffusion cell. *Journal of Power Sources, 196*(2), 674–678.

55. Yu, Z., Carter, R. N., & Zhang, J. (2012). Measurements of pore size distribution, porosity, effective oxygen diffusivity, and tortuosity of PEM fuel cell electrodes. *Fuel Cells, 12*(4), 557–565.

56. Lange, K. J., Sui, P.-C., & Djilali, N. (2012). Determination of effective transport properties in a PEMFC catalyst layer using different reconstruction algorithms. *Journal of Power Sources, 208*, 354–365.

57. Singh, R., et al. (2014). Dual-beam FIB/SEM characterization, statistical reconstruction, and pore scale modeling of a PEMFC catalyst layer. *Journal of the Electrochemical Society, 161*(4), F415–F424.

58. Durst, J., et al. (2015). Hydrogen oxidation and evolution reaction kinetics on carbon supported Pt, Ir, Rh, and Pd electrocatalysts in acidic media. *Journal of the Electrochemical Society, 162*(1), F190–F203.

59. Gasteiger, H.A., J.E. Panels, and S.G. Yan. (2004). Dependence of PEM fuel cell performance on catalyst loading. *Journal of Power Sources, 127*(1–2), 162–171.

60. Kornyshev, A. A., & Kulikovsky, A. A. (2001). Characteristic length of fuel and oxygen consumption in feed channels of polymer electrolyte fuel cells. *Electrochimica Acta, 46*(28), 4389–4395.

61. Jaouen, F., Lindbergh, G., & Sundholm, G. (2002). Investigation of mass-transport limitations in the solid polymer fuel cell cathode. *Journal of the Electrochemical Society, 149*(4), A437–A447.

62. Neyerlin, K. C., et al. (2005). Effect of relative humidity on oxygen reduction kinetics in a PEMFC. *Journal of the Electrochemical Society, 152*(6), A1073–A1080.

63. Shimpalee, S., et al. (2009). Experimental and numerical studies of portable PEMFC stack. *Electrochimica Acta, 54*(10), 2899–2911.

64. Bernardi, D. M., & Verbrugge, M. W. (1991). Mathematical model of a gas diffusion electrode bonded to a polymer electrolyte. *AIChE Journal, 37*(8), 1151–1163.

65. Springer, T. E., Zawodzinski, T. A., & Gottesfeld, S. (1991). Polymer electrolyte fuel cell model. *Journal of the Electrochemical Society, 138*(8), 2334–2342.

66. Kamarajugadda, S., & Mazumder, S. (2008). On the implementation of membrane models in computational fluid dynamics calculations of polymer electrolyte membrane fuel cells. *Computers & Chemical Engineering, 32*(7), 1650–1660.

67. Zawodzinski, T. A., et al. (1991). Determination of water diffusion coefficients in perfluoro-sulfonate ionomeric membranes. *The Journal of Physical Chemistry, 95*(15), 6040–6044.

68. Zawodzinski, J. T. A., et al. (1993). Water uptake by and transport through Nafion[sup [registered sign]] 117 membranes. *Journal of the Electrochemical Society, 140*(4), 1041–1047.

69. Hinatsu, J. T., Mizuhata, M., & Takenaka, H. (1994). Water uptake of perfluorosulfonic acid membranes from liquid water and water vapor. *Journal of the Electrochemical Society, 141*(6), 1493–1498.

70. Jalani, N. H., Choi, P., & Datta, R. (2005). TEOM: A novel technique for investigating sorption in proton-exchange membranes. *Journal of Membrane Science, 254*(1–2), 31–38.

71. Onishi, L. M., Prausnitz, J. M., & Newman, J. (2007). Water−Nafion equilibria. Absence of Schroeder's paradox. *The Journal of Physical Chemistry B, 111*(34), 10166–10173.

72. Kusoglu, A., Kienitz, B. L., & Weber, A. Z. (2011). Understanding the effects of compression and constraints on water uptake of fuel-cell membranes. *Journal of the Electrochemical Society, 158*(12), B1504–B1514.

73. Halim, J., et al. (1994). Characterization of perfluorosulfonic acid membranes by conductivity measurements and small-angle X-ray scattering. *Electrochimica Acta, 39*(8–9), 1303–1307.

74. Weber, A. Z., & Newman, J. (2004). Transport in polymer-electrolyte membranes: II. Mathematical model. *Journal of the Electrochemical Society, 151*(2), A311–A325.

75. Motupally, S., Becker, A. J., & Weidner, J. W. (2000). Diffusion of water in Nafion 115 membranes. *Journal of the Electrochemical Society, 147*(9), 3171–3177.

76. Ye, X., & Wang, C.-Y. (2007). Measurement of water transport properties through membrane-electrode assemblies. *Journal of the Electrochemical Society, 154*(7), B676–B682.

77. Ge, S., Yi, B., & Ming, P. (2006). Experimental determination of electro-osmotic drag coefficient in Nafion membrane for fuel cells. *Journal of the Electrochemical Society, 153*(8), A1443–A1450.

78. Büchi, F. N., & Scherer, G. G. (1996). In-situ resistance measurements of Nafion® 117 membranes in polymer electrolyte fuel cells. *Journal of Electroanalytical Chemistry, 404*(1), 37–43.

79. Buchi, F. N., & Scherer, G. G. (2001). Investigation of the transversal water profile in Nafion membranes in polymer electrolyte fuel cells. *Journal of the Electrochemical Society, 148*(Copyright 2001, IEE), 183–188.

80. Kulikovsky, A. A. (2003). Quasi-3D modeling of water transport in polymer electrolyte fuel cells. *Journal of the Electrochemical Society, 150*(11), A1432–A1439.

81. Zhang, Z., et al. (2008). Spatial and temporal mapping of water content across Nafion membranes under wetting and drying conditions. *Journal of Magnetic Resonance, 194*(2), 245–253.

82. Gebel, G., et al. (2011). The kinetics of water sorption in Nafion membranes: A small-angle neutron scattering study. *Journal of Physics: Condensed Matter, 23*(23), 234107.

83. Tabuchi, Y., et al. (2011). Analysis of in situ water transport in Nafion® by confocal micro-Raman spectroscopy. *Journal of Power Sources, 196*(2), 652–658.

84. Hara, M., et al. (2011). Temperature dependence of the water distribution inside a Nafion membrane in an operating polymer electrolyte fuel cell. A micro-Raman study. *Electrochimica Acta, 58*, 449–455.

85. Hwang, G. S., et al. (2013). Understanding water uptake and transport in Nafion using X-ray microtomography. *ACS Macro Letters, 2*(4), 288–291.

86. Tsushima, S., Teranishi, K., & Hirai, S. (2004). Magnetic resonance imaging of the water distribution within a polymer electrolyte membrane in fuel cells. *Electrochemical and Solid-State Letters, 7*(9), A269–A272.

87. Tsushima, S., et al. (2010). Investigation of water distribution in a membrane in an operating PEMFC by environmental MRI. *Journal of the Electrochemical Society, 157*(12), B1814–B1818.

88. Chen, F., et al. (2004). Transient behavior of water transport in the membrane of a PEM fuel cell. *Journal of Electroanalytical Chemistry, 566*(1), 85–93.

89. Berg, P., et al. (2004). Water management in PEM fuel cells. *Journal of the Electrochemical Society, 151*(3), A341–A353.

90. Majsztrik, P. W., et al. (2007). Water sorption, desorption and transport in Nafion membranes. *Journal of Membrane Science, 301*(1–2), 93–106.

91. Monroe, C. W., et al. (2008). A vaporization-exchange model for water sorption and flux in Nafion. *Journal of Membrane Science, 324*(1–2), 1–6.

92. Ge, S., et al. (2005). Absorption, desorption, and transport of water in polymer electrolyte membranes for fuel cells. *Journal of the Electrochemical Society, 152*(6), A1149–A1157.

93. Adachi, M. (2010). *Proton exchange membrane fuel cells: Water permeation through Nafion (R) membranes*. Department of Chemistry-Simon Fraser University.

94. Adachi, M., et al. (2010). Thickness dependence of water permeation through proton exchange membranes. *Journal of Membrane Science, 364*(1–2), 183–193.

95. Adachi, M., et al. (2010). Water permeation through catalyst-coated membranes. *Electrochemical and Solid-State Letters, 13*(6), B51–B54.

96. Meng, H., & Wang, C.-Y. (2004). Electron transport in PEFCs. *Journal of the Electrochemical Society, 151*(3), A358–A367.

97. Meng, H., & Wang, C. Y. (2005). Multidimensional modelling of polymer electrolyte fuel cells under a current density boundary condition. *Fuel Cells, 5*(4), 455–462.

98. Khandelwal, M., & Mench, M. M. (2006). Direct measurement of through-plane thermal conductivity and contact resistance in fuel cell materials. *Journal of Power Sources, 161*(2), 1106–1115.

99. Ramousse, J., et al. (2008). Estimation of the effective thermal conductivity of carbon felts used as PEMFC gas diffusion layers. *International Journal of Thermal Sciences, 47*(1), 1–6.

100. Sadeghi, E., Djilali, N., & Bahrami, M. (2011). Effective thermal conductivity and thermal contact resistance of gas diffusion layers in proton exchange membrane fuel cells. Part 1: Effect of compressive load. *Journal of Power Sources, 196*(1), 246–254.

101. Teertstra, P., Karimi, G., & Li, X. (2011). Measurement of in-plane effective thermal conductivity in PEM fuel cell diffusion media. *Electrochimica Acta, 56*(3), 1670–1675.
102. Ji, M., & Wei, Z. (2009). A review of water management in polymer electrolyte membrane fuel cells. *Energies, 2*(4), 1057–1106.
103. Bernardi, D. M., & Verbrugge, M. (1992). A mathematical model of the solid polymer electrolyte fuel cell. *Journal of the Electrochemical Society, 139.*
104. Baschuk, J. J., & Li, X. (2000). Modelling of polymer electrolyte membrane fuel cells with variable degrees of water flooding. *Journal of Power Sources, 86*(1), 181–196.
105. Djilali, N., & Lu, D. (2002). Influence of heat transfer on gas and water transport in fuel cells. *International Journal of Thermal Sciences, 41*(1), 29–40.
106. Yan. W.-M., et al. (2004). *Analysis of thermal and water management with temperature-dependent diffusion effects in membrane of proton exchange membrane fuel cells* (Vol. 129, pp. 127–137).
107. Song, D., et al. (2006). Transient analysis for the cathode gas diffusion layer of PEM fuel cells. *Journal of Power Sources, 159*(2), 928–942.
108. Falcão, D. S., et al. (2011). 1D and 3D numerical simulations in PEM fuel cells. *International Journal of Hydrogen Energy, 36.*
109. Gurau, V., Liu, H., & Kakaç, S. (1998). Two-dimensional model for proton exchange membrane fuel cells. *AIChE Journal, 44*(11), 2410–2422.
110. Yi, J., & Nguyen, T. (1998). An along-the-channel model for proton exchange membrane fuel cells. *Journal of the Electrochemical Society, 145,* 1149–1159.
111. Um, S., Wang, C. Y., & Chen, K. S. (2000). Computational fluid dynamics modeling of proton exchange membrane fuel cells. *Journal of the Electrochemical Society, 147*(12), 4485–4493.
112. Siegel, N., et al. (2003). Single domain PEMFC model based on agglomerate catalyst geometry. *Journal of Power Sources, 115,* 81–89.
113. Siegel, N., et al. (2004). A two-dimensional computational model of a PEMFC with liquid water transport. *Journal of Power Sources, 128,* 173–184.
114. Grujicic, M., & Chittajallu, K. M. (2004). Design and optimization of polymer electrolyte membrane (PEM) fuel cells. *Applied Surface Science, 227*(1–4), 56–72.
115. Sui, P. C., & Djilali, N. (2006). Analysis of coupled electron and mass transport in the gas diffusion layer of a PEM fuel cell. *Journal of Power Sources, 161*(1), 294–300.
116. Seddiq, M., Khaleghi, H., & Mirzaei, M. (2006). Numerical analysis of gas cross-over through the membrane in a proton exchange membrane fuel cell. *Journal of Power Sources, 161*(1), 371–379.
117. Lin, H.-H., et al. (2006). Optimization of key parameters in the proton exchange membrane fuel cell. *Journal of Power Sources, 162,* 246–254.
118. Meng, H. (2007). Numerical investigation of transient responses of a PEM fuel cell using a two-phase non-isothermal mixed-domain model. *Journal of Power Sources, 171*(2), 738–746.
119. Beale, S., et al. (2009). Two-phase flow and mass transfer within the diffusion layer of a polymer electrolyte membrane fuel cell. *Computational Thermal Sciences, 1.*
120. Spalding, D. B. (1984). *PHOENICS 84: A multi-dimensional multi-phase general-purpose computer simulator for fluid flow, heat transfer and combustion: 36 lecture panels.*
121. Spalding, D. B. (1981). Numerical computation of multiphase fluid flow and heat transfer. In C. Taylor & K. Morgan (Eds.), *Recent advances in numerical methods in fluids* (pp. 161–191). Swansea: Pineridge Press.
122. Pourmahmoud, N., et al. (2011). Three-dimensional numerical analysis of proton exchange membrane fuel cell. *Journal of Mechanical Science and Technology, 25*(10), 2665–2673.
123. Meng, H., & Wang, C.-Y. (2004). Large-scale simulation of polymer electrolyte fuel cells by parallel computing. *Chemical Engineering Science, 59*(16), 3331–3343.
124. Wu, H., Li, X., & Berg, P. (2009). On the modeling of water transport in polymer electrolyte membrane fuel cells. *Electrochimica Acta, 54*(27), 6913–6927.
125. Chiu, H.-C., et al. (2012). A three-dimensional modeling of transport phenomena of proton exchange membrane fuel cells with various flow fields. *Applied Energy, 96,* 359–370.

126. Zhou, T., & Liu, H. T. (2001). A general three-dimensional model for proton exchange membrane fuel cells. *International Journal of Transport Phenomena, 3,* 177–198.

127. You, L., & Liu, H. (2002). A two-phase flow and transport model for the cathode of PEM fuel cells. *International Journal of Heat and Mass Transfer, 45*(11), 2277–2287.

128. You, L., & Liu, H. (2001). A parametric study of the cathode catalyst layer of PEM fuel cells using a pseudo-homogeneous model. *International Journal of Hydrogen Energy, 26*(9), 991–999.

129. Dutta, S., Shimpalee, S., & Van Zee, J. W. (2000). Three-dimensional numerical simulation of straight channel PEM fuel cells. *Journal of Applied Electrochemistry, 30,* 135–146.

130. Shimpalee, S., & Dutta, S. (2000). Numerical prediction of temperature distribution in PEM fuel cells. *Numerical Heat Transfer, Part A: Applications, 38*(2), 111–128.

131. Sui, P. C., & Djilali, N. (2005). Analysis of water transport in proton exchange membranes using a phenomenological model. *Journal of Fuel Cell Science and Technology, 2*(3), 149–155.

132. Mazumder, S. (2005). A generalized phenomenological model and database for the transport of water and current in polymer electrolyte membranes. *Journal of the Electrochemical Society, 152*(8), A1633–A1644.

133. Um, S., & Wang, C. Y. (2000). Three dimensional analysis of transport and reaction in PEMFC. In *ASME fuel cell division* (pp. 19–25). ASME: Orlando, FL.

134. Um, S., & Wang, C. Y. (2004). Three-dimensional analysis of transport and electrochemical reactions in polymer electrolyte fuel cells. *Journal of Power Sources, 125*(1), 40–51.

135. Wang, Y., & Wang, C.-Y. (2005). Modeling polymer electrolyte fuel cells with large density and velocity changes. *Journal of the Electrochemical Society, 152*(2), A445–A453.

136. Carnes, B., et al. (2013). Validation of a two-phase multidimensional polymer electrolyte membrane fuel cell computational model using current distribution measurements. *Journal of Power Sources, 236,* 126–137.

137. Bvumbe Tatenda, J., et al. (2016). Review on management, mechanisms and modelling of thermal processes in PEMFC. *Hydrogen and Fuel Cells, 1*(1), 1–20.

138. Berning, T., Lu, D. M., & Djilali, N. (2002). Three-dimensional computational analysis of transport phenomena in a PEM fuel cell. *Journal of Power Sources, 106*(1–2), 284–294.

139. Mazumder, S., & Cole, J. V. (2003). Rigorous 3-D mathematical modeling of PEM fuel cells. *Journal of the Electrochemical Society, 150*(11), A1503–A1509.

140. Mazumder, S., & Cole, J. V. (2003). Rigorous 3-D mathematical modeling of PEM fuel cells. *Journal of the Electrochemical Society, 150*(11), A1510–A1517.

141. Al-Baghdadi, M. A. R. S. (2008). *CFD models for analysis and design of PEM fuel cells.* New York: Nova Science Publishers.

142. Schwarz, D. H., & Beale, S. B. (2009). Calculations of transport phenomena and reaction distribution in a polymer electrolyte membrane fuel cell. *International Journal of Heat and Mass Transfer, 52*(17–18), 4074–4081.

143. Strahl, S., Husar, A., & Franco, A. A. (2014). Electrode structure effects on the performance of open-cathode proton exchange membrane fuel cells: A multiscale modeling approach. *International Journal of Hydrogen Energy, 39*(18), 9752–9767.

144. Artemov, V., et al. (2009). A tribute to D. B. Spalding and his contributions in science and engineering. *International Journal of Heat and Mass Transfer, 52,* 3884–3905.

Multiphase Flows

A Review of Computational Models for Falling Liquid Films

Avijit Karmakar and Sumanta Acharya

Nomenclature

C	Species concentration
C_p	Specific heat
d	Tube diameter
D	Mass diffusivity
Fr	Froude number
$\vec{g}$	Gravity vector
Ga	Modified Galileo number
h	Heat transfer coefficient
ΔH	Enthalpy
k	Thermal conductivity
L	Characteristic length for the film
Le	Lewis number
$\dot{m}$	Mass flow rate
$\dot{m}''$	Diffusion mass flux
$\hat{n}$	Unit normal to the interface
Nu	Nusselt number
p	Pressure
Pe	Peclet number
Pr	Prandtl number
q''	Heat flux

A. Karmakar · S. Acharya (✉)
Mechanical, Materials and Aerospace Engineering Department,
Illinois Institute of Technology, Chicago, IL 60616, USA
e-mail: sacharya1@iit.edu

A. Karmakar
e-mail: akarmakar@hawk.iit.edu

© Springer Nature Singapore Pte Ltd. 2020
A. Runchal (ed.), *50 Years of CFD in Engineering Sciences*,
https://doi.org/10.1007/978-981-15-2670-1_16

Q	Volume flow rate
Re	Reynolds number
s	Interfacial line coordinate
Sh	Sherwood number
St	Stanton number
Sc	Schmidt number
t	Time
$\hat{t}$	Unit tangential to the interface
T	Temperature
$\vec{u}$	Velocity vector
u	Streamwise velocity
v	Transverse velocity
W	Domain width
We	Weber number
x	Streamwise coordinate
y	Transverse coordinate
Y	Mass fraction

Greek

ρ	Density
μ	Dynamic viscosity
δ	Film thickness
σ	Surface tension coefficient
κ	Interface curvature
ν	Kinematic viscosity
τ	Shear stress
Γ	Mass flow rate per unit width
ε	Ratio of length scales
λ	Spacing
Λ	Wavelength
ψ	Stream function
ξ	Moving coordinate
Φ	Thermal energy due to viscous dissipation
θ	Dimensionless temperature
β	Mass transfer coefficient
α	Ratio of domain length to width
ζ	Capillary length

Subscript

abs	Absorption
b	Bulk
cr	Critical
d	Diabatic boundary
evap	Evaporation
g	Gas
gp	Gaseous phase
l	Liquid
N	Nusselt
v	Vapor
w	Wall
W	Wave celerity

Superscript

*	Non-dimensional
i	Order
in	Inlet
T	Transpose

Abbreviations

BL	Boundary Layer
CFD	Computational Fluid Dynamics
MAC	Marker and Cell
OpenFOAM	Open-source Field Operation and Manipulation
SIMPLE	Semi-implicit Method for Pressure-Linked Equations
SIMPLER	Semi-implicit Method for Pressure-Linked Equations Revised
UHF	Uniform Heat Flux
UWT	Uniform Wall Temperature
VOF	Volume of Fluid
2D	Two dimensional
3D	Three dimensional

1 Introduction

Liquid films are observed in a variety of applications in nature and engineering practice. In cooling towers, for example, hot water is sprayed onto corrugated plate surfaces, generating a falling liquid film from which evaporation occurs into the counter-flowing air. In nuclear fusion devices, falling liquid films are used to cool the heated chamber walls around the plasma [1]. In modern nuclear reactors, water in the primary circuit flows over fuel elements in the shape of an evaporating film, driven by water vapor [2]. Liquid films are also seen to develop on the outside of horizontal or vertical tubes in power plant condensers [3], or in absorption-cooling devices [4]. In specific gas turbines, liquid fuel is injected as an air-driven film, which atomizes in the combustion chamber [5]. In automobiles, liquid films are seen to develop in reciprocating combustion engines as oil films on the inside of the cylinder wall [6]. In process technology, falling film evaporators are employed for inspissation [7] or distillation [8], including desalination [9]. Such devices consist of vertical or horizontal tubes, along the outside of which a falling film develops, enabling thermal control through the surface of the tube wall. This advantage is relevant for processes requiring the limitation of the product temperature. Thus, the application of falling films is important in many industrial processes, and further examples of it can be found in the comprehensive review article by Fulford [10].

Liquid films are typically thin and generally satisfy the boundary layer assumptions. Broadly, liquid films can be characterized as either shear-driven in the presence of a superimposed external flowfield, or gravity-driven where the falling liquid film is accelerated by the gravitational force. In either case, phase change in the form of vaporization or condensation could be present.

The term *falling liquid films* generally refers to a liquid film flowing under the action of gravity. This flow can be either over a plain flat surface or over tubes of various shapes (round, rectangular, or elliptical) and orientation (vertical, inclined, or horizontal). The review of the literature suggests that the most commonly studied falling liquid film is either over plain flat surfaces or horizontal round tubes. In this chapter, a brief summary is undertaken of the various computational models for falling liquid film over plain flat surfaces (vertical or inclined) and horizontal round tubes. Both the hydrodynamic models as well as the heat and mass transfer models developed recently, and their representative results are reviewed in this context.

2 Falling Liquid Films over Plain Flat Surfaces

One of the most notable characteristics of falling liquid films over plain surfaces is the presence of waves on the liquid–gas interface. Benjamin [11] showed that a smooth vertically falling film is unstable to interfacial perturbations for all values of the Reynolds number (Re). The surface waves arise in these films either due to natural excitation or due to controlled external perturbations. Figure 1 illustrates the typical

Fig. 1 Shadowgraph of a vertical water film with waves and typical wave profile of 2-D wavefronts. Reprinted with permission from Park and Nosoko, AIChE J, 49:2715–2727, John Wiley and Sons, 2003

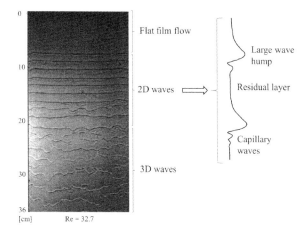

wave topology of a falling liquid film under these conditions at Re = 33 as visualized experimentally by Park and Nosoko [12]. The evolution of these interfacial waves has been divided into distinct wave regimes in the noteworthy works of Chang [13] under the conditions of naturally excited and controlled perturbations on a laminar falling liquid film. Near the inlet of the liquid film, the waves are seen to be short and periodic in nature, and the linearized stability theory predicts an exponential growth rate of the disturbances. For single wavelength or frequency (monochromatic) disturbances, the frequency of the emerging waves is same as the excitation frequency. In contrast to monochromatic disturbances, naturally excited waves assume a non-regular structure consisting of a complex array of large and small waves moving over a surface [14]. As the amplitude of the wave increases downstream, the linear stability theory fails to apply and non-linear effects become prominent, and control the wave amplification process. The amplitude of these waves reach a stage of saturation and thus possess a finite value, which is a function of the film Reynolds number (Re) and liquid properties (Ka). A consequence of this non-linear interaction is the steepening of the sinusoidal shape of the emerging wave. Further downstream in the streamwise direction, subharmonic and sideband instabilities result in coalescence of neighboring waves at intermittent locations. In these particular locations, the distorted waves evolve into localized humps of large size called teardrop humps [13]. The humps have steep fronts which are preceded by a series of front running capillary waves as shown in the profile on the right side of Fig. 1. These solitary humps travel in a nonstationary manner but essentially retain their shapes. Further downstream, these two-dimensional wavefronts with equal spacings break up into three-dimensional structures, which interact with one another and render the interface topology considerably more complex.

2.1 Hydrodynamic Models

The hydrodynamic behavior of falling liquid film over plain surfaces has been addressed by numerous researchers over the past years. This section comprises the summary of the computational approaches used by the researchers to model the flow characteristics of falling liquid film over a plain surface under gravity.

2.1.1 Governing Equations

Assuming Newtonian and incompressible fluids, the flow of a two-dimensional (2D) falling liquid film down a wall in a quiescent gaseous atmosphere (shown in Fig. 2) is governed by the following system of equations for both the liquid and gaseous phases.

The continuity and momentum equations for the flow are respectively given by

$$\nabla \cdot (\rho \vec{u}) = 0 \tag{1}$$

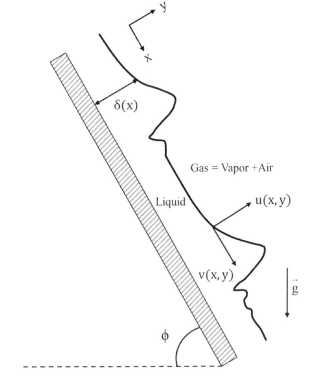

Fig. 2 Schematic diagram of a wavy falling liquid film over a flat plain surface

$$\rho \frac{\partial \vec{u}}{\partial t} + \nabla \cdot (\rho \vec{u}\vec{u}) = -\nabla p + \nabla \cdot (\overline{\overline{\tau}}) + \rho \vec{g} \tag{2}$$

where the stress tensor $\overline{\overline{\tau}}$ for Newtonian fluids is represented by

$$\overline{\overline{\tau}} = \mu \left[(\nabla \vec{u} + \nabla \vec{u}^T) \right] \tag{3}$$

The liquid–gas interface is defined by the scalar equation $y = \delta(x)$. For the solutions to these equations to be well defined in the gas (subscript g) and liquid (subscript l) phases, five inter-phase coupling conditions need to be formulated. Two of the conditions result from the continuity of velocity and shear at the gas–liquid interface, given by

$$u_l|_{y=\delta} = u_g\big|_{y=\delta} \text{ and } \mu_l \frac{\partial u_l}{\partial y}\bigg|_{y=\delta} = \mu_g \frac{\partial u_g}{\partial y}\bigg|_{y=\delta} \tag{4}$$

The next two conditions result from a local interfacial force balances in the normal and tangential directions. According to Brackbill et al. [15], the surface tension force at the fluid interface is balanced by a difference in pressure between the fluids across the interface. In the direction normal to the interface, this boundary condition can be expressed by the following equation:

$$p_l - p_g + \sigma\kappa = 2\mu_l \widehat{n_k} \left(\frac{\partial u_k}{\partial n} \right)_l - 2\mu_g \widehat{n_k} \left(\frac{\partial u_k}{\partial n} \right)_g \tag{5}$$

The interfacial force balance boundary condition in the direction tangential to the interface is given by

$$\mu_g \left(\widehat{t_i} \frac{\partial u_i}{\partial n} + \widehat{n_k} \frac{\partial u_k}{\partial s} \right)_g - \mu_l \left(\widehat{t_i} \frac{\partial u_i}{\partial n} + \widehat{n_k} \frac{\partial u_k}{\partial s} \right)_l = \frac{\partial \sigma}{\partial s} \tag{6}$$

In the above Eqs. (5)–(6), σ is the surface tension coefficient, p is the pressure, κ is the local surface curvature, μ is the viscosity, n_k is the component of the unit normal vector perpendicular to the interface, and t_i is component of the unit normal vector tangential to the interface.

The film thickness δ is related to the velocity field by the kinematic condition and is given as

$$v|_{y=\delta} = \frac{d\delta}{dt} \tag{7}$$

At the wall, assuming no-slip and no-penetration, the following boundary conditions apply:

$$u|_{y=0} = 0 \text{ and } v|_{y=0} = 0 \tag{8}$$

By assuming a constant pressure p_0 in the gaseous phase and a smooth liquid–gas interface, Nusselt [16] derived expressions for the streamwise velocity and the film thickness of a fully developed falling film in terms of liquid flow rate and properties using approximate versions of Eqs. (1–8). This solution is given by

$$\delta_N = \left(\frac{3\mu_l/\Gamma}{\rho_l^2 g_x}\right)^{1/3}; \quad p(y) = p_0 + \rho_l g_y(\delta_N - y); \quad u_N(y) = \frac{g_x}{v_l}\left(\delta_N y - \frac{y^2}{2}\right); \quad v_N = 0 \tag{9}$$

where Γ is the liquid mass flow rate per unit width in the z-direction.

The characteristics length, velocity, and time scales obtained from the Nusselt solution are used for non-dimensionalization of the governing equations and are expressed as follows:

$$\delta_N = \left(\frac{3\mu_l/\Gamma}{\rho_l^2 g_x}\right)^{1/3}; \quad u_N = \frac{g_x \delta_N^2}{3v_l}; \quad t_N = \frac{L}{u_N} \tag{10}$$

Introducing the following dimensionless quantities

$$x^* = \frac{x}{L}; \quad y^* = \frac{y}{\delta_N}; \quad t^* = \frac{t u_N}{L}; \quad p^* = \frac{p}{\rho_l u_N^2}; \quad u^* = \frac{u}{u_N}; \quad v^* = \frac{vL}{u_N \delta_N} \tag{11}$$

the Navier Stokes and continuity equation reduces to the following non-dimensional form for the 2-D case in x-y coordinates with constant fluid properties, [17]

$$\frac{\partial u^*}{\partial t^*} + u^* \frac{\partial u^*}{\partial x^*} + v^* \frac{\partial u^*}{\partial y^*} = -\gamma \frac{\partial p^*}{\partial x^*} + \frac{\chi}{Re}\left(\varepsilon \frac{\partial^2 u^*}{\partial x^{*2}} + \frac{1}{\varepsilon}\frac{\partial^2 u^*}{\partial y^{*2}}\right) + \frac{3\chi}{\varepsilon Re} \tag{12}$$

$$\frac{\partial v^*}{\partial t^*} + u^* \frac{\partial v^*}{\partial x^*} + v^* \frac{\partial v^*}{\partial y^*} = -\frac{\gamma}{\varepsilon^2} \frac{\partial p^*}{\partial y^*} + \frac{\chi}{Re}\left(\varepsilon \frac{\partial^2 v^*}{\partial x^{*2}} + \frac{1}{\varepsilon}\frac{\partial^2 v^*}{\partial y^{*2}}\right) - \frac{1}{\varepsilon^2 Fr^2} \tag{13}$$

$$\frac{\partial u^*}{\partial x^*} + \frac{\partial v^*}{\partial y^*} = 0 \tag{14}$$

where χ and γ have been introduced to distinguish the flow equations in the respective gas and liquid phases. In the gaseous phase $\chi = \frac{\Pi_\mu}{\Pi_\rho}$, $\gamma = \frac{1}{\Pi_\rho}$ and in the liquid phase $\chi = \gamma = 1$ with the density ratio $\Pi_\rho = \frac{\rho_g}{\rho_l}$ and the dynamic viscosity ratio $\Pi_\mu = \frac{\mu_g}{\mu_l}$. As the surrounding atmosphere of the falling liquid film is quiescent gas, the scales in both the liquid and the gaseous phase are imposed by the flow in the liquid phase. The non-dimensionalization of the Navier–Stokes and continuity equations yield the Reynolds number Re, the Froude number Fr, and the ratio of length scales ε, which are defined as follows:

$$\text{Re} = \frac{u_N \delta_N}{v_l}, \quad Fr = \frac{u_N}{(g_y \delta_N)^{0.5}}, \quad \varepsilon = \frac{\delta_N}{L} \tag{15}$$

The inter-phase coupling conditions reduce to

$$p_l^* - p_g^* + \frac{We\varepsilon^2}{\kappa^{1.5}}\frac{\partial^2\delta^*}{\partial x^{*2}} = \frac{2\varepsilon}{\kappa Re}\left[\varepsilon^2\frac{\partial u_l^*}{\partial x^*}\left(\frac{\partial\delta^*}{\partial x^*}\right)^2 + \frac{\partial v_l^*}{\partial y^*} + \frac{\partial\delta^*}{\partial x^*}\left(\frac{\partial u_l^*}{\partial y^*} + \varepsilon^2\frac{\partial v_l^*}{\partial x^*}\right)\right]$$
$$- \frac{2\varepsilon\Pi_\mu}{\kappa Re}\left[\varepsilon^2\frac{\partial u_g^*}{\partial x^*}\left(\frac{\partial\delta^*}{\partial x^*}\right)^2 + \frac{\partial v_g^*}{\partial y^*} + \frac{\partial\delta^*}{\partial x^*}\left(\frac{\partial u_g^*}{\partial y^*} + \varepsilon^2\frac{\partial v_g^*}{\partial x^*}\right)\right], \tag{16}$$

$$\varepsilon^2\frac{\partial\delta^*}{\partial x^*}\left(\frac{\partial u_l^*}{\partial x^*} - \frac{\partial v_l^*}{\partial y^*}\right) + \left(\frac{\kappa}{2} - 1\right)\left(\frac{\partial u_l^*}{\partial y^*} + \varepsilon^2\frac{\partial v_l^*}{\partial x^*}\right)$$
$$= \Pi_\mu\varepsilon^2\frac{\partial\delta^*}{\partial x^*}\left(\frac{\partial u_g^*}{\partial x^*} - \frac{\partial v_g^*}{\partial y^*}\right) + \Pi_\mu\left(\frac{\kappa}{2} - 1\right)\left(\frac{\partial u_g^*}{\partial y^*} + \varepsilon^2\frac{\partial v_g^*}{\partial x^*}\right) \tag{17}$$

where the following definitions were introduced:

$$We = \frac{\sigma}{\rho_l\delta_N u_N^2}, \kappa = 1 + \varepsilon^2\left(\frac{\partial\delta^*}{\partial x^*}\right)^2, \delta^* = \frac{\delta}{\delta_N} \tag{18}$$

and We is the Weber number. The non-dimensional form of the kinematic condition is

$$v^*\big|_{y^*=\delta^*} = \frac{d\delta^*}{dt^*} \tag{19}$$

and the velocity boundary conditions at the wall take the form

$$u^*\big|_{y^*=0} = v^*\big|_{y^*=0} = 0 \tag{20}$$

Equations (12–14) along with the boundary conditions given by Eqs. (16, 17, 19, 20) represent the set of mathematical equations that govern the hydrodynamics of the falling film over plain surfaces. Based on the level of approximations to these governing equations, the computational approaches for the hydrodynamic modeling of vertically falling liquid films can be divided into four groups. In the first group of models, the boundary layer equations are used [13, 18–21]. A second approach is to employ the long-wave equations [22–27], that has been extended to multiple equation models [28–33]. Finally, numerical solutions of the complete Navier–Stokes equations for two-phase flows [17, 34–37] can be obtained, but require the greatest computational effort. A comprehensive overview of available models, as well as characterization of the accuracy of their results, is provided collectively by the works of [13, 20, 33, 38–42]. A brief review is provided on each of the above four approaches in the subsequent sections.

2.1.2 Boundary Layer First-Order Models

The common basis of this mathematical modeling is the boundary layer equations of film flow, which can be derived from the Navier–Stokes equation under the assumption that the wavelength Λ of surface interfacial waves are much larger than the characteristic film thickness δ_N of the flow. This assumption is called the long-wave approximation. Introducing this assumption in the 2D Navier–Stokes and continuity equation for the liquid phase (Eqs. 12–14) and discarding terms containing ε to the second or higher order, provides us the first-order boundary layer equations.

$$\frac{\partial u^*}{\partial t^*} + u^* \frac{\partial u^*}{\partial x^*} + v^* \frac{\partial u^*}{\partial y^*} = -\frac{\partial p^*}{\partial x^*} + \frac{1}{Re}\left(\frac{1}{\varepsilon}\frac{\partial^2 u^*}{\partial y^{*2}}\right) + \frac{3}{\varepsilon Re} \tag{21}$$

$$0 = -\frac{\partial p^*}{\partial y^*} + \frac{\varepsilon}{Re}\left(\frac{\partial^2 v^*}{\partial y^{*2}}\right) - \frac{1}{Fr^2} \tag{22}$$

$$\frac{\partial u^*}{\partial x^*} + \frac{\partial v^*}{\partial y^*} = 0 \tag{23}$$

Under the above assumptions and neglecting the viscous forces in the gaseous phase, the inter-phase coupling conditions (Eqs. 16–17) reduce to

$$p_l^* - p_g^* + We\varepsilon^2 \frac{\partial^2 \delta^*}{\partial x^{*2}} = \frac{2\varepsilon}{Re}\frac{\partial v^{*l}}{\partial y^*};\ \frac{\partial u^{*l}}{\partial y^*} = 0 \tag{24}$$

Importantly, in order to retain the effects of the tensile forces, the term containing the Weber number in the normal coupling boundary condition is retained, although it contains ε^2. The above coupling equation can be further simplified by integrating the crosswise momentum equation from y^* to δ^* using the normal inter-phase coupling conditions. This yields the pressure distribution, which when inserted in the streamwise momentum equation (Eq. 21), takes the form

$$\frac{\partial u^*}{\partial t^*} + u^* \frac{\partial u^*}{\partial x^*} + v^* \frac{\partial u^*}{\partial y^*} = We\varepsilon^2 \frac{\partial^3 \delta^*}{\partial x^{*3}} - \frac{1}{Fr^2}\frac{\partial \delta^*}{\partial x^*} + \frac{1}{Re}\left(\frac{1}{\varepsilon}\frac{\partial^2 u^*}{\partial y^{*2}}\right) + \frac{3}{\varepsilon Re} \tag{25}$$

Introducing the following estimation for the wavelength Λ as:

$$\Lambda = \left(\frac{WeRe^2 v^{l2}}{g_x}\right)^{1/3} \rightarrow \varepsilon = \frac{\delta_N}{\Lambda} = \left(\frac{3}{WeRe}\right)^{1/3} \tag{26}$$

The final form of the two-dimensional first-order boundary layer equations of the film flow is given as

$$\frac{\partial u^*}{\partial t^*} + u^* \frac{\partial u^*}{\partial x^*} + v^* \frac{\partial u^*}{\partial y^*} = \Pi_{BL}\left[\frac{\partial^3 \delta^*}{\partial x^{*3}} + \frac{1}{3}\frac{\partial^2 u^*}{\partial y^{*2}} + 1\right] - \frac{1}{Fr^2}\frac{\partial \delta^*}{\partial x^*}$$

$$\frac{\partial u^*}{\partial x^*} + \frac{\partial v^*}{\partial y^*} = 0, \quad \frac{\partial u^*}{\partial y^*}\bigg|_{y^*=\delta^*} = 0, \quad u^*\big|_{y^*=0} = v^*\big|_{y^*=0} = 0$$

$$\frac{\partial \delta^*}{\partial t^*} = v^*\big|_{y^*=\delta^*} - \frac{\partial \delta^*}{\partial x^*} u^*\big|_{y^*=\delta^*}, \quad \Pi_{BL} = 3^{2/3} \frac{We^{1/3}}{Re^{2/3}} \tag{27}$$

The so-called boundary layer long-wave models rely on the direct numerical solution of the simplified boundary layer equations. Such calculations have been performed by Demekhin [20, 21] and Chang [18, 19] for two-dimensional film flows, and by Chang [13] for three-dimensional situations. Demekhin et al. [20] performed numerical calculations under the same conditions as the experimental measurements of Alekseenko et al. [43], and showed reasonable agreement between numerical and measured data with average deviations of 4.5% and 5.5% in terms of the growth coefficient and the wave number, respectively. The boundary layer equations were able to predict oscillations at the leading front that coincided only quantitatively with the experimental data. Based on the boundary layer equations, Chang et al. [19] developed a numerical model where they studied inlet noise-driven wave dynamics on a falling film at relatively high Reynolds numbers. Experimentally observed phenomena, like wave inception, downstream wave texture coarsening, initial deceleration and subsequent acceleration of wave speeds were quantitatively reproduced by their model. In general, the above-entioned works show that the boundary layer equations of film flow can describe its dynamics with good accuracy. This accuracy, however, bears a considerable amount of computational time and cost, which was in the range of 400–500 h for a 2 m long plate [19].

2.1.3 Long-Wave Film-Thickness Equation Models

Another simplified approach involves the use of the long-wave equation, which consists of deriving a single equation for the evolution of the film thickness. The initial step in this method is the asymptotic expansion of the dimensionless stream function Ψ^* in terms of the length scale ratio ε

$$\Psi^* = \Psi_N^* + \Psi^{*'} = \Psi_N^* + \Psi_0^{*'} + \varepsilon \Psi_1^{*'} + \varepsilon^2 \Psi_2^{*'}$$

$$\frac{\partial \Psi^*}{\partial y^*} = u^*, \quad -\frac{\partial \Psi^*}{\partial x^*} = v^*, \quad \Psi_N^* = \frac{3}{2} y^{*2} - \frac{1}{2} y^{*3} \tag{28}$$

where Ψ_N^* signifies the dimensionless stream function for the developed smooth film and $\Psi^{*'}$ denotes the departure from that solution. This approach must then satisfy Eq. 25 written for the dimensionless stream function. Hence,

$$\frac{\partial^3 \Psi^{*'}}{\partial y^{*3}} = \varepsilon Re \left[\frac{\partial^2 \Psi^{*'}}{\partial y^* \partial t^*} + \frac{\partial^2 \Psi^{*'}}{\partial y^* \partial x^*} \left[u_N^* + \frac{\partial \Psi^{*'}}{\partial y^*} \right] - \frac{\partial \Psi^{*'}}{\partial x^*} \left[\frac{\partial^2 \Psi^{*'}}{\partial y^{*2}} + \frac{\partial u_N^*}{\partial y^*} \right] \right]$$

$$- Re We \varepsilon^3 \frac{\partial^3 \delta^*}{\partial x^{*3}} + \frac{\varepsilon Re}{Fr^2} \frac{\partial^*}{\partial x^*} \tag{29}$$

and the reformulated wall boundary, tangential inter-phase coupling, and kinematic condition

$$\frac{\partial \Psi^{*\prime}}{\partial y^*}\bigg|_{y^*=0} = \frac{\partial \Psi^{*\prime}}{\partial x^*}\bigg|_{y^*=0} = 0 \tag{30}$$

$$\frac{\partial^2 \Psi^{*\prime}}{\partial y^{*2}}\bigg|_{y^*=\delta^*} + \frac{\partial u_N^*}{\partial y^*}\bigg|_{y^*=\delta^*} = 0 \tag{31}$$

$$\frac{\partial \delta^*}{\partial t^*} = -\frac{\partial \Psi^{*\prime}}{\partial x^*}\bigg|_{y^*=\delta^*} - \frac{\partial \delta^*}{\partial x^*}\frac{\partial \Psi^{*\prime}}{\partial y^*}\bigg|_{y^*=\delta^*} - \frac{\partial \delta^*}{\partial x^*}u_N^*\bigg|_{y^*=\delta^*} \tag{32}$$

The different coefficients $\Psi_i^{*\prime}$ of the stream function expansion are successively determined from the Eqs. (29)–(32) by neglecting all terms of orders higher than the considered one (i.e., $< \varepsilon^i$). Therefore, it is important to note that at every step all lower order coefficients Ψ_{i-1}^* have been previously determined and are thus known. Following this approach, Eq. 29 always yields a differential equation for the current coefficient Ψ_i^*.

The result for the zeroth-order approximation $\psi_0^{*\prime}$ is given by

$$\psi_0^{*\prime} = \frac{3}{2}\left(\delta^* - 1\right)y^{*2} \quad \text{and} \quad \frac{\partial \delta^*}{\partial t^*} = -\frac{\partial}{\partial x^*}\left(\delta^{*3}\right) \tag{33}$$

In a similar fashion, using $\psi_0^{*\prime}$, the first-order approximation $\psi_1^{*\prime}$ is given by

$$\psi_1^{*\prime} = \frac{3}{40}y^{*5}\mathrm{Re}\frac{\partial \delta^*}{\partial x^*}\delta^* - \frac{3}{8}y^{*4}\mathrm{Re}\delta^{*2}\frac{\partial \delta^*}{\partial x^*} + \frac{1}{6}y^{*3}\left[\frac{Re}{Fr^2}\frac{\partial \delta^*}{\partial x^*} - \varepsilon^2\mathrm{Re}We\frac{\partial^3 \delta^*}{\partial x^{*3}}\right]$$
$$+ \frac{1}{2}y^{*2}\left[3\mathrm{Re}\delta^{*4}\frac{\partial \delta^*}{\partial x^*} - \frac{Re}{Fr^2}\delta^*\frac{\partial \delta^*}{\partial x^*} + \varepsilon^2\mathrm{Re}We\delta^*\frac{\partial^3 \delta^*}{\partial x^{*3}}\right] \tag{34}$$

which provides the first-order film thickness evolution equation as

$$\frac{\partial \delta^*}{\partial t^*} = -\frac{6}{5}\varepsilon \mathrm{Re}\delta^{*6}\frac{\partial^2 \delta^*}{\partial x^{*2}} - \frac{36}{5}\varepsilon \mathrm{Re}\delta^{*5}\frac{\partial \delta^*}{\partial x^*}\frac{\partial \delta^*}{\partial x^*} - \frac{1}{3}\delta^{*3}\left[\varepsilon^3 \mathrm{Re}We\frac{\partial^4 \delta^*}{\partial x^{*4}} - \varepsilon\frac{Re}{Fr^2}\frac{\partial^2 \delta^*}{\partial x^{*2}}\right]$$
$$- \delta^{*2}\frac{\partial \delta^*}{\partial x^*}\left[\varepsilon^3 \mathrm{Re}We\frac{\partial^3 \delta^*}{\partial x^{*3}} - \varepsilon\frac{Re}{Fr^2}\frac{\partial \delta^*}{\partial x^*} + 3\right] \tag{35}$$

The approach of asymptotic expansion of the stream function described above was first applied to falling liquid films by Benny [22], thus called Benney equation or Benney-type equation. The author did not retain the tensile terms in his first-order approximation, as he assumed $We = \mathcal{O}(1)$. However, Gjevik [23], in opposition to the works of Benny [22], incorporated the tensile effects with $We = \mathcal{O}(\varepsilon^{-2})$ and showed that the model can capture saturated interfacial waves, demonstrating the importance of tensile terms in the model of falling film wave dynamics. This film thickness evolution equation resulting from the asymptotic expansion method is designated as long-wave equation (Eq. 35). It is a non-linear partial differential equation that predicts the instantaneous film thickness at any location over the bounding wall. It can be solved numerically for given initial and streamwise boundary conditions of falling film flow.

Table 1 lists the different long-wave equation models along with the order at which the underlying asymptotic expansion is truncated, the scaling information is employed to simplify the governing equations and the key findings are reported in those works. The studies reported by Benny [22] and Gjevik [23], Lin [24] and Nakaya [25] derived long-wave equations by truncating the asymptotic expansion at first, second, and third order, respectively. In other studies [26], less strict restrictions on the Weber number were imposed in order to retain the tensile terms in the long-wave equations. An evaluation of the accuracy of the solution produced by the various long-wave equations of Gjevik [23], Lin [24], and Nakaya [25] is provided in Panga et al. [44]. The authors compared the bounds of the linear stability and non-linear saturated wave celerities predicted by the respective model equations. The

Table 1 List of long-wave equation models

Author	Order	Scaling	Findings
Benny [22]	1	$We = \mathcal{O}(1), Re = \mathcal{O}(1)$	Wave dispersion effects are not predicted; overprediction of critical wave number compared to linear stability predictions
Gjevik [23]	1	$We = \mathcal{O}(\varepsilon^{-2}), Re = \mathcal{O}(1)$	Identified interfacial wave saturation by inclusion of tensile terms in the LW equation model
Lin [24]	2	$We = \mathcal{O}(\varepsilon^{-1}), Re = \mathcal{O}(1)$	Less strict restrictions on the Weber number in order to retain the tensile terms in the model
Nakaya [25]	3	$We = \mathcal{O}(1), Re = \mathcal{O}(1)$	Inclusion of higher order terms determined the harmonics generated by the non-linear terms in the LW model equation
Takeshi [26]	2	$We = \mathcal{O}(\varepsilon^{-2}), Re = \mathcal{O}(1)$	Improved instability predictions than traditional LW model till $Re/We^{1/3} \leq 1$
Panga and Balakotaiah [27]	6/5	$We = \mathcal{O}(\varepsilon^{-2}), Ka = \mathcal{O}(1)$	Improved accuracy over existing LW equations in stability predictions by inclusion of viscous dissipation and pressure correction terms

results show that stability bounds for all long-wave equations match well with those computed from the Orr–Sommerfeld equation for large values of the Weber number, with average deviations of less than 5%. One of the important characteristics of these long-wave equations is that below a certain Weber number threshold, these equations do not provide saturated wave solutions. This phenomenon was first investigated by Pumir et al. [45] on the basis of the long-wave equation of Gjevik [23]. In that work, the authors showed that below the threshold Weber number value, an unlimited local increase in film thickness arises which leads to unphysical results. This behavior of the long-wave equation is called *"finite time blow up"* and it is attributed to the presence of highly non-linear terms in the film thickness evolution equation. It is perhaps evident that the "finite time blow up" constitutes a major limitation of standard long-wave equations. Since then, researchers [41, 46] have focused their attention toward remedies for this unwanted property so that it can be used to predict falling film wave dynamics.

The effect of the strong non-linear terms in the standard long-wave equations that leads to finite time blow up can be suppressed by considering the weakly non-linear forms of the long-wave equations. This can be achieved under the assumption of small wave amplitudes. Therefore, under this approach, the dimensionless film thickness in the considered long-wave equation is expressed by $\delta^* = 1 + \gamma \eta^*$, whereby $\gamma \eta^*$ designates the departure of the film thickness from the smooth developed state and γ is a small parameter ($\gamma \ll 1$) quantifying the amplitude of that deviation. Inserting this into the first-order long-wave equation derived by Gjevik [23] (Eq. 35) and expressing Re/Fr^2 in terms of the critical Reynolds number (Re_{cr}) obtained from the study of Yih and Guha [47] yields

$$4\frac{\partial \eta^*}{\partial t^*} + 6\gamma \eta^* \frac{\partial \eta^*}{\partial \xi^*} + \frac{6}{5}\varepsilon(Re - Re_c)\frac{\partial^2 \eta^*}{\partial \xi^{*2}} + \varepsilon^3 \gamma Re We \frac{\partial \eta^*}{\partial \xi^*}\frac{\partial^3 \eta^*}{\partial \xi^{*3}}$$
$$+\frac{1}{3}\varepsilon^3 Re We \frac{\partial^4 \eta^*}{\partial \xi^{*4}} + \frac{18}{5}\varepsilon\gamma(2Re - Re_c)\left(\frac{\partial \eta^*}{\partial \xi^*}\right)^2 = 0 \qquad (36)$$

where the moving coordinate ξ^* was introduced and only terms of $\mathcal{O}(\gamma \varepsilon)$ or higher were retained. It can be observed that the strong non-linear terms of the equation are no longer present in this weakly non-linear equation. Further assuming $\mathcal{O}(\gamma) = \varepsilon$ and retaining only terms of the order ε the amplitude equation takes the following form:

$$\frac{\partial H}{\partial T} + H\frac{\partial H}{\partial \Xi} + \frac{\partial^2 H}{\partial \Xi^2} + \frac{\partial^4 H}{\partial \Xi^4} = 0 \qquad (37)$$

where Ξ, T and H are chosen substitutions for the moving coordinate time and amplitude, and are expressed as

$$\Xi = \xi^* \left[\frac{18(Re - Re_{cr})}{5ReWe} \right]^{0.5}, \mathrm{T} = t^* \left[\frac{54(Re - Re_{cr})}{50ReWe} \right]^2,$$

$$H = \eta^* 5 \left(\frac{5}{18} \right)^{0.5} (ReWe)^{0.5} (Re - Re_{cr})^{-1.5} \tag{38}$$

In the literature of falling film wave dynamics, Eq. 37 is designated as *Kuramoto–Shivasinsky* equation. Sivashinsky and Michelson [46] identified its applicability to film flows and used it to simulate falling film wave dynamics. Chang [48] evaluated predictions of the equation on the basis of experimental data provided by [43] with respect to maximal film thickness and non-linear wave celerity, showing reasonable agreement for small wave amplitudes where the deviations were less than 12% of the experimental data. However, the *Kuramoto–Shivasinsky* equation has a serious drawback as it does not describe dispersion, which is primarily the dependence of wave growth rates on the wavelength. Following a similar approach, another weakly non-linear equation evolved on the basis of the third-order long-wave equation [25]. The equation is designated as *Korteweg–de Vries* equation (see [49]) and is given as

$$\frac{\partial H}{\partial T} + H \frac{\partial H}{\partial \Xi} + v \frac{\partial^3 H}{\partial \Xi^3} = 0 \tag{39}$$

The quantities H, T, and Ξ are defined differently in the *Korteweg–de Vries* equation in comparison to *Kuramoto–Shivasinsky* equation. An important property of the equation is its ability to describe the behavior of solitary waves that preserve their properties after interacting with one another. This formulation, however, does not account for the inertial and tensile effects. The three-dimensional form of the equation was employed by Petviashvili and Tsvelodub [50] to construct the first published solution of a three-dimensional solitary wave on a falling liquid film over a vertical surface where Re ~ 1. Later, Chang and Demekhin [41] presented preliminary results on 3D solitary waves at moderate Reynolds numbers obtained on the basis of the generalized *Kuramoto–Sivashinsky* equation, which took into account an additional term for dispersion effects. Their results indicate that for vertically falling film, there exist numerous branches of stationary solutions. The solution belongs to one of the branches, and for other branches, all stationary solutions have the form of multihump solitary waves traveling at the liquid film surface.

2.1.4 Multiple Equation Models

The class of multiple equations evolved as another method of suppressing finite time blow up where the long-wave equations are written in a regularized fashion. This attempt was first performed by Takeshi [26] that based on the second-order long-wave equation of Gjevik [23]. A drawback of the regularized version of the model is the fact that it contains two equations, i.e., film thickness δ^* and instantaneous local flow rate per unit width Q^*. This serves as proof that an accurate prediction of linear and non-linear falling film dynamics is not possible through the modeling of just one kinematic quantity (in this case the film thickness δ^*). Consequently, other kinematic and dynamic quantities such as liquid flow rate and wall shear stress should be considered and modeled using adequate evolution equations. Such so-called multiple equation models consisting of more than one model equation constitute the third class of wave dynamics models and is addressed in the next section.

One of the simplified approach to solve the boundary layer equations can be achieved by applying the general *Von Karman–Pohlhausen* approximate method to the liquid film, i.e., integrating the streamwise momentum continuity in Eq. 27 from the wall ($y^* = 0$) to the liquid film interface ($y^* = \delta^*$). In relevant to flat-plate boundary layer solutions, the critical modeling step of this approach is the assumption of a realistic instantaneous local profile for the streamwise velocity u^*. As mentioned previously, this method leads to two evolution equations for the film thickness δ^* and instantaneous local flow rate per unit width Q^*. The derivation is achieved by applying Leibniz's integration rule and incorporating the wall and interface boundary conditions (as well as the kinematic condition) in Eq. 27, where the integral streamwise momentum and continuity equations take the form

$$\frac{\partial Q^*}{\partial t^*} + \frac{\partial}{\partial x^*} \int_0^{\delta^*} u^{*2} \partial y^* = \Pi_{BL}\left[\delta^* \frac{\partial^3 \delta^*}{\partial x^{*3}} + \delta^*\right] - \frac{\delta^*}{Fr^2}\frac{\partial \delta^*}{\partial x^*} - \frac{\Pi_{BL}}{3}\frac{\partial u^*}{\partial y^*}\bigg|_{y^*=0} \quad (40)$$

$$\frac{\partial Q^*}{\partial x^*} + \frac{\partial \delta^*}{\partial t^*} = 0 \quad (41)$$

The first authors to use this methodology for falling films were Kapitza [28] and Shkadov [29]. In the works of Shkadov [29], the author assumed a self-similar semi-parabolic local instantaneous velocity profile leading to the Shkadov's 2-equation model:

$$u^* = \frac{3Q^*}{\delta^*}\left[y^* - \frac{y^{*2}}{2}\right] \quad (42)$$

In physical significance, the above equation shows that the flow is locally developed at all times, which is clearly not the situation in experimental situations. Shkadov and co-workers (see [20] as an example) later reviewed their model by comparing its results to experimental data and direct solutions of the boundary layer equations. The linearized version of the Shkadov's model provided differing stability bounds compared to those obtained from the Orr–Sommerfeld equation. This model was further analyzed by Demekhin et al. [20], where the authors showed that the model performs reasonably well in prediction of the non-linear wave dynamics for monochromatically excited waves as well as wave saturation. However, with respect to capillary wave dynamics; the model deviates significantly from experimental data and the solution of the boundary layer equations. The author concluded that this is due to the complexity in the velocity field of the capillary wave region. In recent years, significant advances in the modeling of film flows based on Shkadov's approach have been

Table 2 List of multiple equation models

Source	M.O	E.N	V.P.O	Ka	Re	L.P	N.L.P
Shkadov [29]	1	2	2	193	3	>15%	10–15%
Yu et al. [30]	2	4	4	3550	600	5–10%	10–15%
Ruyer-Quil and Manneville [51]	2	3	8	700	<20	2–5%	<1%
Nguyen and Balakotaiah [14]	2	3	2	<100	<100	<2%	5–10%
Ruyer-Quil and Manneville [31]	2	4	14	252	<10	<2%	–
Ruyer-Quil and Manneville [31]	2	2	2	252	<10	<2%	2–5%
Scheid et al. [32]	2	2	6	1106	<60	–	<2%
Mudunuri and Balakotaiah [33]	2	2	2	500	<100	–	<2%

M.O—Model order; E.N—Number of model equations; V.P.O—Velocity profile order; Ka—Kapitza number, Re – Reynolds number; L.P—Linear predictions; N.L.P—Non-linear predictions

performed. The performance of the improved multiple equation models developed by two promising groups [14, 31–33, 51] is summarized in Table 2. The second-order four-equation model by Yu et al. [30] predicted the linear stability bounds and the film thickness statistics reasonably well in comparison to the first-order boundary layer equations. Phenomena such as backflow [32] and negative wall shear stress in the capillary wave region [33] of the film flow were captured under this approach. It can be concluded that by increasing the model order and the number of the equations (for the flow rate per unit width, film thickness, wall shear stress, etc.) and using a velocity profile with higher order terms, the authors were able to capture both the linear and non-linear instabilities observed in the falling film wave dynamics. Therefore, it is clear that the most promising approach to modeling falling liquid film dynamics is based on the multiple equation model. However, it is important to note that the assumption of a physically meaningful crosswise velocity profile is extremely crucial to the predictions of these models.

2.1.5 Navier–Stokes Equations

In order to avoid velocity profile assumptions and solve the non-linearity issues in falling film flows, numerical solutions of the full Navier–Stokes equations for falling film flows have been implemented by several researchers [17, 34–37]. There have been many standard strategies to solve the Navier–Stokes equations such as the MAC method by Harlow and Welch [52], the SIMPLE method and its improved variants by Spalding and his group [53–55], the PISO method of Issa [56] and many others. Over the last four decades, these methods have been successfully adapted to multiphase flow formulations, see Ref. [57, 58] for more details. Depending on the required resolution, there are three main formulations for multiphase flows: Eulerian framework for all phases where the interface between the phases is not explicitly accounted; Eulerian framework for the continuous phase and Lagrangian framework for all the dispersed phases and lastly, interface capturing approaches where a Eulerian framework for both the phases with reformulation of interface forces on volumetric basis is used. The Eulerian–Eulerian approach models the flow of all phases in a Eulerian framework based on the continuum assumption. This approach was implemented in the Inter-phase Slip Algorithm (IPSA) which Spalding pioneered [59] and was successfully adopted by researchers to model dispersed flows [58] such as bubbles in liquid, solid particles in gas or liquid and liquid droplets in gas or other immiscible liquid. However, such a model of multiphase flow does not describe the local distribution of the two phases. It is this property that leads to the simplicity of the formulation and also to its main weaknesses to handle complex phenomena at particle level (such as change in bubble/dispersed phase size due to reaction/evaporation). In the Eulerian–Lagrangian approach, the trajectories of dispersed phase particles are simulated by solving an equation of motion for each dispersed phase particle and the motion of the continuous phase is modeled using a conventional Eulerian framework. The gas–liquid interface is handled by several different approaches proposed, but the most popular approach is the interface capturing techniques such as the Volume of Fluid method [60–62], Level-set method [63, 64], etc. These approaches track the motion of all the phases, from which motion of the interface is inferred. The local instantaneous conservation equations are solved with appropriate jump boundary conditions at the interface. Although these interface capturing approaches require larger computational resources to resolve flow processes around the interface, they can be very useful as research tools and can provide valuable information in developing appropriate closure models for Eulerian–Eulerian approaches. These approaches have been successfully applied to the class of falling films and typical challenges for such simulations include correct implementation of flow boundary conditions on the evolving interface and resolving the different length scales associated with the film thickness and the flow domain. However, these approaches represent improvements over numerical studies [65–68] that assume a waveform for the gas–liquid interface.

The finite element method has been used by several researchers to study the falling film hydrodynamics over plain surfaces. In fact, the first numerical approach to solve both the waveform and flow field simultaneously was carried out by Bach and Villadsen [69], where the authors used a Lagrangian finite element method to solve the governing equations. The wave profiles obtained in their study were quite comparable with the experimental data. Kheshgi and Scriven [70] also used the Lagrangian finite element method and compared their results with the Orr–Sommerfeld linear stability predictions. Malamataris and Papanastasiou [71] used the Galerkin finite element method to solve the equations in a truncated domain with modified outflow boundary conditions. Their study was able to capture the solitary waves and the relation between the wave speed to wavenumber in very short truncated domains. Salamon et al. [72] conducted a numerical study using a similar approach and compared their results with the approximated long-wave and boundary layer theories. They assumed the periodic boundary condition in their problem, which relates to the assumption of a stationary wave. Recently, Ramaswamy et al. [34] did a comprehensive numerical study with and without the periodic condition, where Sommerfeld radiation conditions were imposed at the outflow boundary. They used an Lagrangian–Eulerian finite element framework and simulated the spatial stability with a very long domain and imposing time-periodic disturbances at the inlet. A constant mean film thickness in the whole calculation region was applied in order to modify the temporal film thickness variation. This assumption is not compatible with the experimental conditions corresponding to an open flow condition where the flowrate is conserved. Indeed, both Kapitza [28] and Alekseenko [43] have reported a decrease in the average film thickness downstream.

The finite difference method is also widely used for numerical simulations of falling film flows. Nagasaki and Hijikata [35] employed an adaptive grid with the finite difference approach. The adaptive grid fitted to the free surface has the advantage of providing higher resolution at the free boundary. With periodic boundary conditions in the streamwise direction, they calculated variations from small-amplitude sinusoidal disturbance to fully developed solitary wave which composed of a solitary (big-amplitude) wave and small waves. They recognized the existence of a circulation flow in the big wave of the film flow. Kiyota et al. [73] carried out calculations for variations in wave shape under a similar approach of Nagasaki and Hijikata [35]. They reported the initial stage of wave evolution and compared its characteristics with the linear stability theory. Later, Miyara [36] felt that an advanced technique for the fixed grid might be required in order to consider the gas phase effects and to extend the simulation capability. They solved the two-dimensional Navier–Stokes equations using a finite difference approach for the film velocity distribution and free surface dynamics. Their solution method was based on the MAC algorithm. They neglected the shear stress from the vapor phase generated at the liquid–vapor interface. Under low-frequency disturbances, they observed that the circulation flows generated in the big waves grow downstream, and the scale of the circulation flow is comparable to the wave amplitude. However, no circulation was observed in the capillary waves. This behavior is shown in Fig. 3. With an increase in the frequency of disturbance, the amplitude of the waves becomes small and the capillary waves disappear. The circulation flow repeats the process of generation and collapse due to the wave interaction at a specific frequency.

Gao et al. [37] simulated 2-D falling film flow on a vertical plane by solving the complete equations using finite volume discretization with the Volume of Fluid (VOF) method of Hirt and Nichols [60] as the interface capturing method. They accounted for surface tension effects at the interface using the Continuum Surface Force (CSF) model originally proposed by Brackbill et al. [15]. They generated flow disturbances in their simulations by imposing monochromatic perturbation of velocity at the inlet. At low frequency and high-flow rate, the small inlet disturbance develops into large solitary waves preceded by small capillary bow waves. On the other hand, at high frequency and low Re, small-amplitude waves in nearly sinusoidal shape without forerunning capillary waves are formed on the surface. The variation of velocity and pressure along a wave are strong at the wave trough and capillary wave region, due to the presence of large surface curvature. The pressure variation perpendicular to the wall was found to be negligible.

In comparison to the low-dimensional models presented in the earlier sections, the computational time of the fully resolved simulations of falling films are usually

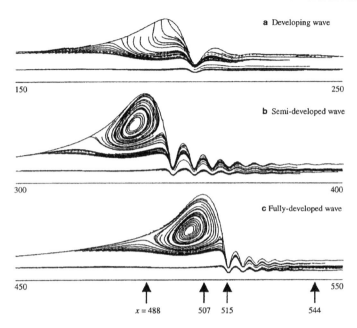

Fig. 3 Streamlines observed from moving coordinates with wave celerity u_w for Re = 100, We = 0.503, Fr = $-\infty$, disturbance amplitude (ε) = 0.03, and frequency (f) = 0.01: **a** Developing wave: u_w = 1.36; **b** Semi-developed wave: u_w = 1.58; **c** Fully developed wave: u_w = 1.72. Reprinted with permission from Miyara, Heat Mass Transf, 35:298–306, Springer nature, 1999

higher and becomes particularly evident when 3-dimensional films are investigated. In this context, simulations using the long-wave or multiple equation models can be conducted with significantly reduced computational power. Moreover, large domain simulations, e.g., interacting wave patterns where symmetry boundary conditions are not applicable, become manageable without huge computational effort. Based on a recent comparative study by Rohlfs et al. [74], the computational effort for the Direct Numerical Simulations (DNS) for 1 s of film evolution over a plate length of 20 mm required nearly 80,000 processor hours on 1024 processors. Respective simulations using the multiple equation model required a computational effort of about 1500 processor hours, which is less than 2% of the time required for DNS. However, in terms of accuracy, these results provided only a qualitative match with the DNS results, and quantitative differences were noted between the two sets of results [75].

2.2   Sensible Heat Transfer Models

For the heat transfer investigation of the falling film flowing over a plain wall (see Fig. 2), the governing equations described in Sect. 2.1 for the flow hydrodynamics is extended by the energy equation

$$\frac{\partial}{\partial t}\left(\rho C_p T\right) + \nabla \cdot \left(\rho C_p \vec{u} T\right) = \nabla \cdot (k \nabla T) + \mu \Phi \tag{43}$$

In addition to the continuity of temperature across the interface, the thermal inter-phase coupling condition is given as

$$-k_l \frac{\partial T_l}{\partial n} = -k_g \frac{\partial T_g}{\partial n} \tag{44}$$

Finally, assuming an isothermal bounding wall, the thermal boundary condition is given as

$$T|_{y=0} = T_w \tag{45}$$

Neglecting thermal dissipation, the non-dimensional form of the above equations is given as

$$\frac{\partial \theta^*}{\partial t^*} + u^* \frac{\partial \theta^*}{\partial x^*} + v^* \frac{\partial \theta^*}{\partial y^*} = \frac{\chi \varepsilon}{Pr\, Re}\left(\frac{\partial^2 \theta^*}{\partial x^{*2}} + \frac{1}{\varepsilon^2}\frac{\partial^2 \theta^*}{\partial y^{*2}}\right) \tag{46}$$

with the inter-phase coupling condition

$$\frac{\partial \theta_l^*}{\partial x^*}\frac{\partial \delta^*}{\partial x^*} + \frac{1}{\varepsilon}\frac{\partial \theta_l^*}{\partial y^*} = -\prod_k \frac{\partial \theta_g^*}{\partial x^*}\frac{\partial \delta^*}{\partial x^*} + \frac{\prod_k}{\varepsilon}\frac{\partial \theta_g^*}{\partial y^*} \tag{47}$$

and boundary condition

$$\theta^*|_{y^*=0} = 1 \tag{48}$$

where in the gaseous phase $\chi = \Pi_\alpha$ and in the liquid phase $\chi = 1$ with the thermal diffusivity ratio $\Pi_\alpha = \alpha_g/\alpha_l$ and thermal conductivity ratio $\Pi_k = k_g/k_l$. The Prandtl number (Pr) and the dimensionless temperature difference (θ^*) are expressed as

$$Pr = \frac{\nu_l}{\alpha_l},\ \theta^* = \frac{T - T_{in}}{T_w - T_{in}} \tag{49}$$

where T_{in} is the liquid inlet temperature and T_w is the wall temperature (depending on the thermal boundary conditions, a different reference temperature may be introduced).

Before addressing the thermal transport in wavy liquid films, it is useful to review the characteristics of thermal transport in smooth films, where computational approaches have been used to obtain approximate solutions. Figure 4 depicts two such films for different heat transfer scenarios, i.e., wall-side heat transfer and interfacial heat transfer. In both cases, the flow is hydrodynamically and thermally developed,

Fig. 4 Schematics of velocity and temperature profiles in smooth hydrodynamically and thermally developed film for wall-side and interfacial heat transfer

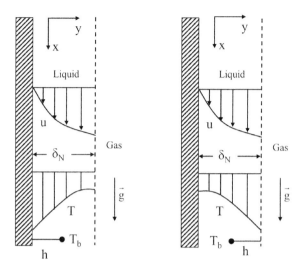

so that velocity profile and film thickness follow Nusselt relations. Introducing a dimensionless temperature

$$\theta^* = \frac{T - T_d}{T_b - T_d}, \; T_b = \frac{\int_0^{\delta_N} \rho_l c_l u_N T \, dy}{\int_0^{\delta_N} \rho_l c_l u_N \, dy} \tag{50}$$

where T_b designates the mean bulk liquid temperature and T_d denotes the temperature at the diabatic boundary. The condition of thermally developed flow can be expressed as $\partial \theta^* / \partial x^* = 0$. The results for different wall temperature boundary conditions such as Uniform Heat Flux (UHF) and Uniform Wall Temperature (UWT) together with the governing equations are summarized in Table 3.

For the uniform heat flux boundary condition, the boundary-value problem can be solved analytically, and the exact Nusselt number is given in Table 3 for wall-side and interfacial heat transfer, respectively. By contrast, no closed-form solution has been

Table 3 Nusselt number for smooth hydrodynamically and thermally developed film

Wall conditions	Governing equations	Wall-side transfer $\left.\frac{\partial \theta^*}{\partial y^*}\right|_{y^*=0} = -Nu, \left.\frac{\partial \theta^*}{\partial y^*}\right|_{y^*=1} = 0$	Interfacial transfer $\left.\frac{\partial \theta^*}{\partial y^*}\right|_{y^*=0} = 0, \left.\frac{\partial \theta^*}{\partial y^*}\right|_{y^*=1} = -Nu$
UHF	$\left[1.5 y^{*2} - 3y^*\right] Nu = \frac{\partial^2 \theta^*}{\partial y^{*2}}$	$Nu \approx 1.88$	$Nu \approx 3.41$
UWT	$\left[1.5 y^{*2} - 3y^*\right] Nu \theta^* = \frac{\partial^2 \theta^*}{\partial y^{*2}}$	$Nu = 35/17$	$Nu = 140/33$

obtained for the case of a uniform temperature boundary condition, although several authors have derived approximate solutions, using iterative computational methods or series expansions for the dimensionless temperature (see e.g., Nusselt [76], Kays and Crawford [77], and Limberg [78] regarding wall-side transfer and Brauer [79] regarding interfacial transfer). Kays and Crawford [77] solved the energy equation without the assumption of thermally developed flow, using an iterative method, and obtained the following approximate relation for the Nusselt number averaged over a considered transfer length x:

$$\overline{Nu_x} = \frac{1}{x} \int_0^x Nu(x)dx \approx 1.88 + 0.0942b, \quad b = RePr\frac{\delta_N}{x} \tag{51}$$

This relation was stated to hold for $b < 1/0.15$ and, for thermally developed flow ($b \rightarrow 0$), converges toward the value 1.88 (see Table 3). Brauer [79] derived a similar relation for the case of interfacial transfer, given as

$$\overline{Nu_x} = 3.41 + \frac{0.276b^{-1.2}}{1 + 0.2b^{-0.7}} \tag{52}$$

It is important to note that the UWT conditions possess a higher Nusselt number than the UHF conditions for fully developed smooth films.

With the understanding of the Nusselt number distribution for smooth films, attention will now be directed toward the numerical approaches for the thermal transport in wavy liquid films.

Hirshburg and Florschuetz [80] analyzed the falling wavy films based on the assumptions of a parabolic velocity profile and a linear temperature distribution. They showed that their heat transfer model incorporating wave effects was in good agreement with the experimental data. Jayanti and Hewitt [81] conducted numerical simulations by assuming sinusoidal and solitary waveforms for the film interface shape and obtained the field variables for periodic laminar flow. They concluded that the overall heat transfer is still dominated by conduction primarily due to a reduction in effective film thickness. The contribution of circulation to the heat transfer enhancement was observed to be minimal. Other studies have solved the energy equation coupled with the Navier–Stokes equation in order to clarify the heat transfer enhancement mechanism by the interfacial waves. The finite difference model by Miyara [82] was also used to explore the flow dynamics and heat transfer of falling liquid films with interfacial waves flowing on a vertical plate. Their studies contended that heat transfer enhancement was due to a combination of both film thinning and convection effects resulting from circulation flow in solitary or roll waves. The dominating effect between the film thinning or convective circulation depends on the Prandtl number, as shown in Fig. 5. For low Pr liquids, convection currents may be neglected, and for high Pr liquids, both effects are seen to be dominant. Kunugi and Kino [83] performed both two- and three-dimensional numerical simulations of falling liquid film flow on the vertical wall in order to investigate the relationship between the interfacial wave behavior and heat transfer under both uniform wall

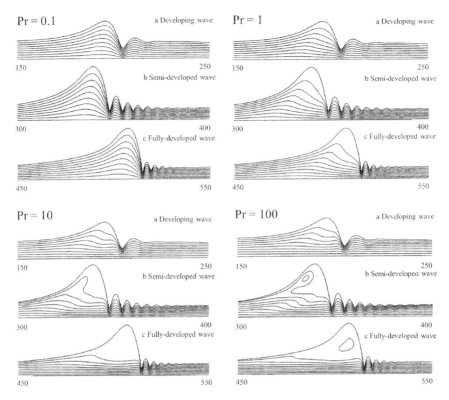

Fig. 5 Temperature contours for Prandtl number (Pr) = 0.1, 1, 10 and 100 with wave celerity u_w for Re = 100, We = 0.503, Fr = $-\infty$, disturbance amplitude (ε) = 0.03, and frequency (f) = 0.01: **a** Developing wave: u_w = 1.36; **b** Semi-developed wave: u_w = 1.58; **c** Fully developed wave: u_w = 1.72. Reprinted with permission from Miyara, Heat Mass Transf, 35:298–306, Springer nature, 1999

temperature and heat flux conditions. They used an interface volume tracking technique based on the Multi-interface Advection and Reconstruction Solver (MARS). The MARS uses a new Piecewise Linear Interpolation (PLIC) algorithm for tracking and reconstructing the surface between neighborhood computational cells and can capture the surface behavior with high accuracy. Their numerical results indicate that the heat transfer of falling film flow is enhanced by small vortices between the solitary and capillary waves in addition to the flow recirculation in the solitary wave. The heat transfer calculations by Dietze et al. [17] using the interface tracking VOF approach also reached similar conclusions on the importance of the small vortices and the flow recirculation on the heat transfer. More recently, Yu et al. [84] performed a numerical investigation of the wave dynamics using a modified version of the Volume Of Fluid (VOF) method for tracking the free surface. The artificial compression term used in this approach leads to compression of the liquid–gas interface and prevents the interface smearing. An important advantage of the modified VOF method is the possibility of capturing the interface region much more sharply than

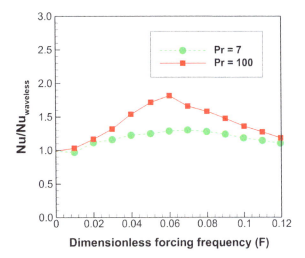

Fig. 6 The relative average Nu as a function of disturbance frequency for Re = 69 on a vertical plate. Reprinted with permission from Yu et al., J Heat Transfer, 135:101010, ASME, 2013

with the conventional VOF approach. They summarized the waveforms dependent on the Reynolds number and disturbance frequency in the form of a regime map. They also proposed a correlation describing the separation curve between the sinusoidal waves regime and solitary waves regime. The thermal transport in the wavy film has been simulated for the uniform wall temperature boundary condition. Figure 6 shows the influence of waves on the Nusselt number, which is quite significant for higher Prandtl numbers in a specific range of disturbance frequencies. For Re = 69 and a dimensionless forcing frequency of 0.06, the maximum enhancement in the Nusselt number amounts to around 85% of the Nusselt number without waves for Pr = 100 and around 24% for Pr = 7.

2.3 Vapor Absorption Models

Absorption of vapor in a falling film on a plain wall has been investigated in a number of studies. The situation is similar to the condition shown in Fig. 2, except that the liquid absorbs vapor from the surrounding through the gas–liquid interface. In this process, the secondary fluid (e.g., lithium bromide solution) falls over the plain surface. The region around the liquid film is filled with refrigerant vapor (e.g., water vapor), which is absorbed into the secondary fluid as it falls down the plain surface. The species concentration equation for vapor absorption in the liquid is given by

$$\frac{\partial}{\partial t}(\rho C_m) + \nabla \cdot \left(\rho \overrightarrow{u} C_m \right) = \nabla \cdot (\rho D_{AB} \nabla C_m) \tag{53}$$

The modified inter-phase coupling condition for the energy transfer between the phases is given by

$$-k_l \frac{\partial T_l}{\partial n} = -k_g \frac{\partial T_g}{\partial n} + \Delta H_{abs} \dot{m}''_{abs} \tag{54}$$

Based on the Fick's law of mass diffusion, the absorbing mass flux from vapor to the liquid is calculated as

$$\dot{m}''_{abs} = \frac{D_{AB}\rho}{C_m} \nabla C_m \widehat{n_\Gamma} \tag{55}$$

In the above Eqs. 53–55, ΔH_{abs} is the heat of absorption, C_m is the species concentration in the liquid and D_{AB} denotes the binary diffusion coefficient. The working fluid for the vapor absorption system is essentially a mixture of two fluids: a refrigerant and a fluid to absorb the refrigerant. At the liquid–vapor interface, the concentration is determined using the vapor pressure equilibrium assumption. The fluid solution is assumed to be in equilibrium with the vapor when the chemical potential of the refrigerant in the fluid solution is equal to the chemical potential of its vapor. The heat and mass transfer coefficients are usually defined at the gas–liquid interface for the absorption processes.

There have been only a few numerical studies [85–94] of the falling film absorption phenomenon over plain walls. Among the early studies, under major assumptions, Kawae et al. [85] presented a finite difference model for a vertical film with a fully developed parabolic velocity profile and the film thickness varying as a function of the absorbed vapor. They also included the effect of a change in liquid properties with changing temperature and concentration. They found that the variable properties have very little effect on the final results with a 5% difference in the total amount of vapor absorbed calculated using constant and variable properties. They also varied the operating conditions to study their effect on absorption. One important conclusion from their study was that the effect of flow rate and inlet temperature depended on the flow length and that the trend actually reverses after a certain downstream distance over the plate surface. Grossman [86] presented a solution for vapor absorption in a constant thickness film falling down a vertical wall for both isothermal and adiabatic wall boundary conditions. The energy and diffusion equations were solved simultaneously to give the temperature and concentration variations at the liquid–gas interface and at the wall. The author presented a Fourier series solution to the heat and mass transfer problem and demonstrated the numerical calculation of the eigenvalues under different wall boundary conditions. The effect of the number of eigenvalues used on the stability and accuracy of the series solution was also discussed in their study where they found that a very high number of eigenvalues were required for the series solution to converge near the fluid inlet. The interfacial distribution of temperature (θ_i) and concentration (γ_i) as a function of plate length for various Lewis number (Le $= D_{AB}/\alpha$, the ratio of mass diffusivity to thermal diffusivity) is shown in Fig. 7. An increase in Le leads to a decrease in θ_i and to an increase in γ_i at $\zeta = 0$ for both wall conditions and at $\zeta = \infty$ for the adiabatic case. In addition to this, the Lewis number has an effect on the development of the concentration boundary layer. The larger Le, the shorter the distance required for the concentration change to reach the

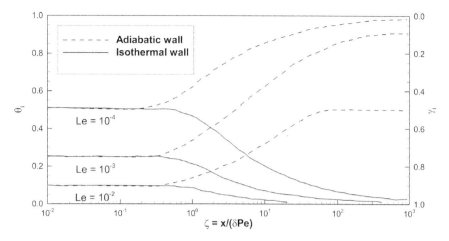

Fig. 7 Dimensionless interface temperature, $\theta_i = (T_i - T_0)/(T_e - T_0)$, and concentration $\gamma_i = (C_i - C_0)/(C_e - C_0)$, versus normalized length, $\xi = x/Pe\delta$ for various Lewis numbers (Le). T_e: equilibrium interface temperature at inlet concentration, C_0; Ce: equilibrium interface concentration at inlet temperature, T_0. Reprinted with permission from Grossman, Int J Heat Mass Transf, 26:357–371, Elsevier, 1983

wall. The Nusselt and Sherwood number (dimensionless mass transfer coefficient) were found to depend on the Peclet number (Pe = RePr) and Lewis numbers, as well as on the equilibrium characteristics of the working materials. In addition, Grossman [87] also presented a finite difference method-based numerical solution approach for the falling film vapor absorption. The author encountered substantial difficulties in obtaining a numerical solution near the inlet region due to a disconnection caused by the simultaneous assumption of a uniform lithium bromide concentration at the inlet and a vapor pressure equilibrium concentration at the liquid–vapor interface. As a remedy, a similarity solution for the inlet region was developed, where the effects of the interfacial boundary layer do not reach the wall. The author concluded that this similarity solution can be used for the inlet region, complemented with the Fourier series solution or the numerical solution to be used downstream.

The study by Ramadane et al. [88] used a wall-side coolant instead of an adiabatic or isothermal wall in their implicit finite difference model of a laminar falling film. The governing equations for the coolant-side heat transfer were included in their numerical model that could accurately predict the heat transfer coefficient on the coolant side. They also presented a model in which the heat transfer coefficient on the coolant side was calculated using correlations available in the literature. They showed the temperature and concentration distribution in the film for various downstream positions on the wall. In addition to this, they also presented the liquid-side interface heat transfer coefficient values for film Reynolds numbers in the range of 15–70. The comparison of their simulated data with the experimental data was found to be quite satisfactory in nature. Ibrahim and Nabhan [89] developed a mixed model that combines the analytical solution proposed by Nakoryakov and Grigor'eva [90] for the inlet regions close to the liquid–vapor interface and a finite difference model for the rest of the solution domain. They claimed that their model predictions are as accurate as the more complex models proposed by Grossman [86] but require less computational resources. They defined the effectiveness of the absorber and presented the relationship between effectiveness, the number of transfer units, and coolant capacity ratio. These results showed that the absorber effectiveness could be increased by increasing the coolant flow rate and/or increasing the contact area. Patnaik and Perez-Blanco [91] presented a numerical model for the absorption process in falling films where they investigated the effect of roll waves. They assumed a parabolic film velocity profile and the diffusion to be equimolar in the formulation of the energy balance at the liquid–vapor interface. Based on image analysis of the experimental data, they selected a wave frequency of 13 Hz as the dominant frequency for their model. With their wavy film model, they observed an enhancement in the Sherwood numbers by around 200% compared to the smooth film predictions. They attributed this enhancement to the hydrodynamics of the wave. They concluded that when the velocity components in the direction normal and parallel to the interface were in phase, it led to a significant enhancement in the mass transfer in the film. Sabir et al. [92] investigated the effect of wave motion on the performance of the falling film absorbers by solving the combined energy and diffusion equations at a low Reynolds number (Re = 100) that employed an Alternating Direction Implicit (ADI) finite difference method. In light of the results obtained for smooth flow and wavy flow, it was found that the interfacial wave motion improved the heat and mass transfer rates despite using a film velocity profile that did not permit mixing. Habib and Wood [93] presented a numerical model using the control volume approach that accounts for the momentum transfer and its effect on the fluid velocity profiles. They also accounted for pressure gradients in the absorber and interfacial shear between the two phases. They found that the rate of absorption was highest just after the inlet when the effect of the wall temperature first reaches the interface. After the peak, they found that the absorption rates decrease exponentially downstream on the wall. More recently, Karami and Farhanieh [94] developed a model for coupled heat and mass transfer process in absorption of refrigerant vapor into a LiBr solution of water-cooled vertical plate absorber in the Reynolds number range of 5–150. The boundary layer assumptions were used for the transport of mass, momentum, and energy equations and the fully implicit finite difference method was employed to solve the governing equations in the film flow. The dependency of lithium bromide aqueous properties to the temperature and concentration and film thickness to vapor absorption was employed. They investigated the effects of various operating conditions such as solution flow rate, inlet solution concentration, etc. on the local as well as on the average heat and mass transfer behavior. To investigate the effect of the flow rate or Re, as shown in Fig. 8, the average mass transfer coefficient shows an increasing trend as the ability of the solution to absorb more vapor increases with an increase in the flow rate. The effect of the solution Re on the Nusselt number is exhibited in two distinct regimes. At low Re numbers, increase in Re decreases the dominating effect of thermal diffusion and hence the Nusselt number decreases. This continues till Re = 10, where the reduction in the thermal diffusion is compensated by the increase

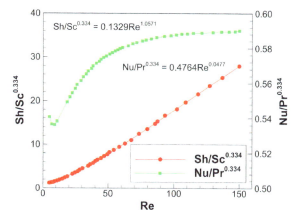

Fig. 8 The effect of film Re on Nusselt and Sherwood number for 5 < Re < 150. Reprinted with permission from Karami and Farhanieh, Heat Mass Transf, 46:197–207, Springer Nature, 2009

in convective heat transfer. Thereafter, Nusselt number shows an increasing trend where the convection effects become a dominant contributor to the heat transfer as the solution Re increases. Correlations for average Nusselt and Sherwood numbers were also provided in their study which can be used to calculate heat and mass transfer for designing the vertical plate absorbers.

2.4 Liquid Evaporation Models

Heat transfer through latent heat exchange connected with the vaporization of a liquid film, the so-called evaporative cooling, is often encountered in several types of thermal equipment. The schematics of the situation are similar to Fig. 2 where interfacial heat and mass transfer take place from the falling liquid film into the surrounding gas phase. Depending on the industrial applications, the surrounding medium may be still or have a finite velocity in any directions (cocurrent or countercurrent). In addition to the hydrodynamic equations presented in Sect. 2.1 and thermal equations presented in Sect. 2.2, a vapor concentration equation is solved for the heat and mass transfer process are given by

$$\frac{\partial}{\partial t}(\rho Y_v) + \nabla \cdot \left(\rho \vec{u} Y_v\right) = \nabla \cdot \left(\rho D_{vg} \nabla Y_v\right) \tag{56}$$

The modified inter-phase coupling condition for the energy transfer between the phases is given by

$$-k_l \frac{\partial T_l}{\partial n} = -k_g \frac{\partial T_g}{\partial n} + \Delta H_{evap} \dot{m}''_{evap} \tag{57}$$

Based on the Fick's law of mass diffusion, the evaporating mass flux from liquid to vapor is given by

$$\dot{m}''_{evap} = \frac{D_{vg}\rho_{gp}}{1 - Y_v}\nabla Y_v \widehat{n_\Gamma} \qquad (58)$$

In the above Eqs. 56-58, ΔH_{evap} is the latent heat of the evaporating fluid, Y_v is the vapor mass fraction, and D_{vg} is the binary diffusion coefficient of vapor in the gas. The vapor pressure at the gas–liquid interface is assumed to be at the saturation pressure. It depends primarily on the interface temperature and is estimated by the Wagner equation. This pressure is subsequently used to calculate the vapor concentration at the interface, assuming it to be at saturation condition.

Similar to the previous sections, modeling strategies for falling film heat and mass transfer (evaporation) differ with respect to the level of approximation to the above governing equations. Numerical investigation of liquid film evaporation over an insulated wall or with heated conditions has already been studied by a number of researchers. Chow and Chung [95] investigated water evaporation processes in the presence of both laminar air and superheated steam. Under constant thermophysical properties, they presented similarity solutions, which were finally solved numerically by the quasi-linearization method. This method has been successfully applied to two-point boundary-value problems. It was shown that below a certain stream temperature, named an inversion temperature, water evaporation rate decreases as air humidity increases; and above the inversion temperature, the rate of water evaporation increases with an increase in the humidity of air. Schröppel and Thiele [96] performed a numerical study on liquid film evaporation in both laminar and turbulent flows of a binary gas mixture flowing over a plane. They employed temperature-dependent thermophysical properties for the gas components in the mixture. Numerical results obtained through finite difference approximations were found to be in good agreement with measurements in terms of flow profiles and Stanton numbers (St = Nu/RePr). Yan et al. [97] numerically investigated the effects of a wetted wall on laminar mixed convection heat and mass transfer processes in vertical channels through the finite difference approach where the thickness of the water film was assumed to be very thin. Their results showed that the heat transfer rate was greatly enhanced under the mass diffusion process connected with water vaporization. However, the later study performed by Yan [98] showed that the validity of the extremely thin-film assumption exists only for low liquid mass flow rate conditions. For conditions with higher liquid mass flow rate, the assumption becomes inapplicable, especially near the plate surface entrance. Studies by Tsay et al. [99] dealt with the ethanol film evaporation along insulated vertical surfaces where they solved the governing equations with the finite difference method. The inertia terms were dropped off, and only the body forces were retained for the liquid film. The respective governing equations for the liquid film and gas stream were solved in a coupled manner under this numerical approach. They observed that the cooling of the liquid film is mainly caused by the latent heat transfer connected with the vaporization process. Significant liquid cooling results for the system with a high inlet liquid temperature or a low liquid flowrate. Tsay and Lin [100] investigated the heat and mass transfer characteristics for a falling liquid film evaporating into a cocurrent gas stream. The liquid water film falls along a vertical plate subjected to uniform heat flux. The numerical predictions were found to be in reasonable agreement with the measured data, where the numerical predictions of the wall temperature were within 10% of the measured data. It was concluded that the latent heat transport connected with the vaporization of the liquid film plays an important role in the transfer processes. Based on their data, they provided the following correlations for the Nusselt number (Nu_l) distribution over the plate surface:

$$
Nu_l = \frac{q_w''\left(v_l^2/g\right)^{1/3}}{k_l(T_w - T_{l,in})} = 0.76 S^{1.1} Re_l^{0.7} Re_g^{0.14}
\begin{cases}
3.7 \times 10^{-4} < S = \dfrac{\rho_g D_{vg} H_{lg}(Y_{v,r} - Y_{v,\infty})}{q_w'' x} < 3.7 \times 10^{-2} \\
100 < Re_l = 4\dot{m}_l/\mu_l < 250 \\
0.93 < Re_g = \dfrac{u_{g,\infty}\left(v_l^2/g\right)^{1/3}}{v_g} < 8.52
\end{cases}
\tag{59}
$$

Debbissi et al. [101, 102] conducted a numerical investigation of coupled heat and mass transfers by natural convection during water evaporation in a vertical channel, where they accounted for the radiative heat transfer between the plates. The governing equations were solved by finite difference methods. In this study, they paid particular attention to explore the effect of ambient conditions on the evaporation process and inversion temperature phenomenon under free and mixed convection. Feddaoui et al. [103] studied the effects of falling water evaporation inside insulated vertical channels without the thin-film approximation. They solved the steady non-linear coupled differential equations by the finite difference numerical scheme. They investigated the effect of gas flow (cocurrent) on the evaporative cooling process and concluded that significant liquid cooling results for the cases with a higher gas flow rate. More recently, Ren and Wan [104] solved the steady 2-D model equations for countercurrent laminar moist airflow inside thermally insulated vertical plate channels with falling water film evaporation using the finite volume method. They assumed the film to be smooth, and the SIMPLE method was adopted in solving the governing equations. They examined the influence of moist air inlet mass flow rate (Re_g), water inlet mass flow rate (Re_l), moist air inlet dimensionless temperature ($\theta_{g,i}$) and the ratio of channel length to half channel width (α_L) on the heat and mass transfer coefficients. In an extension to the previous study, Wan et al. [105] investigated the combined heat and mass transfer characteristics in vertical plate channels with falling film evaporation where the plates are externally subjected to Uniform Heat Flux (UHF) or Uniform Wall Temperature (UWT) conditions. Correlations were proposed for the prediction of average moist air side heat (Nu_M) and mass transfer (Sh_M) coefficients under the specific flow conditions, given by

Uniform Heat Flux (UHF)

$$
\begin{cases}
Nu_M = \frac{4(W-\delta)h_m}{k_g} = 2.762 Re_l^{-0.346} Re_g^{0.883} \theta_{g,i}^{0.757} q_w^{-0.143} \alpha_L^{-0.650} \\
Sh_M = \frac{4(W-\delta)\beta_m}{D_{AB}\rho_g} = 3.345 Re_l^{0.027} Re_g^{0.212} \theta_{g,i}^{0.041} q_w^{0.027} \alpha_L^{-0.136}
\end{cases}
$$

Uniform Wall Temperature (UWT)

$$
\begin{cases}
Nu_M = \frac{4(W-\delta)h_m}{k_g} = 12.876 Re_l^{-0.027} Re_g^{0.136} \theta_{g,i}^{-0.128} \theta_w^{0.092} \alpha_L^{-0.217} \\
Sh_M = \frac{4(W-\delta)\beta_m}{D_{AB}\rho_g} = 10.919 Re_l^{-0.047} Re_g^{0.011} \theta_{g,i}^{0.041} \theta_w^{0.165} \alpha_L^{-0.030}
\end{cases}
\tag{60}
$$

The application ranges for the above correlation are ($20 \leq Re_l \leq 120$, $600 \leq Re_g \leq 2200$, $0.4 \leq \theta_{g,i} \leq 1.4$, $200 \leq \alpha_L \leq 350$, $0.5 \leq q_w \leq 3.0$, $1.5 \leq \theta_w \leq 2.5$).

One of the critical drawbacks of the foregoing studies was that the effect of interfacial waves on the falling film evaporative cooling was neglected. The most rigorous approaches for modeling evaporation involve direct evaluation of the vapor concentration gradient (evaporation in unsaturated conditions) or temperature gradients (evaporation in saturated conditions) at the interface to determine phase change rates. Hence, capturing these effects would require full-scale solutions of the governing equations using the interface tracking technique for three phases (liquid, liquid–vapor, and gas) as developed by Schlottke and Weigand [106]. Under this approach, preliminary CFD simulations (A. Karmakar and S. Acharya, personal communication, 2019) of an evaporating liquid film flowing over an insulated vertical surface were undertaken with a solver implemented for tracking 3 phases using OpenFOAM. The evaporation of the liquid film is driven by the vapor concentration gradient existing in the gas medium. Figure 9 shows the distribution of the time-averaged insulated wall temperature and liquid film bulk temperature obtained from the numerical simulations. Due to computational costs, simulations were undertaken on a limited portion of the experimental configuration. For a plate length of 0.1 m and a solution time of 2 s, the required computational time was in the range of 576 processor hours, consuming 18 h of the supercomputer. The total number of grid points in the simulations was around 1.3 hundred thousand with a minimum grid size of 1×10^{-5} m and the time step was limited to 0.01 ms in order to avoid numerical instabilities at the interface. The results from the numerical simulations are seen to very good agreement with the literature [99], where the maximum deviation was less than 5% of the experimental data. The mass transfer process at the gas–liquid interface is seen to absorb a significant amount of energy, which in turn cools down the liquid film.

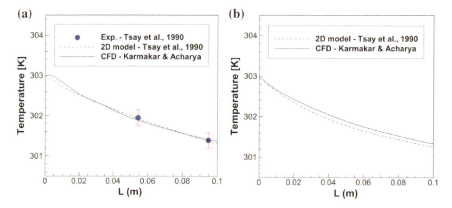

Fig. 9 a Insulated wall temperature and **b** bulk liquid temperature distribution for ethanol mass flowrate $(m_{L,in}) = 0.02$ kg/(ms) and inlet liquid temperature $(T_{L,in}) = 30$ °C. A. Karmakar and S. Acharya, personal communication, 2019

3 Falling Film Over Horizontal Round Tubes

Many researchers have studied the falling film modes or hydrodynamics over horizontal tubes. Several experimental studies have been conducted where they observed that when the liquid flow rate over horizontal tubes is increased, three-basic intertube flow modes are successively observed: the droplet, jet, and sheet modes (shown in Fig. 10) along with two intermediate modes: droplet-jet and jet-sheet [107, 108]. The effects of the liquid flow rate and fluid properties on the flow mode transitions have also been investigated. The flow mode transitions are usually represented as a relationship between the transitional Reynolds and the modified Galileo number [107–110]. Based on the flow mode transition studies, the flow patterns for the jet mode for various liquids (based on Ga) include inline jets, staggered jets, and cases where both inline and staggered configurations are observed. The inline jet mode is observed at a lower mass flow rate in comparison to the staggered jet mode. The transition relations are given as

$$\text{Droplet to droplet - jet} : \text{Re} = 0.074 Ga^{0.302}$$
$$\text{Droplet-jet to jet} : \text{Re} = 0.096 Ga^{0.301}$$
$$\text{Jet to jet-sheet} : \text{Re} = 1.414 Ga^{0.233}$$
$$\text{Jet-sheet to sheet} : \text{Re} = 1.448 Ga^{0.236}$$
$$\text{Inline-jet to staggered jet} : \text{Re} = 20.53 + 0.256 Ga^{0.25} \tag{61}$$

The effects of geometry, such as tube diameter and tube spacing on the flow mode transitions, have also been studied. Experimental studies [107, 109] concluded that the transitional Reynolds numbers showed no dependence on tube spacing in the experimental range studied. The tube diameter has also been shown to have little effect on the mode transitions. The effect of vapor shear on flow mode transitions has also been investigated in the literature [107, 111] and results show that most flow

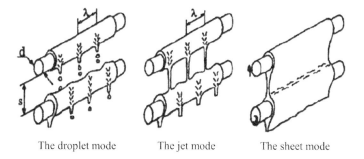

| The droplet mode | The jet mode | The sheet mode |

Fig. 10 The basic intertube flow modes in flow over horizontal tubes. Reprinted with permission from Mitrovic, Chem Eng Technol 28:684–694, John Wiley and Sons, 2005

mode transitions occurred at higher Reynolds numbers as Weber number increased, but that this is not the case for all transitions.

In the droplet and jet falling film modes, the liquid falls from specific sites under the tube surface, which are a fixed distance apart, denoted by λ. The physics behind these formation sites originating from the thin film on the underside of the horizontal tube relates to Taylor instability [112]. For Taylor instability in inviscid incompressible fluids, Richard Bellman [113] found the critical or the shortest unstable disturbance wavelength is $\lambda_{cr} = 2\pi\sqrt{\sigma/g(\rho - \rho_g)}$, and the most amplified wavelength is given by $\lambda_{max} = \sqrt{3}\lambda_{cr}$. However, experimental results have shown that simple Taylor's instability theory and the underlying assumptions do not accurately capture all the underlying physics of the droplet and jet departure sites. Studies by Maron-Moalem et al. [114] showed that λ was dependent on the liquid flow rate, tube spacing, and tube diameter. In the experimental studies of Ganic and Roppo [115], it was reported that the value of λ was lower than that predicted by Taylor instability. Based on the experimental data, Hu and Jacobi [116] developed a correlation (Eq. 62) for these departure-site spacings in terms of liquid properties (Ga) and flow rate (Re).

$$\frac{\lambda}{\zeta} = \frac{0.836Z - 0.836\text{Re}/Ga^{1/4}}{\left\{1 + \left(\frac{0.836Z - 0.836\text{Re}/Ga^{1/4}}{0.75Z - 85/Ga^{1/4}}\right)^2\right\}^{1/12}}, \quad Z = \frac{2\pi\sqrt{3}d}{\sqrt{d^2 + 2\zeta^2}} \qquad (62)$$

In addition to studies on flow mode transitions and site spacings, a number of investigations have focused on characteristics of the falling film flow in between the tubes. It was observed that at very small liquid flow rates, liquid droplets dripped from the underside of the horizontal tube surface at regularly-spaced sites. Early studies by Rayleigh [117] laid the foundation for understanding the formation of droplets. Eggers and Villermaux [118] provide an excellent summary of the experimental, analytical, and computational methods used in the understanding of the droplet behavior including the formation of droplets from axisymmetric jets or capillary tubes and the dynamics of bifurcation which includes droplet detachment; bridge break up, etc. However, owing to the three-dimensional shape of the horizontal tubes, the case of droplet formation and detachment in between the horizontal tubes is somewhat different from most of the literature in terms of the overall shape, size, and internal velocity fields of the droplets, especially in the initial stages of droplet formation [119]. The extent of the droplet in the axial direction of the tube is greater than its extent in the circumferential direction. With a gradual increase in the mass flow rate, the liquid bridge between the horizontal tubes overcomes the bifurcation instabilities and leads to the formation of the jet mode. The diameter of the jet increases with an increase in flow rate. An interesting phenomenon of the pattern of the liquid jets is the surface outline of the liquid film on the tube wall. Experimental studies [107, 120] have observed that when liquid jets fall in a staggered pattern, there is a crest region where a ring of high film thickness is visible between two impinging jets. The lower jet then falls from the bottom of the ring, generating a staggered pattern. On the other hand, when liquid jets fall in an inline configuration, there are two surface curvatures. In the first case, the film is thick around the jet impingement region, and the film thickness decreases in the axial direction, reaching a minimum thickness between the two impinging jets. In the second case, there exists a dimple around the jet impingement region. The film thickness then increases along the axial direction, followed by a decrease in its thickness. The film thickness still reaches its minimum between the two jets. The first case was usually observed at a low Reynolds number, while the second one was observed at a higher Reynolds number. Another interesting phenomenon associated with liquid jets is the geometrical shape of the liquid jets [116] in between the tubes. Depending on the tube spacing and fluid properties, the jets possess a converging shape with some ripples observed toward its base prior to the impingement. The formation of these ripples on the jet surface is attributed to the development of capillary waves under the influence of the impingement surface. At high enough mass flow rates, a stable and uniform sheet mode is established. When the mass flow rate is increased further, the sheet in between the tubes becomes wavy.

3.1 Hydrodynamic Models and Flow Behavior

The review of the literature shows that studies related to hydrodynamic modeling of flow over horizontal tubes are mostly based on interface tracking techniques. A comprehensive overview of different types of interface capturing techniques is available in the landmark book by Tryggvason, Scardovelli, and Zaleski [57]. The complex flow patterns observed between the tubes present serious modeling challenges to the researchers. The governing equations for the flow over round horizontal tubes are similar to that of a plain flat surface, except the fact that it is extended to 3 dimensions. The overall results obtained using the numerical approaches have been compared with experiments, and generally show good agreement.

Killion and Garimella [4] performed a simulation of droplets of aqueous lithium bromide pendant from horizontal tubes by numerically solving the equations of motion on a fixed three-dimensional (3D) grid. The so-called *Volume of Fluid* (VOF) method of Hirt and Nichols [60] was used to handle the interface between the liquid and vapor phase. They assumed two vertical planes of symmetry that allowed the model to be reduced to 1/4 of a full droplet model. Periodic boundary conditions were considered in the gravitational direction. Their model could capture some of the essential characteristics of droplet formation, detachment, and impact (shown in Fig. 11), although some differences were noticed in comparison to their experiments. They also performed a comparative study of their 3D simulations with 2D axisymmetric models where the 2D models underpredicted the primary droplet diameter by around 20%. Their study was limited to the droplet flow mode only.

Jafar et al. [121] performed preliminary simulations using a CFD code (FLUENT) for 2D configurations with the flow over one, two, and three horizontal tubes. They reported the flow modes and film thickness for the Reynolds numbers range of 400 and 3200. Sun et al. [122] performed numerical simulations using FLUENT for 2D configurations with one and two tubes. The VOF methodology is used to track the

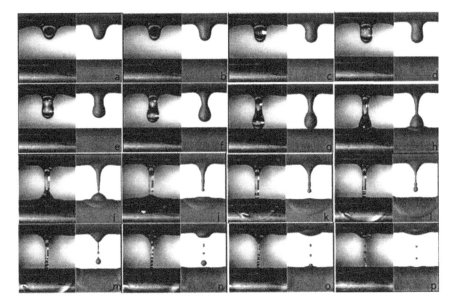

Fig. 11 Visual comparison of the experiment (left) with simulation (right) synchronized at impact: frames **a–g** 10 ms between frames; **g–p** 6 ms between frames. Reprinted with permission from Killion and Garimella, J Heat Transfer 126:1003, ASME, 2004

motion of liquid falling film and the gas–liquid interface. The simulations indicated that (i) the velocity distributions of the upper and lower parts of the tube are not strictly symmetric and non- uniform, (ii) the film thickness depends on flow rate and angular distributions, and (iii) the flow characteristics of the top tube are different from those of the bottom tube. Qiu et al. [123] performed 2D numerical simulations in order to study the film thickness characteristics over a fully wetted horizontal round tube in a falling film evaporator. They used the commercial CFD code Fluent that employed the VOF method to capture the gas–liquid interface. The temporal variation characteristics of film flow process and the steady film thickness distribution over horizontal tubes were analyzed with the help of the qualitative understanding of the forces (surface tension, gravitational, adhesion, inertia) acting on the liquid film over the tube surface. In an extended study, Qiu et al. [124] employed a 3-D model to simulate the falling film flow process over a horizontal tube in jet mode, where the Reynolds number spanned from 171 to 368. The study showed that with a single liquid jet, a saddle-shaped spreading liquid lamella was formed along the axial direction with an elongation distance maximum in the axial direction. In contrast to a single jet, the flow behavior with two adjacent liquid jets changed a lot due to the interaction between the jets at higher Reynolds number. As shown in Fig. 12, a crest was formed between the two jets, and the maximum film thickness over the tube surface was more than three times the minimum in the axial direction.

For the accurate setup of a heat and mass transfer model, Chen et al. [125] performed 3D simulations of the falling film pattern of a horizontal tube bundle. The

Fig. 12 Variation of film thickness at circumferential angles of 45° and 135° with $L^* = L/\lambda$, where λ denotes the spacing between two neighboring jets. Reprinted with permission from Qiu et al., Int J Heat Mass Transf 107:1028–1034, Elsevier, 2017

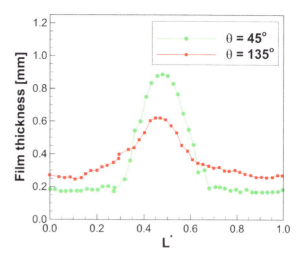

technique was based on the VOF method on Fluent used for the multiphase flow simulations. The experimental data were used to validate the numerical model. The simulated results showed the various flow pattern with different flow rates under which they observed the droplet, droplet-jet, jet, jet-sheet, and sheet flow patterns. The critical value of the flow rate for the transition from one mode to another (droplet to jet and jet to sheet) was close to what they observed in the experimental situations. A three-dimensional, two-phase transient CFD model based on the VOF method was developed by Li et al. [126] focusing on the local hydrodynamic characteristics of falling film over horizontal tubes, both in quiescent surroundings and with a countercurrent gas flow. The film Reynolds number (Re) ranged between 50 and 1500, in which laminar and turbulent flow regimes were present. For different liquid flow rates, the laminar model and k-ε turbulence model were employed to improve the model accuracy. With the introduction of the countercurrent gas flow, they observed that the circumferential bulk film thickness increases with an increase in the countercurrent gas flow rate. They also presented the transient behavior of the axial film thickness distributions for the three-basic intertube flow modes under quiescent surrounding conditions. Fiorentino and Starace [127] presented a two-dimensional numerical model in order to investigate the flow characteristics on horizontal tubes with countercurrent airflow. They characterized the film flow process under different magnitudes of water-to-air mass flow ratio. They also analyzed the effect of the arrangement of the tubes on the flow mode transitions. More recently, Ding et al. [128] presented a 3D CFD model, employing the VOF method, in which they simulated the flow behavior of drop and jet modes (inline only) and provided some direct comparison with the experimental studies of Killion and Garimella [129]. They highlighted the movement of complex saddle-shaped waves on the tube surfaces for the droplet mode. In addition to this, using a static contact angle model, they also investigated the effect of tube wettability on the falling film mode behavior in the droplet and jet mode. Their results indicated that the wetting area, spreading speed, and instability time of the falling film is reduced with an increase in the contact angle, while the first two parameters increase at higher specific flow rates. In order to have a greater understanding of the falling film hydrodynamics, a 3D numerical study (A. Karmakar and S. Acharya, personal communication, 2019) of the falling film encompassing all the six modes of flow patterns with increasing Reynolds number: droplet, droplet-jet, inline jet, staggered jet, jet-sheet, and sheet modes was undertaken. The simulations were performed using an algebraic VOF method available on Open-FOAM. The solver utilizes an interface compression technique in order to capture the gas–liquid interface sharply. This numerical study providing resolved solutions for the flow characteristics (film thickness and velocity) over all the six regimes of flow modes that are currently not available in the literature. Figure 13 represents a qualitative comparison of the complex fluid behavior observed in the droplet and jet mode between the experimental studies and those predicted by our numerical model. Phenomena such as droplet formation and fall, wave formation due to droplet impact, break up of liquid bridge leading to satellite droplets in the droplet mode, formation of capillary waves due to Raleigh–Plateau instability in inline/staggered jet mode and regular stable-crest-stable distribution in the staggered jet mode are well captured as those observed in the experiments. Quantitative agreement of the falling film flow characteristics was also obtained in terms of the departure site spacing of the droplet and jet mode [116] with deviations of around 6% and droplet dimensions [130] with deviations of 4.4% in terms of the droplet diameter. The high-fidelity simulations showed the presence of liquid waves (saddle-shaped or crescent-shaped) that were seen to travel on the tube surface downstream of the impingement region in each of the modes. These waves are created either by the droplet impact at low mass flow rates or by capillary instability of the impinging jet at higher mass flow rates.

The computational effort required for the numerical simulations of falling film over horizontal tubes is usually very high and depends primarily on the grid resolution, time step and the extent of the domain around the tube surface. In the above studies [4, 123, 124, 126, 128], the minimum cell dimension varied from 2×10^{-5} to 1×10^{-4} m and the time step ranged between 0.02 to 0.2 ms. In the CFD studies by our group, for the numerical domain consisting of 3 horizontal tubes of 16 mm, a spacing of 16 mm between the tubes and half of the departure site extent in the tube axial direction, the number of grid points amounted to 3.5 million cells with 2×10^{-5} m as the least cell size. In terms of computational time, the simulations were performed with 96 processors on a supercomputer, and it took around 40 h to simulate 4 secs of flow time for each intertube mode of flow with a time step of 0.02 ms. This point toward the high computational cost associated with these highly resolved simulations.

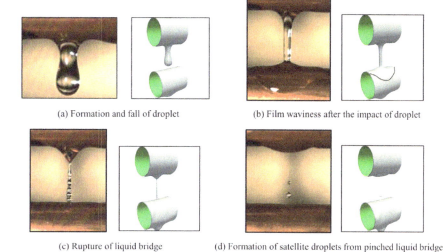

(a) Formation and fall of droplet

(b) Film waviness after the impact of droplet

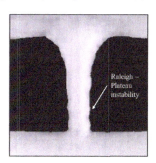

(c) Rupture of liquid bridge

(d) Formation of satellite droplets from pinched liquid bridge

(e) Formation of capillary waves on the free surface of the jet

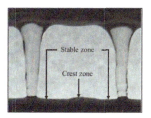

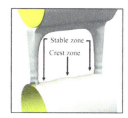

(f) The stable-crest-stable configuration in staggered jet mode

Fig. 13 The left images are from experiments, and the right images are from our numerical simulations. A. Karmakar and S. Acharya, personal communication, 2019; **a–d** Comparison of the evolution of the pendant drop with experiments [119], reprinted with permission from Killion and Garimella S, Int J Heat Mass Transf, 47:4403–4414, Elsevier, 2004. **e** Comparison of jet shape observed in experiments [116], reprinted with permission from Hu and Jacobi, Exp Therm Fluid Sci, 16:322–331, Elsevier, 1998 **f** Comparison of the stable-crest-stable configuration of film thickness with experiments observations [120], reprinted with permission from Chen et al., Int J Heat Mass Transf, 89:707–713, Elsevier, 2015

3.2   Sensible Heat Transfer Models

Several studies have reported measurements of the heat transfer characteristics for falling film flow over the horizontal cylindrical tubes. Prior research reviewed in Ribatski and Jacobi [131], Ribatski and Thome [132] have shown that the intertube flow modes, flow rate, tube spacing, tube diameter, heat flux, vapor flow, and liquid properties have a significant effect on the heat transfer performance of falling films on horizontal round tubes. Inconsistencies in the experimental observations were noticed between the researchers regarding the effect of various parameters on falling film heat transfer. Nevertheless, it is difficult to isolate the effect of the falling film mode on the overall heat transfer behavior, and the problem can be exacerbated by the complexity of the flow characteristics, especially in the jet and droplet modes. Hu and Jacobi [133] pointed out that understanding the role of the falling film mode in sensible heat transfer may help in isolating these effects on evaporation heat transfer. However, numerical models for the heat transfer characteristics for flow around the circular horizontal tubes are somewhat limited. Luo et al. [134] presented a two-dimensional CFD study of the falling film heat transfer on horizontal tubes of three different shapes that are applied in the seawater desalination process. The flow and heat transfer characteristics of the falling water film on a circular shaped, a drop-shaped tube, and an oval-shaped tube were analyzed, respectively. The VOF method was employed to study the influence of the liquid mass flow rate and the feeder height on the film thickness distribution and the heat transfer performance. Their numerical results showed that the minimum film thickness appears approximately at the angular positions of $125°$, $160°$, and $170°$ for the circular, oval-, and drop-shaped tubes, respectively. The film thickness grows with the increase in the mass flow rate and the decrease in the feeder height. It is also concluded that increasing liquid flow rate or Reynolds number results in an increase in the heat transfer coefficients. Moreover, in comparison with the circular tube, the drop-, and oval-shaped tubes have a thinner thermal boundary layer and a lower dimensionless temperature, which means it provides a better heat transfer performance. Jafar et al. [135] presented a numerical investigation of the liquid film falling over three horizontal plain tubes. The flow Reynolds number ranged from 50 to 3000 and tube diameters of 0.022, 0.1, and 0.123 m were employed. The numerical methodology was based on the Volume Of Fluid (VOF) method to account for two phases. Under this investigation, the effect of liquid flow rate (Reynolds number) and the tube diameter on heat transfer coefficient have been studied. Similar conclusions made by Luo et al. [134] was reached in this study regarding the effect of liquid flow rate on the heat transfer coefficient. The study also revealed that decreasing the tube diameter results in an increase in the heat transfer coefficient. Karmakar and Acharya [136] developed a 3D numerical model in which they investigated the flow and sensible heat transfer behavior over three successive horizontal tubes. The simulations were performed in OpenFOAM with three different liquids and under two different tube diameter and tube spacings, respectively. The numerical model could capture the possible basic intertube flow modes (droplet, inline jet, staggered jet, and sheet) using different

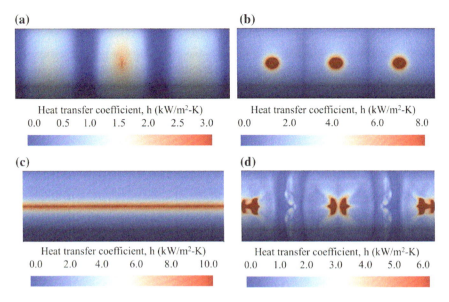

Fig. 14 Contours of the local heat transfer coefficient with the liquid inlet temperature as the reference temperature in **a** droplet mode at Re = 2 **b** inline jet mode at Re = 20 **c** sheet mode at Re = 60 with ethylene glycol and **d** staggered jet mode at Re = 300 with water as the working fluid. Reprinted from Karmakar and Acharya [136]

liquid properties. The heat transfer predictions were validated with the experimental observations presented by Hu and Jacobi [133]. The time-averaged distribution of the local heat transfer coefficient as shown in Fig. 14 depends on the flow features in each mode: it is constant in the axial direction with significant variation in the circumferential direction for the sheet mode and shows substantial variations in both directions for the jet mode (inline and staggered), and droplet mode. The average heat transfer behavior was seen to increase with the liquid mass flow rate for all the liquids. The heat transfer coefficient also increases with an increase in the tube spacing. The increase with spacing is mainly due to the higher velocity of jet impingement and smaller diameter of the jet between the tubes. The heat transfer coefficient is higher in tubes with a smaller diameter. This was due to the flow impingement and the thermal layer development occupying a relatively larger fraction of the heat transfer area over tubes with a smaller diameter.

3.3 Vapor Absorption Models

Falling film absorption over horizontal tubes is one of the complex processes observed in a number of industrial applications. In situations involving horizontal tubes, a coolant flows through the inside of the tubes, which is used to cool the secondary fluid solution as it heats up during the absorption process. Due to the dynamic and locally variable nature of the phenomenon, measurements are typically limited to overall heat and mass transferred in the absorber but do not provide details of the local variations in the absorption process. Nevertheless, experiments conducted in conjunction with analytical or numerical work provide valuable insights into the absorption phenomenon. Several models have been proposed in the literature that captures the heat and mass transfer during absorption over these horizontal tubes. Early studies kicked off by numerical solution of the simplified equations over the tube surfaces. Andberg and Vliet [137] and Andberg [138] were among the first to model the heat and mass transfer phenomenon in films falling over horizontal tubes. They used a finite difference formulation, with a coordinate system fit to the shape of the film around the tube. Some of the major assumptions of this study were the planar jet flow approximation in between the tubes, boundary layer approximations in the transition region from jet to film flow over the tube surface and occurrence of heat and mass transfer only when the liquid is on the tube surface. The tube was cooled with a coolant flowing inside at 30 °C while the solution inlet temperature was varied from 46 to 32 °C. The results were presented in terms of local and bulk variation of velocity, temperature, and concentration in the fluid. Their study reported an 80% increase in the absorption rate as the solution inlet temperature changed from 46 to 32 °C. Choudhury et al. [139] modeled the absorption heat and mass transfer in laminar films falling over horizontal tubes. Boundary layer approximations in the transition region were ignored, and the tube surface was assumed to be isothermal, unlike the model by Andberg [138]. However, they too assumed the flow between the tubes to be in the form of free-falling jets with no heat and mass transfer. They solved the problem using the finite difference method on a grid fitted around the fluid film on the horizontal tube. They varied various flow parameters such as flow rate, tube diameter, etc. and reported the effect of these variations on absorber performance. They found that an optimum absorption rate that depends on the tube diameter occurs at a low solution flow rate. Min and Choi [140] performed a 2D numerical analysis of the absorption of water vapor into a lithium bromide solution flowing over a horizontal tube. They adopted a two-step procedure to solve the problem. Initially, they solved the continuity and momentum equations on an orthogonal grid over the entire domain, to obtain the velocity field and the liquid–vapor interface location and later used these predictions to solve the energy and species equations on a new non-orthogonal grid fitted only over the film region of the flow domain. The governing equations were solved using the SIMPLER algorithm accompanied with the QUICK scheme and the incomplete Cholesky conjugate gradient method. They found that contrary to the common belief that the film is thinner when the flow rate is smaller, for Re < 40, the stagnation film thickness increased with Reynolds number. They attributed this to the presence of recirculation in the region close to the stagnation point. They also found that this recirculation significantly affects the mass transfer process at low flow rates. Jeong and Garimella [141] developed a flow mechanism-based model for the absorption of water vapor into a lithium bromide solution flowing over a bank of tubes. They analytically solved the two-dimensional governing equations using a differential algebraic solver. They included the effect of incomplete wetting of the tubes by introducing a wetting ratio in their model and found that the wetting ratio played an important role in determining the efficiency of the absorption process. Their calculations showed that the vapor absorption occurred primarily in the film and droplet-formation region of the flow and that the absorption was negligible in the droplets between the tubes. They also analyzed the performance of lithium bromide absorbers for different tube diameters, spacing, and pass arrangements and found that smaller diameters provided a significantly better absorber performance than larger diameters. Subramaniam and Garimella [142] numerically analyzed the coupled heat and mass transfer for the absorption of water into lithium bromide solution flowing over a column of tubes. They used the VOF method in order to track the liquid–vapor interface. Their 3D model confirmed with high fidelity the droplet formation, detachment, and impact observed in experiments. Fully resolved 3D computations of the coupled fluid flow and heat and mass transfer revealed the local changes in the concentration and temperature distribution, as shown in Fig. 15. The bulk concentration and temperature of the lithium bromide solution at each of the time steps are also shown, which shows a decreasing trend as the LiBr solution absorbs more water vapor over time. To sustain the absorption, the solution is cooled by the coolant flowing on the inside of the tube. As a result of this, the vapor absorption rates are higher (low concentration) at the interfaces close to the tube, as compared to interfaces on the droplet, which are not in good thermal contact with the coolant in the tube. It is seen that during the droplet formation stages, the solution in the parts of the film immediately above the droplet has a lower concentration than those in parts of the film away from droplet formation. This is primarily due to the high velocity and the low film thickness in the regions above the droplet formation sites. When the droplet falls on the tube, the impact of the droplet mixes the lithium bromide solution in the film, leading to a more uniform concentration distribution.

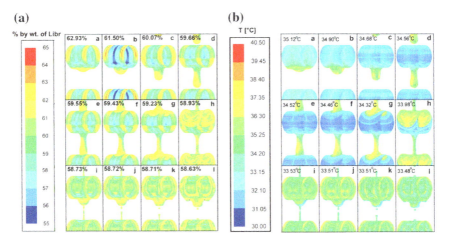

Fig. 15 Falling film and intertube **a** concentration and **b** temperature profiles in the droplet mode. Reprinted with permission from Subramaniam and Garimella, Int J Refrig, 32:607–626, Elsevier, 2009

This exothermic absorption process causes an increase in the temperature of the solution at the liquid–vapor interface, which is compensated by cooling due to the tube-side coolant.

Later, in an extension to their previous work, Subramaniam and Garimella [143] analyzed their numerical results in more detail. They observed significant effects of mixing and the waves caused by droplet impact on heat and mass transfer taking place at the liquid–gas interface over the tube surface. Harikrishnan et al. [144] carried out a 2D numerical study using cosine grid in the direction of film thickness in order to investigate the heat and mass transfer characteristics over a falling film horizontal absorber. The refrigerant R-134a was absorbed by the falling film of the R-134a-DMAC (dimethyl-acetamide) solution. The velocity, temperature, and concentration fields in the domain were computed for various operating conditions. The peak value of absorbed mass flux was found to decrease with increase in the solution flow rate, and for higher solution flow rates, the peak occurred at a further downstream location on the tube surface. The local heat transfer coefficient was seen to be higher for lower solution flow rates and vice versa. Their results showed that the optimum solution flow rate for maximum total mass flux corresponds to a particular tube diameter. Recently, among different falling film regimes, Hosseinnia et al. [145] numerically simulated the heat and mass transfer characteristics of the drop and jet modes with the aid of an in-house CFD code. The well-known VOF method was employed to capture the gas–liquid interface. The authors utilized adaptive mesh refinement, based on the magnitude of volume fraction gradient, that strongly improved the interface capturing and hence increased the accuracy of the simulation. The simulation results revealed the higher mass flux of vapor in the drop mode with respect to the jet mode, which is primarily attributed to the unsteady nature of the drop mode with respect to the jet one. The film thickness around the tubes in the drop mode becomes thinner in several locations and as a result, becomes cooler than the jet mode. The cooler film temperature generates a higher concentration gradient at the interface, and therefore, a higher local mass flux is achieved over the cooler regions. Similar to the observations of Subramaniam and Garimella [143], the impact of the droplets on the absorber surface makes a laminar wave that accompanies the mixing effect. On the other side, the jet mode reaches a nearly steady shape in terms of the hydrodynamics. Figure 16 shows the typical distribution of temperature and concentration observed in the jet mode, revealing the spatial heat and mass transfer variations in this mode. Similar to the droplet mode, the regions in between the jet departure sites have a higher concentration than the regions above the jet departure sites due to a high velocity and low film thickness above the jet departure regions. The water vapor is mostly absorbed in the interface of jets, which was primarily ignored in the previous simplified models. This absorption process causes higher temperature on jets than the liquid film existing around the cooling tubes. The temperature also shows lower values at a midway location between two jets around the cooling tube, where the velocity of the two neighboring impinging jet mixes with each other.

(a)

T [°C]

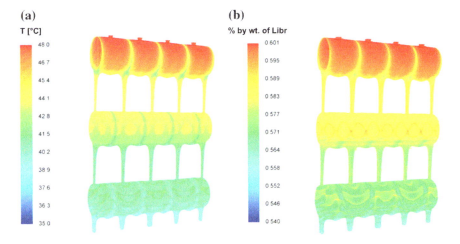

(b)

% by wt. of Libr

Fig. 16 Falling film and intertube **a** temperature and **b** concentration profiles in the inline jet mode. Reprinted with permission from Hosseinnia et al., Appl Therm Eng, 115:860–873, Elsevier, 2017

3.4 Liquid Evaporation Models

Similar to falling film absorption, film evaporation over horizontal tubes is also a very complex phase change heat transfer phenomenon. Few scattered numerical studies are present in the literature, most of which are correlation-based approaches and assumes a saturation temperature at the gas–liquid interface. Kocamustafaogullari and Chen [146] investigated the heat transfer behavior of falling film evaporation on a bank of horizontal tubes. The liquid falling from one tube to the next was considered as a planar liquid jet impinging on the top of the tube with a uniform free-falling velocity. The hydrodynamic and thermal solutions for the rest of the tube surface are obtained by solving the 2D governing equations with a finite difference method. Local and average heat transfer coefficients are obtained for the uniform heat flux and isothermal boundary conditions. In order to compare these solutions with the experimental data, the heat transfer coefficients were modified to account for film waviness. Their predictions were in reasonable agreement with the reported heat transfer data with a maximum discrepancy of ± 20% and the mean deviation is about 10.3%. The. Their study showed that convective heat transfer effects become dominant as the Peclet number increases and the local variation of the heat transfer coefficient shows a decreasing trend along the tube periphery until the fully developed region is reached. Under a similar approach, Liu et al. [147] carried out a numerical simulation for evaporative heat transfer of a falling water film over a smooth horizontal tube bundle evaporator. Laminar as well as turbulence models were used respectively to calculate the heat transfer coefficients of the falling water film on horizontal heated tubes. The periphery of the heated tube surface was divided into two zones: the top stagnation zone and the lateral free film zone. The results calculated from the stagnation zone were fed as the initial boundary conditions for the free film zone.

A comparative study between the experimentally measured data and the numerical solutions by use of two flow models showed that the experimental data lie between the laminar model solutions and the turbulence model solutions and that they are closer to the latter solutions. A simple correction based on the numerical simulations was proposed for predicting the evaporation heat transfer of falling water film for actual engineering applications. Abraham and Mani [148] performed 2D numerical simulations for horizontal tube falling film evaporation that finds wider applications in multi-effect distillation system in recent years. The simulations were performed with the help of two-phase VOF method. The effect of feed rate, tube diameter, wall temperature, etc. on the heat transfer performance was studied. It was observed that convective evaporation heat transfer performance increases with feed rate but decreases with tube diameter. The inline tube configuration was found to perform better compared to staggered tube configuration in their simulations.

Numerical 3D models for the falling film evaporation over horizontal tubes are currently not available in the literature, and hence the local evaporation process over horizontal tubes is not well understood. In this regard, high-fidelity resolved simulations that provide a vast amount of flow and heat transfer data should be undertaken. Its results can then be analyzed to advance the state-of-the-art understanding of the evaporative heat transfer over horizontal tubes. This would also serve as an improvement to the previous heat transfer correlations over horizontal tubes obtained under the experimental uncertainties.

4   Conclusion

In this chapter, a comprehensive review of numerical studies for falling liquid films flows over plain flat surfaces (vertical or inclined) and horizontal tubes have been presented. Experiments provide limited information on the flow characteristics and its effect on the thermal performance, and hence numerical simulations serve as a complementary tool that provides insight into the local flow physics and transport process taking place over the plain or horizontal tube surfaces. A detailed review of the governing equations, boundary conditions, assumptions, solution methods, and results of the reported numerical studies is provided for the flow hydrodynamics as well as for the coupled heat and mass transfer over the respective geometries. The evolution of the computational models has been presented in a hierarchical manner that would facilitate the reader to gain a complete understanding of the building block models involved in the computational process.

The following conclusions can be outlined regarding the computational models employed for the falling liquid films over plain surfaces:

i.   For accurate hydrodynamic simulation of the falling film dynamics the direct numerical solutions of the boundary layer or full Navier–Stokes equations are required which, in turn, requires considerable computational effort.

ii. In order to simplify the modeling approaches, the assumption of small-amplitude waves have been used with the boundary layer equations, and weakly non-linear forms of the first-order boundary layer equations have been derived and shown to capture the solitary wave dynamics of the falling film flow. These models are unable to capture the complex dynamics of smaller capillary structures and wave mergers.

iii. Further simplification is obtained via the long-wave equation models for the wave amplitude (single-equation model). These models have been used to describe the temporal evolution of the liquid film thickness. The single-equation models exhibit finite time blow-up phenomenon due to the presence of the non-linear terms in the equation.

iv. The multiple equation models evolved as an approach toward suppressing *finite time blowup* where the long-wave equations are written in a regularized fashion. Apart from the film thickness, other physical quantities such as liquid flow rate and wall shear stress are modeled through appropriate additional equations. The results show that the multiple equation models turn out to be the most cost-effective model that can capture the essential linear as well as non-linear wave dynamics of the falling liquid films with reasonable accuracy.

v. Full-scale modeling approaches through finite element, finite difference or finite volume solutions of the Navier–Stokes equations have also been attempted by several researchers which either employ waveform assumptions or various interface tracking techniques like the VOF, etc. in order to model the gas–liquid interface. These approaches are, however, computationally expensive compared to the previous models discussed above. Typical challenges for such simulations included correct implementation of flow boundary conditions on the evolving interface and resolving the different length scales associated with the film thickness and the flow domain. Under low-frequency disturbances, the full-scale models revealed the circulation in the big-amplitude solitary waves while the capillary waves are devoid of any circulation effects. The high-frequency disturbances produce small-amplitude waves without any forerunning capillary waves.

vi. For sensible heat transfer models on falling liquid films, it has been found that the interfacial waves have a significant impact on the heat transfer performance. Based on the Pr of the liquids, the interfacial waves at the gas–liquid interface produce film thinning or convective circulation effects, that results in an enhanced heat transfer coefficient. In addition to the above effects, some researchers have also noticed the presence of small vortices between the solitary and capillary waves that leads to this heat transfer enhancement.

vii. The existing models for absorption make many simplifying assumptions and do not account for the transient development of the liquid film thickness on the absorber surface. The models simply assume velocity and thickness profiles within the film rather than solving them. The effect of the fluid recirculation within the wave, or the non-periodicity of actual waves at higher Re on the absorber performance is not yet understood.

viii. The evaporation models for falling liquid films were initiated with the thin-film approximations which were only valid for low liquid mass flow rate conditions. For high liquid flow rate conditions, the governing equations of both the liquid and gas phases are required to be solved in a coupled manner in order to assess the evaporative heat transfer. The presence of a cocurrent or countercurrent surrounding medium significantly affected the vaporization of the liquid film. However, many simplifying assumptions are made for the above models such as steady state, Nusselt film thickness, etc. which neglects the effect of the temporal evolution of the interfacial waves on the coupled heat and mass transfer process under the liquid film evaporation.

Conclusions regarding the computational models employed for the falling liquid films over horizontal tube surfaces can be summarized as

i. The hydrodynamic modeling of falling liquid film over horizontal surfaces are mostly based on interface capturing techniques, due to the complex flow patterns observed in between the intertube regions. The 2D and 3D models developed so far can accurately capture the film flow dynamics in each of the modes, as observed in the experiments. Recent studies have revealed the temporal evolution of the liquid waves that are created either by the droplet impact at low mass flow rate or by the capillary instability of the impinging jet at higher mass flow rates. These highly resolved 3D simulations are essential for the accurate predictions of the heat and mass transfer performance over the tube surfaces.

ii. In addition to the hydrodynamic models, many 2D and 3D sensible heat transfer models have been developed in order to evaluate the heat transfer performance over the tube surfaces. The review shows that although the effect of several operating parameters such as liquid flow rate, tube diameter, and spacings, etc. on the average heat transfer process is addressed numerically, the transient information on the heat transfer process is still not available, pertaining to the transient hydrodynamics on the liquid film over the tube surfaces. It is also to be noted that local variations in velocity and temperature should be incorporated in the evaluation of heat transfer performance, that would help us to analyze the heat transfer mechanism occurring over the tube surfaces.

iii. In the early stages, the vapor absorption models over horizontal tube surfaces were based on simplified assumptions of boundary layer and planar jet approximations where it was assumed that heat and mass transfer takes place only over the tube surfaces. However, in recent years, absorption models on horizontal tubes have been improved significantly through full-scale 3D modeling approaches. The fully resolved simulations of the coupled fluid flow, heat and mass transfer in the intertube modes revealed the local gradients in the concentration and temperature distribution at the gas–liquid interface, that plays a significant role in determining the absorber performance.

iv. In comparison to vapor absorption models, the literature for the liquid film evaporation on horizontal tubes lacks in numerical 3D models, and hence the process is not well understood. In this regard, high-fidelity resolved simulations should be undertaken for evaporation situations, whose results can then be analyzed to advance the state-of-the-art understanding of the evaporative heat transfer over horizontal tubes.

Overall, the review shows that although the effect of the interfacial waves on the sensible heat transfer process is addressed, its effect on coupled heat and mass transfer process of evaporation and absorption, is not yet clear. The presence of cocurrent or countercurrent flow of the gas medium further complicates this process and is not well understood. In the coming years, improved 3D simulations should be undertaken by the researchers, especially for the situations involving evaporation and absorption in order to understand the underlying mechanisms involved in this process. Under film evaporation, the tube surfaces are often subjected to processes such as dryout or liquid maldistribution that occurs under the situations of low mass flow rate, high magnitudes of surface heat flux, or surface wettability effects. Detailed simulations of these phenomena will help in the evaluation of the empirical dryout models proposed in the literature and in making model improvements.

Acknowledgements The authors wish to acknowledge support from a DOE-ARPAE Grant project (DE-AR0000572) under the ARID program through the Electric Power Research Institute (EPRI) as the prime contractor. This financial support is gratefully acknowledged. Numerical simulations undertaken by the authors (Avijit Karmakar and Sumanta Acharya) and reported in this paper used the resources of the Oak Ridge Leadership Computing Facility at the Oak Ridge National Laboratory, which is supported by the Office of Science of the U.S. Department of Energy under Contract No. DE-AC05-00OR22725.

References

1. Abdou, M. A. (1999). Exploring novel high power density concepts for attractive fusion systems. *Fusion Engineering and Design, 45,* 145–167. https://doi.org/10.1016/S0920-3796(99)00018-6.
2. Takase K, Yoshida H, Ose Y, et al (2003) Large-scale numerical simulations of multiphase flow behavior in an advanced light-water reactor core. Annual Report of the Earth Simulator Center, April 2004:
3. Sarma, P. K., Vijayalakshmi, B., Mayinger, F., & Kakac, S. (1998). Turbulent film condensation on a horizontal tube with external flow of pure vapors. *International Journal of Heat and Mass Transfer, 41,* 537–545. https://doi.org/10.1016/S0017-9310(97)00192-0.
4. Killion, J. D., & Garimella, S. (2004). Simulation of Pendant Droplets and Falling Films in Horizontal Tube Absorbers. *Journal of Heat Transfer, 126,* 1003. https://doi.org/10.1115/1.1833364.
5. Gerendas, M., & Wittig, S. (2001). Experimental and Numerical Investigation on the Evaporation of Shear-Driven Multicomponent Liquid Wall Films. *Journal of Engineering for Gas Turbines and Power, 123,* 580. https://doi.org/10.1115/1.1362663.
6. Yilmaz, E. (2003). *Sources and characteristics of oil consumption in a spark-ignition engine* (Ph.D. thesis, Massachusetts Institute of Technology)
7. Jebson, R. S., & Chen, H. (1997). Performances of falling film evaporators on whole milk and a comparison with performance on skim milk. *Journal of Dairy Research, 64,* 57–67. https://doi.org/10.1017/S0022029996001963.
8. Brotherton, F. (2002). Alcohol recovery in falling film evaporators. *Applied Thermal Engineering, 22,* 855–860. https://doi.org/10.1016/S1359-4311(01)00125-9.

9. Uche, J., Artal, J., & Serra, L. (2003). Comparison of heat transfer coefficient correlations for thermal desalination units. *Desalination, 152,* 195–200. https://doi.org/10.1016/S0011-9164(02)01063-9.

10. Fulford, G. D. (1964). The Flow of Liquids in Thin Films. *Adv Chem Eng, 5,* 151–236. https://doi.org/10.1016/S0065-2377(08)60008-3.

11. Benjamin, T. B. (1957). Wave formation in laminar flow down an inclined plane. *Journal of Fluid Mechanics, 2,* 554. https://doi.org/10.1017/S0022112057000373.

12. Park, C. D., & Nosoko, T. (2003). Three-dimensional wave dynamics on a falling film and associated mass transfer. *AIChE Journal, 49,* 2715–2727. https://doi.org/10.1002/aic.690491105.

13. Chang, H.-C. (1994). Wave Evolution on a Falling Film. *Annual Review of Fluid Mechanics, 26,* 103–136. https://doi.org/10.1146/annurev.fluid.26.1.103.

14. Nguyen, L. T., & Balakotaiah, V. (2000). Modeling and experimental studies of wave evolution on free falling viscous films. *Physics of Fluids, 12,* 2236–2256. https://doi.org/10.1063/1.1287612.

15. Brackbill, J., Kothe, D., & Zemach, C. (1992). A continuum method for modeling surface tension. *Journal of Computational Physics, 100,* 335–354. https://doi.org/10.1016/0021-9991(92)90240-Y.

16. Nusselt, W. (1916). *Die oberflachenkondensation des wasserdamphes. VDI-Zeitschrift, 60,* 541–546.

17. Dietze, G. F., Leefken, A., & Kneer, R. (2008). Investigation of the backflow phenomenon in falling liquid films. *Journal of Fluid Mechanics, 595,* 435–459. https://doi.org/10.1017/S0022112007009378.

18. Chang, H. C., Demekhin, E. A., & Kopelevich, D. I. (1993). Nonlinear evolution of waves on a vertically falling film. *Journal of Fluid Mechanics, 250,* 433–480. https://doi.org/10.1017/S0022112093001521.

19. Chang, H. C., Demekhin, E. A., & Kalaidin, E. (1996). Simulation of Noise-Driven Wave Dynamics on a Falling Film. *AIChE Journal, 42,* 1553–1568. https://doi.org/10.1002/aic.690420607.

20. Demekhin, E. A., Kaplan, M. A., & Shkadov, V. Y. (1988). Mathematical models of the theory of viscous liquid films. *Fluid Dynamics, 22,* 885–893. https://doi.org/10.1007/BF01050727.

21. Demekhin, E. A., Demekhin, I. A., & Shkadov, V. Y. (1984). Solitons in viscous films flowing down a vertical wall. *Fluid Dynamics, 18,* 500–507. https://doi.org/10.1007/BF01090610.

22. Benney, D. J. (1966). Long Waves on Liquid Films. *Journal of Mathematical Physics, 45,* 150–155. https://doi.org/10.1002/sapm1966451150.

23. Gjevik, B. (1970). Occurrence of finite-amplitude surface waves on falling liquid films. *Physics of Fluids, 13,* 1918–1925. https://doi.org/10.1063/1.1693186.

24. Lin, S. P. (1974). Finite amplitude side-band stability of a viscous film. *Journal of Fluid Mechanics, 63,* 417–429. https://doi.org/10.1017/S0022112074001704.

25. Nakaya, C. (1975). Long waves on a thin fluid layer flowing down an inclined plane. *Physics of Fluids, 18,* 1407–1412. https://doi.org/10.1063/1.861037.

26. Takeshi, O. (1999). Surface equation of falling film flows with moderate Reynolds number and large but finite Weber number. *Physics of Fluids, 11,* 3247–3269. https://doi.org/10.1063/1.870186.

27. Panga, M. K. R., & Balakotaiah, V. (2003). Low-Dimensional Models for Vertically Falling Viscous Films. *Physical Review Letters, 90,* 4. https://doi.org/10.1103/PhysRevLett.90.154501.

28. Kapitza, P. L. (1948). Wave flow of thin layers of viscous liquid. Part I. Free flow. *Zhurnal Eksp i Teor Fiz, 18,* 3–28.

29. Shkadov, V. Y. (1970). Wave flow regimes of a thin layer of viscous fluid subject to gravity. *Fluid Dynamics, 2,* 29–34. https://doi.org/10.1007/BF01024797.

30. Yu, L. Q., Wasden, F. K., Dukler, A. E., & Balakotaiah, V. (1995). Nonlinear evolution of waves on falling films at high Reynolds numbers. *Physics of Fluids, 7,* 1886–1902. https://doi.org/10.1063/1.868503.

31. Ruyer-Quil, C., & Manneville, P. (2000). Improved modeling of flows down inclined planes. *European Physical Journal B: Condensed Matter and Complex Systems, 15,* 357–369. https://doi.org/10.1007/s100510051137.

32. Scheid, B., Ruyer-Quil, C., & Manneville, P. (2006). Wave patterns in film flows: modelling and three-dimensional waves.

33. Mudunuri, R. R., & Balakotaiah, V. (2006). Solitary waves on thin falling films in the very low forcing frequency limit. *AIChE Journal, 52,* 3995–4003. https://doi.org/10.1002/aic.11015.

34. Ramaswamy, B., Chippada, S., & Joo, S. W. (1996). A full-scale numerical study of interfacial instabilities in thin-film flows. *Journal of Fluid Mechanics, 325,* 163–194. https://doi.org/10.1017/S0022112096008075.

35. Nagasaki, T., Hijikata, K. (1989). A numerical study of interfacial waves on a falling liquid film. In *National Heat Transfer Conference* (pp. 22–30)

36. Miyara, A. (2000). Numerical simulation of wavy liquid film flowing down on a vertical wall and an inclined wall. *International Journal of Thermal Sciences, 39,* 1015–1027. https://doi.org/10.1016/S1290-0729(00)01192-3.

37. Gao, D., Morley, N. B., & Dhir, V. (2003). Numerical simulation of wavy falling film flow using VOF method. *Journal of Computational Physics, 192,* 624–642. https://doi.org/10.1016/j.jcp.2003.07.013.

38. Savage, M. D. (1998). Wave flow of liquid films. Alekseenko, S. V., Nakoryakov, V. E., Pokusaev, B. G., & Fukano, T. (1994). 313 pp. ISBN 1567800 0215. $135. J Fluid Mech 363:348–349. https://doi.org/10.1017/S002211209821874X

39. Frenkel, A. L., & Indireshkumar, K. (1996). Derivations and simulations of evolution equations of wavy film flows. In *Mathematical modeling and simulation in hydrodynamic stability* (pp. 35–81).

40. Ruyer-quil, C. (2012). *Instabilities and modeling of falling film flows* (Ph.D. thesis, Université Pierre et Marie Curie, Paris, France).

41. Chang, H., Demekhin, E. A. (2002). *Complex wave dynamics on thin films.* Elsevier Science.

42. Trifonov, Y. Y. (2008). Trifonov_08. 17:30–52. https://doi.org/10.1134/S1810232808010049.

43. Alekseenko, S. V., Nakoryakov, V. E., & Pokusaev, B. G. (1985). Wave formation on vertical falling liquid films. *International Journal of Multiphase Flow, 11,* 607–627. https://doi.org/10.1016/0301-9322(85)90082-5.

44. Panga, M. K., Mudunuri, R. R., & Balakotaiah, V. (2005). Long-wavelength equation for vertically falling films. *Physical Review E Statistical Nonlinear, Soft Matter Physics* 71. https://doi.org/10.1103/PhysRevE.71.036310

45. Pumir, A., Manneville, P., & Pomeau, Y. (1983). On solitary waves running down an inclined plane. *Journal of Fluid Mechanics, 135,* 27–50. https://doi.org/10.1017/S0022112083002943.

46. Sivashinsky, G. I., & Michelson, D. M. (1980). On irregular wavy flow of a liquid film down a vertical plane. *Progress of Theoretical Physics, 63,* 2112–2114. https://doi.org/10.1143/PTP.63.2112.

47. Yih, C.-S., & Guha, C. R. (1955). Hydraulic Jump in a Fluid System of Two Layers. *Tellus, 7,* 358–366. https://doi.org/10.1111/j.2153-3490.1955.tb01172.x.

48. Chang, H.-C. (1986). Traveling waves on fluid interfaces: Normal form analysis of the Kuramoto-Sivashinsky equation. *Physics of Fluids, 29,* 3142. https://doi.org/10.1063/1.865965.

49. Korteweg, D. J., & de Vries, G. (1895). XLI. On the change of form of long waves advancing in a rectangular canal, and on a new type of long stationary waves. *London, Edinburgh, Dublin Philos Mag J Sci, 39,* 422–443. https://doi.org/10.1080/14786449508620739.

50. Petviashvili, V. I., & Tsvelodub, D. Y. (1978, February). Horseshoe-shaped solitons on a flowing viscous film of fluid. In *Soviet Physics Doklady* (Vol. 23, p. 117).

51. Ruyer-Quil, C., & Manneville, P. (1998). Modeling film flows down inclined planes. *European Physical Journal B: Condensed Matter and Complex Systems, 6,* 277–292. https://doi.org/10.1007/s100510050550.

52. Harlow, F. H., Welch, J. E., et al. (1965). Numerical calculation of time-dependent viscous incompressible flow of fluid with free surface. *Physics of Fluids, 8,* 2182.

53. Patankar S V, Spalding DB (1972) A calculation procedure for heat, mass and momentum transfer in three-dimensional parabolic flows. Int J Heat Mass Transf 15:1787–1806. https://doi.org/10.1016/0017-9310(72)90054-3

54. Van, Doormaal J. P., & Raithby, G. D. (1984). Enhancements of the SIMPLE method for predicting incompressible fluid flows. *Numer Heat Transf, 7*, 147–163. https://doi.org/10.1080/01495728408961817.

55. Patankar, S. V. (1981). A calculation procedure for two-dimensional elliptic situations. *Numer Heat Transf, 4*, 409–425. https://doi.org/10.1080/01495728108961801.

56. Issa RI (1986) Solution of the implicitly discretised fluid flow equations by operator-splitting. J Comput Phys 62:40–65. https://doi.org/10.1016/0021-9991(86)90099-9

57. Scardovelli R, Tryggvason TAR, Zaleski P (2014) Direct Numerical Simulations of Gas – Liquid Multiphase Flows

58. Artemov, V., Beale, S. B., de Vahl, Davis G., et al. (2009). A tribute to D.B. Spalding and his contributions in science and engineering. *International Journal of Heat and Mass Transfer, 52*, 3884–3905. https://doi.org/10.1016/j.ijheatmasstransfer.2009.03.038.

59. Spalding DB (1981) Numerical computation of multi-phase fluid flow and heat transfer. In: In Von Karman Inst. for Fluid Dyn. Numerical Computation of Multi-Phase Flows

60. Hirt, C., & Nichols, B. (1981). Volume of fluid (VOF) method for the dynamics of free boundaries. *Journal of Computational Physics, 39*, 201–225. https://doi.org/10.1016/0021-9991(81)90145-5.

61. D. Y (1982) Them dependent multimaterial flow with large fluid distribution. Numer Methods Fluid D 24:

62. Lafaurie, B., Nardone, C., Scardovelli, R., et al. (1994). Modelling merging and fragmentation in multiphase flows with SURFER. *Journal of Computational Physics, 113*, 134–147.

63. Sussman, M., & Puckett, E. G. (2000). A Coupled Level Set and Volume-of-Fluid Method for Computing 3D and Axisymmetric Incompressible Two-Phase Flows. *Journal of Computational Physics, 162*, 301–337. https://doi.org/10.1006/jcph.2000.6537.

64. Enright D, Fedkiw R, Ferziger J, Mitchell I (2002) A hybrid particle level set method for improved interface capturing

65. Hirshburg, R. I., & Florschuetz, L. W. (1982). Laminar Wavy-Film Flow: Part I. *Hydrodynamic Analysis. J Heat Transfer, 104*, 452. https://doi.org/10.1115/1.3245114.

66. Brauner, N., Moalem Maron, D., & Toovey, I. (1987). Characterization of the interfacial velocity in wavy thin films flow. *Int Commun Heat Mass Transf, 14*, 293–302. https://doi.org/10.1016/0735-1933(87)90030-3.

67. Brauner, N. (1989). Modelling of wavy flow in turbulent free falling films. *International Journal of Multiphase Flow, 15*, 505–520. https://doi.org/10.1016/0301-9322(89)90050-5.

68. Wasden, F. K., & Dukler, A. E. (1989). Numerical investigation of large wave interactions on free falling films. *International Journal of Multiphase Flow, 15*, 357–370. https://doi.org/10.1016/0301-9322(89)90006-2.

69. Bach, P., & Villadsen, J. (1984). Simulation of the vertical flow of a thin, wavy film using a finite-element method. *International Journal of Heat and Mass Transfer, 27*, 815–827. https://doi.org/10.1016/0017-9310(84)90002-4.

70. Kheshgi, H. S., & Scriven, L. E. (1987). Disturbed film flow on a vertical plate. *Physics of Fluids, 30*, 990–997. https://doi.org/10.1063/1.866286.

71. Malamataris, N. T., & Papanastasiou, T. C. (1991). Unsteady Free Surface Flows on Truncated Domains. *Industrial and Engineering Chemistry Research, 30*, 2211–2219. https://doi.org/10.1021/ie00057a025.

72. Salamon, T. R., Armstrong, R. C., & Brown, R. A. (1994). Traveling waves on vertical films: Numerical analysis using the finite element method. *Physics of Fluids, 6*, 2202–2220. https://doi.org/10.1063/1.868222.

73. Kiyota, M., Morioka, I., & Kiyoi, M. (1994). Numerical Analysis of Waveforms of Falling Films. *Trans Japan Soc Mech Eng Ser B, 60*, 4177–4184. https://doi.org/10.1299/kikaib.60.4177.

74. Rohlfs, W., Rietz, M., & Scheid, B. (2018). WaveMaker: The three-dimensional wave simulation tool for falling liquid films. *SoftwareX, 7,* 211–216. https://doi.org/10.1016/j.softx.2018.07.003.

75. Dietze, G. F., Rohlfs, W., Nährich, K., et al. (2014). Three-dimensional flow structures in laminar falling liquid films. *Journal of Fluid Mechanics, 743,* 75–123. https://doi.org/10.1017/jfm.2013.679.

76. Nusselt, W. (1923). *Der wärmeaustausch am berieselungskühler. VDI-Zeitschrift, 67,* 206–210.

77. Kays, W. M., & Crawford, M. E. (1980). *Convective heat and mass transfer.* Limited: McGraw-Hill Ryerson.

78. Limberg, H. (1973). Wärmeübergang an turbulente und laminare rieselfilme. *International Journal of Heat and Mass Transfer, 16,* 1691–1702. https://doi.org/10.1016/0017-9310(73)90162-2.

79. Brauer, H. (1971). *Stoffaustausch einschließlich chemischer Reaktion.* Aarau: Verlag Sauerländer.

80. Hirshburg, R. I., & Florschuetz, L. W. (1982). Laminar Wavy-Film Flow: Part II. *Condensation and Evaporation. J Heat Transfer, 104,* 459. https://doi.org/10.1115/1.3245115.

81. Jayanti, S., & Hewitt, G. F. (1996). Hydrodynamics and heat transfer of wavy thin film flow. *International Journal of Heat and Mass Transfer, 40,* 179–190. https://doi.org/10.1016/S0017-9310(96)00016-6.

82. Miyara, A. (1999). Numerical analysis on flow dynamics and heat transfer of falling liquid films with interfacial waves. *Heat and Mass Transfer, 35,* 298–306. https://doi.org/10.1007/s002310050328.

83. Kunugi, T., & Kino, C. (2005). DNS of falling film structure and heat transfer via MARS method. *Computers & Structures, 83,* 455–462. https://doi.org/10.1016/j.compstruc.2004.08.018.

84. Yu, H., Gambaryan-Roisman, T., & Stephan, P. (2013). Numerical Simulations of Hydrodynamics and Heat Transfer in Wavy Falling Liquid Films on Vertical and Inclined Walls. *Journal of Heat Transfer, 135,* 101010. https://doi.org/10.1115/1.4024550.

85. Kawae, N., Shigechi, T., Kanemaru, K., & Yamada, T. (1980). Water vapor evaporation into laminar film flow of a lithium bromide-water solution (influence of variable properties and inlet film thickness on absorption mass transfer rate). *Heat Transf - Japanese Res, 18,* 58–70.

86. Grossman, G. (1983). Simultaneous heat and mass transfer in film absorption under laminar flow. *International Journal of Heat and Mass Transfer, 26,* 357–371. https://doi.org/10.1016/0017-9310(83)90040-6.

87. Grossman, G. (1986). *No Title.* Houstan: Gulf Publishing Company.

88. Ramadane, A., Aoufoussi, Z., & Le Goff, H. (1992). Experimental investigation and modeling of gas-liquid absorption with a high thermal effect. *Institution of Chemical Engineers Symposium Series, 1,* A451–A459.

89. Ibrahim, G., Nabhan, M. B., & Anabtawi, M. (1995). An investigation into a falling film type cooling tower. *International Journal of Refrigeration, 18,* 557–564. https://doi.org/10.1016/0140-7007(96)81783-X.

90. Nakoryakov VE, Grigor'eva NI (1980) Calculation of heat and mass transfer in nonisothermal absorption on the initial portion of a downflowing film. Theor Found Chem Eng 14:305–309

91. Patnaik, V., & Perez-Blanco, H. (1996). A study of absorption enhancement by wavy film flows. *International Journal of Heat and Fluid Flow, 17,* 71–77. https://doi.org/10.1016/0142-727X(95)00076-3.

92. Sabir, H., Suen, K. O., & Vinnicombe, G. A. (1996). Investigation of effects of wave motion on the performance of a falling film absorber. *International Journal of Heat and Mass Transfer, 39,* 2463–2472. https://doi.org/10.1016/0017-9310(95)00336-3.

93. Habib, H. M., & Wood, B. D. (2001). Simultaneous Heat and Mass Transfer in Film Absorption With the Presence of Non-Absorbable Gases. *Journal of Heat Transfer, 123,* 984. https://doi.org/10.1115/1.1370523.

94. Karami, S., & Farhanieh, B. (2009). A numerical study on the absorption of water vapor into a film of aqueous LiBr falling along a vertical plate. *Heat and Mass Transfer, 46,* 197–207. https://doi.org/10.1007/s00231-009-0557-y.

95. Chow, L. C., & Chung, J. N. (1983). Evaporation of water into a laminar stream of air and superheated steam. *International Journal of Heat and Mass Transfer, 26,* 373–380. https://doi.org/10.1016/0017-9310(83)90041-8.

96. Schröppel, J., & Thiele, F. (1983). On the calculation of momentum, heat, and mass transfer in laminar and turbulent boundary layer flows along a vaporizing liquid film. *Numer Heat Transf, 6,* 475–496. https://doi.org/10.1080/01495728308963101.

97. Yan, W. M., Tsay, Y. L., & Lin, T. F. (1989). Simultaneous heat and mass transfer in laminar mixed convection flows between vertical parallel plates with asymmetric heating. *International Journal of Heat and Fluid Flow, 10,* 262–269. https://doi.org/10.1016/0142-727X(89)90045-3.

98. Yan, W.-M. (1992). Effects of film evaporation on laminar mixed convection heat and mass transfer in a vertical channel. *International Journal of Heat and Mass Transfer, 35,* 3419–3429. https://doi.org/10.1016/0017-9310(92)90228-K.

99. Tsay, Y. L., Lin, T. F., & Yan, W. M. (1990). Cooling of a falling liquid film through interfacial heat and mass transfer. *International Journal of Multiphase Flow, 16,* 853–865. https://doi.org/10.1016/0301-9322(90)90008-7.

100. Tsay, Y. L., & Lin, T. F. (1995). Evaporation of a heated falling liquid film into a laminar gas stream. *Exp Therm Fluid Sci, 11,* 61–71. https://doi.org/10.1016/0894-1777(94)00112-L.

101. Debbissi, C., Orfi, J., & Ben, Nasrallah S. (2001). Evaporation of water by free convection in a vertical channel including effects of wall radiative propeties. *International Journal of Heat and Mass Transfer, 44,* 811–826. https://doi.org/10.1016/S0017-9310(00)00125-3.

102. Debbissi, C., Orfi, J., & Ben Nasrallah, S. (2003). Evaporation of water by free or mixed convection into humid air and superheated steam. *International Journal of Heat and Mass Transfer, 46,* 4703–4715. https://doi.org/10.1016/S0017-9310(03)00092-9.

103. Feddaoui, M., Mir, A., & Belahmidi, E. (2003). Numerical simulation of mixed convection heat and mass transfer with liquid film cooling along an insulated vertical channel. *Heat Mass Transf und Stoffuebertragung, 39,* 445–453. https://doi.org/10.1007/s00231-002-0340-9.

104. Ren, C., & Wan, Y. (2016). A new approach to the analysis of heat and mass transfer characteristics for laminar air flow inside vertical plate channels with falling water film evaporation. *International Journal of Heat and Mass Transfer, 103,* 1017–1028. https://doi.org/10.1016/j.ijheatmasstransfer.2016.07.109.

105. Wan, Y., Ren, C., Xing, L., & Yang, Y. (2017). Analysis of heat and mass transfer characteristics in vertical plate channels with falling film evaporation under uniform heat flux/uniform wall temperature boundary conditions. *International Journal of Heat and Mass Transfer, 108,* 1279–1284. https://doi.org/10.1016/j.ijheatmasstransfer.2016.12.110.

106. Schlottke, J., & Weigand, B. (2008). Direct numerical simulation of evaporating droplets. *Journal of Computational Physics, 227,* 5215–5237. https://doi.org/10.1016/j.jcp.2008.01.042.

107. Hu, X., & Jacobi, A. M. (1996). The Intertube Falling Film: Part 1—Flow Characteristics, Mode Transitions, and Hysteresis. *Journal of Heat Transfer, 118,* 616. https://doi.org/10.1115/1.2822676.

108. Mitrovic, J. (2005). Flow structures of a liquid film falling on horizontal tubes. *Chemical Engineering and Technology, 28,* 684–694. https://doi.org/10.1002/ceat.200500064.

109. Roques, J. F., Dupont, V., & Thome, J. R. (2002). Falling Film Transitions on Plain and Enhanced Tubes. *Journal of Heat Transfer, 124,* 491. https://doi.org/10.1115/1.1458017.

110. Mohamed, A. M. I. (2007). Flow behavior of liquid falling film on a horizontal rotating tube. *Exp Therm Fluid Sci, 31,* 325–332. https://doi.org/10.1016/j.expthermflusci.2006.05.004.

111. Ruan, B., Jacobi, A. M., & Li, L. (2009). Effects of a countercurrent gas flow on falling-film mode transitions between horizontal tubes. *Exp Therm Fluid Sci, 33,* 1216–1225. https://doi.org/10.1016/j.expthermflusci.2009.07.009.

112. Lewis, D. J. (1950). The Instability of Liquid Surfaces when Accelerated in a Direction Perpendicular to their Planes. II. *Proc R Soc A Math Phys Eng Sci, 202,* 81–96. https://doi.org/10.1098/rspa.1950.0086.
113. Richard Bellman, R. H. P. (1953). *Effects of surface tension and viscosity on Taylor instability.*
114. Maron-Moalem, D., Sideman, S., & Dukler, A. E. (1978). Dripping characteristics in a horizontal tube film evaporator. *Desalination, 27,* 117–127. https://doi.org/10.1016/S0011-9164(00)88106-0.
115. Ganic, E. N., & Roppo, M. N. (1980). An Experimental Study of Falling Liquid Film Breakdown on a Horizontal Cylinder During Heat Transfer. *Journal of Heat Transfer, 102,* 342. https://doi.org/10.1115/1.3244285.
116. Hu, X., & Jacobi, A. M. (1998). Departure-site spacing for liquid droplets and jets falling between horizontal circular tubes. *Exp Therm Fluid Sci, 16,* 322–331. https://doi.org/10.1016/S0894-1777(97)10031-0.
117. Rayleigh, L. (1878). On the instability of jets. *Proceedings of London Mathematical Society* s1-10:4–13. https://doi.org/10.1112/plms/s1-10.1.4.
118. Eggers, J., Villermaux, E. (2008). Physics of liquid jets. *Reports Program Physics, 71.* https://doi.org/10.1088/0034-4885/71/3/036601.
119. Killion, J. D., & Garimella, S. (2004). Pendant droplet motion for absorption on horizontal tube banks. *International Journal of Heat and Mass Transfer, 47,* 4403–4414. https://doi.org/10.1016/j.ijheatmasstransfer.2004.04.032.
120. Chen, X., Shen, S., Wang, Y., et al. (2015). Measurement on falling film thickness distribution around horizontal tube with laser-induced fluorescence technology. *International Journal of Heat and Mass Transfer, 89,* 707–713. https://doi.org/10.1016/j.ijheatmasstransfer.2015.05.016.
121. Jafar, F., Thorpe, G., & Turan, O. F. (2007). Liquid Film Falling on Horizontal Circular Cylinders. *16th Australasian Fluid Mechanics Conference* (pp. 1193–1200). Queensland, Australia: Gold Coast.
122. Sun, F., Xu, S., & Gao, Y. (2012). Numerical simulation of liquid falling film on horizontal circular tubes. *Front Chem Sci Eng, 6,* 322–328. https://doi.org/10.1007/s11705-012-1296-z.
123. Qiu, Q., Zhu, X., Mu, L., & Shen, S. (2015). Numerical study of falling film thickness over fully wetted horizontal round tube. *International Journal of Heat and Mass Transfer, 84,* 893–897. https://doi.org/10.1016/j.ijheatmasstransfer.2015.01.024.
124. Qiu, Q., Meng, C., Quan, S., & Wang, W. (2017). 3-D simulation of flow behaviour and film distribution outside a horizontal tube. *International Journal of Heat and Mass Transfer, 107,* 1028–1034. https://doi.org/10.1016/j.ijheatmasstransfer.2016.11.009.
125. Chen, J., Zhang, R., & Niu, R. (2015). Numerical simulation of horizontal tube bundle falling film flow pattern transformation. *Renewable Energy, 73,* 62–68. https://doi.org/10.1016/j.renene.2014.08.007.
126. Li, M., Lu, Y., Zhang, S., & Xiao, Y. (2016). A numerical study of effects of counter-current gas flow rate on local hydrodynamic characteristics of falling films over horizontal tubes. *Desalination, 383,* 68–80. https://doi.org/10.1016/j.desal.2016.01.016.
127. Fiorentino, M., & Starace, G. (2016). Numerical investigations on two-phase flow modes in evaporative condensers. *Applied Thermal Engineering, 94,* 777–785. https://doi.org/10.1016/j.applthermaleng.2015.10.099.
128. Ding, H., Xie, P., Ingham, D., et al. (2018). Flow behaviour of drop and jet modes of a laminar falling film on horizontal tubes. *International Journal of Heat and Mass Transfer, 124,* 929–942. https://doi.org/10.1016/j.ijheatmasstransfer.2018.03.111.
129. Killion, J. D., & Garimella, S. (2003). Gravity-driven flow of liquid films and droplets in horizontal tube banks. *International Journal of Refrigeration, 26,* 516–526. https://doi.org/10.1016/S0140-7007(03)00009-4.
130. Yung, D., Lorenz, J. J., & Ganic, E. N. (1980). Vapor/liquid interaction and entrainment in falling film evaporators. *Journal of Heat Transfer, 102,* 20. https://doi.org/10.1115/1.3244242.
131. Ribatski, G., & Jacobi, A. M. (2005). Falling-film evaporation on horizontal tubes—A critical review. *International Journal of Refrigeration, 28,* 635–653. https://doi.org/10.1016/j.ijrefrig.2004.12.002.

132. Ribatski, G., & Thome, J. R. (2007). Two-phase flow and heat transfer across horizontal tube bundles-a review. *Heat Transfer Engineering, 28,* 508–524. https://doi.org/10.1080/01457630701193898.

133. Hu, X., & Jacobi, A. M. (1996). The intertube falling film: Part 2—Mode effects on sensible heat transfer to a falling liquid film. *Journal of Heat Transfer, 118,* 626. https://doi.org/10.1115/1.2822678.

134. Luo, L. C., Zhang, G. M., Pan, J. H., & Tian, M. C. (2013). Flow and heat transfer characteristics of falling water film on horizontal circular and non-circular cylinders. *Journal of Hydrodynamics, 25,* 404–414. https://doi.org/10.1016/S1001-6058(11)60379-0.

135. Jafar, F. A., Thorpe, G. R., & Turan, Ö. F. (2014). Liquid film falling on horizontal plain cylinders: Numerical study of heat transfer in unsaturated porous media. *International Journal for Computational Methods in Engineering Science and Mechanics, 15,* 101–109. https://doi.org/10.1080/15502287.2013.874056.

136. Karmakar, A., & Acharya, S. (2017). Heat transfer characteristics of falling film over horizontal tubes—A numerical study. In: *55th AIAA Aerospace Sciences Meeting.* American Institute of Aeronautics and Astronautics, Grapevine, Texas

137. Andberg, J. W., & Vliet, G. C. (1983). Design guidelines for water-lithium bromide absorbers. *ASHRAE Trans 89,* 220–232

138. Andberg, J. W. (1986). *Absorption of vapors into liquid films flowing over cooled horizontal tubes,* Ph.D. thesis. Texas University, Austin (USA)

139. Choudhury, S. K., Nishiguchi, A., Hisajima, D., et al. (1993). Absorption of vapors into liquid films flowing over cooled horizontal tubes. *ASHRAE Transaction, 99,* 81–89.

140. Min, J. K., & Choi, D. H. (1999). Analysis of the absorption process on a horizontal tube using Navier-Stokes equations with surface-tension effects. *International Journal of Heat and Mass Transfer, 42,* 4567–4578. https://doi.org/10.1016/S0017-9310(99)00104-0.

141. Jeong, S., & Garimella, S. (2002). Falling-film and droplet mode heat and mass transfer in a horizontal tube LiBr/water absorber. *International Journal of Heat and Mass Transfer, 45,* 1445–1458. https://doi.org/10.1016/S0017-9310(01)00262-9.

142. Subramaniam, V., & Garimella, S. (2009). From measurements of hydrodynamics to computation of species transport in falling films. *International Journal of Refrigeration, 32,* 607–626. https://doi.org/10.1016/j.ijrefrig.2009.02.008.

143. Subramaniam, V., & Garimella, S. (2014). Numerical study of heat and mass transfer in lithium bromide-water falling films and droplets. *International Journal of Refrigeration, 40,* 211–226. https://doi.org/10.1016/j.ijrefrig.2013.07.025.

144. Harikrishnan, L., Maiya, M. P., & Tiwari, S. (2011). Investigations on heat and mass transfer characteristics of falling film horizontal tubular absorber. *International Journal of Heat and Mass Transfer, 54,* 2609–2617. https://doi.org/10.1016/j.ijheatmasstransfer.2011.01.024.

145. Hosseinnia, S. M., Naghashzadegan, M., & Kouhikamali, R. (2017). CFD simulation of water vapor absorption in laminar falling film solution of water-LiBr—Drop and jet modes. *Applied Thermal Engineering, 115,* 860–873. https://doi.org/10.1016/j.applthermaleng.2017.01.022.

146. Kocamustafaogullari, G., & Chen, I. Y. (1988). Falling film heat transfer analysis on a bank of horizontal tube evaporator. *AIChE Journal, 34,* 1539–1549. https://doi.org/10.1002/aic.690340916.

147. Liu, Z. H., Zhu, Q. Z., & Chen, Y. M. (2002). Evaporation heat transfer of falling water film on a horizontal tube bundle. *Heat Transfer Asian Research, 31,* 42–55. https://doi.org/10.1002/htj.10016.

148. Abraham, R., & Mani, A. (2015). Heat transfer characteristics in horizontal tube bundles for falling film evaporation in multi-effect desalination system. *Desalination, 375,* 129–137. https://doi.org/10.1016/j.desal.2015.06.018.

Study of Pool Boiling Through Numerical Approach

Vinod Pandey, Gautam Biswas and Amaresh Dalal

1 Introduction

The processes involving high heat fluxes over a heated surface result in the vapor generation from the cavities and this phenomenon is termed as boiling. In the presence of gravity with the increase in superheat as shown in the boiling curve (Fig. 1), heat transfer prevails through natural convection which subsequently leads to the generation of vapor bubbles at some preferred cavity sites (onset of nucleate boiling). Depending on the heat flux or the superheat at the heated surface, the mode of boiling transforms from nucleate to film boiling with an intermediate stage of transition between the two modes. The nucleate boiling is influenced effectively by various factors like surface superheat, wettability, roughness, liquid-subcooling, etc. As the superheat increases the adjacent bubbles interact and merge to form an enlarged dry-out region resulting in vapor slugs and columns [1]. It is called as the boiling crisis where a sudden drop in heat transfer efficiency is experienced. Contrary to the vapor column theory, an alternative theory of vapor recoil has also been established recently [2] which explains the boiling crisis even under microgravity. The transition mode continues with increase in superheat to ultimately establish a continuous thin layer of vapor film which corresponds to the minimum heat flux point or the Leidenfrost point in the boiling curve. Thereafter, the film boiling regime continues which is dominated by the instabilities at the liquid-vapor interface.

Before approaching to the numerical aspects of boiling study, the following section discusses the overall development in the boiling-research.

V. Pandey · A. Dalal
Department of Mechanical Engineering, Indian Institute of Technology,
Guwahati 781039, India
e-mail: amaresh@iitg.ac.in

G. Biswas (✉)
Department of Mechanical Engineering, Indian Institute of Technology,
Kanpur 208016, India
e-mail: gtm@iitk.ac.in

© Springer Nature Singapore Pte Ltd. 2020
A. Runchal (ed.), *50 Years of CFD in Engineering Sciences*,
https://doi.org/10.1007/978-981-15-2670-1_17

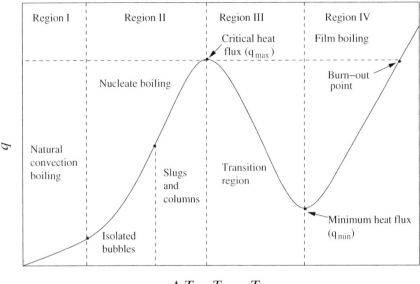

$$\Delta T = T_w - T_{sat}$$

Fig. 1 Typical boiling curve for saturated water

1.1 Brief History of Boiling-Research

Heat transfer correlations During boiling, thermal energy from solid surface is transferred to the fluid that is directly in contact with the surface. Most of the heat is transferred to the liquid side due to higher thermal conductivity. The bubble growth during nucleate boiling is the combined result of evaporation at the liquid-vapor interface and due to the evaporation of the liquid microlayer beneath the bubble. However, the effective contribution of each in the growth process has not been clear so far. Rohsenow [3] provided a correlation for the heat transfer data from the surface during boiling as a function of various physical properties calculated at the saturated condition as

$$\frac{C_l T_x}{h_{lv}} = C_{sf} \left(\frac{q/l}{\mu_l h_{lv}} \sqrt{\frac{g_0 \sigma}{g(\rho_l - \rho_v)}} \right)^{0.33} \left(\frac{C_l \mu_l}{k_l} \right)^{1.7} \tag{1}$$

The two most important parameters which define the characteristics of a boiling curve are the critical heat flux and the minimum heat flux points corresponding to the maximum and the minimum values of heat flux in the boiling curve. As described earlier, the maximum value of heat flux is obtained when the adjacent bubbles merge together to form columns of vapor. As a result, vapor film starts to cover the surface which may result in sudden rise of surface temperature. This is the critical heat flux point beyond which transition regime starts. Once the whole surface is covered by vapor layer, the minimum heat flux point is obtained which is continued by a stable film boiling regime.

Based on the theory of vapor jet formation and instabilities at the interface near the critical point, Zuber [1] deduced the correlation for critical heat flux as

$$q_{crit} = L\rho_v \frac{\pi}{24} \frac{3}{\sqrt{2\pi}} \left[\frac{\sigma m_0}{\rho_v}\right]^{1/2} \left[\frac{\rho_l}{\rho_l - \rho_v}\right]^{1/2} \tag{2}$$

and for the minimum heat flux as

$$q_{min} = L\rho_v \frac{\pi}{24} \left(\frac{\pi 0.4\sqrt{2}}{3^{1/4}}\right) \left[\frac{\sigma g(\rho_l - \rho_v)}{(\rho_l + \rho_v)^2}\right]^{1/4} \tag{3}$$

Later, it has been found by various researchers that the heat transfer coefficients during nucleate boiling is a function of surface superheat as well as the surface properties. Tien's [4] hydrodynamical model considered the inverted stagnation flow of liquid behind the wake of rising vapor column and provided the relation between heat transfer coefficient and thermal boundary-layer over the surface as

$$q = 61.3k Pr^{1/3} n^{1/5} \Delta T \tag{4}$$

Although the influence of liquid agitation caused by bubble departure has been considered to have a significant effect on heat transfer during nucleate boiling, the latent heat transport or the microlayer evaporation has been found to be dominant at high heat fluxes [5]. There is still not a common consensus on the exact contribution due to microlayer evaporation on growth of a bubble and both experimental and numerical investigations are being extensively performed (discussed in Sec. 3.1) in this area.

Nucleation over a surface Zuber [1] was the first to develop a hydrodynamic model for bubble growth in nucleate boiling. He also deduced the criteria for the activation superheat temperature for nucleation at a cavity in a heating surface. The nucleation temperature during heterogeneous boiling is much lower than the temperature obtained from classical nucleation theory or homogenous nucleation temperature. This is due to the presence of potentially active nucleation sites at the surface. The required condition of superheat temperature for any cavity to act as the active site of nucleation [6] is

$$T_w - T_{sat}(P_l) > \frac{2\sigma T_{sat} v_{lv}}{h_{lv} r_{min}} \tag{5}$$

Hsu [7] proposed a thermal boundary-layer model for a minimum and maximum radius of active nucleating cavity as a function of the surface superheat. The model also proposed the minimum superheat or heat flux value required for the growth of bubble from a cavity of a given size. A criterion depending on the waiting period (the time required for the redevelopment of thermal layer after a bubble departs from a cavity) was given as "for any two given cavities which are potential nucleating sites, the bubble would generate at the cavity for which the corresponding waiting period is less". The required nucleation temperature also depends on the surface wettability. The nucleation rate also depends upon the size distribution of the cavities which is a material property. The slope of boiling curve differs with the change in roughness of the same material. Kurihara and Myers [8] found that the number of active nucleation sites is directly proportional to the surface roughness and has direct influence on the heat transfer coefficient. Thus, the characteristic of the surface has been found to be an important parameter while determining the boiling behavior (especially in the nucleate boiling regime) over any surface.

Dynamics of bubbles Nucleate boiling has been considered to consist of two regimes [4, 9]—one, the isolated bubble regime where the nucleating site density is small and another, where the influence between the neighboring sites is large. In the former regime, the flow field around the bubble generates as a result of growth and departure of bubble and the bubble departure diameter has been found to be almost constant. Fritz [10] gave a correlation for the departure diameter in the isolated bubble regime as

$$D_d = 0.02008\psi\sqrt{\sigma/g(\rho_l - \rho_v)}. \tag{6}$$

A balance between buoyant and adhesive forces has been taken into account in the above equation. The contact angle ψ has been measured in degrees.

With an increase in superheat of the surface, the bubble departure frequency increases, and hence the bubbles start to interfere with each other. The interference occurs between the departing bubbles as well as the neighboring bubbles from different nucleating sites. This leads to the formation of multiple continuous vapor columns at various locations over the surface. The bubbles tend to interact with each other when the average spacing between bubbles becomes more or less equal to the departure diameter.

In the film boiling regime, the dynamics of bubble formation completely differs from that in the nucleate boiling regime. In this regime, the interfacial instability governs the bubble generation which is affected by the combined influence of buoyancy and surface tension. The bubble release is periodic and is a function of surface superheat [11].

In the present work, the studies pertinent to the numerical simulation of bubble formation and heat transfer during various regimes of boiling have been discussed. Some important findings and results related to the field have been presented with an intention of explaining the different aspects of boiling.

2 Numerical Approaches

In consideration with the complexity of physical aspects involved in boiling experiments due to the dependence on various factors (surface characteristics, system pressure, heat flux control etc.), the numerical study of the phenomenon has been evolved as an important means to explore the various flow characteristics and mechanisms during boiling. The important features involved in the numerical simulations of boiling process are the interface capturing technique, surface tension formulation, discretization of convective terms and the evaporation model. In addition to these numerical aspects, simulation of nucleate boiling requires to solve the contribution of microlayer evaporation in the bubble growth through a separate numerical model.

Considering a viscous, incompressible and immiscible two-fluid system, the Navier–Stokes equation governing the flow can be written in a single-fluid formulation [12] as

$$\rho[\mathbf{U}_t + \nabla \cdot (\mathbf{U}\mathbf{U})] = -\nabla p + \rho\mathbf{g} + \nabla \cdot \mu(\nabla\mathbf{U} + \nabla\mathbf{U}^T)_v + \sigma\kappa\hat{\mathbf{n}}\delta_s \qquad (7)$$

The effects of discontinuity in fluid properties and the surface tension have been considered in the above equation. The incompressibility criteria satisfy

$$\nabla \cdot \mathbf{U} = 0 \qquad (8)$$

In case of the phase-change processes such as in boiling, the above Eq. 8 is not valid at the interfacial computational cells. Following condition for the jump in velocity needs to be satisfied in all the two-phase cells,

$$\int_{S_C} \mathbf{U} \cdot \hat{\mathbf{n}}dS + \int_{S_I(t)} \left(\frac{1}{\rho_l} - \frac{1}{\rho_v}\right)\frac{\|\mathbf{q}\| \cdot \hat{\mathbf{n}}}{h_{lv}} = 0 \qquad (9)$$

where $\|.\|$ represents the jump in the quantity and $\mathbf{q}/h_{lv}$ is the mass-flux at the interface. S_C is the boundary of the computational cell and S_I is the interfacial area. The detailed procedure to obtain the modified continuity Eq. 9 has been explained in Sect. 2.4.

The energy conservation equation can be expressed in the following form,

$$\frac{\partial T}{\partial t} + \mathbf{U} \cdot \nabla T = \frac{k}{\rho c_p}\nabla^2 T. \qquad (10)$$

2.1 Interface Capturing

Problems involving free boundaries or interface between two fluids possess complexity in terms of extreme deformation of interface and thus require accurate estimation of the interface position and shape as a function of time. Both Lagrangian [13, 14] and Eulerian [15, 16] approaches have been employed successfully in boiling problems. However, methods following Eulerian approach have proved to be much efficient in capturing interface motion and complex topology.

Two most widely used techniques for interface capturing are the Volume of Fluid (VOF) and the Level-Set (LS) approaches. Hirt and Nichols [17] first introduced the VOF method for free boundary problems where the donor-acceptor flux approximation technique has been utilized for mass-flux calculation with time. Based on the interface geometry, the volume fraction is defined as fraction of a fluid component in the computational cell. It is inherently a mass conserving technique and requires a special smoothening procedure to facilitate the calculation of normal and curvature at the interface.

In the level-set technique developed by Osher and Sethian [18], the front evolves with the simultaneous advection of a smooth scalar function (ϕ) with the flow velocity. The interface in this case is implicitly defined as the zero level-set of the function ϕ.

If a bounded domain Ω with boundary $\partial\Omega$ is considered which separates two immiscible fluids, the level-set function is chosen such that it assumes positive values in the region Ω^+ while negative values in the region Ω^-, so that the fluid interface corresponds to the zero level-set of ϕ (see Fig. 2). The level-set function is initialized at $t = 0$ by defining ϕ at each computational cell as a signed distance function from the interface given by

$$\phi(\mathbf{r}, t) = \begin{cases} -d & \text{in the gaseous region,} \\ 0 & \text{at the interface,} \\ +d & \text{in the liquid region} \end{cases} \qquad (11)$$

where $d = d(t)$ is the shortest distance of the interface from point $\mathbf{r}$.

Advection of level-set and volume fraction As the interface propagates with time, it should be ensured that the level-set function be advected in such a way that interface is still represented by $\phi(\mathbf{r}, t)$. Let us consider a point 'p' on the interface whose

Fig. 2 Definition of a
level-set function in a
two-fluid system

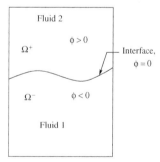

location with time is given by $\mathbf{r}_p(t)$. The prerequisite that the point p lies on the interface with the evolution of ϕ implies that

$$\phi(\mathbf{r}_p, t) = 0 \quad \text{for} \quad t \geq 0 \tag{12}$$

which requires the total derivative of ϕ along the interface to be zero, i.e.,

$$\frac{d}{dt}\left[\phi(\mathbf{r}_p, t)\right] = 0 \tag{13}$$

which further gives

$$\frac{\partial \phi}{\partial t} + \frac{\partial \mathbf{r}_p}{\partial t}.\nabla \phi = 0 \tag{14}$$

or

$$\frac{\partial \phi}{\partial t} + \mathbf{v}_p.\nabla \phi = 0 \tag{15}$$

where $\mathbf{v}_p$ is the interface-velocity. Incorporating the kinematic boundary condition for the interface, the above equation transforms to

$$\frac{\partial \phi}{\partial t} + \mathbf{u}.\nabla \phi = 0 \tag{16}$$

where $\mathbf{u}$ represents the velocity field. The gradient of ϕ, i.e. $\nabla \phi$ has the direction normal to the level-sets of ϕ and towards the increasing direction of ϕ. The calculation of unit normal, $\mathbf{n}$, to the interface can be evaluated as

$$\hat{\mathbf{n}} = \frac{\nabla \phi}{|\nabla \phi|} \tag{17}$$

A coupled level-set and volume of fluid (CLSVOF) approach has been evolved [12, 19] to include the advantages of both level-set method (accurate calculation of normal and curvature) and volume of fluid method (mass-conservation) in the numerical results.

The advection equation for the liquid volume fraction (F) can be written as

$$\frac{\partial F}{\partial t} + \mathbf{U} \cdot \nabla F = 0 \tag{18}$$

Using the continuity equation (Eq. 8), it can be written as

$$\frac{\partial F}{\partial t} + \nabla \cdot (\mathbf{U}F) = 0 \tag{19}$$

which can be discretised by finite differencing methods as

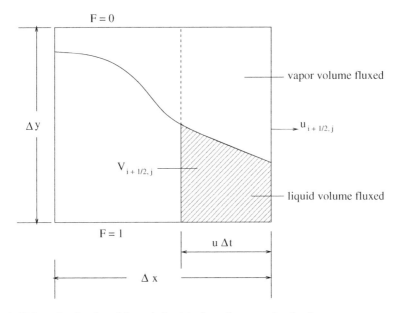

Fig. 3 Volume fraction fluxed through the right face of computational cell

$$F_{i,j}^{n+1} = F_{i,j}^n + \frac{\Delta t}{\Delta x}(\delta V_{i-1/2,j} - \delta V_{i+1/2,j}) + \frac{\Delta t}{\Delta y}(\delta V_{i,j-1/2} - \delta V_{i,j+1/2}) \quad (20)$$

where $\delta V_{i+1/2,j} = (uF)_{i+1/2,j}$ is the amount of liquid volume fraction fluxed through the right cell face as can be referred to Fig. 3. The superscript $(n+1)$ denotes the values at the next time step and Δt, Δx and Δy are the time step and the grid spacing in x and y direction, respectively. The operator split approach due to Puckett et al. [20] is applied to solve Eq. 20 as

$$F_{i,j}^* = F_{i,j}^n + \frac{\Delta t}{\Delta x}(\delta V_{i-1/2,j} - \delta V_{i+1/2,j}) + \frac{\Delta t}{\Delta x}F_{i,j}^*(u_{i+1/2,j} - u_{i-1/2,j}) \quad (21)$$

$$F_{i,j}^{n+1} = F_{i,j}^* + \frac{\Delta t}{\Delta y}(\delta V_{i,j-1/2}^* - \delta V_{i,j+1/2}^*) + \frac{\Delta t}{\Delta y}F_{i,j}^*(v_{i,j+1/2} - u_{i,j-1/2}) \quad (22)$$

Eq. 21 is solved implicitly in the first sweep direction whereas Eq. 22 in the second sweep direction, is solved explicitly. The star (∗) denotes the values obtained after the first sweep. In addition, it can be inferred from Eqs. 21 and 22 that we are basically solving $\partial F/\partial t + \nabla \cdot (\mathbf{U}F) = F\nabla \cdot \mathbf{U}$, rather than Eq. 19 into two equations with the third term on the right hand side of both equations as 'divergence correction' terms.

The volume flux δV in discretized form can be obtained using geometrical-based calculation as shown in Fig. 3. Knowing the interface position and the velocity in the two-phase control volume, the fluxed volume from the right face can be calculated as

$$\delta V_{i+1/2,j} = \frac{u_{i+1/2,j} V_{i+1/2,j}}{u_{i+1/2,j} \Delta t \Delta y} = \frac{V_{i+1/2,j}}{\Delta t \Delta y} \tag{23}$$

Eq. 21 can be solved after determining the value of δV from Eq. 23 and the interface is again reconstructed based on $F_{i,j}^*$. Same method is followed for the vertical fluxes, calculating geometrically the fluxes of the volume fractions crossing the top and bottom cell face. Hence, the new volume fraction field $F_{i,j}^{n+1}$ is obtained. The procedure can be made second-order accurate by alternating the sweep directions at each time step [21].

Smoothening of the fluid properties The presence of free surface brings in abrupt discontinuities of the fluid properties which may bring numerical instabilities to the solution of hydrodynamic equations. Numerical smoothening can be achieved for density (ρ) and viscosity (μ) using Heaviside function as incorporated in Sussman et al. [19] and Gerlach et al. [12] as

$$\rho(\phi) = \rho_l H_\delta(\phi) + \rho_v [1 - H_\delta(\phi)] \tag{24}$$

$$\mu(\phi) = \mu_l H_\delta(\phi) + \mu_v [1 - H_\delta(\phi)] \tag{25}$$

where the Heaviside function, $H_\delta(\phi)$, is defined as

$$H_\delta(\phi) = \begin{cases} 0 & \text{if } \phi < -\delta_{int} \\ \frac{1}{2} + \frac{\phi}{2\delta_{int}} + \frac{1}{2\pi}\left[\sin\left(\frac{\pi\phi}{\delta_{int}}\right)\right] & \text{if } |\phi| \leq \delta_{int} \\ 1 & \text{if } \phi > \delta_{int} \end{cases} \tag{26}$$

where δ_{int} is the numerical interface thickness considered towards both the fluid-phases. Such numerical smoothening removes all the discontinuities across the interface which makes the solution of the discretized equations more accurate. The smoothening effectively provides the interface with an artificial thickness of $2\delta_{int}$ across which the fluid properties changes values corresponding to liquid to vapor as shown in Fig. 4. Although the level-set technique has proved to be efficiently capable of capturing interface morphology accurately [22], it requires reinitialization after every time iteration and is not mass conserving.

A coupled level-set and volume of fluid (CLSVOF) approach has been evolved [12, 19] to include the advantages of both level-set method (accurate calculation of normal and curvature) and volume of fluid method (mass-conservation) in the numerical results.

Fig. 4 Smooth variation of
ρ with level-set function ϕ

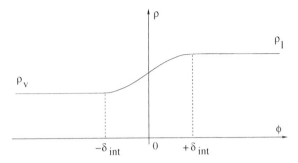

Fig. 5 Transition region
across the interface

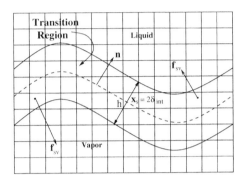

2.2 Surface Tension Model

The continuum surface force (CSF) model by Brackbill et al. [23] is the widely
accepted and applied numerical technique for the calculation of surface tension force
involving free surface. According to this model, the surface tension as a volume force
(f_{sv}) can be represented as

$$f_{sv} = \sigma \kappa \delta(d_n)\hat{\mathbf{n}} \tag{27}$$

where $\delta(d_n)$ is the Dirac delta function, d_n is a normal distance function measured
from the interface and $\hat{\mathbf{n}}$ is the unit normal to the interface. In the level-set framework
close to the interface, $\delta(d_n)$ takes the form

$$\delta(d_n) = |\nabla \phi| \delta(\phi) \tag{28}$$

which when combined with Eq. 17 transform Eq. 27 to

$$f_{sv} = \sigma \kappa \delta(\phi) \nabla \phi \tag{29}$$

The properties are smoothened in a thin region across the interface as shown in
Fig. 5. Similar to the smoothening in Heaviside function (26), the Dirac delta function
can be smeared out (δ_ϵ) near the interface to remove the instabilities as

$$\delta(\phi) = \begin{cases} 0 & \text{if } \phi < -\delta_{int} \\ \frac{1}{2\epsilon}\left[1 + cos\left(\frac{\pi\phi}{\epsilon}\right)\right] & \text{if } |\phi| \leq \delta_{int} \\ 1 & \text{if } \phi > \delta_{int} \end{cases} \tag{30}$$

However, the smearing of Dirac delta function is not required if the normal is calculated utilizing the Heaviside function as

$$\mathbf{n} = \nabla H_\delta(\phi) \tag{31}$$

Thereafter, the curvature can be calculated as

$$\kappa = -\nabla \cdot \mathbf{n}, \tag{32}$$

In the level-set framework,

$$\kappa = -\frac{\phi_y^2 \phi_{xx} - 2\phi_x \phi_y \phi_{xy} + \phi_x^2 \phi_{yy}}{(\phi_x^2 + \phi_y^2)^{3/2}} \tag{33}$$

2.3 ENO Scheme for Convection Terms

To avoid large spurious oscillations or high smearing of the properties across the jump discontinuities, it is required to discretize the convective terms in the Navier–Stokes equations using some high order numerical schemes. The *essentially non-oscillatory* or ENO scheme introduced by Harten et al. [24] produces high accuracy in smooth regions and adequately resolve the steep gradients without creating spurious oscillations. It is based on a class of piecewise polynomial interpolation for reconstructing a discontinuous solution.

The key idea of ENO schemes is the use of an adaptive stencil, which is chosen based on the local smoothness of the solution, measured by the Newton divided differences of the numerical solution. This adaptive polynomial interpolant is constructed to avoid the use of higher order polynomial interpolation across a steep gradient in the data. Thus the order of accuracy of the scheme is never reduced, however, the local stencil automatically avoids crossing discontinuities. The polynomial is also biased to use data by upwinding for physical consistency and stability. ENO schemes have been extremely successful in applications, because they are simple in concept, allow arbitrary orders of accuracy, and generate sharp, monotone solutions together with high-order accuracy in smooth regions of the solution.

The numerical discretization procedure [25] for the second-order accurate ENO scheme has been explained further. The 'minmod' function is defined as

$$\text{Minmod}(u, v) = \begin{cases} sgn(u)\min\left(|a|, |b|\right), & \text{if } u.v > 0 \\ 0, & \text{Otherwise.} \end{cases} \tag{34}$$

where $sgn(u)$ is a sign function defined as

$$sgn(u) = \begin{cases} -1, & \text{if } u < 0 \\ 1, & \text{if } u > 0 \\ 0, & \text{if } u = 0 \end{cases} \tag{35}$$

Now in the 'x' direction, as mentioned in Chang et al. [25], the second-order discretization of the convection term $u\phi_x$ is given by

$$u_{i,j}\left[\frac{\phi_{i,j} - \phi_{i-1,j}}{\Delta x} + \frac{\Delta x}{2}\text{minmod}\left(\frac{\phi_{i+1,j} - 2\phi_{i,j} + \phi_{i-1,j}}{\Delta x^2}, \frac{\phi_{i,j} - 2\phi_{i-1,j} + \phi_{i-2,j}}{\Delta x^2}\right)\right],$$
$$\text{if } u_{i,j} > 0, \tag{36}$$

and

$$\phi_x^+ = \frac{\phi_{i+1,j} - \phi_{i,j}}{\Delta x} - \frac{\Delta x}{2}\text{minmod}\left(\frac{\phi_{i+1,j} - 2\phi_{i,j} + \phi_{i-1,j}}{\Delta x^2}, \frac{\phi_{i+2,j} - 2\phi_{i+1,j} + \phi_{i,j}}{\Delta x^2}\right),$$
$$\text{otherwise.} \tag{37}$$

Similarly, the convection terms in the 'y' direction can be calculated with the discretization of $v\phi_y$.

2.4 Evaporation Model

The phase-change occurring at the liquid-vapor interface needs to be calculated accurately to predict the exact rate of vapor generation. Therefore, an evaporation model is required to be coupled with the hydrodynamical equations which is capable of evaluating the heat flux across the interface. In the present section, a widely accepted mass transfer model is explained which is developed by Welch and Wilson [16] in the volume of fluid framework. The similar model has been utilized in solving various problems related to boiling [26, 27]. The mass-flux across the interface for a two-phase cell (shown in Fig. 6) is given by

$$\dot{m} = \rho_l(\mathbf{U}_l - \mathbf{U}_I) = -\rho_v(\mathbf{U}_v - \mathbf{U}_I) \tag{38}$$

from which the mass jump condition at the interface is obtained as

$$\|\rho\left(\mathbf{U} - \mathbf{U}_I\right)\| \cdot \hat{\mathbf{n}} = 0 \tag{39}$$

and the energy jump condition as

$$\|\rho h_{lv}\left(\mathbf{U} - \mathbf{U}_I\right)\| \cdot \hat{\mathbf{n}} = -\|\mathbf{q}\| \cdot \hat{\mathbf{n}} \tag{40}$$

Fig. 6 A two-phase
computational cell

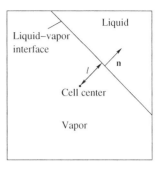

where $\|.\|$ represents the jump in the variable across the interface. From Eqs. 39 and
40, the jump in normal velocity can be calculated as

$$\|(\mathbf{U} - \mathbf{U}_I)\| \cdot \hat{\mathbf{n}} = \left(\frac{1}{\rho_l} - \frac{1}{\rho_v} \right) \frac{\|\mathbf{q}\| \cdot \hat{\mathbf{n}}}{h_{lv}} \tag{41}$$

The kinetic energy term, viscous work term and the viscous dissipation are neglected
in the energy jump condition.

The mass-conservation can be expressed in the following form considering the
mass transfer from each phase boundary,

$$\frac{d}{dt} \int_{V_v(t)} \rho dV + \int_{S_v(t)} \rho \mathbf{U} \cdot \hat{\mathbf{n}} dS + \int_{S_I(t)} \rho(\mathbf{U} - \mathbf{U}_I) \cdot \hat{\mathbf{n}} dS = 0 \tag{42}$$

$$\frac{d}{dt} \int_{V_l(t)} \rho dV + \int_{S_I(t)} \rho \mathbf{U} \cdot \hat{\mathbf{n}} dS + \int_{S_I(t)} \rho(\mathbf{U} - \mathbf{U}_I) \cdot \hat{\mathbf{n}} dS = 0 \tag{43}$$

where $V_l(t)$, $S_I(t)$, $V_v(t)$ and $S_v(t)$ are the volume and surface area of the cell boundary at the liquid and the vapor region, respectively. $S_I(t)$ is the phase interface at
the common boundary of the two regions moving with velocity $\mathbf{U}_I$. The unit normal
vector $\hat{\mathbf{n}}$ points into the liquid phase on $S_I(t)$.

Now, considering the overall volume of the cell as time invariant and taking into
account the incompressibility of each phase, the conservation of mass equation for
the cell volume is given by

$$\int_{S_C} \mathbf{U} \cdot \hat{\mathbf{n}} dS + \int_{S_I(t)} \|(\mathbf{U} - \mathbf{U}_I)\| \cdot \hat{\mathbf{n}} dS = 0 \tag{44}$$

From Eqs. 41 and 44, the conservation of mass equation in modified form can be
written in the form of Eq. 9.

3 Nucleate Boiling Regime

As described in Sect. 1.1, the mechanism of bubble generation, growth and interaction in the nucleate boiling regime is influenced by the various factors like surface superheat, roughness, and wettability. Nucleate boiling initiates with the generation of vapor bubbles over the surface at some preferred locations (imperfections or cavities which may entrap some dissolved gases in the liquid). Starting from the isolated bubble regime, the bubbles interact to form slugs and columns of vapors and thus, vigorous vapor generation and heat transfer occur. Bubble departure diameter, frequency of departure, growth time and waiting time are various useful parameters that decide the nature of boiling in the nucleate regime. Various heat transfer mechanisms occur during the lifetime of a bubble [28]. Due to the agitated motion of liquid as a result of bubble growth and departure, transient conduction results from the surface to the liquid. During the growth period of bubble, the outward motion of liquid results in convection currents which also contributes to the heat transfer. Direct heat transfer from the substrate to the thin liquid microlayer beneath the bubble also contributes substantially to the overall heat transfer rate.

Numerical simulations possess high feasibility due to the advances in computing power in predicting the heat transfer rate during boiling. Complex bubble morphological variations can also be analyzed through simulations which otherwise are difficult to capture through experimental means. Although various efforts have been made through experimental and analytical studies to understand the boiling mechanism, a perfect numerical model which can predict the overall physics of the process, is still lacking. However, some numerical approaches [29, 30] have been able to successfully predicting the growth dynamics of bubbles during nucleate boiling as well as the heat transfer rate during the process. Dhir and coworkers [29, 31, 32] have studied the growth of a single isolated bubble, vertical merger and the lateral merger of the bubbles through their numerical model. The influence of surface wettability and superheat on the bubble growth rate have been determined through numerical simulations; however, it has not been possible yet to observe the exact waiting time period during the bubble growth and departure through numerical techniques.

Incorporating the microlayer-thickness formulation by Cooper [33], Lee and Nydahl [34] modeled the saturated nucleate boiling of water at 1 atm. and 8.5 K wall-superheat. Quantitative analysis has been performed for the heat transfer mechanisms involved in the process and microlayer evaporation has been found to be the dominant micromechanism during the bubble growth. Constant wall-superheat and properties of the fluids were considered and the change in shape of the bubble was not considered. Welch [35] also attempted the simulation of heterogeneous boiling using Lagrangian approach to capture interfacial physics, with a pinned contact line and not considering the effects of microlayer.

Dhir's group [28] has worked extensively in the study of boiling – both experimentally and numerically. Son et al. [29] performed a complete numerical simulation of a single bubble cycle where level-set approach has been adopted for interface capturing. They adopted a microlayer model based on the lubrication approximation as proposed by Lay and Dhir [36]. The simulation domain has been considered to consist of a macroregion where hydrodynamic equations are valid and solved to calculate the flow field and a microregion which assumes a thin layer of liquid beneath the bubble and is separately modeled for determining its contribution in bubble growth. Later, the same model has been utilized to study the vertical bubble merger through axisymmetric calculations [31] and lateral merger through three-dimensional simulations [32] which were supported by their experimental results. The model is extended with the moving mesh technique to study the growth dynamics of a single bubble during subcooled nucleate boiling [37, 38].

The microregion model by Son et al. [29] requires the solution of fourth-order ordinary differential equation to determine the microlayer-thickness; however, the variation in microlayer-thickness at every location was not possible to determine. Further, the calculation assumes arbitrary values of dispersion constant which depends on the contact angle of the interface with solid.

Sato and Niceno [30] developed a microlayer depletion model which calculates the decrease in thickness of the microlayer at every location beneath the bubble with time. The latent heat is directly conducted to the microlayer from the heated surface which contributes in the bubble growth due to its evaporation. The model, thereby, depends only on the initial empirical thickness [39] of the microlayer.

Besides these, there are some other works [40–43] which concentrated on the bubble growth and microlayer evaporation effects using different models. Kunkelmann and Stephen [44] developed the microlayer model in the OpenFOAM two-phase solver and studied the influence of contact line motion on the heat transfer characteristics during nucleate boiling. Recently, Guion et al. [45] performed direct numerical study of the microlayer formation using a high-resolution computational grid around the contact line.

The most critical aspect while simulating the bubble growth in nucleate boiling is incorporating a microlayer model to account for the effect of evaporation of microscale liquid layer on the growth rate.

3.1 Microlayer Model

During nucleate boiling the vapor generation results owing to latent heat exchange either through the bubble interface in the bulk liquid or directly through the liquid microlayer beneath the bubble. Whereas the quantitative contribution of microlayer evaporation in the growth rate of bubble has been reported differently [28, 34, 46], its presence and influence on bubble growth has been established through various experimental studies [33, 39, 47, 48].

A microlayer is a thin liquid layer underneath the growing bubble which is formed as a result of viscous stresses acting at the liquid attached to the wall. The drag experienced by the liquid inside the layer is overcome by the disjoining pressure and the capillary pressure acting as a result of change in curvature of the interface [36]. With the passage of time as the bubble grows, the microlayer starts to evaporate

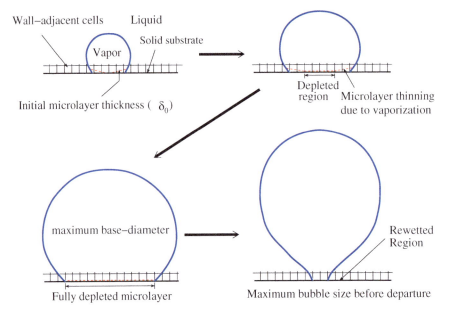

Fig. 7 Microlayer behavior during bubble lifecycle

and decrease in thickness culminating into an adsorbed layer of liquid where the molecular forces are high enough to cease the evaporation process. At this stage the portion of the layer can be considered to be depleted or dried out [30]. When the bubble attains a maximum base-diameter, the microlayer is almost completely depleted. After this stage when the buoyancy overcomes the capillary forces, the interface at the three-phase contact point retracts back towards the cavity centre and the bubble finally departs from the surface. The sequence of the phenomenon has been illustrated in Fig. 7.

The results presented in this section focuses the work presented in Pandey et al. [49] where the microlayer model similar to Sato and Niceno [30] have been utilized with an initial interface thickness as

$$\delta_{ml0} = C_0 r_l \tag{45}$$

where C_0 is an empirical constant depending on the combination of fluid and the surface and surface properties as well. The term r_l is the radial distance of the three-phase contact line from the cavity-center. While solving for the microlayer evaporation rate, it has been assumed that the liquid inside the microlayer is stagnant and heat is transferred purely through conduction. Moreover, the liquid-vapor interface between the microlayer and vapor bubble is considered to be at the saturated condition corresponding to the system pressure (1 atm.).

With the substrate temperature as T_w and the saturation temperature as T_{sat}, the heat flux conducted from the substrate to the microlayer (q_{ml}) can be calculated as

$$q_{ml} = k_l \frac{T_w - T_{sat}}{\delta_{ml}} \tag{46}$$

where k_l is the thermal conductivity of the liquid. The variation in microlayer-thickness with time as a result of evaporation due to latent heat exchange is given by

$$\frac{d\delta_{ml}}{dt} = -\frac{1}{\rho_l} q_{ml}/h_{lv} \tag{47}$$

which is solved explicitly with time as

$$\frac{\delta_{ml}^{n+1} - \delta_{ml}^{n}}{\Delta t} = -\frac{1}{\rho_l} q_{ml}^{n}/h_{lv} \tag{48}$$

The quantity q_{ml}/h_{lv} is basically the mass-flux transferred as a result of microlayer evaporation. The above set of equations (Eq. 46–48) are solved at every wall-adjacent cells to obtain the resulting mass-flux due to microlayer evaporation. The threshold limit of the microlayer-thickness below which it is considered to be depleted is 1×10^{-10} m.

The overall growth of the bubble is the result of mass-flux due to evaporation at the bulk liquid region ($m_{lv} = q_{lv}/h_{lv}$) and at the microlayer ($m_{ml} = q_{ml}/h_{lv}$) which are incorporated in the modified continuity equation (Eq. 9) to calculate the velocity-jump.

3.2   Boundary Conditions

Axisymmetric simulations have been performed in the computational domain (8) considering the left-bottom corner as the center of the nucleating bubble and left boundary as the axis of symmetry (Fig. 8).

The boundary conditions satisfies the following symmetry condition in the left and right sides of the domain

$$u = 0, \ \frac{\partial v}{\partial x} = 0, \ \frac{\partial T}{\partial x} = 0, \ \frac{\partial f}{\partial x} = 0, \ \frac{\partial \phi}{\partial x} = 0. \tag{49}$$

At the top boundary, the outflow condition for fluid follows

$$\frac{\partial u}{\partial y} = \frac{\partial v}{\partial y} = \frac{\partial T}{\partial y} = \frac{\partial f}{\partial y} = \frac{\partial \phi}{\partial y} = 0; \quad P = P_{sat} - \rho g h \tag{50}$$

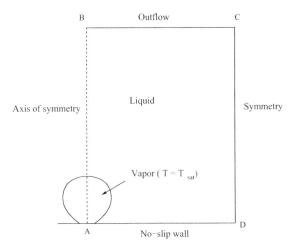

Fig. 8 Schematic of the computational domain

where h is the height of the top boundary from the bottom wall.

The bottom boundary acts as a solid wall with no-slip and impermeability conditions ($u = v = 0$) and a specified static contact and superheat,

$$T_w = T_{sat} + \Delta T_{sup} \tag{51}$$

and

$$\frac{\partial \phi}{\partial y} = -\cos\psi. \tag{52}$$

3.3 Effect of Superheat

The numerical results of an isolated single bubble growth during nucleate boiling presented by Pandey et al. [49] were in good agreement with the experimental results of Siegel and Keshok [50] and Son et al. [29] for the contact angles of 38° and 50°, respectively. During nucleate boiling, the growth rate of bubble increases with an increase in superheat temperature of the substrate due to the enhanced rate of vapor generation. It can be depicted from the variation in equivalent diameter with time for different superheat values as shown in Fig. 9a. The bubble interface profiles (Fig. 9b) at the instants of departure from the surface also illustrate the increase in departure size with the increase in superheat of the surface. The departure time decreases with an increase in superheat as the bubble attains the required volume of vapor more rapidly for higher values of superheat.

Due to the increased surface superheat, the heat transfer rate to the liquid side also increases since the temperature gradient at the surface is higher. Figure 10 shows the variation in heat flux value at the liquid side during the lifetime of a bubble from

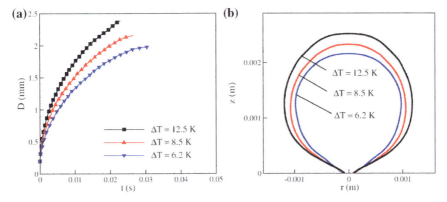

Fig. 9 Effect of superheat on **a** bubble growth rate up to the instant of departure and **b** bubble morphology at the instant of departure for the contact angle of 38° [49]

Fig. 10 Comparison of wall heat flux to the liquid side at different degrees of superheat for $\psi = 38°$ [49]

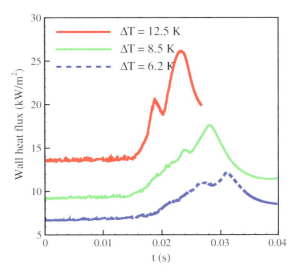

its initiation to its departure. It is evident from the plots that the maximum heat flux value is attained at the instant corresponding to the time of departure. Due to the retracement of the contact line towards the cavity-center, the neighboring liquid rushes towards the bubble-base and thereby enhancing the convective heat transfer.

To understand the dynamics of bubble growth more clearly, the variation of bubble-base radius with time can be observed from Fig. 11. Base-radius increases rapidly during the initial growth period of the bubble which is dominated by the inertia effects and reaches a maximum value where an equilibrium between oppositely acting forces of buoyancy and capillary is attained. As the bubble growth continues, the buoyancy force dominates over the surface tension force which leads to the retraction of the bubble interface towards cavity-center and hence the reduction in bubble-base radius.

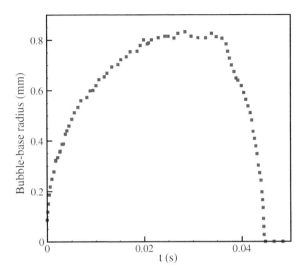

Fig. 11 Variation of base-radius with time for $\psi = 50°$ and $\Delta T_{sup} = 8.5$ K [49]

3.4 Microlayer Parameters

The initial thickness of the microlayer is of the range of few micrometers. As the bubble grows the bubble-base extends outwards from the cavity-center, the geometry of microlayer underneath the bubble modifies. The depletion of the microlayer initiates from the cavity-center where the thickness is minimum and continues toward the radial direction. The maximum thickness prevails at the three-phase contact point. The degree of substrate superheat also affects the evaporation rate of the microlayer which can be observed from Fig. 12a. As the bubble grows, the span of depleted microlayer increases as a result of evaporation. Figure 12b illustrates the geometry of the microlayer at various instants of bubble growth from the initial stage.

At any instant of time since the thickness of the microlayer is minimum toward the cavity-center side. Therefore, the maximum mass-flux transfer occurs at that region and the mass-flux decreases as the microlayer thickness increases.

3.5 Effect of Subcooling

The numerical model is extended and validated in Pandey et al. [49] for subcooled nucleate boiling. A single bubble under subcooled liquid has been studied during its growth and after departure to analyze its morphological and dynamical variation at varying degree of liquid-subcooling. The study is performed for low subcooling range (up to $\Delta T_{sub} = 4.0$ K) and it has been observed that with an increase in degree of liquid-subcooling the bubble departure diameter slightly decreases. After departure, there is an appreciable decrease in diameter of bubble due to the condensation during

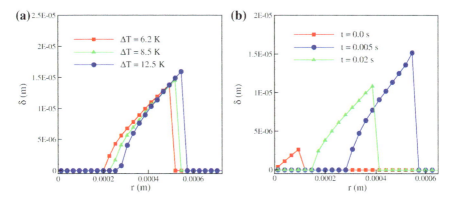

Fig. 12 Variation of microlayer-thickness along the surface **a** at different values of superheat after 0.01 s of bubble initiation and **b** for the superheat value of $\Delta T_{sup} = 6.2$K at different instants of time for $\psi = 38°$ [49]

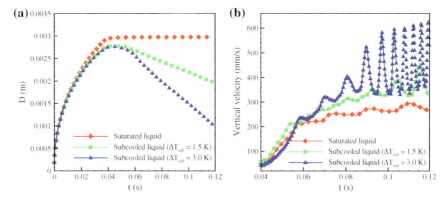

Fig. 13 Comparison of **a** diameter and **b** vertical velocity of bubble among saturated liquid condition and subcooled conditions of $\Delta T_{sub} = 1.5$ K and $\Delta T_{sub} = 3.0$ K [49]

its rise through the subcooled liquid. The rate of decrease in diameter with time has been compared at different subcooling in Fig. 13a.

After the bubble departure from the heated surface, the rise velocity of bubble has been plotted in Fig. 13b. It can be observed from the plots that as the degree of subcooling increases the oscillation of the bubble increases due to the continuous variation in drag force and change in size as a result of condensation at its surface.

The variation in bubble morphology at different degrees of liquid-subcooling during the bubble rise after detachment has been shown in Fig. 14 where the enhancement in condensation-rate of vapor bubble can be clearly observed with an increase in subcooling.

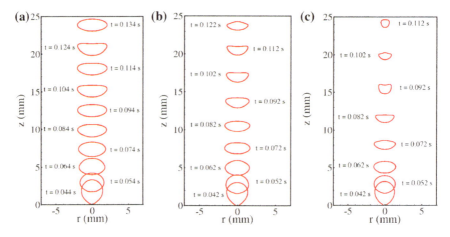

Fig. 14 Profile of a bubble after the instant of departure from the surface in **a** saturated liquid condition, **b** subcooled condition of $\Delta T_{sub} = 1.5$ K and **c** subcooled condition of $\Delta T_{sub} = 3.0$ K for $\psi = 50°$ and $\Delta T_{sup} = 8.5$ K [49]

4 Film Boiling Regime

Factors instigating the bubble incipience and growth during film boiling are very different than in nucleate boiling. Unlike to the surface effects and random bubble generation in nucleate boiling, the bubble formation in film boiling occurs as a result of instabilities at the liquid-vapor interface of the vapor film and the superheat of the surface irrespective of surface characteristics unless liquid contact occurs.

During film boiling, a vapor film is supported over a heated surface against the force of gravity acting on the bulk liquid above it. The film is retained due to the continuous generation of vapor as a result of latent heat transfer from the surface. According to Zuber's [9] hydrodynamic analysis, bubbles emanate from the vapor film alternately from the nodes and antinodes of a wavelength of disturbance which lies between the most critical wavelength ($\lambda_c = 2\pi\sqrt{\sigma/(\rho_l - \rho_v)g}$) and the fastest growing wavelength ($\lambda_0 = 2\pi\sqrt{3\sigma/(\rho_l - \rho_v)g}$). Bubble release is periodic both in space and time. Based on his hypothesis, he provided a dispersion relation for the frequency of bubble release as

$$\omega^2 = \frac{\sigma}{\rho_l + \rho_v}m^3 - \frac{(\rho_l - \rho_v)}{\rho_l + \rho_v}gm \tag{53}$$

where m is the wave number ($m = 2\pi/\lambda$) and the two terms on the right hand side signifies the effects of surface tension and buoyancy.

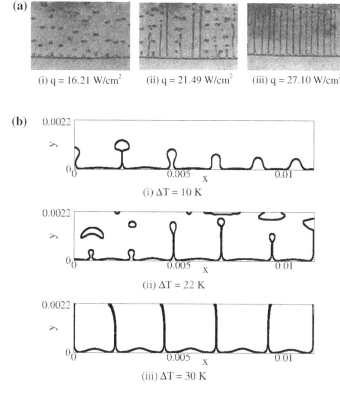

Fig. 15 Hydrodynamic transition in vapor release observed by **a** experimental results of Reimann and Grigull [51] and **b** numerical simulations by Pandey et al. [52]

4.1 Hydrodynamic Transition in Bubble Release Pattern

The phenomena have been observed experimentally by Reimann and Grigull [51] with an increasing value of applied heat flux at the substrate. At lower heat flux values, the bubbles tend to form at their locations and release alternately from nodes and antinodes. As the heat flux is increased, some of the sites transform into vapor columns while others still releasing discrete bubbles. With a further increase in heat flux, every bubble formation site transforms into continuous vapor columns (Fig. 15a).

Numerical validation of the change in vapor release pattern has been observed in a smaller domain by Son and Dhir [53]. Similar transitions in multimode film boiling in a larger domain have been presented and compared with the Reimann and Grigulls's [51] results by Pandey et al. [52] as shown in Fig. 15b.

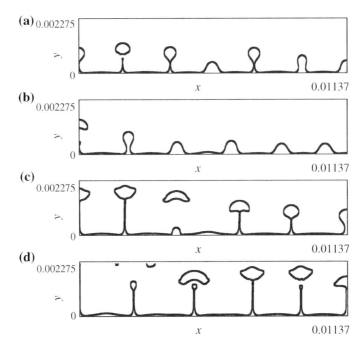

Fig. 16 Interface morphology for wall-superheat of **a** 2 K **b** 5 K **c** 18 K and **d** 22 K [11]

4.2 Change in Instability Mode Due to Superheat

The instability prevailing at the interface depends on the superheat of the heated surface. At the low surface superheat when the lubrication approximation holds appropriate [54], the bubble formation is governed by the Rayleigh–Taylor mode of instability. Whereas at comparatively higher superheat where inviscid flow can be assumed, the phenomenon is dominated by the Taylor–Helmholtz instability [55]. Similar shift in instability mode has been observed through numerical simulations by Tomar et al. [56].

A detailed analysis of the influence of increase in superheat on the bubble morphology and the heat transfer characteristics have been performed by Pandey et al. [11]. Figure 16 shows the effect of increasing superheat on bubble morphology. The increase in bubble separation distance with an increase in superheat temperature signifies the shift in instability mode. Further, the analyses extend to the influence of superheat on the bubble release frequency which can be studied from the variation in heat transfer characteristics with time as shown in Fig. 17. The peaks in the plot represent the instant of bubble release from the vapor film. At that instant, the interface attains a minimum distance from the surface and therefore sudden increase in heat transfer occurs as a result of increase temperature gradient. The increase in superheat enhances the rate of vapor generation, hence, the bubble release frequency increases

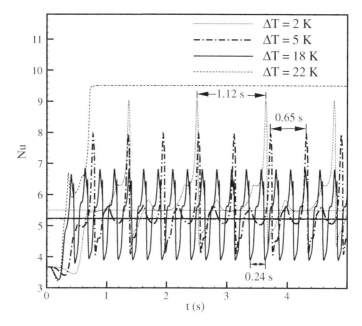

Fig. 17 Variation of space averaged Nu number at various values of applied electric field for different values of superheat [11]

leading to an increase in space-time-averaged value of Nusselt number. With an increase of superheating the time period of bubble release increases, as illustrated in the plots. After a certain degree of superheat, the stable vapor columns are formed above the surface which leads to the constant value of heat transfer rate.

4.3 Influence of Electric Field

The perturbations prevailing at the liquid-vapor interface enhances due to the application of any external force fields such as the electric field. The electrohydrodynamic (EHD) forces when applied across the interface indulge in further destabilizing the dynamical balance which leads to bubble generation from the vapor film. Johnson [57] extended the hydrodynamic dispersion relation for harmonic frequency of bubble release (Eq. 53) to account for the destabilizing eelectric field force as

$$\omega^2 = \frac{\sigma}{\rho_l + \rho_v} m^3 - \frac{(\rho_l - \rho_v)}{\rho_l + \rho_v} gm - \frac{f(E, \epsilon_l, \epsilon_v)}{\rho_l + \rho_v} m^2 \tag{54}$$

The magnitude of electric force depends on the relative dielectric permittivity of the medium considered. The function f is defined as

$$f(E, \epsilon_l, \epsilon_v) = \frac{\epsilon_l(\epsilon_l - \epsilon_v)^2}{\epsilon_v(\epsilon_l + \epsilon_v)} \epsilon_0 E_l^2 = \frac{\epsilon_v(\epsilon_l - \epsilon_v)^2}{\epsilon_l(\epsilon_l + \epsilon_v)} \epsilon_0 E_v^2 \tag{55}$$

The relation for fastest growing wavelength in the presence of an electric field is established as

$$\lambda_J = \frac{6\pi\sigma}{f\left[1 + \sqrt{1 + \frac{3(\rho_l - \rho_v)g\sigma}{f^2}}\right]} \tag{56}$$

Johnson [57] also predicted the ratio of heat flux with electric field, $q(E)$ to heat flux without electric field, $q(0)$ as

$$\frac{q(E)}{q(0)} = \left[1 + \frac{f}{3[\frac{4}{3}(\rho_l - \rho_v)g\sigma]^{1/2}}\right]^{1/2} \tag{57}$$

Welch and Biswas [26] performed numerical simulation to analyze the influence of EHD forces on the film boiling using CLSVOF approach. A modified formulation for electric field forces [58] has been later utilized by Tomar et al. [59] to analyze the effect of electric forces during multimode film boiling. The smoothening in dielectric permittivity across the interface has been performed by harmonic interpolation in their formulation.

Electric field formulation The general form of electric force can be represented by

$$\mathbf{f}_v^E = q_v \mathbf{E} - \frac{1}{2}E^2 \nabla \epsilon_0 \epsilon + \frac{1}{2}\nabla\left[E^2 \rho \left(\frac{\partial \epsilon_0 \epsilon}{\partial \rho}\right)_T\right] \tag{58}$$

The various terms on the right-hand side of the above equation starting from the first terms are the electrophoretic force (depending on the free charge present), force due to gradient in permittivity (acts only near the interface if the bulk properties are homogeneous) and the electrostrictive force (due to the non-uniformity in electric field or due to the volumetric change).

If the medium is considered to be pure dielectric in the presence of uniform electric field, Eq. 58 simplifies to

$$\mathbf{f}_v^E = -\frac{1}{2}E^2 \nabla \epsilon_0 \epsilon \tag{59}$$

In the absence of any dynamic current or induced magnetic field, the irrotationality of eletric field ($\nabla \times \mathbf{E} = 0$) is valid and therefore it can be represented as the gradient of a scalar function (basically the electric potential function, ψ),

$$\mathbf{E} = -\nabla\psi \tag{60}$$

and Gauss law in the absence of any free charge can be reprented as

$$\nabla \cdot (\epsilon \epsilon_0 \mathbf{E}) = 0 \tag{61}$$

where $\epsilon_0 = 8.85 \times 10^{-12}$ C/Vm is the dielectric permittivity of the vacuum. From Eqs. 60 and 61, the electric potential equation in the Laplacian form to be solved is

$$\nabla \cdot (\epsilon \epsilon_0 \nabla \psi) = 0. \tag{62}$$

Modified form of governing equation The momentum Eq. 7 is modified to include the influence of volumetric electric field force ($\mathbf{f}_v^E$) as

$$\rho(\mathbf{U}_t + \nabla \cdot \mathbf{UU}) = -\nabla p + \rho \mathbf{g} + \nabla \cdot (2\mu \mathbf{D}_v) + \mathbf{f}_{sv} + \mathbf{f}_v^E \tag{63}$$

In the subsequent sections where the results associated with the applied electric field from the work performed by Pandey et al. [11, 60] has been discussed, the electric field formulation due to Tomar et al. [58] is utilized.

With the application of an external electric field, the number of bubble formation sites increases verifying the decrease in dominating wavelength of disturbance. Similar conclusion can be drawn from the analytical expression (Eq. 56) given by Johnson [57]. Figure 18 illustrates the bubble morphologies with the application of different intensities of electric field where the decrease in bubble separation distance with an increase in electric field intensity is evidently observed.

Although it is apparent from the bubble-profiles that the spatial periodicity is maintained even with the destabilizing influence of electric field, the temporal periodicity is lost as can be discernible from the variation in Nusselt number with time (Fig. 19a). Without the application of electric field, there is a periodic release of bubbles represented by the peaks in the plot. The periodicity is maintained even after the application of electric field of low intensity. However when the intensity is increased to 2×10^5 V/m, bubble starts to emanate randomly at a rapid rate, thus, disturbing the periodicity. There is a continuous increase in heat transfer rate with the increasing intensity of electric field.

It has also been observed that the enhancement in heat transfer rate is significant only above some threshold value of intensity of electric field and that it further depends on the degree of superheat of the surface. From Fig. 19b, it can be deduced that the increase in heat transfer rate is higher for lower value of superheat compared to the higher value of superheat for almost all values of applied electric field. The increase in heat transfer with applied electric field is significant only above the intensity of 5×10^4 V/m.

The change in bubble separation distance and increase in heat flux value has been found to be consistent with the theoretically calculated values of the critical wavelength by Johnson [57].

Apart from these, a significant numerical verification of the analytical [61] and experimental [62] observation of invariability of bubble departure-height with applied electric field has been performed by Pandey et al. [11]. A comparison of

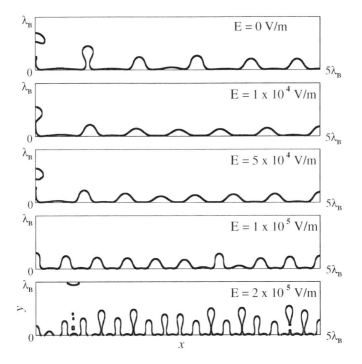

Fig. 18 Interface profiles for the first set of bubble release with different applied electric field intensities at 5 K superheat [11]

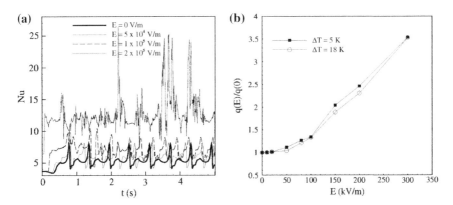

Fig. 19 **a** Variation of space averaged Nusselt number for 5 K wall-superheat and varying electric field intensity and **b** Comparison of the ratio of heat flux values with electric field to without electric field between superheats of 5 K and 18K [11]

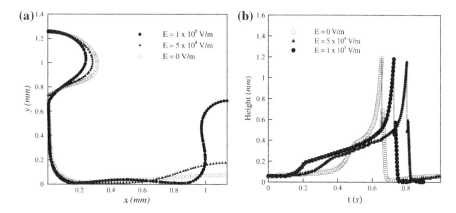

Fig. 20 **a** Comparison of maximum bubble height at various values of applied electric field and superheat of 5 K and **b** Variation of apex height of the bubble with time at 5 K superheat and different values of applied electric field [11]

bubble-profiles at the time of departure (Fig. 20a) and variation of apex height with time (Fig. 20b) shows that even with the application of electric field, the bubble height at the time of departure remains invariant although the radius of bubble decreases significantly. The morphology of bubble is decided by the combined influence of Maxwell stresses and the surface tension force acting at the bubble interface.

4.4 Effect of Gravity

Both, bubble morphology and heat transfer rate are highly affected by the change in gravitational acceleration due to its significant role in the balance of forces across the interface. In almost all heat transfer correlations, gravity has a proved to be an important factor to consider. However, in the low-gravity environment, boiling crisis may result due to the dominant effect of thermocapillary forces [2] instead of gravitational forces. Therefore, the study concerning the influence of gravity on the boiling phenomena is of utmost importance.

Experiments by Siegel and coworkers [50, 63, 64] have been considered to be pioneering in the study of the effects of gravity during boiling. Their major findings include larger bubble departure diameter and decrease in critical heat flux with the reduction in gravity-level. Most of the studies [65–67] concentrating on the influence of gravity have been performed in the nucleate boiling regime i.e. before the critical heat flux is achieved. Most of the researchers observed a deterioration in heat transfer coefficient with the reduction in gravity. Dhir's group performed numerical simulations [68] for single bubble and multiple bubble growth during nucleate boiling regime followed by a series of experiments [69, 70].

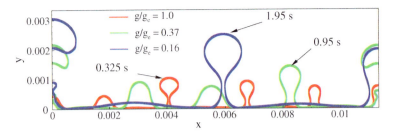

Fig. 21 Comparison of interface profile at the instant of first set of bubble release at three different levels of gravity for R134a with $\Delta T = 20$K [60]

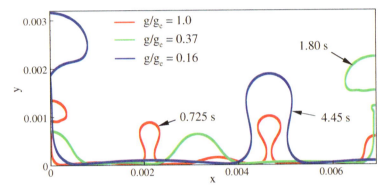

Fig. 22 Comparison of interface profile at the instant of first set of bubble release at three different levels of gravity for water with $\Delta T = 5$K [60]

Straub [71, 72] carried out experiments under low gravity to analyze the influence of controlling factors like surface tension, evaporation and coalescence mechanism during boiling. In addition, the heat transfer data from the experiments signifies less effect in the nucleate regime while a significant reduction in the film boiling regime. Di Marco and Grassi [73] performed low-gravity experiments in the presence of external electric field to determine the variation in bubble generation phenomena under varying physical conditions. They emphasized that through an appropriate combination of electric field and gravity, the bubble size and departure frequency can be controlled.

Numerical simulations on film boiling under varying gravity conditions have been performed by Pandey et al. [60] for water and R134a as the working fluids. Figures 21 and 22 presents the interface profiles of bubbles at the time of departure of first set of bubbles for R134a and water, respectively. Three different levels of gravity have been considered and the time for the first set of bubble departure is mentioned. With a decrease in the level of gravity, both length and time scale increases. The increase in length scale is evident from the increase in bubble separation distance with the reduction in gravity. At reduced gravity the buoyant forces acting on the vapor bubbles are weaker than that in higher gravity, therefore, the larger vapor mass is required for

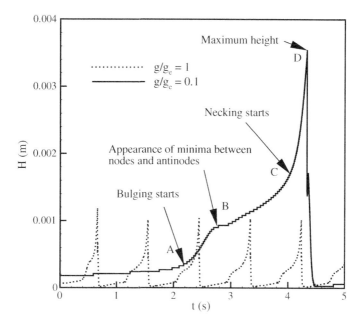

Fig. 23 Variation of apex height of bubble with time at normal and reduced gravity conditions for water at $\Delta T = 5$K [60]

the bubbles to buoyed away which requires larger time too. The increase in bubble size can be observed by bubble-profiles as well as by the variation of bubble apex height with time as shown in Fig. 23.

Due to the decrease in buoyant forces in the reduced gravity, the vapor bubble stays above the heated surface for a longer period than in normal gravity conditions. Further, the increase in most critical wavelength as a result of reduced gravity accommodates less number of bubbles over a given heat transfer area. Although the bubble release maintains the spatial and temporal frequency even in low gravity, the time period of bubble release is much higher. These effects result in the deterioration of heat transfer rate as can be illustrated from Fig. 24.

4.5 Combined Influence of Gravity an Electric Field

In Sect. 4.3 the destabilizing effects of applied electric field at the liquid-vapor interface have been focused. The application of electric field can be utilized in low-gravity conditions to recover the reduction in heat transfer rate and obtain the desired vapor release rate. Figure 25 shows a comparison of the profiles of first set of bubbles at various conditions (gravity-levels and applied electric field). The computational domain in the horizontal direction is equivalent to $3\lambda_B$ where λ_B is the Berenson's

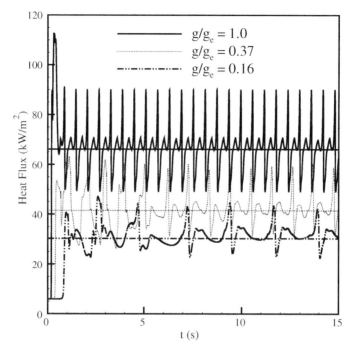

Fig. 24 Comparison of heat flux variation with time at different levels of gravity for water at $\Delta T = 5$K [60]

most dominant wavelength corresponding to normal gravity (g_e). In the reduced gravity condition, the number of emanating bubbles decreases as compared to that in normal gravity for a specified domain. However when an electric field intensity of 1×10^5 V/m is applied across the interface, the number of bubbles again increases which will also restore the heat transfer rate.

The effective percentage increase in heat transfer rate as a result of applied electric field depends on the level of gravity as can be contemplated from Eq. 57. As shown in Fig. 26a, the value of heat flux is higher for normal gravity condition ($g/g_e = 1.0$) than in lower gravity ($g/g_e = 0.16$) for every intensity of applied electric field considered. However at a much higher intensity of electric field, the heat flux values tend to converge (i.e., the difference minimizes as the electric field intensity increases).

Moreover, it can also be observed from Fig. 26b that the ratio of heat flux with applied electric field to without electric field enhances as the applied electric field intensity increases. This emphasizes the increasing trend of dominance of electric field forces with increasing electric field intensity.

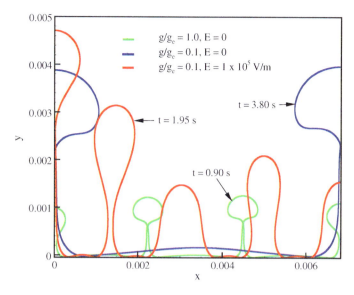

Fig. 25 Comparison of interface profile at the instant of first set of bubble release for water with $\Delta T = 18$K [60]

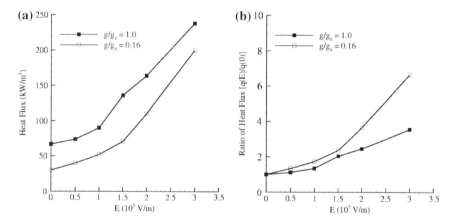

Fig. 26 Variation of **a** heat flux and **b** ratio of heat flux with electric field to that of without electric field in normal and reduced gravity conditions for water at $\Delta T = 5$K [60]

4.6 Self-similarity in Film Boiling

The bubble growth during film boiling is basically a singularity phenomenon exhibited due to the growing instability at the liquid-vapor interface. As discussed in Sect. 4.2, the phenomena is governed by Rayleigh–Taylor mode of instability while Taylor–Helmholtz instability prevails at comparatively higher superheat values.

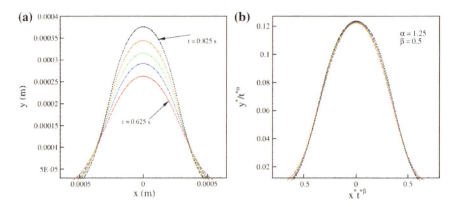

Fig. 27 **a** Interface profiles at different instants of time from t = 0.625 s to t = 0.825 s for $\Delta T = 2$K. **b** Profiles at the same instants of time as in left hand side, rescaled following $y^*/t^* = ax^*t^{*\beta}$ [60]

The bubble-profiles in the initial stage of growth reveal self-similarity behavior as studied by Duchemin et al. [74] in the Rayleigh–Taylor mode of instability of an interface. A self-similarity analyses of the interface profiles during the initial linear stage of bubble growth have been performed by Pandey et al. [60] through a proper scaling of the dimensional parameters and utilizing a fitting function of the form

$$y^*/t^* = ax^*t^{*\beta} \tag{64}$$

as a general solution for the kinetic free surface problem of Hogrefe et al. [75].

At a low surface superheat of $\Delta T = 2$K, the values of α and β are found to be 1.25 and 0.5, respectively, for the scaled bubble-profiles as shown in Fig. 27b. The unscaled profiles have been shown in Fig. 27a. The value of coefficient a in Eq. 64 is considered as unity. The relation between α and β is found to be as $\alpha = 1 + \beta/2$. The characteristic wavelength, $\lambda_B/2$ and the viscous time, $\tau_v = l^2\rho_l/\mu_l$ are used as the scaling parameters.

When the value of superheat is slightly increased to $\Delta T = 5$K and the similar scaling parameters are employed as in $\Delta T = 2$K, the profiles at different instants of time still merge together but for a comparatively smaller time period (Fig. 28).

The values of α and β are not same in the self-similar scaling of the bubble-profiles in case of higher superheat value of $\Delta T = 10$K. The self-similar profiles of bubbles at various instants of time have been shown in Fig. 29. The exponents follows the relation $\alpha = 1 + 6\beta/5$ with $\alpha = 6.4$ and $\beta = 4.5$. The change in scaling parameters with change in superheat indicates the transition in instability mode that governs the bubble growth.

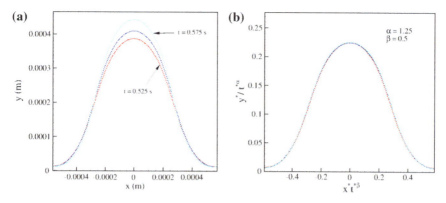

Fig. 28 **a** Interface profiles at different instants of time from t = 0.525 s to t = 0.575 s $\Delta T = 5K$. **b** Profiles at the same instants of time as in left hand side, rescaled following $y^*/t^* = ax^*t^{*\beta}$ [60]

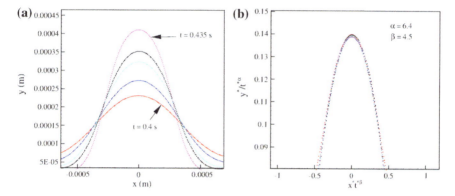

Fig. 29 **a** Interface profiles at different instants of time from t = 0.40 s to t = 0.435 s $\Delta T = 10K$. **b** Profiles at the same instants of time as in left hand side, rescaled following $y^*/t^* = ax^*t^{*\beta}$ [60]

5 Closure

The work presented in this chapter focused on the numerical simulation of boiling phenomena explaining various regimes of bubble formation. Physical aspects influencing the bubble generation and its growth during various modes of boiling have been discussed to provide a conceptual understanding of the mechanism.

The numerical aspects involving interface capturing, surface tension formulation and advection schemes have been discussed along with the formulation for an externally applied electric field force. For nucleate boiling, the present numerical capability is still not capable to simulate the full-scale microlayer dynamics simultaneously with the growth of bubble up to the departure. Therefore, separate microlayer evaporation models have been developed and coupled with the solutions of hydrodynamics equations in order to obtain the exact growth rate. However, the models for microlayer so far developed are either dependent on the empirical values of thickness or the unknown constants (like Hamaker constant). A more efficient and accurate modeling requires further experimental studies to determine the exact shape variation data of the microlayer beneath the bubble due to evaporation.

Numerical investigation of film boiling has been comparatively easier than the nucleate boiling since the interface contact dynamics with the surface is not present. Moreover, the bubble dynamics is highly deterministic owing to its dependence on the instability modes which govern the spatial and temporal frequencies of bubble release. Although film boiling is not desirable due to the reduced heat transfer rate as compared to the nucleate boiling, the bubble release frequency can be enhanced with the application of external forces like the electrohydrodynamic forces. With a specified value of applied electric field, the heat transfer efficiency can be controlled even at varying values of gravity-level.

References

1. Zuber, Z. (1959). Hydrodynamic aspects of boiling heat transfer. Ph.D. thesis, University of California, Los Angeles
2. Nikolayev, V., Chatain, D., Garrabos, Y., & Beysens, D. (2006). *Physical Review Letters*, *97*(18), 184503.
3. Rohsenow, W. M. (1951) A method of correlating heat transfer data for surface boiling of liquids. Tech. rep., Cambridge, Mass.: MIT Division of Industrial Corporation
4. Tien, C. (1962). *International Journal of Heat and Mass Transfer*, *5*(6), 533.
5. Moore, F. D., & Mesler, R. B. (1961). *AIChE Journal*, *7*(4), 620.
6. Carey, V. P. (2008). *Liquid-Vapor Phase-Change Phenomena* (2nd ed.). LLC: Taylor and Francis Group.
7. Hsu, Y. (1962). *ASME Journal of Heat Transfer*, *84*(3), 207.
8. Kurihara, H., & Myers, J. (1960). *AIChE Journal*, *6*(1), 83.
9. Zuber, N. (1963). *International Journal of Heat and Mass Transfer*, *6*(1), 53.
10. Fritz, W. (1935). *Physikalische Zeitschrift*, *36*, 379384.
11. Pandey, V., Biswas, G., & Dalal, A. (2016). *Physics of Fluids*, *28*(5), 052102.
12. Gerlach, D., Tomar, G., Biswas, G., & Durst, F. (2006). *International Journal of Heat and Mass Transfer*, *49*(3–4), 740.
13. Son, G., & Dhir, V. K. (1997). *International Journal of Heat and Mass Transfer*, *1*, 525.
14. Welch, S. W. J. (1995). *Journal of Computational Physics*, *121*, 142.
15. Juric, D., & Tryggvason, G. (1998). *International Journal of Multiphase Flow*, *24*(3), 387.
16. Welch, S. W. J., & Wilson, J. (2000). *Journal of Computational Physics*, *160*(2), 662.
17. Hirt, C. W., & Nichols, B. (1981). *Journal of Computational Physics*, *39*(1), 201.
18. Osher, S., Sethian, J. A. (1988). Journal of Computational Physics *79*, 12.
19. Sussman, M., & Puckett, E. G. (2000). *Journal of Computational Physics*, *162*(2), 301.
20. Puckett, E. G., Almgren, A. S., Bell, J. B., Marcus, D. L., & Rider, W. J. (1997). *Journal of Computational Physics*, *130*(2), 269.
21. Strang, G. (1968). *SIAM Journal on Numerical Analysis*, *5*(3), 506.
22. Sussman, M., Smereka, P., & Osher, S. (1994). *Journal of Computational Physics*, *114*, 146.
23. Brackbill, J. U., Kothe, D. B., & Zemach, C. (1992). *Journal of Computational Physics*, *100*, 335.
24. Harten, A., Engquist, B., Osher, S., Chakravarthy, S.R. (1987). *Upwind and high-resolution schemes* (Springer, pp. 218–290).

25. Chang, Y. C., Hou, T. Y., Meriman, B., & Osher, S. (1996). *Journal of Computational Physics*, *464*(124), 449.
26. Welch, S. W. J., & Biswas, G. (2007). *Physics of Fluids*, *19*(1), 012106.
27. Agarwal, D. K., Welch, S. W. J., Biswas, G., & Durst, F. (2004). *ASME International Journal of Heat and Mass Transfer*, *126*(3), 329.
28. Dhir, V. K., Warrier, G. R., & Aktinol, E. (2013). *Journal of Heat Transfer*, *135*(6), 061502.
29. Son, G., Dhir, V. K., & Ramanujapu, N. (1999). *ASME Journal of Heat and Mass Heat Transfer*, *121*(3), 623.
30. Sato, Y., & Niceno, B. (2015). *Journal of Computational Physics*, *300*, 20.
31. Son, G., Ramanujapu, N., & Dhir, V. K. (2002). *ASME Journal of Heat Transfer*, *124*(1), 51.
32. Mukherjee, A., & Dhir, V. K. (2004). *ASME International Journal of Heat and Mass Transfer*, *126*(6), 1023.
33. Cooper, M., & Lloyd, A. (1969). *International Journal of Heat and Mass Transfer*, *12*(8), 895.
34. Lee, R., & Nydahl, J. (1989). *ASME International Journal of Heat and Mass Transfer*, *111*(2), 474.
35. Welch, S. W. (1998). *International Journal of Heat and Mass Transfer*, *41*(12), 1655.
36. Lay, J., & Dhir, V. K. (1995). *ASME International Journal of Heat and Mass Transfer*, *117*, 394.
37. Wu, J., Dhir, V. K., & Qian, J. (2007). Numerical heat transfer. *Part B: Fundamentals*, *51*(6), 535.
38. Wu, J., & Dhir, V. K. (2010). *ASME International Journal of Heat and Mass Transfer*, *132*(11), 111501.
39. Utaka, Y., Kashiwabara, Y., & Ozaki, M. (2013). *International Journal of Heat and Mass Transfer*, *57*(1), 222.
40. Yoon, H. Y., Koshizuka, S., & Oka, Y. (2001). *International Journal of Multiphase Flow*, *27*(2), 277.
41. Liao, J., Mei, R., & Klausner, J. F. (2004). *International Journal of Heat and Fluid Flow*, *25*(2), 196.
42. Fuchs, T., Kern, J., & Stephan, P. (2006). *Journal of Heat Transfer*, *128*(12), 1257.
43. Hänsch, S., Walker, S., & Narayanan, C. (2017). *Nuclear Engineering and Design*, *321*, 230.
44. Kunkelmann, C., & Stephan, P. (2009). Numerical heat transfer. *Part A: Applications*, *56*(8), 631.
45. Guion, A., Afkhami, S., Zaleski, S., & Buongiorno, J. (2018). *International Journal of Heat and Mass Transfer*, *127*, 1271.
46. Demiray, F., & Kim, J. (2004). *International Journal of Heat and Mass Transfer*, *47*(14), 3257.
47. Voutsinos, C., & Judd, R. (1975). *ASME International Journal of Heat and Mass Transfer*, *97*(1), 88.
48. Zou, A., Chanana, A., Agrawal, A., Wayner, P. C., Jr., & Maroo, S. C. (2016). *Scientific Reports*, *6*, 20240.
49. Pandey, V., Biswas, G., Dalal, A., & Welch, S. W. J. (2018). *Journal of Heat Transfer*, *140*(12), 121503.
50. Siegel, R., & Usiskin, C. (1959). *ASME International Journal of Heat and Mass Transfer*, *81*, 230.
51. Reimann, M., & Grigull, U. (1975). *Wrme-und Stoffbertragung*, *8*, 229.
52. Pandey, V., Biswas, G., & Dalal, A. (2018). Numerical Heat Transfer. *Part A: Applications*, *1*. https://doi.org/10.1080/10407782.2018.1515332.
53. Son, G., & Dhir, V. K. (1998). *International Journal of Heat and Mass Transfer*, *120*, 183.
54. Panzarella, C. H., Davis, S. H., & Bankoff, S. G. (2000). *Journal of Fluid Mechanics*, *402*, 163.
55. Berenson, P. J. (1961). *ASME International Journal of Heat and Mass Transfer*, *83*, 351.
56. Tomar, G., Biswas, G., Sharma, A., & Welch, S. W. J. (2008). *Physics of Fluids*, *20*(9), 092101.
57. Johnson, R. L. (1968). *AIAA Journal*, *6*, 1456.
58. Tomar, G., Gerlach, D., Biswas, G., Alleborn, N., Sharma, A., Durst, F., et al. (2007). *Journal of Computational Physics*, *227*(2), 1267.
59. Tomar, G., Biswas, G., Sharma, A., & Welch, S. W. J. (2009). *Physics of Fluids*, *21*(3), 032107.

60. Pandey, V., Biswas, G., & Dalal, A. (2017). *Physics of Fluids*, *29*(3), 032104.
61. Verplaetsen, F. M., & Berghmans, J. A. (1997). *Rev. Gen. Therm.*, *37*, 83.
62. Verplaetsen, F. M., & Berghmans, J. A. (1999). *Heat and Mass Transfer*, *35*, 235.
63. Usiskin, C., & Siegel, R. (1961). *ASME International Journal of Heat and Mass Transfer*, *83*, 243.
64. Siegel, R., & Keshock, E. G. (1964). *AIChE J.*, *10*(4), 509. http://dx.doi.org/10.1002/aic. 690100419.
65. Oka, T., Abe, Y., Mori, Y. H., & Nagashima, A. (1995). *ASME International Journal of Heat and Mass Transfer*, *117*, 408.
66. Zell, M., Straub, J., & Vogel, B. (1989). PCH, PhysicoChem. *Hydrodyn*, *11*, 812.
67. Lienhard, J. H. (1985). *ASME International Journal of Heat and Mass Transfer*, *107*(1), 262.
68. Aktinol, E., Warrier, G. R., & Dhir, V. K. (2014). *International Journal of Heat and Mass Transfer*, *79*, 251.
69. Dhir, V. K., Warrier, G. R., Aktinol, E., Chao, D., Eggers, J., Sheredy, W., et al. (2012). *Microgravity Science and Technology*, *24*(5), 307.
70. Warrier, G. R., Dhir, V. K., & Chao, D. F. (2015). *International Journal of Heat and Mass Transfer*, *83*, 781.
71. Straub, J. (1994). *Experimental Thermal and Fluid Science*, *9*(3), 253.
72. Straub, J. (2000). *Experimental Thermal and Fluid Science*, *39*(4), 490.
73. Di Marco, P., & Grassi, W. (2002). *International Journal of Thermal Sciences*, *41*(7), 567.
74. Duchemin, L., Josserand, C., & Clavin, P. (2005). *Physical Review Letters*, *94*(22), 224501.
75. Hogrefe, J. E., Peffley, N. L., Goodridge, C. L., Shi, W. T., Hentschel, H. G. E., & Lathrop, D. P. (1998). *Physica D: Nonlinear Phenomena*, *123*(1), 183.
76. Westwater, J., & Santangelo, J. (1955). *Industrial & Engineering Chemistry Research*, *47*(8), 1605.
77. Jerome, B. P. (1960). Transition boiling heat transfer from a horizontal surface. Tech. rep., Cambridge, Mass.: Massachusetts Institute of Technology, Division of Industrial Cooperation.
78. Witte, L., & Lienhard, J. (1982). *International Journal of Heat and Mass Transfer*, *25*(6), 771.
79. Li, J. Q., Mou, L. W., Zhang, Y. H., Yang, Z. S., Hou, M. H., Fan, L. W., et al. (2018). *Experimental Thermal and Fluid Science*, *92*, 103.
80. Raj, R., Kim, J., & McQuillen, J. (2009). *ASME Journal of Heat Transfer*, *131*(9), 091502.
81. Fan, L. W., Li, J. Q., Zhang, L., Yu, Z. T., & Cen, K. F. (2016). *Applied Thermal Engineering*, *109*, 630.
82. Kang, J. Y., Kim, S. H., Jo, H., Park, G., Ahn, H. S., Moriyama, K., et al. (2016). *International Journal of Heat and Mass Transfer*, *93*, 67.
83. Freud, R., Harari, R., & Sher, E. (2009). *Nuclear Engineering and Design*, *239*(4), 722.

Applications and Validation

CFD Modeling of Data Centers

Kailash Karki, Suhas Patankar and Amir Radmehr

1 Introduction

A data center houses racks, or cabinets, containing computer equipment, such as servers, networking equipment, and storage devices. The heat dissipation of this equipment is continually increasing, and its cooling is becoming challenging. To address this cooling challenge, data center designers and owners are turning to the Computational Fluid Dynamics (CFD) modeling.

In this paper, we discuss the salient physical processes associated with airflow and temperature distributions in data centers and describe the contributions of CFD modeling in explaining these processes and facilitating the design of effective and efficient cooling systems.

We note that a large number of experimental studies are also available on data center cooling. These studies improve our understanding of various physical processes. But the quantitative aspects of the results are specific to the data centers studied and cannot be generalized.

1.1 Methods for Delivering the Cooling Air

Normally, cooling air enters a rack through the front face and hot air exits from the rear face. In a large room, with racks spread all over the room, it is not easy to supply cooling air to each rack. A variety of strategies are used to accomplish this task.

Raised-Floor Data Centers. The most common strategy for delivering the cooling air is called the raised-floor data center. The racks are installed on a tile floor that is raised 0.3–1.2 m (12–48 in.) above the real solid floor. Downflow Computer Room

K. Karki (✉) · S. Patankar · A. Radmehr
Innovative Research, Inc., 3030 Harbor Lane N., Suite 201, Plymouth, MN, USA
e-mail: karki@inres.com

© Springer Nature Singapore Pte Ltd. 2020
A. Runchal (ed.), *50 Years of CFD in Engineering Sciences*,
https://doi.org/10.1007/978-981-15-2670-1_18

647

Air-Conditioning (CRAC) units are used to pump cold air into the space below the raised floor. The floor tiles are removable and some of the solid tiles can be replaced by perforated tiles or grilles to permit the cold air to enter the above-floor space. By locating perforated tiles at the feet of the racks, cooling air is delivered to them. The hot air then finds its way back to the CRAC units.

Racks are commonly arranged in rows to form alternate hot and cold aisles. The cold aisles have perforated tiles, and the rack inlet faces are oriented toward these aisles. The hot exhaust emerges in the hot aisles.

If the downflow CRAC units cannot provide adequate cooling, they can be combined with other cooling solutions. These include in-row coolers, rear door heat exchangers, and overhead cooling units. An in-row cooler is usually interspersed in a row of racks. It draws in hot air from its backside (from the hot aisle), internally cools it, and exhausts cold air into the cold aisle. This cold air then enters the inlets of the surrounding racks. A rear door heat exchanger is mounted literally as a rear door for a rack. The normal hot exhaust from the rack is cooled in this heat exchanger so that the exhaust air is at an acceptably low temperature. This eliminates any hot-air streams in the data center and ensures proper cooling. There are also overhead cooling units that draw in hot air from the hot aisle, cool it, and blow cold air downwards into the cold aisle.

Non-raised-Floor Data Centers. In a non-raised-floor data center, the cooling air may come from the upflow CRAC units (as opposed to the downflow units used in a raised-floor environment), overhead ducts, and cooling devices mentioned in the previous paragraph.

1.2 The Cooling Challenge

Each rack in a data center consumes electrical energy and dissipates a large amount of heat in the range of 1–30 kW. For the electronics to function properly, it needs to be cooled and kept at an acceptable temperature level. Overheating of electronics may cause the equipment to malfunction, melt, or burn; but more commonly, safety devices on the servers will detect high temperatures and shut down the equipment. It is this interruption that presents a serious problem for a data center and needs to be prevented.

The objective of the cooling design is to ensure that the air temperatures at the inlet faces of the racks are below the maximum acceptable value. The ASHRAE guidelines [1] specify the recommended maximum temperature as 27 °C (80.6 °F), but a value around 24 °C (75 °F) is more common.

To meet this objective, the cooling design must address many questions. These include: How much cold air do we need? How does it distribute in the whole data center? Are we meeting the *individual demands* of all racks? Does the cold air wash the entire front face of a rack (height is usually close to 2 m)? Can the hot air coming out of one rack enter the inlet of another rack and compromise its cooling?

Another consideration is the temperature of the cooling air. As mentioned earlier, the desirable maximum inlet temperature is around 24 °C. The temperature of the cooling air from the air conditioners could be close to this value if this air could be made to enter all racks (and is sufficient to meet their airflow demands). However, the cooling air may not always enter at all the inlet locations on the racks. Often, the hot air exhausted by a rack finds its way to the inlet of the same rack or some other rack. To compensate for this intake of hot air, the temperature of the cooling air is usually kept lower, in the range 12–20 °C. But still there is no guarantee that the inlet temperatures for all racks will be satisfactory.

An additional complication is the dynamic nature of data centers. The equipment layout changes frequently: New servers and racks are installed, and the old ones are removed. Typically, ten percent of the equipment in a data center is replaced *each month* [2]. The cooling design has to keep pace with this frequent change.

1.3 Methods to Meet the Cooling Challenge

As in any complex engineering situation, the cooling in a data center has been conventionally handled by accumulated experience (including some rules of thumb), field measurements, and ad hoc design changes. These conventional methods are inadequate, time-consuming, and expensive, and they often lead to poor designs. The need for a more scientific approach is quite obvious.

1.4 Role of CFD Simulation

Cooling in a data center is an excellent application for Computational Fluid Dynamics (CFD). It offers a new paradigm for meeting the cooling challenge. One can create a computer model of the whole data center. The CFD simulation then provides a detailed distribution of air velocity, pressure, and temperature throughout the room. The simulation can be used to analyze an existing data center, but more importantly, any *proposed* layout for a new or reconfigured data center. One can detect unacceptably high rack inlet temperatures (hot spots) in a simulation (before they arise in reality) and explore ways of mitigating them. As already mentioned, data centers are dynamic environments; their equipment layout changes frequently. A CFD simulation provides invaluable help in planning the changes and ensuring proper cooling.

1.5 Brief Overview of the Available Literature

The CFD simulations for data centers have become popular only over the last 10–15 years. In the prior years, there is limited archival literature on this topic. The available literature has been reviewed in several publications [e.g., 2–4]. A brief overview is given below. Some of the recent papers will be referred to later when various aspects of the modeling are discussed.

Kang et al. [5], in one of the early efforts to predict the airflow distribution through the perforated tiles, proposed a simplified model that assumes the whole volume under the raised floor to be at a *uniform* pressure. The entire flow system is then represented as a network of flow resistances. The results from this simplified model did agree well with CFD results for a particular small data center. However, the assumption is not valid for most practical configurations.

In a subsequent study, Schmidt et al. [6] presented the results of a depth-averaged model (to convert the three-dimensional problem into a two-dimensional one) that allowed for the pressure variations in the under-floor space. For the cases they considered, a good agreement with measurements was demonstrated. However, later investigations have shown that the depth-averaged model is adequate when the height of the raised floor is small (normally less than 0.15 m, or 6 in.). Practical data centers usually have floor heights in the range of 0.3–1.2 m. Thus, this simplification is not acceptable, and one must perform the full three-dimensional computation.

Karki et al. [7] presented a three-dimensional model for the prediction of airflow rates through the perforated tiles. In this model, the pressure above the raised floor was assumed to be uniform. The calculated flow rates were shown to be in good agreement with measurements. This model has been used to study the effect of various factors that affect the distribution of flow rates and to propose techniques to modify or control the distribution [8, 9]. VanGilder and Schmidt [10] consider the factors that govern the uniformity of airflow through the perforated tiles. Patankar [2] and Bhopte et al. [11] have studied the effect of under-floor obstructions on the airflow distribution.

Abdelmaksoud et al. [12], Arghode et al. [13], and Arghode and Joshi [14] have proposed models to account for the air entrainment just above the perforated tile.

Patankar [2], Samadiani et al. [15], and Alkharabsheh et al. [16, 17] discuss the effect of external pressure losses on the flow rate of a CRAC unit.

Radmehr et al. [18] and Karki et al. [19] have addressed the issue of distributed leakage through the raised floor; they report airflow measurements and corresponding CFD analysis.

Schmidt [20] and Schmidt and Cruz [21–24] have studied the inlet temperatures to racks in a data center. Guggari et al. [25] describe ways of optimizing data center layout. Bhopte et al. [26] have used multivariable analysis to optimize the data center layout.

The interaction of the air stream emerging from a perforated tile with the internal fans in a front-to-rear rack is analyzed by Radmehr et al. [27]. They determine the conditions for which the inlet flow is essentially unaffected by the pressure variations in the inlet stream.

Patel et al. [28, 29] describe application of CFD modeling to data centers and evaluate some specific overhead cooling solutions. Sorell et al. [30] have compared the performance of a raised-floor system with that of a non-raised-floor system with overhead delivery. Shrivastava et al. [31] have compared various combinations for supplying the cooling air and extracting the hot air.

1.6 Scope of the Paper

The purpose of this paper is to describe important physical behavior in the airflow and cooling in a data center and summarize studies, mainly involving CFD simulations, that explain the behavior or enhance our understating of it. A large portion of the material is taken from Patankar [2]; this publication and the references cited there should be consulted for full details.

The primary focus is on raised-floor data centers. But a large portion of the discussion, especially that related to the above-floor space, or the computer room, is applicable to non-raised-floor data centers.

A complete consideration of the flow and heat transfer in a raised-floor data center would include both the under-floor and above-floor spaces. This paper initially focuses on the flow in the under-floor space and the resulting airflow rates through the perforated tiles. Although the under-floor space appears to be insignificant, it is here that the cooling battle is primarily won or lost. If we are able to deliver the required amount of airflow at the foot of each rack, proper cooling is possible. If we fail to satisfy this requirement, any attempted remedy in the above-floor space is usually ineffective.

The flow in the under-floor space is influenced by various factors such as the layout of perforated tiles, their open area, the height of the raised floor, and under-floor obstructions. A number of simple case studies are described to illustrate these effects.

The above-floor space brings its own unique behavior and surprises. These are explained through several examples. It is shown that the main challenge in the above-floor space is to ensure that the hot exhaust air from a rack does not enter the inlet of the same rack or some other rack. Various strategies used for ensuring this are described.

Much of the available work on the CFD simulations is limited to the steady-state behavior. The simulation of transient conditions requires input that is difficult to provide or not readily available, and it is very time consuming. However, a limited number of studies are available on this topic and will be included in this review.

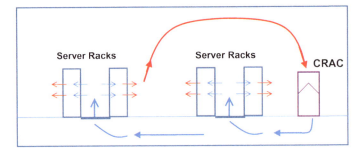

Fig. 1 A raised-floor data center

2 A Raised-Floor Data Center

Figure 1 shows an outline of a raised-floor data center. On the right, a downflow CRAC unit is placed on the raised floor. It draws the hot air in the room into its top (inlet) face and supplies cold air from the openings on its bottom face into the under-floor space. The cold air enters the above-floor space through perforated tiles that are placed at the feet of the racks. The server fans, in turn, draw in this air through the front face of the rack and exhaust hot air from the rear face. The hot air finally returns to the top of the CRAC unit.

To minimize the possibility that the hot air exhausted by one rack would enter the inlets of servers in another rack, data centers are often laid out in the "hot aisle/cold aisle" arrangement, which is shown in Fig. 2. This arrangement was suggested by Sullivan [32] and has now become a standard practice. The so-called cold aisle has the perforated tiles. The racks are placed on both sides of the cold aisle such that their inlet faces are toward the cold aisle. As a result, the exhausts from two neighboring rows of racks emerge into the hot aisle. The hot air collected there returns to the CRAC units. Figure 3 shows a photograph of a cold aisle in a data center. (The object at the back is a CRAC unit.)

The raised-floor arrangement gives unlimited flexibility. If the layout of the racks is changed, all that is needed is to rearrange the perforated tiles so that cooling air is delivered at the new locations of the racks. Since there is no permanent ducting, no elaborate dismantling or construction is necessary.

3 Representation of Basic Components

A raised-floor data center has three essential components: perforated tiles, CRAC units, and racks. In this section, the representation of these components is discussed.

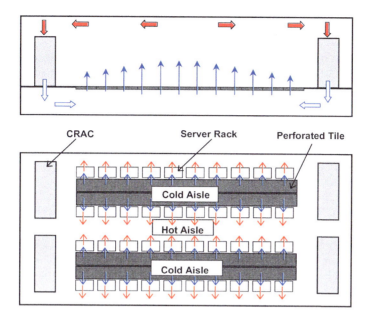

Fig. 2 The hot aisle/cold aisle arrangement

Fig. 3 The cold aisle in a data center

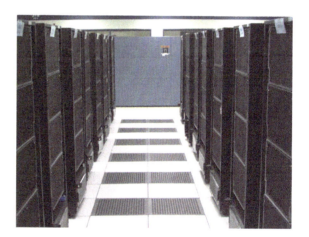

3.1 Perforated Tiles

The perforated tiles used in a data center have a large number of small holes. Such a tile is usually characterized by the percentage open area; the most common perforated tiles have 25% open area. The flow resistance of the perforated tiles can be obtained from well-known pressure-drop correlations for such plates. The pressure drop Δp across a perforated tile is expressed as

$$\Delta p = K \left(0.5 \, \rho V^2 \right)$$

where V is the velocity approaching the perforated tile, ρ is the density of air, and K is the flow resistance factor (the "K factor"). An empirical formula for K, based on a large number of measurements, is given by Idelchik [33]:

$$K = \frac{1}{F^2} \left(1 + 0.5(1 - F)^{0.75} + 1.414(1 - F)^{0.375} \right)$$

where F is the fractional open area of the perforated tile. For a 25% open tile, this formula gives $K = 42.8$.

Figure 4 [2] shows the variation of the pressure drop Δp, given by the above formula for the K factor, with the volumetric airflow rate through the perforated tile for tiles of 6, 11, and 25% open area. For the case of 25% open tile, the figure also shows the experimental data. The pressure drop values given by the formula are in good agreement with the data. (This figure assumes that the tile size is 2 ft × 2 ft, which is the common size used in the USA. The size in other countries is 0.6 m × 0.6 m, which is not very different.)

Table 1 gives, for the 25% open tile, the pressure drop values for several airflow rates through the tile. Practical flow rates through perforated tiles are of the order of 0.25 m³/s, for which the pressure drop is about 12 Pa. That this value of the pressure drop is rather small is relevant for the discussion on the behavior of the CRAC units (see the next section).

In the above discussion, a perforated tile is treated as a porous plate with a given flow resistance. Some recent studies have focused on the details of the flow exiting a perforated tile. Abdelmaksoud et al. [12] have proposed a body force model to

Fig. 4 Pressure drop as a function of airflow rate for perforated tiles [2]

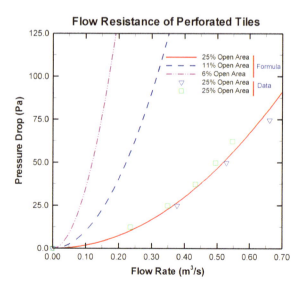

Table 1 Pressure drop for a 25% open tile at different airflow rates	Airflow rate (m³/s)	Pressure drop (Pa)
	0.05	0.47
	0.10	1.87
	0.15	4.21
	0.20	7.48
	0.25	11.7
	0.30	16.8

account for the increase in the momentum of the air as it accelerates through the small holes. It adds a momentum source in a volumetric region of a preselected height just above the perforated tile. The source term is calculated using the average pore velocity. Arghode et al. [13] have suggested a modified model in which the average pore velocity is replaced by the velocity at the jet vena contracta. Recently, Arghode and Joshi [14] have presented an analysis for obtaining the dimensions of the momentum source region and related input needed for this model.

The treatment so far is concerned with standard perforated tiles, which are essentially plates with a large number of holes. To manage the airflow rates through the perforated tiles, the manufacturers have introduced complex tiles, which could have dampers, directional vanes, or even fans. These tiles usually have a complex understructure. They cannot be characterized simply by their open area, and the flow rate versus pressure drop data must be used. This data is usually available from the tile manufacturer. See Fig. 5 for sample data for 25% open tiles with dampers [34].

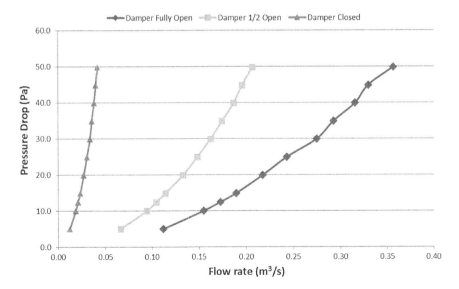

Fig. 5 Pressure drop as a function of flow rate for 25% open perforated tiles with dampers [34]

3.2   CRAC Units

The flow emerging from the perforated tiles originates at the CRAC units. Usually, the flow from a CRAC unit is directed vertically downwards from one or more blowers. It impinges on the solid floor (usually called the subfloor), turns 90°, moves horizontally in the under-floor space, and emerges from the perforated tiles wherever they are placed.

The amount of airflow from a CRAC unit depends on the characteristics of its blowers and the pressure drop within and outside the unit. However, for all practical purposes, it can be regarded as a constant-flow device that delivers that same amount of airflow rate for different numbers, layouts, and open areas of the perforated tiles [2].

To understand this conclusion, we need to examine the various pressure drops that the CRAC unit blower is required to overcome. These comprise the pressure drop in

- The filter and cooling coils in the CRAC unit. (This is known as the internal static and is usually in the range 250–500 Pa.)
- The impingement and turning on the subfloor. (This is known as the external static and is usually in the range 100–200 Pa.)
- The perforated tiles (about 12 Pa as seen above).

It is now obvious that the pressure drop across the perforated tiles represents only a small fraction of the flow resistance experienced by the CRAC blower. Therefore, for all practical purposes, the CRAC unit gives *nearly the same flow rate* for different numbers, layouts, and open areas of the perforated tiles.

The flow rate of a CRAC unit may deviate from the rated value if the pressure loss outside the unit is significantly different than the rated external static. The flow rate may reduce, for example, if there are large under-floor obstructions under the blowers. However, the sensitivity of the flow rate to the external pressure drop is rather small. Reference [35] presents an analysis of this sensitivity for a typical CRAC unit and concludes that, even when the external pressure loss is twice the rated external static, the flow rate reduces by less than 10% of the rated value. This behavior is to be expected because the internal static makes up a large fraction of the total pressure drop the CRAC unit blower is required to overcome. The effect of external pressure drop on the CRAC unit flow rate has also been addressed by Samadiani et al. [15], and Alkharabsheh et al. [16, 17]. The findings are specific to the data center considered, and the deviation of the flow rate from the rated value depends on the open area of the tiles, the details of under-floor obstructions, and presence of containments.

The above discussion focuses on the flow behavior of the CRAC unit. The calculation of temperature distribution requires the temperature of air exiting the CRAC unit. If the exit temperature is known (either maintained by the control mechanism or measured), it is used as the boundary condition; no other information about the thermal characteristics of the CRAC unit is needed.

However, often, the control mechanism is set to maintain a certain return (inlet) temperature. Under this condition, the exit temperature of the CRAC unit is not known a priori; it depends on the cooling load of the unit, which, in turn, depends on the airflow pattern and temperature distribution in the computer room. The calculation of the exit temperature now requires the relationship between the inlet and exit temperatures.

The CRAC unit manufacturer provides exit temperatures corresponding to various inlet temperatures at different operating conditions (e.g., water flow rates and inlet temperatures). Since the inlet temperatures and operating parameters in the data center could be different than those included in the manufacturer data, interpolation or extrapolation is required. For this, the common approach is to treat the CRAC unit as a heat exchanger and use the manufacturer data to deduce the heat exchanger effectiveness as a function of the operating parameters, such as the air and water flow rates [e.g., 36–38].

The treatment described here is also applicable to other cooling devices, such as in-row coolers and rear-door heat exchangers.

3.3 Racks

A rack, or cabinet, houses servers and other computer equipment. The electronic components within a server dissipate a certain amount of heat. The cooling air enters a server from the front face, picks up the heat, and is expelled from the rear face. A rack usually has screen doors at the front and rear faces. Some specialized racks have plenums behind their rear faces to collect the hot air, which exits through an opening at the top of the plenum. A rack may not be completely filled with servers; the empty spaces could be closed using blanking panels or left as open spaces. Thus, it can be considered a collection of servers, blanking panels, and open spaces. In addition, racks could have cable management systems.

Even when the entire rack is occupied by servers and blanking panels, small gaps remain above and below a server and a blanking panel. Gaps are also present between the sides of a server and the rack sidewalls. Additional gaps could be associated with cable management systems. These gaps present air leakage paths; the amount and direction of air leakage depends on the pressure difference across the rack.

The representation of a rack involves two components: (a) representation of the servers, and (b) representation of the rack structure (frames, doors, and mounting rails, etc.). For servers, a variety of options are available. The simplest is to treat a server as a black box with one heat load and airflow demand. The mass and energy sink and source terms are introduced at the inlet and outlet faces, and no calculations are performed inside the server. This representation has been used extensively [e.g., 21–24, 39–42].

An improvement of the black box approach is to treat a server as a porous object and specify its flow resistance and the characteristics of the server fans [e.g., 16, 17, 27]. This representation can distinguish between different types of devices, such as, servers, storage devices, and networking devices. Now the server flow rate depends on the external pressure difference across the rack. This dependence, of flow rate on the pressure difference, is important in data centers with containments. This porous object approach allows the possibility for directional flow resistance or permeabilities, which can be used to make the flow inside a server three dimensional [43].

Another approach is to fully describe the internal details of the servers. This approach is, however, impractical, especially for data centers with large number of racks.

For the rack structure, the options are to ignore the details or include them. The latter approach requires a fine computational mesh for the rack.

Zhang et al. [44] have studied three approaches for representing a rack: (a) a black box, (b) detailed representation of the rack (frames, doors, and mounting rails), and the black box approach for the servers, and (c) detailed representation of both the rack and the servers. They found that all three approaches gave similar results for the temperatures in the computer room. Zhai et al. [45] also have compared the results of the black box approach with one that allows flow through the rack and involves detailed representation of servers. They also found that the two approaches produce similar results, confirming the conclusions of Zhang et al. [44].

A number of researchers have studied other aspects of racks, such as the effect of screen doors, cable management arm, and air leakage. Coxe [46] has estimated that the combination of screen doors and cable management arm reduces the flow rate of a 1U server by 6%. North [47] has concluded that doors with open area ranging from 64 to 80% cause a small reduction in the server flow rate. Rubenstein [48] has performed tests to study the impedance of the cable management arm and reported it to be about 2% of the server impedance.

The air leakage through the racks is usually not of concern if an open arrangement is used for the racks, that is, both their inlet and outlet faces are exposed to the room conditions. In this arrangement, the pressures on the two faces are nearly the same (negligible external pressure difference). The external pressure difference, and hence the air leakage, can become important when containments are used. Tatchell-Evans et al. [36] have presented laboratory measurements and predictions using a system model for air leakage in the presence of cold-aisle containment. They report that, under typical conditions, up to 20% of the airflow entering the cold aisle may bypass the servers, leaking through the gaps in the racks. Kennedy [49] has reported tests, which indicate similarly large levels of air leakage when containments are used.

Here it is important to note that these conclusions are specific to the operating conditions, layout, servers, and racks considered and cannot be generalized.

4 Airflow in the Under-Floor Space

As discussed below, the airflow rate emerging from each perforated tile holds the key to successful cooling. We now turn our attention to the factors that control the distribution of airflow through the perforated tiles. Interestingly, it is not what happens *above* the raised floor but what happens *below* the raised floor that determines the flow through the perforated tiles. Thus, the fluid mechanics of the tiny (and usually invisible) space below the raised floor controls the success or failure of cooling in a raised-floor data center. From a computational point of view, this is good news. If the CFD simulation is limited to the under-floor space, the calculation domain is small and a fast solution is possible. Yet, this small computational effort leads to the most valuable information needed for the cooling of the data center.

4.1 A Necessary Condition for Proper Cooling of Racks

A rack has internal fans that draw a known amount of airflow rate. A necessary condition for proper cooling is to supply the *required amount of cold airflow* at the foot of each rack. If this is done, satisfactory cooling is possible. If this cannot be done, cooling difficulties arise and they are usually very difficult to overcome.

4.2 Surprising Airflow Distribution Through the Perforated Tiles

As seen in Fig. 3, different perforated tiles are at different distances from the CRAC unit. How does the airflow distribute through the perforated tiles? At first sight, we may conclude that, as we go away from the CRAC unit (the source of air), the airflow through the perforated tiles diminishes. However, the flow distributes in a surprising and counter-intuitive manner. The perforated tiles that are farthest way from the CRAC unit get the largest flow. For the tiles close to the CRAC unit, the airflow is quite small.

Figure 6 shows the so-called maldistribution of airflow in a schematic manner. The reason for this maldistribution is also explained in the figure. The mechanism is similar to the maldistribution that occurs in manifolds [50, 51]. In Fig. 6, if we consider the velocity of the under-floor air in the horizontal direction, this velocity must decrease as air exits through the perforated tiles. The Bernoulli equation would then imply that the pressure increases as we go away from the CRAC unit. The airflow rate through a perforated tile depends on the pressure drop across it. Since the tiles on the right experience a greater pressure drop, they deliver more airflow.

Fig. 6 Maldistribution of airflow and its cause

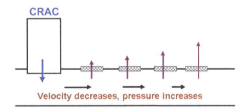

(This conclusion assumes that the pressure above the raised floor is nearly uniform. This is a valid assumption and has been verified by measurements and computations.)

4.3   A Sample Validation

The above discussion explains the maldistribution in a qualitative sense. To verify its quantitative accuracy, a simple experiment [6] was conducted at IBM Corporation in Poughkeepsie, NY. The experimental setup is shown in Fig. 7. It is a small data center with two CRAC units and four rows of perforated tiles. For the initial tests, the CRAC unit on the left was turned off. Then, only the unit on the right supplies cold air to all the perforated tiles. The airflow from each perforated tile was measured and compared with the airflow rate predicted by a CFD simulation of the under-floor space. The comparison is shown in Fig. 8. The maldistribution is clearly visible.

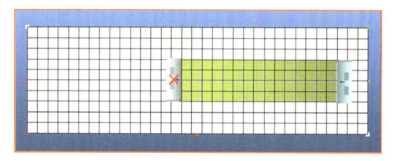

Fig. 7 Layout for a small test data center

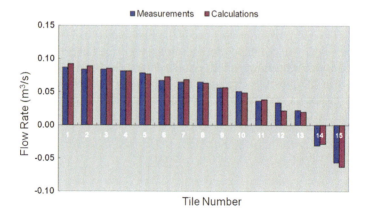

Fig. 8 Comparison of measured and calculated airflow rates (The tiles are numbered from left to right.)

For tiles 14 and 15, which are closest to the working CRAC unit, the flow rate is actually negative. The flow rate can be seen to increase rapidly for the tiles that are further away from the CRAC unit. The agreement between the measurements and calculation is very satisfactory.

4.4   Additional Validations

The prediction of airflow rates through the perforated tiles by CFD simulation has been validated by comparison with measurements in a number of data centers. Karki et al. [7] have presented the comparison for a real-life data center (floor area close to 900 sq m) and shown good agreement. Bhopte et al. [11] have presented this comparison for a data center with significant under-floor obstructions. Other studies include Radmehr et al. [18, 52, 53], Samadiani et al. [15], and Athavale et al. [54]; in the last publication, the focus is on active tiles (i.e., tiles with fans).

4.5   Factors Affecting the Airflow Distribution

As we have seen before, the flow rate through a perforated tile depends on the pressure drop across the tile, that is, the difference between the plenum pressure just below the tile and the ambient pressure above the raised floor. Pressure variations above the raised floor are generally small compared to the pressure drop across the perforated tiles. Thus, relative to the plenum, the pressure just above the perforated tiles can be assumed to be uniform. The flow rates, therefore, depend primarily on the pressure levels in the plenum, and the nonuniformity in the airflow distribution is caused by the pressure variations in the horizontal plane under the raised floor.

For the nonuniformity in the airflow distribution to be significant, the horizontal pressure variations (or change in velocity heads) must be comparable to the pressure drop across the perforated tiles. This condition is satisfied if the area available for horizontal flow in the plenum is comparable to or less than the total open area of the perforated tiles.

Parameters Considered. The key to controlling the airflow distribution is the ability to influence the pressure variation in the plenum. For specified (horizontal) floor dimensions and total flow rate, the effect of the following parameters is significant:

- Plenum height
- Open area of perforated tiles

Plenum height. The plenum height has a major influence on the horizontal velocity and pressure distributions in the plenum. As the plenum height increases, the velocities decrease and the pressure variations diminish, leading to a more uniform airflow distribution.

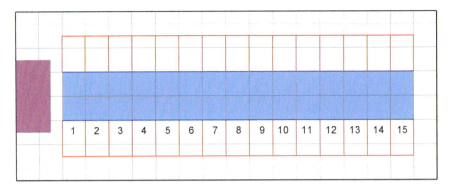

Fig. 9 The base case

Open area of perforated tiles. As the open area of perforated tiles is reduced, the pressure drop across the tiles increases and, at some point, becomes much larger compared to the horizontal pressure differences under the raised floor. Under these conditions, all perforated tiles experience essentially the same pressure drop and the airflow distribution becomes nearly uniform.

The Base Case. The effect of these two parameters on the airflow distribution will be illustrated with reference to the simple configuration shown in Fig. 9. This layout uses the conventional hot aisle/cold aisle arrangement. The CRAC units are located in the cold aisles. The racks are arranged on both sides of the cold aisles, with their intake sides facing the cold aisles. The hot aisles are formed between the back ends of two rows of racks. The configuration is symmetric; thus, only a portion of the data center around one CRAC unit needs to be considered; this portion is shown in Fig. 9.

Thus, the base configuration consists of a CRAC unit and two rows of perforated tiles, each containing 15 tiles with 25% open area. The flow rate of the CRAC unit is 10,000 cfm (4.72 m³/s). The under-floor plenum height is 12 in. (0.3048 m). The tile size is 2 ft × 2 ft. The overall dimensions of this part of the data center are 36 ft × 14 ft (10.97 m × 4.27 m).

The distribution of airflow rates for this configuration is shown in Fig. 10. The flow rates are smaller near the CRAC unit and increase toward the opposite wall. There is reverse flow through the perforated tiles next to the CRAC unit.

Figure 11 shows the velocity vectors and the pressure distribution on the horizontal plane just under the raised floor. The cold air exiting the CRAC unit impinges on the subfloor and expands horizontally. In the impingement region, the pressure levels are high, and they decrease rapidly as the air rushes out of this region. As we move away from the CRAC unit, since the cold air is exiting the plenum, the horizontal velocity diminishes and the pressure rises. Note that the pressure under the perforated tiles next to the CRAC unit is negative (that is, *below* the pressure in the above-floor space) and produces a reverse flow through these tiles.

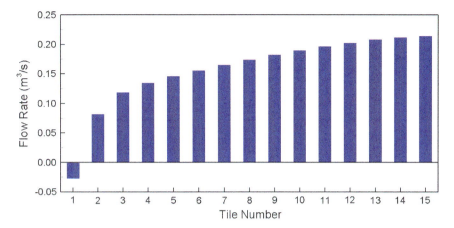

Fig. 10 Flow rates through perforated tiles for the base case

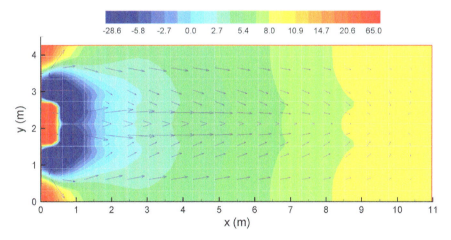

Fig. 11 Pressure distribution and velocity vectors under the raised floor for the base case (Plenum height = 12 in.; the pressure values are in Pa)

Effect of the Plenum Height. To illustrate the effect of plenum height on the airflow rates, the height for the base configuration is varied from 6 in. (0.1524 m) to 24 in. (0.6096 m).

The flow rates for different plenum heights are shown in Fig. 12. The nonuniformity in flow rates is most pronounced for plenum height of 6 in. and diminishes as the height is increased. As the plenum height is increased, the intensity of reverse flow through the perforated tiles next to the CRAC unit also weakens. Note that the curves for the plenum heights of 18 and 24 in. are almost coincident. This implies that, once the plenum height is large enough, any further increase does not affect the flow distribution. In Fig. 12, even for the 24-in. height, the flow distribution is not

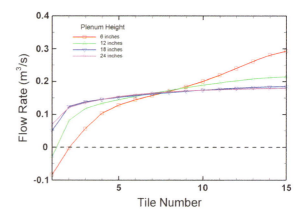

Fig. 12 Effect of plenum height on the airflow distribution

quite uniform. This is due to the complex flow and lateral spreading in the vicinity of the CRAC unit.

Figure 13 shows the velocity vectors and pressure distribution on a horizontal plane just under the raised floor for plenum height of 6 in. For this smaller plenum height, the pressure variations are more significant (compared with Fig. 11), with an extensive region of negative pressure near the CRAC unit. Figure 14 shows the same plot for plenum height of 24 in. Here, the pressure distribution is much more uniform; the variations are limited to a small region near the CRAC unit. These pressure plots explain the airflow distribution shown in Fig. 12.

Effect of the Open Area of Perforated Tiles. To illustrate the effect of the open area of perforated tiles, the open area in the base configuration is varied from 10% to 60%.

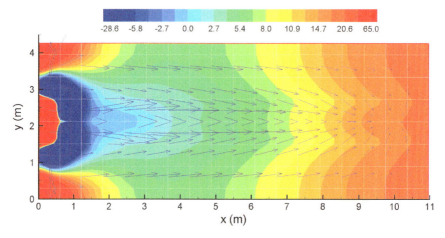

Fig. 13 Pressure distribution and velocity vectors under the raised floor for plenum height = 6 in. (The pressure values are in Pa)

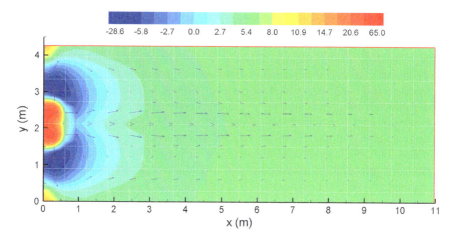

Fig. 14 Pressure distribution and velocity vectors under the raised floor for plenum height = 24 in. (The pressure values are in Pa)

The airflow distributions for different open areas are shown in Fig. 15. For a fixed layout and plenum height, the nonuniformity in flow rates diminishes as the open area is reduced. A reduction in the open area also reduces the likelihood of reverse flow near the CRAC units. Note that there is no reverse flow for open area of 15% and 10%.

It may appear that using highly restrictive tiles (such as 10% open) is a good way of making the airflow distribution uniform. However, there is an undesirable side effect. At smaller open areas, the pressure levels in the plenum increase and a large proportion of cold air escapes through extraneous openings on the floor, e.g., openings around cables and pipes and other leakage paths. (The flow resistance of these openings now becomes comparable to the flow resistance of the perforated tiles.) This wasted air will not be available for cooling of equipment.

Fig. 15 Effect of open area of perforated tiles on the airflow distribution

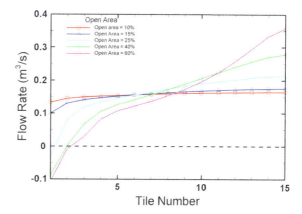

4.6   Leakage Through the Raised Floor

In practical data centers, the raised floor is not an impermeable surface; there are always small gaps between individual floor tiles. The amount of the leakage area depends on how well the raised floor has been installed. Radmehr et al. [18] have used airflow measurements in a prototype data center to estimate the leakage area. Their measurements indicate that it is around 0.2% of the floor area and that the air leakage can easily be 10–15% of the total airflow entering the under-floor plenum.

If we ignore the leakage area, then we would expect that, as the number of perforated tiles decreases, the plenum pressure would increase and the average airflow rate through each perforated tile would be larger. These expectations are qualitatively in agreement with actual measurements, but quantitatively the pressure rise and the increase in airflow rate are not found to be as high. The reason for the discrepancy is the leakage area. As the number of perforated tiles decreases, the leakage area becomes comparable to the open area of the perforated tiles. As a result, the leakage airflow must be considered while estimating the plenum pressure and the average airflow rate through each perforated tile.

The gaps that constitute the leakage area are very narrow (of the order of 1 mm), and it is not realistic to resolve them in the computational mesh. The preferred—and practical—approach is to assume that the leakage area is dispersed throughout the floor except for the area covered by solid objects, such as the CRAC units and support columns. This approach is described by Karki et al. [19].

A detailed study of the effect of leakage is presented by Radmehr et al. [18] and Karki et al. [19]. The study includes careful measurements performed in a data center with and without distributed leakage from the floor. The measurements are also compared with the results of CFD simulations. Athavale et al. [54] have studied the distributed leakage in the presence of active tiles.

4.7   Effect of the Under-Floor Obstructions

In the under-floor space, usually, there are pipes, cable trays, cables, structural beams, and other objects. Their presence reduces the area available for the airflow and creates non-uniformities in the pressure distribution. These are substantial effects, which are illustrated in this section (taken from Patankar [2]).

As a demonstration of how an under-floor obstruction influences the airflow distribution through the perforated tiles, a modification to the base case is considered. As shown in Fig. 16, a circular pipe with diameter of 6 in. (0.15 m) is placed in the under-floor space as an obstruction. (Only the centerline of the pipe is shown in the figure. In the 12-in. height of the raised floor, the pipe is placed on the solid subfloor leaving a 6-in. clearance above it.) Figure 17 shows the resulting airflow distribution through the perforated tiles. The corresponding velocity vectors and pressure distribution just under the raised floor are displayed in Fig. 18. In general, the pipe

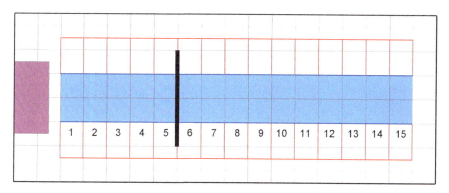

Fig. 16 A circular pipe as an under-floor obstruction (Only the centerline of the pipe is shown.)

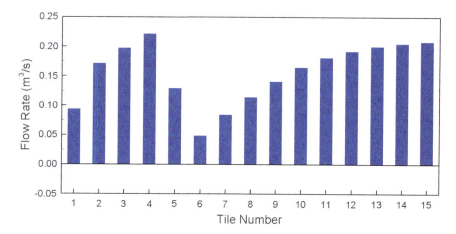

Fig. 17 Flow rates through perforated tiles as affected by the circular-pipe obstruction

allows the pressure to build up in the upstream region. So, the flow rates close to the CRAC unit are not as small as in the base case. In the immediate upstream region of the pipe, air velocity increases (due to the blockage caused by the pipe) and pressure falls. Then, there is a further lowering of the pressure on the downstream side of the pipe. Thereafter, the pressure gradually builds up as the horizontal flow velocity diminishes. These pressure changes are reflected in the airflow distribution through the perforated tiles.

The above example illustrates the substantial effect of large under-floor obstructions. This effect of obstructions has been corroborated by Bhopte et al. [11] and Samadiani et al. [15].

In an existing data center, there is limited flexibility in the placement of the obstructions and one has to accept the consequences. In a new design, however, they can be judiciously placed to minimize the detrimental effect on the airflow distribution. Bhopte et al. [11] have combined measurements and simulations to

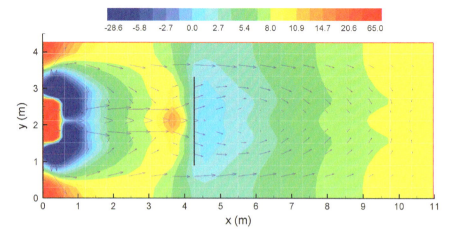

Fig. 18 Pressure distribution and velocity vectors under the raised floor for the case of the circular-pipe obstruction (The pressure values are in Pa)

propose the concept of critical and safe flow paths and used this concept to give guidelines for placing the obstructions.

4.8   Using Under-Floor Partitions to Modify the Airflow Distribution

In an ideal design, the airflow rates through the perforated tiles must vary in accordance with the rack heat loads (or airflow demands). To obtain the desired distribution of airflow rates, one could change the plenum height or open areas of the perforated tiles. However, in certain situations, these changes may not be enough to yield the desired distribution or may not even be feasible. This is especially true in an existing data center, where it is not practical to change the plenum height and there may be restrictions on the open areas of the perforated tiles. Karki et al. [9] and Patankar [2] have proposed the use of under-floor partitions, solid and perforated, for modifying the airflow distribution. This section discusses their work.

Use of Inclined Partitions. The flow maldistribution in the base case occurs because the horizontal flow velocity decreases as the air emerges from the perforated tiles. If this velocity can be kept uniform (by decreasing the area available for the under-floor flow), then the pressure would remain uniform, making it possible to achieve a uniform airflow distribution through the perforated tiles. One method to reduce the flow area is to use vertical partitions.

Figure 19 shows the base case modified by the placement of two inclined partitions in the under-floor space. The partitions attempt to reduce the available area for the horizontal flow in a linear fashion. However, in order to avoid intersecting the final

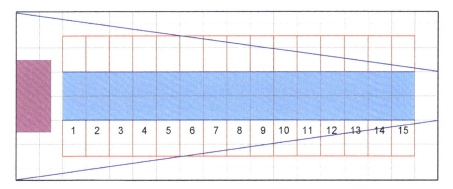

Fig. 19 Use of inclined partitions in the under-floor space

perforated tiles, the final area at the right end is kept finite and not made zero. The resulting airflow distribution through the perforated tiles is presented in Fig. 20, along with the distribution for the base case (no partitions). The use of the inclined partitions leads to a much more uniform airflow distribution compared to the base case. The only non-uniformities are in the region close to the CRAC unit (because of rather complex flow there) and at the right end (because of our inability to make the area reduce to zero).

Although the solid impermeable partitions can improve the airflow distribution, they have a serious limitation. In a data center with many CRAC units, if a CRAC unit fails, the other CRAC units would supply some of the air to the perforated tiles that are normally supplied by the failed CRAC unit. The presence of the solid partitions prevents this cross transfer and may turn a normal failure into a catastrophic one.

From this point of view, a perforated plate can be a more suitable partition. Such a plate does not completely stop the flow through it; however, it offers extra flow

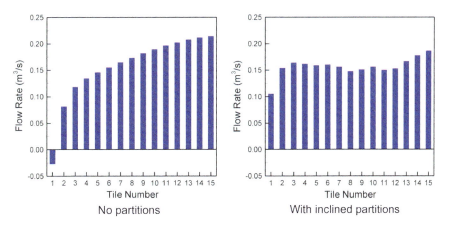

Fig. 20 Airflow distribution with and without inclined partitions

resistance, which can be used to discourage the flow in one place and consequently to encourage the flow in other places. The use of perforated partitions is discussed next.

Use of Perforated Partitions. The flow maldistribution in the base case delivers large airflow rates through the perforated tiles located far away from the CRAC unit. The flow distribution can be made more uniform by placing perforated partitions normal to the horizontal flow so that the flow will be somewhat discouraged from moving fast to the downstream region. This concept is shown in Fig. 21 with two proposed locations for the partitions. The required percent open area of the partitions will be determined by computational experiments.

In Fig. 22, the results of airflow distribution are shown for the case of the two perforated partitions set at 70% and 30% open, respectively. Also shown for reference are the results for the base case (no partitions). The use of the partitions has increased the airflow in the tiles near the CRAC unit. At each partition location, there is a drop in the airflow rate due to the pressure drop across the partition. In fact, the "discouragement" of the flow is so strong that the six tiles furthest away from the CRAC unit have a rather small flow through them. To get a uniform flow, we need to adjust the percent open area of the partitions. However, if we do desire a large airflow for tiles 1–9 and a small airflow for tiles 10–15, then the current arrangement is quite satisfactory. In general, the perforated partitions can be used to get any *desired* distribution of airflow rates, not just a uniform distribution.

Figure 23 shows the results with partitions at 80 and 65% open area. The resulting airflow distribution is quite uniform. In fact, except for the tile closest to the CRAC unit (which is affected by the very complex flow in the vicinity of the CRAC unit), all other tiles give nearly the same airflow.

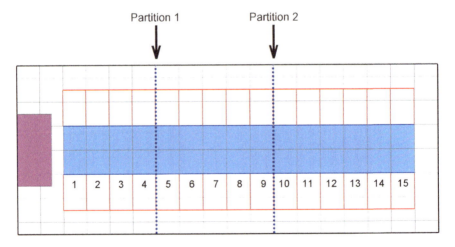

Fig. 21 Use of perforated partitions in the under-floor space

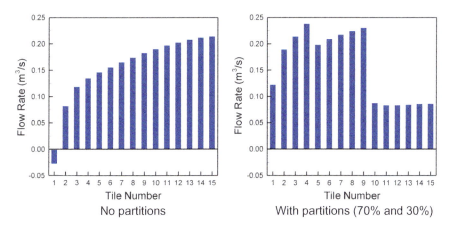

Fig. 22 Airflow distribution without and with perforated partitions (70 and 30%)

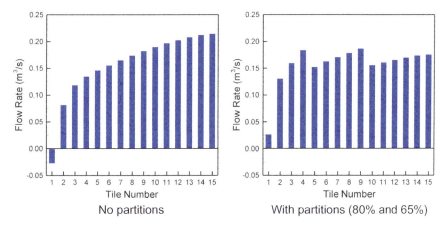

Fig. 23 Airflow distribution without and with perforated partitions (80 and 65%)

5 Airflow and Temperatures in the Above-Floor Space

As already mentioned above, in a data center, the goal of cooling is to ensure that the maximum inlet temperature to any rack does not exceed the maximum allowable temperature. The primary purpose of a simulation in the above-floor space (computer room) is the calculation of the temperature field, which, in turn, requires the calculation of velocity and pressure fields.

5.1 Some Common Reasons for Unsatisfactory Cooling

The primary cause of high rack inlet temperatures is insufficient airflow supplied through the perforated tile at the foot of the rack. If the airflow demand of the rack cannot be met by the perforated tile, the upper part of the rack draws in hot air. To illustrate this, a small section of a data center is shown in Fig. 24, in which insufficient cooling airflow is supplied. There are 22 racks. The heat load and airflow demand per rack are 4 kW and 0.223 m³/s. (These parameters imply a temperature rise of 11.1 °C, or 20 °F, across the rack.) The total airflow demand is 4.91 m³/s and the airflow rate of the CRAC unit is 3.54 m³/s. For this situation, the calculated inlet temperatures for the racks are displayed in Fig. 25. (The color scale used for plotting the temperature contours is included in Fig. 25. The same color scale is used for all the remaining figures that show a temperature distribution and is not repeated in those figures.) Since the cooling airflow is insufficient, all racks have hot air entering the tops of their inlets. Figure 26 shows where this hot air originates. It is the hot exhaust of the rack that gets recirculated into the inlet when the cooling air is insufficient.

If sufficient cooling airflow is supplied at the foot of the rack, usually there is no reason for unsatisfactory cooling. However, there are special locations and circumstances in which hot air can enter the inlets even when sufficient cold airflow is supplied at the perforated tiles. Some of these are described below.

End Effects. Suppose that in Fig. 24, the airflow was increased to meet the demands of the racks. The corresponding calculated inlet temperatures are displayed in Fig. 27. Now, most of the racks in the middle have acceptable low inlet temperatures over the whole inlet face. However, inlet temperatures are higher for the racks near the

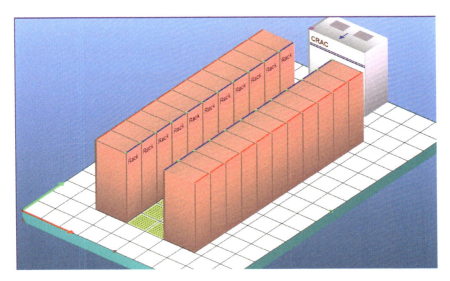

Fig. 24 A simple data center model with one CRAC, several racks, and perforated tiles (insufficient cooling airflow)

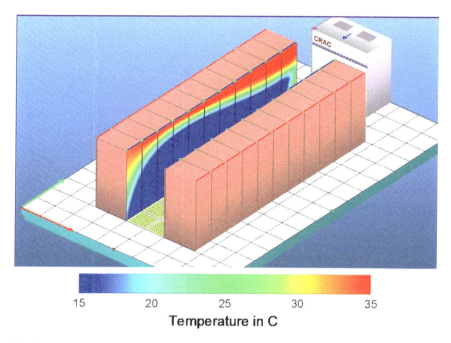

Fig. 25 Temperature distribution on the inlet faces of the racks (insufficient cooling airflow)

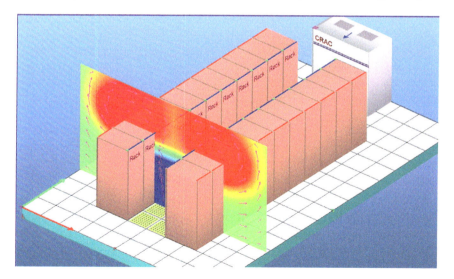

Fig. 26 Temperature distribution and velocity vectors on a plane (insufficient cooling airflow)

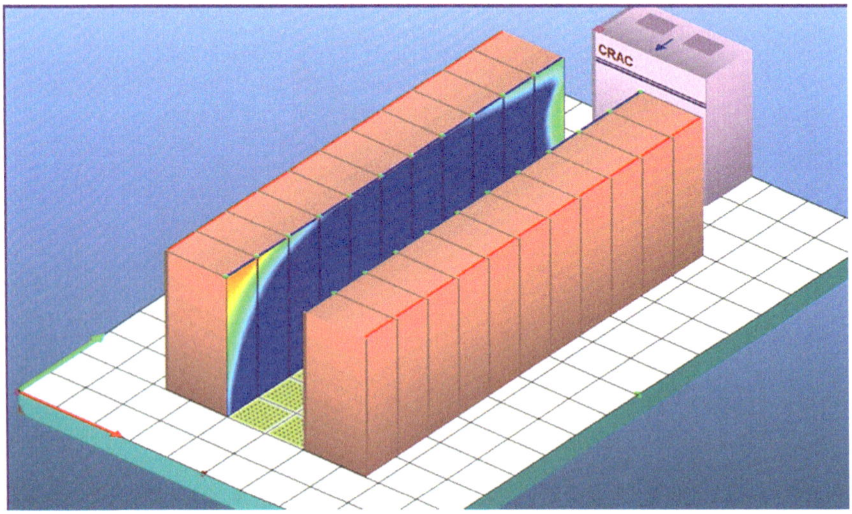

Fig. 27 Temperature distribution on the inlet faces of the racks with increased cooling airflow

CRAC unit and for the racks furthest away from the CRAC unit. The behavior near the CRAC unit is easy to understand. As we have seen before, the usual maldistribution of airflow leads to small airflow rates at the perforated tiles near the CRAC unit. This insufficient cooling flow leads to the higher inlet temperatures for the racks in that region.

Far away from the CRAC unit, the perforated tiles deliver the highest airflow rates. In this end region, the perforated tiles deliver sufficient flow, but the hot exhaust air finds its way around the last rack and enters the rack inlets. This penetration of hot air is evident from the temperature distribution on a horizontal plane near the top of the racks, shown in Fig. 28.

One quick remedy for this effect is shown in Fig. 29. Here, additional perforated tiles are placed in the end region to create an air curtain of cold air. The corresponding inlet temperatures, shown in Fig. 29, are an improvement over those in Fig. 27. Figure 30 shows how the air curtain is effective in preventing the recirculation of hot air into the rack inlets.

Gaps Between the Racks. Normally, the racks are placed in a row in a contiguous manner. However, occasionally, there may be gaps between them. For example, in practice, gaps are created by *removing* a rack from a row. It is easy to see that the gaps provide additional places where the "end effects" can be observed. Hot air from the back of the racks can enter the cold aisle through the gaps and influence the inlet temperatures of the racks. An obvious remedy is to close the gaps by using impermeable plates or partitions.

High-Velocity Flow Through the Perforated Tiles. The heat loads of modern racks can be very high (10–30 kW) and the corresponding airflow demand may be of the order of $1.0\,\mathrm{m^3/s}$. At these flow rates, air emerges from the perforated tile at a velocity

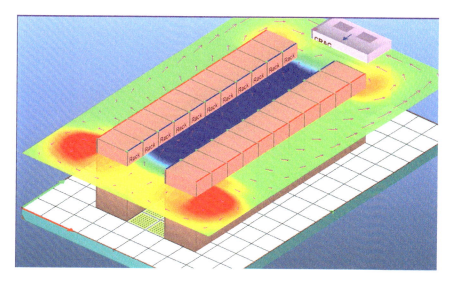

Fig. 28 Temperature distribution and velocity vectors on a horizontal plane

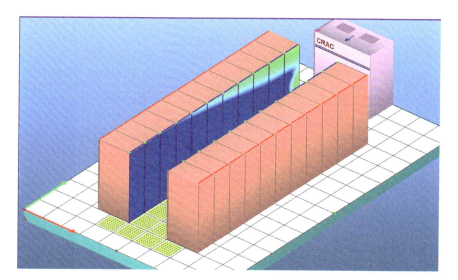

Fig. 29 Temperature distribution on the inlet faces of the racks with air curtain

of about 3 m/s. When this high-velocity stream flows over the inlet face of the rack, it may not enter the rack. Instead, it may flow past it. Radmehr et al. [27] considered this issue and performed a detailed analysis of the situation. They have shown that the high-velocity airflow does create a low-pressure region at the bottom of the rack. This means that the server fans in the bottom region deliver a lower flow rate compared to the uniform-pressure environment. However, this flow reduction is not large. The

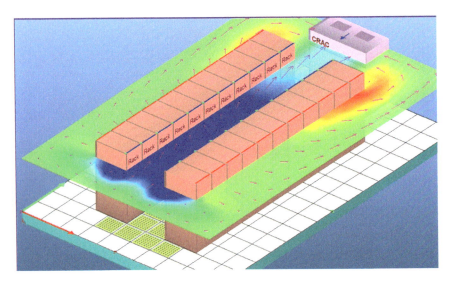

Fig. 30 Temperature distribution and velocity vectors on a horizontal plane (with air curtain)

results in the paper indicate that, for realistic values of the flow resistance inside a rack and for common fan curves, it is less than 15%.

5.2   Use of Above-Floor Partitions to Improve Cooling

The success of cooling in a data center depends on keeping the hot air away from the inlets of the racks. This can be partially arranged by placing solid partitions in appropriate places. Figure 31 shows some possible arrangements. In the two configurations at the left, the top or the sides of the cold aisle are closed. This would prevent hot air entering the cold aisle from the side or from the top. The arrangement in the middle tries to prevent both top and side recirculation. The arrangement at the right of the figure is an attempt to deal with the effects of a bad original layout. Here, the hot exhaust from one set of racks blows into the inlet faces of the next set of racks. (The hot aisle/cold aisle arrangement is not used.) The partition between the two rows of racks helps to keep the hot air away from the next set of inlets.

An ultimate use of partitions is the arrangement known as the ceiling plenum or the use of a drop ceiling. Figure 32 shows a schematic of the drop-ceiling arrangement, in which a false ceiling is created below the real ceiling. Above the hot aisles, vents are placed on the drop ceiling to take the hot air to the space above the drop ceiling. From this space, the hot air is ducted to the inlet faces of the CRAC units. This arrangement creates a near-perfect separation between the hot and cold air and ensures satisfactory cooling.

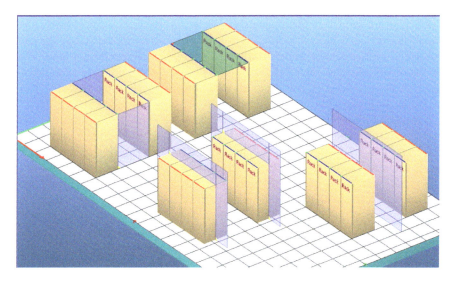

Fig. 31 Use of partitions to prevent hot air from entering the inlets of the racks

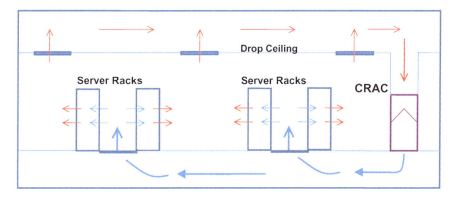

Fig. 32 A schematic of the drop-ceiling arrangement

5.3 *Containments*

As discussed previously, even when sufficient cooling air is available in front of the racks, the mixing of cold and hot air streams can cause the rack inlet temperatures to exceed the allowable maximum value. A popular remedy to prevent this mixing is to deploy containments. The containments could be installed over cold aisles or hot aisles. For a discussion on these two arrangements see, for example, Patankar et al. [55] and Niemann et al. [56].

The use of containments, although a seemingly simple solution, introduces several complexities. These are discussed with reference to the contained cold aisles, but the conclusions are valid even for contained hot aisles. With a contained cold aisle, the relationship between the total cooling airflow entering the aisle through the perforated tiles and the airflow demand of the racks becomes important. Depending on how these two flow rates compare, there is likely to be a pressure difference across the racks. Thus, one has to consider the effect of this pressure difference on the rack flow rates and on the air leakage through the containment walls and racks.

If the total airflow rate from the perforated tiles exceeds the total airflow demand (over-provisioned condition), the cold aisle is pressurized relative to the rest of the computer room. Thus, the pressure at the inlet faces of racks is higher than that at the outlet faces. The higher pressure at the inlets faces will increase the airflow rate through the server fans and cause the cooling air to leak from the cold aisle into the computer room. Although the increased server flow rates represent wastage of cooling resources, the higher pressure in the cold aisle ensures that the rack inlet temperatures are close to the temperature of air through the perforated tiles. Thus, satisfactory cooling is assured. The rack inlet temperatures for this condition are shown in Fig. 33; they are all equal to the supply-air temperature. (Note that if the pressure variation within the cold aisle is large, pressure in certain regions of the aisle could be below the room pressure, leading to leakage of hot air into the cold aisle.)

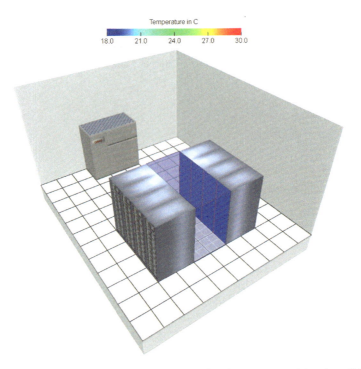

Fig. 33 Rack inlet temperatures for a contained cold aisle for an over-provisioned condition

However, if the total airflow rate from the perforated tiles is less than the total airflow demand (under-provisioned condition), the cold aisle is at a lower pressure than the room. Under this condition, the rack/server flow rates will be less than the design values and the air leakage will bring room (hot) air into the cold aisle. For a given inlet temperature, the reduced server flow rate will cause the electronics to be at a temperature higher than that achieved with the design flow rate. Thus, to ensure proper cooling, the acceptable inlet temperature must be reduced. (This derating of acceptable inlet temperature, to compensate for reduced flow rate, is similar to that used to account for the effect of altitude.) The air leakage into the cold aisle increases the inlet temperature and exacerbates the problem caused by reduced server flow rates. The rack inlet temperatures under this condition are shown in Fig. 34. Now the rack inlet temperatures are higher than the supply-air temperature. In addition, the rack/server flow rates are less than the design values.

In recent years, the subject of containments has been studied extensively. Arghode et al. [57] have presented an experimental and computational study of airflow and temperatures in contained cold aisle. For the under-provisioned condition, the leakage of hot air into the cold aisle increases the rack inlet temperatures. For over-provisioned conditions, the server inlet temperatures are nearly ideal. These findings confirm the general conclusions drawn above. Arghode and Joshi [58] have presented a computational study with similar conclusions.

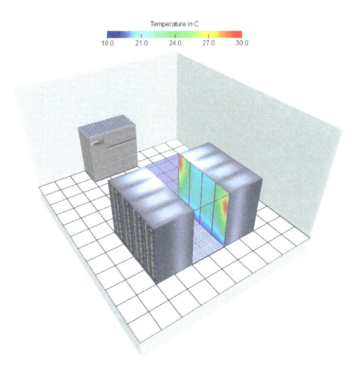

Fig. 34 Rack inlet temperatures for a contained cold aisle for an under-provisioned condition

In most of the studies devoted to containments, the subject of air leakage through racks has received little attention. Air leakage could be significant if the external pressure difference across the racks is large. Tatchell-Evans et al. [36] have presented measurements and theoretical results for air leakage when cold aisles are contained and have reported that it can be up to 20% of the air supplied to the cold aisle, depending on the pressure difference and leakage area in the racks.

The external pressure difference across the racks also brings into focus the flow through turned-off servers, which behave like porous bodies. If the pressure difference is large, this flow could be significant [49]. If the cold aisle is at a lower pressure than the room, hot (room) air will enter the cold aisle and increase the inlet temperatures.

It is thus clear that a cold-aisle containment can be effective only if the amount of the cooling air entering the aisle is sufficient to meet the airflow demand of the racks. When this condition is met, the temperature of the cooling air can be increased. In the ideal case, it can be equal to the maximum acceptable inlet temperature.

The discussion so far has focused on fully contained cold aisles. If the total airflow from the perforated tile meets the airflow demand of the racks, partial containments are also effective. Two possibilities are open top (only side walls) and open sides (only top wall). These are shown in Fig. 31. The containment with only sidewalls eliminates the end effects by preventing the hot air from entering the cold aisle. The containment with only top wall prevents the cooling air to escape from the top, thus eliminating the flow starvation near the bottom of the racks. As with the full containment, a partial containment helps only if the flow rate through the perforated tile(s) in front of a rack is sufficient to meet the airflow demand of the rack.

5.4 Chimney Racks

A chimney rack has a plenum at the back and a duct, or chimney, over the plenum. The hot air exiting the servers is collected in the plenum and directed upwards via the chimney.

Under certain situations, such as the presence of blockages above the chimney, the pressure in the plenum could be significantly higher than the pressure at the inlet face of the rack, reducing the rack flow rate and causing the hot air, from the plenum, to leak through the inlet face, raising the rack inlet temperatures.

These racks are commonly used in conjunction with a ceiling plenum. The chimneys are connected to drop ceiling. Thus, the hot air enters the ceiling plenum, from where it returns to the CRAC units, which are also usually ducted to the ceiling plenum. Such as arrangement is schematically shown in Fig. 35. It guarantees complete separation between hot and cold air and is a special case of hot-aisle containment. As long as the *total* amount of required airflow is supplied through the perforated tiles and leakage of hot air from the rack plenums is prevented, proper cooling of all racks is assured. The flow distribution through the perforated tiles is

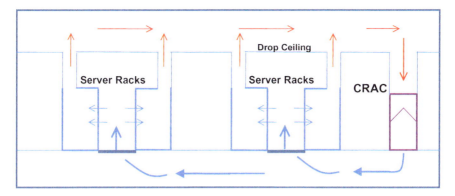

Fig. 35 A schematic of ducted racks with drop ceiling

no more important. The cold air can be supplied anywhere in the room; it will find its way to the inlets of the racks without getting "diluted" by any hot air.

5.5 *Validation of Above-Floor Simulation in Real-Life Data Centers*

As a sample of how the computed results agree with measurements in real-life data centers, a comparison with the measurements is shown here. For full details of the data center, simulations, and measurements, refer to Radmehr et al. [52]. Figure 36 shows the layout of the data center. Figure 37 gives a plot of computed and measured flow rates through the perforated tiles in one row. Figure 38 presents the comparison of computed and measured temperatures at the rack inlets at heights of 0.9 m (3 ft) and 1.8 m (6 ft) from the floor (for selected rows). Finally, Fig. 39 compares the computed and measured values of the return temperatures for the CRAC units in the data center. In general, the agreement is good. Comparisons are also presented by Patankar [2], Shrivastava [40], Radmehr et al. [53], and Schmidt et al. [59].

6 Transient Simulations

A variety of events in a data center can cause the rack inlet temperatures to become unacceptably high. Such events include interruption of cooling, for example, because of power failure; change in the server/rack heat load; and change in the supply temperature. Normally, the effect of these events cannot be studied by making measurements in operating data centers, because they are likely to cause the computer equipment to fail. Transient CFD simulations offer a safer and practical approach.

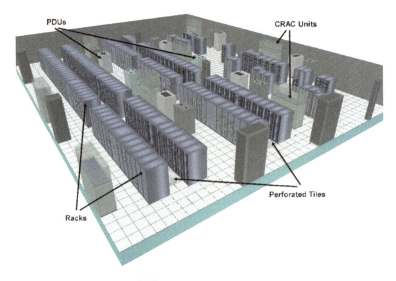

Fig. 36 Layout of the data center [52]

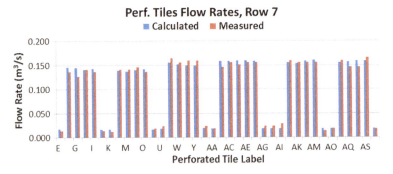

Fig. 37 Comparison of measured and computed airflow rates through perforated tiles [52]

Beitelmal and Patel [60], Gondipalli et al. [61], and Ibrahim et al. [62] have reported some of the early transient simulations. They have considered small data centers. These studies ignore the heat storage capacity of the CRAC units and computer equipment in the racks. The lack of the heat capacity leads to unrealistic results.

There are several challenges in performing transient simulations for real-life data centers. These include prohibitively long calculation times, uncertainties in the input, and difficulties in representing the details of racks/servers and CRAC units (and other cooling devices). As a result, the focus in many of the studies is on developing simplified models, guided by measurements.

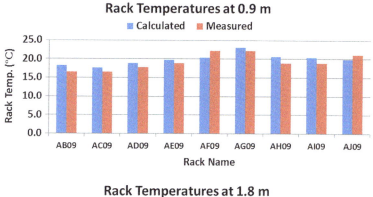

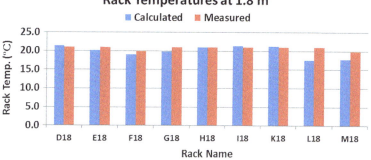

Fig. 38 Comparison of measured and computed temperatures at the rack inlets [52]

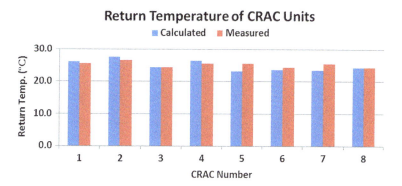

Fig. 39 Comparison of measured and computed return temperatures for the CRAC units [52]

The most common approach is to treat the servers as black boxes with specified lumped-capacitance [e.g., 63, 64]. The server characteristics are deduced from limited experiments. A similar approach has been proposed for representing the cooling devices [65].

7 Concluding Remarks

This paper has dealt with a number of issues pertaining to the airflow and cooling in data centers and summarized various studies involving CFD modeling that provide insight into these issues and help in devising effective cooling strategies. The focus is on raised-floor data centers, but the material is applicable to other designs. For a raised-floor data center, the flow field in the under-floor space holds the key to the distribution of airflow through the perforated tiles. If the airflow demand of each rack is met by supplying the required airflow at the foot of the rack, proper cooling is, in general, assured. The airflow distribution through the perforated tiles is governed by the pressure variation under the raised floor. This is affected by the height of the raised floor, the locations of the CRAC units, the layout of the perforated tiles, their open area, and the presence of under-floor obstructions. Whereas some obstructions are present as a practical necessity, deliberate placement of obstructions (such as perforated partitions) can be used to influence the flow field in a desirable way.

In the above-floor space, the goal is to prevent any hot air stream from reaching the inlets of the racks. This is primarily achieved by supplying sufficient cold air at the perforated tiles. In addition, air curtains, partitions, drop ceiling, containments, and ducted racks can be used to achieve a separation between the hot and cold air streams.

Cooling in a data center is a topic of enormous practical importance. The design of an effective and efficient cooling system requires a good understanding of the airflow and temperature distributions. This understanding is best achieved through CFD modeling.

References

1. ASHRAE (2008). *Thermal guidelines for data processing environments*, ASHRAE, Atlanta.
2. Patankar, S. V. (2010). Airflow and cooling in a data center. *Journal of Heat Transfer, 132,* 073001-1–073001-17.
3. Joshi, Y. & Kumar, P. (2012). *Energy efficient thermal management of data centers*, Springer.
4. Alkharabsheh, S., Fernandes, J., Gebrehiwot, B., Agonafer, D., Ghose, K., Ortega, A., et al. (2015). A brief overview of recent developments in thermal management of data centers. *ASME Journal of Electronic Packaging, 137,* 040801-1–040801-19.
5. Kang, S., Schmidt, R. R., Kelkar, K. M., Radmehr, A., & Patankar, S. V. (2001). A methodology for the design of perforated tiles in raised floor data centers using computational flow analysis. *IEEE Transactions on Components and Packaging Technologies, 24,* 177–183.
6. Schmidt, R. R., Karki, K. C., Kelkar, K. M., Radmehr, A., & Patankar, S. V. (2001). Measurements and predictions of the flow distribution through perforated tiles in raised-floor data centers. In *Proceedings of InterPack'01, The Pacific Rim/ASME International Electronic Packaging Technical Conference and Exhibition, Paper No. IPACK2001-15728.*
7. Karki, K. C., Radmehr, A., & Patankar, S. V. (2003). Use of computational fluid dynamics for calculating flow rates through perforated tiles in raised-floor data centers. *International Journal of Heating, Ventilation, Air-Conditioning, and Refrigeration Research, 9,* 153–166.
8. Patankar, S. V., & Karki, K. C. (2004). Distribution of cooling airflow in a raised-floor data center. *ASHRAE Transactions, 110,* 629–635.

9. Karki, K. C., Patankar, S. V., & Radmehr, A. (2003). Techniques for controlling airflow distribution in raised-floor data centers. *Paper No. IPACK 2003-35282, Proceedings of IPACK'2003, The Pacific Rim/ASME International Electronics Packaging Technical Conference and Exhibition.*

10. Van Gilder, J., & Schmidt, R. R. (2005). Airflow uniformity through perforated tiles in a raised-floor data center, *IPACK2005–73375. ASME InterPack'05.*

11. Bhopte, S., Sammakia, B., Iyengar, M., & Schmidt, R. (2011). Numerical and experimental study of the effect of underfloor blockages on data center performance. *Journal of Electronic Packaging, 133,* 011007–1–011007-7.

12. Abdelmaksoud, W. A., Khalifa, H. E., Dang, T. Q., Elhadidi, B., Schmidt, R. R., & Iyengar, M. (2010). Experimental and computational study of perforated floor tile in data centers. In *Intersociety Conference on Thermal and Thermomechanical Phenomena in Electronic Systems* (ITherm).

13. Arghode, V. K., Kumar, P., Joshi, Y., Weiss, T., & Meyer, G. (2013). Rack level modeling of air flow through perforated tile in a data center. *ASME Journal of Electronic Packaging, 135,* 030901-1–030902-7.

14. Arghode, V. K., & Joshi, Y. (2016). Modified body force model for air flow through perforated floor tiles in data centers. *ASME Journal of Electronic Packaging, 138,* 031002-1–031002-11.

15. Samadiani, E., Rambo, J., & Joshi, Y. (2010). Numerical modeling of perforated tile flow distribution in raised-floor data center. *ASME Journal of Electronic Packaging, 132,* 021002-1–021002-8.

16. Alkharabsheh, A., Sammakia, B., Shrivastava, S., & Schmidt, R. (2013). Utilizing practical fan curves in CFD modeling of a data center. In *29th IEEE Semi-Therm Symposium.*

17. Alkharabsheh, A., Sammakia, B., & Shrivastava, S. (2015). Experimentally validated computational fluid dynamics model for a data center with cold aisle containment. *ASME Journal of Electronic Packaging, 137,* 021010-1–021010-9.

18. Radmehr, A., Schmidt, R. R, Karki, K. C. & Patankar, S. V. (2005). Distributed leakage flow in raised-floor data centers. *IPACK2005–73273, ASME InterPack'05.*

19. Karki, K. C., Radmehr, A., & Patankar, S. V. (2007). Prediction of distributed air leakage in raised-floor data centers. *ASHRAE Transactions, 113,* 219–226.

20. Schmidt, R. R. (2001). Effect of data center characteristics on data processing equipment inlet temperatures. *Paper No. IPACK2001–15870, Proceedings of InterPack'01, The Pacific Rim/ASME International Electronic Packaging Technical Conference and Exhibition.*

21. Schmidt, R. R. & Cruz, E. (2002). Raised floor computer data center: effect on rack inlet temperatures of chilled air exiting both the hot and cold aisles. In *IEEE 2002 Inter Society Conference on Thermal Phenomena* (pp. 580–594).

22. Schmidt, R. R., & Cruz, E. (2004). Cluster of high-powered racks within a raised-floor computer data center: Effect of perforated tile flow distribution on rack inlet temperatures. *ASME Journal of Electronic Packaging, 126,* 510–518.

23. Schmidt, R., & Cruz, E. (2002). Raised floor data center: Effect on rack inlet temperatures when high powered racks are situated amongst lowered powered racks. *ASME Paper No. IMECE2002-39652.*

24. Schmidt, R., & Cruz, E. (2003). Raised floor data center: Effect on rack inlet temperatures when rack flow rates are reduced. *Paper No. IPACK2003-35241.*

25. Guggari, S., Agonafer, D., Belady, C., & Stahl, L. (2003). A hybrid methodology for the optimization of data center room layout. *Paper No. IPACK2003-35273, Proceedings of IPACK'03, The Pacific Rim/ASME International Electronics Packaging technical Conference and Exhibition.*

26. Bhopte, S., Aganofer, D., Schmidt, R., & Sammakia, B. (2006). Optimization of data center room layout to minimize rack inlet temperatures. *ASME Journal of Electronic Packaging, 128,* 380–387.

27. Radmehr, A., Karki K. C., & Patankar, S. V. (2007). Analysis of airflow distribution across a front-to-rear server rack. *Paper No. InterPack2007-33574, Proceedings of IPACK2007.*

28. Patel, C.D., Bash, C. E., Belady, C., Stahl, L., & Sullivan, D. (2001). Computational fluid dynamics modeling of high compute density data centers to assure system air inlet specifications. *Paper No. IPACK2001-15622, Proceedings of InterPack'01, The Pacific Rim/ASME International Electronic Packaging Technical Conference and Exhibition.*
29. Patel, C. D., Sharma, R., Bash, C. E. & Beitelmal, A. (2002). Thermal considerations in cooling large scale high compute data centers. In *ITHERM 2002, The Eighth Intersociety Conference on Thermal and Thermomechanical Phenomena in Electronic Systems.*
30. Sorell, V., Escalante, S., & Yang, J. (2005). Comparison of overhead and underfloor air delivery systems in a data center environment using CFD modeling. *ASHRAE Transaction, 111*, 756–764.
31. Shrivastava, S., Sammakia, B., Schmidt R., & Iyengar, M. (2005). Comparative analysis of different data center airflow management configurations. *Paper No. IPACK2005-73234.*
32. Sullivan, R. F. (2002). *Alternating cold and hot aisles provides more reliable cooling for server farms*, a White Paper from the Uptime Institute, Inc., Santa Fe, NM, USA.
33. Idelchik, I. E. (1994). *Handbook of hydraulic resistance*. Florida: CRC Press.
34. Tate Access Floors Inc. (2008). PERF1250 Air Flow Panel-24.
35. Innovative Research, Inc. (2004). *Effect of backpressure on the flow rate delivered by a CRAC unit*, Technical Note.
36. Tatchell-Evans, M., Kapur, N., Summers, J., Thompson, H., & Oldham, D. (2017). An experimental and theoretical investigation of the extent of bypass air within data centers employing aisle containment, and its impact on power consumption. *Applied Energy, 186*, 457–469.
37. Erden, H. S., Khalifa, H. E., & Schmidt, R. R. (2014). A hybrid lumped capacitance-CFD model for the simulation of data center transients. *HVAC&R Research, 20*, 688–702.
38. Alkharabsheh, S., Sammakia, B., Shrivastava, S., & Schmidt, R. (2014). Dynamic models for server rack and CRAH in a room level CFD model of a data center. In *IEEE Intersociety Conference on Thermal and Thermomechanical Phenomena in Electronic Systems (ITHERM)* (pp. 1338–1345).
39. Patel, C. D., Bash, C. E., Belady, C., Stahl, L., & Sullivan, D. (2001). Computational fluid dynamics modeling of high compute density data centers to assure system inlet air specifications. In *Pacific Rim Technical Conference and Exposition of Packaging and Integration of Photonic Systems, Paper No. IPACK2001-15622.*
40. Shrivastava, S. K., Iyengar, M., Sammakia, B. G., Schmidt, R., & vanGilder, J. W. (2006). Experimental-numerical comparison for a high-density data center: hot spot fluxes in excess of 500 W/ft². In *IEEE Intersociety Conference on Thermal and Thermomechanical Phenomena in Electronics Systems (ITHERM)* (pp. 402–411).
41. Tan, S. P., Toh, K. C., & Wong, Y. W. (2007). Server-rack air flow and heat transfer interactions in a data center. *Paper No. IPACK2007-33672.*
42. Bash, C. E., Patel, C. D. & Sharma, R. K. (2006). Dynamic thermal management of air cooled data centers. In *IEEE Intersociety Conference on Thermal and Thermomechanical Phenomena in Electronics Systems (ITHERM)* (pp. 445–452).
43. Almoli, A. (2013). *Airflow management inside data centres*, Ph.D. Thesis, University of Leeds.
44. Zhang, X., VanGilder, J. W., Iyengar, M., & Schmidt, R. R. (2008). Effect of rack modeling detail on the numerical results of a data center test cell. In *IEEE Intersociety Conference on Thermal and Thermomechanical Phenomena in Electronics Systems (ITHERM)* (pp. 1183–1190).
45. Zhai, J. Z., Hermansen, K. A., & Al-Saadi, S. (2012). The development of simplified rack boundary conditions for numerical data center models. *ASHRAE Transactions, 118*, 436–449.
46. Coxe, K. C. (2009). *Rack infrastructure effects on the thermal performance of a server*, Dell White Paper.
47. North, T. (2011). Understanding how cabinet door perforation impacts airflow. *BICSI News Magazine*, September/October 2011 (pp. 36–42).
48. Rubenstein, B. (2008). Cable management arm airflow impedance study. In *IEEE Intersociety Conference on Thermal and Thermomechanical Phenomena in Electronics Systems (ITHERM)* (pp. 577–582).

49. Kennedy, D. (2012). *Ramifications of server airflow leakage in data centers with aisle containment*. White Paper, Tate Access Floors, Inc.

50. Bajura, R. A., & Jones, E. H. (1976). Flow distribution in manifolds. *Transactions on ASME Journal of Fluids Engineering, 98,* 654–666.

51. Majumdar, A. K. (1980). Mathematical modelling of flows in dividing and combining flow manifolds. *Applied Mathematical Modelling, 4,* 424–432.

52. Radmehr, A., Fitzpatrick, J., & Karki, K. (2018). Optimizing cooling performance of a data center using CFD simulations and measurements. *ASHRAE Journal, 60,* 22–30.

53. Radmehr, A., Noll, B., Fitzpatrick, J., & Karki, K. (2013). CFD modeling of an existing raised-floor data center. In *29th IEEE Semi-Term Symposium*.

54. Athavale, J., Joshi, Y., & Yoda, M. (2018). Experimentally validated computational fluid dynamics model for data center with active tiles. *ASME Journal of Electronic Packaging, 140,* 010902–1–010902-10.

55. Patankar, S., Karki, K., & Radmehr, R. (2012). Cold-aisle and hot-aisle containment, 7 × 24 Magazine. *Fall,* 52–58

56. Niemann, J., Brown, K., & Avelar, V. *Hot-Aisle vs. Cold-Aisle Containment for Data Centers*, White Paper 135, APC by Schneider Electric

57. Arghode, V., Sundaralingam, V., Joshi, Y., & Phelps, W. (2013). Thermal characteristics of open and contained data center cold aisle. *ASME Journal of Heat Transfer, 135,* 061901–1–061901-11.

58. Arghode, V., & Joshi, Y. (2014). Room level modeling of air flow in a contained data center aisle. *ASME Journal of Electronic Packaging, 36,* 011011–1–011011-10.

59. Schmidt, R., Iyengar, M., & Caricari, J. (2010). Data center housing high performance super-computer cluster: above floor thermal measurements compared to CFD analysis. *ASME Journal of Electronic Packaging, 132,* 021009–1–021009-8.

60. Beitelmal, A., & Bash, C. D. (2004). Thermo-fluids provisioning of a high performance high density data center. *Technical Report No. HPL-2004-146. Hewlett Packard Laboratories, Palo Alto, CA*.

61. Gondapalli, S., Ibrahim, M., Bhopte S., Sammakia, B., Murray, B., & Ghose, K. (2010). Numerical modeling of data centers with transient boundary conditions. In *12th IEEE Intersociety Conference on Thermal and Thermomechanical Phenomena in Electronic Systems (ITherm)* (pp. 262–268).

62. Ibrahim, M., Gondapalli, S., Bhopte S., Sammakia, B., Murray, B., & Ghose, K. (2010). Numerical modeling approach to dynamic data center cooling. In *12th IEEE Intersociety Conference on Thermal and Thermomechanical Phenomena in Electronic Systems (ITherm)* (pp. 1262–1268).

63. Erden, H. S., Khalifa, H. E., & Schmidt, R. R. (2014). Determination of the lumped-capacitance parameters of air-cooled servers through air temperature measurements. *ASME Journal of Electronic Packaging, 136,* 031005–1–031005-9.

64. VanGilder, J, Pardey, Z., Healey, C., & Zhang, X. (2013). A compact server model for transient data center simulations. *ASHRAE Transactions*, Paper No. DE-13-032.

65. VanGilder, J. W., Healey, C. M., Condor, M., Tian, W., & Menusier, Q. (2017). A compact cooling-system model for transient data center simulations. In *17th IEEE ITHERM Conference*.

Validation Problems in Computational Modelling of Natural Convection

Victoria Timchenko⬤ and John A. Reizes⬤

1 Introduction

Commercial computer programs for solving Computational Fluid Dynamics (CFD) problems have been available for 40 years since PHOENICS, the first commercial CFD code, created by Spalding [1], to whom this volume is dedicated, was released. It would therefore appear that CFD is a well-established field. In many ways, it is as it is applied now widely in industry. Yet, despite the continuing improvement of numerical methods, there appears to be no reliable method of ensuring that the computer code is solving your particular problem correctly. That is, the thorny problem of verification and validation remains.

CFD programs are large and complex, so the first step is to ensure that the set of partial differential equations has been correctly discretised, that the resulting algebraic equations, the boundary and initial conditions as well as the solution procedure have been correctly coded and, finally, that the resulting program provides the correct solution. This is the so-called verification problem. Thus, verification is an attempt to identify and if possible to quantify errors which result from the transformation of a mathematical model into a computer program. It should be noted therefore, that verification in no way ensures that the mathematical model is appropriate for the problem being modelled.

Unfortunately, verification is by no means simple. In their excellent report, Oberkampf and Trucano [2] carefully outline the necessary steps in verifying a CFD code. Although written more than ten years ago, they present an excellent review of the unsatisfactory situation as far as the verification of CFD codes is concerned. They state "Analytical or formal error analysis is inadequate in numerical algorithm verification because it is the code itself that must demonstrate the analytical and formal results of the numerical analysis" and that as a consequence "Numerical algorithm

V. Timchenko (✉) · J. A. Reizes
School of Mechanical and Manufacturing Engineering, UNSW Sydney, 2052 Kensington, NSW, Australia
e-mail: v.timchenko@unsw.edu.au

© Springer Nature Singapore Pte Ltd. 2020
A. Runchal (ed.), *50 Years of CFD in Engineering Sciences*,
https://doi.org/10.1007/978-981-15-2670-1_19

verification is usually conducted by comparing computational solutions with highly accurate solutions".

The "highly accurate solutions" are often referred to as "benchmark solutions". A number of such solutions are presented by de Vahl Davis [3] for the steady natural laminar convection flow in a two-dimensional cavity with the vertical boundaries being isothermal at different temperatures. Twenty years later Le Quere et al. [4] developed reference solutions for three test cases of steady natural laminar convection flows of gases with temperature-dependent properties in cavities with large temperature differences. Two benchmark solutions are also presented by Nicolas et al. [5, 6] for a three-dimensional mixed convection flow. Many others are to be found in the literature and are often used for verification purposes.

However, the benchmark solutions mentioned above are based on the assumption of steady flow, and as programs involving new physical models are developed there is a need to establish new benchmarks for each type of problem; a gigantic undertaking!

Interestingly, Oberkampf and Trucano [2] write: "Although we have not surveyed all the major commercial CFD codes available, we have not found extensive, formally documented verification or validation benchmark sets for those codes we have examined. As an indication of the poor state of maturity of CFD software, a paper by Abanto et al. [7] tested three unnamed commercial CFD codes on relatively simple verification test problems. The poor results of the codes were shocking to some people, but not to the authors of the paper and not to us".

This is surely a worrying situation and it may be because commercial programs are designed to always converge. What is unknown to the user is that if the program detects that the solution is diverging, it changes the discretisation schemes to stabilise the solution process whilst at the same time it may be lowering its accuracy. Indeed, commercial codes use strategies which help ensure that the program converges by, for example, making some form of upwinding the default option.

Many workers in CFD who use commercial codes assume that because the codes have been run so many times that should there have been "bugs", they would have been discovered and fixed. Therefore, it is often assumed when using commercial codes that verification is not necessary. This unfortunately is not always true, although bugs are rare. On a number of occasions assessors of our PhD students who were supposed to have a considerable knowledge of CFD queried the need for the thorough verification which the candidates had performed on the "well known and respected" commercial codes which they had used to generate their results. They claimed that all that was required was to demonstrate that mesh convergence had been achieved. It should be noted that OpenFoam®, an open source code, which is capturing much of the market previously serviced by commercial suppliers, advises that verification is essential to ensure reliable results. As it seems unlikely that a benchmark solution can be available for all possible physics and chemistry models verification and validation of the program sometimes may not be possible and the code user or writer is reduced to attempting the validation of the code against experimental data.

Another very important aspect of CFD modelling is validation of numerical results. Validation is defined as "The process of determining the degree to which a model is an accurate representation of the real world from the perspective of the intended uses of the model" [2, 8]. "Real world" is understood to mean "experimental results". Since, even now, computer resources are limited, a numerical model is often only a partial representation of the physical equipment, containing only the parts that the modeller considers essential. This together with simplified models of the physics means that, as Kowalewski [9] notes, "there are multiple sources of uncertainties and serious errors".

An issue with validation of numerical results in CFD is that that boundary value problems need to be solved. Unfortunately, it has long been known that small changes in boundary and initial conditions can lead to large changes in the resulting rate of heat transfer and the flow topology. Further, in many instances, because of computer resource limitations, the problem needs to be simplified and the boundary conditions idealised. For example, the boundary conditions assumed for the two-dimensional cavity benchmark solution by de Vahl Davis [3] are that the vertical boundaries are isothermal surfaces at different temperatures and that the horizontal surfaces are adiabatic. The results produced with adiabatic boundary conditions when compared with experimental data indicated the necessity to impose heat losses at the side, "inactive" walls of the cavity [9–11]. This issue will be discussed in detail later in the paper. Whilst this benchmark solution is a valuable tool for verifying incompressible laminar convective flows with constant transport properties in closed cavities, it cannot be used for validation as the boundary conditions are not realistic and the flow is different if all properties are functions of temperature [12].

Kowalewski [9] proposed a strategy to overcome the problem of validating a code against experimental data. This approach is based on a "model experiment including all expected physical ingredients" and suggests that this is usually a laboratory experiment. Once the experimental and numerical data is available a sensitivity analysis needs to be performed of the numerical model to the "inevitable experimental errors". This sensitivity analysis may indicate a redesign of the experimental apparatus, or that the model may need to cover more of the details of the physical environment and the experiment to include monitoring this extended region. He stresses the need for there being sufficient data to ensure that the comparison between the numerical and experimental data sets "to quantify the degree of agreement and to evaluate the improvement of generated results with different physical models.

The extent to which a particular problem is modelled is one of the difficulties in determining the boundary condition which should be applied. To verify boundary conditions for the natural convection flow in a "typical" benchmark case [3] full 3D modelling of heat conduction through the insulating side walls was undertaken by Leonardi et al. [10]. Including heat transfer in side walls in this 3D study provided numerical results not only valid in terms of global parameters (heat fluxes, velocity extremes), but also allowed the accurate prediction of the motion of fluid particles [9]. Kowalewski [9] illustrates this in his paper and shows the surprisingly large effects of extending the calculation domain. Experimentally it is often difficult to control and determine the exact boundary conditions being applied, whereas, numerically one needs to know how complicated the model should be to adequately represent the physical situation being modelled; that is, which equations, in which geometric configuration with which boundary conditions are required. He shows that the inclusion of sidewall heat losses results in the merging of the cross-flow spirals which occur with idealised boundary conditions. This is similar to the results obtained by Leonardi and Reizes [12] who obtained excellent agreement with experiment when the heat transfer through the top of the cavity was also included.

Kowalewski [9] then used the approach proposed in the paper to obtain the actual boundary conditions at the interface of the containing enclosure of the experimental apparatus and the fluid to validate the CFD computer code. However, he warns that the experimental uncertainty might be too large and "that [a] validation is valid only if the experimental uncertainty is below [the] deviation from the numerical result". It should be emphasised that all problems presented in Kowalewski [9] relate to laminar flow. If the flow is turbulent, in many cases it might not be practical to perform such experiments to validate computer codes and the comparison of numerical and experimental results becomes more problematic.

The first difficulty is that unless a Direct Numerical Simulation (DNS) is sought, a turbulence model needs to be used if the Reynolds Averaged Navier–Stokes (RANS) equations are solved or alternatively should a Large Eddy Simulation (LES) approach be taken, a subgrid model has to be used. In both the latter cases, as there is no turbulence or subgrid model that is able to satisfactorily model all flows, the resultant flow and thermal fields are crucially dependent on the model adopted. The computer code may have been validated against a benchmark which uses the particular turbulence model, but does not guarantee that the results for a different physical situation will be modelled correctly.

Secondly, when a RANS steady-state solver is used, it is difficult to ensure that mesh convergence has been achieved whenever a CFD code is used to model turbulent flows. This is usually easily achieved if a steady solution is sought. However, if Large Eddy Simulation (LES) is employed or the solution is unsteady, the determination of a mesh and time step size which meet the criteria of "mesh convergence" are more problematic. This is mainly due to the fact that such problems are large and take a great amount of CPU time to achieve quasi-steady conditions needed to obtain reliable averages and other statistics. Very often, although attempts are made to obtain mesh convergence criteria, the fact that this can take a very long time is a severe disincentive, so that a "reasonable" mesh and time step are chosen. This of course, creates another possible source of error.

However, in order to be able to make rational comparisons between experimentally and numerically generated data, one needs to be sure that the computational model represents the experimental situation sufficiently well for whatever purposes it is intended to use the results; a task which is not trivial. Further, since the equations are highly non-linear there is no guarantee that the solution found is unique or steady. This is something that is rarely questioned. The interesting consequences of these difficulties will be illustrated by considering a number of numerical and experimental studies.

2 Laminar Natural Convection in Cavities

Extensive numerical studies [12–15] have been performed on the effects that vari-able fluid properties have on the calculated details of thermal, velocity and pressure fields in differentially heated vertical two-dimensional square cavities in which the vertical walls are isothermal and the horizontal boundaries are adiabatic. As a result, we have had occasion to compare our numerical solutions with experimental results and have had other researchers compare their numerical work with ours. Interesting problems have arisen in attempting such evaluations. These difficulties have been related to a number of factors: initial conditions assumed by other numericists in attempting comparisons with our numerical results; reasons given by experimen-talists for explaining differences between their results and those from theoretical studies and the boundary conditions which need to be provided by experimentalists so that the numerical calculations adequately represent the physical situation being modelled. These are the very problems discussed by Kowalewski [9].

2.1 Flow of Fluids with Properties Dependent on the Temperature

A non-dimensional approach is generally adopted for two reasons, one being that the number of parameters is reduced and the second that the results can be generalised. For example, when the Boussinesq approximation (all transport properties are con-stant except for the density—expressed usually, as a linear function of temperature in the buoyancy term in the momentum equation) is used in the numerical study of natural convection, it does not matter how a particular Rayleigh number, Ra, and Prandtl number, Pr, are obtained. Thus, any combination of parameters which gives the same value of Ra, will lead to the same solution at the same Pr. Further the two parameters Ra and Pr can be varied independently.

It is true that when all properties are variable the number of parameters is reduced by non-dimensionalisation, but in this case, unfortunately, they cannot be varied independently. For instance, as mentioned above, once the fluid and the reference temperature have been chosen, the Prandtl number and other parameters, which depend only on the temperature, are also allocated values. Moreover, it is interesting to note that once the non-dimensional problem is fully defined, all the values of the physical system represented by the non-dimensional parameters can be evaluated [13]. Thus, contrary to normal practice, no generalisation is obtained by using a non-dimensional approach.

Therefore, when variable properties are used in numerical calculations, the vari-ation in non-dimensional values is limited by the constraints of dealing with real fluids, that is only the same manipulations of non-dimensional parameters is pos-sible as would be experienced in "real life" experiments and the interpretation of results has to be performed with great assiduousness. As will be discussed in some detail, the choice of the reference, in this case initial conditions, has an important impact on the interpretation of the results.

Let us begin the discussion by examining what appears to be a counter-intuitive result when variable properties are considered. Since the functional relation between the transport properties determines how the thermally driven flows will behave, the results will only apply to a particular fluid. Traditionally, when modelling such flows numerically, the Boussinesq Approximation, has often been, and continues to be used. However, when the flow and thermal fields in differentially heated vertical cavities were calculated allowing all properties to vary as functions of temperature and pressure [12–17], results were obtained which are contrary to expectation.

It was found that when the cold wall temperature, of 15 °C, was used as the reference temperature, the numerical solutions in which the Boussinesq approximation has been invoked overestimated the convection obtained for air-filled cavities whilst underestimating it for water-filled cavities. At the time there was no way of validating this result and therefore it was necessary to find an explanation for the phenomenon. Since the pressure in the cavities investigated in this paper was approximately constant throughout the cavity, the reasons for the differences between the solutions in which the Boussinesq approximation was invoked and results obtained when variable properties were used, were determined by examining the variation of density with temperature at constant pressure.

It may be seen in Fig. 1a that the use of the Boussinesq approximation instead of the perfect gas law in air-filled cavities must always lead to an overestimate of the density change for a given temperature. For example, for air, with a temperature difference between the hot and cold walls of 288 K, the Boussinesq approximation overestimates the maximum density change by 50%. Hence, stronger convective flows are always predicted with the use of the Boussinesq approximation in air-filled cavities than when variable properties are used.

On the other hand, in water-filled cavities, as may be seen from Fig. 1b, the employment of the Boussinesq approximation leads to a serious underestimate of the density change. For example, with a 57.6 K temperature difference between the hot and cold walls, the Boussinesq approximation underestimates the density change by over 200%. This leads to a considerably increased convection when variable properties are used in numerical calculations, rather than the Boussinesq approximation.

2.2 Comparison with Experimental Data

Morrison and Tran [18] stated that their experimental flow pattern in an air-filled vertical cavity with isothermal vertical walls at different temperatures and adiabatic horizontal walls were not adequately represented by the numerical solution obtained with the Boussinesq Approximation, particularly at the centre of the cavity. They attributed the difference between the two flows to the fact that property variations were not taken into account in the numerical calculations. Further, they wrote that the position of the cell centre could not have been ascribed to the "different heat

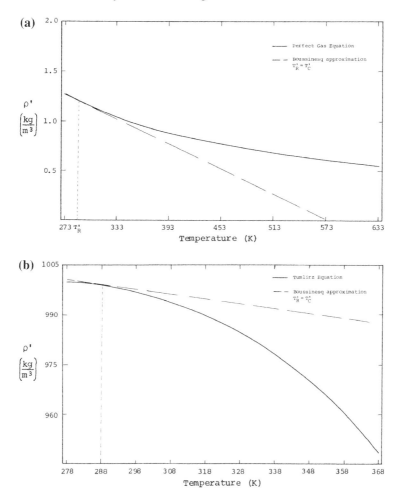

Fig. 1 Density as a function of temperature at a pressure of 101 kPa for **a** air, **b** water [11]

losses through the ends of the walls" [18]. Yet the cell centre moved downwards and towards the cold wall in calculations performed when properties were allowed to vary as a function of temperature whereas the cell centre moved upwards and towards the hot wall in Morrison and Tran's experiments.

The fact that the cell centre moved towards the cold wall is in agreement with the fact that since the density of the cooler fluid flowing downwards on the cold wall side of the cavity is greater than that of the warmer fluid flowing upwards on the hot wall side of the cavity, and since the average velocities are approximately the same on both sides, the area occupied by the cold fluid should be less than that occupied by the hot fluid; that is the cell centre should be on the cold wall side of the centre of the cavity. Similarly, it may be argued that the cell centre should move towards the lower horizontal wall. There had to be another reason for the experimental result.

The position of the cell centre must therefore have been due to the only effect not taken into account, namely, heat losses through the "well insulated walls". A very simple model of the insulation was developed [13] but there was insufficient information in the paper [18] to fully define the model. Morrison and Tran were therefore approached to obtain the relevant data. Fortunately, they were at the same institution and the details of the insulation could be readily obtained. However, the temperature in the laboratory could not be obtained as it had not been recorded. Since the date of the experiment was known, it was possible to estimate the temperature in the laboratory with reasonable accuracy. The excellent agreement between the measured and experimental upward velocity distributions at the mid height of the cavity after the heat transfer through the end walls had been included in the calculations was a proof that the heat transfer through the end is a very important effect which cannot be neglected. The cell centre now moved upward in the numerical calculations by 5.6 mm and this was in agreement with a shift of 6 mm in the experiment.

Again as Kowalewski [9] notes, this means that ideal boundary conditions are unlikely to lead to satisfactory agreement between experimental and numerical results and that whilst ad hoc assumptions might be useful in determining what needs to be done, either the actual boundary conditions can be used for analysis of sensitivity for validation so that the actual boundary conditions for steady flow natural convection problems can be established from experiment or the region used in the numerical calculations needs to be extended until the boundary conditions can be established with reasonable certainty. However, if an engineer uses CFD to model a natural convection problem, for say design purposes, the first course of action is not practical, so it follows that boundaries of computational domain need to incorporate much more of the environment than has generally been the case in the past.

2.3 Generalisation from Non-dimensionalisation

As mentioned above when all properties are variable no generalisation is possible. In some cases this may lead to confusion and complete misinterpretation of the meaning of the results of a calculation and initial conditions can also be equally important in determining the final result. This was brought to the second author's attention when Chenoweth and Paolucci [19] claimed Leonardi and Reizes [12] were wrong in the data presented on their evaluation of the corner pressure in the air-filled cavity in which variable properties were taken into account.

In the in-house computer code used only the dynamic pressure was calculated, so that the thermodynamic pressure can be evaluated by determining the reference pressure which was the pressure in the corner of the hot wall and the top wall. The calculation was based on the fact that the mass of air, m, trapped in the cavity was constant. The mass of air, m, in the cavity is given by,

$$\int_A \rho \, dS = m \tag{1}$$

in which ρ is the density and A is the area of the cavity. The density can be evaluated from the equation of state so that Eq. (1) becomes,

$$\int_A \frac{p_o + \Delta p}{RT} \, dS = m \tag{2}$$

in which $\Delta p = p - p_o$, R is the specific gas constant and p and T are the absolute pressure and temperature, respectively. Since the mass in the cavity remains constant and p_o is a known pressure at a particular point, it follows that the non-dimensional pressure is

$$\frac{p_o}{p_a} = \frac{m - \int_A \frac{\Delta p}{RT} \, dS}{p_a \int_A \frac{1}{RT} \, dS} \tag{3}$$

in which p_a is the standard atmosphere.

Clearly the corner pressure depends on the mass, m, initially in the enclosure. Leonardi and Reizes [12] assumed that the temperature in the cavity had been the cold wall temperature, whereas Chenoweth and Paolucci [19] assumed that the initial temperature had been the average temperature between the hot and cold walls and both assumed that the initial pressure was the standard atmosphere. Further, Leonardi and Reizes [12] showed that if the cold wall was used as the initial temperature $(p_o/p_a) > 1$, whereas if the mean temperature between the two isothermal wall was used $(p_o/p_a) < 1$. Because of the different assumptions the two sets of pressure data could not have been the same, whether dimensional or non-dimensional; yet Chenoweth and Paolucci [19] did not seem to have been aware of that fact. When validating one code against another, if the initial conditions used in the two codes are different, a careful check should be made to ensure that the initial conditions do not affect the results. The initial as well as the boundary conditions must be identical in the data submitted to the two codes.

As discussed above, a number of difficulties arise in validating computer codes for relatively simple laminar flows in laminar natural convection. The numerical modelling of turbulent flows introduces an additional level of difficulty to the validation problem.

3   Three-Dimensional Turbulent Convection in Tall Cavities

There are two aspects of the validation process as discussed by Kowalewski [9]. One is the validity of the equations being solved and the other is appropriateness of the boundary conditions to be applied.

In the case of the laminar flow discussed above, once all property variations had been included, properly modelled, correctly solved and validated, the question of appropriate boundary conditions remains. Since in general the numericist tries to simplify the region of interest as much as possible and that may be complicated, as demonstrated by Kowalewski [9], the determination of the actual boundary conditions is required. In modelling turbulent flows, unless a direct numerical simulation is employed, something still very computationally expensive, a model is required to overcome the closure problem. As a result, the validation of CFD codes dealing with turbulent flows becomes more difficult.

Turbulent natural convection in a tall rectangular air-filled differentially heated cavity was carefully studied experimentally by Cheesewright et al. [20], so as to generate data suitable for validation of computer codes. The details of the cavity, schematically shown in Fig. 2, are given in King's Ph.D. thesis [21]. As may be seen in Fig. 2 the hot and cold walls were each 2.5 m high (H) by 1 m wide (W) and the cavity was 0.5 m deep (D). They used two temperature differences between the hot and cold wall, the one used for here is 45.8 K, giving a Rayleigh number Ra based on the height of 4.6×10^{10}. King [21] took all possible steps to ensure that the temperature distribution was nearly constant on the hot and cold walls, particularly on the vertical centre lines of these walls and that the heat losses were reduced as much as possible on the other surfaces of the cavity. The implication was given that in validating CFD codes the boundary conditions to be used are: hot and cold walls could be treated as isothermal and the remaining walls as adiabatic. These therefore were the boundary conditions initially adopted in the simulations.

The three-dimensional computational model developed by Lau et al. [22] was based on the conservation equations for variable-property Newtonian fluids. The dynamic viscosity and thermal conductivity of air were modelled with the Sutherland [23] equation and the specific heat capacity at constant pressure is taken to be a

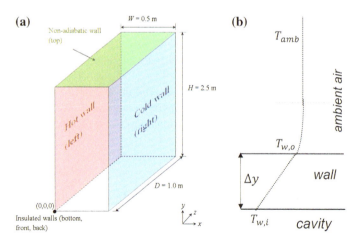

Fig. 2 a Computational domain **b** heat loss model at the top wall (polystyrene)

polynomial function of temperature used by Lau et al. [22]. To adequately model transitional flow structures, a Large Eddy simulation was adopted, in which large-scale eddies were resolved explicitly and small-scale eddies were filtered and a subgrid-scale (SGS) model was therefore necessary. The Vreman [24], VM, formulation was used as the SGS model as it has been found to be less dissipative for simulations of wall-bounded natural convective flows than the Smagorinsky SGS model used by Barhaghi [25]. In contrast to the Smagorinsky model, the Vreman model was found to predict zero subgrid dissipation in regions in which the flow is laminar, hence giving more accurate predictions of the location of transition [22].

Although the Vreman model has been shown to be successful in predicting many turbulent and transitional flows [24], the model coefficient of VM is non-optimal [26] and still needs to be adjusted in different flow configurations. You and Moin [27] utilised the energy equation on the basis of internal energy to derive an expression for the subgrid Prandtl number. Following this approach, the present co-workers [28] adapted a "global equilibrium" dynamic procedure into the Vreman model, called DVME SGS, for the flows at low Mach numbers in which compressibility effects due to propagation of acoustic waves may be ignored. It was demonstrated that this formulation leads to a reasonably accurate capture of the location of transition in boundary layers in natural convection flows. Details of this formulation as well as of the equations and numerical procedures used in the in-house code employed for the solutions presented below can be found in [28].

3.1 Comparison of Numerical Solutions with Experiment

Since King [21] performed measurement on the vertical centre plane between the hot and cold plates, of the mean velocity, velocity fluctuation statistics, mean temperature, turbulent fluctuation and turbulent heat flux statistics. Some comparison between King's experimental data and the numerically generated results were generated in [29] and are shown in Figs. 3 and 4. All comparisons presented below are for the vertical centre plane between the hot and cold isothermal walls. The temperature distributions obtained when either VM or the DVME SGS models were used are not in good agreement with the experimental data as shown in Fig. 3a. Indeed there was little change between the two SGS models.

Following [9, 12] heat transfer through the top of the enclosure was introduced. A one-dimensional model, shown in Fig. 2b, was used to calculate the heat transfer through the top face of enclosure. Although King [21] presented details of the construction of the equipment, he did not give the temperature of the laboratory in which the equipment was located. King wrote that the temperatures of both the hot and cold walls were maintained above the ambient temperature of the laboratory and that the cold wall was kept at 31.4 °C. Since the laboratory was located in London in Great Britain, the ambient temperature was therefore assumed to be 15 °C, some 16 °C below the cold wall temperature. Two values of the external convective heat transfer

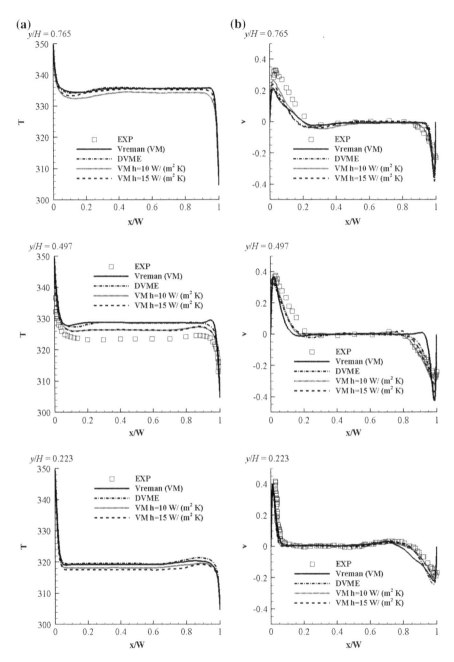

Fig. 3 Comparison of the time-averaged **a** temperature profiles and **b** v-velocity profiles with experimental data at the central plane $z/D = 0.5$

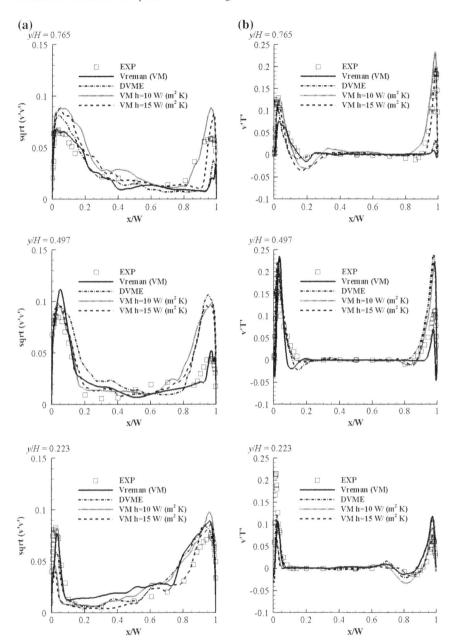

Fig. 4 Comparison of the **a** rms *v*-velocity fluctuations and **b** of the vertical turbulent flux with experimental data at the central plane $z/D = 0.5$

coefficients, h, were assumed, namely 10 and 15 W/m^{2}K (cases VM10 and VM15) representing typical values for natural convection on a flat horizontal surface.

As mentioned above, there is little difference in the mean temperature distributions obtained with either the VM or DVME SGS models in Fig. 3a. However, once heat losses are introduced through the top of the cavity with the VM SGS model there is a marked improvement in the agreement between the numerical and experimental data for the temperature distribution on the centre plane at a height of y/H = 0.223 and y/H = 0.497 with little difference between the two values of h. It should be noted that the agreement is almost exactly at the lower mid-height, namely y/H = 0.223.

Higher in the cavity, at y/H = 0.765 also shown in Fig. 3a, again the VM and the DVME SGS model solutions for the temperature distribution at y/H = 0.765, are not close to the experimental data, and there is very little difference between them. However, the solutions with the VM SGS model when $h = 10$ W/m^{2}K or $h = 15$ W/m^{2}K are used bring the numerically obtained temperature values closer to the experimental data and both are in better agreement with the experiment than the solutions with either the VM or DVME SGS model.

As is shown in Fig. 3b, the numerical predictions for the mean velocity distributions are in better agreement with experimental data than those for the mean temperature, mainly at the lowest level. Here there is a great improvement with the use of the DVME SGS and the VM SGS with either value of h. Similarly, in the region $0.2 \leq x/D \leq 0.8$ at $y/H = 0.497$, there is an almost exact agreement between all solution and experimental values. Near the hot and cold walls, at $y/H = 0.497$, the introduction of heat transfer at the top surface leads to approximately the same improvement for the vertical velocity as the use of the DVME SGS.

The agreement between the numerical and experimental velocity data sets, given in Fig. 3b, is similar at $y/H = 0.765$ to that at $y/H = 0.497$. However, the difference between the velocity distributions obtained numerically and experimentally in the boundary layer adjacent to the hot wall is larger here than at $y/H = 0.497$. Once again the velocity distribution obtained with heat transfer through the top surface is closest to the experimental values particularly through the boundary layer near the cold wall as well as having the lowest overshoot in the magnitude of the maximum velocity.

It is clear from comparison of Fig. 3a, b that all solutions for the velocity are closer to the experimental results than the solutions for the temperature. Less clear is whether the introduction of DVME SGS model or heat transfer through the top of the enclosure leads to the numerical data being closer to the experimental results. A quantitative evaluation of the improvement in a solution relative to the VM SGS model case with adiabatic top condition is therefore required. Let E_m, be the relative improvement over VM SGS solution, a solution using approach m predicts the experimental date for quantity f. E_m can then be defined as,

Table 1 Uncertainty estimates in the experimental data [21]		T (K)	V (m/s)
	Time-averaged values	0.01–0.16	0.002–0.007
	Time-averaged root mean square values	0.013–0.139	0–0.005

$$E_m = \frac{\int\limits_0^1 |f_{VM} - f_{\exp}|\mathrm{d}x - \int\limits_0^1 |f_m - f_{\exp}|\mathrm{d}x}{\int\limits_0^1 |f_{VM} - f_{\exp}|\mathrm{d}x} \tag{4}$$

in which f is the value of a quantity, and the subscripts VM and exp, refer to the Vreman SGS solution and the experimental data, respectively. A positive value of E_m, indicates an improvement, whereas a negative value indicates deterioration in outcome comparing with the VM SGS solution. Kowalewski [9] indicates that a proposed benchmark experiment is only valid if the errors are small. This appears to be true in the case of the King [21] experiments as indicated in Table 1. So that $f_{\exp}$ values can be used with confidence in Eq. (4) and need not have any error bounds.

Values for the E_m, expressed as a percentage, are presented in Table 2 for the time-averaged temperature and velocity distributions at $y/H = 0.223$, the middle of the lower part, at $y/H = 0.497$, the middle, and at $y/H = 0.765$, the middle of the upper part of the cavity.

There appears to be no improvement in the solution for the time-averaged temperature when the DVME SGS model is used instead of the VM SGS as is seen in Table 2a. However, there is a substantial improvement in the solutions for the average temperature at all levels of the cavity when VM SGS is used with heat transfer through the top of the cavity. The relative improvement in the mean temperature is approximately the same with $h = 15\,\mathrm{W/m^2K}$ as with $h = 10\,\mathrm{W/m^2K}$.

It may also be seen in Table 2b, that the use of DVME SGS model yields solutions closer to the experimental data for the time-averaged vertical velocity than the VM solution up to the mid-height of the cavity. Unfortunately, a deterioration of -11% is found in the middle upper part of the cavity. However, as is shown in Table 2b,

Table 2 Relative percentage improvement of different solutions types—mean values

(a) Average temperature $\overline{T}$				(b) Average vertical velocity $\bar{v}$			
Case	Lower middle (%)	Middle (%)	Upper middle (%)	Case	Lower middle (%)	Middle (%)	Upper middle (%)
DVME adiabatic	−20	7	−1	DVME adiabatic	24	36	−11
VM15	73	43	9	VM15	39	48	−3
VM10	49	43	27	VM10	40	57	17

there is much larger relative improvement when the Vreman SGS model is used with heat transfer through the top of the cavity with either value of h than when with the DVME SGS The relative improvement is a maximum of 57% in the middle of the cavity with $h = 10\,\mathrm{W/m^2 K}$ and it is more uniform than with $h = 15\,\mathrm{W/m^2 K}$ and the VM SGS.

It follows from Table 2 that when heat transfer through the top of the cavity is introduced, solutions with the VM SGS are closer to the experimental data for mean values than solutions in which a better subgrid model, namely the DVME SGS, is used.

A comparison is presented in Fig. 4 for the various numerical solutions with the experimental data of the root mean square (rms) vertical velocity, v, fluctuations, $\sqrt{\overline{v'^2}}$, and the turbulent vertical heat flux, $\overline{v'T'}$. It is difficult to determine from Fig. 4 whether there is any improvement in the numerically generated data relative to the VM SGS model in the fluctuating statistics as there are such large variations depending on the vertical and horizontal location in the cavity. For example at $y/H = 0.223$, near the hot wall the values of $\sqrt{\overline{v'^2}}$ are in good agreement with the experimental data when the DVME SGS is used, and are significantly underestimated when heat transfer through the top is introduced. However, at this vertical position the VM SGS with $h = 10\,\mathrm{W/m^2 K}$ matches the experimental results $0.15 \leq x/W \leq \sim 0.09$ and has the least overshoot. Similar remarks apply to the $\overline{v'T'}$ numerical results, but this time all the numerical solutions match the experimental data in the $0.1 \leq x/W \leq 0.7$ range, with the solution with $h = 10\,\mathrm{W/m^2 K}$ matching the experimental values up to the cold wall.

At $y/H = 0.497$, all models overshot at the hot boundary and with the exception of the VM SGS, wildly overshoot the experimental values at the cold boundary, where once again VM SGS does not overshoot by much. At the mid cavity height, all models either significantly overshoot or undershoot the experimental data for near the boundaries, but match the experimental data between in the $0.2 \leq x/W \leq \sim 0.7$ range with the solutions with heat transfer matching the experimental $0.2 \leq x/W \leq \sim 0.9$ range.

At the upper middle of the cavity the experimental values of $\overline{v'T'}$ are best matched by the solution with $h = 15\,\mathrm{W/m^2 K}$ in the boundary layer at the hot wall, the DVME solution in the centre and solution with $h = 10\,\mathrm{W/m^2 K}$ in the boundary layer at the cold wall. It should be noted that both the solutions with the VM and the DVME SGS underpredict the turbulent heat flux values in the boundary layer near the cold wall.

The data presented in Table 3 can be interpreted to indicate that there is an improvement in the numerically predicted vertical component of the turbulent energy, $\sqrt{\overline{v'^2}}$, and the vertical turbulent heat flux, $\overline{v'T'}$ when heat transfer through the top of the cavity is included. In the middle of the cavity, as may be seen in Table 3a, all schemes lead to poorer predictions than the VM SGS for $\sqrt{\overline{v'^2}}$. Only the VM SGS with $h = 15\,\mathrm{W/m^2 K}$ produced a relative improvement on the average in both $\sqrt{\overline{v'^2}}$ and $\overline{v'T'}$.

Table 3 Relative percentage improvement of different solutions—Fluctuating quantities

(a) RMS Vertical velocity $\sqrt{\overline{v'^2}}$				(b) Mean Vertical turbulent heat flux $\overline{v'T'}$			
Case	Lower middle (%)	Middle (%)	Upper middle (%)	Case	Lower middle (%)	Middle (%)	Lower middle (%)
DVME adiabatic	31	−34	−5	DVME adiabatic	1	−15	−5
VM15	50	−7	11	VM15	25	10	20
VM10	11	−30	−2	VM10	−8	26	−27

3.2 Discussion

Unless DNS, to which "the closure problem" does not apply, is used to simulate turbulent flows numerically, some "turbulence model" needs to be used when solving turbulent flow problems. Unfortunately, DNS requires a very fine grid and a very small time step, which despite the gigantic developments in computer hardware, make DNS too computationally expensive to use in most research. Therefore, whether this involves a subgrid model to represent the effects of averaging over a finite time step and a finite grid in an LES code, or a turbulence model in a RANS code, some approaches lead to better results in different situations. The word "better" is, we believe, normally simply defined as "more closely approximating experimental results" as no other metric exists.

In the sections above we have demonstrated that when modelling laminar flows, validation is very difficult to achieve, even if all property variations are correctly represented in the code. Unless boundary conditions are known accurately, a numerical solution cannot yield the same results as an experiment [9]. The problem of validating a turbulence model therefore is doubly problematic and as demonstrated in Sect. 3.1 immediately above, a change in boundary conditions led to a closer agreement between numerical solutions and experimental results than an "improvement" in the SGS.

Since CFD simulations of turbulent flows often require long runs to achieve "quasi static" conditions and further long runs to achieve reliable averages and flow statistics, there is often a tendency to simplify the model geometry to contain the "crux of the problem". For example, the boundaries of the two and three-dimensional cavities were initially taken as the inside of the walls enclosing the fluid, well-insulated boundaries were treated as adiabatic, no conduction of heat was assumed to occur along the walls of the cavity and the effects due to the necessity of observing and taking measurements by experimentalists completely neglected. However, it is shown above that each time when some of the adiabatic boundaries are replaced by very simple one-dimensional models of walls and insulation, the numerical solution is brought into better agreement with experimental data. It follows that unless accurate boundary conditions are used in the numerical solution, it will be very difficult in problems which involve heat transfer to validate turbulence models.

The quest for the development of the simplest numerical model which should be used is resolved once boundary conditions are accurately determined. This means that the boundaries of the region of interest need to be extended outwards from the equipment that is being simulated until this condition is met without the need for experimental input to the numerical model. In the next section, this is examined in some detail.

4   Open-Ended Heated Vertical Channel

For a number of years we have collaborated on the research of natural convection in heated open-ended vertical channels with colleagues from the CNRS-INSA Lyon-University of Lyon, France, who have mainly performed the experiment work, whereas our team at UNSW in Sydney mainly carried the numerical simulations. The initial numerical study of natural convection in an open-ended channel, run as a validation exercise, was to simulate the experiments of Miyamoto et al. [30]. The computational domain was simplified to a rectangular vertical channel with rectangular regions above and below the channel representing the regions through which air flowed in and out of the channel. A Large Eddy Simulation was used with both Smagorinsky and Vreman subgrid schemes. The results obtained with the Vreman SGS were in very good agreement with the experimental data obtained by Miyamoto et al. [30], particularly in the regions in which the flow is laminar. However, the results obtained with the Smagorinsky SGS were in very poor agreement with the experimental data [22, 31, 32]. This gave us the confidence to attempt the simulations of the experiments being performed by our French colleagues.

The computational domain for the simulation of the experiments performed by our French colleagues [33–38] was simplified to reduce the computational burden, as shown in Fig. 5. The artificial ceiling located above the channel was included as was the floor beneath. Since the interest in convective flows in channels open at both ends stemmed from building integrated photovoltaic (BIPV) systems, the heating was applied to one of the wide sides of the rectangular channel as shown in Fig. 5. The effect of the insulation was approximated by a heat loss to the surroundings calculated with a one-dimensional approximation and the properties of the insulation. Since the experimental data were restricted to the narrow vertical centre plane of the channel, the numerical effort was exclusively directed towards matching that data. As a consequence, only numerical results on the centre plane were initially examined in detail.

Surprisingly the LES numerical results using the Vreman SGS on the geometry in Fig. 5 were in very poor agreement with experimental data, yet the same model has produced excellent agreement with Miyamoto et al. [30] experimental results.

Fedorov and Viskanta [40] used two-dimensional Reynolds averaged Navier–Stokes equations with the k-ε turbulence model to model the Miyamoto et al. [30] experiments and found that their solution was very sensitive to the value of the turbulence kinetic energy at the inlet to the computational domain. Indeed they had

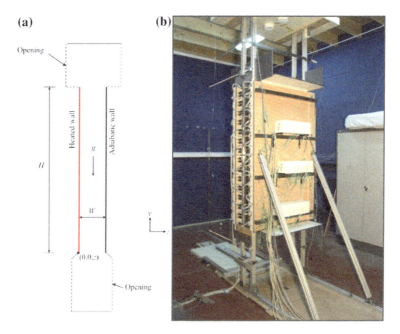

(a)

Opening

Heated wall

Adiabatic wall

g

H

W

$(0,0,z)$

Opening

v

(b)

Fig. 5 **a** Schematic diagram of the computational region used in LES simulations. **b** Experimental apparatus built in CETHIL, Lyon, France (reprinted from [39] with permission of the Begell House)

to use unexpectedly high values of turbulence intensity to obtain any semblance of agreement with experimental data of Miyamoto et al. [30]. Thus, although it was not clear at the time why disturbances should be introduced at the inlet to the computational domain, following Fedorov and Viskanta [40], velocity fluctuations were introduced in LES simulations on the outward facing surface of each cell at the inlet to the domain. The components of the velocity on the outward face of the cell at the inlet of the domain were given by

$$u_i^{n+1} = u_i^n + r\varepsilon, \quad i = 1, 2, 3 \tag{5}$$

in which the superscripts n and n + 1 indicate the end of the time step n and the beginning of the next, $n + 1$, time step, respectively, ε is the coefficient of disturbance and r is a random number in the range $1 \geq \varepsilon \geq -1$. Various values of ε were tried with $0.02 \, \text{ms}^{-1}$ yielding the best agreement with experimental data. Sanvicente et al. [37] and Daverat et al. [41], respectively, reported experimental turbulent intensity of 25 and 35% at the inlet of the open-ended channel, which was in agreement with the large fluctuations assumed in a numerical study by Lau et al. [39].

In order to attempt to answer the reason for disturbances in both the Miyamoto et al. [29] and Lyon experiments [42], calculations were undertaken with the inclusion of the laboratory in the computational domain for the Lyon case. A two-dimensional study of the flows within the laboratory was therefore performed. It immediately became clear that unless the laboratory was vented to the outside, both the temperature and the pressure within it would rise due to the heat input from the open channel. Since the physical laboratory was not hermetically sealed, air would flow out if the pressure within the laboratory increased and heat would also be transferred to the wall as well to the outside if there were any windows In order to partially simulate the flows in and out of the laboratory, a "chimney" was added with an opening near the floor to represent the gap beneath door through which air may move in and out of the laboratory.

It was not intended to exactly determine the flows in the room, but to "schematically" represent it so as to understand the reasons for the need to introduce fluctuations. The walls of the laboratory, the artificial ceiling and the horizontal projections from the open channel were treated as adiabatic. Air in the room was modelled as a dry gas with variable properties. The insulation on either side of the heated open channel had a thermal conductivity of 0.033 W/mK with the specific heat capacity of 1300 J/kgK and the channel was heated on both sides with a uniform heat flux of $230\,\mathrm{Wm^{-2}}$. Simulations were run with the commercial code CFD-ACE+. The total number of mesh points used was 850,000 with 17,000 in the channel. An Implicit LES scheme was used with a second-order upwinding discretisation instead of a SGS [43]. A 0.005 s time step was used. The code was run for 1063.15 s model time which took approximately 10 days on a 20 core Intel Xeon CPU E5-2650 v4 @ 2.20 GHz.

The initial conditions for the numerical simulation including the laboratory were 25 °C. The first image in the series of images presented in Fig. 6a, was obtained at 859.75 simulation seconds after the start of the calculations. At this time the temperature stratification in the room had begun to develop, but the insulation, as could have been expected due to its low coefficient of thermal conductivity remained at room temperature except for a thin layer adjacent to the heating strips. Indeed, even at nearly 18 min (1063.15 s), the time at which the last image had been generated, the temperature of the insulation had barely changed; however, the temperature stratification had developed much further. The high thermal gradients generated at the interface between the insulation and the heating surface mean that a significant proportion of the heat flux enters the insulation, thereby reducing the thermal energy available to heat the air. As a result, the rise in temperature of the heating surfaces is quite slow. When the quasi-steady-state temperature distribution has been reached in the insulation, the heat flux into the insulation would be reduced to a small value and nearly all the thermal energy would be used to heat the air in the channel. Thus the flow rates present in Fig. 6 are far from those expected when quasi-steady flow is eventually reached.

Soon after the beginning of the calculations the plume issuing from the heated channel began to break up into large coherent structures which cooled as they moved in the air surrounding the channel. Although not clearly seen in these images the fact that the outside of the insulation is at the initial room temperature the hot air emanating from the channel in the vicinity of the insulation is cooled by it and is therefore rapidly convected downwards towards the inlet of the channel. Other structures as may be seen in Fig. 6a–c meander around the room whilst being cooled by the surrounding colder air which is heated as the result. The temperature stratification is thereby

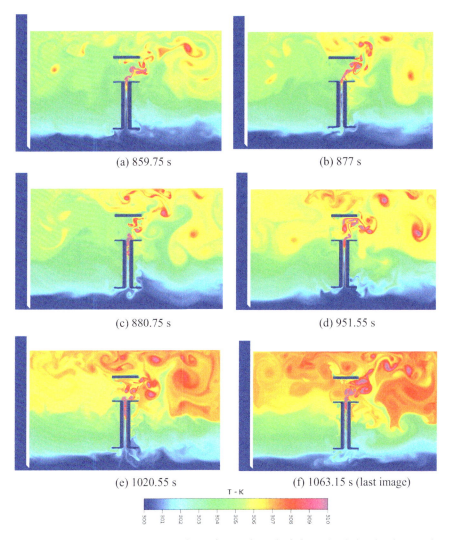

(a) 859.75 s   (b) 877 s

(c) 880.75 s   (d) 951.55 s

(e) 1020.55 s   (f) 1063.15 s (last image)

Fig. 6 Instantaneous images at various times after starting calculations: simulation time in seconds, colour indicate temperature

gradually developed. The development of the temperature stratification is clearly seen in Fig. 6e, f at the same time there is a leakage of hot air out the "chimney" and an inflow of cold air from outside so that the lower region below the intake of the channel remains quite cold.

The lower pressure at the inlet to the channel eventually results in these large structures being "drawn" into the channels, as may be seen in Fig. 6b–f. Each time when such a structure is "drawn" into the channel, the flow is significantly disrupted and each time the disruption is different. The results presented in Fig. 6 are far from quasi-steady flow which is only achieved when the temperature stratification in the laboratory and the temperature distribution in the insulation are steady. There is no reason to suppose that the large flow structures would not occur when quasi-steady conditions have been reached. Thebault [42] showed experimentally that large structures are present at the entrance of the channel at Lyon, thereby validating the above conclusion.

Lau et al. [39] demonstrated that turbulent flow did not develop within the channel when numerical simulations were performed without large disturbances at the lower faces of the channel and these disturbances needed to be maintained for the whole duration of the simulations. It appears therefore that the turbulent flow developing in the channel is the result of factors external to the channel rather than in conventionally developing natural convective flow. Further it is clear that Miyamoto et al. [30] also had disturbances at the inlet to their channel similar to those experienced in the Lyon channel. Katoh et al. [44] attempted to reduce the disturbances by introducing a bell-mouth shape at the entrance of the channel, but since the results of the later study by Katoh et al. [44] were very similar to those of Miyamoto et al. [30] it does not appear that they were successful in significantly reducing these disturbances. Neither Miyamoto et al. [30], nor Katoh et al. [44] mentioned the source of the disturbances.

4.1 Discussion

The two-dimensional numerical study of the heated channel in the large laboratory was only performed in order to *qualitatively* establish the cause of the disturbances at the inlet to the heated channel. It is clearly shown above that large coherent structures which meander through the space and the flow moving down the insulation are the cause of the disturbances. It follows that the very simplified model of the channel is not adequate for the task of numerically simulating the convective flow in a vertical channel in a laboratory, unless it is "tuned" as Lau [39] and Fedorov and Viskanta [40] had done. Were a tuning approach to be used for design, an experimental model would be needed to tune the model. It is therefore not clear that such a model could be used to develop a new design.

An alternative approach would be to simulate the entire laboratory in three dimensions, with all its heat sources such as computers, lasers and personnel as well as other equipment and furniture; the heat and mass exchange with the atmosphere outside the laboratory as well as the heat exchanges between the ambient in the room and at least the walls, ceiling and floor of the laboratory. Such a numerical model would possibly capture all, or nearly all the physical exchanges occurring in the laboratory. Unfortunately, the mesh required to model all the details in three dimensions would need to be very large and it is not clear that all the exchanges with the external atmosphere could be easily established. Such a task might possibly be performed for research purposes, however, the time to set up the mesh and the boundary conditions together with the expected simulation wall clock time of many weeks, possibly months makes this approach prohibitively long for design purposes.

Fig. 7 Experimental double façade-roof configurations used by Gaillard at al. [46]

Experimenters in Lyon could not reproduce experimental results from one day to the next despite taking great care to set up identical conditions [36, 42]. In particular, Thebault [45] could not control the gradient of the temperature stratification in the laboratory. It follows that since in both the above cases boundary and initial conditions cannot be established with a great level of confidence, validation can only be approximately established.

The experimental program performed in the Lyon laboratory was designed to obtain data for the much more complex problem of flow and temperature distributions in building integrated photovoltaic (BIPV) systems. The scale of the experiments in Lyon was too small to allow extrapolation to full-scale experimental installations such as those described by Gaillard et al. [46] and shown in Fig. 7. The experimental data were presented in the form of graphs of results taken over 133 days.

Before proceeding with attempting to simulate the full-scale system shown in Fig. 7, a quarter scale combined façade/roof arrangement on the right of the house in Fig. 7 was simulated using LES as described above. It was suggested that the heating surface, representing the photovoltaic arrays on the roof section of the duct could either be mounted on the top of the duct or directly on the roof. In a like manner to the Lyon physical model, the computational model was assumed to be insulated on all external surfaces. Only internal flows in the combined configuration were therefore studied. One significant conclusion of this study was that the photovoltaic arrays should be located directly on the roof since they were about 10 K cooler than when they are placed as the upper surface of the duct.

When numerically modelling the full size building shown in Fig. 7, the parameters which must be included as boundary conditions include: the magnitude and direction of the solar heat flux; wind direction, velocity distribution and its turbulence property distributions; ambient temperature and pressure; the radiation properties of the terrain, all surfaces of the house exposed to solar radiation, its attached double façade ducts together with the radiation properties of the internal surfaces of these ducts. The list of initial conditions is similarly long. The computational domain must include a large region around the building in all directions to ensure that the location of boundaries does not affect the flow and thermal fields around the building and the ducts.

In the initial study of the full-scale experiments the wind direction was restricted to being perpendicular to the face of the building incorporating the two BIPV systems (Fig. 7). The computational domain was taken as $30H_B$ in the direction of wind, $10H_B$ wide in the direction perpendicular to the wind and $5H_B$ high, in which H_B is the height of the building.

Further many of the boundary conditions such as, at least, the magnitude and direction of the solar heat flux and wind velocity and direction are time-dependent. This means that were the diurnal variations to be simulated, the numerical solution would have to be unsteady, creating a numerical problem requiring very large computer resources and extremely long run times. In fact the computing problem is of such a magnitude that it is not possible to resolve with present-day computers. It follows that only a steady-state numerical solution was attempted.

In order to attempt to approximately validate the computer program, a particular moment in time needs to be selected for the steady-state solution. Wind would appear to be an important parameter and since wind direction was presented in the data for the building in Fig. 7, a period with zero wind needed to be selected for comparison purposes. Approximately 6×10^6 elements of which approximately 3×10^6 elements were internal to the two channels were created to mesh the computational domain. The k-ω Shear Stress Transport (SST) turbulence model and the ANSYS CFX software were used to solve the equations. One of the difficulties encountered was that each of the channels shown in Fig. 7 had a floor made from perforated plates which were modelled as a porous medium.

The experimental mass flow rate was ~0.9 kg/s through ETNA A and ~0.7 kg/s through ETNA B [45] whereas the numerically calculated mass flow rates were 0.95 and 1.4 kg/s, respectively. Experimental mass flow rates were estimated from readings obtained from rakes consisting of five anemometers mounted on the centre plane of each of the inclined ducts; one rake near the inlet and the other near the outlet of each of the inclined ducts [45]. The values of the experimental mass flow rates are therefore an approximation of the actual experimental flow rates and would depend on the velocity distribution within the ducts. Numerically generated flow structures through both ducts with zero wind as indicated by the iso-surfaces of λ_2 criterion are shown in Fig. 8. The flow through the inclined portion of ETNA A is more or less uniform and the centre plane could be reasonably representative of the average flow. However, the flow in ETNA B is particularly complex with most of the flow not being near the central plane. It follows that the experimental flow rate could be greatly under-predicted in ETNA B and explain the larger flow rate in ETNA B; opposite to the experimental results. Indeed the numerically generated results indicate that the mass flow rate in ETNA B is larger than that in the smaller duct. It follows that validation, could only be approximately achieved.

It is interesting to note that the experimental mass flow rates presented by Gaillard et al. [45] do not vary greatly with wind which is surprising. With no wind, the mean velocity in ETNA A is less than 0.5 m/s. The calculated flows through the ducts and around the house are shown in Fig. 9 for a wind of 1 m/s at a height of 3 m. It appears

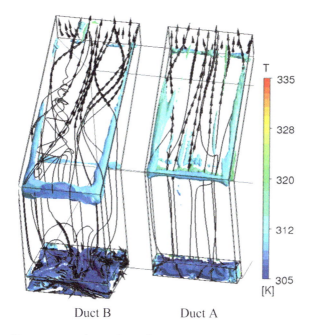

Fig. 8 Coloured by temperature iso-surfaces of $\lambda_2 = -0.125$ for ETNA A and ETNA B—no wind condition

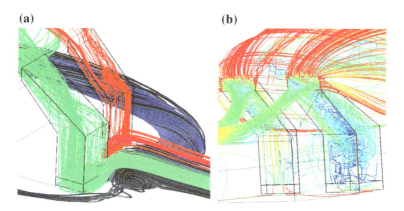

Fig. 9 Effect of 1 m/s wind on the flow through the ducts A and B **a** positive wind **b** negative wind

that whether the flow is towards the face of the building containing the ducts, hereby termed positive, or in the opposite direction, hereby termed negative, the flows in the channels in Fig. 9 differ greatly from those in Fig. 8. The numerically determined mass flow rates through the ETNA A and B are now 1.1 and 1.6 kg/s, respectively, for a positive 1 m/s wind and 0.6 and 0.5 kg/s for a negative 1 m/s wind. Since such small wind velocities can lead to such large changes in mass flow rate, the validation problem is made significantly more difficult.

The equipment in the Lyon laboratory experiments was well insulated, so that it was assumed that the quarter scale combined façade/roof configuration was well insulated in the discussion above. One of the conclusions of that study, mentioned above, was that the PV arrays should be placed directly on the roof. However, it is shown numerically that since a PV array on the outer face of the channel is cooled both by internal and external flows and radiation exchanges, it is much cooler than if it were placed directly on the roof. Whilst this may not seem relevant to the validation of CFD codes, it shows that if an inappropriate model is used whether experimental or numerical the results may lead to the wrong conclusions.

5   Conclusion

It is shown that the use of experimental data to validate computer codes for use in natural convection simulations is particularly problematic. The difficulty arises because the boundary conditions to be used in numerically modelling the flow and thermal fields are difficult to determine. This first became clear in the apparently simple problem of laminar flow in a well-insulated differentially heated square air-filled cavity in which the experimental and numerical results differed. Since the numerical code had been written with the transport properties dependent on temperature and pressure, the only difference between numerical and experimental data sets had to be due to the boundary conditions. It was shown numerically that if heat transfer occurred through the top of the enclosure, differences between the numerical results and the experimental data vanished. This indicated that, in this case, the numerical solution should include an interaction between the atmosphere in the laboratory and the equipment included. However, since the conditions in the laboratory are imperfectly known, it is only possible to approximately validate the code even if the flow in the experimental apparatus is laminar.

Validation becomes even more difficult if the flow is turbulent. Since DNS solutions are still too computationally expensive, it follows that either a turbulence model needs to be chosen if steady state is sought or a subgrid model in the case of LES solutions. This introduces another level of complexity. Again, experimental data obtained in a differentially heated enclosure, but this time with the flow turbulent, were used as the basis for validation. Although an improved Vreman subgrid model gave numerical results which were closer to the experimental data, the introduction of heat transfer through the top of the cavity and the use of the standard Vreman subgrid scheme, resulted in a closer agreement between the experimental data and the numerical solution. Once again the interaction between the experiment and its environment needs to be included in the numerical model. Unless this is correctly performed it is not possible to validate the turbulence or subgrid models. Since the ambient temperature and flow distributions probably cannot be simply determined or evaluated, the computer code can only be approximately validated.

This is vividly illustrated by the simulation of the heated open-ended channel in the Lyon laboratory. Unless perturbations were introduced in the numerical simulations at the inlet of the Lyon channel, it was not possible to obtain any match between the numerical and the experimental data sets. The clue for the reason that this was necessary was given by the fact that it was not possible to repeat the results experimentally from one day to the next, despite having the identical setting of heat input to the channel. It follows that the laboratory environment played an important role in the flow inside the channel. The modelling of the entire laboratory room including the channel revealed that the hot plume emanating from the channel broke up and the resulting large structures meandered around the laboratory. Some of these structures enter the channel and disrupt the flow. Similar large disturbances were also detected experimentally. The three-dimensional transient simulation of the channel and the laboratory, including its interaction with its external surroundings, is a major computational effort with no guarantee that all interactions would be captured. It follows that validation of a reduced computational model is not possible, only tuning is available.

Validation is only possible if the boundary conditions of the numerical model have been well identified and all the discussions of validation are based on the assumptions that the boundary conditions to be used in numerical simulations are known. Indeed in all the above numerical simulations, validation was not possible precisely because the problematic aspect in natural convection is the determination of the boundary conditions. For laminar flow in the experimental apparatus, one approach used to overcome the problem was to establish the boundary conditions by measuring what they are in the equipment. This is an iterative approach and therefore expensive. Further, when dealing the next problem, once again the whole procedure needs to be repeated to establish the boundary conditions to be used in the numerical solution. Thus this approach is not suitable for engineering design problems.

Another method is to extend the domain of the numerical simulation, as was done above to establish the origin of the disturbances in the Lyon channel flow. Extending the numerical domain where the boundary conditions are better known is done in the expectation that small differences between the actual boundary conditions and those used in the simulation would have "minimal" effects on the flow and thermal fields occurring in the equipment being investigated. Such a methodology may result in a physically large domain-containing regions of very different turbulence scales thereby making an LES solution quite difficult. In any case the numerical simulation may become so computationally expensive, that steady solutions is all that can be sought.

This was the case of a house with BIPV ducts. In this case, it should be noted that despite the complexity of the numerically generated steady flows and the reasonably fine mesh, details such as vortex shedding were inevitably lost. Despite the existence of a large body of experimental data, validation was not possible as the instrumentation was not sufficiently detailed to allow meaningful comparisons between numerical and experimental results. Despite this, it was possible to show that the wind has an overwhelming effect on the flows and thermal fields in the ducts and that it acts to reduce the temperature of the PV cells.

It would therefore appear that in natural convection validation of computational models can only be partially achieved. In some cases, such as in the heated channel in the Lyon laboratory, tuning is possible if sufficient experimental data are available. However, it is not clear that the same tuning can be used in another similar situation. In general, it is recommended that the boundaries of the computational domain be set as far as possible from the equipment being numerically modelled and where boundary conditions can be determined with some certainty.

Acknowledgements The authors would like to thank Svetlana Tkachenko, Oksana Tkachenko and Peter Brady for allowing us to use their work in this chapter.

References

1. Spalding, D. B. (1981). A general purpose computer program for multi-dimensional one-and two-phase flow. *Mathematics and computers in simulation, 23*(3), 267–276.
2. Oberkampf, W. L., & Trucano, T. G. (2007). *Verification and validation benchmarks*. 2007: Albuquerque, New Mexico 87185 and Livermore, California 94550.
3. de Vahl Davis, G. (1983). Natural convection of air in a square cavity: a bench mark numerical solution. *International Journal for Numerical Methods in Fluids, 3*(3), 249–264.
4. Le Quéré, P., et al. (2005). Modelling of natural convection flows with large temperature differences: A benchmark problem for low Mach number solvers. Part 1. Reference solutions. *ESAIM: Mathematical Modelling and Numerical Analysis, 39*(3), 609–616.
5. Nicolas, X., et al. (2011). Benchmark solution for a three-dimensional mixed-convection flow, part 1: Reference solutions. *Numerical Heat Transfer, Part B: Fundamentals, 60*(5), 325–345.
6. Nicolas, X., et al. (2011). Benchmark solution for a three-dimensional mixed-convection flow, part 2: Analysis of richardson extrapolation in the presence of a singularity. *Numerical Heat Transfer, Part B: Fundamentals, 60*(5), 346–369.
7. Abanto, J., et al. (2005). Verification of some commercial CFD codes on atypical CFD problems. In *43rd AIAA Aerospace Sciences Meeting and Exhibit*.
8. Roache, P. J. (1998). *Fundamentals of computational fluid dynamics* (Vol. 2). NM: Hermosa Publishers Albuquerque.
9. Kowalewski, T. A. (2011). Validation problems in computational fluid mechanics. *Computer Assisted Mechanics and Engineering Sciences, 18*(1/2), 39–52.
10. Leonardi, E., et al. (1994). Effects of finite wall conductivity on flow structures in natural convection. *International journal of Heat transfer, 1999*(4), 1–6.
11. Timchenko, V. (2012). Eddie Leonardi Memorial Lecture: "Natural Convection From Earth to Space". *Journal of Heat Transfer, 134*(3), 031014-undefined.
12. Leonardi, E., & Reizes, J. (1981). Convective flows in closed cavities with variable fluid properties. *Numerical Methods in Heat Transfer*, 387–412.
13. Leonardi, E. (1984). *A numerical study of the effects of variable properties on natural convection*. The University of New South Wales.
14. Leonardi, E., & Reizes, J. (1979). Numerical convection in compressible fluids with variable properties. In Lewis, R. W., & Morgan, K. (Eds.) *Numerical Methods in Thermal Problems*. Pineridge Press, Swansea, United Kingdom, pp. 297–306
15. Leonardi, E., & Reizes, J. Natural convection in inclined cavities filled with a compressible fluid with variable properties: A numerical study. In *7th Australasian Conference on Hydraulics and Fluid Mechanics 1980: Preprints of Papers*. 1980. Institution of Engineers, Australia.
16. Davis, G. D. V. (1986). *Private Communication*, The University of New South Wales.

17. Reizes, J. A., Leonardi, E., & Davis, G. D. V. (1985). Natural convection near the density extremum of water. In *Fourth International Conference on Numerical Methods in Laminar and Turbulent Flow*. Pineridge Press, Swansea, UK.

18. Morrison, G. L., & Tran, V. Q. (1978). Laminar flow structure in vertical free convective cavities. *International Journal of Heat and Mass Transfer, 21*(2), 203–213.

19. Chenoweth, D. R., & Paolucci, S. (1986). Natural convection in an enclosed vertical air layer with large horizontal temperature differences. *Journal of Fluid Mechanics, 169,* 173–210.

20. Cheesewright, R., King, K. J., & Ziai, S. (1986). Experimental data for the validation of computer codes for the prediction of two-dimensional buoyant cavity flows. In *ASME Winter Annual Meeting, HTD-60, Anaheim.*

21. King, K. J. (1989). *Turbulent natural convection in rectangular air cavities.* Queen Mary College, University of London, London.

22. Lau, G., et al. (2011). Large-eddy simulation of turbulent natural convection in vertical parallel-plate channels. *Numerical Heat Transfer, Part B: Fundamentals, 59*(4), 259–287.

23. Sutherland, W., *LII. The viscosity of gases and molecular force.* The London, Edinburgh, and Dublin Philosophical Magazine and Journal of Science, 1893. **36**(223): p. 507–531

24. Vreman, A.W., *An eddy-viscosity subgrid-scale model for turbulent shear flow: Algebraic theory and applications.* Physics of Fluids (1994-present), 2004. **16**(10): p. 3670–3681

25. Barhaghi, D. G., Davidson, L., & Karlsson, R. (2006). Large-eddy simulation of natural convection boundary layer on a vertical cylinder. *International Journal of Heat and Fluid Flow, 27*(5), 811–820.

26. Park, N., et al. (2006). A dynamic subgrid-scale eddy viscosity model with a global model coefficient. *Physics of Fluids, 18*(12), 125109.

27. You, D. and P. Moin, *A dynamic global-coefficient subgrid-scale model for compressible turbulence in complex geometries.* Center for Turbulence Research Annual Research Briefs, 2008: p. 189–196

28. Lau, G. E., et al. (2012). Application of dynamic global-coefficient subgrid-scale models to turbulent natural convection in an enclosed tall cavity. *Physics of Fluids, 24,* 094105.

29. Timchenko, V., et al. *Is comparison with experimental data a reasonable method of validating computational models?* in *Journal of Physics: Conference Series.* 2016. IOP Publishing

30. Miyamoto, M., et al. (1986). *Turbulent free convection heat transfer from vertical parallel plates.* in *Proceedings of the eighth international heat transfer conference, Volume 4.* 1986

31. Lau, G., et al. (2012). Large-eddy simulation of natural convection in an asymmetrically-heated vertical parallel-plate channel: Assessment of subgrid-scale models. *Computers & Fluids, 59,* 101–116.

32. Lau, G., et al. (2012). Numerical investigation of passive cooling in open vertical channels. *Applied Thermal Engineering, 39,* 121–131.

33. Ménézo, C., M. Fossa, and E. Leonardi. *An experimental investigation of free cooling by natural convection of vertical surfaces for building integrated photovoltaic (bipv) applications.* in *Thermal Issues in Emerging Technologies: Theory and Application, 2007. THETA 2007. International Conference on.* 2007. IEEE

34. Vareilles, J., *Étude des transferts de chaleur dans un canal vertical differentiellement chauffé: application aux enveloppes photovoltaïques thermiques.* 2007

35. Fossa, M., Mènèzo, C., & Leonardi, E. (2008). Experimental natural convection on vertical surfaces for building integrated photovoltaic (BIPV) applications. *Experimental Thermal and Fluid Science, 32*(4), 980–990.

36. Giroux-Julien, S., et al. (2009). Natural convection in a nonuniformly heated channel with application to photovoltaic facades. *Computational Thermal Sciences, 1*(3), 231–258.

37. Sanvicente, E., et al. (2013). Transitional natural convection flow and heat transfer in an open channel. *International Journal of Thermal Sciences, 63,* 87–104.

38. Saadon, S., et al. (2016). Simulation study of a naturally-ventilated building integrated photovoltaic/thermal (BIPV/T) envelope. *Renewable Energy, 87,* 517–531.

39. Lau, G.E., et al., *Numerical and experimental investigation of unsteady natural convection in a vertical open-ended channel.* Computational Thermal Sciences: An International Journal, 2012. **4**(5)

40. Fedorov, A. G., & Viskanta, R. (1997). Turbulent natural convection heat transfer in an asymmetrically heated, vertical parallel-plate channel. *International Journal of Heat and Mass Transfer, 40*(16), 3849–3860.
41. Daverat, C., et al. (2013). Experimental investigation of turbulent natural convection in a vertical water channel with symmetric heating: Flow and heat transfer. *Experimental Thermal and Fluid Science, 44,* 182–193.
42. Thebault, M., *Coherent structures and impact of the external thermal stratification in a transitional natural convection vertical channel.* 2018, University of Lyon and UNSW Sydney
43. Brady, P., *An investigation of free surface hydraulic structures using large eddy simulation and computational fluid dynamics.* 2011
44. Katoh, Y., et al., *Turbulent free convection heat transfer from vertical parallel plates: effect of entrance bell-mouth shape.* JSME international journal. Ser. 2, Fluids engineering, heat transfer, power, combustion, thermophysical properties, 1991. **34**(4): p. 496–501
45. Thebault, M., et al. (2018). Impact of external temperature distribution on the convective mass flow rate in a vertical channel–A theoretical and experimental study. *International Journal of Heat and Mass Transfer, 121,* 1264–1272.
46. Gaillard, L., et al. (2014). Experimental study of thermal response of PV modules integrated into naturally-ventilated double skin facades. *Energy Procedia, 48,* 1254–1261.

Integral Transform Benchmarks of Diffusion, Convection–Diffusion, and Conjugated Problems in Complex Domains

Renato M. Cotta, Diego C. Knupp, João N. N. Quaresma, Kleber M. Lisboa, Carolina P. Naveira-Cotta, José Luiz Z. Zotin and Helder K. Miyagawa

1 Introduction

Integral transforms have been recognized [1] to be introduced by Leonhard Euler (1707–1783) for the solution of second-order linear differential equations, as presented in his acclaimed compendium *Institutiones Calculi Integralis* [2]. In general terms, his proposition introduces the integral transformation of a function $f(x)$ in the form [1]:

$$F(n) = \int_{x_0}^{x_1} K(x, n) f(x) dx \tag{1}$$

R. M. Cotta (✉) · K. M. Lisboa · C. P. Naveira-Cotta
Mechanical Engineering Department POLI & COPPE, CT, Federal University of Rio de Janeiro, UFRJ, Rio de Janeiro, Brazil
e-mail: cotta@abc.org.br; cotta@mecanica.coppe.ufrj.br

R. M. Cotta
General Directorate of Nuclear and Technological Development, DGDNTM, Brazilian Navy, Rio de Janeiro, RJ, Brazil

D. C. Knupp
Mechanical Engineering Department, Instituto Politécnico, IPRJ/UERJ, State University of Rio de Janeiro, Nova Friburgo, Brazil

J. N. N. Quaresma · H. K. Miyagawa
Chemical Engineering Department and Graduate Program, Natural Resource Engineering in the Amazon, Institute of Technology, University of Pará, UFPA, Belém, PA, Brazil

J. L. Z. Zotin
Centro Federal de Educação Tecnológica Celso Suckow da Fonseca, CEFET/RJ, Itaguaí, RJ, Brazil

© Springer Nature Singapore Pte Ltd. 2020
A. Runchal (ed.), *50 Years of CFD in Engineering Sciences*,
https://doi.org/10.1007/978-981-15-2670-1_20

with the aid of a transformation kernel $K(x, n)$, yielding a transformed dependent variable $F(n)$ that involves a parameter n, which should in principle be a posteriori obtainable from a simpler transformed algebraic (or differential) problem, upon integral transformation of the original ordinary (or partial) differential system. Then, an inverse transformation is required to recover the original function $f(x)$ from the transformed one, $F(n)$. If additional independent variables are considered in Eq. (1), multiple integral transformations, with respect to each variable, can be adopted to reduce the original differential system by eliminating these variables in the resulting integral transformed system. The choices of kernels and integration bounds in Eq. (1), and consequently of the required inverse transformation, have been historically identified with the names of their particular proposers, such as the best-known Laplace and Fourier transforms.

The finite Fourier transform, of major relevance in the present context, can be said to have been introduced in a series of papers by Joseph Baptiste Fourier (1768–1830) and consolidated in his famous heat conduction treatise of 1822 [3]. It can be interpreted that from the application of separation of variables to the linear transient heat conduction equation in Cartesian coordinates, Fourier reached the eigenvalue problem, later on generalized as the so-called Sturm–Liouville problem, that naturally provided the transformation kernel and inverse formulae in terms of the associated eigenfunctions, essentially sines and cosines in the particular case of heat conduction in a slab. A number of contributions followed that dealt with the proposition of integral transforms of linear partial differential equations in different coordinates systems, such as the Hankel and Legendre transforms, respectively, for cylindrical and spherical geometries. In the popular book by Academician Nikolai Sergeevich Koshlyakov (1891–1958), published with the translated title of "Basic Differential Equations of Mathematical Physics" [4], the finite integral transform technique was formalized to handle nonhomogeneous problems, already in terms of eigenfunction expansions obtained from the generalized Sturm–Liouville problem. A number of classical works have then followed, such as in [5–17], expanding and consolidating this knowledge from the mathematical point of view and providing exact analytical solutions to various physical applications, notably in heat and mass diffusion, as systematically illustrated for seven different classes of problems [17].

Despite the usefulness of exact analytical solutions for such linear diffusion problems, the limitation of the classical integral transform approach was soon recognized when dealing with a priori non-transformable problems, such as in the case of time-dependent equation and/or boundary conditions coefficients [18, 19], yielding coupled infinite transformed ordinary differential systems with variable coefficients, of unknown analytical treatment. Approximate analytical solutions were yet made available in [18, 19] by considering only the diagonal of the coupled transformed system, until a hybrid numerical–analytical framework was introduced in [20] for the solution of a moving boundary mass diffusion problem, combining the integral transform method with the error-controlled numerical solution of ordinary differential systems. This hybrid approach was coined as the Generalized Integral Transform Technique (GITT) [20], following the terminology previously proposed for the approximate solution of non-transformable linear problems [18, 19]. In a natural development sequence, the GITT was then gradually extended to handle different classes of problems, including the first considered class of non-transformable linear problems with time-dependent coefficients [21]. It would not take long until this methodology was challenged to handle irregular geometries [22] and nonlinear formulations [23], followed by the solution of the boundary layer and Navier–Stokes formulations of heat and fluid flow problems [24, 25]. The various applications and extensions that were then pursued through the GITT, led to the first compilation of such developments as a monograph [26] and to a couple of invited articles [27, 28], that allowed for a broader dissemination of the methodology among the thermal sciences community.

During the participation as a keynote lecturer at the International Heat Transfer Conference in Brighton, UK [27], the first author had the unique opportunity of being properly introduced to Prof. Brian Spalding, who then kindly commented on the GITT development and provided suggestions for future work. The possibility of obtaining independent reference benchmark results in different classes of problems was one of the positive aspects, as pointed out by Prof. Spalding, to be further pursued, and he also referred us to some of the test cases that were then being tackled by the CHAM development team, such as the IAHR diverging channel test case [29], that would have numerical results compiled in electronic format through Phoenics version 2.2.1 in 1996. The handling by GITT of the suggested test case dealing with the Navier–Stokes formulation in an irregular domain was then pursued and first provided in [30], and later on presented as a dedicated article [31].

Since then, the GITT hybrid approach has been further extended and widely applied in different physical contexts, as compiled in various sources up to recent reviews [32–39]. The present chapter provides a general description of the methodology, that is, here focused on the solution of diffusion, convection–diffusion, and conjugated problems in irregular domains and/or heterogeneous media. In terms of formalism, first general solutions are provided for nonlinear diffusion problems, either considering linear or nonlinear eigenvalue problems. Second, convection–diffusion problems are discussed, either as a direct application of the previous formal solution of diffusion problems in the total transformation scheme or by skipping the integral transformation along the coordinate of predominant convective effects, through the so-called partial transformation scheme. Alternatively, the total transformation of convection–diffusion problems adopting convective eigenvalue problems is also described. Third, the treatment of eigenvalue problems by GITT is briefly reviewed, including the direct integral transformation for irregular domains, followed by the description of a single domain reformulation strategy, which markedly facilitates the handling of both heterogeneous domains and irregular regions, including the class of conjugated heat transfer problems here emphasized. Finally, selected test cases for laminar flow and convection in corrugated channels and conjugated heat transfer are described. The GITT approach results are examined in terms of convergence rates and critically compared to results from general-purpose numerical computer codes. The chapter is concluded with a discussion on the progress achieved within the last fifty years and on the next steps toward the full establishment of this computational–analytical (CAFD) approach in heat and fluid flow.

2 Diffusion Problems

Let us consider a fairly general nonlinear transient diffusion problem for the potential $T(\mathbf{x}, t)$, defined in the arbitrary region V, with all equation and boundary conditions coefficients written as functions of the independent and dependent variables, including the respective nonlinear source terms, $P(\mathbf{x}, t, T)$ and $\phi(\mathbf{x}, t, T)$, given as

$$w(\mathbf{x}, t, T)\frac{\partial T(\mathbf{x}, t)}{\partial t} = \nabla.k(\mathbf{x}, t, T)\nabla T - d(\mathbf{x}, t, T)T + P(\mathbf{x}, t, T), \quad \text{in } \mathbf{x} \in V, \ t > 0 \qquad (2a)$$

with initial and boundary conditions

$$T(\mathbf{x}, 0) = f(\mathbf{x}), \quad \mathbf{x} \in V \qquad (2b)$$

$$\alpha(\mathbf{x}, t, T)T(\mathbf{x}, t) + \beta(\mathbf{x}, t, T)k(\mathbf{x}, t, T)\frac{\partial T}{\partial \mathbf{n}} = \phi(\mathbf{x}, t, T), \quad \mathbf{x} \in S, \ t > 0 \quad (2c)$$

where α and β are the nonlinear boundary condition coefficients that allow for recovering the three most usual kinds of boundary conditions, and $\mathbf{n}$ is the outward-drawn normal vector to surface S. Although more general situations could be considered, such as accounting for elliptic- and hyperbolic-type formulations, through appropriate t-operators, and including coupled multiple potentials [38], Eq. (2a) provides enough information to fully illustrate the methodology.

The most usual formal integral transform solution of problem (2b) involves the selection of a linear eigenvalue problem, which offers the basis for the eigenfunction expansion that represents the potential, as introduced in [23]. This is in fact equivalent to rewriting problem (2c) with characteristic linear coefficients that have only $\mathbf{x}$ dependence, i.e. $w(\mathbf{x})$, $k(\mathbf{x})$, $d(\mathbf{x})$, $\alpha(\mathbf{x})$, and $\beta(\mathbf{x})$, while the nonlinear source terms then incorporate the remaining nonlinear portions of the equation and boundary conditions operators. This more traditional approach is thoroughly documented in previous reviews [32–39], and therefore is not repeated here. Instead, a more general formalism is presented, as introduced in [40], which adopts a nonlinear eigenvalue problem, with all the nonlinear coefficients present in the formulation. Thus, consider the following nonlinear eigenvalue problem:

$$\nabla.k(\mathbf{x}, t, T)\nabla\psi_i(\mathbf{x}; t) + \left[\mu_i^2(t)w(\mathbf{x}, t, T) - d(\mathbf{x}, t, T)\right]\psi_i(\mathbf{x}; t) = 0, \quad \mathbf{x} \in V \tag{3a}$$

with boundary conditions

$$\alpha(\mathbf{x}, t, T)\psi_i(\mathbf{x}; t) + \beta(\mathbf{x}, t, T)k(\mathbf{x}, t, T)\frac{\partial\psi_i(\mathbf{x}; t)}{\partial\mathbf{n}} = 0, \quad \mathbf{x} \in S \tag{3b}$$

The solution for the associated time-dependent eigenfunctions, $\psi_i(\mathbf{x}; t)$, and eigenvalues, $\mu_i(t)$, has been presented in [40], as will be discussed in what follows. Thus, the following integral transform pair is defined from problem (3a):

$$\bar{T}_i(t) = \int_V w(\mathbf{x}, t, T)\ \psi_i(\mathbf{x}; t)\ T(\mathbf{x}, t)\ dv, \quad \text{transform} \tag{4a}$$

$$T(\mathbf{x}, t) = \sum_{i=1}^{\infty} \frac{1}{N_i(t)} \psi_i(\mathbf{x}; t)\bar{T}_i(t), \quad \text{inverse} \tag{4b}$$

with the normalization integral given as

$$N_i(t) = \int_V w(\mathbf{x}, t, T)\psi_i^2(\mathbf{x}; t)dv \tag{4c}$$

After application of the integral transformation procedure through the operator $\int_V (-)\psi_i(\mathbf{x}; t)dv$, the resulting ODE system for the transformed potentials, $\bar{T}_i(t)$, is written as

$$\frac{d\bar{T}_i(t)}{dt} + \sum_{j=1}^{\infty} A_{i,j}(t, \bar{\mathbf{T}})\bar{T}_j(t) = \bar{g}_i(t, \bar{\mathbf{T}}), \quad t > 0, i, j = 1, 2 \ldots \tag{5a}$$

with initial conditions

$$\bar{T}_i(0) = \bar{f}_i \tag{5b}$$

where

$$A_{i,j}(t, \bar{\mathbf{T}}) = \delta_{ij}\mu_i^2(t) + A_{i,j}^*(t, \bar{\mathbf{T}}) \tag{5c}$$

$$A_{i,j}^*(t, \bar{\mathbf{T}}) = -\frac{1}{N_j(t)} \int_V \frac{\partial}{\partial t}[w(\mathbf{x}, t, T)\psi_i(\mathbf{x}; t)\ ]\psi_j(\mathbf{x}; t)dv \tag{5d}$$

$$\bar{g}_i(t, \bar{\mathbf{T}}) = \int_V \psi_i(\mathbf{x}; t)\ P(\mathbf{x}, t, T)dv + \int_S \phi(\mathbf{x}, t, T)\left(\frac{\psi_i(\mathbf{x}; t) - k(\mathbf{x}, t, T)\frac{\partial \psi_i}{\partial \mathbf{n}}}{\alpha(\mathbf{x}, t, T) + \beta(\mathbf{x}, t, T)}\right)ds \tag{5e}$$

$$\bar{f}_i = \int_V w(\mathbf{x}, 0, T(\mathbf{x}, 0))\tilde{\psi}_i(\mathbf{x}; 0) f(\mathbf{x})dv \tag{5f}$$

This more general solution path provides a formal solution that encompasses the usual formalism with a linear eigenvalue problem and has been shown to result in improved convergence rates [40, 41]. On the other hand, it has the drawback of requiring that the eigenvalue problem be solved simultaneously with the transformed ODE system, yielding time-dependent transformed potentials, eigenvalues, and eigenfunctions. The GITT solution of the nonlinear eigenvalue problem (3) has been presented in [40], and will be later on reviewed, by considering an auxiliary linear eigenvalue problem of known solution to offer an eigenfunction expansion for the nonlinear eigenfunctions, leading, upon integral transformation, to a nonlinear algebraic eigenvalue problem that needs to be solved simultaneously with the ODE system (5). Alternatively, the algebraic eigenvalue problem can be differentiated with respect to the t variable and solved as a larger coupled ODE system jointly with the transformed potentials, Eq. (5a).

It should be noted that the above formal solution derivation did not account for the employment of a filtering solution, either explicit or implicit, so as to reduce the importance of the source terms in the convergence behavior, which essentially leads to the same system (2) but with redefined source terms and initial conditions. Also, only the total transformation scheme of the GITT has been so far described, when all space variables in the position vector $\mathbf{x}$ are eliminated through integral transformation. Alternatively, one may also consider the partial transformation scheme [42, 43], when one of the spatial variables is left out of the transformation process, thus leading to a transformed partial differential system with only time and one spatial coordinate as independent variables, as will be briefly described in the next section for convection–diffusion problems.

3   Convection–Diffusion Problems

Now consider an also fairly general convection–diffusion problem, which is essentially the nonlinear formulation of Eq. (2a) plus a nonlinear convective term, defined for a nonlinear velocity vector $\mathbf{u}(\mathbf{x}, t, T)$:

$$w(\mathbf{x}, t, T)\frac{\partial T(\mathbf{x}, t)}{\partial t} + \mathbf{u}(\mathbf{x}, t, T).\nabla T = \nabla.k(\mathbf{x}, t, T)\nabla T - d(\mathbf{x}, t, T)T + P(\mathbf{x}, t, T), \quad \text{in } \mathbf{x} \in V, t > 0 \tag{6}$$

with similar initial and boundary conditions as in Eq. (2b, 2c).

The most usual formalism in dealing with the integral transformation of Eq. (6), similarly to the above diffusion problem (2), is to consider a linear diffusive eigenvalue problem with space dependent coefficients only, while incorporating the above nonlinear convection term and the remaining nonlinear terms, into the nonlinear source term, as introduced in [44] and widely employed throughout the development of the GITT approach. In addition, one may again merge the convection term, $\mathbf{u}(\mathbf{x}, t, T).\nabla T$, into the nonlinear equation source term, $P(\mathbf{x}, t, T)$, in Eq. (6), while keeping the remaining nonlinear coefficients in the equation to be transformed, thus being accounted for by the nonlinear diffusive eigenvalue problem. This derivation is not repeated here, since it is essentially a direct application of the methodology in Sect. 2. However, two alternative solution paths for convection–diffusion problems are here described, that have also been successfully employed in different classes of applications.

3.1 Partial Transformation

In the treatment of transient convection–diffusion problems with a preferential convective direction, one possible approach with relative merits is to consider the integral transformation in all but this one space coordinate, yielding an infinite coupled system of partial differential equations for the transformed potentials, to be solved numerically. This partial integral transformation scheme offers an interesting combination of advantages between the eigenfunction expansion approach and the selected numerical method for handling the coupled system of one-dimensional partial differential equations that results from the transformation procedure. To illustrate this procedure, a transient convection–diffusion problem is considered, separating the preferential direction that is not to be integral transformed. The vector $\mathbf{x} = \{x_1, x_2, x_3\}$ is then formed by the space coordinates that will be eliminated through integral transformation, here denoted by $\mathbf{x}^* = \{x_1, x_2\}$, as well as by the space variable to be retained in the transformed partial differential system, here denoted by x_3. In addition, a linear eigenvalue problem is here preferred, by selecting characteristic $\mathbf{x}^*$ dependent coefficients and incorporating the remaining terms in the source terms, including the nonlinear convection term and all the remaining nonlinear terms. The problem to be solved is now written in the following form:

$$w(\mathbf{x}^*)\frac{\partial T(\mathbf{x}, t)}{\partial t} = \nabla^* \cdot \left(k(\mathbf{x}^*)\nabla^* T(\mathbf{x}, t)\right) - d(\mathbf{x}^*)T(\mathbf{x}, t) + P(\mathbf{x}^*, x_3, t, T), \quad \mathbf{x} \in V, t > 0 \quad \text{(7a)}$$

where the operator ∇^* refers only to the coordinates to be integral transformed, $\mathbf{x}^*$, and with initial and boundary conditions given, respectively, by

$$T(\mathbf{x}, 0) = f(\mathbf{x}), \; \mathbf{x} \in V; \; \left[\alpha(\mathbf{x}^*) + \beta(\mathbf{x}^*)k(\mathbf{x}^*)\frac{\partial}{\partial \mathbf{n}^*}\right]T(\mathbf{x}, t) = \phi(\mathbf{x}^*, x_3, t, T), \quad \mathbf{x}^* \in S^*, t > 0$$

$$\text{(7b,c)}$$

$$\left[\lambda(x_3) + (-1)^{l+1}\gamma(x_3)\frac{\partial}{\partial x_3}\right]T(\mathbf{x}, t) = \varphi(\mathbf{x}^*, x_3, t, T), x_3 \in S_3 = \{x_{3,l}\}, \quad l = 0, 1, t > 0 \quad \text{(7.d)}$$

where $\mathbf{n}^*$ denotes the outward-drawn normal to the surface S^* formed by the coordinates $\mathbf{x}^*$ and S_3 refers to the boundary values of the coordinate x_3. The coefficients in Eq. (7a) inherently carry the information to the auxiliary eigenvalue problem that will be chosen for the eigenfunction expansion, and all the remaining terms from this rearrangement are collected into the source terms, $P(\mathbf{x}^*, x_3, t, T)$ and $\phi(\mathbf{x}^*, x_3, t, T)$, including the existing nonlinear terms and diffusion and/or convection terms with respect to the dimensional variable x_3. By performing the integral transformation of Eq. (7a) with respect to the selected space coordinates $\mathbf{x}^* = \{x_1, x_2\}$, one obtains the following transformed system dependent on the remaining variables t and x_3:

$$\frac{\partial \bar{T}_i(x_3, t)}{\partial t} + \mu_i^2 \bar{T}_i(x_3, t) = \bar{g}_i(x_3, t, \overline{\mathbf{T}}), i = 1, 2, \ldots, \quad x_3 \in V_3, t > 0, \qquad (8a)$$

$$\bar{g}_i(x_3, t, \overline{\mathbf{T}}) = \int_{V^*} \psi_i(\mathbf{x}^*) P(\mathbf{x}^*, x_3, t, \overline{\mathbf{T}}) dv^* + \int_{S^*} \phi(\mathbf{x}^*, x_3, t, T) \left[\frac{\psi_i(\mathbf{x}^*) - k(\mathbf{x}^*) \frac{\partial \psi_i(\mathbf{x}^*)}{\partial \mathbf{n}^*}}{\alpha(\mathbf{x}^*) + \beta(\mathbf{x}^*)} \right] ds^*$$

$$(8b)$$

$$\bar{T}_i(x_3, 0) = \bar{f}_i(x_3) \equiv \int_{V^*} w(\mathbf{x}^*) \psi_i(\mathbf{x}^*) f(\mathbf{x}) dv^* \qquad (8c)$$

$$\left[\lambda(x_3) + (-1)^{l+1} \gamma(x_3) \frac{\partial}{\partial x_3} \right] \bar{T}_i(x_3, t) = \overline{\varphi}_i^*(x_3, t, \overline{\mathbf{T}}) \equiv \int_{v^*} w(\mathbf{x}^*) \psi_i(\mathbf{x}^*) \varphi(\mathbf{x}^*, x_3, t, \overline{\mathbf{T}}) dv^*,$$

$$x_3 \in S_3 = \{x_{3,l}\}, l = 0, 1, t > 0 \qquad (8d,e)$$

Equation (8a–e) form an infinite coupled system of nonlinear partial differential equations for the transformed potentials, $\bar{T}_i(x_3, t)$. After truncation to a sufficiently large finite order, this PDE system can be numerically solved. For instance, the *Mathematica* system provides the routine NDSolve, which implements the Method of Lines in the numerical solution of this problem, under automatic absolute and relative error control.

3.2   Convective Eigenvalue Problems

Quite recently, an alternative solution was proposed adopting a convective eigenvalue problem, again either linear or nonlinear, that through a coefficient transformation could allow to rewrite Eq. (6) as a generalized diffusion problem [45]. Consider that the convective term coefficient vector $\mathbf{u}$ can be represented in the three-dimensional situation by the three components $\{u_x, u_y, u_z\}$, here illustrating the transformation in the Cartesian coordinates system, $\mathbf{x} = \{x, y, z\}$. Then, Eq. (6) can be rewritten in the generalized diffusive form as

$$w^*(\mathbf{x}, t, T) \frac{\partial T(\mathbf{x}, t)}{\partial t} = \frac{1}{\hat{k}_x(\mathbf{x}, t, T)} \frac{\partial}{\partial x} [\hat{k}_x(\mathbf{x}, t, T) \frac{\partial T(\mathbf{x}, t)}{\partial x}]$$

$$+ \frac{1}{\hat{k}_y(\mathbf{x}, t, T)} \frac{\partial}{\partial y} [\hat{k}_y(\mathbf{x}, t, T) \frac{\partial T(\mathbf{x}, t)}{\partial y}]$$

$$+ \frac{1}{\hat{k}_z(\mathbf{x}, t, T)} \frac{\partial}{\partial z} [\hat{k}_z(\mathbf{x}, t, T) \frac{\partial T(\mathbf{x}, t)}{\partial z}]$$

$$- d^*(\mathbf{x}, t, T) T(\mathbf{x}, t)$$

$$+ P^*(\mathbf{x}, t, T), \quad \mathbf{x} \in V, \quad t > 0 \qquad (9a)$$

where

$$w^*(\mathbf{x}, t, T) = w(\mathbf{x}, t, T)/k(\mathbf{x}, t, T); \quad d^*(\mathbf{x}, t, T) = d(\mathbf{x}, t, T)/k(\mathbf{x}, t, T);$$

$$P^*(\mathbf{x}, t, T) = P(\mathbf{x}, t, T)/k(\mathbf{x}, t, T); \quad \mathbf{u}^*(\mathbf{x}, t, T) = \frac{1}{k(\mathbf{x}, t, T)}[\mathbf{u}(\mathbf{x}, t, T) - \nabla k(\mathbf{x}, t, T)];$$

$$\hat{k}(\mathbf{x}, t, T) = \hat{k}_x(\mathbf{x}, t, T)\hat{k}_y(\mathbf{x}, t, T)\hat{k}_z(\mathbf{x}, t, T)$$

$$\hat{k}_x(\mathbf{x}, t, T) = e^{-\int u_x^*(\mathbf{x}, t, T)dx}; \quad \hat{k}_y(\mathbf{x}, t, T) = e^{-\int u_y^*(\mathbf{x}, t, T)dy}; \quad \hat{k}_z(\mathbf{x}, t, T) = e^{-\int u_z^*(\mathbf{x}, t, T)dz}$$

$$(9b\text{-}i)$$

In the special simpler case when the transformed diffusion coefficients are functions of only the corresponding space coordinate, or $\hat{k}_x(\mathbf{x}, t, T) = \hat{k}_x(x)$, $\hat{k}_y(\mathbf{x}, t, T) = \hat{k}_y(y)$; $\hat{k}_z(\mathbf{x}, t, T) = \hat{k}_z(z)$, with the consequent restrictions on the related coefficients k and $\mathbf{u}$, a diffusion formulation is constructed which leads to a self-adjoint eigenvalue problem [45]. Alternatively, one may seek an adjoint eigenvalue problem that allows for the construction of a biorthogonal eigenfunctions set. This convective eigenvalue problem solution path was recently implemented in the analysis of conjugated heat transfer problems, also with significant convergence rates improvement [46].

4   Vector Eigenfunction Expansion

Although flow problems governed either by the boundary layer or full Navier–Stokes equations formulations can be cast into the general form of Eq. (6), with corresponding initial and boundary conditions, their GITT treatment deserves some special considerations that are here briefly reviewed, while the most recent developments are pointed out.

As mentioned before, the first GITT solution of the boundary layer equations was proposed in [24], in the primitive variables formulation; while the Navier–Stokes equations were first solved by GITT in [25], but preferring instead the streamfunction-only formulation, which eliminates the pressure field and automatically satisfies the continuity equation, while introducing a fourth-order differential eigenvalue problem. A number of contributions then followed extending the applicability of the GITT to different classes of flow problems, also considering the GITT solution for the Navier–Stokes equations in the primitive variables formulation [47], including, for instance, transient problems, compressible flow, three-dimensional formulations, variable physical properties, non-Newtonian fluids, porous and partially porous media, irregular regions, MHD flows, among others, as reviewed in [39].

Recently, a unified framework was proposed, based on a vector eigenfunction expansion [48], which includes the streamfunction formulation treatment as a special case, while generalizing the GITT in dealing with heterogeneous media and three-dimensional flow problems. The vector eigenfunction expansion represents all velocity components with one set of transformed potentials and an appropriately chosen vector eigenfunction basis, while the velocity vector field can be interpreted as the result of the influence of an infinite number of vortices disturbing a base flow. Consider the transient Navier–Stokes equations for incompressible flow, in dimensionless vector form:

$$\nabla \cdot \mathbf{u} = 0, \quad \mathbf{x} \in V \tag{10a}$$

$$\frac{\partial \mathbf{u}}{\partial t} + \nabla \cdot (\mathbf{u} \otimes \mathbf{u}) = -\nabla p + \frac{1}{\mathrm{Re}} \nabla^2 \mathbf{u} + \mathbf{b}, \quad \mathbf{x} \in V \tag{10b}$$

where V represents the domain occupied by the Newtonian fluid, $\mathbf{u}$ is the dimensionless velocity vector, p is the dimensionless pressure field, Re is the Reynolds number, $\mathbf{b}$ is a volumetric source term.

The first step in the GITT solution, as usual, is the proposition of a filtering solution to reduce the importance of source terms, specially to homogenize the boundary conditions, in the form:

$$\mathbf{u}(\mathbf{x}, t) = \hat{\mathbf{u}}(\mathbf{x}, t) + \mathbf{u}_{\mathbf{f}}(\mathbf{x}; t) \tag{11}$$

The vector eigenfunction expansion for the filtered velocity field is then proposed as

$$\hat{\mathbf{u}}(\mathbf{x}, t) = \sum_{i=1}^{\infty} \bar{u}_i(t) \left(\nabla \times \tilde{\boldsymbol{\Phi}}_i \right) \tag{12}$$

where Eq. (12) warrants mass conservation, as in the streamfunction-only formulation, dropping the need to further deal with Eq. (10a). A self-adjoint fourth-order vector eigenvalue problem, extracted from the analytical solution for the limiting linear situation of Re $\rightarrow$ 0 (Stoke's flow), is given by [49]:

$$\nabla^2 \left(\nabla \times \nabla \times \tilde{\boldsymbol{\Phi}}_i \right) + \lambda_i^2 \left(\nabla \times \nabla \times \tilde{\boldsymbol{\Phi}}_i \right) = \mathbf{0} \tag{13}$$

The orthogonality property of the eigenfunction from Eq. (13) allows for the proposition of a transformed velocity, in the form:

$$\bar{u}_i(t) = \int_V \left(\nabla \times \tilde{\boldsymbol{\Phi}}_i \right) \cdot \hat{\mathbf{u}}(\mathbf{x}, t) dv \tag{14}$$

Equations (12) and (14) thus provide the inverse-transform pair required for the integral transformation process, following the same formalism as in the usual application of the GITT above described. The integral transformation of the Navier–Stokes equations, with the curl of the solution of Eq. (13) as kernel, then proceeds, leading to the transformed problem below, as detailed in [49]:

$$\frac{d\bar{u}_i}{dt} + \frac{\lambda_i^2}{\text{Re}}\bar{u}_i(t) + \sum_{k=1}^{\infty}\sum_{j=1}^{\infty} A_{ijk}\bar{u}_j(t)\bar{u}_k(t) + \sum_{j=1}^{\infty} B_{ij}(t)\bar{u}_j(t) = \bar{g}_j(t) \qquad (15a)$$

with integral coefficients given by

$$A_{ijk} = -\int_V \nabla\left(\nabla\times\tilde{\mathbf{\Phi}}_i\right)\cdot\left[\left(\nabla\times\tilde{\mathbf{\Phi}}_j\right)\otimes\left(\nabla\times\tilde{\mathbf{\Phi}}_k\right)\right]dv \qquad (15b)$$

$$B_{ij}(t) = -\int_V \nabla\left(\nabla\times\tilde{\mathbf{\Phi}}_i\right)\cdot\left[\left(\nabla\times\tilde{\mathbf{\Phi}}_j\right)\otimes\mathbf{u_f} + \mathbf{u_f}\otimes\left(\nabla\times\tilde{\mathbf{\Phi}}_j\right)\right]dv \qquad (15c)$$

$$\bar{g}_i(t) = \int_V \left(\nabla\times\tilde{\mathbf{\Phi}}_i\right)\cdot\left\{\mathbf{b} + \frac{1}{\text{Re}}\nabla^2\mathbf{u_f} - \frac{\partial\mathbf{u_f}}{\partial t} - \nabla\cdot(\mathbf{u_f}\otimes\mathbf{u_f})\right\} \qquad (15d)$$

It is noteworthy that the automatic elimination of the pressure gradient term from the transformed problem is achieved through the use of the proper integral transform kernel, thus dropping the need to directly deal with the pressure term. Notwithstanding, the pressure field can be determined a posteriori from the original Navier–Stokes equations, once the velocity field is known.

5 Eigenvalue Problems and Irregular Domains

As seen in the previous sections, the accurate solution of the associated eigenvalue problems is a crucial step in the application of the GITT approach. Except for those simpler cases in which an exact analytical solution is available for the Sturm–Liouville problem, it is necessary to implement a more general and automatic procedure for its computational–analytical solution. The GITT itself can be used for this purpose, including the treatment of nonlinear eigenvalue problems and irregular domains, as now reviewed. Thus, consider the nonlinear eigenvalue problem defined in region V and boundary surface S:

$$L\psi(\mathbf{x}; t) = \mu^2(t)w(\mathbf{x}, t, T)\psi(\mathbf{x}; t), \quad \mathbf{x} \in V \qquad (16a)$$

$$B\psi(\mathbf{x}; t) = 0, \quad \mathbf{x} \in S \qquad (16b)$$

where the operators L and B are given by

$$L = -\nabla\cdot(k(\mathbf{x}, t, T)\nabla) + d(\mathbf{x}, t, T) \qquad (16c)$$

$$B = \alpha(\mathbf{x}, t, T) + \beta(\mathbf{x}, t, T)k(\mathbf{x}, t, T)\frac{\partial}{\partial\mathbf{n}} \qquad (16d)$$

The problem given by Eq. (16a–d) can be rewritten as

$$\hat{L}\psi(\mathbf{x}; t) = \left(\hat{L} - L\right)\psi(\mathbf{x}; t) + \mu^2(t)w(\mathbf{x}, t, T)\psi(\mathbf{x}; t), \quad \mathbf{x} \in V \tag{17a}$$

$$\hat{B}\psi(\mathbf{x}; t) = (\hat{B} - B)\psi(\mathbf{x}; t), \quad \mathbf{x} \in S \tag{17b}$$

where $\hat{L}$ and $\hat{B}$ are simpler operators with linear coefficients that define an auxiliary eigenvalue problem of known solution for the eigenvalues, λ, and corresponding eigenfunctions, $\Omega(\mathbf{x})$, given by

$$\hat{L}\Omega(\mathbf{x}) = \lambda^2 \hat{w}(\mathbf{x})\Omega(\mathbf{x}), \quad \mathbf{x} \in V \tag{18a}$$

$$\hat{B}\Omega(\mathbf{x}) = 0, \quad \mathbf{x} \in S \tag{18b}$$

where

$$\hat{L} = -\nabla \cdot \left(\hat{k}(\mathbf{x})\nabla\right) + \hat{d}(\mathbf{x}) \tag{18c}$$

$$\hat{B} = \hat{\alpha}(\mathbf{x}) + \hat{\beta}(\mathbf{x})\hat{k}(\mathbf{x})\frac{\partial}{\partial \mathbf{n}} \tag{18d}$$

Problem (18) thus allows definition of the following integral transform pair:

$$\bar{\psi}_i(t) = \int_V \hat{w}(\mathbf{x})\tilde{\Omega}_i(\mathbf{x})\psi(\mathbf{x}; t)dv, \quad \text{transform} \tag{19a}$$

$$\psi(\mathbf{x}; t) = \sum_{i=1}^{\infty} \tilde{\Omega}_i(\mathbf{x})\bar{\psi}_i(t), \quad \text{inverse} \tag{19b}$$

where the normalized auxiliary eigenfunctions and corresponding norms are given by

$$\tilde{\Omega}_i(\mathbf{x}) = \frac{\Omega_i(\mathbf{x})}{\sqrt{N_{\Omega_i}}}, \text{ with } N_{\Omega_i} = \int_V \hat{w}(\mathbf{x})\Omega_i^2(\mathbf{x})dv \tag{19c, d}$$

Problem (17) is now operated on with $\int_V \tilde{\Omega}_i(\mathbf{x})(\cdot)dv$, to yield the transformed nonlinear algebraic system, truncated to the Mth order, in matrix form, as

$$(\mathbf{A}(t) + \mathbf{C})\{\bar{\%}(t)\} = \mu^2(t)\mathbf{B}(t)\{\bar{\%}(t)\} \tag{20a}$$

with the elements of the $M \times M$ matrices and vector $\mu(t)$ given by

$$a_{ij}(t) = -\int_S \gamma_i(\hat{B} - B)\tilde{\Omega}_j(\mathbf{x})ds - \int_V \tilde{\Omega}_i(\mathbf{x})(\hat{L} - L)\tilde{\Omega}_j(\mathbf{x})dv \qquad (20b)$$

$$c_{ij} = \lambda_i^2 \delta_{ij} \qquad (20c)$$

$$b_{ij}(t) = \int_V w(\mathbf{x}, t, T)\tilde{\Omega}_i(\mathbf{x})\tilde{\Omega}_j(\mathbf{x})dv \qquad (20d)$$

$$\boldsymbol{\mu}(t) = \{\mu_1(t), \mu_2(t), \dots, \mu_M(t)\} \qquad (20e)$$

$$\gamma_i = \frac{\tilde{\Omega}_i(\mathbf{x}) - \hat{k}(\mathbf{x})\frac{\partial \tilde{\Omega}_i(\mathbf{x})}{\partial \mathbf{n}}}{\hat{\alpha}(\mathbf{x}) + \hat{\beta}(\mathbf{x})} \qquad (20f)$$

The nonlinear algebraic eigenvalue problem, Eq. (20a–f), should now be solved simultaneously with the transformed system, Eq. (5a–f), or the equivalent for convection–diffusion problems, yielding the time evolution of the transformed potentials, eigenvalues and eigenfunctions. For linear eigenvalue problems, system (18) is solved only once, prior to the numerical solution of the transformed system (5).

The expressions here derived are valid for any arbitrary region V and corresponding surface S, and essentially require the evaluation of the volume and surface integrals defined in the transformed coefficients and source terms of the transformed ODE system, Eq. (5d–f), and in the coefficients of the transformed eigenvalue problem, Eq. (20b, d). In the more general situation of complex geometric configurations, domain decomposition techniques can be handy in the automatic computational evaluation of such integral transformations. However, in a fairly wide class of problems for which the domain bounding surfaces, in one coordinate, can successively be expressed as functions of the remaining space variables, the volume integral can be organized so as to permit a direct integration of the irregular region [50, 51], including the generalization to nonlinear moving boundaries, i.e. $V(t)$ and $S(t)$ [52]. For instance, considering such a region in the Cartesian coordinates system, the bounding surfaces in each spatial coordinate, may be written as

$$x_0(t) \le x \le x_1(t), \quad y_0(x, t) \le y \le y_1(x, t), \quad z_0(x, y, t) \le z \le z_1(x, y, t) \tag{21a}$$

Then, the auxiliary eigenfunction (or directly the original eigenfunction or the potential), can be expressed as an eigenfunction expansion of the product of one-dimensional eigenfunctions in each coordinate, as

$$\tilde{\Omega}_n(\mathbf{x}; t) = \tilde{X}_j(x; t)\tilde{Y}_k(y; x, t)\tilde{Z}_m(z; x, y, t) \tag{21b}$$

and the corresponding integral transform pair, in terms of this auxiliary eigenfunction basis, would be

$$\bar{\psi}_n(t) = \int_{V(t)} \hat{w}(\mathbf{x}) \tilde{\Omega}_n(\mathbf{x}; t) \psi(\mathbf{x}; t) dv, \quad \text{transform} \tag{21c}$$

$$\psi(\mathbf{x}; t) = \sum_{n=1}^{\infty} \tilde{\Omega}_n(\mathbf{x}; t) \bar{\psi}_n(t), \quad \text{inverse} \tag{21d}$$

while the volume integrals are undertaken in the appropriate sequence as

$$\int_{V(t)} (-) \tilde{\Omega}_n(\mathbf{x}; t) dv \equiv \int_{x_0(t)}^{x_1(t)} \{ \int_{y_0(x,t)}^{y_1(x,t)} | \int_{z_0(x,y,t)}^{z_1(x,y,t)} (-) \tilde{Z}_m(z; x, y, t) dz | \bar{Y}_k(y; x, t) dy \} \bar{X}_j(x; t) dx \tag{21e}$$

The special case of boundary surfaces mapped as functions of the space coordinates, as discussed above, is particular advantageous in the analytical or semi-analytical evaluation of the transformation integrals [53], and an appropriate choice of the positioning of the coordinates system may allow for this direct integration in many situations. In any case, as discussed above, domain decomposition with numerical or semi-analytical integration provides a more general-purpose algorithm for determination of the transformed coefficients and source terms [42, 54].

6   Single Domain Formulation

In dealing with heterogeneous media, either defined in regular or irregular subregions, the derivation task of the integral transform process can become tedious, especially for multiple regions. Besides, the computational task itself can become cumbersome, since many transformed subregions will lead to a large coupled transformed system to be numerically solved for. Therefore, a single domain formulation strategy was proposed in [55], originally aimed at solving conjugated heat transfer problems when solid and fluid subregions would require separate integral transformations or solving coupled eigenvalue problems. The strategy is based in rewriting the diffusion or convection–diffusion equations for each subdomain, with their respective physical properties and source terms, as one single formulation for the whole region, with spatially variable coefficients and functions that vary abruptly at the interfaces of the subregions, representing the original heterogeneities. This approach was then employed in various classes of conjugated heat transfer problems [56–58], heat conduction applications [59], natural convection in partially porous media [60], and convective mass transfer problems [61].

To illustrate the single domain formulation strategy, consider a nonlinear diffusion problem, defined in a multi-region configuration that is formed by n_V subregions of volumes V_l, $l = 1, 2, \ldots, n_V$, with potential and flux continuity at the interfaces, as illustrated in Fig. 1a, in the form

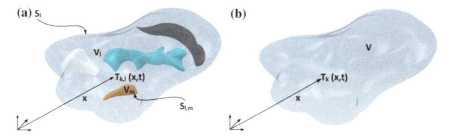

Fig. 1 a Diffusion or convection–diffusion problem in a complex multidimensional configuration with n_V sub-regions; **b** Single domain representation keeping the original overall domain

$$w_l(\mathbf{x}, t, T_l)\frac{\partial T_l(\mathbf{x}, t)}{\partial t} = \nabla \cdot [k_l(\mathbf{x}, t, T_l)\nabla T_l(\mathbf{x}, t)] - d_l(\mathbf{x}, t, T_l)T_l(\mathbf{x}, t)$$

$$+ P_l(\mathbf{x}, t, T_l), \quad \mathbf{x} \in V_l, t > 0, l = 1, 2, \ldots, n_V \qquad (22a)$$

with initial, interface and boundary conditions given, respectively, by

$$T_l(\mathbf{x}, 0) = f_l(\mathbf{x}), \quad \mathbf{x} \in V_l \qquad (22b)$$

$$T_l(\mathbf{x}, t) = T_m(\mathbf{x}, t); \quad k_l(\mathbf{x}, t, T_l)\frac{\partial T_l(\mathbf{x}, t)}{\partial \mathbf{n}} = k_m(\mathbf{x}, t, T_m)\frac{\partial T_m(\mathbf{x}, t)}{\partial \mathbf{n}}, \quad \mathbf{x} \in S_{l,m}, t > 0$$

$$(22c, d)$$

$$\left[\alpha_l(\mathbf{x}, t, T_l) + \beta_l(\mathbf{x}, t, T_l)k_l(\mathbf{x}, t, T_l)\frac{\partial}{\partial \mathbf{n}}\right]T_l(\mathbf{x}, t) = \phi_l(\mathbf{x}, t, T_l), \quad \mathbf{x} \in S_l, t > 0$$

$$(22e)$$

where $\mathbf{n}$ denotes the outward-drawn normal to the interfaces among the different subregions, $S_{l,m}$, and at the external surfaces, S_l.

The idea in the single domain reformulation is, as illustrated in Fig. 1.b, to merge all the Eq. (22a) into one single equation, for the whole region V, and to represent the various properties and source terms in each subregion as one single set of nonlinear space variable coefficients accounting for the abrupt variations at the interfaces. Then, problem (22) is simply rewritten in identical form to problem (1), with the appropriate reformulation of the coefficients, thus requiring only a single integral transformation process, yielding a single transformed ODE system, and carrying the information on the domain heterogeneity to the associated single region eigenvalue problem with spatially variable coefficients.

7 Test Cases

The chosen test cases to illustrate the hybrid GITT methodology are closely related to the suggestions of Prof. Spalding, which motivated the present review. First, laminar flow inside a corrugated microchannel is analyzed [62], followed by the convective heat transfer analysis. This flow problem has been previously analyzed through the GITT methodology [63], but recently it has been observed that the domain singularity at the corrugated duct inlet could introduce some numerical disturbances, which are here corrected for. Also, the heat transfer problem was also previously considered, including upstream and downstream axial diffusion effects, but employing an analytical approximate velocity field representation for very low Reynolds numbers [64]. More recently, the substrate conjugation in the micro-system thermal behavior was also accounted for [37, 65], again by considering a simplified velocity field for low Reynolds number. Here, the full set of Navier–Stokes equations for the fluid flow are solved for and employed in the solution of the corresponding energy equation for the fluid. Second, the analysis of conjugate heat transfer problems is considered to illustrate the single domain reformulation approach in dealing with heterogeneous and irregular regions. A multi-stream perfused substrate configuration with channels of polygonal cross section is then considered and handled through the partial transformation scheme [58]. The adoption of a convective eigenvalue problem in handling conjugated heat transfer is also demonstrated, considering a transient two-dimensional formulation in the total transformation scheme [66].

7.1 Steady Heat Transfer and Fluid Flow in Corrugated Channel

The first test case deals with the steady forced convection heat transfer in a wavy wall channel. The geometrical duct configuration and the physical aspects of this problem are similar to that analyzed in the work of Wang and Chen [62]. Thus, we consider incompressible laminar flow of a Newtonian fluid within a wavy irregular channel in simultaneous hydrodynamic and thermal developments, with fully developed velocity and uniform temperature profiles at the inlet. Viscous dissipation is disregarded and constant physical properties are imposed. The channel walls are maintained at a uniform dimensional temperature T_w^*. Figure 2 shows a schematic representation of the problem and the respective boundary conditions. This problem is governed by the continuity, Navier–Stokes and energy equations in two dimensions, which in terms of the streamfunction-only formulation is written in dimensionless form as

$$\frac{\partial \psi}{\partial y}\left(\frac{\partial^3 \psi}{\partial x^3} + \frac{\partial^3 \psi}{\partial x \partial y^2}\right) - \frac{\partial \psi}{\partial x}\left(\frac{\partial^3 \psi}{\partial x^2 \partial y} + \frac{\partial^3 \psi}{\partial y^3}\right) = \frac{1}{\text{Re}}\left(\frac{\partial^4 \psi}{\partial x^4} + 2\frac{\partial^4 \psi}{\partial x^2 \partial y^2} + \frac{\partial^4 \psi}{\partial y^4}\right) \tag{23a}$$

$$\frac{\partial \psi}{\partial y}\frac{\partial T}{\partial x} - \frac{\partial \psi}{\partial x}\frac{\partial T}{\partial y} = \frac{1}{\text{Re Pr}}\left(\frac{\partial^2 T}{\partial x^2} + \frac{\partial^2 T}{\partial y^2}\right) \tag{23b}$$

subjected to the following inlet, outlet, and boundary conditions:

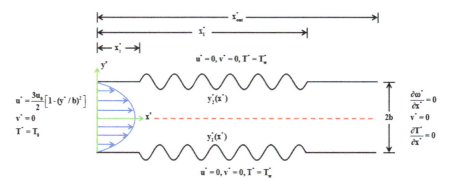

Fig. 2 Geometric configuration and boundary conditions for the problem of forced convection heat transfer within a wavy walls channel

$$\psi(0, y) = k_1 + \frac{3}{2}\left(y - \frac{y^3}{3}\right) + 1; \quad \frac{\partial \psi(0, y)}{\partial x} = 0; \quad T(0, y) = 1 \quad (23\text{c–e})$$

$$\frac{\partial \omega(x_{out}, y)}{\partial x} = \frac{\partial^3 \psi(x_{out}, y)}{\partial x^3} + \frac{\partial^3 \psi(x_{out}, y)}{\partial x \partial y^2} = 0; \quad \frac{\partial \psi(x_{out}, y)}{\partial x} = 0; \quad \frac{\partial T(x_{out}, y)}{\partial x} = 0$$
$$(23\text{f–h})$$

$$\psi(x, -y_1(x)) = k_1; \quad \frac{\partial \psi(x, -y_1(x))}{\partial \mathbf{n}} = 0; \quad T(x, -y_1(x)) = 0 \quad (23\text{i–k})$$

$$\psi(x, y_2(x)) = k_2; \quad \frac{\partial \psi(x, y_2(x))}{\partial \mathbf{n}} = 0; \quad T(x, y_2(x)) = 0 \quad (23\text{l–n})$$

where $\mathbf{n}, k_1$, and k_2 represent the outward-drawn normal vector to the channel wall and the values of the streamfunction at the duct walls, respectively, which are associated through the overall mass balance with the volumetric flow rate per unit length Q, as $k_2 = Q + k_1$. In the streamfunction-only formulation the continuity equation is automatically satisfied, and the velocity components are related to the streamfunction through its definition:

$$u = \frac{\partial \psi}{\partial y}; \quad v = -\frac{\partial \psi}{\partial x} \quad (24\text{a, b})$$

Also, the following dimensionless groups were employed in Eqs. (22a–e) to (23a–n):

$$x = \frac{x^*}{b}; \quad y = \frac{y^*}{b}; \quad y_1(x) = \frac{y_1^*(x^*)}{b}; \quad y_2(x) = \frac{y_2^*(x^*)}{b};$$
$$u = \frac{u^*}{u_0}; \quad v = \frac{v^*}{u_0}; \quad T = \frac{T^* - T_w^*}{T_0^* - T_w^*}; \quad Re = \frac{u_0 b}{\nu}; \quad Pr = \frac{\nu}{\alpha} \quad (25\text{a–h})$$

where b is the half-spacing between the plates at the straight sections, u_0 is the average velocity at the channel inlet, $y_1(x)$ and $y_2(x)$ are the functions that describe the wall contours, Re and Pr are the Reynolds and Prandtl numbers, respectively.

The functions that define the channel walls are here taken as

$$y_2(x) = -y_1(x) = \begin{cases} 1, & \text{for } 0 \le x \le x_s \\ 1 + \alpha \sin[\pi(x - x_s)], & \text{for } x_s \le x \le x_l \\ 1, & \text{for } x_l \le x \le x_{out} \end{cases} \qquad (26)$$

where $\alpha = a/b$ is the dimensionless channel amplitude, and $x_s = x_s^*/b$ and $x_l = x_l^*/b$ are the dimensionless lengths for the beginning and the end of the wavy walls, respectively, and $x_{out} = x_{out}^*/b$ is the dimensionless channel length, which is taken equal to 20. The geometry selected for illustration has the parameters $x_s = 3$ and $x_l = 15$, yielding six complete sinusoidal waves in the corrugated part of the channel.

In order to avoid incorrect solutions stemming from the discontinuity of the derivatives of the function defined in Eq. (26), an approximate continuous unit step function is introduced to smooth the transition between the straight and sinusoidal sections of the channel depicted in Fig. 2. The modified geometry is then given by

$$y_2(x) = -y_1(x) = 1 + \alpha \sin[\pi(x - x_s)][U_s(x, x_s) - U_s(x, x_1)] \qquad (27a)$$

with a continuous approximate unit step written as,

$$U_s(x, x') = \frac{1}{1 + \exp[-\beta(x - x')]} \qquad (27b)$$

where β is an adjustable parameter.

Simplifying Eq. (13) for two-dimensional problems in the Cartesian coordinate system, there is only one component of $\tilde{\Phi}_i$ different from zero, resulting in a scalar eigenvalue problem [49]. For that case, the curl of the vector base automatically reproduces the streamfunction-only formulation [48]. Therefore, the problem defined by Eqs. (23a–n) is solved via the GITT approach by eliminating the transversal coordinate through integral transformation, considering a biharmonic-type fourth-order eigenvalue problem for the flow problem and a classical Sturm–Liouville problem for the temperature problem, both yielding x-variable eigenvalues and eigenfunctions due to the irregular contours of the wavy walls.

7.2 Conjugated Heat Transfer

The second application is aimed at illustrating the single domain reformulation, the partial transformation scheme, and the convective eigenvalue problem alternative in handling conjugated heat transfer. Consider the thermally developing flow inside one or multiple straight channels of arbitrarily irregular cross section, perfusing a rectangular prismatic substrate. The single domain dimensionless formulation for the energy balance is given by

$$W(X,Y)\frac{\partial\theta}{\partial\tau} + U(X,Y)\frac{\partial\theta}{\partial X} = \frac{K(X,Y)}{Pe^2}\frac{\partial^2\theta}{\partial X^2} + \frac{\partial}{\partial Y}\left(K(X,Y)\frac{\partial\theta}{\partial Y}\right) + \frac{\partial}{\partial Z}\left(K(X,Y)\frac{\partial\theta}{\partial Z}\right).$$

$$0 < X < L_X, \quad 0 < Y < L_Y, \quad 0 < Z < L_Z, \quad \tau > 0 \qquad (28a)$$

where

$$U(X,Y) = \begin{cases} U_f(X,Y), & \text{in fluid region} \\ 0, & \text{in solid region} \end{cases} : \quad K(X,Y) = \begin{cases} 1, & \text{in fluid region} \\ k_s/k_f, & \text{in solid region} \end{cases}$$

$$(28b, c)$$

$$W(X,Y) = \begin{cases} 1, & \text{in fluid region} \\ w_s/w_f, & \text{in solid region} \end{cases} \qquad (28d)$$

with the following dimensionless groups:

$$X = \frac{x/L_{ref}}{RePr} = \frac{x}{L_{ref}Pe}: \quad Y = \frac{y}{L_{ref}}: \quad Z = \frac{z}{L_{ref}}:$$

$$U = \frac{u}{4u_{av}}: \quad \theta = \frac{T - T_{in}}{T_w - T_{in}}: \quad K = \frac{k}{k_f}: \quad W = \frac{w}{w_f}Re = \frac{u_{av}4L_{ref}}{\nu}:$$

$$Pr = \frac{\nu}{\alpha}: \quad Pe = RePr = \frac{u_{av}4L_{ref}}{\alpha}: \quad \alpha = \frac{k_f}{w_f}: \quad \tau = \frac{\alpha t}{L_{ref}^2} \qquad (28e)$$

where x is the longitudinal coordinate, y and z are the transversal space coordinates (height and width, respectively). In order to more closely illustrate the conjugated heat transfer application, two examples are analyzed. The first example considers multiple parallel fluid streams perfusing a substrate through channels with polygonal cross sections, illustrating the handling of arbitrary geometries and the partial integral transformation procedure, while the second one considers heat and fluid flow inside parallel plates, illustrating the total integral transformation procedure and the convergence gains achieved with the convective eigenvalue problem alternative previously discussed.

7.2.1 Multiple Irregular Regions

The transversal view of the multi-stream perfused substrate with parallel channels of irregular cross section is illustrated in Fig. 3.

The substrate and fluid inlet region is assumed to be at the dimensionless temperature $\theta_{in} = 0$, the outlet surface is considered adiabatic, and the four lateral surfaces are considered at the prescribed dimensionless temperature $\theta_w = 1$. In order to obtain

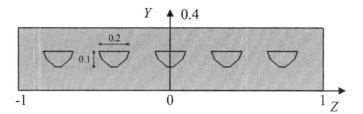

Fig. 3 Schematic representation of the multi-stream perfused substrate cross section

the velocity field inside the fluid flow regions, $u_f(Y, Z)$, the single domain formulation was also employed for the momentum equation in the longitudinal direction (X), assuming that the flow is fully developed and governed by

$$\frac{\partial}{\partial Y}\left(v(Y, Z)\frac{\partial u_f(Y, Z)}{\partial Y}\right) + \frac{\partial}{\partial Z}\left(v(Y, Z)\frac{\partial u_f(Y, Z)}{\partial Z}\right) - \frac{C(Y, Z)}{\rho(Y, Z)} = 0 \qquad (29a)$$

with

$$C(Y, Z) = \begin{cases} \frac{dp}{dX} = \Delta p/L_X, & \text{in the fluid region} \\ 0, & \text{in the solid region} \end{cases};$$

$$v(Y, Z) = \begin{cases} v_f, & \text{in the fluid region} \\ v_s \to \infty, & \text{in the solid region} \end{cases} \qquad (29b, c)$$

$$\rho(Y, Z) = \begin{cases} \rho_f, & \text{in the fluid region} \\ \rho_s, & \text{in the solid region} \end{cases} \qquad (29d)$$

where v_f and ρ_f stand for the kinematic viscosity and density of the fluid, and v_s and ρ_s for the solid. For v_s it suffices to choose a sufficiently large value, when the value for ρ_s will no longer affect the final result and the calculated velocities in the solid region will recover the no slip condition, as physically expected. The dimensionless velocity is calculated from $U_f = u_f/u_{av}$. The solution of problem (29a) is readily handled through GITT based upon an eigenvalue problem with space variable coefficients, as detailed in [58]. The energy equation is solved through the partial transformation scheme, described in Sect. 3.1, in which the longitudinal variable, X, is not integral transformed, leading to a PDE system to be numerically solved, as described in [58].

7.2.2 Convective Eigenvalue Problem

Consider a laminar incompressible internal flow between parallel plates, as illustrated in Fig. 4. In this case, the lateral coordinate Z can be neglected in Eq. (28a), and the space variable coefficients are function of the Y-coordinate only, leading to a 2D transient problem.

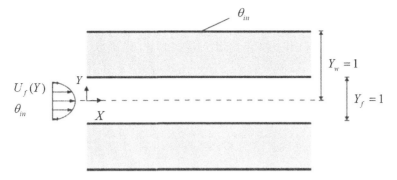

Fig. 4 Schematic representation of the transient conjugated heat transfer example

The flow is considered fully developed, with known parabolic velocity profile, the inlet surface and the external walls are considered at a prescribed dimensionless temperature $\theta_{in} = \theta_w = 0$, and the initial condition is taken as $\theta_0 = 1$. Adopting the procedure described in Sect. 3.2, the energy equation can be rewritten in the following generalized diffusive form:

$$\hat{W}\frac{\partial \theta}{\partial \tau} = \frac{\hat{K}_Y}{Pe^2}\frac{\partial}{\partial X}\left(\hat{K}_X\frac{\partial \theta}{\partial X}\right) + \hat{K}_X\frac{\partial}{\partial Y}\left(\hat{K}_Y\frac{\partial \theta}{\partial Y}\right), \quad 0 < Y < L_Y, \quad 0 < X < L_X, \tau > 0$$

$$(30a)$$

$$\hat{K}_X(X,Y) = e^{-\int Pe^2 U_X^* dX}; \quad \hat{K}_Y(Y) = e^{-\int U_Y^* dY} \qquad (30b, c)$$

$$U_X^* = \frac{U(Y)}{K(Y)}; \quad U_Y^* = -\frac{1}{K(Y)}\frac{dK(Y)}{dY}; \quad \hat{W}(X,Y) = \frac{W(Y)\hat{K}_x(X,Y)\hat{K}_Y(Y)}{K(Y)}$$

$$(30d\text{--}f)$$

Separation of variables is then applied to problem (28a), yielding the following non-classical eigenvalue problem:

$$\frac{\hat{K}_Y}{Pe^2}\frac{\partial}{\partial X}\left(\hat{K}_X\frac{\partial \psi}{\partial X}\right) + \hat{K}_X\frac{\partial}{\partial Y}\left(\hat{K}_Y\frac{\partial \psi}{\partial Y}\right) + \mu_i^2\psi = 0, \quad 0 < Y < L_Y, \quad 0 < X < L_X$$

$$(31)$$

with boundary conditions analogous to the original problem. This eigenvalue problem is non-self-adjoint, meaning the eigenfunctions $\psi_i(X, Y)$, $i = 1, 2, 3 \ldots$, do not follow the same orthogonality property as for the classical Sturm–Liouville problem. Also, the corresponding eigenvalues spectrum is not known a priori and eventually complex quantities may be present. This eigenvalue problem does not allow for explicit analytic solution, but the integral transforms procedure described in Sect. 5 can be employed in its solution. The solution to problem (30) can be written as

$$\theta(X, Y, \tau) = \sum_{i=1}^{\infty} A_i \psi_i(X, Y) e^{-\beta_i^2 t} \tag{32}$$

where the expansion coefficients A_i must be determined from the initial condition. Hence, operating on Eq. (30a) with $\int_0^1 \int_0^{L^*} \psi_j(X, Y)(\cdot) dX dY$ at $\tau = 0$ yields the following system:

$$\int_0^{L_Y} \int_0^{L_X} \psi_j(X, Y) dX dY = \sum_{i=1}^{\infty} A_i \int_0^{L_Y} \int_0^{L_X} \psi_i(X, Y) \psi_j(X, Y) dX dY, \tag{33}$$

$$i = 1, 2, 3 \ldots, \quad j = 1, 2, 3 \ldots$$

which, after truncated to a finite order N, can be solved for the coefficients A_i, and the expansion given by Eq. (33) can be readily used to calculate the dimensionless temperature θ at any position (X, Y) and time τ.

8   Results and Discussion

8.1   *Steady Heat Transfer and Fluid Flow in Corrugated Channel*

Numerical results are presented for the forced convection heat transfer problem in the wavy walls channel described in Sect. 7.1. For the sake of reporting numerical results, it is considered the case of Re $= 400$, Pr $= 6.93$, and $\alpha = 0.1$. Results for the streamfunction and isotherms along the wavy walls channel are presented. In Fig. 5.a, the streamfunction isolines show that recirculation zones are present in almost all cavities along the duct walls. Figure 5b shows the isotherms, and it can be noticed that the temperature distribution is affected mainly in the vicinity of the channel walls and it remains practically unchanged in the central regions of the channel.

Figure 6 shows the comparison of the present GITT results with those generated by using the software COMSOL Multiphysics, and one may observe a good agreement between the two sets of results. Figure 6a shows the results for the product of the skin-friction coefficient by the Reynolds number, $C_f \text{Re} = -\left(\frac{\partial u}{\partial y} + \frac{\partial v}{\partial x}\right)\Big|_{y=y_2(x)}$, and one can readily observe the oscillatory behavior for this parameter along the channel. In Fig. 6.b, it is shown the axial velocity component at the centerline, U_c, and it is also noted the expected oscillatory behavior accompanying the distribution of peaks and valleys along the channel, slightly increasing until the flat outlet region of the duct is reached. Finally, Fig. 6.c shows the distribution of the local Nusselt number, Nu, along the channel. One can observe that the Nusselt number increases in the constricted regions and decreases in the diverging ones. This behavior can be

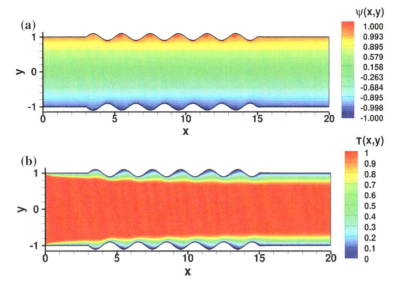

Fig. 5 Patterns of **a** streamlines and **b** isotherms for Re = 400, Pr = 6.93, and $\alpha = 0.1$

explained due to an increase in the average flow velocity and temperature gradients in such constricted areas.

8.2 Conjugated Heat Transfer

Numerical results are now presented for the conjugated heat transfer examples described in Sect. 7.2. As test case, an application with water as the working fluid and acrylic as the channel substrate is here considered, leading to the adopted dimensionless values $k_s/k_f = 0.25$ and $w_s/w_f = 0.35$.

8.2.1 Multiple Irregular Regions

The flexibility of the single domain approach in handling arbitrary domains is here illustrated by considering the multi-stream perfused substrate with the five microchannels shown in Fig. 7, which can be modeled as a single domain by properly defining the space variable coefficients so as to capture the five microchannels geometries. The wall and fluid temperature profiles are plotted in Fig. 7a, b in steady-state regime ($\tau \rightarrow \infty$) for Pe = 1, comparing the GITT and COMSOL solutions for different longitudinal positions, (a) along Z, at $Y = 0.2$, and (b) along Y at $Z = 0$. Both across the thickness of the micro-system and along its width, taking just half of the width due to symmetry, one may observe the perfect adherence to the graphical

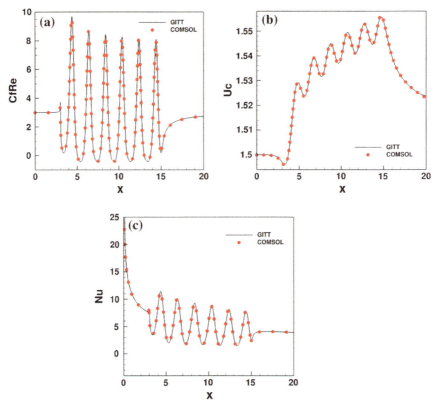

Fig. 6 Parameters related to the velocity and temperature fields for Re = 400, Pr = 6.93, and α = 0.1: **a** Product $C_f Re$; **b** centerline axial velocity; **c** local Nusselt number

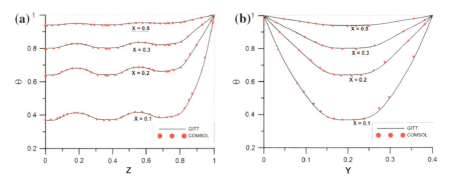

Fig. 7 Comparisons between GITT and COMSOL solutions for the fluid and wall temperature profiles: **a** along Z, at $Y = 0.2$, and **b** along Y at $Z = 0$, for different longitudinal positions (X), for the multiple irregular regions situation

Table 1 Convergence behavior of the steady-state temperature profile along Y with respect to the truncation order of the expansion (N), with fixed $M = 120$ terms in the eigenvalue problem solution (multiple irregular regions), at $X = 0.1$

N	$\theta(0.1, Y, 0, \tau \to \infty)$		
	$Y = 0.1$	$Y = 0.2$	$Y = 0.3$
N = 75	0.5151	0.3703	0.5010
N = 85	0.5147	0.3700	0.5007
N = 95	0.5150	0.3699	0.5013
N = 105	0.5146	0.3695	0.5025
N = 115	0.5145	0.3688	0.5022

Table 2 Convergence behavior of the steady-state temperature profile along Y with respect to the truncation order of the expansion (N), with fixed $M = 120$ terms in the eigenvalue problem solution (multiple irregular regions), at $X = 0.2$

N	$\theta(0.2, Y, 0, \tau \to \infty)$		
	$Y = 0.1$	$Y = 0.2$	$Y = 0.3$
N = 75	0.7504	0.6424	0.7382
N = 85	0.7505	0.6421	0.7384
N = 95	0.7505	0.6421	0.7386
N = 105	0.7502	0.6418	0.7393
N = 115	0.7501	0.6411	0.7391

scale between the two sets of temperature results. Clearly, the transitions between the solid and fluid regions are accurately accounted for by the single domain formulation and its corresponding eigenfunctions.

Besides the remarkable adherence between the hybrid and numerical solutions observed in Figs. 7, Tables 1 and 2 illustrate the convergence behavior of the GITT solution with respect to the number of terms employed in the temperature field expansion, N, where it can be observed a convergence of three significant digits to within the truncation orders considered. In these results, the eigenvalue problem was also solved using GITT, as described in Sect. 5, keeping the truncation order constant with $M = 120$ terms.

8.2.2 Convective Eigenvalue Problem

In order to demonstrate the convergence rate gains in adopting the convective eigenvalue problem formulation, three representative situations are considered in the analysis, with Pe = 1, 10, and 100. In all cases, it was employed M = 75 as the truncation order in the eigenvalue problem solution via GITT (Sect. 5). Figure 8a–c graphically illustrate the convergence behaviors for Pe = 1, 10, and 100, respectively, by presenting some transversal temperature profiles calculated with different truncation orders ($N = 3, 6, 9$) together with purely numerical solutions calculated with COMSOL (automatically generated mesh with the "extremely fine" option), demonstrating that a truncation order as small as $N = 9$ is enough to provide curves fully converged to the graph scale and in full agreement with the numerical solutions.

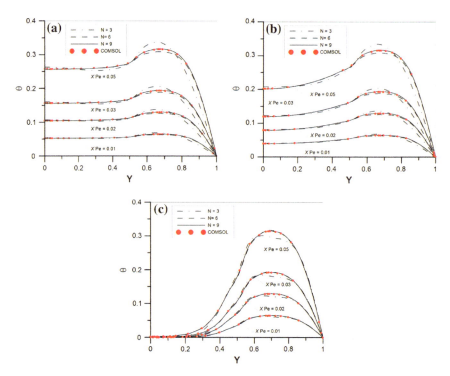

Fig. 8 Transversal temperature profiles at. XPe $= 0.01, 0.02, 0.03$ and 0.05 and $\tau = 0.01$, calculated employing the convective eigenvalue problem and the truncation orders $N = 3$ (green dots-dashes), $N = 6$ (blue dashes), $N = 9$ (solid black), and COMSOL (red dots). **a** Pe $= 1$, **b** Pe $= 10$, **c** Pe $= 100$

Table 3 Convergence behavior of the calculated temperatures at $XPe = 0.05$, $\tau = 0.01$, for Pe $= 1$, employing the convective eigenvalue problem

	$Y = 0.2$	$Y = 0.3$	$Y = 0.4$	$Y = 0.6$	$Y = 0.7$	$Y = 0.8$
$N = 3$	0.256079	0.252575	0.256287	0.328441	0.330669	0.269604
$N = 6$	0.257835	0.259193	0.263403	0.304034	0.307831	0.284545
$N = 9$	0.257596	0.259685	0.264093	0.310626	0.316237	0.290713
$N = 12$	0.259063	0.260886	0.265352	0.310019	0.315476	0.290549
$N = 15$	0.259085	0.260927	0.265346	0.309965	0.315517	0.290559
$N = 18$	0.259077	0.260928	0.265353	0.309960	0.315529	0.290546
$N = 21$	0.259077	0.260928	0.265353	0.309961	0.315527	0.290547
$N = 24$	0.259077	0.260928	0.265353	0.309964	0.315531	0.290551

Table 4 Convergence behavior of the calculated temperatures at $X\text{Pe} = 0.05$, $\tau = 0.01$, for Pe = 10, employing the convective eigenvalue problem

	$Y = 0.2$	$Y = 0.3$	$Y = 0.4$	$Y = 0.6$	$Y = 0.7$	$Y = 0.8$
N = 3	0.210106	0.217457	0.234580	0.323458	0.328714	0.272011
N = 6	0.211044	0.223159	0.241558	0.302504	0.307638	0.284495
N = 9	0.210796	0.223583	0.242205	0.309022	0.316087	0.290680
N = 12	0.212114	0.224704	0.243447	0.308494	0.315314	0.290522
N = 15	0.212132	0.224737	0.243437	0.308459	0.315344	0.290520
N = 18	0.212124	0.224738	0.243443	0.308454	0.315355	0.290508
N = 21	0.212125	0.224737	0.243444	0.308457	0.315358	0.290512
N = 24	0.212125	0.224737	0.243444	0.308458	0.315358	0.290512

Table 5 Convergence behavior of the calculated temperatures at $X\text{Pe} = 0.05$, $\tau = 0.01$, for Pe = 100, employing the convective eigenvalue problem

	$Y = 0.2$	$Y = 0.3$	$Y = 0.4$	$Y = 0.6$	$Y = 0.7$	$Y = 0.8$
N = 3	0.003741	0.021366	0.084100	0.296500	0.295620	0.280568
N = 6	0.008122	0.029257	0.087402	0.289878	0.316661	0.288770
N = 9	0.008837	0.028693	0.086863	0.292425	0.314506	0.290178
N = 12	0.008819	0.028761	0.087426	0.292684	0.313462	0.290293
N = 15	0.008837	0.028817	0.087474	0.292616	0.313595	0.290154
N = 18	0.008853	0.028836	0.087467	0.292643	0.313572	0.290172
N = 21	0.008854	0.028835	0.087468	0.292644	0.313575	0.290177
N = 24	0.008853	0.028835	0.087469	0.292644	0.313574	0.290177

Tables 3, 4, 5 further illustrate the convergence behavior, by presenting the results in tabular form for Pe = 1, 10, and 100, respectively, demonstrating that with only $N = 18$ terms, a full convergence of five to six significant digits is observed at the selected positions, despite the increase in the Péclet number, clearly demonstrating the importance of incorporating the convective term into the eigenvalue problem.

9 Closing Remarks

Fifty years ago, roughly by the late 60s and early 70s, the classical integral transform method for the solution of linear diffusion problems reached a maturity level that is evident from the seminal contributions published in this period, as here reviewed [10–14]. Also around this period, the limitations on the classical approach were faced, in the pioneering works of Ozisik and Murray [18] and Mikhailov [19], that provided the first approximate analytical solutions for non-transformable linear problems, and would plant the seeds for the development of the hybrid numerical–analytical approach [20–28], nowadays known as the Generalized Integral Transform Technique, GITT, that has been extended to various classes of linear and nonlinear diffusion and convection–diffusion problems along about thirty years, as here briefly described. Also, fifty years ago, Prof. Spalding and his collaborators were breeding a sequence of academic contributions on the finite volume method, that would lead to the very first version of the Phoenics code, launched in 1981, inaugurating the era of modern CFD & HT. This achievement was soon followed by developments on other classical numerical approaches that also led to the establishment of general-purpose computational tools based on the finite element, finite differences, and boundary element methods. The need for independent benchmarks of classical heat and fluid flow test cases, toward the verification and critical comparison of competing numerical schemes, was the original motivation in the parallel development of the hybrid GITT approach along this period. The analytical nature behind the GITT, which concentrates most of the numerical work in one single independent variable, has proved to offer error-controlled solutions with mild computational costs that, once derived and implemented for a certain class of problems, become an interesting alternative path for computational simulation, especially when associated with very computer intensive tasks such as optimization, inverse problem analysis, and simulation under uncertainty.

More recently, with the aid of mixed symbolic-numerical systems [67], the construction of a Unified Integral Transforms (UNIT) algorithm has been advanced [42, 54], offering both an automatic symbolic-numerical open source solver and a development platform for researchers and practitioners interested in this class of hybrid methods for partial differential equations. In parallel to the attempt of offering a more general-purpose hybrid simulation tool, the method has been advanced through both mathematical and computational novel aspects, as here partially described, to challenge applications from now on that, even nowadays pose difficulties to the well-established numerical methods, such as in dealing with the wide classes of unstable nonlinear problems and multiscale phenomena.

Acknowledgements The authors are grateful for the financial support offered by the Brazilian Government agencies CNPq (projects no. 401237/2014-1 and no. 207750/2015-7), CAPES-INMETRO, and FAPERJ. RMC is also grateful to the Leverhulme Trust for the Visiting Professorship (VP1-2017-028) and to the kind hospitality of the Department of Mechanical Engineering, University College London (UCL), UK, along 2018.

References

1. Deakin, M. A. B. (1985). Euler´s Invention of the Integral Transforms. *Archive for History of Exact Sciences, 33,* 307–319.

2. Euler, L. (1769). *Institutiones Calculi Integralis* (Vol. 2 (Book 1, Part 2, Section 1)). St. Petersburg Imp. Acad. Sci. Reprinted in https://archive.org/details/institutionescal020326mbp/page/n1.

3. Fourier, J. B. (2007). *The analytical theory of heat (Unabridged)*. Cosimo Inc., New York. Translation of Original: (1822) Théorie Analytique de la Chaleur. Paris: Firmin Didot Père et Fils.

4. Koshlyakov, N. S. (1936). *Basic differential equations of mathematical physics (In Russian)*. ONTI, Moscow, 4th edition.

5. Titchmarsh, E. C. (1946). *Eigenfunction expansion associated with second order differential equations*, Oxford University Press.

6. Grinberg, G. A. (1948). *Selected problems of mathematical theory of electrical and magnetic effects (In Russian)*. Nauk SSSR: Akad.

7. Koshlyakov, N. S., Smirnov, M. M., & Gliner, E. B. (1951). *Differential equations of mathematical physics*. North Holland, Amsterdam: Translated by Script Technica, 1964.

8. Eringen, A. C. (1954). The finite Sturm-Liouville transform. *The Quarterly Journal of Mathematics, 5*(1), 120–129.

9. Olçer, N. Y. (1964). On the theory of conductive heat transfer in finite regions. *International Journal of Heat and Mass Transfer, 7,* 307–314.

10. Mikhailov, M. D. (1967). *Nonstationary temperature fields in skin*. Moscow: Energiya.

11. Luikov, A. V. (1968). *Analytical heat diffusion theory*. New York: Academic Press.

12. Ozisik, M. N. (1968). *Boundary value problems of heat conduction*. New York, Int: Textbooks Co.

13. Sneddon, I. N. (1972). *Use of integral transforms*. New York: McGraw-Hill.

14. Mikhailov, M. D. (1972). General solution of the heat equation of finite regions. *International Journal of Engineering Science, 10,* 577–591.

15. Ozisik, M. N. (1980). *Heat conduction*. New York: John Wiley.

16. Luikov, A. V. (1980). *Heat and mass transfer*. Moscow: Mir Publishers.

17. Mikhailov, M. D., & Özisik, M. N. (1984). *Unified analysis and solutions of heat and mass diffusion*. John Wiley: New York; also, Dover Publications, 1994.

18. Ozisik, M. N., & Murray, R. L. (1974). On the solution of linear diffusion problems with variable boundary condition parameters. *ASME J. Heat Transfer, 96c,* 48–51.

19. Mikhailov, M. D. (1975). On the solution of the heat equation with time dependent coefficient. *International Journal of Heat and Mass Transfer, 18,* 344–345.

20. Cotta, R. M. (1986). Diffusion in media with prescribed moving boundaries: Application to metals oxidation at high temperatures. *Proc. of the II Latin American Congress of Heat & Mass Transfer*, Vol. 1, pp. 502–513, São Paulo, Brasil, May.

21. Cotta, R. M., & Ozisik, M. N. (1987). Diffusion problems with general time-dependent coefficients. *Rev. Bras. Ciências Mecânicas, 9*(4), 269–292.

22. Aparecido, J. B., Cotta, R. M., & Ozisik, M. N. (1989). Analytical Solutions to Two-Dimensional Diffusion Type Problems in Irregular Geometries. *J. of the Franklin Institute, 326,* 421–434.

23. Cotta, R. M. (1990). Hybrid Numerical-Analytical Approach to Nonlinear Diffusion Problems. *Num. Heat Transfer, Part B, 127,* 217–226.

24. Cotta, R. M., & Carvalho, T. M. B. (1991). Hybrid Analysis of Boundary Layer Equations for Internal Flow Problems. *7th Int. Conf. on Num. Meth. in Laminar & Turbulent Flow*, Part 1, pp. 106–115, Stanford CA, July.

25. Perez Guerrero, J. S., & Cotta, R. M. (1992). Integral Transform Method for Navier-Stokes Equations in Stream Function-Only Formulation. *Int. J. Num. Meth. in Fluids, 15,* 399–409.

26. Cotta, R. M. (1993). *Integral Transforms in Computational Heat and Fluid Flow*. Boca Raton, FL: CRC Press.

27. Cotta, R. M. (1994). The Integral Transform Method in Computational Heat and Fluid Flow. Special Keynote Lecture. *Proc. of the 10th Int. Heat Transfer Conf.*, Brighton, UK, SK-3, Vol. 1, pp. 43–60, August.

28. Cotta, R. M. (1994). Benchmark Results in Computational Heat and Fluid Flow: - The Integral Transform Method. *Int. J. Heat Mass Transfer,* Invited Paper, *37,* 381–394.
29. Napolitano, M., & Orlandi, P. (1985). Laminar Flow in a Complex Geometry: A Comparison. *Int. Journal for Numerical methods in Fluids, 5,* 667–683.
30. Perez Guerrero, J. S., & Cotta, R. M. (1995). A Review on Benchmark Results for the Navier-Stokes Equations Through Integral Transformation. *Revista Perfiles de Ingenieria,* (Invited Paper), no.4, pp.C.30–33, Peru, July.
31. Perez Guerrero, J. S., Quaresma, J. N. N., & Cotta, R. M. (2000). Simulation of Laminar Flow inside Ducts of Irregular Geometry using Integral Transforms. *Computational Mechanics, 25*(4), 413–420.
32. Cotta, R. M., & Mikhailov, M. D. (1997). *Heat Conduction: Lumped Analysis, Integral Transforms, Symbolic Computation.* Chichester, UK: Wiley.
33. Cotta, R. M. (1998). *The Integral Transform Method in Thermal and Fluids Sciences and Engineering.* New York: Begell House.
34. Cotta, R. M., & Mikhailov, M. D. (2006). Hybrid Methods and Symbolic Computations. In W. J. Minkowycz, E. M. Sparrow, & J. Y. Murthy (Eds.), *Handbook of Numerical Heat Transfer, 2nd edition, Chapter 16.* New York: John Wiley.
35. Cotta, R. M., Knupp, D. C., & Naveira-Cotta, C. P. (2016). *Analytical Heat and Fluid Flow in Microchannels and Microsystems.* Mechanical Eng. Series. New York: Springer.
36. Cotta, R. M., Knupp, D. C., & Quaresma, J. N. N. (2018). Analytical Methods in Heat Transfer. In *Handbook of Thermal Science and Engineering,* F. A. Kulacki et al., Eds., Chapter 1. Springer.
37. Cotta, R. M., Naveira-Cotta, C. P., Knupp, D. C., Zotin, J. L. Z., Pontes, P. C., & Almeida, A. P. (2018). Recent Advances in Computational-Analytical Integral Transforms for Convection-Diffusion Problems. *Heat & Mass Transfer,* Invited Paper, *54,* 2475–2496.
38. Cotta, R. M., Su, J., Pontedeiro, A. C., & Lisboa, K. M. (2018). Computational-Analytical Integral Transforms and Lumped-Differential Formulations: Benchmarks and Applications in Nuclear Technology. *Special Lecture, 9th Int. Symp. on Turbulence, Heat and Mass Transfer, THMT-ICHMT,* Rio de Janeiro, July 10th–13th. In *Turbulence, Heat and Mass Transfer 9,* pp. 129–144, Eds. A. P. Silva Freire et al., Begell House, New York.
39. Cotta, R. M., Lisboa, K. M., Curi, M. F., Balabani, S., Quaresma, J. N. N., Perez-Guerrero, J. S., et al. (2019). A Review of Hybrid Integral Transform Solutions in Fluid Flow Problems with Heat or Mass Transfer and under Navier-Stokes Equations Formulations. *Num. Heat Transfer, Part B - Fundamentals, 76,* 1–28.
40. Cotta, R. M., Naveira-Cotta, C. P., & Knupp, D. C. (2016). Nonlinear Eigenvalue Problem in the Integral Transforms Solution of Convection-diffusion with Nonlinear Boundary Conditions. *Int. J. Num. Meth. Heat & Fluid Flow,* Invited Paper, 25th Anniversary Special Issue, 26, 767–789.
41. Pontes, P. C., Almeida, A. P., Cotta, R. M., & Naveira-Cotta, C. P. (2018). Analysis of Mass Transfer in Hollow-Fiber Membrane Separator via Nonlinear Eigenfunction Expansions. *Multiphase Science and Technology, 30*(2-3), 165–186.
42. Cotta, R. M., Knupp, D. C., Naveira-Cotta, C. P., Sphaier, L. A., & Quaresma, J. N. N. (2014). The Unified Integral Transforms (UNIT) Algorithm with Total and Partial Transformation. *Comput. Thermal Sciences, 6,* 507–524.
43. Ozisik, M. N., Orlande, H. R. B., Colaço, M. J., & Cotta, R. M. (2017). *Finite Difference Methods in Heat Transfer* (2nd ed.). Boca Raton, FL: CRC Press.
44. Serfaty, R., & Cotta, R. M. (1992). Hybrid Analysis of Transient Nonlinear Convection-Diffusion Problems. *Int. J. Num. Meth. Heat & Fluid Flow, 2,* 55–62.
45. Cotta, R. M., Naveira-Cotta, C. P., & Knupp, D. C. (2017). Convective Eigenvalue Problems for Convergence Enhancement of Eigenfunction Expansions in Convection-diffusion Problems. *ASME J. Thermal Science and Eng. Appl., 10*(2), 021009 (12 pages).
46. Knupp, D. C., Cotta, R. M., Naveira-Cotta, C. P., & Cerqueira, I. G. S. (2018). Conjugated Heat Transfer via Integral Tranforms: Single Domain Formulation, Total and Partial Transformation, and Convective Eigenvalue Problems. *Proc. of the 10th Minsk International Seminar "Heat*

Pipes, Heat Pumps, Refrigerators, Power Sources", pp. 171–178, Minsk, Belarus, September 10th–13th.

47. Lima, G. G. C., Santos, C. A. C., Haag, A., & Cotta, R. M. (2007). Integral Transform Solution of Internal Flow Problems Based on Navier-Stokes Equations and Primitive Variables Formulation. *Int. J. Num. Meth. Eng., 69,* 544–561.

48. Lisboa, K. M., & Cotta, R. M. (2018). Hybrid Integral Transforms for Flow Development in Ducts Partially Filled with Porous Media. *Proc. Royal Society A - Mathematical, Physical and Eng. Sciences, 474,* 1–20.

49. Lisboa, K. M., Su, J., & Cotta, R. M. (2019). Vector Eigenfunction Expansion in the Integral Transform Solution of Transient Natural Convection. *Int. J. Num. Meth. Heat & Fluid Flow, 29,* 2684–2708.

50. Sphaier, L. A., & Cotta, R. M. (2000). Integral Transform Analysis of Multidimensional Eigenvalue Problems Within Irregular Domains. *Numerical Heat Transfer, Part B-Fundamentals, 38,* 157–175.

51. Sphaier, L. A., & Cotta, R. M. (2002). Analytical and Hybrid Solutions of Diffusion Problems within Arbitrarily Shaped Regions via Integral Transforms. *Computational Mechanics, 29*(3), 265–276.

52. Monteiro, E. R., Quaresma, J. N. N., & Cotta, R. M. (2011). Integral transformation of multidimensional phase change problems: Computational and physical analysis. *21st International Congress of Mechanical Engineering, COBEM-2011,* ABCM, pp.1–10, Natal, RN, Brazil, October.

53. Cotta, R. M., & Mikhailov, M. D. (2005). Semi-analytical evaluation of integrals for the generalized integral transform technique. *Proc. of the 4th Workshop on Integral Transforms and Benchmark Problems – IV WIT,* pp. 1–10, CNEN, Rio de Janeiro, RJ, August.

54. Cotta, R. M., Knupp, D. C., Naveira-Cotta, C. P., Sphaier, L. A., & Quaresma, J. N. N. (2013). Unified Integral Transforms Algorithm for Solving Multidimensional Nonlinear Convection-Diffusion Problems. Num. *Heat Transfer, part A - Applications, 63,* 1–27.

55. Knupp, D. C., Naveira-Cotta, C. P., & Cotta, R. M. (2012). Theoretical Analysis of Conjugated Heat Transfer with a Single Domain Formulation and Integral Transforms. *Int. Comm. Heat & Mass Transfer, 39*(3), 355–362.

56. Knupp, D. C., Naveira-Cotta, C. P., & Cotta, R. M. (2014). Theoretical–experimental Analysis of Conjugated Heat Transfer in Nanocomposite Heat Spreaders with Multiple Microchannels. *Int. J. Heat Mass Transfer, 74,* 306–318.

57. Knupp, D. C., Cotta, R. M., Naveira-Cotta, C. P., & Kakaç, S. (2015). Transient Conjugated Heat Transfer in Microchannels: Integral Transforms with Single Domain Formulation. *Int. J. Thermal Sciences, 88,* 248–257.

58. Knupp, D. C., Cotta, R. M., & Naveira-Cotta, C. P. (2015). Fluid Flow and Conjugated Heat Transfer in Arbitrarily Shaped Channels via Single Domain Formulation and Integral Transforms. *Int. J. Heat Mass Transfer, 82,* 479–489.

59. Almeida, A. P., Naveira-Cotta, C. P., & Cotta, R. M. (2018). Integral Transforms for Transient Three-dimensional Heat Conduction in Heterogeneous Media with Multiple Geometries and Materials. Paper # IHTC16–24583. *Proc. of the 16th International Heat Transfer Conference – IHTC16,* Beijing, China, August 10th–15th.

60. Lisboa, K. M., Su, J., & Cotta, R. M. (2018). Single Domain Integral Transforms Analysis of Natural Convection in Cavities Partially Filled with Heat Generating Porous Medium. *Num. Heat Transfer, Part A – Applications, 74*(3), 1068–1086.

61. Lisboa, K. M., & Cotta, R. M. (2018). On the Mass Transport in Membraneless Flow Batteries of Flow-by Configuration. *Int. J. Heat & Mass Transfer, 122,* 954–966.

62. Wang, C. C., & Chen, C. K. (2002). Forced Convection in a Wavy-Wall Channel. *Int. Journal of Heat and Mass Transfer, 45,* 2587–2595.

63. Silva, R. L., Santos, C. A. C., Quaresma, J. N. N., & Cotta, R. M. (2011). Integral Transforms Solution for Flow Development in Wavy-Wall Ducts. *Int. J. Num. Meth. Heat & Fluid Flow, 21*(2), 219–243.

64. Castellões, F. V., Quaresma, J. N. N., & Cotta, R. M. (2010). Convective Heat Transfer Enhancement in Low Reynolds Number Flows with Wavy Walls. *Int. J. Heat & Mass Transfer, 53,* 2022–2034.
65. Zotin, J. L. Z., Knupp, D. C., & Cotta, R. M. (2017). Conjugated Heat Transfer in Complex Channel-Substrate Configurations: Hybrid Solution with Total Integral Transformation and Single Domain Formulation. *Proc. of ITherm 2017 - Sixteenth Intersociety Conference on Thermal and Thermomechanical Phenomena in Electronic Systems,* Paper #435, Orlando, FL, USA, May 30th–June 2nd.
66. Knupp, D. C., Cotta, R. M., & Naveira-Cotta, C. P. (2020). Conjugated Heat Transfer Analysis via Integral Transforms and Convective Eigenvalue Problems. *J. Eng. Physics & Thermophysics,* (in press).
67. Wolfram, S. (2017). *Mathematica, version 11.* Champaign, IL: Wolfram Research Inc.

Alternatives and Future of CFD

A Comparison Between FEM and FVM via the Method of Weighted Residuals

Darrell W. Pepper⊙, S. Pirbastami⊙ and David B. Carrington⊙

1 Introduction

The method of weighted residuals (MWR) is a ubiquitous approach that establishes a functional form for dependent variables commonly deployed in PDEs. The procedure produces a set of equations consistent with the number of unknowns, i.e., the problem domain is subdivided into intervals of integration, e.g., elements or volumes, that coincide with the number of unknowns (or node points). The finite volume method (FVM) and the finite element method (FEM) fall within the family of weighted residual methods [1].

1.1 Short History of the FVM

The FVM has been used for many years and follows from early applications of the FDM for approximating ODEs and PDEs. While FDMs are based on point discretizations and Taylor series expansions, the unique quality of the FVM is its natural ability to balance quantities (or fluxes) across control volumes. Application of the FVM began in the late 1950s, and its popularity has continued to the present time.

As in the FDM and FEM, a mesh must be constructed of the problem domain. These individual partitions that surround node points are actually control volumes, a procedure commonplace in analyzing thermodynamic processes. Each individual PDE defined at each node is then integrated over the control volume, and the integral expressions summed globally to establish the set of discretized PDEs. This operation

D. W. Pepper (✉) · S. Pirbastami
University of Nevada Las Vegas, Las Vegas, NV 89154, USA
e-mail: darrell.pepper@unlv.edu

D. B. Carrington
T-3, Los Alamos National Laboratory, Los Alamos, NM 55555, USA

© Springer Nature Singapore Pte Ltd. 2020
A. Runchal (ed.), *50 Years of CFD in Engineering Sciences*,
https://doi.org/10.1007/978-981-15-2670-1_21

results in a balanced set of equations containing unknown variable values that are then solved numerically. The important concept lies with the fluxes at the boundaries of each control volume, i.e., what flows into the control volume is balanced by what flows out of the volume. This results in the FVM being classified as naturally conservative, i.e., a flux entering a control volume from an adjacent control volume must be balanced by an opposite flux leaving the control volume. As in the FDM, as the mesh decreases in size, the discretized FVM relations tend toward the exact mathematical relation—however, levels of truncation of the series produce various levels of error. As shown later, simple expressions of the FVM result in equivalent FDM expressions. However, the discretization employed in the FVM is based on local control volume balancing, and not on the PDEs as in the FDM. It is the fluxes that are discretized in the FVM, and that distinction becomes quite important if not critical in applications to nonlinear PDEs arising in the numerical analysis, for example, in the mechanics of incompressible and compressible fluids.

A system of equations resulting from the integral operations is solved on the set of discrete points established at the nodal points defining the control volume. There are actually two choices in establishing the control volumes in 2D, either cell centered or cell vertex (staggered grid), as shown in Fig. 1. In the cell-centered volumes, the quantities are stored at the centroidal nodes; in the vertex-based volumes, quantities are stored at the corner grid points. Discussions regarding the pros and cons of each approach are numerous in the literature as well as on the web. More recent absorption of the FEM mesh generation techniques for unstructured grids into the FVM (including polygonal volumes) has enabled the method to compete with the meshing advantages of the FEM.

The choice of solution technique, whether it be iterative or direct (Gauss elimination based) resides with the developer. These equations create banded symmetric matrices for Laplace and Poisson equations, and sparse nonsymmetric matrices when dealing with nonlinear problems, e.g., fluid flow. These constraints also affect FEM, but the FVM has a bit more leeway in how the discretized relations can be solved whereas the FEM is inherently implicit.

Avid development and application of the FVM essentially accelerated in the 1960s, with a fervent push from Spalding and later by his students at Imperial College [2, 3] with particular focus on fluid flow (CFD), heat transfer, and species transport. At

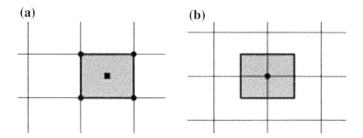

Fig. 1 Common 2D control volumes using **a** vertex and **b** centroid

the same time, similar interest and development in FVM began at the Los Alamos National Laboratory (LANL). Today, one of the most popular commercial FVM codes that spun from Imperial College and Spalding's students is FLUENT (ANSYS) [4].

1.2 Short History of the FEM

The history of the finite element method stems back to the mid-1950s. The use of FDM and FVM approaches prior to the formal development of the FEM could be particularly frustrating when attempting to model complex, geometrically irregular problems. Many of these early numerical solutions were based on simple geometries that could be easily created using orthogonal meshes. Coordinate transformations (also requiring the governing equations to be transformed) were typically required—and could be tricky to implement while yielding low-order results.

Early efforts using FEM generally dealt with structural problems to resolve deformations and stress/strain. Using small, discrete elements (analogous to volumes), Argyris [5] and Turner et al. [6] were the first to publish results stemming from this approach for problems occurring in the aircraft industry. Clough [7] actually coined the name "finite element." The popularity of the FEM approach rapidly gained acceptance for structural analysis and became the defacto standard in the industry.

It did not take long to realize the versatility of the FEM and the MWR for problems other than structural analysis, and the method began to be applied to field problems related to heat transfer and potential flow, i.e., solution of Laplace and Poisson equations [1]. There are numerous references and textbooks discussing the development of the finite element approach available in libraries and on the web, including vigorous mathematical treatments.

The development of mesh generators that can discretize actual physical problem geometry utilizing elements began in earnest as the FEM began to be pushed into nonlinear and fluid-related areas. One of the earliest finite element commercial codes was PATRAN [8], which is still in existence today, although many newer packages are now available. In fact, most of the current commercial FEM software have their own built-in mesh generators, e.g., ANSYS [9] and COMSOL [10] are well-known and popular programs with many thousands of users. CAD commercial codes such as SOLIDWORKS, AUTOCAD, and PRO/E likewise rely on the generation of surface-oriented unstructured finite elements.

As the FEM has continued to expand into more areas, the method has become a serious competitor to the FVM for CFD, particularly with its ability to easily couple multiphysics components. The COMSOL FEM code is a good example of the ability to link different physics into one overall simulation [10].

1.3   The Method of Weighted Residuals (MWR)

There are many variations of the MWR. In the Galerkin procedure, the approximating function, N, for the dependent variable is chosen to be the same as the weighting function, W. When the weighting function is not equal to the approximating function, other weighted residual methods are obtained. For example, least squares, collocation, and spectral methods are common alternatives. The Galerkin approach is the most popular, and we will employ it in this comparison study. A discussion of other methods employed in the MWR can be found in [11]. We briefly describe the Galerkin method and then apply it to the one- and two-dimensional equation for heat transfer. The general advection–diffusion transport equation for various 2D FEM configurations is listed in the Appendix. The best method depends upon user preference and the class of problems to be solved. Factors to consider include whether there are singularities or discontinuities, regular or irregular boundaries, the necessity for mesh refinement, and computer platform [12].

A very expedient method for one-dimensional problems may not be appropriate for two- or three-dimensional problems [13–15]. For example, three-dimensional FEMs become very costly and storage intensive compared to FVMs; however, for one- or two-dimensional problems, FEM is competitive with conventional FVMs. Low-order methods such as linear finite elements or simple finite volumes are best for singular problems; high-order methods, e.g., quadratic finite elements, are better suited for problems with smooth solutions. Once the mesh, or nodal discretization of the problem domain, is established, the manner in which the equivalent algebraic equations are solved (iteratively versus direct) also influences the choice of a method.

The underlying principle of the FEM is its ability to easily solve problems described by complex boundary shapes. The FEM was initially developed to calculate stress in irregularly shaped objects and analyze structural problems in aircraft. Since its inception, the FEM has been found to be equally effective in nonstructural problems, particularly those in heat transfer and fluid dynamics. More recently, the FVM has incorporated the nonstructural meshing strategy of the FEM, enabling polygonal volumes to be utilized, but ensuring that orthogonality of the fluxes is maintained through the volume edges.

Like many analytical methods, the FEM is based on the series expansion of the functions themselves. In a typical series expansion, an *infinite* number of global basis functions (sines, cosines, etc.) span the entire domain [16, 17]. However, in the FEM, only a *finite* number of basis functions that are *local* in nature (nonzero over only a small segment of the domain) are employed. With the use of local basis functions, the coefficient matrices that result from the approximations of the governing equation are banded and sparse. FVMs also produce banded sparse matrices. However, the coefficients in the FEM vary from node to node, since nodal locations are arbitrary; hence, the resulting matrices become considerably fuller due to the greater degree of coupling among nodes. For example, the Laplacian operator generates 3-, 5-, and 7-nodal point coupling for one-, two-, and three-dimensional domains in a second-order FVM. In the FEM, the simplest basis function generates a 3-, 9-, or 11-point coupling [18, 19]. In addition, time derivative terms are coupled in the FEM, whereas they are not necessarily in the FVM.

In the FEM a partial differential equation is reduced to a finite system of ordinary differential equations, which is then solved by matrix solution techniques. The reduction of the governing equation is typically performed using either (1) the Rayleigh–Ritz method or (2) the MWR. In the Rayleigh–Ritz method, variational calculus is used to formulate a variational statement of the problem. In order to use the variational method, an appropriate integral over the problem domain must possess an extremum. In most instances, this extremum is based on energy concepts, that is, Lagrangian multipliers. For simple problems, such as potential flow or conduction heat transfer, the variational formulation is easily established. However, for most practical problems, particularly when advection terms or odd derivatives are present, variational principles cannot be developed, and one has to resort to pseudo-variational formulations. In the MWR, the governing equation is multiplied by a weighting function, W, and the product is integrated over space (hence the term weighted residual). The MWR can be applied naturally to any differential equation that contains even and odd derivatives. FVMs are simpler to formulate, can easily be extended to 2- or 3-dimensions, and require considerably less computational work than finite elements (for equivalent nodes).

In both FEM and FVM methods, irregular geometries are easily handled; small mesh sizes can be used in those regions where the solution changes significantly. Derivative or flux-type boundary conditions are automatically incorporated in Galerkin methods without diminishing the accuracy of the overall solution. In the FVM, boundary conditions are generally resolved following solution of the domain interior, often as simple one-sided approximations of lower order accuracy. In some flow situations, lower order boundary approximations must be used with the FVM to achieve stable solutions. Higher order approximations for the overall solution are more easily generated in the FEM and do not require special treatment near boundaries. Finally, the FEM generates numerical approximations that satisfy certain global conservation laws independent of the boundary conditions and geometry, which lead to computationally well-behaved results.

The biggest drawback to the FEM is that it is inherently an implicit method (since the time derivatives are coupled in the mass matrix). Inversion of the mass matrix, if an explicit scheme is used, is not practical except for small problems. Hence, the trapezoidal rule is usually employed to advance over one time step; banded Gaussian elimination or iterative techniques are generally used to solve the linear system of algebraic equations. Although the mass matrix terms can be lumped to permit explicit integration, loss of accuracy can become significant.

Large FEM codes were historically troublesome to contain totally in the CPU memory of even the most massive computers. However, more recent advances in computer hardware and capabilities now permit rather large FEM programs to be run without difficulty, depending on the number of available computer cores. PCs today have multiple cores, allowing many commercial FEM codes to be run on laptops. Large supercomputers have massive numbers of cores and RAM capabilities, including peripheral storage employing high-speed tapes and disks, permitting

$>10^9$ nodal problems to be addressed. The solution of nonlinear problems normally requires that many of the Galerkin integrals (matrices) be recalculated at each time step; in contrast, for linear problems, they can be calculated once and stored for later use.

2   The MWR

In order to illustrate the MWR formulation, consider a steady-state heat transfer problem, where the temperature, $T(x, y)$, is governed by the heat conduction equation and subjected to either Dirichlet (prescribed temperature) or Neumann (flux) boundary conditions:

$$\nabla^2 T(x, y) + Q = 0 \tag{1}$$

where

$$T(x_s, y_s)|_{\Gamma_T} = T_s$$
$$\left.\frac{\partial T}{\partial n}\right|_{(x_s, y_s)}\Big|_{\Gamma_q} = q_s,$$

as shown in Fig. 2. We chose this problem as an illustrative example, and the procedure can be applied to any other governing scalar or vector, linear or nonlinear equation, as shown in Fig. 2.

The basic concept of the MWR is to approximate the temperature using a set of trial functions, $N_j(x, y)$, as

$$\tilde{T}(x, y) = \sum_{j=1}^{N} \alpha_j N_j(x, y) \tag{2}$$

Fig. 2 Method of weighted residuals applied to heat conduction equation [11]

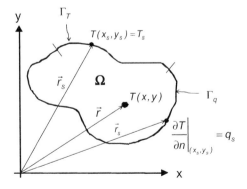

We are free to choose localized or global support, with the only requirement being that the trial functions must be linearly independent. The expansion coefficients, α_j, represent the nodal temperatures.

Introducing Eq. (2) into the governing equation leads to a domain residual, $R_\Omega(x, y)$:

$$R_\Omega(x, y) = \nabla^2 \tilde{T}(x, y) + Q \tag{3}$$

where Ω denotes the problem domain. Likewise, introducing Eq. (2) into the boundary conditions leads to boundary residuals, i.e., $R_{\Gamma_T}(x, y)$, on the Γ_T portion of the boundary, where the Dirichlet boundary condition is imposed:

$$R_{\Gamma_T}(x, y) = \tilde{T}(x_s, y_s) - T_s \tag{4}$$

and to a residual, $R_{\Gamma_q}(x, y)$, on the Γ_q portion of the boundary, where a flux boundary condition is imposed

$$R_{\Gamma_q}(x, y) = \frac{\partial \tilde{T}(x_s, y_s)}{\partial n} - q_s \tag{5}$$

Depending on the selection of the trial function, as shown in Fig. 3, any of these residuals may be zero. The choices are typically described in an MWR perspective as:

1. Interior methods: Trial functions satisfy the boundary conditions
2. Boundary methods: Trial functions satisfy the governing equation
3. Mixed methods: Trial functions satisfy neither governing equation nor boundaries.

The FVM and FEM are mixed methods with trial functions that have local support.

The next step in using MWR is to determine the unknown expansion coefficients by minimizing the residual. Weighting functions are subsequently introduced:

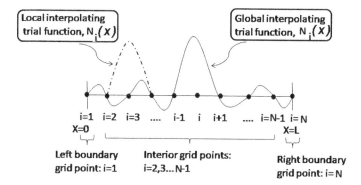

Fig. 3 Illustration of local and global trial functions, $N_i(x)$ [11]

(a) A weighting function, $W_\Omega(x, y)$, for the domain residual $R_\Omega(x, y)$;
(b) A weighting function, $W_{\Gamma T}(x, y)$, for the boundary residual, $R_{\Gamma T}(x, y)$, on the section Γ_T; and
(c) A weighting function, $W_{\Gamma q}(x, y)$, for the boundary residual, $R_{\Gamma q}(x, y)$, on Γ_q of the boundary.

A weighted residual statement is then established to solve for the expansion coefficients. The resulting integral equation is

$$\oiint_\Omega R_\Omega(x, y)W_{\Omega,j}(x, y)d\Omega + \int_{\Gamma_T} R_{\Gamma_T}(x, y)W_{\Gamma_T,j}(x, y)d\Gamma$$

$$+ \int_{\Gamma_q} R_{\Gamma_q}(x, y)W_{\Gamma_q,j}(x, y)d\Gamma = 0 \quad j = 1, 2 \ldots N \tag{6}$$

The choice of the weighting function dictates the type of minimization technique. The most common choices are (1) collocation, (2) subdomain, (3) Galerkin, (4) constant weights, and (5) least squares. We focus on the subdomain (FVM) and Galerkin (FEM) methods.

2.1 Subdomain Method

The domain Ω is subdivided into N-subdomains, Ω_j, as shown in Fig. 4. The weighting function is chosen to be:

$$W_j(\vec{r}) = 1 \; if \; \vec{r} \in \Omega_j$$
$$= 0 \; if \; \vec{r} \notin \Omega_j \tag{7}$$

Fig. 4 The Subdomain method, which subdivides domain Ω into N-subdomains, Ω_j

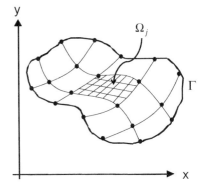

Fig. 5 Subdomain
method—finite volume

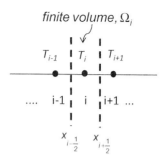

The FVM sits within the MWR subdomain with local shape functions, typically taken as polynomials. The subdomains, known as finite volumes, are generated automatically by mesh generation techniques. The expansion coefficients, α_j, are the FVM nodal temperatures.

In a one-dimensional FVM, a subdomain (finite volume), $\Omega_i \in [x_{i-\frac{1}{2}}, x_{i+\frac{1}{2}}]$ surrounds each grid point where x_i is defined to extend from $x_{i-\frac{1}{2}}$, the halfway mark between i and i − 1, and $x_{i+\frac{1}{2}}$, located at the halfway mark between i and i + 1, as shown in Fig. 5.

To illustrate the simplicity of the procedure, we will use the one-dimensional heat transfer equation over the domain shown in Fig. 6,

$$\frac{d^2 T(x)}{dx^2} + Q = 0 \quad x \in [0, L] \tag{8}$$

$$T(0) = T_o$$
$$T(L) = T_L$$

where Dirichlet (fixed) values are set for the end points. We now integrate over the domain, Ω,

$$\int_{\Omega} \left(\frac{d^2 \tilde{T}(x)}{dx^2} + Q \right) dx = 0, \quad or \quad \int_{\Omega} R(x)dx = 0 \tag{9}$$

The residual equation, R(x), is then multiplied by the weighting function, W(x).

$$\int_{\Omega} \left(\frac{d^2 \tilde{T}(x)}{dx^2} + Q \right) W_i(x)dx = 0, \quad or \quad \int_{\Omega} R(x)W_i(x)dx = 0 \tag{10}$$

Fig. 6 Simple
one-dimensional problem

Setting $W_i(x) = 1$ in each subdomain $\Omega_i \in [x_{i-1/2}, x_{i+1/2}]$, we see that

$$\int\limits_{x_{i-\frac{1}{2}}}^{x_{i+\frac{1}{2}}} \left(\frac{d^2\tilde{T}(x)}{dx^2} + Q \right) dx = 0 \quad for\ i = 2, 3, \ldots N - 1 \tag{11}$$

Note that $W_i(x) = 0$ outside the subdomain, Ω_i. If Q is constant and integrating the second derivative, we obtain the relation

$$\left. \frac{d\tilde{T}}{dx} \right|_{x_{i+\frac{1}{2}}} - \left. \frac{d\tilde{T}}{dx} \right|_{x_{i-\frac{1}{2}}} + Q \int\limits_{x_{i-\frac{1}{2}}}^{x_{i+\frac{1}{2}}} dx = 0 \quad for\ i = 2, 3 \ldots N - 1 \tag{12}$$

The first two terms are related to the flux in and out of the subdomain (finite volume) Ω_i; the integral of the source term requires separate treatment over the finite volume. The FVM expresses a conservation principle on the grid, which becomes important in nonlinear and multidimensional problems.

What remains now is to introduce the approximation for $\tilde{T}(x)$ to arrive at the FVM relation. In FVM, various local interpolations are utilized. Interpolation is generally used between grid points to evaluate $\tilde{T}(x)$ at the finite volume faces. For example,

$$\tilde{T}(x) = \alpha_1 + \alpha_2 x = \begin{cases} \frac{T_{i+1}+T_i}{2} + \left(\frac{T_{i+1}-T_i}{\Delta x} \right) x & for\ x \in [x_i, x_{i+1}] \\ \frac{T_i+T_{i-1}}{2} + \left(\frac{T_i-T_{i-1}}{\Delta x} \right) x & for\ x \in [x_{i-1}, x_i] \end{cases} \tag{13}$$

This results in the following expressions for the derivatives in Eq. (11):

$$\left. \frac{d\tilde{T}}{dx} \right|_{x_{i\pm\frac{1}{2}}} = \begin{cases} \left. \frac{d\tilde{T}}{dx} \right|_{x_{i+\frac{1}{2}}} = \left(\frac{T_{i+1}-T_i}{\Delta x} \right) \\ \left. \frac{d\tilde{T}}{dx} \right|_{x_{i-\frac{1}{2}}} = \left(\frac{T_i-T_{i-1}}{\Delta x} \right) \end{cases} \tag{14}$$

We can employ a simple difference to evaluate the source term integral over the finite volume, i.e.,

$$Q \int\limits_{x_{i-\frac{1}{2}}}^{x_{i+\frac{1}{2}}} dx = Q(x_{i+\frac{1}{2}} - x_{i-\frac{1}{2}}) \tag{15}$$

The final discretized expression becomes

$$\left(\frac{T_{i+1} - T_i}{\Delta x} \right) - \left(\frac{T_i - T_{i-1}}{\Delta x} \right) + Q\Delta x = 0 \tag{16}$$

Dividing by Δx, we arrive at the FVM relation:

$$\left(\frac{1}{\Delta x^2}\right)T_{i-1} - \left(\frac{2}{\Delta x^2}\right)T_i + \left(\frac{1}{\Delta x^2}\right)T_{i+1} = -Q \tag{17}$$

Defining the FVM coefficients, $a_i = \left(\frac{1}{\Delta x^2}\right)$, $b_i = \left(-\frac{2}{\Delta x^2}\right)$, $c_i = \left(\frac{1}{\Delta x^2}\right)$, $d_i = -Q$, and applying the Dirichlet boundary conditions at $x = 0$ and $x = L$, we "assemble" (a common FEM term) the relation over a sequence of adjacent finite volumes, a tri-diagonal matrix is formed that can be solved using the Thomas algorithm, i.e.,

$$\begin{bmatrix} 1 & 0 & 0 & 0 & \cdots & 0 \\ a_2 & b_2 & c_2 & 0 & \cdots & 0 \\ 0 & a_3 & b_3 & c_3 & \cdots & 0 \\ \vdots & \vdots & \ddots & \ddots & \ddots & \vdots \\ 0 & \cdots & 0 & a_{N-1} & b_{N-1} & c_{N-1} \\ 0 & 0 & 0 & 0 & 0 & 1 \end{bmatrix} \begin{Bmatrix} T_1 \\ T_2 \\ T_3 \\ \vdots \\ T_{N-1} \\ T_N \end{Bmatrix} = \begin{Bmatrix} d_1 \\ d_2 \\ d_3 \\ \vdots \\ d_{N-1} \\ d_N \end{Bmatrix} \tag{18}$$

2.2 1D FEM

In the FEM, the weighting function is chosen to be the expansion function itself, that is, $W_j = N_j$. Most often, the FEM is formulated using the Galerkin MWR by using local shape functions, typically taken as polynomials defined over a set of subdomains, Ω_j, called finite elements. The expansion coefficients, α_j, are the FEM nodal temperatures.

In the FEM, the same discretization can be used over each region (or finite element) $\Omega_i \in [x_i, x_{i+1}]$, as shown in Fig. 7.

Fig. 7 Two adjacent linear finite elements

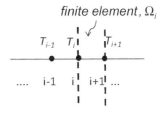

Applying Galerkin minimization:

$$\int_0^L \left(\frac{d^2 \tilde{T}(x)}{dx^2} + Q \right) W_k(x) dx = 0 \tag{19}$$

and given that $W_i(x) = N_i(x)$, where trial (or shape) functions $N_i(x)$ are local and are zero outside the finite element, Ω_i, then

$$\int_{x_i}^{x_{i+1}} \left(\frac{d^2 \tilde{T}(x)}{dx^2} + Q \right) N_k(x) dx = 0 \tag{20}$$

This is called the **strong form** since the approximation $\tilde{T}(x)$ needs to be twice differentiable; hence, linear trial functions cannot be used. In other words, the approximation $\tilde{T}(x)$ is required to be C^2—continuous in the strong form. Consequently, integration by parts is used on the highest order derivative; in two dimensions, this is equivalent to applying Green's first identity. This results in the so-called **weak form** statement:

$$\left(N_k(x) \frac{d\tilde{T}(x)}{dx} \right) \Bigg|_{x_i}^{x_{i+1}} - \int_{x_i}^{x_{i+1}} \left(\frac{d\tilde{T}(x)}{dx} \frac{dN_k}{dx} \right) dx + \int_{x_i}^{x_{i+1}} Q N_k(x) dx = 0 \tag{21}$$

As only first-order derivatives appear, the weak statement admits linear interpolation. Consequently, using linear interpolating functions $N_1(x)$ and $N_2(x)$, we set

$$\tilde{T}(x) = T_i N_1^{(i)}(x) + T_{i+1} N_2^{(i)}(x) \tag{22}$$

The interpolating functions are defined as linear Lagrange interpolating functions within each element, i.e.,

$$\begin{aligned} N_1^{(i)} &= \begin{cases} 1 \; if \; x = x_i \\ 0 \; if \; x = x_{i+1} \end{cases} \\ N_2^{(i)} &= \begin{cases} 1 \; if \; x = x_{i+1} \\ 0 \; if \; x = x_i \end{cases} \end{aligned} \tag{23}$$

Moreover, these functions are defined to be zero outside the finite element i, as seen in Fig. 8.

Fig. 8 Linear shape functions

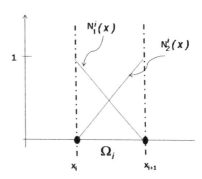

Notice that

$$
\begin{aligned}
N_1^{(i)} &= \begin{cases} \left(\dfrac{x-x_{i+1}}{x_i-x_{i+1}}\right) & if \ x \in [x_i, x_{i+1}] \\ 0 & if \ \notin [x_i, x_{i+1}] \end{cases} \\
N_2^{(i)} &= \begin{cases} \left(\dfrac{x-x_i}{x_{i+1}-x_i}\right) & if \ x \in [x_i, x_{i+1}] \\ 0 & if \ \notin [x_i, x_{i+1}] \end{cases}
\end{aligned} \tag{24}
$$

By setting i = 1, 2 in each element *i*, we arrive at the FEM equation for each finite element:

$$
\begin{bmatrix} \int\limits_{x_i}^{x_{i+1}} \left(\frac{dN_1^i}{dx}\frac{dN_1^i}{dx}\right)dx & \int\limits_{x_i}^{x_{i+1}} \left(\frac{dN_1^i}{dx}\frac{dN_2^i}{dx}\right)dx \\ \int\limits_{x_i}^{x_{i+1}} \left(\frac{dN_2^i}{dx}\frac{dN_1^i}{dx}\right)dx & \int\limits_{x_i}^{x_{i+1}} \left(\frac{dN_2^i}{dx}\frac{dN_2^i}{dx}\right)dx \end{bmatrix} \left\{ \begin{array}{c} T_i \\ T_{i+1} \end{array} \right\} = \left\{ \begin{array}{c} \int\limits_{x_i}^{x_{i+1}} QN_1^i dx \\ \int\limits_{x_i}^{x_{i+1}} QN_2^i dx \end{array} \right\} + \left\{ \begin{array}{c} -\frac{dT}{dx}\big|_{x_i} \\ \frac{dT}{dx}\big|_{x_{i+1}} \end{array} \right\} \tag{25}
$$

In matrix form, the algebraic finite element equation is:

$$
\begin{bmatrix} K_{11}^i & K_{12}^i \\ K_{21}^i & K_{22}^i \end{bmatrix} \left\{ \begin{array}{c} T_i \\ T_{i+1} \end{array} \right\} = \left\{ \begin{array}{c} F_i \\ F_{i+1} \end{array} \right\} \tag{26}
$$

where {F} is termed the load vector and

$$
K_{mn}^i = \int\limits_{x_i}^{x_{i+1}} \left(\frac{dN_m^i}{dx}\frac{dN_n^i}{dx}\right)dx \tag{27}
$$

$$
F_1^i = Q \int\limits_{x_i}^{x_{i+1}} N_1 dx - \frac{dT}{dx}\bigg|_{x_i} = f_1^i - q_i \quad and \quad F_2^i = Q \int\limits_{x_i}^{x_{i+1}} N_2 dx + \frac{dT}{dx}\bigg|_{x_{i+1}}
$$

$$
= f_2^i + q_{i+1} \tag{28}
$$

In the FEM, the integrals in Eqs. (27) and (28) are numerically evaluated using Gauss-Legendre quadratures. Given that at each node, there is an influence from the interpolating function N_1 from element i and from the interpolating function N_2 from element $i - 1$, a nodal equation can be assembled, assuming that the derivatives are continuous at common nodes as:

$$K_{21}^i T_i + (K_{22}^i + K_{11}^{i+1})T_{i+1} + K_{12}^{i+1} T_{i+2} = f_2^i + f_1^{i+1} \quad for \quad i = 2, 3 \ldots N - 1 \tag{29}$$

For the first node, the flux at the first node does not cancel out and:

$$K_{11}^1 T_1 + K_{12}^2 T_2 = f_1^1 - q_1 \quad for \quad i = 1 \tag{30}$$

Similarly, for node N, the flux at node N does not cancel out and:

$$K_{21}^N T_{N-1} + K_{22}^N T_N = f_2^N + q_N \quad for \quad i = N \tag{31}$$

This process is automatically accomplished by loading the local matrix equation into a global matrix equation (known as assembly). The result is again a tri-diagonal matrix set of equations in the form:

$$\begin{bmatrix} a_1 & b_1 & 0 & 0 & \cdots & 0 \\ a_2 & b_2 & c_2 & 0 & \cdots & 0 \\ 0 & a_3 & b_3 & c_3 & \cdots & 0 \\ \vdots & \vdots & \ddots & \ddots & \ddots & \vdots \\ 0 & \cdots & 0 & a_{N-1} & b_{N-1} & c_{N-1} \\ 0 & 0 & 0 & 0 & a_N & b_N \end{bmatrix} \begin{Bmatrix} T_1 \\ T_2 \\ T_3 \\ \vdots \\ T_{N-1} \\ T_N \end{Bmatrix} = \begin{Bmatrix} d_1 \\ d_2 \\ d_3 \\ \vdots \\ d_{N-1} \\ d_N \end{Bmatrix}$$

where $\{d\}$ denotes the right-hand side vector of known values.

The final step in the FEM is to rearrange the above matrix set according to the imposed boundary conditions, moving unknowns to the left and knowns to the right, when necessary. For example, since we imposed the temperatures at nodes $i = 1$ and $i = N$, then q_1 and q_N are unknowns; consequently, the first and last unknowns are switched with corresponding column switches.

$$\begin{bmatrix} 1 & b_1 & 0 & 0 & \cdots & 0 \\ a_2 & b_2 & c_2 & 0 & \cdots & 0 \\ 0 & a_3 & b_3 & c_3 & \cdots & 0 \\ \vdots & \vdots & \ddots & \ddots & \ddots & \vdots \\ 0 & \cdots & 0 & a_{N-1} & b_{N-1} & c_{N-1} \\ 0 & 0 & 0 & 0 & 0 & -1 \end{bmatrix} \begin{Bmatrix} q_1 \\ T_2 \\ T_3 \\ \vdots \\ T_{N-1} \\ q_N \end{Bmatrix} = \begin{Bmatrix} -a_1 T_1 \\ -a_2 T_2 \\ 0 \\ \vdots \\ 0 \\ 0 \end{Bmatrix} + \begin{Bmatrix} 0 \\ 0 \\ 0 \\ \vdots \\ -c_{N-1} T_N \\ -b_N T_N \end{Bmatrix} + \begin{Bmatrix} d_1 \\ d_2 \\ d_3 \\ \vdots \\ d_{N-1} \\ d_N \end{Bmatrix} \tag{32}$$

At this point, this tri-diagonal matrix set of equations can be solved. The integrals in the FEM are typically carried out automatically using Gauss-type quadratures.

Let's examine the following 1D ODE equation and produce the equivalent discretized expressions for each method. Assume a generic variable $\phi(x)$ and the relation

$$\frac{d^2\phi}{dx^2} + \frac{d\phi}{dx} + \phi = 0$$

FVM:

$$\int\limits_{i-1/2}^{i+1/2} \frac{d^2\phi}{dx^2} dx + \int\limits_{i-1/2}^{i+1/2} \frac{d\phi}{dx} dx + \int\limits_{i-1/2}^{i+1/2} \phi dx = 0$$

$$\frac{d\phi}{dx}\Big|_{i+1/2} - \frac{d\phi}{dx}\Big|_{i-1/2} + \phi_{i+1/2} - \phi_{i-1/2} + \int\limits_{i-1/2}^{i+1/2} \phi dx = 0$$

$$\frac{\phi_{i+1}-\phi_i}{\Delta x} - \frac{\phi_i-\phi_{i-1}}{\Delta x} + \frac{\phi_{i+1}+\phi_i}{2} - \frac{\phi_i+\phi_{i-1}}{2} + \int\limits_{i-1/2}^{i+1/2} \phi dx = 0$$

$$\int\limits_{i-1/2}^{i+1/2} \phi dx \rightarrow \text{ use } \tfrac{1}{3}\text{Simpson's Rule}$$

$$\frac{\phi_{i+1}-\phi_i}{\Delta x^2} - \frac{\phi_i-\phi_{i-1}}{\Delta x^2} + \frac{\phi_{i+1}+\phi_i}{2\Delta x} - \frac{\phi_i+\phi_{i-1}}{2\Delta x} + \frac{1}{3}(\phi_{i-1} + \phi_i + \phi_{i+1}) = 0$$

Combining terms,

$$\frac{\phi_{i+1} - 2\phi_i + \phi_{i-1}}{\Delta x^2} + \frac{\phi_{i+1} - \phi_{i-1}}{2\Delta x} + \frac{1}{3}(\phi_{i-1} + \phi_i + \phi_{i+1}) = 0 \qquad (33)$$

FEM:

$$\int\limits_\Omega \frac{d^2\phi}{dx^2} W dx + \int\limits_\Omega \frac{d\phi}{dx} W dx + \int\limits_\Omega \phi dx = 0$$

$$or$$

$$-\int\limits_\Omega \frac{d\phi}{dx}\frac{dW}{dx} dx + W\frac{d\phi}{dx}\Big|_\Gamma + \int\limits_\Omega \frac{d\phi}{dx} W dx + \int\limits_\Omega \phi W dx = 0$$

For the elements spanning nodes $i - 1$ to i

$$\hat{\phi} = N_{i-1}\phi_{i-1} + N_i\phi_i$$
$$\text{with } N_{i-1} = \tfrac{x_i-x}{\Delta x}, N_i = \tfrac{x-x_{i-1}}{\Delta x}$$
$$\text{now let } W_i = N_i$$

$$-\int\limits_\Omega \frac{dN_{i-1}}{dx}\frac{dN_i}{dx} dx \left\{\begin{matrix} \phi_{i-1} \\ \phi_i \end{matrix}\right\} + N_i\frac{d\phi}{dx_\Gamma} + \int\limits_\Omega \frac{dN_{i-1}}{dx} N_i dx \left\{\begin{matrix} \phi_{i-1} \\ \phi_i \end{matrix}\right\}$$

$$+ \int\limits_\Omega N_{i-1} N_i dx \left\{\begin{matrix} \phi_{i-1} \\ \phi_i \end{matrix}\right\} = 0$$

$$\text{now use } \int N_1^a N_2^b dx = \Delta x \frac{a!b!}{(a+b+1)!} \rightarrow \text{ exact integration}$$

In this simple case, we have been able to use exact integration, producing the following generic expression

$$-\frac{1}{\Delta x^2}\int\left\{\begin{matrix}-1\\1\end{matrix}\right\}\underline{[-1\ \ 1]}dx\left\{\begin{matrix}\phi_{i-1}\\\phi_i\end{matrix}\right\}+\frac{1}{\Delta x}\int\left\{\begin{matrix}N_{i-1}\\N_i\end{matrix}\right\}\underline{[-1\ \ 1]}dx\left\{\begin{matrix}\phi_{i-1}\\\phi_i\end{matrix}\right\}+\int\left\{\begin{matrix}N_{i-1}\\N_i\end{matrix}\right\}\underline{[N_{i-1}\ \ N_i]}dx\left\{\begin{matrix}\phi_{i-1}\\\phi_i\end{matrix}\right\}=0$$

$$-\frac{1}{\Delta x^2}\int\begin{bmatrix}1&-1\\-1&1\end{bmatrix}dx\left\{\begin{matrix}\phi_{i-1}\\\phi_i\end{matrix}\right\}+\frac{1}{\Delta x}\int\begin{bmatrix}-N_{i-1}&N_{i-1}\\-N_i&N_i\end{bmatrix}dx\left\{\begin{matrix}\phi_{i-1}\\\phi_i\end{matrix}\right\}+\int\begin{bmatrix}N_{i-1}^2&N_{i-1}N_i\\N_{i-1}N_i&N_i^2\end{bmatrix}dx\left\{\begin{matrix}\phi_{i-1}\\\phi_i\end{matrix}\right\}=0$$

$$-\frac{1}{\Delta x}\begin{bmatrix}1&-1\\-1&1\end{bmatrix}\left\{\begin{matrix}\phi_{i-1}\\\phi_i\end{matrix}\right\}+\frac{1}{2}\begin{bmatrix}-1&1\\-1&1\end{bmatrix}\left\{\begin{matrix}\phi_{i-1}\\\phi_i\end{matrix}\right\}+\frac{\Delta x}{6}\begin{bmatrix}2&1\\1&2\end{bmatrix}\left\{\begin{matrix}\phi_{i-1}\\\phi_i\end{matrix}\right\}=0$$

Notice how the integral terms are evaluated, e.g.,

$$\int_\Omega N_{i-1}N_i\,dx=\Delta x\begin{bmatrix}\frac{2!0!}{(1+2+0)!}&\frac{1!1!}{(1+1+1)!}\\\frac{1!1!}{(1+1+1)!}&\frac{0!2!}{(1+2+0)!}\end{bmatrix}=\frac{\Delta x}{6}\begin{bmatrix}2&1\\1&2\end{bmatrix}$$

$$\int_\Omega\frac{dN_{i-1}}{dx}\frac{dN_i}{dx}\,dx=\frac{1}{\Delta x}\left\{\begin{matrix}-1\\1\end{matrix}\right\}[-1\ \ 1]=\frac{1}{\Delta x}\begin{bmatrix}1&-1\\-1&1\end{bmatrix}$$

Assembling over two adjacent elements involving the nodal values and performing the assembly over each element produces the following two relations,

for $\Delta x_- = x_i - x_{i-1}$

$$-\frac{1}{\Delta x_-}\begin{bmatrix}1&-1\\-1&1\end{bmatrix}\left\{\begin{matrix}\phi_{i-1}\\\phi_i\end{matrix}\right\}+\frac{1}{2}\begin{bmatrix}-1&1\\-1&1\end{bmatrix}\left\{\begin{matrix}\phi_{i-1}\\\phi_i\end{matrix}\right\}+\frac{\Delta x_-}{6}\begin{bmatrix}2&1\\1&2\end{bmatrix}\left\{\begin{matrix}\phi_{i-1}\\\phi_i\end{matrix}\right\}=0$$

for $\Delta x_+ = x_{i+1} - x_i$

$$-\frac{1}{\Delta x_+}\begin{bmatrix}1&-1\\-1&1\end{bmatrix}\left\{\begin{matrix}\phi_i\\\phi_{i+1}\end{matrix}\right\}+\frac{1}{2}\begin{bmatrix}-1&1\\-1&1\end{bmatrix}\left\{\begin{matrix}\phi_i\\\phi_{i+1}\end{matrix}\right\}+\frac{\Delta x_+}{6}\begin{bmatrix}2&1\\1&2\end{bmatrix}\left\{\begin{matrix}\phi_i\\\phi_{i+1}\end{matrix}\right\}=0$$

Combining both equations produces a *global* 3 × 3 matrix (known as the stiffness matrix) which is used to establish the recursion relation

$$\frac{1}{6}\begin{bmatrix}2\Delta x_-&\Delta x_-&0\\\Delta x_-&2\Delta x_-+2\Delta x_+&\Delta x_+\\0&\Delta x_+&2\Delta x_+\end{bmatrix}\left\{\begin{matrix}\phi_{i-1}\\\phi_i\\\phi_{i+1}\end{matrix}\right\}+\frac{1}{2}\begin{bmatrix}-1&1&0\\-1&0&1\\0&-1&1\end{bmatrix}\left\{\begin{matrix}\phi_{i-1}\\\phi_i\\\phi_{i+1}\end{matrix}\right\}$$

$$+\frac{1}{6}\begin{bmatrix}\frac{1}{\Delta x_-}&-\frac{1}{\Delta x_-}&0\\-\frac{1}{\Delta x_-}&\frac{1}{\Delta x_-}+\frac{1}{\Delta x_+}&-\frac{1}{\Delta x_+}\\0&-\frac{1}{\Delta x_+}&\frac{1}{\Delta x_+}\end{bmatrix}\left\{\begin{matrix}\phi_{i-1}\\\phi_i\\\phi_{i+1}\end{matrix}\right\}=0$$

Extracting only the central terms (those involving the middle node—see [19]),

$$\frac{1}{6}[\phi_{i-1}\cdot\Delta x_-+\phi_i\cdot(\Delta x_-+\Delta x_+)+\phi_{i+1}\cdot\Delta x_+]+\frac{1}{2}[-\phi_{i-1}+\phi_{i+1}]$$

$$+ \left[\frac{\phi_{i-1}}{\Delta x_-} - \phi_i \left(\frac{1}{\Delta x_-} + \frac{1}{\Delta x_+} \right) + \frac{\phi_{i+1}}{\Delta x_+} \right] = 0$$

If $\Delta x_- = \Delta x_+$, one obtains a relation that looks similar to the FVM.

$$\frac{\Delta x}{6} \left[\phi_{i-1} + 2\phi_i + \phi_{i+1} \right] + \frac{\phi_{i+1} - \phi_{i-1}}{2} + \frac{\phi_{i+1} - 2\phi_i + \phi_{i-1}}{\Delta x} = 0 \qquad (34)$$

The FEM yields the same second-order truncation error as the FVM (which is Δx^2). So why is it that the FEM is normally rated as being Δx^4? It turns out that this occurs when the time-dependent term appears in the equation. A more thorough discussion on this issue is described in Pepper and Baker (1979).

The popularity of the FEM approach is based on the assumption that the Galerkin method yields more accurate results. This belief stems from the fact that the Galerkin method seeks to render the set of residues orthogonal to the set of trial functions, $N(x)$, by using the shape functions as the weighting functions. If the domain is subdivided into an infinite number of elements, then the set of shape functions forms a complete set of functions. Any function that is orthogonal to every member of a complete set of functions must be identically zero, i.e., implying that the set of residuals must be zero, or the approximate solution close to exact. However, when the set of trial functions is finite, this is not true. In this case, the type of trial function and the geometry of the element can influence the accuracy. It should be noted that conservation equations represent flux balances over the control volumes—as the control volume approaches zero, the approximation approaches the original differential equation, also common to finite difference methods utilizing Taylor series expansions.

3 The FVM-FEM 2D Analogy

We now examine the 2D equivalent discretization for the FVM and FEM. There are several options open for the choice of FEM element—triangle, quadrilateral (bilinear), and quadrilateral (Lagrangian). These are listed in the Appendix. Here we are looking at the 9-node molecule established by the 2D mesh, as shown in Fig. 9.

Assume a simple Poisson equation where $\phi(x, y)$ represents a field variable,

$$K \nabla^2 \phi(x, y) + Q = 0 \qquad (32)$$

where K denotes a constant exchange (diffusion) coefficient and Q is a source/sink term. Also assume a rectangular domain over which Eq. (32) is to be solved, as shown in Fig. 10. The domain is discretized into a set of linear, 3-node triangles (dotted lines) along with rectangular bilinear elements (solid lines) overlaid on the FVM control volume (dashed lines).

In the FVM, the corners of the discretized volume are typically denoted as North (N), South (S), East (E), West (W), Northwest (NW), etc. Focusing on point P, both

Fig. 9 2D Nodal molecule

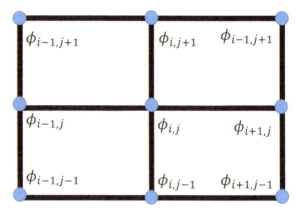

Fig. 10 2D element
discretization of the finite
volume

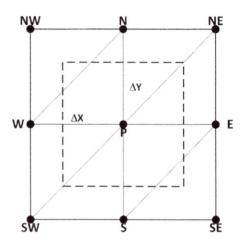

the Galerkin and the FVM produce the same relations,

$$2\left(\frac{\Delta y}{\Delta x} + \frac{\Delta x}{\Delta y}\right)\phi_P = \frac{\Delta x}{\Delta y}(\phi_N + \phi_S) + \frac{\Delta y}{\Delta x}(\phi_E + \phi_W) + Q\Delta x \Delta y/K$$

If $\Delta x = \Delta y = \Delta$, the equation reduces to

$$4\phi_P = \phi_N + \phi_S + \phi_E + \phi_W + Q\Delta^2/K$$

This is the typical 5-node molecule also commonly attributed to the finite difference method. Both the FVM and FEM produce the same accuracy.

Examining Fig. 10, the bilinear trial function for the FEM is defined as

$$N_P = a_P + b_P x + c_P y + d_P xy$$

which applies over each element. In this case, the FVM and the FEM differ from each other. For the bilinear quadrilateral FEM for node P,

$$8\left(\frac{\Delta y}{\Delta x} + \frac{\Delta x}{\Delta y}\right)\phi_P = \left(4\frac{\Delta x}{\Delta y} - 2\frac{\Delta y}{\Delta x}\right)(\phi_N + \phi_S) + \left(4\frac{\Delta y}{\Delta x} - 2\frac{\Delta x}{\Delta y}\right)(\phi_E + \phi_W)$$
$$+ \left(\frac{\Delta y}{\Delta x} + \frac{\Delta x}{\Delta y}\right)(\phi_{NE} + \phi_{NW} + \phi_{SE} + \phi_{SW}) + 6Q\Delta x \Delta y / K$$

If $\Delta x = \Delta y = \Delta$,

$$8\phi_P = \phi_N + \phi_S + \phi_E + \phi_W + \phi_{NE} + \phi_{NW} + \phi_{SE} + \phi_{SW} + 3Q\Delta^2/K$$

The control volume method for the FVM produces

$$12\left(\frac{\Delta y}{\Delta x} + \frac{\Delta x}{\Delta y}\right)\phi_P = 2\left(3\frac{\Delta x}{\Delta y} - \frac{\Delta y}{\Delta x}\right)(\phi_N + \phi_S) + 2\left(3\frac{\Delta y}{\Delta x} - \frac{\Delta x}{\Delta y}\right)(\phi_E + \phi_W)$$
$$+ \left(\frac{\Delta y}{\Delta x} + \frac{\Delta x}{\Delta y}\right)(\phi_{NE} + \phi_{NW} + \phi_{SE} + \phi_{SW}) + 8Q\Delta x \Delta y / K$$

which for $\Delta x = \Delta y = \Delta$,

$$12\phi_P = 2\phi_N + 2\phi_S + 2\phi_E + 2\phi_W + \phi_{NE} + \phi_{NW} + \phi_{SE} + \phi_{SW} + 4Q\Delta^2/K$$

We see that the Galerkin method, when employing bilinear quadrilateral elements, produces equal importance to all the neighbors of point P. In the FVM, the nearer side neighbors N, S, E, and W produce stronger influence coefficients than the corner neighbors, and maybe the more plausible relation. Simple test cases show that the Galerkin method can produce errors about twice as large as those produced by the FVM, with the finite difference method essentially equal in magnitude to the Galerkin method [3, 8]. However, the FEM has the added flexibility of being able to incorporate p-refinement, i.e., enhanced trial function selection. If one elects to use quadratic quadrilaterals or higher interpolation, the accuracy increases several orders of magnitude, but at the expense of additional storage and compute times.

Other FEM trial function choices for the discretization of the domain include the 9-node quadratic Lagrangian quadrilateral and the 4-triangular element discretization about the nodes P, N, S, E, and W. The aspects of using these elements is discussed in [20].

4 Conclusion

The FVM and FEM are essentially cousins within the method of weighted residuals. The FVM system can also be formulated directly from evaluating the fluxes through a control volume, making it perhaps more intuitively developed, unlike the MWR and the FEM system. The idea of a residual formulation is intriguing, and doesn't require a leap of faith regarding conservation laws being applied.

We simply understand that a discrete approximation of the continuum must have some associated error at the outset of a methods development. When the two methods are formulated by similar constructions within the method of weighted residual framework and employing subdomain weighting, the intuitive conservation laws are not as easily identified. Being formed directly from the control volume upon which they are often directly developed, the FVM system of equations is easy to apply to the physics being modeled, i.e., there is little calculus required between FVM and the physics. The FVM, in its more modest development as presented here, is easily related to finite difference methods employing first- and second- order expansions of derivatives by Taylor series.

While easy to apply to linear and nonlinear continuum equations, such as the Navier–Stokes equations for fluid dynamics, the FVM has some advantages in accuracy over simple FEM methodology. However, the FEM finds advantages in the more robust mathematical underpinning of Hilbert inner product spaces of the weighted residual, allowing for measures of error in the solution, and knowing where those error bounds exist. At the nodes of the FEM system, the solutions are exact for linear Laplacian and Poisson systems—known as super convergence—yet there exists error within the element. Usual formulations for the FVM have piecewise continuous solutions, e.g., constant dependent variable properties within the cell. The FEM's material properties and dependent variables are continuously approximated to the order of the basis or shape function that is applied to the elements. Note that constructing higher order FVM approximations is more complicated than in the FEM; simply adding higher order orthogonal polynomials that increases the accuracy of FEM.

FVM methods were nearly the first numerical method on the scene for solving fluid dynamics problems thanks to the great insight of Professor Spalding. Another method arrived at about the same time as FVM for the solution of fluid dynamics was Frank Harlow's marker and cell system, or MAC [21]. However, MAC was not nearly as popular as the more ubiquitous FVM approach now found in most commercial CFD products. As more use of the FVM method has continued over the years, its ability to resolve many of the continuum problems for material motion has evolved with increased accuracy [22]. The FEM found its way into CFD once the advection terms were properly handled to adjust or mitigate the numerical dispersion created by what is essentially a Taylor series central difference expansion that occurs when using linear basis functions. This enhanced advection development for the FEM, occurring nearly 20 years after the advent of FVM, has led to larger popularity of the method in recent years for CFD.

Appendix

1D FEM Formulation with Advection

For simplicity, we define $\phi(x, t)$ as a 1D scalar transport variable. Linear shape functions are used for two consecutive adjacent linear elements. After assembly (creating a 3×3 global matrix), the recursion relation for the ith node can be established.

Setting $W(x) = N(x)$ and using the integral relation for the 1D element spanning nodes i-1 to i, the governing equation for time-dependent transport and diffusion,

$$\frac{\partial \phi}{\partial t} + U \frac{\partial \phi}{\partial x} - K \frac{\partial^2 \phi}{\partial x^2} = 0$$

can be written numerically as

$$\int N_i N_{i-1} dx \begin{Bmatrix} \dot{\phi}_{i-1} \\ \dot{\phi}_i \end{Bmatrix} + \int U N_j N_i \frac{dN_{i-1}}{dx} dx \begin{Bmatrix} \phi_{i-1} \\ \phi_i \end{Bmatrix} + K \int \frac{dN_i}{dx} \frac{dN_{i-1}}{dx} dx \begin{Bmatrix} \phi_{i-1} \\ \phi_i \end{Bmatrix} = 0 \tag{33}$$

where $\dot{\phi}(x, t)$ denotes time dependence and U(x) is velocity. The integral terms are similarly evaluated for elements spanning i to $i + 1$. Thus, for the two adjacent linear elements,

For $\Delta x_- = x_i - x_{i-1}$

$$\frac{\Delta x_-}{6} \begin{bmatrix} 2 & 1 \\ 1 & 2 \end{bmatrix} \begin{Bmatrix} \dot{\phi}_{i-1} \\ \dot{\phi}_i \end{Bmatrix} - \frac{U}{2} \begin{bmatrix} -1 & 1 \\ -1 & 1 \end{bmatrix} \begin{Bmatrix} \phi_{i-1} \\ \phi_i \end{Bmatrix} - \frac{K}{\Delta x_-} \begin{bmatrix} 1 & -1 \\ -1 & 1 \end{bmatrix} \begin{Bmatrix} \phi_{i-1} \\ \phi_i \end{Bmatrix} = 0 \tag{34}$$

For $\Delta x_+ = x_{i+1} - x_i$

$$\frac{\Delta x_+}{6} \begin{bmatrix} 2 & 1 \\ 1 & 2 \end{bmatrix} \begin{Bmatrix} \dot{\phi}_i \\ \dot{\phi}_{i+1} \end{Bmatrix} - \frac{U}{2} \begin{bmatrix} -1 & 1 \\ -1 & 1 \end{bmatrix} \begin{Bmatrix} \phi_i \\ \phi_{i+1} \end{Bmatrix} - \frac{K}{\Delta x_+} \begin{bmatrix} 1 & -1 \\ -1 & 1 \end{bmatrix} \begin{Bmatrix} \phi_i \\ \phi_{i+1} \end{Bmatrix} = 0 \tag{35}$$

For simplicity, the velocity is assumed constant. The use of hypermatrix formulation is needed to deal with variable velocity—more details are described in [19].

Assembling the two adjacent elements into a 3×3 global matrix,

$$\frac{1}{6} \begin{bmatrix} 2\Delta x_- & \Delta x_- & 0 \\ \Delta x_- & 2\Delta x_- + 2\Delta x_+ & \Delta x_+ \\ 0 & \Delta x_+ & 2\Delta x_+ \end{bmatrix} \begin{Bmatrix} \dot{\phi}_{i-1} \\ \dot{\phi}_i \\ \dot{\phi}_{i+1} \end{Bmatrix} + \frac{U}{2} \begin{bmatrix} -1 & 1 & 0 \\ -1 & 0 & 1 \\ 0 & -1 & 1 \end{bmatrix} \begin{Bmatrix} \phi_{i-1} \\ \phi_i \\ \phi_{i+1} \end{Bmatrix}$$

$$+ \frac{K}{6} \begin{bmatrix} \frac{1}{\Delta x_-} & \frac{-1}{\Delta x_-} & 0 \\ \frac{-1}{\Delta x_-} & \frac{1}{\Delta x_-} + \frac{1}{\Delta x_+} & \frac{-1}{\Delta x_+} \\ 0 & \frac{-1}{\Delta x_+} & \frac{1}{\Delta x_+} \end{bmatrix} \begin{Bmatrix} \phi_{i-1} \\ \phi_i \\ \phi_{i+1} \end{Bmatrix} = 0 \tag{36}$$

Setting $\Delta x_- = \Delta x_+ = \Delta x$, the recursion relation in 1D is obtained by stripping out the central expression within the global matrix, i.e.,

$$\frac{1}{6}\left(\dot{\phi}_{i-1} + 4\dot{\phi}_i + \dot{\phi}_{i+1}\right) + \frac{U}{2\Delta x}(\phi_{i+1} - \phi_{i-1}) - \frac{K}{\Delta x^2}(\phi_{i+1} - 2\phi_i + \phi_{i-1}) = 0 \tag{37}$$

which appears similar to the central difference FDM (or FVM). The improved accuracy of the FEM lies in the time-dependent terms, $\dot{\phi}$. When employing Crank–Nicolson time averaging, Eq. (37) yields $O(\Delta x^4)$ versus $O(\Delta x^2)$ in space.

2D FEM Formulation with Advection

The general transport equation in 2D can be discretized using either 2D triangular elements or quadrilateral elements and then assembled over eight triangular elements, four quadrilateral elements, or one quadratic Lagrangian quadrilateral element, consecutively, to establish a set of recursion relations. The resulting set of recursion relations can then be solved using the Strongly Implicit Procedure (SIP), or with a Modified SIP [20, 22]. A more detailed discussion on the formulation of FEM and FVM methods that includes advection and upwinding is provided in Idelsohn and Onate [23].

The matrix equivalent equation of the transport relation (assume $\phi(x, y) \equiv T(x, y)$ for simplicity) can be written as

$$[M]\{\dot{T}\} + [A]\{T\} - [K]\{T\} = 0 \tag{38}$$

where [M] is the mass matrix, [A(V)] is the advection matrix, and [K] is the diffusion matrix

$$[M] = \iint N_i N_j dx dy \tag{39}$$

$$[A(V(u, v))] = \iint \left(u \cdot N_i N_j \frac{\partial N_k}{\partial x} + v \cdot N_i N_j \frac{\partial N_k}{\partial y} \right) dx dy \tag{40}$$

$$[K] = \iint K \left(\frac{\partial N_i}{\partial x} \frac{\partial N_j}{\partial x} + \frac{\partial N_i}{\partial y} \frac{\partial N_j}{\partial y} \right) dx dy \tag{41}$$

where the i, j, k indices denote row and column vectors [20]. After establishing the global matrices for a patch consisting of 8 bilinear triangular elements, 4 bilinear quadrilateral elements, and one Lagrangian quadratic quadrilateral element, a set of recursion relations for the $T_{i,j}$ node can be established.

2D BASIS Functions

Bilinear triangular and quadratic elements are used for the 2D trial functions [19, 20]. The general 2D bilinear element configurations are shown in Figs. 11 and 12. Notice that the triangular element array permits either a 9-node configuration or a 5-node configuration.

The set of finite element expressions, defined over a rectangular subspace and formulated using chapeau basis functions, can be interpreted as integrated averaged difference approximations. After applying Galerkin's method, integrating by parts, and using isoparametric transformations, we obtain the equations as follows:

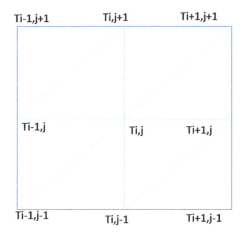

Fig. 11 Basis function-triangular element

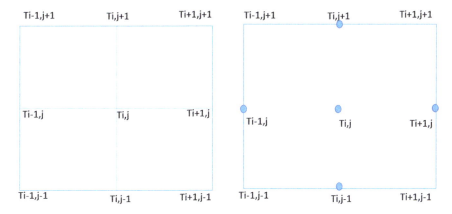

Fig. 12 Basis function-quadrilateral element

2D bilinear triangular elements:

$$\frac{1}{36}\Big[\big(3\big(\dot{T}_{i+1,j+1} + \dot{T}_{i-1,j-1} + \dot{T}_{i+1,j} + \dot{T}_{i-1,j} + \dot{T}_{i,j+1} + \dot{T}_{i,j-1}\big) + 18\dot{T}_{i,j}\big)\Big]$$
$$+ \frac{K}{\Delta x^2}\big[T_{i-1,j} - 2T_{i,j} + T_{i+1,j}\big] + \frac{K}{\Delta y^2}\big[T_{i,j+1} - 2T_{i,j} + T_{i,j-1}\big]$$
$$+ \frac{u}{6\Delta x}\big[+T_{i,j+1} - 2T_{i+1,j} + 2T_{i-1,j} - T_{i,j-1}\big]$$
$$+ \frac{v}{6\Delta y}\big[T_{i-1,j} - T_{i-1,j-1} + 2T_{i,j+1} - 2T_{i,j-1} + T_{i+1,j+1} - T_{i+1,j}\big] = 0 \quad (42)$$

2D bilinear quadrilateral elements:

$$\frac{1}{9}\left[\left(\begin{array}{c}\left(\dot{T}_{i-1,j+1}+\dot{T}_{i+1,j+1}+\dot{T}_{i-1,j-1}+\dot{T}_{i+1,j-1}\right)\\+4\left(\dot{T}_{i-1,j}+\dot{T}_{i+1,j}+\dot{T}_{i,j+1}+\dot{T}_{i,j-1}\right)+16\dot{T}_{i,j}\end{array}\right)\right]$$

$$+\frac{K}{6\Delta x^2}\left[\begin{array}{c}-T_{i-1,j+1}-4T_{i-1,j}-T_{i-1,j-1}+2T_{i,j+1}\\+8T_{i,j}+2T_{i,j-1}-T_{i+1,j+1}-4T_{i+1,j}-T_{i+1,j-1}\end{array}\right]$$

$$+\frac{K}{6\Delta y^2}\left[\begin{array}{c}-T_{i-1,j+1}+2T_{i-1,j}-T_{i-1,j-1}-4T_{i,j+1}\\+8T_{i,j}-4T_{i,j-1}-T_{i+1,j+1}+2T_{i+1,j}-T_{i+1,j-1}\end{array}\right]$$

$$+\frac{u}{6\Delta x}\left[\begin{array}{c}-T_{i-1,j+1}-4T_{i-1,j}-T_{i-1,j-1}\\+T_{i+1,j+1}+4T_{i+1,j}+T_{i+1,j-1}\end{array}\right]$$

$$+\frac{v}{6\Delta y}\left[\begin{array}{c}T_{i-1,j+1}-T_{i-1,j-1}\\+4T_{i,j+1}-4T_{i,j-1}+T_{i+1,j+1}-T_{i+1,j-1}\end{array}\right]=0 \qquad (43)$$

2D Quadratic Lagrangian quadrilateral element:

$$\frac{1}{225}\left[\left(\begin{array}{c}4\left(\dot{T}_{i-1,j+1}+\dot{T}_{i-1,j-1}+\dot{T}_{i+1,j+1}+\dot{T}_{i+1,j-1}\right)\\+32\left(\dot{T}_{i-1,j}+\dot{T}_{i,j+1}+\dot{T}_{i,j-1}+\dot{T}_{i+1,j}\right)+256\dot{T}_{i,j}\end{array}\right)\right]$$

$$+\frac{K}{45\Delta x^2}\left[\begin{array}{c}-8T_{i-1,j+1}-64T_{i-1,j}-8T_{i-1,j-1}+16T_{i-1,j}+128T_{i,j}\\+16T_{i,j-1}-8T_{i+1,j+1}-64T_{i+1,j}-8T_{i+1,j-1}\end{array}\right]$$

$$+\frac{K}{45\Delta y^2}\left[\begin{array}{c}-8T_{i-1,j+1}+16T_{i-1,j}-8T_{i-1,j-1}-64T_{i-1,j}\\+128T_{i,j}-64T_{i,j-1}-8T_{i+1,j+1}+16T_{i+1,j}-8T_{i+1,j-1}\end{array}\right] \qquad (44)$$

$$+\frac{u}{45\Delta x}\left[\begin{array}{c}-4T_{i-1,j+1}-32T_{i-1,j}-4T_{i-1,j-1}\\+4T_{i+1,j+1}+32T_{i+1,j}+4T_{i+1,j-1}\end{array}\right]$$

$$+\frac{V}{45\Delta y}\left[\begin{array}{c}4T_{i-1,j+1}-4T_{i-1,j-1}+32T_{i-1,j}\\-32T_{i,j-1}+4T_{i+1,j+1}-4T_{i+1,j-1}\end{array}\right]=0$$

References

1. Zienkiewicz, O. C., & Taylor, R. L. (1989). The finite element method. *Basic Formulation and Linear Problems*, 4th Ed., McGraw-Hill, London, UK.
2. Patankar, S. V. (1980). *Numerical heat transfer and fluid flow*. Hemisphere, NY.
3. Gosman, A. D., Pun, W. M., Runchal, A. K., Spalding, D. B., & Wolfshtein, M. (1969). *Heat and mass transfer in recirculating flows*. London: Academic Press.
4. Ansys Fluent (2019). Ansys Inc., Canonsburg, PA.
5. Argyris, J. H. (1954). *Recent advances in matrix methods of structural analysis*. Pergamon Press, Elmsford, NY.
6. Turner, M., Clough, R. W., Martin, H., & Topp, L. (1956). Stiffness and deflection of complex structures. *Journal of Aerosol Science, 23*, 805–823.

7. Clough, R. W. (1960). The finite element method in plane stress analysis. In *Proceedings of the 2nd Conference on Electronic Computation*, ASCE, Pittsburgh, PA (pp. 345–378).
8. Patran 2019 Installation and Operations Guide, 2019, MSC Software Corporation, Newport Beach, CA, 132 p.
9. Ansys 19.2, 2019, Ansys Inc., Canonsburg, PA.
10. COMSOL 5.4, 2019, User's Manual, COMSOL, Inc., Burlington, MA.
11. Pepper, D. W., Kassab, A. J., & Divo, E. A. (2014). *Introduction to finite element, boundary element, and meshless methods*, ASME Press, NY.
12. Roache, P. J. (1998). *Fundamentals of computational fluid dynamics*. Albuquerque, New Mexico: Hermosa Press.
13. Fletcher, C. A. J. (1991). *Computational techniques for fluid dynamics*, Vol. I and II, Springer, New York.
14. Finlayson, B. A. (1972). *The method of weighted residuals and variational principles*. New York: Academic Press.
15. Pepper, D., & Heinrich, J. (2017). *The finite element method: basic concepts and applications with MATLAB, Maple, and COMSOL* (3rd ed.). Boca Raton, FL: CRC Press.
16. Gottlieb, D., & Orzag, S. A. (1977). *Numerical analysis of spectral methods: theory and applications*. Bristol, England: Society for Industrial and Applied Mathematics.
17. Maday, Y., & Quateroni, A. (1982). Spectral and pseudo-spectral approximations of the Navier–Stokes equations. *SIAM Journal of Numerical Analysis, 19*(4), 761–780.
18. Gray, W. G., & Pinder, G. F. (1976). On the relationship between the finite element and finite difference methods. *International Journal for Numerical Methods in Engineering, 10*, 893–923.
19. Pepper, D. W., & Baker, A. J. (1979). A simple one-dimensional finite element algorithm with multidimensional capabilities. *Numerical Heat Transfer, 2*, 81–95.
20. Pirbastami, S., & Pepper, D. W. (2018). *Two-dimensional chapeau function recursion relations*, WCCM XIII, July 22–27, NY.
21. Harlow, F. H., & Amsden, A. A. (1971). *Fluid Dynamics*, A LASL Monograph, LA-4700, June 1971.
22. Schneider, G. E., & Zedan, M. (1981). A modified strongly implicit procedure for the numerical solution of field problems. *Numerical Heat Transfer, 4*, 1–19.
23. Idelsohn, S. R., & Onate, E. (1994). Finite volumes and finite elements: two 'Good Friends'. *International Journal for Numerical Methods in Engineering, 37*, 3323–3341.

CFD of the Future: Year 2025 and Beyond

Akshai Kumar Runchal and Madhukar M. Rao

1 Introduction

1.1 Background

In 2012, the first author [1] talked about the future of CFD and the CFD of the future. There, it was "predicted" that developments in CFD would result in CFD being "embedded" inside specific applications and integrated with EVR (Engineering Virtual Reality) software and design tools. As of today, it appears that that day is still in the future. Indeed one could even say that developments in CFD have stagnated to some extent—or at least, not shown any significant revolutionary changes from what the situation was a decade ago. To be sure, there have been advances in many aspects; computer systems are more capable, software is more user-friendly, and the industry has grown substantially. According to the imarc market research group [2], the global CFD market was worth about US$ 1.6 Billion in 2018 and is projected to grow to about US$ 3.1 Billion in 2024, registering a compound annual growth rate (CAGR) of about 11%. However, the fundamental limitations of the last decade still continue to plague us today. In fact, an early annual review discussing the prospects for computational fluid dynamics [3] mentions similar limitations. One of the major drawbacks is the fact that, for many practical applications, the computational time is excessive, if not impractical. Yet, the solution algorithms are still variants of what we used in the eighties and the early nineties, albeit with some improvements.

However, we believe that a fundamental driving force for change has appeared in the last four or five years—machine learning. This technology is already transforming our lives and we believe that it will have a substantial impact on almost every

A. K. Runchal (✉) · M. M. Rao
CFD Virtual Reality Institute, Analytic & Computational
Research, Inc, 1931 Stradella Road, Bel Air, CA 90077, USA
e-mail: runchal@ACRiCFD.com
URL: http://www.ACRiCFD.com; http://www.CFDVRinstitute.org

© Springer Nature Singapore Pte Ltd. 2020
A. Runchal (ed.), *50 Years of CFD in Engineering Sciences*,
https://doi.org/10.1007/978-981-15-2670-1_22

aspect of life. Of course, that includes CFD (and structural FEM, thermal sciences, computational reacting flows, and other simulation sciences). Here, we will explore the potential impact of these developments in CFD and the computational sciences.

1.2   Current Limitations

The limitations that were discussed in [1] are still alive and well today. For convenience, we reproduce them below:

1. Translating Geometry to Grid

    a. Time consuming and difficult to automate
    b. Difficult to produce good quality grids
    c. Non-optimal and unnecessarily large grids for complex geometries

2. Ease of Use

    a. Still needs Ph.Ds
    b. Complex Initial and Boundary conditions
    c. Accommodating disparate Physical processes

3. Speed of Computations

    a. Most codes of today require solution of large matrices
    b. Inefficient codes for highly parallel and grid computing systems

4. Robustness

    a. Can you leave your simulation unattended?
    b. Large matrices and iterative solvers
    c. Subject to truncation errors and non-convergence

5. Lack of Knowledge of Physics

    a. Turbulence
    b. Chemistry
    c. Nano-scale physics
    d. Nonlinearity, non-equilibrium, etc.
    e. Thermal radiation
    f. Multiphase flow

6. Uncertainty and Stochastic Processes

    a. What is the nature of the uncertainty?
    b. How to account for it?

7. Confidence in Predictions

    a. Verification and Validation
    b. Design input considerations

Each of these influences, as well as limits, the practice of CFD of today. That remark is as true today as it was a decade or so ago.

2 The Future of CFD: Impact of AI and Machine Learning

2.1 Resurgence in Artificial Intelligence/Machine Learning Technology

The past few years have seen a lot of interest in statistical machine learning and related applications. In particular, artificial neural networks have been enjoying a resurgence since 2012, especially in the field of deep neural networks and machine vision [4, 5]. Most of us have experienced firsthand, the profound impact of AI and big data in our everyday lives. Self-driving cars, natural language interfaces, speech recognition, handwriting recognition, machine translation between human languages, medical imaging, are, but, a few examples. It is, then, logical to examine its impact on CFD and computational simulation technology. In the following, we will offer a brief description of the profound ways in which AI is beginning to have an impact on CFD (for example, physics informed deep neural networks [6, 8]) and then we will try to "see" where things are heading over the next several years.

The artificial neural network (ANN) appears to be the most powerful machine learning technique, at present. Figure 1 shows the schematic of an artificial neural network. A typical feedforward ANN consists of an input layer, one or more hidden layers and an output layer. Each layer takes its input from the preceding layer, performs a linear transformation with a set of weights followed by a nonlinear transformation via an activation function and delivers its output to the next layer. This is shown in Eq. (1),

$$O_i = \sigma \left(\sum_j w_{ij} q_i + b_i \right) \tag{1}$$

where σ is a suitable nonlinear function such as the sigmoid, hyperbolic tanh, rectified linear unit, etc. Artificial neural networks with sufficiently large hidden layers are universal function approximators. ANNs with a large number of hidden layers are typically referred to as deep neural networks.

Figure 2 is a schematic of a "physics informed" neural network [6]. This is a recent innovation which incorporates the residuals of the Navier–Stokes equations and other conservation laws into the objective function/loss function that is used while training the network. These residuals are,

$$\begin{aligned} R_1 &\equiv \tfrac{\partial \rho}{\partial t} + \nabla \cdot (\rho u) \\ R_2 &\equiv \tfrac{\partial (\rho u)}{\partial t} + \nabla \cdot (\rho u u) - \mu \nabla^2 u + \nabla P - f \end{aligned} \tag{2}$$

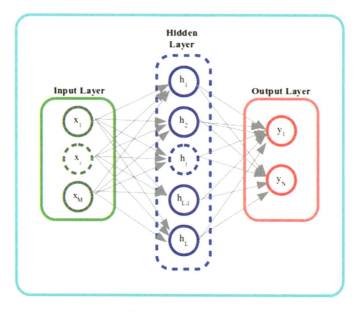

Fig. 1 Schematic of feed forward, artificial neural network (ANN) with one input layer, one hidden layer, and one output layer. Any number of hidden layers may be present

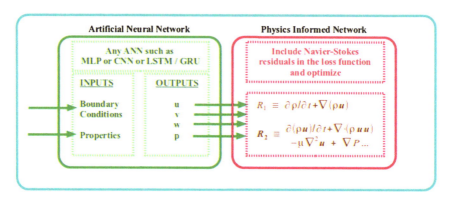

Fig. 2 Schematic of physics informed neural networks. Here, a feed forward, artificial neural network is enhanced by a physics informed layer that computes the residuals of the governing equations from the output of the ANN

with the spatial and temporal gradients usually evaluated by automatic differentiation. Open source libraries for machine learning, such as TensorFlow [7], usually provide this capability. This has been shown to yield better accuracy for systems governed by differential equations [8] and is a very promising approach. Another recent innovation is the use of data-driven neural networks to directly discretize the partial differential equations instead of the usual finite differences [9].

2.2 Reduced Order Models

The last several decades have seen a steady increase in computing power. However, even on present generation hardware, CFD calculations require large memory and CPU resources. This becomes even more challenging when repeated solutions need to be undertaken for real-time contexts [10] such as control, inverse problems, optimal design, etc. Hence, there is a need for suitable reduced order models that can substantially reduce storage and CPU requirements. Popular model order reduction techniques include proper orthogonal decomposition (POD) [11–14] and/or dynamic mode decomposition (DMD) [15–19]. Reference [20] uses a combination of POD and a deep neural network to develop a surrogate model for the stress distribution in patient-specific model of the aorta. Here, the POD is used to characterize the geometry of the aorta and the deep neural network is trained on stresses predicted by structural FEM.

2.3 Digital Twin Technology

A digital twin is a virtual replica of a physical product or process or service [21]. Digital twins incorporate multi-physics simulations, data analytics, machine learning, and Internet of Things to ensure accurate modeling over the lifetime of a product or process. Digital twin was named as one of the top ten strategic technology trends by Gartner in 2019 [22]. According to Refs. [23], model order reduction is a core component of digital twin technology, since it allows for real-time simulation as well as model exchange with substantially reduced intellectual property risk.

2.4 Turbulence Modeling

As may be expected, turbulence modeling is a fertile area for the application of deep learning. Duraisamy and coworkers [24–29] have developed a framework called field inversion and machine learning (FIML). Their strategy is to introduce a data-informed modifier to the production term of a standard RANS model (such as the Spalart–Allmaras) for improved predictions in specific classes of problems (such as airfoil computations). The data (ground truth) has to be available from experimental measurements and/or from high fidelity computations (DNS or LES). It is noted that the production term modifier is multiplicative in nature and is a field variable. The production term modifier/correction is obtained by solving an inverse problem which minimizes the discrepancy between the RANS result and the high fidelity/experimental data (ground truth). In Ref. [26], they train a neural network to output the correction given some flow features as the input. The input flow features include an adverse pressure gradient indicator, ratio of production to dissipation, ratio of rate of strain (magnitude) to the vorticity (magnitude), etc. In [28], they have considered other flow features such as mean velocity gradients and turbulent time scales as inputs to the model.

In contrast to the approach of Duraisamy, Ling, and coworkers have used deep neural networks with focus on preserving Galilean invariance in their data-informed corrections to RANS models [30, 32]. In particular, they preserve rotational invariance. There are at least three ways of preserving invariance in neural networks. One is to generate additional training data by rotation/translation of the given data by incrementally rotating/translating the coordinate system. Another way is to build equivariance into a convolutional neural network [31] by using equivariant filters. The third way is to ensure that the inputs to the model are themselves invariant. Reference [30] uses 10 isotropic basis tensors formed from the strain rate and the rotation rate tensors to determine scalar coefficients which are then used to compute the Reynolds stress anisotropy tensor. The Reynolds stress anisotropy tensor is further used to enhance the predictive capability of a quadratic eddy viscosity model. The inputs to the deep neural network are always the results from the RANS model, however, the training data are based on results from high fidelity LES or DNS computations. They used a total of six canonical flows for training, one case for optimizing the neural network hyper parameters, and two cases for testing. They have obtained improved accuracy over the standard quadratic eddy viscosity model. Other researchers, such as Xiao and coworkers, have followed a similar approach. In Refs. [33] and [34], the input features have been augmented by including the pressure gradient, the gradient of the turbulent kinetic energy, and the wall distance.

A somewhat different approach has been taken by Weatheritt [37] and related works [35, 36], which have focused upon algebraic Reynolds stress models. These studies have, also, proposed decomposition of the anisotropy tensor in terms of tensor bases and scalar invariants. However, the algebraic form of the expressions has been obtained via symbolic regression and gene expression programming (GEP). This role of the training data is to evaluate the candidate expressions and find the best possible match. In particular, in Ref. [35], they also integrate CFD calculations into the evaluation step because Reynolds stress expressions that are trained on the high fidelity (LES or DNS) data may yield RANS models that are inconsistent or even ill-conditioned. They have tested the new RANS models against turbine wake mixing calculations with significant improvements in prediction accuracy. Reference [36] also compares the results of GEP + symbolic regression against the performance of deep neural network in the context of a jet in cross-flow with improved results over a baseline simulation with a linear eddy viscosity model.

Other applications of deep learning to turbulence modeling includes generation of synthetic turbulence (time-dependent) at inflow boundaries for LES/DNS [38], sub-grid scale modeling [39], modeling of turbulent combustion, especially sub-grid scale reaction rates [40, 41] and generative approach to turbulence by learning the underlying attractor [42].

2.5 Multiphase Flow

Artificial neural networks have found various applications in the simulation of multiphase flows. In Ref. [47], an ANN is trained to compute the curvature of an interface between two immiscible fluids (typically gas-liquid) computed by the volume-of-fluid (VOF) method. Usually, the curvature is computed by computing the gradient of the normal to the interface. This normal is itself computed using the gradient of a smoothed volume fraction field. The expense of this operation can be substantially reduced by training an ANN to estimate the curvature. In Ref. [48], a physics informed deep recurrent neural network is combined with a POD-Galerkin projection-based ROM to model two-phase oil-water flow in reservoirs. Deep neural networks have been trained to predict bubble sizes in two-phase bubbly flows [49]. In Ref. [50], deep neural network models for bubble column reactors have been trained using data from CFD simulations. A neural network model for boiling heat transfer has been developed in Ref. [51].

2.6 CFD Compression

Deep neural networks have been used for compressing the results of CFD simulations. Reference [43] employs a particularly innovative approach to compress the results of high order discontinuous Galerkin computations. A neural network-based autoencoder is used in combination with proper orthogonal decomposition to form a compact representation of the computed results. The dynamics are learned using vectorial kernel orthogonal greedy algorithm using radial basis functions. Thereafter, the CFD solution can be recovered at any time instance from the compressed state as required. Reference [44] uses neural network-based autoencoders to form a compressed representation of lattice Boltzmann simulations.

2.7 Convergence Acceleration and Numerical Stability

It may be possible to accelerate the solution of the discretized system and/or the linearized discrete system using deep neural networks. Reference [45] uses a convolutional neural network to accelerate the solution of the pressure Poisson equation used to enforce the divergence constraint in incompressible flows. In Ref. [46], neural networks are used to control numerical oscillations in the vicinity of discontinuities by tuning the artificial viscosity.Generally, CFD solvers have many tunable parameters such as relaxation parameters, inner iterations, etc. which are used to ensure convergence of the iterative schemes. Deep neural networks can be used to learn the optimum values of these parameters to yield the fastest convergence while retaining stability.

2.8   Geometry, Mesh Generation, Visualization, and Post-processing

Deep neural networks (DNN) are already being used to classify geometry, especially in areas such as medical imaging. Convolutional neural networks were devised to aid in processing images which are 2D or 3D structured data. However, recently, deep neural networks have been extended to learn unstructured data and/or "raw" point clouds [54, 55]. DNNs are also being used for CAD model processing and repair [56]. They are also expected to play an increasingly important role in PLM [57]. While their use in meshing is still relatively undeveloped [58], DNNs have been used in visualization and post-processing of CFD and FEM simulation results. A typical use is in identification of vortices [59, 60] in 3D and 4D (3D space + time). This application will continue to see more interest and developments in the years to come.

2.9   Uncertainty Quantification

There are several sources of uncertainty in numerical simulations. These include uncertainties in geometry (including those due to manufacturing tolerances), physics, material properties, initial conditions, boundary conditions, etc. For instance, although saturated groundwater flow can be computed in a deterministic manner [61, 62], the medium itself is subject to substantial variability. The hydraulic conductivity of the soil, recharge rates, etc. need to be considered as random variables [63]. Reference [64] presents a framework for uncertainty quantification (UQ). A review of polynomial chaos techniques has been presented in Ref. [65]. However, UQ is quite expensive computationally because of the need to evaluate the model many times (hundreds or even thousands of times). Thus, the use of a cheap to evaluate surrogate model arises and has been treated in Ref. [66] using deep neural networks. The extension of UQ to utilize physics informed neural networks has been considered in Refs. [67] and [68] and this seems to be a promising approach in the future.

3   The CFD of the Future: Expected Trends

3.1   Higher Order Schemes

According to Jameson [69], most commercial and academic CFD codes are at the most 2nd order accurate in space and handle turbulent flow via RANS models. However, problems involving complex physics, such as vortical flows, aeroacoustics,

LES, etc. require numerical schemes with lower dissipation and higher order accuracy, while retaining the geometrical flexibility of unstructured meshes [70]. Also, curved elements become possible which makes it easier to discretize curved geometry [71]. Reference [70] outlines some practical strategies for effecting wider adoption of higher order schemes in practice. Additionally, the penetration of deep learning into CFD practice will increase the demand for higher accuracy solutions, over a wide range of parameters, and at a rate that will outstrip the increases in computing power by ever wider margins. Therefore, increasing adoption of higher order schemes can be expected in the future.

3.2 Discovery of New Models from Data

Many natural physical systems are based on physical principles expressed as differential equations (both, ODE and PDE) or integro-differential equations such as the population balance equations [72]. However, for many complex systems (such as neuroscience) the physical principles and the governing equations are still unknown. It is expected that the future will call for the integration of such systems with fluid dynamics and CFD. With rapid advances in data acquisition, sensors, data storage, Internet-of-Things (IoT), huge quantities of data (big data) can be easily collected. Then, it is logical to try and "learn" the governing equations and closure models from the vast data that will be available [73, 74]. It is expected that this trend will accelerate in the near future.

3.3 User Interfaces

Significant improvements are possible to the current generation of CFD and structural FEM software especially in the area of user interfaces. Recent advances in natural language processing, speech and hand writing recognition, speech synthesis, etc. make it possible to endow CFD and structural FEM software with natural language interfaces [52] and also in multiple languages (human languages). Deep learning can also be used to flag probable user errors in problem setup, incompatible boundary and initial conditions, and guide the user through the entire simulation process. It may also be feasible for the user interface to adapt itself to the user [53], enhancing the user experience and the usability of the software.

3.4 ASIC/FPGA Chips (CFD-on-a-Chip)

In Ref. [1], it was predicted that ASIC (application-specific integrated circuits) and SoC (system-on-a-chip) type of architectures would be developed for CFD and similar disciplines. Whereas CFD ASICs are not yet commonplace, yet the topic has drawn some interest from researchers. The MDGRAPE-4 system [75] is a SoC specially designed for molecular dynamics computations. Reference [76] describes the design and implementation of two CFD solvers on a FPGA (field-programmable gate array) and demonstrates a factor of 17 improvements in execution performance while reducing the power consumption over conventional CPU-based hardware, under some conditions. Reference [77] also demonstrates the implementation of a Laplace equation solver on a FPGA with a speed-up of about 20 times over a conventional CPU-based architecture. It is safe to conclude that "CFD-on-a-Chip" is feasible and capable of delivering performance benefits.

3.5 Expanded Range of Applications

It is certain that CFD will be applied to an increasing range of applications, many of which are in a nascent stage today. A major application area will be in medicine. Patient-specific cardio-vascular modeling is already an active area of research, whereas one dimensional models of unsteady, pulsatile blood flow have been in use for decades [78, 79]. Computational hemodynamics is a useful tool in disease research, diagnostics development, medical device development and evaluation (e.g., stents), and even cardio-vascular surgical planning [80, 81]. It is already possible to extract the geometry of the larger blood vessels from medical images using toolkits such as the vascular modeling toolkit (vmtk) [82] and perform high-quality meshing. Another important potential application is in aneurysm research, diagnosis, and treatment planning. An aneurysm is a bulge in a blood vessel that is likely to prove fatal if it ruptures. CFD is already used as a tool for understanding aneurysms [83–86]. It is quite likely that CFD, perhaps in combination with machine learning, could become an important tool in surgical planning and treatment of aneurysms. CFD is being used to study air flow in the respiratory system [87] and is also finding applications in micro-fluidics [15] and in the study of the locomotion of biological cells and microorganisms [89]. Strook et al. [90] have presented a comprehensive review of the role of fluid mechanics in the vascular physiology of plants and associated transport processes. Other notable subjects of investigation, where fluid mechanics plays a major role, include the eye [91] and the inner ear [92]. Thus, the role of CFD in biological and biomedical studies will increase substantially in the future.

The food processing industry has started to make use of CFD [93]. Processes such as spray drying and spray freezing of food use CFD to analyze and improve the product quality. CFD is also being used to analyze heat transfer and air flow during thermal processing of canned solid and liquid foods, and for the analysis of processes such as baking, pasteurization, sterilization, refrigeration, crystallization, and so on. Another possibility is to use CFD for the analysis of the digestion process.

Competitive sport started making use of CFD in the early nineties. Motor racing was among the first to utilize CFD for competitive advantage [94]. Apart from external aerodynamics of the vehicle, CFD has been applied to engines, intakes, exhausts, radiators, wheels, etc. Future applications will probably include maneuvers overtaking and complex motion such as aquaplaning, etc. CFD is also used in sailing yacht design [95–97]. Apart from sail design, CFD can be used to analyze the hull shape, the interaction between the various underwater and above water components, sailing tactics, etc. CFD has also been used to analyze the aerodynamics for ball sports and this trend will continue in the future [98]. Wei et al. [99] have provided a detailed discussion of the fluid mechanics of competitive swimming.

Other applications of current and future importance include buildings ventilation and energy efficiency [100] and fire safety [101–103].

3.6   Vision for the Year 2025 and Beyond

We expect that computational fluid dynamics will find application in almost every sphere of human activity. Apart from the conventional application in the aerospace, automotive, and other industrial sectors, we believe that CFD will have a profound impact in biomedical and healthcare fields, with beneficial impact on human health, longevity, and quality of life. The combination of more capable hardware, software, and deep machine learning will make CFD more efficient and easier to use. It will become common practice to routinely generate digital twin/reduced order models from the results of high fidelity CFD simulations. This will make it possible to utilize CFD directly in CAD and PLM systems. To facilitate this, standards will evolve for embedding digital twin models into PLM and design systems. It may even become possible to embed the digital twin generated from CFD simulations into spreadsheets (MS Excel, OpenOffice calc, gnumeric) for design calculations. In turn, this implies that engineers will be able to use CFD in design without needing the assistance of a CFD expert. Also, since digital twins/ROMs are orders of magnitude faster to evaluate than full order CFD models, it will become substantially easier to solve inverse problems [104], thus allowing for better designs [105], and also facilitating control, typically by minimization of suitable objective functions. All these developments are likely to result in an explosion of CFD usage since deep learning requires copious amounts of high-quality training data that needs to be generated via high fidelity simulations.

4 Conclusion

The CFD of the future will be profoundly affected by ongoing advances in deep machine learning. A typical CFD simulation of the future will, in addition to the usual static pictures and video animations, be used to generate a digital twin/reduced order model of the system under investigation. The digital twin will be an accurate replica of the system, but, orders of magnitude faster to evaluate. These digital twins can then be "plugged" into CAD/PLM software tools or even into spreadsheets. We assume that suitable software interfaces will be developed and standardized for this purpose. These can then be used for design without requiring the services of a CFD expert. A clear separation between the tasks of generating a digital twin and their use in design will expedite the future growth of CFD and render it substantially easier to use. We expect that this will result in the application of CFD to a wide range of problems and have an impact on virtually every aspect of human activity. For instance, expanded use of CFD and digital twins in biomedical applications will substantially improve the quality and effectiveness of healthcare and keep it affordable too. The social benefits are obvious.

Acknowledgements The authors thank Dr. M. Damodaran of Temasek Laboratories, National University of Singapore, for very enlightening discussions and valuable input during the preparation of this contribution.

References

1. Runchal, A. K. (2012). The future of CFD and the CFD of the future. *Computational Thermal Sciences: An International Journal, 4*(6), 517–524.
2. IMARC Group (2019). *Computational Fluid Dynamics Market: Global Industry Trends, Share, Size, Growth, Opportunity and Forecast 2019–2024.* https://www.imarcgroup.com/computational-fluid-dynamics-market (last accessed 22 June 2019)
3. Patterson, G. S., Jr. (1978). Prospects for computational fluid mechanics. *Annual Review of Fluid Mechanics, 10*(1), 289–300.
4. Krizhevsky, A., Sutskever, I., & Hinton, G. E. (2012). Imagenet classification with deep convolutional neural networks. In *Advances in Neural Information Processing Systems* (pp. 1097–1105).
5. Goodfellow, I., Bengio, Y., & Courville, A. (2016). *Deep learning.* MIT Press.
6. Raissi, M., Perdikaris, P., and Karniadakis, G. E. (2017). *Physics informed deep learning (part i): Data-driven solutions of nonlinear partial differential equations.* arXiv:1711.10561.
7. Abadi, M., Agarwal, A., Barham, P., Brevdo, E., Chen, Z., Citro, C., & Ghemawat, S. (2016). *Tensorflow: Large-scale machine learning on heterogeneous distributed systems.* arXiv:1603.04467.
8. Kani, J. N., & Elsheikh, A. H. (2017). *DR-RNN: A deep residual recurrent neural network for model reduction.* arXiv:1709.00939
9. Bar-Sinai, Y., Hoyer, S., Hickey, J., & Brenner, M. P. (2019). *Learning data-driven discretizations for partial differential equations.* arXiv:1808.04930
10. Lassila, T., Manzoni, A., Quarteroni, A., & Rozza, G. (2014). Model order reduction in fluid dynamics: Challenges and perspectives. In *Reduced Order Methods for Modeling and Computational Reduction* (pp. 235–273). Springer.

11. Narasimha, R. (2011). Kosambi and proper orthogonal decomposition. *Resonance, 16*(6), 574–581.
12. Berkooz, G., Holmes, P., & Lumley, J. L. (1993). The proper orthogonal decomposition in the analysis of turbulent flows. *Annual Review of Fluid Mechanics, 25*(1), 539–575.
13. Bui-Thanh, T., Damodaran, M., and Willcox, K. (2003). Proper orthogonal decomposition extensions for parametric applications in compressible aerodynamics. In *21st AIAA Applied Aerodynamics Conference* (p. 4213).
14. Bui-Thanh, T., Damodaran, M., & Willcox, K. E. (2004). Aerodynamic data reconstruction and inverse design using proper orthogonal decomposition. *AIAA journal, 42*(8), 1505–1516.
15. Schmid, P. J. (2010). Dynamic mode decomposition of numerical and experimental data. *Journal of Fluid Mechanics, 656,* 5–28.
16. Tu, J. H., Rowley, C. W., Luchtenburg, D. M., Brunton, S. L., & Kutz, J. N. (2013). *On dynamic mode decomposition: Theory and applications.* arXiv:1312.0041.
17. Kutz, J. N., Brunton, S. L., Brunton, B. W., & Proctor, J. L. (2016). Dynamic mode decomposition: data-driven modeling of complex systems. *SIAM.*
18. Proctor, J. L., Brunton, S. L., & Kutz, J. N. (2016). Dynamic mode decomposition with control. *SIAM Journal on Applied Dynamical Systems, 15*(1), 142–161.
19. Williams, M. O., Kevrekidis, I. G., & Rowley, C. W. (2015). A data–driven approximation of the koopman operator: Extending dynamic mode decomposition. *Journal of Nonlinear Science, 25*(6), 1307–1346.
20. Liang, L., Liu, M., Martin, C., & Sun, W. (2018). A deep learning approach to estimate stress distribution: a fast and accurate surrogate of finite-element analysis. *Journal of the Royal Society, Interface, 15*(138), 20170844.
21. Grieves, M., & Vickers, J. (2017). Digital twin: Mitigating unpredictable, undesirable emergent behavior in complex systems. In *Transdisciplinary Perspectives on Complex Systems* (pp. 85–113). Springer.
22. Kerremans, M., Burke, B., Cearley, D., & Velosa, A. (2019). *Top 10 strategic technology trends for 2019: Digital Twins.* Gartner Research, Document ID: G00377678. https://www.gartner.com/en/documents/3904569/top-10-strategic-technology-trends-for-2019-digital-twin. Retrieved 11 July 2019.
23. Hartmann, D., Herz, M., & Wever, U. (2018). Model order reduction a key technology for digital twins. In *Reduced-Order Modeling (ROM) for Simulation and Optimization* (pp. 167–179). Springer.
24. Duraisamy, K., Iaccarino, G., & Xiao, H. (2019). Turbulence modeling in the age of data. *Annual Review of Fluid Mechanics, 51,* 357–377.
25. Duraisamy, K., & Durbin, P. (2014). Transition modeling using data driven approaches. In *CTR Summer Program* (p. 427).
26. Holland, J. R., Baeder, J. D., & Duraisamy, K. (2019). Towards integrated field inversion and machine learning with embedded neural networks for RANS modeling. In *AIAA Scitech 2019 Forum* (p. 1884).
27. Singh, A. P., Medida, S., & Duraisamy, K. (2017). Machine-learning-augmented predictive modeling of turbulent separated flows over airfoils. *AIAA Journal,* 2215–2227.
28. Singh, A. P., & Duraisamy, K. (2016). Using field inversion to quantify functional errors in turbulence closures. *Physics of Fluids, 28*(4), 045110.
29. Parish, E. J., & Duraisamy, K. (2016). A paradigm for data-driven predictive modeling using field inversion and machine learning. *Journal of Computational Physics, 305,* 758–774.
30. Ling, J., Kurzawski, A., & Templeton, J. (2016). Reynolds averaged turbulence modeling using deep neural networks with embedded invariance. *Journal of Fluid Mechanics, 807,* 155–166.
31. Thomas, N., Smidt, T., Kearnes, S., Yang, L., Li, L., Kohlhoff, K., & Riley, P. (2018). *Tensor field networks: Rotation-and translation-equivariant neural networks for 3d point clouds.* arXiv:1802.08219
32. Ling, J., Jones, R., & Templeton, J. (2016). Machine learning strategies for systems with invariance properties. *Journal of Computational Physics, 318,* 22–35.

33. Wu, J. L., Xiao, H., & Paterson, E. (2018). Physics-informed machine learning approach for augmenting turbulence models: A comprehensive framework. *Physical Review Fluids, 3*(7).
34. Wang, J. X., Wu, J. L., & Xiao, H. (2017). A Physics-informed machine learning approach for reconstructing Reynolds stress modeling discrepancies based on DNS data. *Physical Review Fluids, 2*(3).
35. Zhao, Y., Akolekar, H. D., Weatheritt, J., Michelassi, V., & Sandberg, R. D. (2019). *Turbulence model development using CFD-driven machine learning.* arXiv:1902.09075
36. Weatheritt, J., Sandberg, R. D., Ling, J., Saez, G., & Bodart, J. (2017). A comparative study of contrasting machine learning frameworks applied to RANS modeling of jets in crossflow. GT2017-63403. In *Proceedings of the ASME Turbo Expo 2017: Turbomachinery Technical Conference and Exposition, GT2017*, June 26–30, 2017, Charlotte, USA.
37. Weatheritt, J. (2015). *The development of data driven approaches to further turbulence closures* (Doctoral dissertation, University of Southampton).
38. Fukami, K., Kawai, K., & Fukagata, K. (2018). A synthetic turbulent inflow generator using machine learning. arXiv:1806.08903
39. Maulik, R., San, O., Jacob, J. D., & Crick, C. (2019). Sub-grid scale model classification and blending through deep learning. *Journal of Fluid Mechanics, 870,* 784–812.
40. Lapeyre, C. J., Misdariis, A., Cazard, N., Veynante, D., & Poinsot, T. (2019). Training convolutional neural networks to estimate turbulent sub-grid scale reaction rates. *Combustion and Flame, 203,* 255–264.
41. Nikolaou, Z. M., Chrysostomou, C., Vervisch, L., & Cant, S. (2018). *Modelling turbulent premixed flames using convolutional neural networks: application to sub-grid scale variance and filtered reaction rate.* arXiv:1810.07944
42. Mohan, A., Daniel, D., Chertkov, M., and Livescu, D. (2019). *Compressed convolutional LSTM: An Efficient deep learning framework to model high fidelity 3D turbulence.* arXiv: 1903.00033
43. Carlberg, K. T., Jameson, A., Kochenderfer, M. J., Morton, J., Peng, L., & Witherden, F. D. (2018). *Recovering missing CFD data for high-order discretizations using deep neural networks and dynamics learning.* arXiv:1812.01177.
44. Hennigh, O. (2017). *Lat-net: Compressing lattice Boltzmann flow simulations using deep neural networks.* arXiv:1705.09036
45. Tompson, J., Schlachter, K., Sprechmann, P., & Perlin, K. (2017). Accelerating Eulerian fluid simulation with convolutional networks. In *Proceedings of the 34th International Conference on Machine Learning* (Vol. 70, pp. 3424–3433).
46. Discacciati, N., Hesthaven, J. S., & Ray, D. (2019). *Controlling oscillations in high-order Discontinuous Galerkin schemes using artificial viscosity tuned by neural networks.* https://infoscience.epfl.ch/record/263616/files/Artificial_viscosity_nn.pdf. Retrieved 23 June 2019.
47. Qi, Y., Lu, J., Scardovelli, R., Zaleski, S., & Tryggvason, G. (2019). Computing curvature for volume of fluid methods using machine learning. *Journal of Computational Physics, 377,* 155–161.
48. Kani, J. N., & Elsheikh, A. H. (2019). Reduced-order modeling of subsurface multi-phase flow models using deep residual recurrent neural networks. *Transport in Porous Media, 126*(3), 713–741.
49. Montes-Atenas, G., Seguel, F., Valencia, A., Bhatti, S. M., Khan, M. S., Soto, I., et al. (2016). Predicting bubble size and bubble rate data in water and in froth flotation-like slurry from computational fluid dynamics (CFD) by applying deep neural networks (DNN). *International Communications in Heat and Mass Transfer, 76,* 197–201.
50. Mosavi, A., Shamshirband, S., Salwana, E., Chau, K. W., & Tah, J. H. (2019). Prediction of multi-inputs bubble column reactor using a novel hybrid model of computational fluid dynamics and machine learning. *Engineering Applications of Computational Fluid Mechanics, 13*(1), 482–492.
51. Liu, Y., Dinh, N., Sato, Y., & Niceno, B. (2018). Data-driven modeling for boiling heat transfer: Using deep neural networks and high-fidelity simulation results. *Applied Thermal Engineering, 144,* 305–320.

52. Goldberg, Y. (2016). A primer on neural network models for natural language processing. *Journal of Artificial Intelligence Research, 57,* 345–420.
53. Soh, H., Sanner, S., White, M., & Jamieson, G. (2017, March). Deep sequential recommendation for personalized adaptive user interfaces. In *Proceedings of the 22nd International Conference on Intelligent User Interfaces* (pp. 589–593). ACM.
54. Wu, Z., Song, S., Khosla, A., Yu, F., Zhang, L., Tang, X., & Xiao, J. (2015). 3d shapenets: A deep representation for volumetric shapes. In *Proceedings of the IEEE conference on computer vision and pattern recognition* (pp. 1912–1920).
55. Qi, C. R., Su, H., Mo, K., & Guibas, L. J. (2017). Pointnet: Deep learning on point sets for 3d classification and segmentation. In *Proceedings of the IEEE Conference on Computer Vision and Pattern Recognition* (pp. 652–660).
56. Danglade, F., Pernot, J. P., & Véron, P. (2014). On the use of machine learning to defeature CAD models for simulation. *Computer-Aided Design and Applications, 11*(3), 358–368.
57. Dekhtiar, J., Durupt, A., Bricogne, M., Eynard, B., Rowson, H., & Kiritsis, D. (2018). Deep learning for big data applications in CAD and PLM–Research review, opportunities and case study. *Computers in Industry, 100,* 227–243.
58. Yao, S., Yan, B., Chen, B., & Zeng, Y. (2005). An ANN-based element extraction method for automatic mesh generation. *Expert Systems with Applications, 29*(1), 193–206.
59. Deng, L., Wang, Y., Liu, Y., Wang, F., Li, S., & Liu, J. (2019). A CNN-based vortex identification method. *Journal of Visualization, 22*(1), 65–78.
60. Rajendran, V., Kelly, K. Y., Leonardi, E., & Menzies, K. (2018). Vortex detection on unsteady CFD simulations using recurrent neural networks. AIAA 2018–3724.
61. Analytic and Computational Research, Inc. (2019) The PORFLOW® Reference Manual. https://www.acricfd.com/
62. Harbaugh, A. W., Banta, E. R., Hill, M. C., & McDonald, M. G. (2000). MODFLOW-2000, *The U.S. geological survey modular ground-water model-user guide to modularization concepts and the ground-water flow process.* Open-file Report. U. S. Geological Survey, 92, 134
63. Hill, M. C., & Tiedeman, C. R. (2006). *Effective groundwater model calibration: With analysis of data, sensitivities, predictions, and uncertainty.* ISBN 9780471776369, Wiley.
64. Roy, C., & Oberkampf, W. (2010). A complete framework for verification, validation, and uncertainty quantification in scientific computing. In *48th AIAA Aerospace Sciences Meeting Including the New Horizons Forum and Aerospace Exposition*, AIAA (pp. 2010–124).
65. Najm, H. N. (2009). Uncertainty quantification and polynomial chaos techniques in computational fluid dynamics. *Annual Review of Fluid Mechanics, 41,* 35–52.
66. Tripathy, R. K., & Bilionis, I. (2018). Deep UQ: Learning deep neural network surrogate models for high dimensional uncertainty quantification. *Journal of Computational Physics, 375,* 565–588.
67. Yang, Y., & Perdikaris, P. (2018). *Adversarial uncertainty quantification in physics-informed neural networks.* arXiv:1811.04026
68. Zhu, Y., Zabaras, N., Koutsourelakis, P. S., & Perdikaris, P. (2019). Physics-constrained deep learning for high-dimensional surrogate modeling and uncertainty quantification without labeled data. *Journal of Computational Physics, 394,* 56–81.
69. Jameson, A. (2015). *Computational fluid dynamics: Past, present, and future.* http://aero-comlab.stanford.edu/Papers/CFD_Past,Present,Future-SciTech2015.pdf (last accessed on May 29, 2019)
70. Vincent, P. E., & Jameson, A. (2011). Facilitating the adoption of unstructured high-order methods among a wider community of fluid dynamicists. *Mathematical Modelling of Natural Phenomena, 6*(3), 97–140.
71. Heinrich, C., & Simeon, B. (2012). A finite volume method on NURBS geometries and its application in isogeometric fluid–structure interaction. *Mathematics and Computers in Simulation, 82*(9), 1645–1666.
72. Ramakrishna, D. (2000). *Population balances.* San Diego, CA, USA: Academic Press.
73. Raissi, M., Perdikaris, P., & Karniadakis, G. E. (2017). *Physics informed deep learning (part ii): Data-driven discovery of nonlinear partial differential equations.* arXiv:1711.10566

74. Long, Z., Lu, Y., Ma, X., & Dong, B. (2017). *Pde-net: Learning PDES from data*. arXiv: 1710.09668

75. Ohmura, I., Morimoto, G., Ohno, Y., Hasegawa, A., & Taiji, M. (2014). MDGRAPE-4: a special-purpose computer system for molecular dynamics simulations. *Philosophical Transactions of the Royal Society A: Mathematical, Physical and Engineering Sciences, 372* (2021).

76. Bin Abu Talip, M. S. (2013). Partial *Reconfiguration Implementation on Fluid Dynamics Computation Using an FPGA*. Doctoral dissertation, Keio University.

77. Ebrahimi, A., & Zandsalimy, M. (2017). Evaluation of FPGA hardware as a new approach for accelerating the numerical solution of CFD problems. *IEEE Access, 5,* 9717–9727.

78. Ku, D. N. (1997). Blood flow in arteries. *Annual Review of Fluid Mechanics, 29*(1), 399–434.

79. Van de Vosse, F. N., & Stergiopulos, N. (2011). Pulse wave propagation in the arterial tree. *Annual Review of Fluid Mechanics, 43,* 467–499.

80. van Bakel, T. M., Lau, K. D., Hirsch-Romano, J., Trimarchi, S., Dorfman, A. L., & Figueroa, C. A. (2018). Patient-specific modeling of hemodynamics: Supporting surgical planning in a Fontan circulation correction. *Journal of cardiovascular translational research, 11*(2), 145–155.

81. Taylor, C. A., & Figueroa, C. A. (2009). Patient-specific modeling of cardiovascular mechanics. *Annual Review of Biomedical Engineering, 11,* 109–134.

82. Antiga, L., Piccinelli, M., Botti, L., Ene-Iordache, B., Remuzzi, A., and Steinman, D. A. (2008). An image-based modeling framework for patient-specific computational hemodynamics. *Medical & Biological Engineering & Computing, 46*(11), 1097. http://www.vmtk. org/.

83. Sforza, D. M., Putman, C. M., & Cebral, J. R. (2012). Computational fluid dynamics in brain aneurysms. *International Journal for Numerical Methods in Biomedical Engineering, 28*(6–7), 801–808.

84. Turjman, A. S., Turjman, F., & Edelman, E. R. (2014). Role of fluid dynamics and inflammation in intracranial aneurysm formation. *Circulation, 129*(3), 373–382.

85. Roi, D. P., Mueller, J. D., Lobotesis, K., McCague, C., Memarian, S., Khan, F., et al. (2019). Intracranial aneurysms: looking beyond size in neuroimaging: The role of anatomical factors and haemodynamics. *Quantitative Imaging in Medicine and Surgery, 9*(4), 537.

86. Sforza, D. M., Putman, C. M., & Cebral, J. R. (2009). Hemodynamics of cerebral aneurysms. *Annual Review of Fluid Mechanics, 41,* 91–107.

87. Basri, E. I., Basri, A. A., Riazuddin, V. N., Shahwir, S. F., Mohammad, Z., & Ahmad, K. A. (2016). Computational fluid dynamics study in biomedical applications: A review. *International Journal of Fluids and Heat Transfer, 1*(2).

88. Stone, H. A., Stroock, A. D., & Ajdari, A. (2004). Engineering flows in small devices: microfluidics toward a lab-on-a-chip. *Annual Review of Fluid Mechanics, 36,* 381–411.

89. Shyy, W., Francois, M., Udaykumar, H. S., N'dri, N., & Tran-Son-Tay, R. (2001). Moving boundaries in micro-scale biofluid dynamics. *Applied Mechanics Reviews, 54*(5), 405–454.

90. Stroock, A. D., Pagay, V. V., Zwieniecki, M. A., & Michele Holbrook, N. (2014). The physicochemical hydrodynamics of vascular plants. *Annual Review of Fluid Mechanics, 46,* 615–642.

91. Siggers, J. H., & Ethier, C. R. (2012). Fluid mechanics of the eye. *Annual Review of Fluid Mechanics, 44,* 347–372.

92. Obrist, D. (2019). Flow phenomena in the inner ear. *Annual Review of Fluid Mechanics, 51,* 487–510.

93. Anandharamakrishnan, C. (2013). *Computational fluid dynamics applications in food processing*. New York, NY: Springer.

94. Hanna, R. K. (2012). CFD in Sport-a retrospective; 1992–2012. *Procedia Engineering, 34,* 622–627.

95. Hedges, K. L., Richards, P. J., & Mallinson, G. D. (1996). Computer modelling of downwind sails. *Journal of Wind Engineering and Industrial Aerodynamics, 63*(1–3), 95–110.

96. Shyy, W., Udaykumar, H. S., Rao, M. M., & Smith, R. W. (2007). *Computational fluid dynamics with moving boundaries*. Taylor & Francis, Washington DC (1996); Dover, New York.
97. Peters, M. (2009). *Computational fluid dynamics for sport simulation*. Springer.
98. Jalilian, P., Kreun, P. K., Makhmalbaf, M. M., & Liou, W. W. (2014). Computational aerodynamics of baseball, soccer ball and volleyball. *American Journal of Sports Science, 2*(5), 115–121.
99. Wei, T., Mark, R., & Hutchison, S. (2014). The fluid dynamics of competitive swimming. *Annual Review of Fluid Mechanics, 46*, 547–565.
100. Lawson, T., & Lawson, T. V. (2001). *Building aerodynamics* (Vol. 10). London: Imperial College Press.
101. Yeoh, G. H., and Yuen, K. K. (2009). *Computational fluid dynamics in fire engineering: Theory, modelling and practice*. Butterworth-Heinemann.
102. McGrattan, K., Hostikka, S., McDermott, R., Floyd, J., Weinschenk, C., & Overholt, K. (2013). *Fire dynamics simulator user's guide*. NIST Special Publication, 1019 (6).
103. Wang, C. J., Wen, J. X., & Chen, Z. B. (2014). Simulation of large-scale LNG pool fires using FireFoam. *Combustion Science and Technology, 186*(10–11), 1632–1649.
104. Stoecklein, D., Lore, K. G., Davies, M., Sarkar, S., & Ganapathysubramanian, B. (2017). Deep learning for flow sculpting: Insights into efficient learning using scientific simulation data. *Scientific reports, 7*, 46368.
105. Jiang, Z. (2018). *Distributed optimization for control and learning*. Ph.D. Dissertation, Department of Mechanical Engineering, Iowa State University, Ames, Iowa, USA.

Appendix I

Appendix A

Brian Spalding: January 9, 1923–November 27, 2016
Scientist, Polymath, Poet, Family Man and Believer in Right.
by Colleen I Spalding

Brian was born on January 9 1923 in the apartment; his family then occupied over Lloyds Bank, 94–96 High Street, New Malden, Surrey. His father, Harold Spalding, who was awarded the Military Cross during WWI, was the local bank manager. Harold had 3 siblings and married Kathleen Davey from Jersey (known as Kitty). They had two children; Brian, and his older sister Katie who was born on December 2, 1919.

Brian as a Baby

His Parents pre 1920

A. Runchal (ed.), *50 Years of CFD in Engineering Sciences*,
https://doi.org/10.1007/978-981-15-2670-1

Lloyds Bank New Malden

The Bank is still there and Brian's account remained there, nominally at least, until his death. I am not sure how long he lived over the Bank but at some point in his childhood, the family moved to Glenthorne, Woodside Road, New Malden. He went to a "local preparatory school" but, when asked, could not recall the name. The photo below is, he thought, from 1927 or 1928. The picture below right is with his mother, Kitty, and sister, Katie, from about the same time.

Brian said, often, that he had a happy childhood. His mother encouraged him to respect his father and his father (whom I did not meet) seems to have been liked by all. Brian had a dog, of indeterminate breed, called Chum and they spent many hours wandering Wimbledon Common, Richmond Park and the banks of the

Beverley Brook. He got on with things, was not pushed to be academic and cannot recall his parents ever being involved in his education. He was bought a chemistry set and endeavoured, unsuccessfully, to blow up the garden shed. He started life with a belief in a superior being and travelled from that, via communism, to scepticism and atheism.

Brian liked the area of his birth sufficiently to remain in close proximity for most of his life other than when he was absent in pursuit of education. From New Malden he moved to Oxford, then the outskirts of Kingston, to Cambridge then Wimbledon and thence to Putney Heath. His mental agility more than made up for his perhaps somewhat limited geographical perambulations. He was proud that New Malden appeared in the Domesday Book a fact which he oft times quoted. When his parents retired, he bought them a flat in New Malden but his mother preferred the seaside so they moved to Paignton.

At 8 he went to King's College School Wimbledon where he remained until he was 18. He thrived. His first report card advised that he had made a "capital start" for what the master foresaw as a "successful and useful career", he was described as a "most delightful little boy". His report cards continued to be positive, he did as he was told and "obeyed the rules" though remembered feeling somewhat aggrieved with the French teacher who told him "not to be silly" when Brian (in a lesson learning the names of cities in French) postulated, correctly, that London would be called Londres.

He was in the CCF and whilst on exercises had to rewind lengths of wire. The process had a long term (indeed permanent) impact on his future in that it taught him the value of tenacity, patience and working methodically to achieve an end. It also seemed to have had an impact on his choice of study at Oxford in that he decided to study engineering and was not put off that choice despite being advised by one of the master's, as he sat in the Headmaster's Garden after being offered his place, that they had educated him to get a double first in classics not waste his time on some "new-fangled" subject.

Brian and the cricket team at KCS 1934

Brian, a Prefect at KCS 1944

The Great Hall at KCS

He liked cycling and cycled through France just before war broke out. Rowing took him to The Wye Valley and a submerged boat. Walking took him to what he always described as his favourite Christmas; he was to meet a group of friends and stay in Snowdonia. He was the only one to make it due to inclement weather, spent the holiday in splendid isolation, and infinitely preferred it to the more commercial occasions which followed.

He went up to The Queen's College, Oxford in October 1941 as a Styring Scholar and obtained a BA, 1st Class, in Engineering Science in July 1944. Whilst there he was Captain of the Boat Club, represented the University at Ju Jitsu, rowed for the Summer VIII's and was awarded a Cronshaw Scholarship in 1943. His connexion with The Queen's was important. He went back for a gathering of the Engineering Class of 1944 in 1994 (see below) and was scheduled to attend a Gaudy in December 2016. He and Ian Hutcheon, a close friend who was at KCS and Oxford with Brian, found time for non-academic activities including climbing the Sheldonian to put a policeman's helmet on one of the flag poles!

Matriculation Photograph 1944
provided by Oxford University

The Engineering Class of 1944 in 1994
Brian standing back left, Ian standing back right

The Sheldonian

After Oxford, Brian joined RPE Westcott, part of the Ministry of Supply (MoS), in October 1945 specialising in rocket research. He was sent to Germany, probably under Colonel Colbrook, to research activities at a Rocket Station at Volkenrohde near Brausschweig where he met Eda Ilse-Lotte Goericke whom he subsequently married. Before the marriage, Brian rescued his sister-in-law (to be) from East Berlin. As I heard it, he "borrowed" a laundry van and drove to the Russian sector with a laundry list, in Russian, as his "authorization". He collected mother and (sedated) baby. Near the exit point, a troop of Russian soldiers hove into view across the road. Had they stopped his vehicle they would have recognized a laundry list! Fortunately they moved to either side; Brian drove on.

Brian returned to England and his work was at RPE. Eda joined him, they were married in November 1947 (he was 24) and he adopted Eda's daughter from her first marriage. Also that November (14) there was a fatal accident at RPE associated with experiments on the use of hydrogen peroxide as rocket fuel oxidisers. Brian said he was scheduled to be there that day but was either ill or on holiday due to his marriage–he was unsure. Had he been present his history would have been a deal shorter and the scientific world poorer. As it was two British technicians were killed along with Dr. Johannes Schmidt, the leader of the German rocket team.

Brian 1945

Brian & Eda with Brian's Parents, November 1947

Brian enjoyed his work at RPE Westcott until, shortly after his marriage, it was brought abruptly to a close at the instigation of the Ministry of Supply. He, and his friend Ian Hutcheon, had both joined the Communist Party whilst at Oxford, Brian remained a member for only a year rapidly becoming disaffected with party policy. At this point after the war, the government undertook a purge of fascists and communists; Brian and Ian were effectively "dismissed" from their posts. Their dismissal made the front page of the Daily Express on Saturday, January 24, 1948.

Whether his removal from RPE Westcott affected Brian's decision to return to academe I am unsure but, for whatever reason, he joined Pembroke College, Cambridge in 1948 and obtained his Ph.D. in October 1952. His thesis was on *The Combustion of Liquid Fuels*. It was, and is, generally considered to have been a ground-breaking work.

Whilst in Cambridge, he and his family (which now numbered 4) lived in a house which, evidently, had no electricity and only basic heating and plumbing. Milk was delivered in a churn by a milkman who let himself into the kitchen and filled jugs left out for that purpose. Living conditions may have played a part in his moving from a role as a Demonstrator at Cambridge University to Imperial College London when he was "head-hunted" and invited to take up the post of Reader in Heat Transfer.

Brian and Family circa 1960

During his period at Imperial, Brian progressed from Reader to Professor, holding the Chair of Heat Transfer which he retained until his retirement in 1988. He also headed up a new unit—the Computational Fluid Dynamics Unit—which reflected his ever enquiring mind exploring new areas of interest. He was a founding father of Computational Fluid Dynamics (CFD) and also worked in, and was creatively passionate about, Combustion, Heat Transfer, Fluid Flow, Boundary Layer and Unified Theory, the SIMPLE algorithm, Finite Difference and Finite-Volume Methods, Turbulence Modelling, Multiphase Flows, Heat Exchangers and other aspects of his varied and beloved scientific work.

Brian in the Computer Room at Imperial College (Mechanical Engineering Department)

I met Brian, at Imperial, in 1973 and, having little experience of the higher echelons of academe, assumed that all Professors lectured, interacted with undergraduate and postgraduate students as well as post-doctoral researchers, were invited to be keynote lecturer at many and varied conferences, wrote books, wrote and presented papers, participated in research, and, in addition, liaised with MPs, held press conferences, got scientists out of Russia and helped publishers regain control of their companies, wrote computer software—oh and ran a company carrying out commercial consulting work. It took a while to realize that this was not always the case but it was a fascinating introduction.

Brian had a highly developed, and personal, sense of right and wrong. It did not always agree with the mores of the day because *his* beliefs were his. He *would* change them but only after thought and deliberation and NOT because any external force wanted him so to do. Brian acted according to his own, strongly held, principles because "*it was the right thing to do*". This phrase represented him in so many aspects of his life. It made him principled but not always popular.

Brian's passion for doing what he perceived as the right thing (whether it endeared him to his fellow man or not) meant that whilst I was with him in Moscow throughout what proved to be his last (and only real) illness I was supported by Jana Levich. We have been friends for 40 years because Brian decided, in the early 1970s, to take on the USSR and fight for the right of Academician Levich and his family to move to Israel. This alienated the bureaucracy of the time which was a sorrow to him but did not prevent him from succeeding, against all odds. In the process of obtaining freedom for Ben and Tanya, their sons and daughters-in-law, Brian decided that, in addition to petitions and pressures, he would pursue a scientific route. Ben had founded the field of PhysicoChemical Hydrodynamics (PCH) so Brian decided to hold a PCH Conference at Somerville College Oxford in 1977 to mark Ben's 60th birthday. It was attended by many scientists of international repute as well as Ben and Tanya's sons (Evgeny and Alexander) and daughter-in-law Jana who had been allowed out in 1975 and had settled in Israel. A festschrift was published followed by the PCH Journal of which Brian was Editor-in-Chief. The initial Conference did not achieve the desired result so a second was held in Washington DC in 1978. Edward Kennedy came to the Conference dinner, albeit briefly, and advised that he was using his influence in the Levich case. Not long afterwards we were able to fly to Vienna and meet Ben and Tanya.

Brian with Benjamin and Tanya Levich, Vienna 1978

Also in the 1970s, Brian supported Robert Maxwell in his endeavours to regain control of Pergamon Press. Maxwell was not always a popular figure but he was popular with the editors of various scientific Journals published by Pergamon—he treated them well and they responded accordingly.

Back to science. In the early 1970s, Brian and Suhas Patankar published *Heat and Mass Transfer in Boundary Layers* which contained Genmix–Brian's first foray into software I think. Prior to publication, Brian set up CHAM (Concentration, Heat & Momentum Limited) within Imperial College offering consulting services to industry and, by so doing, founded commercial CFD. As computers became more prevalent and acceptable, Brian and his CHAM/IC team developed innovative application-specific codes including ESTER, HESTER, CORA, FLASH, TACT, PLANT and others. Brian moved CHAM out of IC to perform independently in the 1970s—it was a busy decade.

Also in the 1970s, Brian was appointed as a Reilly Professor of Combustion at Purdue University, Indiana. This was crucial. Why? Whilst at Imperial, Brian had colleagues to code for him. At Purdue, this was not the case so, rather than suspend his activities, he honed his Fortran skills and started to develop PHOENICS, his Parabolic, Hyperbolic, Or Elliptic Numerical Integration Code Series, the first general-purpose CFD code. PHOENICS came to the market in 1981 and continues today as, I think, probably the only *independent* CFD Code available commercially.

Time at Purdue did not last. Brian was archetypally English and missed home so we returned and settled. We had two sons (William and Jeremy) bringing Brian's family, at the time he died, to 6 children, 8 grandchildren and 5 great grandchildren (now 7). There were times when his passion for science seemed greater than his passion for family. Indeed, given that he was not a great one for small talk, it could be more difficult for him to relax socially than to create a new equation, re-solve an old one, or extend his beloved PHOENICS. He loved, supported and was proud of his family. Amongst the many and varied celebrations of his 90th Birthday, the one *he requested* was a family weekend.

Brian's 90th Birthday Celebration with his Family

He was conscious of family history and enjoyed discussions with his cousin Frances who created the Spalding family tree. When we were in Jersey, he went to find where his mother had been born and successfully located the house as it was being demolished. He did not have a camera with him so there is no record. He was particularly interested in the fate of Grandfather William (after whom our son was named) who had been "banished" to Australia, besought his family to provide him with funds to "buy a coat" (which action was forbidden by the family member who banished him) and who succumbed to illness in Vancouver Canada. Brian was delighted when a search for Grandfather William's resting place proved successful, contributed to the plaque which marks his resting place, and attended a conference on Vancouver Island in order to visit and have a photograph taken.

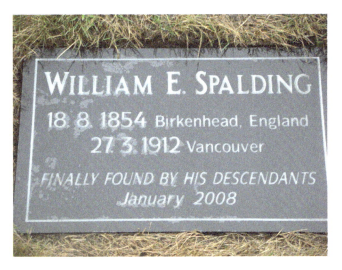

What else? There were conferences and parties, there were awards aplenty. He received the Global Energy Prize from President Medvedev in 2009. He was awarded the Benjamin Franklin Medal and sat on the same platform as Bill Gates; he was a Fellow of the Royal Society; a Fellow of the Royal Academy of Engineering; an Honorary Fellow of King's College (the School he interacted with throughout his life and sent his sons to); he received an Honorary Doctorate from Imperial College in 2014; and, in 2018, Professor Bill Jones organized a wonderful day of remembrance to be held there where I learned things including that Brian had been heavily involved in trade union matters—that was before my time.

Brian & President Medvedev 2009

Brian and Family at the Franklin Awards 2011

He went jogging every morning until forbidden by doctors (aging knees), he rowed, played squash (and created something called table squash which got lost in one of our moves), swam every Friday and during all holidays including our last one in Sicily in September 2016. He walked to CHAM across Wimbledon Common every Saturday and Sunday until he went to Moscow on October 29, 2016 (there is now a bench in his memory).

He took up pilates on his 90th birthday and was proud that he managed all exercises set. He once went white water rafting and had the most wonderful time until, on the second time around a whirlpool (the first having been a success), our younger son turned to discover (as he put it) that he had become an orphan with both Brian and I heading down the river at great speed without benefit of boat.

Brian back right

Brian's scientific legacy will live on as long as his academic scientific family continues, there is software based on PHOENICS, the textbooks and papers he authored and co-authored are used; and, of course, through his not-inconsiderable reputation. Physically, he will not meet the next generation who will question his ideas and move them forward in the same, or a different, direction but he would hope that it continues to occur via the written materials he has bequeathed to future generations.

One of Brian's strengths was that he was not bound by an idea, precept or concept —even one of his own. If he saw a better way, or had a better idea, he unhesitatingly explored it and, if it worked, moved forward with it. Two of his favourite precepts were "never let the best be the enemy of the good" and "all decisions are for the future". Thus, having decided in his sixties that individual computer codes would be better combined under one umbrella (PHOENICS), in his nineties, he decided that individual codes (industry-specific front ends) had a place as they enabled those without CFD knowledge to use software relating to their particular field without having to consider, or understand, the underlying physics. He was designing Simulation Scenes with his teams in London and Moscow up to his death.

Brian was a polymath; he "invented" turbulence models, was a "father of CFD", ran into trouble with those who believed computers "new-fangled". He was still having new ideas and, in 2016, was trying, passionately, to put over his "discretised population model" to audiences he was invited to address. He would be delighted if it is taken up and progressed—though would have preferred to do this himself.

Brian was human, intelligent, intellectual, whimsical, humorous, indestructible, brilliant, kind, caring, he twinkled, was a gentleman and never boring—I could go on. He could also be somewhat irritable, stubborn, once set on a path was almost immovable, believed in his scientific abilities in a way which may have driven others mad (yet was humble about his achievements), had little time for fools and less for small talk. He was set in his ways (as anyone who tried to intervene between him and work realized). He loved to have the last word in any argument or discussion which, given his intellectual capacity and command of English, was not difficult.

Brian loved literature, Shakespeare, theatre, music, languages (he taught himself Russian, spoke German, could "get by" in French and read Latin), sport, nature, walking, discussions over a glass of whisky, sitting in the garden in the sun (writing equations or doing the Times Crossword). He loved words particularly those which formed poetry. He read poetry voraciously and four volumes of poems he wrote have been published. His poems reflect his attitude to life, and to death.

We led independently co-dependent lives. Brian was not a great one for social activities and I am so I (or we the family) would socialize and "report back"—he always wanted to share experiences and know what was going on. I attended his lectures, understanding not a word. He attended my concerts despite a dislike of "church" music which he considered "gloomy".

Rome 2014

Sicily September 2016, Brian's last Conference

For me Brian was unique. He was brilliant, articulate, passionate, a great scientist, a lover of life, a lover of justice, a lover of poetry and so much more. He was 93 and had an amazingly productive, rich and full life. Above all, he loved science. He would be honoured that there have been so many articles written about him—though he would have preferred to be able to read, comment, correct, contribute—and, undoubtedly, describe some of the stories as apocryphal!

Brian refused to write his autobiography. I, and others, tried, unsuccessfully, to get him to write, dictate, make notes—anything—for some 15 years. When the subject was raised after his 90th Birthday celebration hosted by Imperial College he responded as follows—Brian, here, has the last word.

"Dear Friend

Colleen has told me that several of those present at Saturday's dinner suggested that I should write my autobiography. "Oh dear!" I thought. "Don't they recognize that I have written it? And that I have just read it to them?" Since you may have been among those to whom I did not make myself clear, I have re-titled my poem and added some explanatory footnotes. Here it is. With thanks to you for your attendance and for your so-kindly-expressed good wishes,

Brian"

My Three Lady Friends[1,]   by Brian Spalding

I have three kindly lady friends
And honour every one;
Each her unique persona lends
To all I've ever done.

The first tells me each week: "*I'm bored;*
"*Invent a newer game:*
"*Fresh goals, fresh rules bring fresh reward.*"
Restlessness[2] is her name.

The next sweeps mental store rooms clear;
So, if by any chance
New thoughts float in, they find space there.
Her name is **Ignorance**[3].

The third, who nothing does all day,
Has brought me most success
By urging: "*Find **some simpler way**.*"
Her name is **Laziness**[4].

That they let other men enjoy
Their charms I do not mind,
But, all the same, some care employ
So they're to **me** inclined.

To show I share the attitudes
Restlessness advocates,
I shun cliches and platitudes;
So now we are soul mates.

To **Ignorance** my gratitude
I show by letting new
Unlikely notions be reviewed;
And sometimes keep a few.

Dear **Laziness** I gratify
By proving **she is right**:
There **is** a simpler way. That's why
She visits me each night.

When DNS[5], then LES[6],
Wasted my energies,
"To discretise;" said **Laziness**;
"Gets pdf's[7] with ease".

Prandtl and Karman, clever guys,
Linked **u**plus with **y**plus.[8]
Too hard for me; so I devise
Yplus as f(**u**plus)[9].

"You want the distance from the wall?
"Use your imagination",
Says **L**; *"It isn't hard at all.*
'Just solve Poisson's equation'.[10]

Which I did later useful find
For transfer (radiative)
Of heat **and noise**[11]. A vacant mind
Can be, by chance, creative.

I studied finite elements[12]
Till stopped by **Laziness**;
For F**VM** (it's common sense!)
Can model **any**[13] stress.

These friends say none more ignorant
Nor lazy[14] they have found;
So of their aid I'll nothing want
Till I lie underground.

"E'en then", wise **Restlessness** has said,
"You'll not have long to wait;
"For I can help the restless dead
"Quickly re-incarnate".

A genius she has, or two,
Now waiting in the womb
For me; so, to **our** precepts true,
My life I can resume[15],

If I shall promise, **as I do**,
Daily all three to bless:
Kind **Restlessness**; **Ignorance** too;
Last ... lovely **Laziness**.

1. For me, innovation means imitation with a touch of originality. Here I am imitating Rudyard Kipling's "Six honest serving men". http://www.kipling.org. uk/poems_serving.htm.

2. "Restlessness" is not of course an unqualified virtue; and too often, alas, I have moved on to the next subject before properly completing the last; but "fresh rewards" have always been for me like the "fresh woods and pastures new" of Milton's Lycidas: http://www.dartmouth.edu/ ∼ milton/reading_room/lycidas.

3. The same can be said of "ignorance". But sometimes it is best **not** to know that the *cognoscenti* have declared to be impossible what one is just about to attempt. Or that someone else has already succeeded. Thus I was unaware that the failure of others to solve the two-phase-flow Navier–Stokes equations had led them to determine, perhaps as an excuse, that the solution problem was "ill-posed"; nor that Frank Harlow had displayed an equal lack of knowledge http://www3.nd.edu/ ∼ gtryggva/CFD-Course/JCP-Harlow-2004.pdf. (I'll skip the references from now on. This isn't a Ph.D. thesis.)

4. Just as "telling lies" is sometimes euphemized as "being economical with the truth", so can "laziness" be defined as "being economical with the effort". And this truly **is**, I believe, a virtue. Moreover, if to restlessness and ignorance there is added lack of practical skill, searching for "a simpler way" is the **only possible** way. Here are some examples.

   • In the 1950s, gas-turbine designers wanted to know the burning rate of kerosine droplets; which G.A.Godsave was skilled enough to measure. I could create only a clumsy one-inch-diameter copper sphere, covered with liquid fuel. Fortunately measurements on that enabled a more general theory to be developed and tested.

   • Others wanted to know the propagation speed of an adiabatic flame, *i.e.* one far enough from boundaries to lose no heat. Felix Weinberg's cleverly devised flat-flame burner could measure it. Mine could measure only that of a heat-**losing** one; but it was easier to construct and to operate. That how-ever also turned out to yield unexpected information of a more far-reaching kind. Laziness had struck lucky again.

   • Hirschfelder and Curtiss used mathematical methods to **calculate** the flame speed which were far beyond my reach. So I had to extend Schmidt's easy-to-understand **graphical** method (for unsteady heat conduction) to chemically reacting mixtures. It is true that I was studying the wrong pro-cess, i.e. **transient** not steady flame propagation. But transients become steady in the end; one just has to use a large-enough number of time steps.

   • Years later, it was the difficulty of finding polynomial or other formulae to fit boundary-layer temperature **profiles**, which led Patankar and me to choose the simplest-of-all **piece-wise-linear** ones, refinement of which led us to invent, or rather re-invent, the "finite-volume method". "Profile methods" we thenceforth abandoned.

- Looking back on my career, I have to admit that "finding the easy way out" has been my constant practice. Although this is regarded by some people as morally reprehensible, I am not ashamed of it.
- Oh! yes; and sometimes I have "moved the goal-posts", too. Equally reprehensible; and profitable.

5. DNS stands for Direct Numerical Simulation of turbulence. This is a practice which continues to consume enormous amounts of computer time, with, so far, little practical advantage; so my friend Laziness believes.
6. LES stands for Large Eddy Simulation, about which she has similar doubts.
7. Pdf's are "probability-density functions", which are what engineers would need if they were to turn the results from DNS and LES to practical account, for example for predicting combustion rates within, or thermal radiation from, turbulent burning gases. Curiously, DNS and LES practitioners rarely present their results in pdf terms, having perhaps lost sight of, or never been interested in, practical outcomes. By following the advice of Laziness to "discretise" has enabled me to procure such information at much less computational expense. On the left below is a contour diagram of one ordinate element of the pdf of concentration, when a bluff body is placed in an approach stream having non-uniform concentration and velocity. On the right is the corresponding diagram computed by means of LES for the same circumstances.

 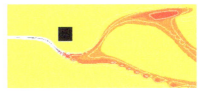

They differ. So the question is which is the more correct? **Nobody knows**, even though the experiments would be easy to perform; but one can be sure that the discretised model gives the better representation of the mixing **upstream** of the obstacle (on the left); while it is possible that LES represents better the **downstream** region (on the right). If only discretisers, LES practitioners and experimentalists would talk together! But **they have no common tongue**. I am trying to provide one.

8. The reference here is to the so-called "law of the wall". Experiments had shown that the velocity profile in a turbulent stream near a wall had a linear shape at small distances and a logarithmic one at large ones; and attempts had been made, by Prandtl and von Karman, to represent its total shape via 2- and 3-part equations. These had the dimensionless velocity, uplus, to the left of the = sign and the dimensionless distance, yplus, to the right.
9. Probably other clever researchers have obtained greater accuracy by formulating 4- and 5-part expressions. But it occurred to me that if one placed **yplus** on the **left** and **uplus** on the **right**, it was easy to devise a 1-part formula which would cover the whole range. I had "moved the goal-posts"; after which kicking the goal was so easy that even I could not miss.

10. In order to **use** the law of the wall, one needs to know how far away the wall is from the point in question. If the space is cluttered with solid objects of various shapes and sizes, it is far from clear how one should proceed. Indeed even how to **define** the "distance from the wall" is questionable. I cannot now either recall or understand what prompted my guess, namely that solving a certain Poisson-type equation would yield distances which were exact very close to the nearest wall, and never unreasonable when many objects were almost equally distant. Inspiration by Lady Laziness, coupled with the vacancy of mind provided by her sister, Lady Ignorance, furnish the most plausible explanation.

11. As luck would have it, the Poisson equation yielded not only the distance from the wall but also the "distance **between** walls". This is exactly what one needs when creating a mathematical model of radiative heat transfer that is simple enough to be used in engineering calculations. So Lady Laziness had turned once more into Lady Luck, presenting me with another easy-to-exploit simulation tool. What is more, in **just the last few days** I have seen how it can be extended to simulating the spread of noise also. If Twelfth-Night Maria's formula ("some are born …", "some achieve …", and "some have … thrust upon …") were applied to scientists, I should have to place myself in the third category.

12. When the finite-element-method (*i.e.* FEM) furore first broke out, I was too busy with CFD to give it attention; but I did take notice when its promoters recommended FEM as being good for solving fluid flow too. However, my mathematical knowledge did not suffice; so Laziness turned into Leave-it-alone; which saved me much time.

13. Later, I saw how the finite-**volume** method (*i.e.* FVM), if it treated displacements like velocities, could solve solid-stress problems as easily as **fluid**-stress ones. So I built the capability into PHOENICS; but I did no proselytising. Now however I recognize the FVM/FEM dichotomy as being comparable with those between Protestant and Catholic, or Sunni and Shia. That is to say that it promotes needless conflict; and expense; and loss of opportunity. I have characterized the conflict as being between N-UWFists and UWFists, because the FEM camp has needlessly carried over from pre-computer solid-stress-analysis days the practice of multiplying the differential equations by Non-Unity Weighting Factors (N-UWFs). The FVM camp, by contrast, uses Unity Weighting Factors (UWFs); which is to say that they use **no weighting at all**. They are all the better for it.

14. Nor "**more opinionated**" either, I can imagine some of you saying.

15. Re-incarnation is a widely believed-in but not-yet-sufficiently-proven way of prolonging one's contribution to mankind. I'll give it a go; but meanwhile I take any opportunity of transmitting what I have learned *via* a written record; like this, my autobiographical poem.

Appendix II

Tributes to an Exceptional Life
D. Brian Spalding
9 January 1923–1927 November 2016

D. Brian Spalding
Sc.D., Ph.D., FRS, FREng, FIMechE, FInstF

Compiled & Edited by:
Prof. Akshai K. Runchal
Ph.D. Student of Prof. Spalding—1965–1968
runchal@gmail.com

Foreword

On 30 May 2017, a Personal Recollection and two Technical Sessions were held at the CHT-2017 Conference in Naples, Italy. A number of former students and colleagues paid tributes to Professor Spalding and talked about his influence on their personal and professional careers. The Conference was co-sponsored by the International Center for Heat and Mass Transfer (ICHMT) and the American Society for Thermal and Fluids Engineering (ASTFE).

Prof. Spalding was the prime mover and a co-Founder of the ICHMT. He had a profound impact on the practice of Fluids and Thermal Engineering through his lifelong contributions in Combustion, Fluid Flow, Heat and Mass Transfer and his innovations in Computational Fluid Dynamics (CFD). Though some core technologies that gave rise to CFD existed earlier, the CFD as we know it today was nurtured in the Heat Transfer Section (also called HTS and ThermoFluids) of the Mechanical Engineering at Imperial College London in the decade between 1965 and 1975 by Professor Spalding and his students. In the late 1960, Patankar and Spalding developed the Integral Profile and Finite Difference methods for Parabolic Boundary Layer equations. Working in parallel, Runchal, Wolfshtein and Spalding developed the Finite Volume Method (FVM) and the high-Reynolds number stabilization schemes for two-dimensional Stream Function–Vorticity form of the Navier–Stokes equations. By the end of 1968, this core technology that was to become the cornerstone of CFD practice was in place. Later, in early 1970, a number of researchers, including Caretto and Curr, worked on developing a computational solution of the three-dimensional Navier–Stokes equations. These efforts, together with some key developments by the group led by Frank Harlow at Los Alamos, finally led to the SIMPLE algorithm by Patankar and Spalding. Throughout these developments, Professor Spalding was the prime mover and the head of this HTS Section.

Professor Spalding's contributions are many and varied. Some of these have already been captured comprehensively in publications including "*Brian Spalding —CFD and Reality: A Personal Recollection*", (Runchal, IJHMT **52**, 2009.), "*A Tribute to D. B. Spalding and his Contributions to Science and Engineering*" (*Artemov et al. IJHMT* **52**, *2009) and "Emergence of Computational Fluid Dynamics at Imperial College-1965–1975*" (*Runchal, Journal of Heat Transfer, 2012*). I also understand that the Royal Society has commissioned a biographical memoir on his professional life that should appear in due time.

A tribute to Brian Spalding has been published in *Int J Heat & Fluid Flow* by Gatski et al. (*Tribute to Professor D. Brian Spalding, FRS, FREng*). The tributes that follow here are of a more personal nature. These are by his students, associates and colleagues whose life was profoundly influenced by Spalding either at a personal or professional level. As a matter of fact, anyone who practices the art and science of CFD today is doing so because of the seeds sown by Brian Spalding.

In 2013, Prof. Spalding's 90th Birthday was celebrated on the occasion of the ASME Summer Heat Transfer Conference meeting in Minneapolis. A Dinner Event was organized at the Hilton Hotel in Minnesota on 13 July and a number of persons either paid a tribute in person or sent messages to be read. These tributes are included in Appendix A of this Monograph. Appendix B contains some of the contributed Photographs.

Professional Spalding, Brian to me, was an intellectual giant and an exceptional human being—a towering intellect, a founder of scientific institutions that endure to the day, a fighter for human rights who risked his professional career for the freedom of Soviet scientists, a poet, a linguist and a family man—and for me an intellectual father and a dear friend.

Most people today remember Brian for his technical contributions. He himself wanted to be remembered as a poet. Poetry was his lifelong passion and almost an obsession during his later years. This monograph therefore starts with the last known poem written by Brian and ends with his poetic rendering of his own autobiography.

Today is exactly one year since Brian passed away. On this day, this collection of tributes is being released to honour and remember him. Brian, we will miss you but generations will remember you for your contributions to science and engineering.

Prof. Akshai Kumar Runchal
McLeod Ganj & Los Angeles
November 27, 2017

A Last Poem by Brian Spalding

Found November 28 2016

I shall have no regrets when I am dead
Of deadlines, none will matter but my own.
Unwritten papers? Hopelessly misled
Inheritors? All claimants I'll disown.
Yet hope, while still alive, there'll be but few
Who think: I was a fool to trust him.
Now that he's gone, what am I going to do?
None I would hope; but guess the chance is slim.
Yet, in that soon-to-close window of time,
There's much I want to do; and think I can.
Always too optimistic is what I'm
Dismissed as. To disprove it is my plan.
"After such labours", I would have it said,
"It must be truly blissful to be dead".

Brian by Colleen

Choir Trip to Rome 2014. Brian was unused to being the "accompanying person" but managed with aplomb.

For me Brian was unique. He was brilliant, articulate, passionate, a great scientist, a lover of life, a lover of justice, a lover of poetry and so much more. He was 93 and had an amazingly productive, rich and full life; the end of it came about unexpectedly and tragically swiftly. I thought that he was immortal—as did many others.

The tragedy is not that he died—we knew, logically, that time was diminishing. The tragedy is that he was ill so far from home and that, at his end, a man who lived through speech and writing could communicate via neither. Enough. Brian would disapprove of my "raging"; he did not believe in it. He was a pragmatist, a logician and a philosopher; he would have said "move on". We try.

Brian's life revolved around, influenced and was influenced by, **families**. He had a large **biological** family; he and Eda (his first wife) had 4 children, 8 grandchildren and 4 great-grandchildren. We had 2 boys, William and Jeremy (now men). There *were* times when his passion for science seemed greater than his passion for family. Indeed, given that he was not a great one for small talk, it could be more difficult for him to relax socially than to create a new equation, re-solve an old one or extend his beloved PHOENICS. He was proud of his family, loved and supported them. The celebration *he* requested for his 90th Birthday was a family weekend. His family are his legacy. Our sons and I miss him every day.

He had a HUGE **scientific** family: children, grandchildren, great-grandchildren and, maybe, even great-great-grandchildren in the world of Computational Fluid Dynamics (of which Brian was a founding father), Combustion, Heat Transfer, Fluid Flow, Boundary Layer and Unified Theory, the SIMPLE algorithm, Finite Difference and Finite Volume Methods, Turbulence Modelling, Multiphase Flows, Heat Exchangers, and other aspects of his varied and beloved scientific work.

One of the things Brian most missed when he retired from his Chair at Imperial College was the day-to-day interaction with young and enquiring scientific minds. When we attended Conferences, he enjoyed young people "brave" enough to ask for a photograph or, even better, for a conversation. He was always happy to oblige. The last such requests came in Cologne in October and in Palermo in September 2016.

His scientific legacy will live on as long as his academic scientific family continues, there is software based on PHOENICS, the textbooks and papers he authored and co-authored are used; and through his not-inconsiderable reputation. Physically, he will not continue to meet the next generation who will question his ideas and move them forward in the same, or a different, direction but I hope it keeps happening.

In addition to his academic scientific family, he had his **CHAM** scientific family. In the early 1970s, Brian and Suhas Patankar published "Heat and Mass Transfer in Boundary Layers" which contained Genmix—Brian's first foray into coding I think. Prior to publication, Brian set up CHAM (Concentration, Heat & Momentum Limited) within Imperial College offering consulting services to industry and, by so doing, founding *commercial* CFD. As computers became more prevalent and acceptable, Brian and his CHAM team developed innovative application-specific codes including ESTER, HESTER, CORA, FLASH, TACT, PLANT and others. Brian moved CHAM out of IC to perform independently in the 1970s—it was a busy decade.

Also in the 1970s, Brian was appointed Reilly Professor of Combustion at Purdue University, Indiana. This was crucial. Why? Whilst at Imperial Brian had colleagues who coded for him, at Purdue this was not the case so, rather than suspend his activities, he honed his Fortran skills and started to develop PHOENICS, his **P**arabolic, **H**yperbolic, **O**r **E**lliptic **N**umerical **I**ntegration **C**ode **S**eries the first general-purpose CFD code. PHOENICS came to the market in 1981 and continues today as, I think, pretty well the only *independent* CFD Code available commercially.

One of Brian's strengths was that he was not bound by an idea, precept or concept—even one of his own. If he saw a better way, or had a better idea, he unhesitatingly explored it and, if it worked, moved forward with it. One of his favourite precepts was "all decisions are for the future". Thus, having decided that individual computer codes would be better combined under one umbrella (PHOENICS); in his 90s, he decided that individual codes (designed as SimScenes or industry-specific front ends) had a place as they enabled those without CFD knowledge to use software relating to their particular field without having to consider the underlying physics. He was designing SimScenes with his

Development teams in London and Moscow up to his death. We at CHAM are attempting to bring them to completion and offer them to industry.

Over the years many Engineers, and non-technical staff, passed through CHAM's portals. Brian was, justifiably, proud of CHAM; he loved it and had every intention of working his standard seven-day week for as long as he lived. He managed pretty well. We at the company will try to move Brian's vision forward and maintain a CHAM legacy.

There was his **Friends and Neighbours** family who joined in our celebrations from children's birthday parties to Brian's 93rd at the start of last year—all held in the home we loved. I close my eyes and see him, in his chair, in the garden room debating the meaning of life, and the political activities of the day, with glasses of single malt assisting the conversation. I see him at our Book Club listening, absorbing and then commenting—clearly, concisely, with humour but..... We still gather but now the talk is about him rather than with him. He is remembered fondly by *this* family and will be for many years.

There was his **Russian** family. After perestroika, he went to Russia often. In St. Petersburg, he met an academic who translated his poetry from English into Russian and had four volumes of it published. That gave Brian *such* pleasure. In Moscow, he loved working with his research group. He was in "his" suite in "his" hotel with "his" group when he was taken ill. It is where he wanted to be.

Brian had a highly developed, and personal, sense of right and wrong. It did not always agree with the mores of the day because *his* beliefs were *his*. He *would* change them but only after thought and deliberation and NOT because any external force wanted him so to do. Brian acted according to his own, strongly held, principles because "*it was the right thing to do*". This phrase represented him in so many aspects of his life.

Brian's passion for doing what he perceived as the right thing (whether it endeared him to his fellow man or not) meant that whilst I was with him in Moscow throughout what proved to be his last (and only real) illness I was supported by Jana Levich. We have been friends for 40 years because Brian decided, in the early 70s to take on the USSR and fight for the right of Academician Levich and his family to move to Israel. This alienated the bureaucracy of the time which was a sorrow to him but sorrow did not prevent him succeeding against all odds. *It was the right thing to do.*

Doing the "right thing" could have been problematic when, as a young man, he rescued his sister-in-law from East Berlin after the War. As I heard it, he "borrowed" a laundry van and drove to the Russian sector with a laundry list, in Russian, as his "authorization". He collected mother and (sedated) baby. Near the exit point, a troop of Russian soldiers hove into view strung across the road. Had they stopped the vehicle *they* would have seen the laundry list as—a laundry list! Fortunately, they peeled off to either side and Brian drove on. *It was the right thing to do.*

When at a younger age, he chose to do Engineering at Oxford and was informed, by a Master at King's College School that they had not educated him so that he could take a degree in the sciences rather than a double first in the classics it influenced the young Brian not a jot. He started the way he went on—his, right, way. He "invented" turbulence models, he was a "father of CFD", he ran into trouble with those who believed computers were a "newfangled notion", etc. He was still having new ideas and, in his last years, was passionately trying to put over his "discretized population model" to any audience to which he was invited to speak. I hope someone will take it up?

It typifies Brian that we spent what was to be our last anniversary in Cologne so he could lecture. It could not be any other way. We led independently co-dependent lives. Brian was not a great one for social activities and I am so I (or we the family) would socialize and "report back"—he always wanted to share experiences and know what occurred. I attended his lectures, understanding not a word. He attended my concerts despite a dislike of "gloomy" church music.

To me, Brian was human, intelligent, intellectual, whimsical, humorous, inde-structible, brilliant, kind, caring, he twinkled, was a gentleman and never boring—I could go on. He could also be somewhat irritable, stubborn, once set on a path was almost immovable, believed in his scientific abilities in a way which may have driven others mad (yet was humble about his achievements), had little time for fools and less for small talk. He was set in his ways (as anyone who tried to intervene between him and work, or the ten o'clock news, realized). He loved to have the last word in any argument or discussion which, given his intellectual capacity and command of English, was not difficult.

Brian loved literature, Shakespeare, theatre, music, languages (he taught himself Russian, spoke German and could "get by" in French) sport, nature, walking, discussions over a glass of whisky, sitting in the garden in the sun (writing equa-tions or doing the Times Crossword), he started pilates on his 90th birthday. He loved words particularly those which formed poetry. He read poetry voraciously and he wrote it. His poems reflect *his* attitude to life and to death. He confronted death in poetry; I found a handwritten poem on the subject the day after he died (see page 5) which I found difficult to read them, and probably always will, but it summed him up in so many ways.

Above all, he loved **science**. He would be honoured that articles are being written about him and that colleagues have gathered in his memory—though he would have preferred to be present to listen, comment, contribute—and, undoubt-edly, describe some of the stories as apocryphal!

Words cannot say how much I, and others who loved him, miss him; he touched many lives across continents and areas of interest. Life will never be the same now he is gone.

Colleen I Spalding
Director, CHAM, cik@cham.co.uk

Brian at 90 with his Family

Should this be read out at my funeral
And some feel tempted to let fall a tear,
Thank you, I say but life's ephemeral.
Be glad: no more I'll grunting fardels bear;
Nor try to show off my Shakespearean knowledge;
Tell tales so trite table-companions groan:
How William said:" Dad works at JAM and PORRIDGE".
And Jeremy: "I'll do it by my own".
To (plural) wives and families: Farewell.
To friends, acquaintances, fond lovers too;
And co-believers in my "What-the-hell"
Philosophy; to students old and new:
A sonnet ends when its last couplet's read.
Finished with this, I'm satisfied; soon dead.

by Brian, from "My Sail Hoistings" published 2011.

Professor Brian Spalding: Brilliant, Inspiring, Peerless—Rabi Baliga

It is an honour for my students and me to be considered as members of Professor D. Brian Spalding's extended academic family. Even though I was not fortunate enough to be one of his students, I have had the privilege and pleasure of having Professor Suhas Patankar as my Ph.D. supervisor (at the University of Minnesota, Minneapolis) and Professor Akshai Runchal as my instructor of undergraduate fluid mechanics (at the Indian Institute of Technology, Kanpur); and they have been my mentors ever since. Many years back, it was Akshai who kindly suggested that I should apply to the University of Minnesota to do my Ph.D. studies with Suhas; and that was one of the best things that ever happened to me.

At the University of Minnesota, Suhas taught me the subject of computational fluid dynamics (CFD) and many other things. His unbounded enthusiasm and passion for CFD and teaching were infectious. The brilliant seminal works in CFD of Professor Spalding and his group from the late 1960s and early 1970s inspired and brought me great joy. Even today, my face lights up every time I think of the following three groundbreaking works, for example, from those early days of CFD and introduce them to my own students: Professor Spalding's proposal of the general form of the governing equations (consisting of the unsteady, advection, diffusion and source terms); the upwinding schemes proposed by Runchal and Spalding; and the semi-implicit method for pressure-linked equations (SIMPLE) of Patankar and Spalding, and its variants.

Over the years, I have had several direct interactions with Professor Spalding and his lovely wife (Colleen). Even though these interactions were brief, they were always wonderfully inspirational and enriching in many ways. Suhas and his beautiful wife (Rajani) have treated me as a member of their family, and the doors of their home have always been open to me. Akshai has been and continues to be a marvellous mentor to me. I have developed very rewarding friendships with several of Professor Spalding's former students, in particular, Pratap Vanka, Andrew Pollard and Steven Beale.

In summary, Professor D. Brian Spalding was and will remain truly peerless. He has influenced my life and the lives of my students in numerous wonderful ways. We will always be very grateful to him, and we wish him eternal peace and happiness.

B. Rabi Baliga
Professor
Department of Mechanical Engineering
McGill University, Montreal, Quebec H3A 0C3, Canada

Working with Brian Spalding—Steven Beale

I joined the computational fluid dynamics unit (CFDU) at Imperial College in 1986, following a recommendation from Prof. Chang-Lin Tien of the University of California at Berkeley. At that time, I was in my early 30s. My main interest at the time was in numerical heat transfer and heat exchangers. Previously, I had worked in the Solar Energy Program at the National Research Council of Canada (NRC). At the time the CFDU was perhaps 20–30 people: professors/lecturers, visiting scientists, professionals, post-doctoral fellows and post-graduates. I recall in Spring 1987, Prof. Spalding taking the entire group up to the Physicochemical Hydrodynamics Conference at Oxford University. This was in honour of his friend Prof. V.I. Levich, the Russian-Jewish "refusenik" on whose behalf Spalding had activated, politically and socially. It was at the PCH conference that I got to appreciate the amazing breadth of CFD, with numerous applications in a vast array of applications: metal casting, magnetohydrodynamics, membrane science, bio-medical applications, etc. Mr. G. Liao and I gave a poster presentation on some side work we were doing for the Health and Safety Executive in developing CFD models of industrial goggles and respirators.

My Ph.D. topic was flow and heat transfer in tube banks and other heat exchangers such as offset fins, but somehow the work kept expanding. At one point, Brian Spalding walked into the office and suggested (insisted) that I derive an analytical solution for potential flow in tube banks, from complex variable theory, for code verification. This seeming distraction leads to several interesting sidelines such as three-dimensional stream functions and the solution to stress analysis problem using the finite volume method, which of course Brian Spalding and others subsequently contributed to, in no small way.

For the latter part of the thesis work, I was back in Canada at the NRC, and I used to regularly communicate with Prof. Spalding via email, rather than by facsimile (fax) which was the norm at the time. When it came to publishing the work, I became nervous about the timeline, and in response to my anxious inquiries, Prof. Spalding sent me an email apologizing for the delay and lamenting as to how he wished that "real life conformed to the SIMPLE algorithm whereby as the pressure rises—the rate of output increases". Alas, the email is locked in a file somewhere on a VAX computer.

In the biographical memoir of the physicist Richard Feynman, the author Jagdish Mehra writes about two types of genius: the "ordinary" genius and the "magician". Brian Spalding was surely in the latter category. A magician of the highest order, he conjured up an entire subject, like a spell, out of nowhere, and then turned it into a practical business, relevant to almost every engineering application, from which countless ex-students and associates have subsequently profited. The  scope of his knowledge was truly unique. It has been a very great privilege to have known and worked with him over these last 30 years.

Steven B. Beale
Professor, Institut für Energie- und Klimaforschung (IEK)
IEK-3: Elektrochemische Verfahrenstechnik
Forschungszentrum Jülich GmbH
52425 Jülich, Germany
s.beale@fz-juelich.de

Memories of Working with Brian Spalding—Larry Caretto

I first met Brian Spalding in the late 1960s when I was an Acting Assistant Professor in the Mechanical Engineering Department at UC Berkeley and Brian was a Visiting Professor there. I often found Brian in a keypunch room where he sat patiently preparing Fortran program cards in a room largely occupied by junior faculty and graduate students. Later, Brian wrote a letter to our Division Chair, Ernie Starkman, asking if Ernie could recommend someone like myself to visit Imperial College for a year to work on CFD. I am not sure why Brian asked for someone like me, but I do remember that he was grateful when I showed him how to read the sort field on binary cards to properly rearrange a dropped binary card deck. I've deliberately used the split infinitive in the last sentence in memory of the first paper I wrote with Brian. I did not realize how much my American English allowed the use of the split infinitive until I counted the number of such grammatical errors in British English that Brian corrected in my draft.

I spent the 1970–1971 academic year working with Brian's group at IC. I remember several memos that Brian prepared for our group with ideas and suggestions usually followed by assigning tasks to individuals working on codes that became known as SIVA and SIMPLE. Brian's knowledge of the developing field of CFD, and his intuitive explanations of fluid phenomena and how to model them on computers, was an inspiration to our group. I also appreciated his willingness to change his mind when another member of the group convinced him that there was a better way to solve a problem.

I remember one time when Brian came into a room where I was working on a teletype terminal, then a new way to communicate with computers. I've forgotten what I was working on, but I remember that I was having problems, and in a discussion with Brian we both agreed to try a solution which we both thought had a dubious theoretical basis and little or no chance of success. Our failed try proved that our initial thoughts were correct. As I deleted the file, Brian commented favourably on the ability the terminal provided to try silly ideas with no worries that there would be any written records of their failure.

Brian was not only a font of ideas, but he also created a strong group of individuals who worked together on various projects related to CFD. When I started at IC, the main team members working on three-dimensional parabolic flow project were David Gosman, Bob Curr, Devraj Sharma and David Tatchell. Later that year I shared an office with Suhas Patankar and Aki Runchal. I remember spending most of one afternoon with Suhas helping him find the words for the SIMPLE acronym that he was determined to use for the algorithm he had developed for three-dimensional parabolic flows. I also discovered the meaning of serendipity when Aki found the solution to the convergence problem in applying SIMPLE to elliptic flows. A coding error produced no correction pressure adjustment for continuity. Although this meant there was no correct solution, it did show that the divergence in the correction pressure that had been plaguing the use of SIMPLE in elliptic flows had vanished. Adding an under-relaxation to the correction pressure algorithm provided the desired solutions.

After a sidetrack working on applying SIVA to free convection, I began work on the application of SIVA and SIMPLE to steady, three-dimensional elliptic flows. This work continued after I left IC in September 1971. An initial publication of this work, modelling the wind flow around buildings, presented at the Third International Conference on Numerical Methods in Fluid Mechanics, was

co-authored with David Gosman, Suhas and Brian. (Brian's policy of placing the names of authors in alphabetical order often made me the first name on the list of authors.).

I also spent two summers at IC working on projects for CHAM modelling rocket exhaust flow (1972) and solution mining (1974).

I remember the time that I spent with Brian and his group as the most intellectually stimulating time of my career and I have always appreciated the opportunity Brian gave me to be a part of that group.

Larry Caretto, Professor Emeritus
Mechanical Engineering Department
California State University, Northridge
Northridge, CA 91330-8348
lcaretto@csun.edu

My Interactions with Spalding—Graham de Vahl Davis

Brian Spalding was the creator of CFD/HT, a solution method many of us have spent our lives using. My first contact with Brian was around 1970—sorry, I can't remember the exact year—when I spent a sabbatical with him when he was at Imperial College. We have been in contact ever since. In May 2008, I had the privilege of organizing an ICHMT Conference in Brian's honour in Marrakech Morocco. And of course, many of us contributed to "A tribute to D.B. Spalding and his contributions in science and engineering", which was published in IJHMT in 2009. He will be sorely missed.

Graham de Vahl Davis
Honorary Professor of Mechanical
and Manufacturing Engineering
UNSW Sydney
g.devahldavis@unsw.edu.au

One Note in Memory of Professor
Brian D. Spalding—Milorad Dzodzo

During my work on M.Sc. Thesis at University of Belgrade, in 1982, I was struggling with some numerical method implementation details. These were my first attempts at creating a CFD program which could simulate, converge and even produce some logical and meaningful results. At that time, the number of papers related to CFD details was limited.

Fortunately, Professors Zoran Zaric and Naim Afgan from Vinca Nuclear Institute, in Belgrade, had informed me that the XIV Symposium of the International Centre of Heat and Mass Transfer (ICHMT) devoted to Heat and Mass Transfer in Rotating Machinery was scheduled in Dubrovnik that year and that Professor Spalding, one of the ICHMT founders, will most likely attend. I seized the opportunity to meet Professor Spalding and had a short but very useful chat about some numerical issues I was facing. At that time, I was not aware of how lucky I was to get an advisory and consultation session during two relaxing Symposium coffee breaks, instead of scheduling an appointment in his office with his secretary one or two weeks in advance. The right words at the right time aided me in overcoming the outstanding issues and the program finally converged.

Written correspondence with Professor Spalding had continued after I obtained an M.Sc. and resulted in a privileged British Council Fellow Scholarship that allowed me to spend the 1985/86 school year, and 1986 fall semester as a research associate, at the Imperial College, Computational Fluid Dynamic Unit, where I studied CFD and observed the very beginning of commercial CFD software creation.

Professor Spalding's unique style of efficient knowledge transfer was at that time adjusted to the dynamics of the Unit and the needs of each already highly specialized visitor who had their own arrival time, planned duration of stay and specific interests. Well-organized and prepared printed lectures with Figures, diagrams and short notes, which were in fact predecessors of today's popular PowerPoint presentations, were available and recommended to the newcomers first, and followed by an assigned list of papers and reports. After that, one-to-one office consultations, scheduled at week in advance, were the most prized learning opportunities.

The mood in the department was very calm except when somebody in the group would report success and obtained good results, or figured out how to generate proper input. I was asking myself if this was a result of proper work environment, or just a difference in culture and climate. Perhaps, this was the British way of doing business, which was drastically different from the Mediterranean norms.

However, everywhere else in the cosmopolitan London the world-renowned British calm demeanour was not as prominent. Everything was more akin to the Mediterranean culture that I was accustomed to. In some cases, emotions and humour were at an even higher elevation, especially during the soccer games, after a variety of TV humour series, or longer Friday's stays in pubs.

Finally, one day I had a chance to understand how a collected and determined spirit is nurtured when I attended an unofficial out of the office lecture by Professor Brian Spalding. Fortunately, it was at the very begging of my stay at IC, so I had a chance to adjust early.

Visitors from the Navy were scheduled to visit IC CFD Unit, and colleagues were preparing poster presentations for two weeks. Each poster was discussed several times and carefully reviewed by Professor Spalding. At the day of the visit in the morning, the posters were nicely positioned near each other in the entrance hall of Tizard dorm just across the street from the IC Mechanical Engineering Department.

The poster session was scheduled in the afternoon after lunch. The whole Unit and visitors were in good mood but after entering the dorm entrance hall everybody had noticed something strange. Lots of papers littered the floor. The posters were damaged and atop of some that were left intact, pacifist protesters wrote messages against the military complex. Everybody was surprised and looked to each other without saying anything but the obvious question was—what are we going to do now? The standstill was broken when Professor Spalding calmly took a piece of paper from the floor, figured out to which poster it belonged, placed it back and kept it at the right spot and asked the planned presenter to start their presentation with an apology to the visitors and an explanation that we needed to change the order of presentations due to the unplanned circumstances. The rest of the poster presenters followed in the same manner, and all of the planned posters were presented. Once back in the department, Professor Spalding calmly asked the secretary to prepare one letter addressed to the Dean's office and that was most likely the only action and reaction to the incident in the dorm.

Later on, from time to time in my life and career, I encountered similar situations where a sudden loss needed to be recovered, or when work needed to start again from the scratch. The difference was that now I was on the spot to lead by example. Several times, having in mind that scene in the entrance hall of Tizard dorm, I was able to convince colleagues to calm down, disallow emotions to waste our time and energy and concentrate on recovering what could be recovered and then proceed ahead. Sometimes when crisis is over, I just tell my colleagues that in spite of my Mediterranean background I was able to keep calm, thanks to the fact that I had spent some time in Britain. On rare occasions, I would disclose to my closest colleagues the whole story which inspired my perspective. I would like to thank Professor Brian Spalding for that unexpected and perhaps most valuable unofficial out of the office lecture that served me well and that I always attempt to pass to the younger generation.

Milorad B. Dzodzo
Fellow Engineer
Westinghouse Electric Company
Global Technology Development
1000 Westinghouse Dr., Suite 305
Cranberry Twp., PA 16066, USA
dzodzomb@westinghouse.com

Memories of DBS—Marcel Escudier

My first encounter with Brian was in 1960 when he taught thermodynamics to first-year undergraduates at Imperial College. In the academic year 1964/5, he was my supervisor on a postgraduate (DIC) project involving a simple theory for the calculation of the Reynolds analogy factor and the temperature recovery factor for a turbulent boundary layer. This was an enjoyable year, not least because Brian was using the results of my calculations in his postgraduate lectures, which I attended, and we met far more often than he did with most of his students. As a consequence, in October 1964 when I started my Ph.D. under his supervision, we knew each other pretty well and had already developed a good working relationship. My Ph.D. was concerned primarily with analysing published experimental data to extract information about the entrainment rate, develop a correlation for the entrainment function, central to the Unified Theory of Friction, Heat Transfer and Mass Transfer in the Turbulent Boundary Layer and Wall Jet, and carry out numerical integration of the ordinary differential equations Brian had developed within the theory. At some stage, he informed me that every well-rounded Ph.D. had to include an experimental aspect. He suggested a hot-wire anemometer investigation of a turbulent boundary layer with blowing upstream and subjected to a severe adverse pressure gradient. I was the first to use hot-wire anemometry in the heat transfer laboratory (according to someone, it was the problems I encountered, and detailed in my thesis, that led Jim Whitelaw to develop LDA with Franz Durst). A major problem in the recirculating wind tunnel I was using was dust build up on the heated wires causing major calibration drift. When I went to see Brian about this, he just said "flypapers", expecting me to collect the dust by installing flypapers ahead of the working section. I didn't find this to be the most helpful advice! However, he arranged for me to discuss my problems with Peter Bradshaw, at that time at the National Physical Laboratory in Teddington, which proved to be very useful.

Brian had been asked to review a paper submitted by a well-respected academic from Southampton University to the Proceedings of the IMechE. It was about hot-wire anemometry about which Brian said he knew very little (see above!) and so asked me to do the review, with an explanation to the editor saying that he felt I was an appropriate choice as being "suitably imbued with disrespect for his seniors".

Personally, I found Brian very easy to work with, and not in the least condescending or intimidating. The latter was probably due to my naivety and unawareness of him as a major international figure in our community. I should have been alerted to this by the number of high profile visitors to the Thermofluids Group, including Bill Kays and Bill Reynolds from Stanford, Roger Eichhorn from Kentucky, and Samson Kutateladze and Aleksandr Leont'ev from Novo Sibirsk. I recall other research students being incredibly nervous about meeting with Brian. One (no names) would empty his pockets beforehand of anything, such as keys and coins, that might cause a distracting noise during their meeting. Arranging such a meeting to discuss day-to-day progress was tricky if you followed the official procedure of arranging the meeting through his secretary, Marjorie Steele. I discovered the trick was to knock on his door after she'd left for the day: he'd always see me.

Kutateladze and Leont'ev had written a monograph with the title "Turbulent boundary layers in compressible gases" which Brian had translated from Russian into English. A remarkable feature of the book is that Brian's comments as footnotes more than doubled its length. Shortly after its publication (in 1964), Kutateladze and Leont'ev came to visit. A meeting was arranged at which each of Brian's research students (I believe Suhas was also there) was asked to say what they were working on. Brian acted as translator. After every presentation, Kutateladze said they'd already done whatever it was. Towards the end, when it came to my turn, I just said "there's no point saying anything, they've done it already". With a straight face, Brian translated. There was a shout in Russian from Kutateladze which Brian translated into English as "sacrilege" to laughter all round.

I fell out with Brian on only one occasion when he said I should leave out of a paper the lengthy appendix giving all the details of the analysis I'd been doing on a drag law correlation. I actually threw a pencil across his office I was so angry. He didn't react and although the incident was never mentioned, it may go some way to explaining the following.

In spite of his legendary insight, Brian wasn't infallible. On one occasion, he asked me to read the draft of a paper he'd written. He'd shown that the quantity $\partial\tau/\partial u$ at the edge of a shear layer is equal to the entrainment rate, an interesting result. He'd then correctly evaluated $\partial\tau/\partial u$ as the quotient of the spatial gradients of τ and u. I pointed out that both gradients were identically zero at the edge of the shear layer but became non-zero when approximate expressions were adopted for the spatial gradients of τ and u. I only found out during the Stanford Conference in 1968 that he agreed with my assessment that this made no sense and so he binned it.

On another occasion, on a Friday afternoon, he gave me a photocopy of a manuscript he was working on, about 200
handwritten pages, and said he'd like my
comments by Monday morning. I duly
obliged with lots of suggestions in red ink
which he largely accepted.

After finishing my Ph.D., I went to MIT
which I found to be culturally very similar to
Imperial College. It was only after I left MIT
that I realized it wasn't like that everywhere.
That made me even more appreciative of
Brian and Imperial, and what a privilege it
had been to work so closely with him.

Marcel Escudier
Emeritus Harrison Professor of Mechanical Engineering
University of Liverpool
Supervised by DBS 1963–64 (DIC), 1964–67 (Ph.D.)
sqda@btinternet.com

The Value of Having a Reference—Norberto Fueyo

While I was his Ph.D. student at the CFD Unit of
Imperial College in the late 80s, I do not think I met
with Brian more than perhaps five times.

In the pre-email era, our communications with
(then) *Professor* Spalding would be primarily via
memos—quite a novelty for a Spaniard who had
never seen such a format before.

I immediately became mesmerized by his writing
style: concise and sharp, yet elegant; eminently technical, but interspersed with cultural references and
witticisms; and often humorous in unpredictable ways.

To this day, I cannot think of any scientist who
can write with the accurate, elegant, colourful,
timeless style that Brian mastered.

A hot London afternoon in July I passed my viva; Professor Spalding went out
of the examination room for some brief minutes and came back with two cans of

Coke ("American Champagne"); we celebrated with the frugality that he professed; I joined his company CHAM; and Professor Spalding became simply Brian.

Working with him at CHAM, I came to witness first-hand the many attributes that no doubt many others have also reported; notably, Brian's ability to search for the essence of any problem, and to find unusual, ingenious solutions. Brian was often as demanding of others, but not any more than he was of himself; he was persistent but he was fair, and possessed an intellectual integrity that does not always go hand-in-hand with outstanding talent.

Brian's scientific stature was indeed paired with his ability to influence by virtue of his example. As his student, I am as grateful for the latter as I am for the former.

Norberto Fueyo
Ph.D. Imperial College 1990
Professor of Fluid Mechanics
University of Zaragoza

Some Recollections on Brian Spalding—Kemo Hanjalić

"Professor Spalding, Imperial College!". That was the instant response of P.O.A.L. Davies from Southampton early in 1965 during his short visit to Sarajevo when I had asked him if he could recommend me a place in the UK to do a Ph.D. in heat transfer. That was my first encounter with the name of Brian Spalding. As a research assistant in the R&D Centre of "Energoinvest", a rising company in process and power engineering, I had just been awarded a nine-month scholarship from the British Fund for Yugoslav Scholars, which I saw as an opportunity to begin doctoral studies. But before I could reach Spalding (no web around then!), a letter from my stipend fund informed me that I had been enrolled in the Postgraduate School in

Thermodynamics at the University of Birmingham with no alternatives offered. Discovering that my Dipl.Ing. degree was not sufficient to enlist in a Ph.D. programme and that I wouldn't have an opportunity to work with Spalding was disappointing. But I didn't give up. As soon as I passed the M.Sc. exams in June 1996, I wrote to Spalding expressing my enthusiasm for pursuing a Ph.D. in his group. He replied within few days that right then he had no vacancies but that he had passed my letter to his young colleague Dr. Brian Launder from whom I ought to hear soon. That indeed happened, and after a short interview in London, I enrolled for the M.Phil. degree to work on Launder's project sponsored by the CEGB, but formally under the supervision of Spalding—at least for a year or so until Launder was granted the status of a "recognized teacher of the University of London". Thus, apart from group seminars and semi-annual meetings of my thesis committee, throughout the whole stay I had almost no direct interaction with Brian Spalding.

Hence, I cannot say that I was a student of Spalding in the true sense. Did I lament that? Well, I never thought of it in such a way, since very soon I established cordial relation with Brian Launder that emerged into genuine friendship that still lasts. With Launder, I and a few fellow Ph.D. students, notably Bill Jones, discovered an opportunity niche in turbulence modelling that became our prime research focus. However, just being around in the Heat Transfer Section in that period was already a great privilege. Spalding's spirit permeated the whole section, filling our minds with his constructive criticisms and visionary ideas. To a great extent, he shaped the future careers of just about every one of us! His intellectual dominance and self-confidence, his great striving for innovation and his passion for the then-emerging CFD had an overwhelming influence on the whole generation at the time and many afterwards. I did interact with his students, followed the developments in CFD, but, in parallel with my experiments, focused on turbulence modelling where Launder's group gradually took over the initiative emerging as a major hub in that field.

After defending my thesis in 1970 and before leaving the UK, I requested an audience with Professor Spalding to express my thanks and to say goodbye. That was my only visit ever to his office after which for years I had no contact except seeing him occasionally at some conferences.

About a decade later, our roads crossed again. After joining the Executive Committee of the *International Centre for Heat and Mass Transfer* (ICHMT), based in the "Boris Kidrič" Institute for Nuclear Sciences in Vinča near Belgrade, I began meeting Spalding more often. Even in the mid-60s, he had established a broad international network but wanted to extend eastbound to enhance scientific communications within the global thermo-fluids community, especially in the Soviet Union. Together with several colleagues from the USA, France, Germany and Soviet Union, he became a powerful driver for the creation of the ICHMT based in Yugoslavia—one of the few European countries in the 60s and 70s that was easily accessible both to East and West. Indeed, it was Spalding, together with colleagues from the USA who pushed for an Adriatic location over Paris promoted by others. I still recall Spalding's constructive input in shaping the ICHMT profile and its mission. In fact, the very first global events organized by ICHMT, "*Heat and Mass Transfer in Turbulent Boundary Layers*" (1968) and "*Heat and Mass Transfer in Flows with Separated Regions and Measurements Techniques*" (1969), both held in Herceg Novi, were inspired by Spalding reflecting his main research focus of the time.

In 1989, I organized in Sarajevo an ICHMT Conference on *Mathematical Modelling and Computer Simulations in Energy Systems* (Hemisphere Publ. Corp. 1990). Spalding kindly accepted my invitation to give a keynote lecture. At a small reception for keynote speakers in my apartment, I vividly recall a sparkling conversation (half in English and half in Russian) between Spalding and Russian academician A. A. Samarskii from the Keldysh Institute for Applied Mathematics in Moscow, that went far beyond the research matters, opening the rivalry between the West and Russia, politics in general and "perestroika" in particular, on the very eve of the dissolution of Soviet Union. Prior to leaving, Spalding discretely inquired if I ever contemplated leaving Sarajevo, and if I would be interested in joining his company CHAM.

A few years later, in 1992, I invited him again to be a keynote speaker in another ICHMT event, a Forum on *Expert Systems and Computer Simulation in Energy Engineering* (Begell House Inc. 1994) held in Erlangen, Germany. He accepted and sent me a title "Friendly face of CFD". Acknowledging the receipt, I inquired if we should replace CFD by Computational Fluid Dynamics, arguing that many participants may not recognize what CFD implies. He promptly responded (rewording): no, keep it; many know, and those who don't, ought to learn.

We continued bumping into each other at various events. In May 2007, at a conference in St. Petersburg dedicated to the 80th birthday of academician A.I. Leontiev, I was scheduled to talk after Brian's keynote. Before the session was to begin, I walked into the lecture hall to upload my presentation and saw two groups of people gathered around two laptops facing two screens. Apparently, Brian wanted to give his presentation simultaneously in English and in Russian, as he did before on some occasions, but he now wanted also some sound effects that posed problems. In the end, everything worked well to great excitement of the predominantly Russian audience.

The last time I met Brian Spalding was again in Sarajevo, in September 2015 at the *Turbulence, Heat and Mass Transfer Symposium* (*THMT-15*). Among abstracts submitted, we noticed one authored by Spalding, which we immediately upgraded to a special plenary lecture. The lecture hall was full: many, especially the young participants, were excited with a unique opportunity to see Brian Spalding in person and to listen about his new turbulence model for fully stirred reactors. Sadly, this was one of his last conference appearances.

Turbulence, Heat and Mass Transfer 9 (THMT'15), Sarajevo Sept 2015

Kemal Hanjalić, FREng.
Professor Emeritus
Delft University of Technology, Nl.
K.Hanhjalic@tudelft.nl

Professor Spalding: A True Genius—Yogesh Jaluria

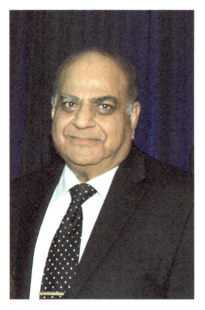

When I first started working after my education, I had not even heard of Professor Spalding. Then, one day, while discussing the leaders in heat transfer, a colleague said, "the only true genius in this field is Professor Spalding". That comment encouraged me to pursue the work done by Professor Brian Spalding and his group. The more I read, the more impressed I got. A few years later, when I learned that he was editor of a book series, I immediately sent my manuscript on Natural Convection for consideration to be included in the series. He accepted it and sent extremely useful and insightful comments that I followed to improve the book. For much of my initial work on recirculating flows, in heat rejection, enclosure fires and solar ponds, I found great inspiration and information in the many papers written by him and his students. Similarly, the work on boundary layers and separated flows was valuable in my research.

Much later, in 1985, I had the great pleasure of spending about 10 days with Brian in Cesme, Turkey, for a workshop on Natural Convection. I distinctly remember his brilliant talks and comments. But what I remember most fondly is that he had checked everything, including his slides, on his flight to Turkey. The bags did not show up for a while and he was seen in the same clothes day after day. He was also forced to use his memory and imagination to prepare viewgraphs for his talks. It was an impressive achievement.

At another conference, where we were both invited to present our work, I was impressed when Brian presented a list of outstanding topics that critically needed further research. I was even more impressed when he said that the list was also presented about a decade earlier and almost nothing had changed, stressing that tough questions were not being addressed. Over the years, I interacted with him at various venues and always came away with clearer and deeper understanding of the problems at hand.

In summary, I go back to my friend's comment on Professor Spalding and fully agree that he has been one of the giants in heat and mass transfer, particularly in computational fluid dynamics and heat transfer. Like many of my contemporaries, I learnt a lot from his work and owe him much gratitude.

Yogesh Jaluria
Board of Governors Professor & Distinguished Professor
Rutgers, The State University of New Jersey
jaluria@jove.rutgers.edu;

My Brief Encounter with Professor
Brian Spalding—Jerry Jones

I was not a student of Prof. Spalding, but I know many who were. I was educated in the MEAM Department at the University of Pennsylvania starting in the mid-1970s, receiving my MSME and Ph.D. degrees there in 1975 and 1981, respectively. Among my thesis advisors was Prof. Stuart Churchill, who was starting to explore more and more the use of computers to solve the differential equations governing fluid flow and convection heat transfer. After hearing his brief talk on the solution of the Navier–Stokes and energy equations to simulate wave behavior emanating from a lightning strike, I was convinced I wanted to coax a computer to solve my problem on forced convection. However, a computer does not solve these equations by itself; it needs to be told how to do it. This is where Prof. Spalding and his students entered my life. I began to read the fluids and heat transfer literature authored by names like Patankar, Launder and Runchal. The thread of commonality that ran alongside these names was a person named D. B. Spalding—who? Though it took me a while to figure out, I discovered in time that he was the intellectual source of so many grand ideas that spilled off the journal pages. With no Google or Wikipedia at that time, I struggled to learn who he was. This came later.

Fast forward to 2009. My career had taken me through the Los Alamos National Laboratory (LANL) and back to the U.S. east coast to Villanova University, not far from UPenn in Philadelphia. I was also a member of the Franklin Institute Committee on Science and the Arts in Philadelphia, which awards the prestigious Franklin Medals each year for outstanding contributions in six areas, including mechanical engineering. Was there ever a better opportunity to recognize the genius and outstanding contributions of Prof. Spalding? With the extensive help of Akshai Runchal, whom I finally met while taking a short course on CFD while at LANL, I advanced a case that awarded Prof. Spalding the 2010 Franklin Medal in Mechanical Engineering "for his seminal contributions to the science and art of Computational Fluid Dynamics, including integrating flow, heat, and mass transport into a single unified mathematical and computational framework, the Finite-Volume methodology, Exponential method, Staggered grid, SIMPLE algorithm, and the standardization of the k-e turbulence model among many others, and creating the practice of CFD in industry including developing the first commercial CFD code (PHOENICS) and company, thus paving the path for widespread application of CFD in industry". I had finally discovered the genius of Prof. Spalding.

Along with the awards ceremony in late April 2010, a sumptuous black-tie affair under the impressive 82-foot-tall dome of Franklin Hall at the Institute, a CFD symposium was held at Villanova University that featured world-renowned experts speaking on CFD topics from the nanoscale to planets. Prof. Spalding spoke on his novel representation of combustion (some photos included in Appendix II). Several of his former students attended and a few presented. It was a truly great event. I spent perhaps less than a day's worth of time with him over this period and, above all, what impressed me about Prof. Spalding was his simplicity. He was a true gentleman who used words sparingly but could easily be coaxed into heated language if moved to do so. The story he told me about his experiences with the Royal Society was one such example. I asked him about the title of one of his books, which I found curious. It was "Some Fundamentals of Combustion". What I asked was what he would title a new book once he discovered additional science in this area. I should have anticipated his answer. It was "Some More Fundamentals of Combustion"—simplicity.

He was a wonderful man.

Gerard F. "Jerry" Jones
Professor, Mechanical Engineering
Sr. Assoc. Dean, Graduate Studies and Research
Villanova University
Villanova, Pennsylvania
gerard.jones@villanova.edu

A Tribute to Brian—Emma Jureidini

I started at Cham in July 1988, Reception had just moved into the front of Bakery House and Brian was about to retire from Imperial College, and at the time only worked three days a week in Wimbledon. It felt like a new start for everyone when Brian started full time at Cham and I was excited to be part of it. I was just 22 at the time and never could have imagined how much influence working at Cham would have over the next 30 years on my life. My husband worked at Cham, and my dearest friends have all worked at Cham over the years, so looking back it has never just been a place of work, it has been a huge part of my life.

One of my lasting memories of Brian will be a more recent memory; he always came to work dressed in a suit and tie and was always first in and last out, seven days a week. During the last few years, once a week he would leave the office an hour early, having changed into his grey tracksuit, laptop in hand and ready for his pilates class; this always made me smile.

So although my contribution to the Cham family has been small (meeting, greeting and answering the telephone), I feel blessed to have had Brian in my life and shall miss him.

Emma Jureidini
Concentration, Heat and Momentum Ltd
ercj@cham.co.uk

Affectionate Memories of DBS—Brian Launder

My first fleeting encounter with Brian Spalding was in early 1961. He was the new Professor of Heat Transfer at IC while I was a final-year undergraduate who had come along to hear about research opportunities within the Thermo-Fluids Division that he led. Only it didn't work out that way. The college was then in the lengthy process of rebuilding on site and, with stunning (though later, I was to appreciate, entirely characteristic) honesty, Brian announced to the 30 or so students gathered to hear him: "If you stay here for a Ph.D. you'll lose at least a year from the building works; so, if you want a speedy doctorate, go elsewhere".

On the basis of his advice, I applied to half-a-dozen US universities along the east coast and, to my surprise, offers of admission with financial support started to arrive: a named scholarship at Princeton and an assistantship at MIT … but which should I choose? I requested an appointment with "the Professor" and five minutes were graciously allocated. Briefly, I explained my agreeable dilemma and, in 10 seconds flat, got his reply: "Go to MIT with Rohsenow. Princeton has Robert M. Drake but he's never done anything original!"

So, I went to MIT (but to the Gas Turbine Lab as they offered an RA post rather than a TA with Rohsenow). Three years later, thanks to a reverse-brain-drain initiative by the British Council, I had a range of UK post-doc positions to choose from at government and industrial research labs. Before deciding, however, I wrote to Brian wondering whether there might be a position as lecturer available at IC—indeed, I even suggested that one day I might hope to become a professor. In ten days (rapid given the leisurely pace of two-way trans-Atlantic airmail) his reply came: yes, an offer of a position would shortly be sent from the registrar's office though he regretted that he could not immediately meet "the full extent of [my] ambition"!

My first task at Imperial was to seek research funding and, since my doctorate at MIT had been mainly experimental, that was the area of the initial proposals. Two projects came through which largely occupied my research time and, as DBS had asked me to process research applications to the group, I was in an excellent position to scoop up as research students Bill Jones and Kemo Hanjalic (both of whom have had stellar research careers and elected Fellows of the Royal Academy of Engineering). DBS also allocated me a soft teaching duty lecturing to the M.Sc. class, sparing me the rigours of the rebellious undergrads. However, when his parabolic solver, GENMIX, was undergoing testing, he asked me if I would like to join him in a consultancy project for the UKAEA using that software to predict condensation in steam flow through tubes. It was a struggle for me (at the time an experimentalist) to get to grips with the method and the code … but eventually we got there.

I remain grateful to this day for the opportunity that the project gave me. But the code only had a rudimentary mixing-length model of turbulent transport. So, after the consultancy was completed, Brian encouraged me to redirect my research to improve the model of turbulence, which I did with my students, Kemo and Bill, re-directed from their initially experimental projects. Indeed, Brian and his students (Wolfgang Rodi and K. H. Ng) also pursued closely parallel research in friendly competition with us. Yes, that half-decade from 1969–1974 was an intensely interesting and happy time to be researching in the group. Brian's charismatic leadership stimulated major advances in the numerical solution of the equations of motion for two- and, subsequently, three-dimensional flow as well as the modelling of turbulence.

Well, this idyllic state couldn't—or, at least, didn't—last. Key students and post-docs moved on to careers outside the UK; several academics from the group decided to pursue their own visions independently while the writer moved to California for four years. Yet, while it is sometimes said that this marked the end of the group's influence, in many respects it was the reverse. Brian, released from running a major research team and all the academic problems of university life, had more time to devote to the development of his CFD software company, CHAM, which he put to very great effect through the development of the first code that could credibly claim to be an all-purpose CFD solver, PHOENICS (an acronym for Parabolic, Hyperbolic or Elliptic Numerical Integration Code Series). Moreover, university CFD research centres and software companies led by former students or staff from the Imperial group emerged at a dozen locations around the globe. Thus, the current, relatively mature state of CFD applicable to complex, industrial problems owes an enormous debt of gratitude to Brian Spalding.

Since my return to the UK, I've been regularly reminded of Brian's continuing presence and influence. Long past the time when even feisty academic warriors normally hang up their spurs, Brian would be engaging in what seemed a perpetual round of conferences and trips to receive honorific awards (all richly deserved). I particularly recall a dinner at Imperial College to celebrate his 90th birthday where he regaled the company with his observations on life—delivered in poetic form, of course! My final meeting with him was at a conference in Sarajevo in autumn 2015 where we both delivered invited lectures. At the end of mine, on a topic outside the main theme of the meeting—hurricanes—in the absence of interjections from elsewhere, he posed a sequence of thought-provoking questions.

Brian Launder
Professor B. E. Launder, FREng, FRS
University of Manchester
brian.launder@umist.ac.uk

Professor Brian Spalding—Sadly Missed, Fondly Remembered—Millie Lyle

When I joined CHAM in January 1992, I certainly did not expect to still be working here in 2017, some 25 years later. Most certainly Brian created a company that has given a true sense of family to its staff, both past and present. I have made many wonderful friends being part of that CHAM family. Brian is greatly missed at 40 High Street but I still feel his presence everywhere around the building. His passion for learning and life will remain with me always.

Farewell Brian and thank you for everything.

Mrs Millie Lyle
Concentration Heat and Momentum Ltd (CHAM)
ml@cham.co.uk

What Can I say about Brian?—Mike Malin

What can I say about Brian? He had a massive influence on my engineering career, and I will always be grateful for the assistance, training and knowledge he imparted to me during our forty-year association together at Imperial College and CHAM. It was a privilege to have known and worked with him. I witnessed first-hand many examples of his technical brilliance and creativity, his deep physical insight and vision and his encyclopedic knowledge of thermo-fluids. It seems to me that Brian's approach was to seek generic and economical solutions to practical engineering problems by making approximations based on physical intuition combined with analytical flair. He also had a remarkable ability to resolve convergence problems by devising novel algorithmic changes or linearization practices.

I think it is true to say that Brian created the CFD industry with the formation of CHAM in 1969. Another landmark achievement occurred during the late 1970s when Brian conceived and executed the revolutionary idea of creating the first ever general-purpose CFD code for simulating all thermo-fluid problems. This code was called PHOENICS, an acronym for Parabolic, Hyperbolic Or Elliptic Numerical Integration Code Series. It was launched commercially in 1981, and by this means Brian founded a worldwide industry based on the selling of CFD software and services into an ever-widening field of applications.

My own involvement in CFD began in 1976 when I joined Brian's research group at Imperial College to undertake an M.Sc. in Heat Transfer Engineering. It is now easy to see that this was essentially the world's first postgraduate degree course in CFD. Brian, who was known to everyone in the group as DBS, delivered the course almost single-handedly, starting off with his brilliant lecture series entitled "Mathematical Modelling of Fluid-Mechanics, Heat Transfer and Chemical-Reaction Processes: A Lecture Course". These lectures provided an all-encompassing, everything-you-need-to-know description of the basic equations of fluid flow, heat and mass transfer, turbulence, combustion, multiphase mixtures and radiation; and how to present them in a unified and generic way, and then solve them by means of the CFD finite volume method. At the time, these lectures were an absolute revelation to me, and I can see that even now they still provide a valuable reference source.

   Evidence of Brian's impressive intuition and creative talent was abundant through our long association together. A few examples will suffice. In 1993, one of CHAM's engineers implemented a generic search algorithm to compute the nearest-wall distances for the low-Reynolds-number turbulence models I had just implemented in PHOENICS. When we briefed Brian, his response was that the method was far too expensive and cumbersome for use in industrial applications with arbitrary complex geometries. A day later he produced a technical note describing an ingenious and generic differential-equation method for calculating approximate values of the nearest-wall distance. Brian wrote: "I have devised a method of computing the turbulence length scale which will be more economical and more realistic than any other which we know of". This certainly proved to be the case when it was immediately applied to large industrial applications. The "wall-distance trick", as Brian liked to call it, was published obscurely in a poster session at the 1994 International Heat Transfer Conference. Thereafter, it was exploited, refined and published by other researchers, and eventually the method or some variant of it found its way into most commercial CFD codes.

   The invention of the wall-distance method emerged almost simultaneously with Brian's proposal for a zero-equation low-Reynolds-number turbulence model for situations in which fluid flows through spaces cluttered with many solid objects. Here, the grid density between nearby solids is often too coarse for any more advanced turbulence model to be meaningfully employed. The key elements of the so-called LVEL turbulence model were Brian's 1962 law of the wall and the "wall-distance trick". Brian said the model didn't offer any new scientific insight, but it would enable PHOENICS to be used with great practical advantage for the electronics cooling applications we were carrying out for IBM. This proved to be the case, and following publication in 1996 the usefulness of the model was demonstrated by many other workers in the field. The consensus was that the model was as effective as the k-ε model, but with the big advantage of significant savings in computer time, sometimes by factors of three or higher. This inevitably led to the model appearing in commercial CFD codes dedicated to the thermal design of electronic equipment.

   Brian's ability to recognize the need for engineers to get an adequate solution within reasonable computing time also led to the development of the IMMERSOL (i.e. immersed solid) radiation model in 1994. I was engaged on a contract for Radian Corporation, who wanted an economical radiation model for use on curvilinear meshes in a participating medium with complex geometry. Radian had previous experience of using the discrete transfer model in another commercial CFD code, but they found the model prohibitively expensive. I proposed we implement Rosseland's model, but Brian expressed the view that while this was a reasonable suggestion, he had something much more generic in mind. A few days later, Brian sent me a technical note outlining his proposal for a diffusional model to represent radiation in arbitrary geometries for both optically thick and thin media. The model involved the solution of a diffusion equation for the radiosity with a diffusivity proportional not only to the medium's absorption and scattering coefficients, but also to the distance between solid walls (needed for transparent media).

The radiosity equation was derived by simplifying the six-flux model, and the inter-wall distances were computed from the "wall-distance trick". Further notes in 1996 elaborated on how the model could be employed in conjugate-heat-transfer situations by reformulating the radiosity equation in terms of a radiant temperature. Subsequent application of IMMERSOL showed that whilst it doesn't necessarily procure close agreement with experiments, it always produces physically reasonable predictions with computer times that are only a fraction of those expended by more sophisticated models. Even so, the model appears to have made little impact outside of CHAM, possibly because Brian didn't publish the model formally until an eloquent description appeared in a 2013 volume of Advances in Heat Transfer.

As a final example of Brian's technical brilliance on algorithmic work, I worked very closely with him during 1998 on a contract for S&C Thermofluids. This involved the extension of the PHOENICS parabolic solver to handle purely supersonic flow (hyperbolic) and under-expanded jets issuing into stagnant surroundings (transonic). The idea was to effect massive savings in computer time by switching from elliptic computations to the marching integration procedure offered by the parabolic solver. Brian devised and implemented very rapidly several

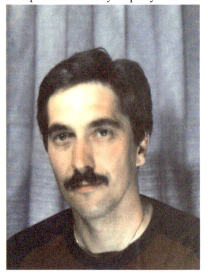

algorithmic changes to the pressure–velocity coupling, and it was my role to create numerous cases to test and validate these enhancements, and then report on the results. This work was never published in the open literature, but the outcome was very successful and pleasing to the customer.

Brian was a remarkable engineer and scientist, and he had a profound influence on the field of engineering through his pioneering work in thermo-fluids and computational fluid dynamics. It is also a testament to Brian's ingenuity that numerous innovations from Imperial College and CHAM are employed in many commercial and open-source CFD codes being marketed today.

Michael R. Malin
M.Sc. (1976–1977) Student of Prof. Spalding
Ph.D. (1983–1986) Student of Prof Spalding
Technical Support Manager
CHAM Limited, Bakery House
40 High Street, Wimbledon,
London SW19 5AU, June 2017.
mrm@cham.co.uk

A Tribute to Professor Brian Spalding—Adrian Melling

As a young undergraduate in Toronto, I had developed a strong interest in thermo-fluid dynamics. Through a particular fascination with the boundary layer concept, I had first found references to the name Spalding. For three years following my graduation in 1966, I worked as a teacher in Zambia and that posed two questions with respect to further studies. Would my interests have changed during this time and how could I identify the best university in UK for my purpose? I decided to minimize the first difficulty by choosing to re-enter academia through an M.Sc. program rather than starting Ph.D. research directly. Living in Zambia, I relied on a subscription to the Manchester Guardian Weekly to maintain contact with the wide world. There in 1968 I found a timely advertisement for the HMSO publication "Scientific Research in British Universities and Colleges" which I promptly ordered. Working my way meticulously through this hefty document, I identified the M.Sc. program "Thermal Power and Process Engineering" run by Spalding's group at Imperial College as my first choice. I was extremely pleased to be offered a place on this course starting in September 1969. My first meeting with the famous professor himself was, of course, an exciting prospect. I did, however, have some reservations. Would I have forgotten too much from my undergraduate studies? Would I have to struggle to keep up with freshly graduated fellow students? Both fears proved to be groundless!

As well as the anticipated boundary layers, the M.Sc. course introduced me to exciting concepts such as the Navier–Stokes equations, turbulence modelling and convective heat and mass transfer, ably and enthusiastically instructed by Brian Spalding and his colleagues. We were also granted insight into something very new at the time, computational fluid dynamics (CFD), presented in the case of boundary layer flows by Professor Spalding himself. His capable explanations of the computer program GENMIX helped me greatly in deciding with confidence to take up experimental rather than computational fluid mechanics! I willfully ignored Professor Spalding's assertion that sooner or later CFD would render experiments superfluous. 50 years later Brian's expectations have in many respects been fulfilled, but parallel developments in experimental techniques, especially those made possible by laser technology, have opened up new fields of experimental investigation.

Back at Imperial College in the 1970s, CFD doctoral workers were commendably required to undertake experimental work related to their project. Accordingly, it seemed only proper that I should demonstrate my limited ability to apply the CFD techniques learned in the M.Sc. program: no complicated geometry, no turbulence model, just a laminar creeping flow over a double backward step. My measurements with laser Doppler anemometry (LDA) had established that velocity profiles far upstream and far downstream of the step were parabolic, while those near the step clearly showed zones of recirculating flow. My calculations of stream function and vorticity using a CFD program developed at Imperial College (Gosman et al.) confirmed these features. In CFD jargon "the agreement was encouraging", at least until the flow rate was doubled. Flow visualization and LDA measurements then indicated a pronounced but stable asymmetry in the flow field. I left the predictions this time in more capable hands!

In the 1970s "we" (the experimentalists) had our "headquarters" in room 200. "They" (the computationalists) were frequently to be found nearby in a sparsely furnished annex of the Fluids laboratory: a couple of card punchers, a card reader, a line printer, one or two chairs and (most importantly) a large waste paper bin. The proximity of the two bases encouraged a degree of rivalry. "We" worked hard but by midnight we were normally on our way home. Two or three hours later, some of "them" would still be standing near the printer, anxiously awaiting for the printout of their most recent jobs on the pile of seemingly endless paper. When each job finally appeared, a glance at the first page would often suffice before the disgusted researcher threw the whole pile into the bin. "Our" reaction varied from sympathy to contempt for the unhappy computationalist, whose latest tweaking of various parameters and "constants" in his program had been unsuccessful in bringing predicted results closer to the reference data.

Sometimes, I regret the passing of such uncomfortable aspects of research. On this occasion, I regret the passing of a great and inspiring man!

Adrian Melling
Department of Chemical and Bioengineering
University of Erlangen-Nuremberg
melling.adrian@gmail.com

Recollections of Professor Brian Spalding—Dubravka Melling

I first met Brian, or Professor Spalding as I then knew him, at the cocktail party in 1971 for the International Conference on Heat and Mass Transfer in Trogir (Croatia). It was a surprise to me to see a famous professor dancing without a pause on the hotel terrace, instead of engaging in the academic work which I thought was a professor's sole occupation. Brian was the last one to leave the dance floor, yet the next morning he was fresh and fit while the rest of us were trying to shake off a hangover. The secret I discovered later: an early morning swim at 6 o'clock!

Another episode which I would like to mention dates back to December 1972, soon after starting my Ph.D. studies at Imperial College. Brian had invited me to a party for students and staff and took the opportunity to ask how I would spend the Christmas period. When I told him that I intended to stay in London and work, he warned me that at Christmas London, the College and the hostel would all be very quiet and depressing. So Brian recommended to me to go home, but as an alternative he generously invited me to spend Christmas with his family. I was very grateful for the invitation, but in the end I took Brian's best advice and returned to my family in Belgrade for Christmas and the New Year.

Many years after I left Imperial College, our paths crossed again in 1987 at the 19th International Conference on Heat and Mass Transfer in Dubrovnik. Brian was approaching formal retirement but looked so well and so fit that my first question to him was: " When are you going to get old?". With two young boys to care for, there would be no retirement yet, Brian, and academic life in its new form was likely to remain busy. Around this time, I was living in UK between periods of residence abroad, so it was easy for our families to get together at the Spalding's home or ours. We were also able to attend Brian's retirement party in 1988 in South Croydon before we moved overseas again. Thereafter, we still met from time to time, notably when Brian was awarded an honorary doctorate from the University of Erlangen-Nuremberg. As I had expected, Brian did not stop working after retirement. I was able to confirm this in 2003 at the celebration of his 80th birthday in Richmond Hill and in 2013 at a dinner for his 90th birthday at Imperial College when I met him sadly for the last time.

Dubravka Melling
SAOT—Erlangen Graduate School in Advanced Optical Technologies
University of Erlangen-Nuremberg
dubravka.melling@gmail.com

Profound Impact of Spalding—Moscow Team

**From Left to Right
Valery, Brian, Alexey, Nickolay**

Elena

Six months passed but the soul still refuses to believe that he is no longer with us. Still are alive the feelings of night prayers for Brian's health.

Everything happened quickly enough. Three weeks of worries, anxiety, hope and strong desire to give ourselves to help him feel better. The words of physicians at times led to despair, at times—gave hope for recovery. At some point, there was a feeling that we prevailed: Brian started to recover, even spoke Russian with medical staff when the lung ventilation system was temporarily switched off. And we sent him from Moscow with full confidence that doctors in his homeland would ensure the continuation of the process of recovering. Therefore, we were all stunned to receive 2 days after his departure a message telling that he died. It was a great shock and we are still under its influence.

It is not important now whether it happened because of somebody's mistake or fatal coincidence of circumstances. The outcome is sad. Let us accept it as is destined by destiny.

What we remember about him? He was a bright and spiritual man, sincere, modest, infinitely wise and taking care of people. It was extremely interesting to deal with him. Sometimes, he was persistent in stubbornness but also knew how to get out of any situation with dignity attracting new volunteers for implementation of his ideas. His enormous capacity for work forced us to work constantly, without regard for time and strength.

It was understandable: for each of us he was a TEACHER whom we immensely respected and loved. We tried to make a new breakthrough in development for his each visit, for our each meeting to see in his eyes notes of encouragement and glint from anticipation of new results. His each visit to Moscow was accompanied by a powerful impetus to continuous development. For us, he was the one who led and pointed the way. In this, he was unsurpassed Master. He always tried to solve any complex problem by splitting it into multiple simple and clear tasks. He knew how to see far ahead and anticipate the direction of further progress. In recent years, he used to say that he could see the solution of problems that had not been solved earlier. Not all his thoughts and ideas are realized. We'll try to realize them to the best of our abilities.

We were lucky to communicate with him for the last 20 years, so it seems that we are partly responsible for conveying to the scientific community the ideas that were of concern to him in recent years.

First of all, it is probably worth mentioning the direction which Brian called the population theory. His first ideas were used in development of the multi-fluid model of turbulence. Professor was not very happy with implementation of this model in PHOENICS. He waited for the occasion to improve it. However, he considered the population theory as being much wider than just a model of turbulence. He intuitively felt that in hydrodynamics, particularly with chemical reactions, there was a special place for random processes.

Another idea was related to awareness that only few could understand CFD methods, whereas CFD results should be used by all engineers. It is important, therefore, to enable any engineer to make calculations even if he does not know the CFD features. In this direction, a separate program, PHOENICS-Direct, was created, which Professor Spalding again spoke of at the OpenFOAM Conference one month before his decease.

Once again thoughts come back to reality in which he is no longer with us. When people in Russia want to commemorate a deceased person with kind words, they say: "May your memory be eternal". So we, too, say: "May your memory be eternal, Brian Spalding. Rest in peace".

Moscow Research Group:

Elena Pankova (worked with Professor Spalding from 2000 to 2016)
Nikolai Pavitsky (worked with Professor Spalding from 1997 to 2016)
Valery Artemov (worked with Professor Spalding from 1999 to 2014)
Alexey Ginevsky (worked with Professor Spalding from 1999 to 2016)
alexeyginevsky@gmail.com

Remembering Brian Spalding—Jayathi Murthy

I am one of Prof. Spalding's academic grandchildren—one of his sprawling "family of the mind". The name of my adviser, Suhas Patankar, is forever entwined with that of Brian Spalding, and so, in a strangely intimate way, I too am connected with this far-flung family. I came to the University of Minnesota to work on my Ph.D. with Suhas in 1981, and though I knew practically nothing about CFD, already the words "Patankar–Spalding" could send a thrill of excitement down my spine. In time, I came to understand and appreciate the sheer creativity of their work together, the clarity of thought, its rootedness in physics, and most of all, its beautiful simplicity. And of course, I heard from Suhas and Rajani the fabled tales of their days at Imperial College, and about the great man himself. I felt like a part of a great legend.

I remember the first time I met Prof. Spalding. I believe it was in 1982, at a summer conference in Seattle, if I remember correctly, and a number of graduate students from Suhas's group were attending. I remember us milling together, out of place and a bit intimidated, looking across the room at Prof. Spalding, and finally, the introduction by Suhas, and Prof. Spalding's extraordinary kindness and interest in each of us. We talked about it for days afterwards.

My own professional life is inextricably intertwined with Prof. Spalding's work. Our work on unstructured solution-adaptive finite volume methods could not exist without SIMPLE and the class of pressure-based finite volume methods; it goes without saying. And it is fascinating to see the evolution of CFD through the years, from a research tool to be used by the enlightened few to a design tool well integrated into the product cycle. I feel fortunate to have been closely involved with both the intellectual and commercial evolutions of this fascinating subject. But more than the specific methods themselves, the most valuable thing I have inherited is a way of thinking, and for this, I will forever be grateful. Thank you, Brian Spalding.

Jayathi Y. Murthy
Ronald and Valerie Sugar Dean
Henry Samueli School of Engineering and Applied Science
University of California, Los Angeles
jmurthy@gmail.com

What I Learned from Professor Brian Spalding—Siva Parameswaran

I worked with Prof. Brian Spalding from September 1977 to July 1983. When I joined Imperial College in September 1977, Prof. Spalding (his students affectionately called DBS) was trying to extend his famous GENMIX computer program to compressible flows with shocks. GENMIX is a general-purpose computer program, developed to solve two-dimensional, boundary layer flows for incompressible or subsonic flows with low Mach numbers. In order to solve high-speed compressible flow in a boundary layer, DBS correctly figured out that he had to let the pressure field to vary inside the boundary layer (for incompressible flows, free stream pressure is impressed inside the boundary layer, and thus pressure does not vary in the direction normal to the streamlines). He came up with an algorithm to correct the vertical pressure field (normal to the marching direction) downstream if the flow is supersonic and upstream if the flow is subsonic. The algorithm worked very well if the flow is fully supersonic or if it is fully subsonic. However, the algorithm failed if the flow goes from supersonic to subsonic through a shock. DBS maintained that we made some programming error, and there was nothing wrong with the algorithm. It took me two semesters and lot of writing back and forth with DBS that I finally convinced him the algorithm is flawed because it created more equations than the unknown pressures. I demonstrated this by solving the transonic flow through a 1D, divergent duct but it cost me additional 5 years to finish my doctoral degree!! My struggles with the compressible flow algorithm made me resilient and prepared me for an academic career. I owe this to Professor Brian Spalding.

Dr. George Carroll was my mentor at Imperial College, and we shared the room 484 in the 4th floor of the Mech. Engineering Department. Over the years, George has become a good friend of mine and helped me in many ways unknown to him. During my struggle with the compressible flow algorithm, George always encouraged me not to give up.

Siva Parameswaran
Professor, Texas Tech University
siva.parameswaran@ttu.edu

Meeting at Texas Tech University October 2013

Brian Spalding: A Kind and Generous Man—Rajani Patankar

I am not Professor Spalding's student. I know nothing about engineering. So I cannot comment on Brian's groundbreaking contributions to heat transfer. However, I got to experience the kindness and generosity of this exceptional individual. This is why I am paying this tribute to him.

My husband, Suhas Patankar, worked with Brian for his Ph.D. After finishing the degree, he returned to India and soon thereafter we got married. Suhas started working at IIT, Kanpur in India. However, Brian kept urging Suhas to return to England to work with him. So, after three years, we came to London with our little daughter. Brian came to the airport to receive us and waited for three hours before we came out. To my surprise, he had made arrangements for us to stay in their home in Wimbledon. Actually, we ended up staying there for two months before we moved into our apartment. I was new to England and to the western culture. So, I was observing the norms of this new culture. Brian and Eda (his first wife) were very kind, supportive and understanding. They went out of their way to make us feel comfortable in their home. Their tremendous generosity and kindness touched my heart. The experience of staying with the Spalding has made a lasting impact in our life. We felt a mutual bond of affection with Brian and Eda. That same closeness has continued with Brian and Colleen. Over the years, we have met many times at conferences, we have visited them a few times in London, and they have come to our house in Minneapolis. Every time, we experienced their closeness and affection towards us. We feel that this has been the greatest gift to us. We will cherish that feeling for the rest of our life. I want to pay my respects to Brian Spalding, who has shown us how to be kind and generous towards others. We will miss him very much.

Rajani Patankar
patan001@umn.edu

Professor Brian Spalding: My Guiding Star—Suhas Patankar

No person has made a more profound and lasting impact on my professional and personal life than Brian Spalding. In 1964, I came to London to work as a Ph.D. student under his guidance. Soon we became good friends and that friendship lasted forever. He was a brilliant scientist and a skillful teacher. Moreover, he was very kind and generous to me and my family.

Brian's scientific work is noteworthy not only for the variety of subjects he covered but also for the strong impact of his inventions. None of his work made just an incremental contribution. It was always a significant breakthrough, opening the door to many scientific opportunities that did not exist before. His vision and creativity provided a quantum increase in our scientific understanding and predictive capability. His specific scientific contributions represent lifelong fireworks of his creative ideas. He was not only successful in creating scientific methods of practical utility but also in establishing them in industry. Today's phenomenal advance in engineering can be traced to the vision and early actions of one man, Brian Spalding.

For me, he was a great source of inspiration. He taught me how to grasp complex problems and pursue grand ideas. I learned from him how to teach and communicate effectively. He is largely responsible for my professional success and personal joy.

I am associated with two scientific breakthroughs. Both were accomplished when I got to spend a lot of time with Brian. The first happened in 1966 when we both travelled to the USA to attend the International Heat Transfer Conference in Chicago. We attempted a finite volume method for two-dimensional boundary layers. This work led to a very popular book and a computer program for this application. In 1971, the Spalding hosted my family for two months. At that time, Brian and I would take the train every day from Wimbledon to South Kensington (where Imperial College was). During our train ride, we would discuss possible methods for three-dimensional boundary layers and recirculating flows. Often, I would propose a method and Brian would express his objections. Next day, I would describe a modified method and hear new objections. However, one day, he tentatively approved what I proposed. I quickly implemented the method in a computer program and solved a substantial problem. With great excitement, I went to show him the results. I distinctly remember the joy and the glow on his face. We both knew that we had achieved something big. That was the birth of the SIMPLE algorithm.

Over the years, I have enjoyed my interactions with him and his family. It has been a pleasure to host them at our house in the USA. I have marvelled at his intelligence and sharpness right up to the final moments in his life. For his teachings and friendship, I remain forever grateful to him.

Suhas V. Patankar
Ph.D. (1964–1967) Student of Prof Spalding
Professor Emeritus, University of Minnesota
President, Innovative Research, Inc.
patankar@inres.com

Remembering Professor Brian Spalding—Darrell Pepper

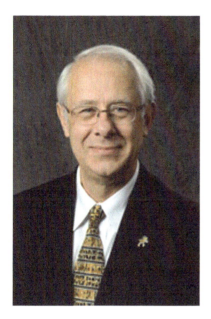

In my early days of CFD model development, I spent many hours reading about the fluids modelling efforts of Brian Spalding and his students, A. K. Runchal, S. V. Patankar, A. D. Gosman and M. Wolfshtein. Their work, under the tutelage of Brian, essentially exploded the application of computational fluid dynamics and made the solution of the Navier–Stokes equations actually feasible. The implementation of the finite volume technique for discretizing these nasty, highly nonlinear equations of fluid motion, and especially combustion processes, were insightful and created interesting and exciting solutions (even the non-real ones). My first interaction with Brian, and later with many of his legacy students, began in the mid-1970s, and I became even more adamant about exploring CFD after listening to one of his talks at a conference—he made modelling fluid flow seem so straightforward. Later, when I began to venture into finite elements (or as Brian would tell me—the dark side of CFD), I began to appreciate even more the magnitude of his work and creativity, and using his work to verify and justify our efforts to employ an alternative numerical method. Over the ensuing years, we would have friendly jousts about finite element versus finite volume (especially during conference presentations). Some years ago, when I introduced the use of meshless methods in lieu of mesh-based techniques, he thought I was at first jesting, but then realized it was an interesting approach. I later boasted that the next logical evolution of numerical modelling would be the absence of points—or the pointless method. He turned to his neighbour in the audience, and asked if I was serious—then he began to smile at me as he quickly grasped the jest of my statement—a priceless moment.

Darrell W. Pepper
University of Nevada Las Vegas
Department of Mechanical Engineering
Las Vegas, NV 89154, USA
darrell.pepper@unlv.edu

Reflections on My Academic Father Professor Brian Spalding—Andrew Pollard

In the annals of human endeavour, few people can measure up to DBS; he was unique to his time with us. In no order of preference, he defined uniqueness as a person, as a scientist, as an engineer, as a fighter for freedom and justice, as a leader in many forms of communication be they scientific or literary and as a role model. He packed so much into his 70 plus years in academe and business it makes one's head swoon.

I have been, and continue to be, influenced by him in all those above-mentioned parts that made up this extraordinarily accomplished man. First introduced to him through an undergraduate lecture on the law of the wall and DBS' novel reformulation of it, and a perchance acquisition of the Patankar–Spalding book on heat and mass transfer in boundary layers, I knew my future was to be influenced by him. I did not appreciate at the time, as a third-year undergraduate student, the magnitude of that influence. Now as a professor emeritus, I know I have immeasurably benefitted from standing on the shoulder of an academic giant. He and his work are historically significant in the evolution of our subject matter and his impact has become legendary.

I have had the fortune to be present and participate in many milestones and events in his life. From his retirement as Chair of Heat Transfer, through to his various birthday celebrations, both he and Colleen have been integral to my journey. How do you thank someone who has influenced me from a very earnest young Ph.D. student to an acknowledged expert in turbulence, computational fluid dynamics, heat and mass transfer and a successful mentor of others? How do you thank someone who, even in death, continues to give: his population model of turbulence, which I shall continue to work on, his lessons on coping from having consumed a little too much whiskey and his critical, often twinkling eye, now seen through mine, on everyday activities: the appreciation of novel science, the appreciation of well-written verse and prose, his exuberance for life and the nurturing of promising new ideas that challenge students and practitioners of physico-chemical hydrodynamics?

To those who know him only through his written works, take heed for there you shall find clarity of thought and exquisitely argued and presented scientific and engineering principles. His books on combustion, mass transfer, turbulence and

CFD are models for communicating complex ideas simply and elegantly. His papers, all 350 of them, are each a significant contribution with *The Numerical Computation of Turbulent Flows* having been cited over 12,000 times! His physical insight was legendary and he encapsulated this through simple experiments, both physical and numerical, which elucidated key phenomena. I particularly continue to enjoy Ricou and Spalding 1961 JFM paper on jet entrainment being a case in point.

He left a not insignificant legacy in print and in his academic children, grandchildren and great and emerging great-great-grandchildren in whom he instilled directly and through generational diffusion his clarity of thought for unifying and insightful ideas and who each have gone on to make their own significant impacts both scientifically and personally in the world.

Brian Spalding received many awards and accolades including the Franklin Institute Medal and so I close with this quote from Benjamin Franklin: "If you would not be forgotten as soon as you are dead, either write something worth reading or do something worth writing". Brian Spalding succeeded admirably in both.

Andrew Pollard
Ph.D. (1975–1978) Student of Prof Spalding
FCAE, FAPS, FASME
Professor and Queen's Research Chair,
Queen's University, Kingston, CANADA
pollard@me.queensu.ca

Tribute to My Boss, Professor D. Brian Spalding—Jill Rayss

I first met Professor Brian Spalding at my interview on 27 April 1989, and started work as his Personal Assistant on 12 June 1989. I remember it being a short interview, with Brian passing me over to Colleen as soon as possible. I would later come to understand this; it took his time away from his work.

Over the years, I have a lot of memories of working closely with Brian on a daily basis. Some days I would arrive at the office to find an enormous amount of correspondence that he had written, his handwriting getting more and more illegible. Other days I wouldn't see him at all, as he would be in his office concentrating on coding or lecture writing, and did not wish to be disturbed. Frequently, he would say "have I not said good morning to you", when I would knock on the door to say "goodnight"; he always looked at his watch whenever I said I was leaving. I remember many times tidying up his office to find snippets of poetry that he had written, something he enjoyed doing.

Brian was courteous as a boss, and always expressed gratitude, that is not to say, he couldn't be difficult, if he made up his mind to do things his way, he wouldn't be moved. I admit to being irritated with him on occasion, but if he ever was with me, he didn't show it. He was thoughtful and nearly always brought me back a bottle of Russian Vodka from his frequent trips to Moscow, which I would find waiting for me on my desk.

Brian disliked taking time out of the office for mundane things, like taking his car to the garage, paying bills,

Spalding & Jill Rayss

going to the bank or organizing the servicing of his caravan, and he was always ready to pass this on to me. I helped him with numerous personal tasks, and I am grateful for the trust he had in me. One of the last things I did was go with him to the florist to purchase flowers for Colleen's birthday.

Brian was not fond of small talk, and any CHAM social event, usually held late in the afternoon, always ended up with our engineers and Brian in one corner of the room discussing technical issues. After one such event in 2011, I did manage to get a photograph with him, the only one I have of the two of us together. I share this with you here. Unfortunately, I did not manage to get a photograph of Brian, doing an impression of Mick Jagger on the dance floor at a CHAM Christmas party, no doubt aided by the amount of whisky he had consumed.

In later years, Brian would ask me not to come into his office while he took a short rest, or if he was changing into his tracksuit to go to Pilates, but I caught him a number of times in his stockinged feet, working out on his treadmill.

The final email I received from Brian was just after he arrived in Moscow on his last ill-fated trip. It was a "thank you"; for organizing his travel, and hotel accommodation, and said that his journey had been uneventful.

I knew Brian in his role as a well-respected and eminent scientist, as well as a businessman who would discuss everyday issues and as a family man who would share thoughts with me. He was a kind and generous man whom I felt very privileged to know and thought of with a great deal of affection, and would describe as a "true gentleman". I miss him very much.

Jill Rayss
Personal Assistant to
Professor D. Brian Spalding
Concentration Heat & Momentum Limited (CHAM)
Tel: +44 20 8947 7651, jpr@cham.co.uk

My Encounters with Brian Spalding—Wolfgang Rodi

My first encounter with Brian Spalding was in 1967 when I spent a year at the University of Minnesota in the Heat Transfer Laboratory of Prof. Eckert. During that time, Brian came by and gave a seminar on Numerical Flow Calculations. I was so fascinated by that topic that I applied to him to pursue a Ph.D. study on the topic. I got accepted and he put me on the development and testing of turbulence models, and that is what I did at Imperial College under his supervision for 5 years during 1968—1972. I was very fortunate to have this opportunity as I could participate in one of the most exciting and momentous developments in fluid mechanics at the time—the dawn of Computational Fluid Dynamics (CFD), i.e. the early stages of numerical flow calculations and sophisticated turbulence modelling. It was indeed an exciting and in fact turbulent time, sometimes a bit tough, as it was not easy to follow the pace of Brian. It laid the foundations for my later career as Brian put me on the increasingly important subject of turbulence modelling and taught me a lot on how to work and how to write. He was not too happy with the first draft of my Ph.D. thesis and gave very clear and helpful instructions on how to improve the writing. I followed these and he was then much happier with the second draft which inspired him to write a poem on the margin. I had abbreviated "equation" as "equat", which reminded Brian of nineteenth-century comic verse and stimulated him to write:

> *The Akhund of Swat*
> *Said "Please tell me what*
> *(For I've quite forgot)*
> *Is an equat?*
> *Does it stand or squat,*
> *Smoke hash or pot?*
> *If it drinks a lot*
> *I would much rather not*
> *Receive it in Swat."*

After I finished my Ph.D., Brian wanted me to work for a newly established software company in Germany, I think related to Prof. Argyris, but I had an attractive offer from the University of Karlsruhe (now Karlsruhe Institute of Technology), which I accepted and stayed there for good. After my return to Germany, I kept in contact with Brian and we exchanged scientific ideas and publications, and we met at various conferences.

In 1988, Brian helped with the birth of ERCOFTAC (European Research Community on Flow, Turbulence and Combustion) and with choosing its name. He became the first chairman of its Scientific Programme Committee and put ERCOFTAC scientifically on track. At that time, he got me into ERCOFTAC and I became very active in this organization, an involvement that continues until now.

Then we met at various celebrations and events: 1988 at Brian's retirement dinner in South Croydon and in January 2003 at his 80th birthday celebration in Richmond Hill. In September 2007, he came to my retirement event in Karlsruhe and gave a philosophical talk on my work in turbulence modelling. In 2008, we had a very enjoyable conference in Marrakesh held on the occasion of Brian's 85th birthday, and we celebrated his 90th birthday in 2013 with a dinner at Imperial College, where he delivered a very interesting and entertaining autobiographical poem summarizing the key elements in his life. I was then fortunate to be with him at his last conference, namely ETMM11 in September 2016 in Palermo, where at the age of 93 he attended many lectures (first picture) and asked pertinent questions and delivered a lecture on combustion modelling (second picture). The last picture shows him at the conference dinner, which he clearly enjoyed and where he was given a small gift. At the dinner, I had the opportunity to say a few words on this very special conference participant and expressed the hope to have Brian back at the next ETMM conference in 2018—but this unfortunately cannot come true and we are all extremely sad and sorry that we have to continue without this great man.

(Photographs Courtesy of Easy Conferences)

Wolfgang Rodi
Karlsruhe Institute of Technology
Karlsruhe, Germany
Ph.D. (1968–1972) Student of Prof. Spalding
rodi@kit.edu
Brian Spalding at ETMM11 Conference, Palermo, Sept. 2016

A Tribute to an Exceptional Life:
D. Brian Spalding—Akshai Runchal

It is tough deciding what to say about Brian Spalding. Where do I begin? That he was an exceptional man. That he made exceptional contributions to science. That he was a multi-dimensional personality who had expertise in multiple branches of science and spoke multiple languages fluently. That he was a man who would risk valuable professional relationships for the sake of a principle—that of right to freedom of a scientist from a dictatorial regime. That in spite of an age difference of 20 years, I could sit with him on a verandah in Kanpur, have a beer, admire the sunset and watch the world go by. That because a flight was cancelled he would risk a drive of 250 miles overnight in a beat-up old taxi in an unknown land—from Delhi to Kanpur—so he could keep an appointment. That he would trust a young undergraduate enough to say "you decide which courses you want to take". Or that he was a man who would jot down some quick incisive thoughts (Spalding Missives) on the train from Wimbledon to South Ken that could take days for us, mere mortals, to unravel. Brian was all that and much more. An accomplished poet. An excellent squash player. A very good swimmer. Whatever Brian decided to do—and he decided quickly—he did with gusto.

Brian said he would want to be remembered for his poetry. That may well turn out to be so. But surely, Brian's contribution to science will be remembered for generations to come. Most scientists make their mark in a niche and distinct field. Not so with Brian. Although he is most recognized today for his contribution to the field of CFD, what is less known is that this was just the icing on the cake because, by the time he turned his attention to CFD, he had already made his mark in multiple fields of science and engineering. First, he made seminal contributions to Combustion; this led him to the theory of Mass Transfer where he has the rare distinction of having a non-dimensional number (Spalding Number) named after him. He then unified the basic theories of fluid flow, heat and a mass transfer which led him, finally, to CFD.

For me personally, Brian was not just an Icon but a living legend. His nurture and mentorship led me to where I am today. His advice to me, to concentrate on what needs to be done rather than on the obstacles to the doing of it, has been a guiding beacon. Though he was born in a land far away, I found his advice to be so much in tune with the teachings of Bhagavad Gita which—for me—is the book for living a conflict-free life. I considered him my intellectual father and I will miss him in so many ways.

Brian once made a memorable statement that has stayed with me. Man inhabits Fluid and Fluid inhabits Man. Now Spalding permeates this fluid space. Alas, he is no more but his legacy will endure for a long time.

Akshai Runchal
Ph.D. (1965–68) Student of Prof D.B. Spalding
Professor, CFD Virtual Reality Institute
Founder, ACRi Group of Companies
Los Angeles, Nice & Dharamsala
runchal@gmail.com

Tribute to Brian Spalding—Yelena Shafeyeva

My contact with Prof. Brian Spalding started through Begell House. Dr. William (Bill) Begell, first met Prof. Spalding at an International Conference of Heat and Mass Transfer in early 60s. Many years of professional interactions, friendship and joint projects started and continued first through Hemisphere and later with Begell House Publishers.

I first met Prof. Spalding during my visit to Moscow with Academician Aleksandr Leontiev. For me, it was very surprising to meet a British professor who translated A. S. Pushkin from Russian to English. I was intrigued and naturally, there were many things I wanted to discuss with him. Bill and I kept a close relationship with Brian and met him on many occasions. After Bill passed away, I had the great pleasure to interact with Professor Spalding during my regular visits to Imperial College and to CHAM headquarters in Wimbledon. Over the years, Hemisphere and Begell House have published many of the important contributions of Prof. Spalding in the areas of CFD and Heat Exchanger Design. I had very good interaction with Prof. Spalding during Minnesota meeting in 2013 as well as ICHMT Rutgers conference in 2015th. Spending good time with great memories of friends and beautiful Colleen—and of course many jokes during reception parties".

Many of Begell House editors and executive staff had a pleasure to work on various writings of Professor Spalding and I am much honoured to be among close circle of considered to be friends.

Yelena Shafeyeva
Publisher, Begell House
elena@begellhouse.com

Brian in My Life—V. Siddhartha

"—- Take a neeewww thowght, and make it be-atter! be-atter! be-atterrr!" —might have been his brand line, but Brian didn't waste his time—he told me—with the common-room addictions of students in the late 60s—The Beatles, and colour TV.

Brian took me in for a Ph.D. on the recommendation of Arthur Lefevbre at Cranfield, under whom I got a PG qualification in Rocket Propulsion in 1966. But I was probably the only student whom he directly supervised who disappointed him in that I did not continue with CFD and Heat Transfer in my professional life.

When still a student under Brian, he came to our flat in Putney Heath to a very South Indian vegetarian dinner cooked by my mother (my father was then posted as *The Hindu* newspaper's correspondent in London). Asked questions about the difference between what he was partaking of, and "curry" as common noun then in use for all Indian food. When Brian came to Bangalore, in 1978 or '79, my wife thanked him for "training my mind". I was with the Indian Space Research Organization (ISRO), and I had the opportunity of taking-him around the then-nascent ISRO Satellite Centre located then in a large shed in Peenya Industrial Estate on the outskirts—as it then was—of Bangalore. Years ago, for one of his birthdays which was being celebrated in London, his (then new) wife asked many of his students for memories of their experiences. I did write a piece I recall in which I mentioned that Devraj Sharma and I became converts to the fountain pen with the broad italics nib, Brian's "signature" style; our handwritings started to match! I have become a Lamarkist: The discipline and habits of mind that Prof. Brian Spalding inculcated into his students will be inherited by their progeny—and theirs.

V. Siddhartha
Adjunct Faculty, NIAS, Bengaluru, India
Ph.D. (1971) Student of Prof D.B. Spalding
scatopsa@gmail.com

Personal Tribute to Prof. Brian Spalding—Ashok Singhal

Brian was truly a legend who touched lives of so many in several ways. He will be missed and remembered fondly by many of us.

I had the privilege of working closely with Brian for 12 years (1974–1986). This included 3 years as a Ph.D. student and 9 years as the Technical Director of CHAM of North America. Prior to this, during my Master's degree at IIT Kanpur (1968-1970) with Prof. Suhas Patankar, I started admiring Brian as a prolific writer with a unique ability to intersperse analogies for various life lessons in his technical writings. Then, in 1972, I met Brian and his wife during their visit to Bombay, India, and got the glimpses of Bran's interests in history, philosophy and various mythologies.

In addition to his lifelong contributions to Combustion, Heat & Mass Transfer, and finally Computational Fluid Dynamics, I also admired and benefited from his perseverance for pursuing and advocating new ideas till these were adopted in scientific communities and then in the business world.

Ashok K. Singhal, Ph.D.
Ph D. student: 1974–1977
Chairman, CFD Research Corporation
Huntsville, AL 35806
aks@cfdrc.com

Close Encounters with the Life of a Genius—Roy Singham

FIRST ENCOUNTERS: I first met Brian Spalding in 1954. We joined the staff of Imperial College, Mechanical Engineering Department, at the same time, he as a Reader, I as a lecturer. At first, I felt somewhat intimidated by this self-assured character—I had no idea of the vast hidden depths that justified this self-assurance. Before long he was made Professor of Heat Transfer and encouraged me to assist him by taking charge of the Heat Transfer Laboratory. It was a time of expansion for Imperial College, and there was no shortage of funds for equipment and we felt under an obligation to spend without due consideration. I remember thinking at the time that it would be better to buy gold sovereigns for spending wisely at a later date but this was not a permissible option.

COOLING TOWERS: One day during our early years I received a phone call from a Mr. L, asking if he could speak to Professor Spalding. Brian was away at the time so I asked if I could help him. He told me that he was speaking for a firm who were in the business of building large natural draft cooling towers. They wanted to bid for a contract offered by the Central Electricity Generating Board for eight such towers. But the firm needed consultants who could prepare designs to the Board's specification. I said I would tell Professor Spalding about this when he returned. So, a few days later I told Brian. His first reaction was to say that he did not want to take the job away from me. I pointed out that I had simply taken the call for him. After some discussion, we entered a blunt agreement: he would tell me how to do it; I would do it; we would share the fees 50/50. Neither of us was skilled in the detailed civil engineering side of the matter so Brian recruited a colleague who would take care of that side of it. After a long period of unsuccessful bidding for several such contracts, Mr. L's firm gained a contract and Brian and I found ourselves in receipt of consultancy fees that lifted our personal finances on to a different plane. The work also generated some ideas for research projects.

START OF COMPUTER PROGRAMMING: I recall a moment in the late 1950s. Brian had taken on a new research student, a Mr. G. As was often the case, I was helping with day-to-day supervision and we were in my office with the student. He had been given a task which required a good deal of computation and was reporting on his progress. Much to our surprise, he had finished the task in what seemed record time. The college had recently been given an (apartment-sized) IBM computer which the student had made use of. He was conversant with "computer programming" which Brian and I knew nothing about. Brian concluded that the student had outreached him and that he could no longer act as his supervisor: it was time for him (and me) to become programmers. The college was running courses in the new art and he and I attended one and gradually became "programmers" ourselves.

START OF CHAM: There was an occasion, early in the 1960s I think, when I was chatting with Brian in his office. He was feeling happy and excited and shared his thoughts with me. He had had a discussion with the Rector (the head of Imperial College) and had been given permission to set up his own research and consultancy company. He duly went ahead and founded the company to which he gave the name CHAM (Combustion, Heat And Mass Transfer). This was gradual to become the great international company that we have been familiar with for many years. I remember joking with him about the Chinese sound of the name.

COLLEEN KING AS SECRETARY OF CHAM: Brian's first secretary at Imperial college was a Miss S. She was loyal and efficient but as the years went by she became over-devoted to the point of possessiveness and Brian found this troublesome. Somehow, he managed to break the bond and appoint a younger and rather stunning replacement, Miss Colleen King. She occupied an office outside that of Professor Spalding and you had to confront her before gaining access to him. She would greet you with a warm and devastating smile but at the same time would manage to indicate that you had to have a good reason to gain access to him. As CHAM grew so did the business of handling the company and in due course separate premises were established in Wimbledon. Colleen King became company secretary and supervisor of CHAM's secretarial business. She became his wife.

The IJHMT: I recall Brian happily telling me one day that he had now got his own journal: The International Journal of Heat and Mass Transfer. It was one of several prestigious publications sponsored by Robert Maxwell, a well-known publisher, based in Oxford. Later Mr. Maxwell spread his net rather wider and acquired the Daily Mirror, a paper with a wide circulation. There was a memorable occasion when Mr. Maxwell decided to give a dinner party, ostensibly in honour of Professor Saunders, head of the Mechanical Engineering Department at Imperial. Brian was invited and asked me to accompany him. We drove up to Oxford in Brian's car, wondering who would be in a fit state to drive back. The dinner was a lavish affair with plenty to eat and drink. I remember how at one point Mr. Maxwell was rather patronizing to Professor Saunders while holding a phone conversation with the Mirror editor (I presume) about the next day's issue. I did not keep an eye on Brian—I was too busy enjoying the food and drink followed by a best Havana. But I know that he drove us safely home.

Most of the papers in the new journal were beyond my ken, but from time to time Brian would send one to me to assess prior to publication.

PARTIES AT RIDINGS: When I retired from Imperial College in 1990, I thought that would be the end of my relationship with Brian Spalding but that turned out not to be the case. I met him by chance after our mutual attendance at the funeral of an old and much-loved colleague, Peter Moore. I had not intended to attend the "wake" after the funeral but Brian persuaded me to go along with him.

From then on, our relationship was on a personal, rather than academic, basis. Colleen was giving wonderful garden parties at their house, Ridings, attended by the many young friends of their two grown-up boys. My wife and I were invited. Neither Brian nor I had much in common with these young people and so we tended to gravitate towards each other. Both he and I enjoyed a glass or two of whisky, usually blended but sometimes Laphroaig single malt, and it loosened our tongues: who would have thought that he was as familiar with the ballad-song "Begin the Beguine" as I was?

DEEPER DISCUSSIONS: I have always been in the habit of questioning my fundamental beliefs but in more recent years I got the idea of setting down in a couple of pages "my credo". I thought there must be a website where prominent people were doing the same but I was wrong. So, I decided to create my own primitive website, calling it "mycian.com", an acronym for <u>M</u>y <u>C</u>redo <u>I</u>n <u>A</u> <u>N</u>utshell. On it I invented the credos of five prominent figures from the past, adding as an aside my own. Brian, it turned, out had never thought of writing his credo and was very interested in mine, including my conclusion that our minds are distinct from our brains. The latter is purely physical and can be weighed, probed and measured; the mind is non-physical and cannot be measured, only discussed. Of course, better minds than mine, e.g. Karl Popper's, have discussed this idea, but not "in a nutshell".

RUSSIA: Although I had nothing to do with it, I am aware that for many years up to his last Brian had a prestigious appointment in Russia, overseeing technical matters at a high level (I believe he was introduced to Russia's leader, Vladimir Putin). He visited Russia frequently, sometimes accompanied by his wife Colleen. I remember seeing a fun video of a party in which the two of them were being, separately, entranced and beguiled in a kind of fairyland. On his last and final visit in 2016, he was taken ill, flown home and never recovered.

John Roy Singham
Professor Emeritus
Mechanical Engineering Department
Imperial College
jroysok@gmail.com

Brian—William Spalding

Brian meant a great deal to a lot of people, and in that sense he has always been more than just our father.

If, as I assume, this collection will be mainly for those of his science family it may be better to give the main voice to those "children" rather than overshadowing it with a more personal vibe that they may or may not provide.

Therefore, as part of me does, indeed, wish to include something, in a tribute to Brian, I will just say that we will always and forever have our memories and an ability to share them.

William Spalding
Concentration, Heat and Momentum Limited (CHAM)
Bakery House
40 High Street
Wimbledon Village
London SW19 5AU, England

Brian, with Jeremy and I
Relaxing after being made a Global Energy Laureate by President Medvedev in Moscow, 2009

My Memories of Prof. D. B. Spalding—Pratap Vanka

Guru Brahma Guru Vishnu, Guru Devo Maheswara, Guru Sakshat Para Brahma, Tasmai Sri Guravenamah

I salute to my teacher who is my Brahma (the creator), the Vishnu (the caretaker), and Shiva (the caretaker of final days).

I had the great fortune of being one of Prof. Spalding's students. I learned a great deal of technical, speaking and writing skills which greatly influenced my academic and personal lives. I first met Professor Spalding when we received him at Bombay airport on 3 December 1971. This was a day to remember in every way. It was not only the day that essentially started my future academic life, but also was a day to remember because of the start of the India–Pakistan war. Between the blacked-out nights, flight delays and tea stallers suspicious of foreigners, we entertained Prof. and Mrs. Spalding for two days. This subsequently led to my joining the Imperial College group as a research assistant on an SRC sponsored project on 3D duct flows. My doctoral research was to study the effects of curvature and rotation on flows through ducts in which the flow is predominantly one way. We had several publications from this effort, published in reputed journals. I subsequently worked for CHAM for two years prior to coming to the United States in 1978.

Professor Spalding was a visionary, a deep thinker and a strong advocate of precision in thinking, writing and speaking. He was very sharp throughout his life, until his final illness. I very fondly remember celebrating his 60th, 80th, 85th and 90th birthdays with celebrations, books and conferences. We were all looking forward to celebrating his 100th birthday with a great conference with him giving the plenary talk. We may still celebrate it with his spirit looking over us!

Professor Spalding will be missed most not only by his immediate family but by the entire CFD and CHT communities globally. I will always remember him for the rest of my life.

Pratap Vanka (V. S. Pratap)
Research Professor and Professor Emeritus
University of Illinois at Urbana-Champaign, Urbana, IL. USA
spvanka@illinois.edu

Tributes at 90th Birthday Celebrations on 17 July 2013 in Minneapolis

At
The Windows Restaurant, Marquette Hotel, Minneapolis
During
The ASME Summer Heat Transfer Meeting

D. Brian Spalding, FRS, FREng by Colleen Spalding

Brian in 1923

Brian was born on 9 January 1923 in New Malden, Surrey, son of Harold and Kitty Spalding the local Lloyds Bank Manager and his wife. He was their second child as they had a daughter, Katie, born three years before.

He spent his youth, and most of his life, within ten miles of the place in which he was born and made up for his lack of geographical movement by his technical accomplishments which moved him, and the scientific fields with which he was, and is, associated, forward further than his colleagues, and often he himself, thought would be the case.

After attending a local prep school, Brian moved to Kings College School Wimbledon where he remained until he was 18. He has always maintained that one of the major causes of any academic success was that he adhered to the rules and obeyed the Masters. Whilst not wishing to take issue with this statement, he does also point out that he had a discussion with a French master who insisted that whilst the French for Paris was Paree the

Brian, Kitty & Katie, late 1920s

French for London was not Londres but…London. He also recalls sitting in the headmaster's garden after being offered a place to read Engineering at Oxford and being told by a senior master that they had not educated him to read some new-fangled science but, rather, to read the Classics. Why did he read Engineering? Brian maintains that many stories about him are apocryphal but he himself has said

Prep School

that one of the reasons was because, during the war, KCS pupils had to play out a lot of wire during CCF (Cadet Corp) activities which then had to be wound tidily onto a spool sometimes when pupils had to repair to bomb shelters. Something about caught his attention and he proceeded with the course, and the interest, which is still occupying his life—unravelling puzzles, rescuing and finding logical solutions.

He moved from KCS to The Queen's College Oxford where he obtained a BA(Hons) in Engineering Science in 1944 and an MA, also from Oxford, in 1948.

At KCS 1934

KCS Prefect 1941

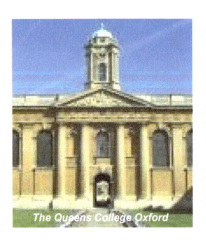

The Queens College Oxford

During the war years, he spent time afloat making a contribution to the war effort which included nearly sinking the ship he was on when, having cleaned the bilges, he forgot to secure them and water made its way in. The war, or rather the period after it, was also meaningful in that Brian was sent to Germany and was allocated a translator/ photographer who was to become his wife, Eda.

From Oxford, his natural progression was to Pembroke College Cambridge where he obtained a Ph.D. in the Combustion of Liquid Fuels in 1952. He was later, in 1966, awarded an ScD from Cambridge.

After completing his Ph.D. Brian and Eda remained at Cambridge from 1948 to 1954 until, whilst a Research Fellow and Engineering Demonstrator there he was invited to join Imperial College of Science & Technology as Reader in Applied Heat.

The family—Brian, Eda, Eddi, Michael, Sylvia and Peter, moved

The Engineering Class of 1944 in 1994

to London and he and Eda lived, initially, off Kingston Hill and then in Vineyard Hill Road in Wimbledon.

1945

Brian was awarded the Chair of Heat Transfer at Imperial in 1958, held the post of Head of the Heat Transfer Section from 1958 until 1981 when he became, Head of the Computational Fluid Dynamics Unit a position he held until his retirement in 1988. As part of his responsibilities at IC, he was the Staff Orator in 1983.

His last connexion with the College was when, in 2014, he was made Doctor of Science (Engineering) Honoris Causa in a ceremony at the Royal Albert Hall. A memorial day will be held at Imperial in 2018 where he will be remembered at the institution where he spent so much of his working life.

Brian's interest in computers and their use in modelling fluid flow led to his pioneering, and being one of the founding fathers of Computational Fluid Dynamics (CFD) at Imperial College and worldwide. As part of this activity, Brian set up a company in 1970 to expand his scientific interests into the world of business. Concentration, Heat & Momentum

Pembroke College, Cambridge

Staff Orator 1983

Limited (CHAM) was the first commercial CFD Company. It originally operated within Imperial College but, after a few years, moved back to Brian's roots in New Malden and, from thence, to Wimbledon Village where it still operates as a software and consultant engineering house and was, until his death, run by its founder and Managing Director. CHAM specializes in computer modelling of fluid flow, heat transfer and combustion in industry and the environment. It has a Branch in Japan and Agents in many parts of the world and had a branch in North America of which Brian was the Founder, Chairman and Director, from *1977 to 1991.*

CHAM's major product was, and is, the PHOENICS computer code, which was the first commercial CFD code, which Brian created, and continued to develop, with his team until 2016. From 1988, he ran the company and promoted applications of CFD in industry worldwide, on a full-time basis. There has been a PHOENICS Journal, and PHOENICS User Meetings held across the world.

From January 1978 to April 1979, Brian was Reilly Professor of Combustion at Purdue University, West Lafayette, Indiana, USA. In addition to the Reilly Professorship, he held positions as Visiting Professor at MIT, the University of California Berkeley and the University of Minnesota over the years.

In 1983, he was made a Fellow of the Royal Society and, over the years, received a number of honours including an Honorary Doctorate (2013), Moscow

View of the Mechanical Engineering Department of Imperial College

Brian & Eda with their children around 1960

HTS Party in 1977

CHAM as it was

Energy Institute, Russia; the Huw Edwards Award (2011) the Physics Institute, England; the A V Luikov Prize (2010), Academy of Sciences of Belarus, Minsk; appointment as a Franklin Laureate (2010) and recipient of the Franklin Institute Medal for Mechanical Engineering, The Franklin Institute, Philadelphia, Pennsylvania, USA; appointment as Global Energy Laureate (2009), by the Global Energy Foundation, Russia; the subject of a Festschrift Issue (2009) of the International Journal of Heat

and Mass Transfer; the holding of an ICHMT Conference to honour and celebrate his 85th birthday in 2008, in Marrakech, Morocco.

He was made an Honorary Member (1999) of the Scientific Council of Institute of Thermophysics, Siberian Branch of Russian Academy of Sciences; a Dr-Ing Eh (1997), Technical Faculty of the Friedrich Alexander University

Brian rescued a local book shop which now operates within CHAMís shop front area.

PHOENICS User Meeting, Cannizaro December 2007

of Erlangen-Nuremberg; became a Member (1994) of the Russian Academy of Sciences and (1994) of the Ukrainian National Academy of Sciences. He is a Fellow (1989) of the Fellowship of Engineering—now The Royal Academy of Engineering and (1983) of the Royal Society of London and a member (1979), Royal Norwegian Society of Sciences and Letters.

He received the Luikov Medal (1986) from the International Centre for Heat and Mass Transfer, the Bernard Lewis Metal (1982) from the Combustion Institute, the Medaille d'Or (1980) from the Institute

Official Photo FRS 1983

Brian Receiving the Global Energy Award from President Medvedev 2009

Francais de l'Energie, the Max Jakob Award (1978) for Services to Heat Transfer, the James Clayton Prize (1970) from the Institution of Mechanical Engineer.

Brian was not much of a Committeeman. He much enjoyed discussing any aspect of his scientific interests —Computational Fluid Dynamics, Turbulence Modelling, Heat Transfer, Two-Phase & Multiphase Flow, Chemical

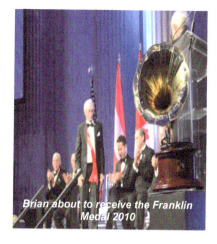

Brian about to receive the Franklin Medal 2010

Brian with Naim Afgan 1988

Reaction and Stresses in Solids amongst others, with colleagues, past students, young people who share his interests today but he was adamant he did not much enjoy sitting on committees.

He did, however, sit on various British Aeronautical Research Council Committees from 1952 to 1979 and has been the Honorary President of ERCOFTAC (European Research Centre on Flow Turbulence and Combustion), Steering Committee Member & Chairman Scientific Program Sub-Committee of ERCOFTAC (1988–1991), an Investigator on the Hermes Space Program, a member of the Executive Committee of the International Centre for Heat and Mass Transfer, a Member of Scientific Council —International Centre for Heat & Mass Transfer (1994–1998) and Chairman, Technical Activities Committee, International Centre for Heat and Mass Transfer.

The ICHMT was important to Brian. He was one of those responsible for founding it to facilitate scientists from the East and the West being able to meet and share their knowledge during the days of the Cold War. The Headquarters were in Belgrade in what was then Yugoslavia and were run by Zoran Zaric and Naim Afgan. Conferences were held in idyllic places such as Dubrovnik, and we have fond memories of attending them accompanied by small children.

Brian has been associated with various scientific Journals. He was the Founding Editor of PhysicoChemical Hydrodynamics: an International Journal (1978-1989) published by Pergamon Press. This came about as part of an ongoing campaign to persuade those in power that Benjamin Levich, an

Brian with Zoranis daughter Milana

Brian with Ben & Tanya Levich in Vienna in 1978

Associate Member of the Russian Academy of Sciences, should be allowed to leave Russia. Brian worked tirelessly to achieve this aim visiting Ben in Moscow, working with various institutions in London and abroad, starting a series of PCH Conferences the first and the last of which were held in Oxford. The first was held before Ben and Tanya came to the west but was attended by their sons Alexander and Yevgeny. The last was meant to be a celebration of the 10 years Ben had spent in Israel and New York but, in the event, became a celebration of his life as it occurred just after his death.

Brian had close links with Pergamon Press, was a founding Editor of the IJHMT (International Journal of Heat and Mass Transfer) and successfully involved with Robert Maxwell's successful efforts to reclaim Pergamon Press during the 1970s.

Other scientific publications with which he was involved include acting as Editor-in-Chief of the HMT Book Series: The Science and Applications of Heat and Mass Transfer; Editor, International Communications in Heat and Mass Transfer; Editorial Board Member, Applied Mathematical Modelling; Editorial Board Member, Computer Methods in Applied Mechanics and Engineering; Editorial Board Member, International Journal for Numerical Methods in Fluids; Member, Conseil Scientifique International Revue General de Thermique; Editorial Advisory Board Member, Numerical Heat Transfer; Editorial Advisory Board Member, Combustion and Flame; Editorial Advisory Board Member, Advances in Transport Processes; Editorial Board Member, International Journal of Turbo and Jet Engines; Editorial Advisory Board Member, Heat and Technology.

Starting in 1949, Brian has published some hundreds of papers and many books including papers on thermodynamics, combustion, heat transfer, boundary layer theory and computational fluid dynamics. Many of his papers can be accessed via the CHAM website (www.cham.co.uk).

His published books are listed below:

1. Spalding D B (1955). *'Some Fundamentals of Combustion'*, Butterworths, London
2. Spalding D B (1963) *'Convective Mass Transfer'*, Edward Arnold, London(and under McGraw Hill imprint)
3. Gosman A, Pun W M, Runchal A K, Spalding D B and Wolfshtein M (1969) *'Heat and Mass Transfer in Recirculating Flows'*, Academic Press, London
4. Patankar S V and Spalding D B (1970) *'Heat and Mass Transfer in Boundary Layers' (2nd Editio*n), Intertext Books, London

5. Launder B E and Spalding D B (1972) *'Lectures on Mathematical Models of Turbulence',* Academic Press, London and New York

6. Cole E H and Spalding D B (1974) *'Engineering Thermodynamics' (3rd Edition),* Edward Arnold, London

7. Afgan N and Spalding D B (Editors) (1977) *'Heat Transfer and Turbulent Convection' vol I and II,* Hemisphere Publishing, Washington DC

8. Spalding D B (1978) *'GENMIX: A General Computer Program for Two-Dimensional Parabolic Phenomena'* HMT Book Series, Number 1, Pergamon Press, Oxford

9. Spalding D B (Editor) (1978) *'PhysicoChemical Hydrodynamics' Proceedings 1st International Conference, vols, I & II,* Advance Publications, London

10. Spalding D B (1979) *'Combustion & Mass Transfer: A Textbook with Multiple Choice Exercises for Engineering Students'* Pergamon Press, Oxford

11. Kakac S and Spalding D B (Editors) (1979) *'Turbulent Forced Convection in Channels and Bundles'* Volumes I and II, Hemisphere Publishing, Washington DC

12. Spalding D B and Afgan N (Editors) (1979) *'Heat and Mass Transfer in Metallurgical Systems',* Hemisphere Publishing and McGraw Hill, Washington DC

13. Schlunder E, et al (Editors) (1982) *'Heat Exchanger Design Handbook',* Hemisphere Publishing, Washington DC

14. Spalding D B (1983) *'Numerical Prediction of Flow, Heat Transfer, Turbulence and Combustion' Collected Works* Editors: Patankar S V, Pollard A, Singhal A K, Vanka S P, Pergamon Press Oxford & New York

15. Afgan N and Spalding D B (Editors) (1989) *Heat and Mass Transfer in Gasoline and Diesel Engines,* Hemisphere Publishing and McGraw Hill, Washington DC

There are, undoubtedly, many aspects of Brian's life missing from this brief synopsis. He worked a 7-day week at CHAM. He attended conferences, wrote papers and had a research group in Moscow which he visited regularly. Outside of work he walked across Wimbledon Common every weekend, used his rowing machine at home each morning, had a treadmill in his CHAM office, swam every Friday night and took up Pilates on his 90th birthday.

He comes from a family which lives to quite an age. His sister in South Africa will be 98 in December 2017 and one of my favourite pictures is of Brian with 2 cousins in 2007 when Brian was the "baby" at 84, Frances was 88 and Cyril 92 (I think!). He was also delighted to track down the grave of his grandfather who had been banished to Australia and died in Vancouver.

His 90th birthday, the actual day, was celebrated with his family in South Africa but he would have preferred to be working at CHAM to visiting the Botanical Gardens in Johannesburg, however, beautiful.

There was then a family party in February followed by a most pleasant evening at Imperial College in March.

Despite these celebrations, or perhaps because of them, I think Brian did

A family Gathering, February 2013

not much like being 90. However, I am sure that events such as that in Marrakesh to mark his becoming 85, and in Minneapolis to celebrate 90, both of which were scientific and gave him the opportunity to interact with colleagues, past students and to meet those who will take "his" subject further brought more pleasure than "just another party".

Brian speaking at a most enjoyable evening at Imperial College & a greatly appreciated gift.

Perhaps, to end, I should mention that Brian had a second family. I like to think that helped to keep him young. Our sons, William and Jeremy, and William's wife, Ashley, joined me in wishing Brian happiness when he was halfway through the first year of his ninth decade.

This piece was written in somewhat of a hurry but I hope that, whilst there were sins of omission and commission, it gave somewhat of a flavour of a life which was long and fruitful. The owner of it accomplished much.

One of the things which I had hoped would be accomplished was HIS version of a life extraordinary—it was not to be.

Tribute to Brian Spalding by Larry Caretto

Although I knew the name of Brian Spalding from my graduate-student research, I met Brian when he was a visiting professor in the Mechanical Engineering Department at UC, Berkeley in the late 1960s. At that time, I was a member of the research staff and an acting assistant professor at Berkeley; Brian had recently started his work on general computer codes. I was surprised to meet him in a room with keypunch machines and teletypes, which was generally used by graduate students, research staff members and junior faculty. From that meeting and subsequent conversations, he invited me to be a visiting member of his research team at Imperial College.

At the time I joined his research group in September 1970, I was assigned to a project which sought to develop computational methods for three-dimensional parabolic flows. At that time, the term CFD had not yet been invented, and there were only a few groups in the world working on numerical methods in fluid dynamics. However, Brian had a vision that future advances in computing power would bring to engineers the ability to model complex flow phenomena using numerical tools.

Brian was not a distant research supervisor, but an active participant in the research. He would write detailed memoranda outlining his ideas and suggesting specific tasks that members of the research team might undertake. Although his memoranda would give forceful statements of his ideas, he was always willing to entertain debate and would willingly write another memorandum correcting his previous one if he found a team-member's argument to be persuasive.

Brian's approach to the numerical solution of the Navier–Stokes equations was an elegant combination of mathematical reasoning and physical insight. He taught us to regard the numerical grid as a collection of cells with the computations modelling a physical flow from one cell to another. I still remember the term "false inertia". I have forgotten what that means, but I remember that working on algorithms using this concept by Brian for one or two weeks and quickly abandoning this when it proved not to be fruitful.

I still regard the time that I spent at Imperial College with Brian and my fellow researchers as one of the most stimulating research experiences of my career.

Happy 90th Birthday, Brian!

Larry Caretto, Professor Emeritus
Mechanical Engineering Department
California State University, Northridge
Northridge, CA 91330-8348
lcaretto@csun.edu

A Message to Brian Spalding from Anil Date

Dear Prof Spalding,

Many many congratulations on your 90th Birthday and to know that you have agreed to go to Minnesota for the felicitation. I met you in 2007 in Xian, China, and it would have been my honour and privilege to meet you again now. But circumstances have prevented me from attending the conference.

On this occasion, I am reminded of your first visit to India in 1971 for the National Heat Transfer Conference. On return to IC, you agreed to meet the Indian students who were keen to know of your impressions. What I distinctly remember is the surprise you expressed about almost everyone (you met at that conference) posing you the question " what was latest in Heat transfer?". And, your answer was: whatever Indian industry and society needed should be the latest! How true.

You are of course world famous for the pioneering contributions in both academic and industrial CFD. But I have found your very early contribution of "Reynolds Flow Model—Reynolds Flux Hypothesis and the Spalding number B" in the book "Convective Mass Transfer" (1963) very useful in a variety of applications involving Heat and Mass Transfer in analysing simple "People's Technologies" from India's rural areas. What is heartening is that papers analysing H & M Transfer in a Clay-Pot Refrigerator (IJHMT, vol 55, p 3977–3983, 2012) and Wood Burning Cook Stove (Comb Sc & Tech, vol 183, p 321–346, 2011) are now found acceptable in peer-reviewed international journals. From the reviewer's comments, I could of course deduce that acceptance was greatly influenced by my reference to the work of D. B. Spalding.

Yes, I have benefitted and been enabled to lead a satisfying professional life—thanks to my 4-year association with IC-Heat Transfer Section and the rich writings of D. B. Spalding.

With respectful regards, as a cricket fan, I am certain that you will cross 100.

Anil Date
Professor Emeritus, IIT Bombay
awdate@me.iitb.ac.in

Message for Brian Spalding from Graham De Vahl Davis

I wrote to Brian on the occasion of his actual birthday back in March, apologizing for my inability to attend at that time. I was in fact listening to a performance of the bawdy and irreverent Carmina Burana at the time, and I raised a glass to Brian during the interval. The opening and closing *O Fortuna* of that music is the most-played classical music of the past 75 years in the United Kingdom. I suspect that PHOENICS is the most-used CFD code of the past 30 years in the world.

There is nothing bawdy or irreverent about the father (or perhaps we should say grandfather) of computational fluid dynamics. You are, on the contrary, a SIMPLE man. I pay a humble tribute to you and treasure the memory and privilege of six months working with you some time ago, and the honour of your friendship (and that of Colleen) over many years.

As we say in the Jewish community: "ad mea v'esrim"—you should live to be 120.

Graham de Vahl Davis
Honorary Professor of Mechanical
and Manufacturing Engineering
UNSW Sydney
g.devahldavis@unsw.edu.au

Best Wishes from Said Elgobashi

Dear Professor Spalding,
All my best wishes for a Happy Birthday and many happy returns. I wish I were able to join the July celebration in Minneapolis.

Said Elghobashi
Professor, UC Irvine
selghoba@uci.edu

Tribute to Brian Spalding from Marcel Escudier

I have nothing but positive memories of Brian. He was an excellent teacher (lecturer) and research supervisor who always found time to discuss progress. I attended classes he taught in 1963/4 as part of the DIC course during which he also supervised my project. From 1964 to 1967, I was one of Brian's many research students, Earl Baker, Suhas Patankar, Peter Duffield and Bill Nicoll, with whom I worked very closely, being among my contemporaries. Jim Whitelaw and Brian Launder were young lecturers at the time, both working with Brian, who also contributed to a wonderful group spirit. The feeling that we were part of a special group was enhanced by visits from eminent researchers such as Bill Reynolds, Bill Kays and Roger Eichhorn from the USA as well as Samson Kutateladze and Aleksandr Leont'ev from Russia. I feel greatly privileged to have worked with Brian and am greatly indebted to him. The recent dinner in London to celebrate his 90th birthday was a warm and memorable occasion. I'm sure the dinner in Minnesota will be just as successful.

Marcel Escudier
Emeritus Harrison Professor of Mechanical Engineering
University of Liverpool
*Supervised by DBS 1963–64 (DIC), 1964–67 (*Ph.D.*)*
sqda@btinternet.com

A Message from Sir John Horlock

I first knew Brian when we were lowly demonstrators together in Cambridge. That group included several future vice-chancellors and fellows of the Royal Society. Brian was senior to me by several years, and it was clear he should have been promoted to a lectureship. But before this happened he was stolen by Professor Saunders at Imperial College who immediately found a readership for him.

My memory of Brian in those days is of his incredible work rate, coupled with the succession of books and papers which kept appearing at a great rate. His concentration was intense—callers at his office were faced with a notice outside his door, which simply said "Will a note do?"

We both enjoyed publishing much of our work through the old Aeronautical Research Committee, where he chaired the Combustion Sub-committee and I chaired the Propulsion Committee. But I did not have the pleasure of working with him in more recent years, when he worked at Imperial and CHAM with such staggering success.

Not the least of Brian's achievements has been the mentoring of a series of research students and associates who subsequently attained considerable distinction. I am delighted that so many have gathered together tonight to honour such a distinguished, bat modest engineer. My only regret is that I am unable to join you all this evening because of my immobility.

Sir John Horlock,
Formerly Vice-Chancellor of the Open University.

A fragment from my many memories of DBS by Brian Launder

In 1971, while I was a lecturer at IC, DBS suggested that a case should be made for my promotion to a readership in fluid mechanics. My pertinent data and publications were forwarded to him, and these were duly sent off to the external assessor, Professor John Horlock. A few weeks later DBS told me that the external assessor had felt the case was premature … but invited me to provide a persuasive rebuttal that he might send to Horlock. Well, I did indeed prepare a highly inflated draft letter for DBS, greatly exaggerating the importance of my contributions. I've no idea how much, if any, of my draft was used but a short time later DBS sent me a note saying that the external was now fully satisfied that my promotion should go ahead. Brian then commented that he knew whom he should contact if he ever wanted a letter of recommendation for himself!

In reality, of course, Brian's standing in the world has been so outstanding and widely appreciated that he has hardly needed letters of support, even from people of far greater eminence than me.

The DBS 90-Birthday Dinner at IC in March was one of the most memorable and fun occasions of the year for my wife and me and I trust the celebratory dinner in Minneapolis is equally as enjoyable.

Brian Launder
Professor B. E. Launder, FREng, FRS
University of Manchester
brian.launder@umist.ac.uk

My Association with Professor Brian Spalding by Alok Majumdar

I first met Professor Spalding at IIT Kanpur in 1971 when I was a doctoral student at Central Mechanical Engineering Research Institute (CMERI), Durgapur, India. In my dissertation, I used Stream Function–Vorticity Method that was developed at Imperial College under his guidance. I received many useful advises from him through mail correspondence during my Ph.D. work. After completion of my Ph.D., he offered me a post-doctoral position at Imperial College. I worked with Professor Spalding from 1975 to 1977 on Numerical Methods for Rotating Flows. I consider those years in Heat Transfer Section at Imperial College as the most productive and educative period of my career. I had the opportunity to meet many distinguished research workers in a vibrant and active academic environment that made a very deep impact on the development of my professional career. I had the opportunity to publish several papers with him on Turbo-machinery during this period. My daughter, Bipasha, was born in London while I was doing my post-doctoral work.

On completion of my fellowship at Imperial College, I returned to CMERI and continued to collaborate with him. In 1980, I joined CHAM of North America in Huntsville, Alabama, a sister organization of CHAM which he founded in London during 1970s. I had the opportunity of writing my first CFD code on Cooling Tower Performance while working at CHAM. Since 1985, I started working for Aerospace Industries and for last twenty years I have been working on the development of a Generalized Fluid System Simulation Program (GFSSP), a thermo-fluid system analysis program for Propulsion System Analysis for Liquid Rocket Engine (https://gfssp.msfc.nasa.gov/). GFSSP is a finite volume based network flow analysis code for analysing complex thermo-fluid system. Two very important things that I learned from Professor Spalding that have influenced my work are (a) Finite

Imperial College (1976)

Volume Formulation of Conservation Equation and (b) concept of separating solver from the user interface which is a key for developing a general-purpose code.

Professor Spalding conceived the power of Computational Thermo-Fluid Dynamics back in 1960s and laid the foundation of this emerging field that has transformed the design of thermo-fluid systems around the world. I congratulate Professor Spalding on his 90th Birthday and wish for his good health and many happy returns of his birthday.

Alok Majumdar
NASA/Marshall Space Flight Center
Huntsville, Alabama, USA

Best Wishes from Nicolas Markatos

Dear Brian,

I send you a few words from Heart. Brian Spalding is certainly the man who "discovered" Computational Fluid Dynamics, at least in practical terms, i.e. for solving real technological problems. It is amazing that people are proud today for doing things which we were doing, because of Brian, in the late 70s! I thank you Brian for directing me into a most interesting and sometimes puzzling but always exciting discipline.

Long ago, during dinner in Brussels, you were in very good mood and you told us "I have a very good family hereditary, my grandfather reached the age of 102, I intend to reach that age too, so do not hope to get rid of me soon!" I am certain that the grandfather part was not true but I am also certain that you will be here at 102 celebrating; and I sincerely hope that everybody with you today will also be here with you in 2025!

My sincere Best Wishes. HAPPY 90th BIRTHDAY and many happy returns of the day!

Incidentally, do not forget to come to my 90th Birthday party at the Rose and Crown, in 2034, around 19.00 please!

Nicolas Markatos
Professor Emeritus, Former Rector
National Technical University of Athens
n.markatos@ntua.gr

A Message to Brian Spalding from Suhas Patankar

Dear Brian:

Congratulations on your 90th birthday. We are delighted that you will be honoured in Minneapolis. We are eagerly looking forward to the event.

You have made the most significant impact on my professional life. I clearly remember two distinct periods in which we spent a lot of time together. One was in 1966 when we shared an apartment in the MIT graduate dormitory and worked at NREC. During this time, our intense technical interaction led to the creation of the boundary layer calculation scheme. The other period was December 1970–February 1971. During this time, my family and I stayed at your house in Wimbledon. Again, the day-to-day interaction between you and me produced the SIMPLE calculation scheme. These periods were not only very productive but also very enjoyable. I am very grateful for the opportunity.

Since I began working with you in 1964, you have been a role model for brilliant ideas, creative thinking and lucid writing. You have been a constant source of inspiration. Your professional contributions to Heat Transfer and CFD have set a new standard for what is truly possible.

On the occasion of your 90th birthday, my wife Rajani joins me in wishing you good health, happiness, and many more productive years.

Sincerely,
Suhas V. Patankar
Professor Emeritus, University of Minnesota
President, Innovative Research, Inc.
Ph.D. (1964–1967) Student of Prof Spalding
patankar@inres.com

What goes around comes around! by Andrew Pollard

Andrew joined Brian's group in September 1975, arriving from the University of Waterloo in Canada with recommendations from Suhas Patankar and George Raithby. Andrew's thesis concerned three-dimensional laminar and turbulent flow in tee junctions, in both Cartesian and cylindrical polar coordinates, which included wall mass transfer as an analogy for wall shear stress. He performed both computations and experiments and has continued throughout his career to enjoy the interplay between these approaches to unravel the underlying physico-chemical processes in many other complex thermo-fluid problems. Because of Brian's broad interests, Andrew, too, embraced combustion, turbulence and heat and mass transfer: he graduated in 1978 and returned to Canada.

Today, Andrew holds a University Research Chair in Fluid Dynamics and Multi-scale Phenomena at Queen's University at Kingston, Canada and looks back on a career that has produced over 250 publications, a number of edited volumes, graduated over 50 graduate students and post-doctoral fellows. The open, inquisitive and unified approach to transport processes engendered by Brian instilled the confidence to tackle myriad problems and most recently spontaneous combustion of re-useable low-carbon fuels and biomass densification and torrefaction technologies. He continues to probe the intricacies of fundamental turbulence using direct and large eddy simulations and detailed optical and probe-based experimental techniques. These include drug delivery to the lungs, quantification and destruction of air emboli during cardiac surgery, non-homogeneous wall roughness modifications to the "log-law" and the effects of mean shear, intermittency and passive flow control in jets.

He has been happily married for over 40 years, with two children Richard (Ph.D. Cantab, 2009) and Michael (CPGA).

In 2013, we celebrate Brian as he reached the age of 90 and Andrew is 65: Andrew had the pleasure to be in attendance at Brian's 65th retirement from IC in 1988……what goes around, comes around!

Andrew Pollard
FCAE, FAPS, FASME
Professor and Queen's Research Chair,
Queen's University, Kingston, CANADA
Ph.D. (1975–1978) Student of Prof Spalding
pollard@me.queensu.ca

Happy Birthday Wishes from Heqing Qin

生日快乐

Happy Birthday
永远 健康,长寿, 智慧永存！
Healthy, Long life forever,
May your genius remain undimmed!

Regards,
Heqing Qin
hqq@cham.co.uk

A Tribute to My Intellectual Father by Akshai Runchal

It is rare to come across an individual who is a good researcher, a good teacher and fun to be with socially and at the pub. Brian, you are that rare individual. Your numerous and outstanding scientific achievements are well known. After all how many engineers get a non-dimensional number named after them or get a Franklin Memorial Prize. Today, I will talk of you as a person.

When I first met you in 1965 as your Ph.D. student, I asked you what courses I should take and your surprising answer was "none". You instead assigned me to do an in-depth literature survey over the next 3 months and get familiar with your Unified Theory. That demonstrated your hands-on approach to life and your essence as an engineer. Do something practical and useful—theory will fit in as needed. Not to talk of your risk-taking—trusting an undergraduate from a foreign university of unknown standing!

Over the next 10 years, I worked closely with you. When I would bring a problem to you, I found that you understood the essence of it before I did and your questions clarified the cobwebs in my mind. Working closely with you was taxing by any account. More often than not, there will be a missive from you that you had just compiled on your trip from Wimbledon to South Ken that might take days for me to unravel.

After I joined IIT Kanpur, I invited you to India. Kanpur is somewhat of a remote place and the only flight of the day from Delhi got cancelled. This is before the days of mobile phones and I started preparing to postpone your seminar. To everyone's utter surprise, you landed late at night (or was it early morning?). You had hired a Taxi to drive the odd 300 miles from Delhi to Kanpur. Given the drivers in India, the road conditions and a foreigner at that, nobody would consider that well advised! Even today I would certainly prohibit anyone—including myself—from doing that. It shows your single-minded determination to focus on the problem at hand and find a solution.

After I was done with the ψ-ω Navier–Stokes solver in 1967, Micha, my partner in crime, was writing his thesis, and since the main thrust of our work was common, it would have been better if both theses came out together. So I asked for a meeting with you. To my utter consternation, you agreed that I had done enough for the thesis but informed me that there was a University of London provision that a Ph.D. thesis in engineering must have an experimental component. So, I spent a year doing experiments and completed my thesis in 1968 instead of 1967. Brian, you then had to explain to the external advisor that my numerical work was "original" and not boot-legged from Micha. It was only in 2011 that you told me—with an enviable smile—that you had made up that rule! I guess I had not spent enough time with you! In my defense, there were no websites of university regulations in those days! But on the positive side, I must say having done those experiments gave me a very different perspective that has come very handy in my professional life.

I must say I am more than glad for the time I spent with you. You gave me excellent intellectual training and provided personal support and advice when most needed. In 1969, I had an offer from IIT Kanpur but was reluctant to go due to personal considerations. You advised me to go and said if I don't I may regret having missed an opportunity. It would have been easy for me to stay and for you to have me around. Yet that advice changed and enriched my professional and personal life.

I consider you to be my "intellectual" father. My father lived to be 101 and in good health—I wish so would you. And need I remind you that I fully expect you to attend my 90th Birthday! Best wishes for a long and healthy life from Chanchal and me. Here is to you—the next drink for your good health!

Akshai Runchal
President & CEO
Analytic & Computational Research, Inc.
Runchal@gmail.com

Prof. Brian Spalding, FRS by Ashok Singhal

Health and Happiness in coming years!
Happy Birthday

We admire and appreciate your legacy of over 60 years of contributions to combustion and CFD.

Your bold initiatives for industrial use of CFD have lead tens of thousands of practitioners around the world.

It has been my privilege to learn and work with you for over 12 years.

Along with Sangeeta, Andrzej Przekwas and all other employees of CFDRC, I wish you the very best of Health and Happiness in coming years!

Thanks to Pratap Vanka and Akshai Runchal for organizing the 90th Birthday.

Ashok K. Singhal, Ph.D.
Chairman, CFD Research Corporation

Researcher

Loss of Hair

Entrepreneur

Best Wishes on 90th Birthday by Pratap Vanka

Dear Professor Spalding,

I consider it a distinct privilege to have been your student. I very much remember the nice comments by Sir Lighthill on our helical coil papers to JFM and the comments by Sir Hawthorne on the partially parabolic computations of flows in curved ducts that you presented at an ARC meeting. I have learned a lot from you on effective writing, speaking and identifying key scientific issues.

The SIMPLE algorithm has made many professors, millionaires and executives and has resulted in a culture of its own, as a consequence of your vision and intelligence.

Raj and I wish you a Happy 90th birthday, and many, many happy returns of such birthdays.

Pratap Vanka (V. S. Pratap)
Research Professor and Professor Emeritus
University of Illinois at Urbana-Champaign,
Urbana, IL. USA
spvanka@illinois.edu

Best Wishes by Nikos Vlachos

Please convey my warm greetings to all. I thank you for your kind invitation to celebrate the 90th Birthday of Prof. Brian Spalding. As I could not, regretfully, join you, let me take this opportunity to refer to some experiences with this great scientist and research leader.

I met DBS in August 1970 as a postgraduate student at Imperial College. I was fascinated by his seminars and those of Suhas Patankar in Heat Transfer. In April 1971, I was asked to modify a CFD code (with its endless STRIDE routine!) in order to handle curved boundaries with Cartesian grids. Thus, I entered the fascinating world of CFD with the help of my classmate, Jim McGuirk. Indeed, the results from a flow in a bend of a circular glass tube matched very well my LDA measurements, taken under the supervision of the late Jim Whitelaw.

Looking for employment with CHAM Ltd in June 1976, DBS asked me to show a listing of TEACH3D which we had developed with my colleague, the late Pepe Humphrey, under the supervision of David Gosman. The next day DBS returned the listing with a handwritten comment: *"Nick, thank you for the listing. The code is very well organized, but surely it is not in a form for commercial use!"* Incidentally, Pepe introduced many of us to developing our codes electronically rather than using the bulky punch cards. We spend endless hours at the *Terminal Room* with Pratap Vanka, Alan Morse, Paul Watkins, Essam Khalil, and young Steve Pope and Alex Taylor! I also remember Dave Gosman, Fred Lockwood and Aki Runchal coming to run their codes!

In July 1976, I joined Nikos Markatos and Stathis Ioannides at CHAM Ltd in Wimbledon, working with curvilinear grids on the flow around a ship for Admiralty Works. We had an interesting time in Bakery House, helped very much by the skillful office management of Colleen King. I recall David Tatchell asking how things were going. Things were not going because we were stuck in handling the free boundary. And then… success! DBS proposed a simple solution that left Nikos and me fascinated by the ingenuity of the man.

On another occasion, I had been working for almost 2 days to finish a project on the mixing of exhaust gases of a GE gas turbine for Boeing. After we prepared the card deck and the user manual, I rushed from Wimbledon to Imperial College. Entering his office, DBS walked towards me saying: *"Is it all done, Nick?"* I replied: *"Yes Sir!"* Then with his usual smile he offered: *"Well done Nick, you will be rewarded in heavens!"*

In August 1988, I left the U of Illinois at Urbana-Champaign to join CHAM North America in Huntsville—Alabama, where I enjoyed 3 years of interesting work for NASA Marshal (global model for space shuttle main engine) and the US Department of Energy (model for flue gas desulfurization). For reasons of loyalty to DBS, the late Toni Mukerjee and I resigned the day after we received his letter referring to the difficulties he was facing with CHAM NA, I guess because of the cuts in the NASA program by the Reagan administration.

Dear Colleagues - Transport phenomena are in the heart of energy, environment, transport and fusion, but also circulation of blood in our arteries and of air in our lungs. Fluid flow is a nonlinear phenomenon that produces unique images, but also gives difficulties. Although the Navier–Stokes equations were formulated in 1840–1850, there exist only few analytical solutions. Unfortunately, experiments are expensive and provide only point measurements, while we are far from plane or volume measurements. Thus, CFD has provided useful results and, undoubtedly, DBS has been a pioneering authority in its development.

Having started my engineering education at the Technical U of Athens, I am indebted to my colleagues at Imperial College for 12 years of innovative research. Whatever I achieved is due to the excellent supervision of Jim Whitelaw and Dave Gosman, and to the sharp thinking of DBS. Indeed, I am grateful to Professor Spalding for his support on two difficult occasions, and I hope to be able to join you all in celebrating the 100th birthday of this unique and gentleman, wishing him wholeheartedly good health and longevity.

Nicholas S. Vlachos
Professor Emeritus U of Thessaly
Meg. Alexander 94, 15124 Marousi-Athens, Greece
vlachos@uth.gr

My life with Brian by Micha Wolfshtein

In 1962, I decided that I want to take some graduate studies. This was already 8 years after graduation; my math was very rusty. I had some practical experience in Heat Transfer and I decided to enrol in the M.Sc. degree at the Technion. At the end of the academic year, I gained some confidence in my ability and I decided to enrol for a Ph.D. I was already 31 years old, married with two children, and I suspected that I may have difficulties getting some financial support. Therefore, I wrote to about 30 Heat Transfer experts in American universities. My recognition of the international heat transfer community was very limited, and the most important name I knew was that of Eckert of "Eckert and Drake" textbook. Naturally, I was hoping for an assistantship at the University of Minnesota. Surprisingly, it took me 51 years to reach this destination on DBS 90th birthday.

Sometime toward the end of summer 1962, somebody gave me a paper on analytical solution of laminar boundary layers. The author was an anonymous man for me, a D. B. Spalding from Imperial College in London. I read the paper and was very impressed. The next day I added yet another application to my records and started waiting. I knew that answers to such applications are due in the spring, say February to April, and I was surprised to get a proposal from Professor Spalding in November. I had some queries and we exchanged a couple of letters, and by the end of December the moment of truth came. Brian sent me a complete proposal, and I had to decide: take it or leave it. The decision was not simple. If I reject the proposal, I may lose what appeared as a very good proposal, although all I knew of Professor Spalding was some journal publications. On the other hand, accepting the offer would mean rejecting all the unknown promises (?) of the other 30 applications.

This was one of the most difficult decisions of my life. I decided to jump into these uncharted waters and accepted. Looking back, it was a very successful decision, and one of the most important decisions of my life. Thus I found myself in London in the autumn of 1964. The first year was very difficult. My English was very poor, and my theoretical background was still limited. Living in London with a family proved more expensive than I estimated, and the family had some social difficulties as well. On top of all this, Brian asked me to write a literature survey on the chosen topic of my research, but when I submitted it he returned it to me because the English was poor. I thought that I shall have to pack up and return to my industrial practice. To my surprise, Brian was willing to wait until I improved my English (British Council course for foreign students), and eventually he read the paper and let me continue into what turned out, after two more years, into a Ph.D. Thesis from London University and a DIC from Imperial College. This was the first (but not the last) time that Brian gave me a hand when I was in real need of help.

Years have passed. My relations with Brian became more and more friendly, but his ability to penetrate the very heart of a problem and to offer a solution which was always simple and convincing, but somehow invisible, is still surprising me every time we meet.

Thank you, Brian. As much as you are a friend, you are still my teacher, and so you will always remain.

Best wishes and regards to you and Colleen from

Aviva and Micha Wolfshtein
Professor Emeritus, Technion
Haifa, Israel
aer0902@aerodyne.technion.ac.il

Happy Birthday! by Jeremy Wu

Dear Professor Spalding,

Whenever I think of you, I feel so grateful. I want to thank you from the bottom of my heart for your supervision, support and the kindness you have shown to me and my family for over 30 years of my life since 1982 when I became your student.

May this birthday bring you all the happiness you deserve! Wish you a good health and a long life!

Happy Birthday!

Jeremy Wu
jzwwu@yahoo.co.uk

Best Wishes from Michael Yianneskis

Dear Brian,

With my best wishes to a most inspiring teacher, colleague and friend.

For a truly wonderful and memorable 90th birthday.

I hope to have an opportunity to give my wishes in person soon.

Michael

Professor Michael Yianneskis
CEng CSci FRSA FIMechE FIChemE FKC
Honorary Professor, University College London
Emeritus Professor of Fluid Mechanics, King's College London

Appendix III: Photo Gallery

Brian 1923

DBS 1927-28

A. Runchal (ed.), *50 Years of CFD in Engineering Sciences*,
https://doi.org/10.1007/978-981-15-2670-1

Sylvia, Peter, Eda, Edd, Brian, Michael in the 1960s

DBS 1974

DBS 5th Heat Transfer Conf. Tokyo, 1974

DBS FRS 1983

Staff Orator 1983

DBS CHAM 1985

DBS St. Petersburg
2005

DBS Franklin Awards CFD Conf., Villanova, 2010

DBS Franklin Awards CFD Conf., Villanova, 2010

Brian Robed at IC, 2014

90th Birthday, London, 2013

Palermo Univ. Lecture, 2016

Palermo Univ.
Reception, 2016

Duomo Piazza, Syracuse,
2016

Thinking in Syracuse, the home town of Archimedes, Sep 2016

Brian carving a Roast
DBS Home London July 2007

Brian & Colleen
DBS Home London July 2007

Chanchal, Brian & Colleen
DBS Home London July 2007

Spalding, Colleen and Family

Rome, 2014

DBS 90th Dinner, Imperial College, London, March 2013

Rajini, William, Colleen, Jeremy & Brian London

Dressed for a 30ís Charity Event, London,

Family, June 2016, The Petersham

Jeremyís Graduation, Leeds, 2008

White Water Rafting, Yosemite, 2005
Brian (back right)

Sister Katie's 90th Birthday, South Africa, 2009

Brian's Family 90th
February 2013

Brian One on One

Runchal & Spalding
Franklin Awards 29 April 2010

Spalding & Patankar 2013

Spalding & Steven Beale, California

Spalding & Heqing Qin
London, 2011

Spalding & Prof. Churchill
Sydney - CHT August 2006

Gerald Jones &
Spalding
Franklin Awards
29 Apr 2010

Jayathi Murthy & Spalding
Villanova, 2010

Patankar & Spalding
London 1999

Spalding & Steven Beale

Spalding & Bill Jones,
Royal Albert Hall, 2014

Brian, Colleen & Rajni:
Marrakesh ó 13 May 2008

Spalding, Vivian de Vahl Davis & Jaluria
Marrakech CHT-2008

CHT Marrakech 2008 Brianis 85th Birthday

Colleen, Brian, & Yogesh
CHT Bath 2012

Gerald Jones, Runchal & Spalding
Franklin Awards 29 Apr 2010

Artumov, Spalding, Ginevsky & Pavitsky
Moscow

Spalding with Leontiev, Irvine, Hatnett and others
Moscow, 1984

Jeremy Spalding, Runchal, Baliga, Spalding, Beale, Colleen
CHT, Sydney , 2006

Colleen and Brian Spalding, Yelena Shafeyeva & Bill Begell, Sydney 2006

DBS and Colleagues at the NATO Natural Convection Workshop, Cesme, Turkey, 1985

Jayathi Murthy, Luigi Martinelli, Gerald Jones, Brian Spalding, Gretar Tryggvason, Phillip. Marcus, Thomas Gatski & Akshai Runchal, Villanova, 2010

Suhas, Brian & Rajni Rutgers, 2015

Spalding, Runchal & Patankar
Rutgers CHT 2015

Rajni, Suhas, Brian & Colleen
Rutgers CHT 2015

Oronzio Manca, Pepper, Spalding & Jaluria
Rutgers 2015

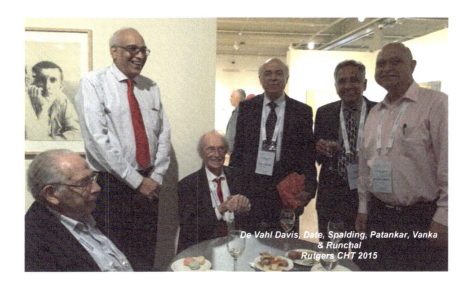

De Vahl Davis, Date, Spalding, Patankar, Vanka & Runchal
Rutgers CHT 2015

Manca, Pepper, Spalding, Colleen, Jaluria & Jeanie Pepper
Rutgers 2015

Heat Transfer Section Party
London, 1977

DBS Birthday Party
London

CHAM Wear Red Day
London

Franklin Awards April, 2010

Doctor of Science (Engineering) Honoris Causa, Imperial College, May, 2014